HEAT TRANSFER

HEAT TRANSFER

JAMES SUCEC

UNIVERSITY OF MAINE AT ORONO
ORONO, MAINE

wcb
WM. C. BROWN PUBLISHERS
DUBUQUE, IOWA

Book Team
Robert B. Stern *Senior Editor*
Nova A. Maack *Assistant Editor*
Lisa K. Bogle *Designer*
Kevin P. Campbell *Production Editor*

wcb
group
Wm. C. Brown *Chairman of the Board*
Mark C. Falb *President and Chief Executive Officer*

wcb

Wm. C. Brown Publishers, College Division
Lawrence E. Cremer *President*
James L. Romig *Vice-President, Product Development*
David A. Corona *Vice-President, Production and Design*
E. F. Jogerst *Vice-President, Cost Analyst*
Bob McLaughlin *National Sales Manager*
Marcia H. Stout *Marketing Manager*
Craig S. Marty *Director of Marketing Research*
Marilyn A. Phelps *Manager of Design*
Eugenia M. Collins *Production Editorial Manager*
Mary M. Heller *Photo Research Manager*

DEDICATED TO MY WIFE

Judith Ann Sucec

CONTENTS

3   UNSTEADY-STATE CONDUCTION   184

5 FORCED CONVECTION HEAT TRANSFER 414

6   FREE CONVECTION   612

9   ADDITIONAL TOPICS IN HEAT TRANSFER   781

APPENDIXES

PREFACE

THIS book is grounded upon, and is an outgrowth of, my experience teaching a one-semester course in heat transfer to students in an engineering curriculum. It is suitable for an engineering course in heat transfer at the junior or senior level and can be used as a reference book for practicing engineers.

The reader's background should include courses in thermodynamics, calculus, differential equations, and fluid mechanics. However, a fluid mechanics course could be taken concurrently with the heat transfer course, since some of the material in Chapter 5 develops the relations in fluid flow that the reader needs to know in order to understand the nature and the solution of forced and free convection heat transfer processes. Chapters 2 and 3 contain a detailed method for separation of variables to solve the partial differential equations governing multidimensional steady-state conduction and unsteady conduction. This will allow the reader with a background in ordinary differential equations and Fourier series to understand the exact solutions presented for conduction, and to use this knowledge to solve other problems.

Detailed solutions to more than one hundred examples supplement the theory, mathematical solution techniques, and experimental information needed to understand heat transfer phenomena and to make quantitative predictions of temperatures, energy transfer rates, and equipment sizes. These examples show some of the more significant and interesting applications of heat transfer, not only to arouse the reader's interest, but also to demonstrate solution techniques, engineering idealizations and approximations, and the order of magnitude of numbers to be expected in practical situations. Each chapter also contains a large number of homework problems, some quite simple, others difficult and challenging.

Chapter 1 discusses the three fundamental modes of heat transfer—conduction, radiation, and convection—as well as the mechanisms of energy transfer in these modes and the laws governing the energy transfer rates. The chapter emphasizes the use of the law of conservation of energy and its importance in solving heat transfer problems. For control volumes fixed in space, the rate form, in words, of the law of conservation of energy is used. Throughout the text, this form of the law is applied to control volumes, differential in extent and finite in extent in one or more directions, to derive fundamental equations and to solve practical problems in engineering heat transfer. I feel that heat transfer students should be very confident in their ability to apply the energy conservation law to any heat transfer problem that they might encounter. Hence, to detail its proper use, a fairly large number of examples are included in Chapter 1.

Chapter 2 devotes much attention to an exposition of the boundary conditions in conduction heat transfer. It is my experience that this is a point of special difficulty for students, particularly the convective-type boundary condition, which is the most realistic and the one which occurs most often in engineering conduction problems. Chapter 2 also emphasizes quasi-one-dimensional conduction, such as may occur in fins or thin rods. This section carefully develops criteria which must be satisfied in order to achieve results of acceptable engineering accuracy when employing this approximation in the solution of what are actually multidimensional conduction problems.

In Chapter 3, "Unsteady-State Conduction," the practical importance of the solutions to some one-dimensional transients (partly embodied by the easy-to-use Heisler charts) is emphasized, not only because they are valuable in their own right, but also because they lead to the solution of certain, even more realistic, two- and three-dimensional transients. Another focal point in this chapter is the lumped parameter method. A wide range of complicated unsteady-state problems can be handled fairly easily by this method when the relevant criteria, such as the Biot number criterion, are satisfied. This is also an area which further develops and tests the student's ability to apply the law of conservation of energy, since almost every problem requires an appropriate energy balance to yield the ordinary differential equation to be solved for the lumped temperature as a function of time.

Since many of the complex conduction problems in industry are solved by digital computers using the finite difference method, the chapters on steady-state and transient conduction contain detailed developments and applications of this invaluable tool. Particular stress is put on the derivation of the algebraic finite difference equations at both interior nodes and surface nodes, the development of stability criteria for transient finite difference equations, and the need to make successive computer solution runs to refine the lattice spacings in space and time to insure that the final values at the lattice points are independent of the lattice spacing.

Chapter 4 introduces a classification scheme for radiation problems, which helps the student choose which of the three radiation calculation methods (reflection method, electric network method, or absorption factor method) will minimize the time and effort needed to calculate the net radiant loss in any specific problem.

In dealing with analytical predictions of surface coefficients of heat transfer, the text uses the powerful approximate integral method, in which an approximating sequence for the temperature profile is made to satisfy a number of conditions which the unknown exact solution satisfies, by proper choice of adjustable functions in the approximate profile. Attention is focused on the reasons why this method often works well and also on the application of the method, such as the derivation of the integral form of the relevant conservation law and the construction of the approximate profiles to be used.

Chapter 5 on forced convection includes a review of the relevant fluid mechanics. Because of the emphasis on boundary layer theory and integral methods of solution, this may be much more than a review for some. For those students who have previously received a good dose of this material, this section can be skipped completely or assigned to be read quickly as a refresher. Almost all students, however, should certainly benefit from studying the subsection on the integral method, since it is developed and explained in detail for velocity boundary layer problems before it is subsequently used as a tool in the analysis of thermal boundary layers in forced convection, free convection, and film condensation heat transfer.

Another feature of Chapter 5 is the presentation of Ambrok's method, which easily allows one to approximately account for the effects of variable free stream velocity and for surface temperature variation in both laminar and turbulent forced convection situations.

Included for the student's convenience at the ends of Chapters 5, 6, and 7 are summary tables of the experimental correlations and analytical and semianalytical results for the surface coefficient of heat transfer in forced convection, free convection, and boiling and condensation, respectively. Having mastered the chapter material, students can rapidly select from the tables the most appropriate expression for the surface coefficient for each problem.

In Chapter 9, "Additional Topics in Heat Transfer," a thorough presentation of differential similarity allows one to arrive at the nondimensional groups appropriate to a particular phenomenon by making the governing partial differential equations (or other types of equations) and the boundary and initial conditions nondimensional. Thus, one can see the source of the nondimensional groups which appear in experimental correlations and analytical expressions in Chapters 5 and 6. Depending upon the instructor's preference, Chapter 9 could be assigned as a prelude to Chapters 5 and 6.

Because of the current change to the International System of Units (SI), both the SI system and the English Engineering System are incorporated in the text, with half of the illustrative examples in the SI system and the remaining half in the English system. Similarly, the end-of-chapter problems are presented with half in each system. There are a sufficient number of end-of-chapter problems to assign homework in both unit systems or, if the instructor prefers, to assign all homework problems in one system or the other.

My experience indicates that most of the first six chapters and Chapter 8 on heat exchangers, along with a light study of Chapter 7 and a selected topic or two from Chapter 9, can be adequately covered in a one-semester course at the senior engineering level.

I wish to thank many students over the years for the pleasure of presenting this material to them and for their questions concerning the aspects of heat transfer which they found most difficult to grasp. These questions, I hope, have resulted in better explanations. I have paid careful attention to the points that students find most bothersome.

The support provided by the editorial and technical staff at Wm. C. Brown Publishers is gratefully acknowledged. I would also like to thank the reviewers of the original manuscript for their valuable suggestions: John R. Biddle, California State Polytechnic University, Pomona; John R. Howell, University of Texas at Austin; Frank J. Lahey, University of Hartford; Gus Plumb, Washington State University; and Larry C. Witte, University of Houston.

Finally, special thanks and deep appreciation are extended to my wife, Judith Ann, for typing the original manuscript and for her support of this writing endeavor.

James Sucec   *Orono, Maine*

INTRODUCTION TO HEAT TRANSFER

1.1 INTRODUCTION

Heat transfer is the science which predicts temperature distributions, which may be functions of both spatial coordinates and time, within regions of matter. Heat transfer also predicts the *rate* at which energy is transferred across a surface of interest due to temperature gradients at the surface and temperature differences between different surfaces. Thus, two examples of the application of heat transfer principles are its importance in the calculation of thermal stresses within a turbine blade in a jet engine, since these are determined by the temperature distribution, and in the determination of the size of a heat exchanger needed to produce a given heat transfer rate between the fluids separated by the heat exchanger walls.

Additional applications of heat transfer include the design of combustors in internal combustion engines, heat treatment processes, the calculation of building energy loads needed to size the furnace or air conditioner, and the determination of cooling water requirements for electric power plants. Many other illustrations of physical situations which require a heat transfer analysis occur throughout the book, both in the examples and in problems for the student to solve.

Heat transfer differs from classical thermodynamics in that thermodynamics involves temperatures and even solves for heat transfer rates from the steady flow energy equation, but actual temperature distributions (except for special cases where the temperature has the same value everywhere within the region of interest and for all time) are not determined in thermodynamics. For example, in a problem involving a heat exchanger, the surface area required to achieve a particular heat transfer rate between the two fluids of the exchanger cannot be calculated using the methods of thermodynamics, but can be determined by the methods of heat transfer. To further illustrate the differences between heat transfer and thermodynamics, consider a bar of metal taken out of a heat treating oven at a relatively high temperature and placed in a large room on an insulating mat to cool. The only information obtainable from classical thermodynamics is that at *thermodynamic equilibrium* (i.e., after some time has elapsed) the temperature of the bar will come to the temperature of the surrounding air (as long as the other surfaces in the room are also at the air temperature). On the other hand, the methods of heat transfer can also predict the temperature in the bar as a function of time and position within the bar. That is, at any time, the temperature at all points of interest within the bar (the temperature will vary from point to point until thermodynamic equilibrium is reached) and the temperature at any point (such as at the center of the bar where it is highest) at any time can be predicted. Also, the instantaneous heat transfer rate can be predicted from all of or any part of the surface of the bar at any time, or even the total internal energy lost by the bar during a time period rather than simply between the initial and final states as can be predicted by equilibrium thermodynamics.

The succeeding sections of this chapter continue the introduction to heat transfer by discussing the mechanisms of, and the laws governing, the energy transfer rates of the three modes of heat transfer: conduction, convection, and radiation. In addition, the form of the law of conservation of energy, the First Law of Thermodynamics, most appropriate to heat transfer problems is reviewed. These topics are illustrated, as will also be the case for the topics in all chapters, by a number of solved examples. Since most scientific and engineering work carried out in other countries uses the International System of Units (SI system), while the United States uses the English Engineering System with a gradual change to the SI system occurring, students should be comfortable using either unit system. Toward this end, about half of the examples and the supplementary problems employ the SI system, and the remaining ones are in the English system. A selection of units in these two systems, with abbreviations and conversion factors, is supplied in Appendix D.

1.2 MODES AND BASIC LAWS OF HEAT TRANSFER

1.2a Conduction

Generally three different modes of heat transfer are delineated: *conduction, convection,* and *radiation.* Transfer by conduction arises from temperature gradients within a material. Ordinarily, this is considered the only mode of heat transfer within opaque solids. In Fig. 1.1, a solid bar is surrounded by perfect insulation except at the left and right faces, where temperatures T_1 and T_2 exist, with T_1 being greater than T_2. Experimentally it is found that if the condition of the bar is not changing with time and if there are no sources of energy within the bar, there is a net energy transfer rate (expressed either in Btu/hr or in watts, with symbol W) from the left face to the right face. This is called *conduction heat transfer,* and the direction of the net energy transfer is in accordance with the Second Law of Thermodynamics.

The actual mechanism of conduction, the means by which the energy transfer takes place, is complex and imperfectly understood. Simply stated, conduction in a solid is due to energy transferred during collisions of adjacent molecules and the migration of free electrons. Conduction in a gas is due to collisions of the molecules which are in continuous random motion.

If an experiment is carried out with the bar shown in Fig. 1.1 under *steady-state conditions* (i.e., no changes in temperature or other variables at any point in the bar as time passes), it is found that the heat transfer rate q in Btu/hr or W is directly proportional

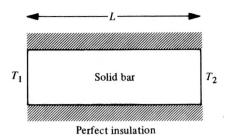

Figure 1.1 Conduction heat transfer through a solid bar due to the temperature difference $T_1 - T_2$.

to the cross sectional area A in m² or ft², is directly proportional to the temperature difference $T_1 - T_2$ in Celsius or Fahrenheit degrees, and is inversely proportional to the path length L, in m or ft, in the direction of heat flow. The proportionality constant k is called the *thermal conductivity* of the material, with the units of k being either Btu/hr ft °F, or W/m °C.

The mathematical expression for the conduction heat transfer rate q implied by our experiment is called *Fourier's simplified law of conduction* and is given as

$$q = kA \frac{(T_1 - T_2)}{L}.$$

(1.1)

The thermal conductivity k in Eq. (1.1) for homogeneous materials which are *isotropic* (i.e., no directional characteristics) is a thermodynamic property of the material, and as such is a function of other thermodynamic properties, such as pressure and temperature. Thermal conductivity is only very weakly dependent upon pressure for solids, and for liquids as well, if they are not near the critical point. Additionally, the thermal conductivity of most gases is essentially independent of pressure if the gases are at pressures near standard atmospheric. (Actually, for the simplest model of a gas, kinetic theory predicts no dependence of k upon the pressure, and a dependence upon temperature which is directly proportional to $\sqrt{T}$.) Hence, the selected values of the thermal conductivity presented in Appendix A are not functions of pressure, but only indicate the variation of k over a particular temperature range. Table 1.1 shows selected values of k from Appendix A.

From Table 1.1, we note that air, a gas, has the lowest value of thermal conductivity shown, while silver, a metal, has the highest. Note that k for silver is about 17,000 times that of air at the same temperature. Thus, the thermal conductivity of real materials varies through a wide range of values. The thermal conductivity values for air are typical of gases, which as a class tend to have the lowest values of k; liquids tend to have higher values than gases; nonmetals tend to have higher values than liquids (although there is some overlap here—compare asbestos and water); metals and alloys tend to have the highest values of thermal conductivity. The additional conduction mechanism possessed by metals and metal alloys, namely the migration or diffusion of the free or valence electrons, is thought to be the reason that metallic solids generally exhibit higher values of thermal conductivity than do nonmetallic solids.

Introduction to Heat Transfer

Table 1.1
Thermal Conductivity of Some Selected Materials

Material Description	Temperature °F	Thermal Conductivity k		Temperature °C
		Btu/hr ft °F	W/m °C	
Gas				
Air	50	0.0143	0.0247	10
	200	0.0181	0.0313	93
	600	0.0265	0.0459	316
Liquids				
Engine oil (unused)	32	0.085	0.147	0
	212	0.079	0.137	100
Water (saturated)	50	0.334	0.578	10
	200	0.392	0.678	93
	400	0.384	0.664	204
Nonmetals				
Asbestos	32	0.087	0.151	0
	392	0.12	0.208	200
Fire-clay brick	392	0.58	1.00	200
	1832	0.95	1.64	1000
Metals and Alloys				
Stainless steel,	32	8.0	13.8	0
Type 304 (18-8)	212	9.4	16.3	100
	572	10.9	18.9	300
Aluminum	32	117.0	202.0	0
	572	133.0	230.0	300
Copper, pure	32	224.0	388.0	0
	212	218.0	377.0	100
	572	212.0	367.0	300
Silver	32	242.0	419.0	0

The value of thermal conductivity k increases with temperature for gases and non-metallic solids, while it tends to decrease with temperature for most liquids, water being a notable exception. The behavior of k for metals and metal alloys is more varied, decreasing with temperature for most pure metals and either increasing or decreasing with temperature for alloys, depending upon the specific alloy.

By a simple rearrangement, Eq. (1.1) can be written as follows:

$$q = \frac{T_1 - T_2}{L/kA}$$

In this form, Eq. (1.1) has the same structure as does Ohm's law for a simple DC electric network element and hence admits of an analogous interpretation: namely that q is the current caused by the driving potential difference, $T_1 - T_2$, across a resistor of magnitude L/kA. Thus it is natural to refer to L/kA as being the thermal resistance, r_t, to heat transfer caused by the finite thermal conductivity k of the material. In certain conduction problems, particularly those involving more than one material, it may be advantageous to write Eq. (1.1) as shown next.

$$q = (T_1 - T_2)/r_t \qquad \text{(1.1a)}$$

$$r_t = L/kA \qquad \text{(1.1b)}$$

(Note that the so-called "R" values, commonly employed by manufacturers in the United States to characterize the insulating quality of their products, are not the same as r_t in Eq. [1.1b]. These refer instead to the thermal resistance per unit area, L/k, in English units of ft^2 °F hr/Btu.)

The *heat flux* q'' is the heat transfer rate per unit area q/A, and is given by

$$\frac{q}{A} = q'' = k \frac{(T_1 - T_2)}{L}. \qquad \text{(1.2)}$$

When there is heat transfer in more than one direction, or temperatures are changing with time, or there is generation, or the thermal conductivity is variable, a form more general than Eq. (1.2) is required. Experiments indicate that for isotropic solid media, Eq. (1.2) can be generalized to *Fourier's law of conduction* given by

$$q'' = -k \frac{\partial T}{\partial n}, \qquad \text{(1.3)}$$

where n is any coordinate direction of interest. The minus sign is inserted so that when energy is transferred in the positive n direction, which requires a negative temperature derivative $\partial T/\partial n$, the heat transfer rate per unit area is positive.

Note that a generalization of Eq. (1.3) for *anisotropic* solids (i.e., those having directional characteristics) is

$$q = -K\nabla T,$$

where q is the heat flux vector, ∇T is the gradient of the scalar temperature distribution also a vector, and K is the thermal conductivity tensor which maps the negative of the temperature gradient vector into the heat flux vector.

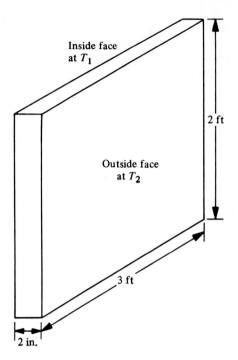

Inside face
at T_1

2 ft

Outside face
at T_2

3 ft

2 in.

Figure 1.2 The heating chamber wall of Ex. 1.1.

EXAMPLE 1.1

The vertical wall of a small heating chamber is 2 ft high, 3 ft long and 2 in. thick. Its inside face is known to be at 500°F, while its outside face is known to be at 100°F. The thermal conductivity of the wall material is 0.10 Btu/hr ft °F, and steady-state conditions prevail. Compute (a) the heat transfer rate across the wall and (b) the heat flux through the wall.

Solution
A sketch of the physical situation is shown in Fig. 1.2.

If it is assumed that the edges are insulated, then, since steady-state conditions prevail with no generation, Fourier's simplified law of conduction, Eq. (1.1), is valid in this situation. Hence, since

$$k = 0.10 \text{ Btu/hr ft } °F, \quad A = 2(3) = 6 \text{ ft}^2, \quad L = \frac{2 \text{ in.}}{12 \text{ in./ft}} = \frac{1}{6} \text{ ft},$$
$$T_1 = 500°F, \quad T_2 = 100°F,$$

from Eq. (1.1), the heat transfer rate across the wall is

$$q = kA \frac{(T_1 - T_2)}{L} = \frac{0.1(6) \ (500 - 100)}{1/6} = 1440 \text{ Btu/hr.} \qquad \textbf{(a)}$$

The heat flux is

$$q/A = q'' = \frac{1440}{6} = 240 \text{ Btu/hr ft}^2.$$ **(b)**

Based on the information given in this example, only conduction heat transfer has to be dealt with. However, the other two modes of heat transfer (which will be discussed shortly in this chapter) are present as well. Energy is arriving at the inside of the wall by free convection from the heating chamber air and by radiation from the source of energy within the heating chamber, perhaps a burner or electric resistor heater, while energy is leaving the outside of the wall by convection with the surrounding air and radiant exchange with other solid surfaces.

EXAMPLE 1.2

The thermal conductivity of a material is measured using a large plane plate of the material which is 0.5 cm thick. It is found, under steady-state conditions, that when a temperature difference of 55°C is maintained between the two surfaces of the plate, a heat transfer rate per unit area near the center of either face (where edge effects are negligible and the heat transfer is approximately one-dimensional) of 9460 W/m² is measured. Determine the thermal conductivity of the material.

Solution
Since the process described is one-dimensional steady-state conduction with no generation, Fourier's simplified law of conduction in the form of Eq. (1.2) can be used to calculate the thermal conductivity k. Since $q'' = 9460$ W/m², $T_1 - T_2 = 55°C$, and $L = (0.5/100)$ m, substitution in

$$q'' = k \frac{T_1 - T_2}{L}$$ **[1.2]**

yields

$$9460 = k \frac{55}{0.5/100}.$$

Hence, $k = 0.86$ W/m °C.

EXAMPLE 1.3

The temperature distribution within a metal bar is

$$T = e^{-0.02\tau} \sin(\pi x/2L),$$

where τ is time in hours, x is a coordinate measured from one end of the bar, and L is the total length of the bar. If the thermal conductivity of the bar material is 25 Btu/hr ft °F and L is 2 ft, find the heat flux in the bar at its center after 10 hours have passed.

Solution

It is surmised that this is a conduction process, since the actual temperature distribution in a metal bar is given; but Eq. (1.2) *cannot* be used because the temperature is not steady, but is instead a function of time. However, Eq. (1.3) can be used after identifying the direction of interest n with the space coordinate x.

Differentiating T with respect to x while holding time constant,

$$\left(\frac{\partial T}{\partial x}\right)_\tau = \frac{\pi}{2L} e^{-0.02\tau} \cos\left(\frac{\pi x}{2L}\right).$$

At $x = \frac{1}{2} L$,

$$\left(\frac{\partial T}{\partial x}\right)_\tau = \frac{\pi}{(2)(2)} e^{-0.02\tau} \cos\left(\frac{\pi}{4}\right) = 0.555 e^{-0.02\tau}.$$

Now at $\tau = 10$ hr,

$$\left(\frac{\partial T}{\partial x}\right)_{\tau\,=\,10\,\text{hr}} = 0.555 e^{-(0.02)\,(10)} = 0.454°\text{F/ft}.$$

Hence, from Eq. (1.3),

$$q'' = -(25)(0.454) = -11.35 \text{ Btu/hr ft}^2.$$

The minus sign indicates that the heat transfer is in the negative x direction.

1.2b Convection

Convection is heat transfer between a solid surface and the adjacent moving medium, which is usually a fluid. For example, consider the horizontal roof shown in Fig. 1.3 with a wind blowing at an appreciable velocity U_s measured far above the roof. Assume that the roof temperature T_w is higher than the wind temperature T_s, so there is heat transfer between the roof and the air. It is not anticipated that this rate will be correctly predicted by Eq. (1.1), since the motion of the air would be expected to influence the temperature distribution through the air and, hence, the heat transfer rate. Energy to the air from the roof is carried away by the moving fluid, thus increasing the heat transfer rate over that which would be expected if the process were one of pure conduction throughout the air. *Newton's law of cooling* is used to determine convection heat transfer rates and is given by

$$q = hA\Delta T, \tag{1.4}$$

where A is the area of the interface between fluid and solid, ΔT is the "appropriate" temperature difference, which in this case is $T_w - T_s$, and h is the *surface coefficient of heat transfer* (or *unit surface conductance*, or *film coefficient*, or simply *convection coefficient*) and is a complicated function of the surface geometry, fluid velocity, and various

Chapter One

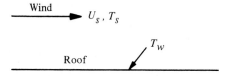

Figure 1.3 Energy transfer by forced convection from a roof to air blown over it.

fluid properties. One of the reasons for choosing the form given by Eq. (1.4) is that h, in many cases, is essentially independent of the temperature difference, and hence q is linear in the temperature difference. The "appropriate" temperature difference, ΔT, is that ΔT which if possible causes h to be independent of the ΔT. This is not always possible as witnessed by the fact, discussed in Chapter 6, that in free convection the value of h does depend on ΔT.

Whenever the fluid motion is induced by an external source, such as a pump, fan, or blower, the process is termed *forced convection,* as in the example of wind blowing over a roof. *Free* or *natural convection* is a process in which the fluid motion is due to density gradients within the fluid usually caused by the temperature field itself. (Density gradients can also be due to concentration differences or to a number of layers of different types of fluids.) Referring to Fig. 1.3, if there were no wind, the air nearest the roof would be heated to a temperature greater than T_s, and since the pressure would be about the same as for air at temperature T_s and the same elevation, this heated air would be less dense than the unheated air, leading to a net upward buoyant force causing the heated air to rise. Therefore, energy is being physically transported away from the roof. Newton's law of cooling, Eq. (1.4), is used to calculate heat transfer rates in both forced and free convection, although the expressions for the surface coefficient of heat transfer h are different. Until heat transfer by convection is studied more formally in Chapters 5–6, numerical values of the surface coefficient of heat transfer h will be given as required.

Although convective heat transfer is characterized by fluid motion, the layer of fluid in contact with the solid surface has no motion. In continuum fluid flow, the velocity of the fluid in contact with a stationary impermeable solid surface is zero. Since the first few molecular layers of fluid against the solid surface are stationary, the energy transferred from the solid to the fluid (or vice versa, depending upon which temperature is higher) must go across those first few molecular layers of stationary fluid by a pure conduction process before moving fluid is encountered. This fact is used in Chapter 5, where surface coefficients of heat transfer are analytically predicted.

In Chapter 5, it will be seen that even in a fairly simple forced convection situation, the surface coefficient of heat transfer h depends upon the density, thermal conductivity, viscosity, and specific heat of the fluid, all of which are thermodynamic properties. It also depends on a characteristic velocity of the fluid and a characteristic length of the solid surface, both of which are mechanical properties. Hence, h is not a thermodynamic property like k, and therefore will not be found tabulated in property tables the way the thermal conductivity k and the mass density ρ are given. In order to establish some idea of roughly what value of h might be expected in some selected circumstances, Table 1.2 has been prepared using the information provided in Chapters 5, 6, 7, and 9 for calculating values of the surface coefficient of heat transfer, h. It is to be stressed that the entries for

Table 1.2
Typical Values of h in Selected Circumstances*

Physical Situation	Possible Values of h		
	Substance	*W/m^2 °C*	*Btu/hr ft² °F*
Free convection over various shapes	Air	3–23	0.5–4
	Water	300–1700	50–300
	Liquid metals	3000–9000	500–1500
Turbulent forced convection in tubes and over various shapes	Air	6–1400	1–240
	Water	1100–9000	200–1600
Film boiling	Water	170–350	30–60
Nucleate boiling	Water	6000–90,000	1000–15,000
Film condensation	Water	4500–17,000	800–3000

*These values are NOT to be employed in calculations.

h in Table 1.2 are not to be used in any calculations and do not even represent the possible range of values of h. Almost any numerical value of h could occur in any of the categories shown, depending upon the value of the characteristic length of the body under consideration or on the fluid velocity's value. For example, even though the values of h typically are between 0.5 and 4 Btu/hr ft² °F or 3 and 23 W/m^2 °C for free convection of air over surfaces of many different shapes, much larger values are possible, such as when one is dealing with a very thin horizontal wire in air. Table 1.2 displays some of the general trends exhibited by the surface coefficient of heat transfer. For instance, of the two most common heat transfer fluids, air and water, usually much lower values of h are found with air. This is due in great part to its much lower thermal conductivity k. Forced convection gives rise to much larger surface coefficients than does free convection for the same fluid, and film boiling leads to rather small values of h, while nucleate boiling causes very high values of h.

EXAMPLE 1.4

A 0.15-m-diameter steam pipe which is 12 m long crosses a room in which the air is at 20°C. The temperature of the outside pipe wall is 65°C and the surface coefficient of heat transfer is estimated to be about 7.4 W/m^2 °C in this free convection situation. Determine the convective heat transfer rate from the surface of the pipe.

Solution
The convective heat transfer rate from the pipe to the surrounding air is given by Newton's cooling law, Eq. (1.4). Since $A = \pi DL$,

$$q = 7.4(3.14)(.15)(12)(65 - 20) = 1883 \text{ W}.$$

As shown in Chapter 6, the net radiant loss from the pipe of Ex. 1.4 is probably of the same order of magnitude as the convective loss, and must be taken into account when the total heat transfer rate from the pipe surface is desired. For most surfaces of ordinary shape and size, and for typical temperature differences between the air and the surface, the free convection coefficient for the surface is usually fairly close to 6 W/m² °C or 1 Btu/hr ft² °F. This relatively low value of h results in a net radiant loss which is of the same order of magnitude as the free convection loss in the type of situation in Ex. 1.4.

1.2c Radiation

When two objects at different temperatures are placed a finite distance apart in a perfect vacuum, a net energy transfer occurs from the higher temperature object to the lower temperature object, even though there is no medium between them to support heat transfer by either conduction or convection. This net energy transfer results from the third mode of heat transfer called *thermal radiation,* or simply, *radiation.* Any surface at an absolute temperature above zero degrees absolute is found to continually emit energy-carrying electromagnetic waves. The rate at which any given surface emits radiant energy per unit area of surface is a complex function of the surface temperature, type of material, and surface condition. However, for the class of surfaces defined as *black bodies,* which absorb all incident radiant energy, the emission rate is given by a simple expression called the *Stefan-Boltzmann law,* which states that

$$W_b = \sigma T^4, \tag{1.5}$$

where W_b is the emission rate of a black surface per unit area of surface with units of Btu/hr ft² or W/m², T is the absolute temperature in degrees Rankine or degrees Kelvin, and σ is a universal constant, the Stefan-Boltzmann constant, whose value is

$$\sigma = 0.1714 \times 10^{-8} \text{ Btu/hr ft}^2 \text{ °R}^4 \quad \text{or}$$
$$\sigma = 5.669 \times 10^{-8} \text{ W/m}^2 \text{ °K}^4 \tag{1.6}$$

in a system of units consistent with the ones chosen for W_b and T. Real surfaces emit at a rate lower than a black surface at the same temperature, although some, such as graphite and soot, come fairly close to the black surface emission rate given by Eq. (1.5). For an actual surface at temperature T with an emissive power W which is less than W_b at temperature T, the expression for the emission rate is

$$W = \epsilon \sigma T^4, \tag{1.7}$$

where ϵ is the total hemispheric emissivity and depends upon the type of material, surface temperature, and surface condition; ϵ is a number between zero and unity. Some values of ϵ for different materials are given in Appendix B. Although Eqs. (1.5) and (1.7) are available to calculate the rate at which a surface continually emits radiant energy, these emission rates are not the primary concern in engineering heat transfer. What is of concern to engineers is the *net* radiant loss from a surface—that is, the *difference* between the rate at which a surface *emits* radiant energy and the rate at which it *absorbs* radiant energy from all sources. This type of calculation depends on the emission rates of all surfaces involved and their spatial relationships, the distribution of the emitted radiant energy above them, and other surface characteristics related to the portion of incident radiant energy which is absorbed and reflected by the surfaces. A full discussion of this problem is contained in Chapter 4. Hence, radiation examples and problems in this chapter are relatively straightforward.

EXAMPLE 1.5

The sun is thought to emit radiant energy as a black body at about 10,000°R. Calculate the heat flux at the surface of the sun.

Solution
The heat flux at the surface is the rate at which radiant energy leaves the surface per unit area. Since the sun is black, this is given by the Stefan-Boltzmann law, Eq. (1.5), as

$$W = \sigma T^4 = 0.1714 \times 10^{-8} (10,000)^4 = 17.14 \times 10^6 \text{ Btu/hr ft}^2.$$

As a result of the directions in which radiant energy is emitted in conjunction with the distance between the sun and the earth, the radiant flux from the sun has a value of 429 Btu/hr ft² at the outer edge of the earth's atmosphere, and its value at the surface of the earth is further reduced by absorption of the radiant energy within the atmosphere.

1.3 LAW OF CONSERVATION OF ENERGY

Many heat transfer problems can be treated most easily by using the concept of a control volume which includes the region or object of interest (such as a pipe). A control volume is *a region of space across whose surface, called the control surface, both mass and energy may pass.* A general control volume may move through space and have a variable volume. Herein, however, discussion will be restricted for the most part to stationary non-variable-volume control volumes. It is advantageous to have a word statement of the *law of conservation of energy* for such a control volume. This statement is derived in thermodynamics by reinterpreting the law of conservation of energy applied to a *system* of fixed mass in terms of a fixed control volume. The final statement is

$$
\begin{bmatrix}
\text{THE RATE AT} \\
\text{WHICH} \\
\text{ENERGY OF} \\
\text{ALL FORMS} \\
\text{ENTERS THE} \\
\text{CONTROL} \\
\text{VOLUME}
\end{bmatrix}
+
\begin{bmatrix}
\text{THE RATE AT} \\
\text{WHICH} \\
\text{ENERGY IS} \\
\text{GENERATED} \\
\text{WITHIN THE} \\
\text{CONTROL} \\
\text{VOLUME} \\
\text{ITSELF}
\end{bmatrix}
=
\begin{bmatrix}
\text{THE RATE AT} \\
\text{WHICH} \\
\text{ENERGY OF} \\
\text{ALL FORMS} \\
\text{LEAVES THE} \\
\text{CONTROL} \\
\text{VOLUME}
\end{bmatrix}
+
\begin{bmatrix}
\text{THE TIME} \\
\text{RATE OF} \\
\text{CHANGE OF} \\
\text{STORED} \\
\text{ENERGY} \\
\text{WITHIN THE} \\
\text{CONTROL} \\
\text{VOLUME} \\
\text{ITSELF}
\end{bmatrix}
$$

which can be rewritten in shorthand form as

$$R_{in} + R_{gen} = R_{out} + R_{stor}. \qquad (1.8)$$

Except perhaps for the R_{gen} term in Eq. (1.8), these terms will be familiar to all who have studied thermodynamics.

We recall from thermodynamics that energy can enter or leave control volumes because of heat interactions, because of work interactions, and because the mass which passes across a control surface carries with it the energy it possesses—in general, internal, kinetic, and gravitational potential energy, as well as the flow work done on or by it as the surface is crossed. Most courses in thermodynamics are concerned with steady-state problems, those in which properties at fixed space points both inside and on the surface of the control volume are not changing with time. When this is the case, the last term in Eq. (1.8), the storage term R_{stor}, will be zero. However in heat transfer this term is often not zero, and Chapter 3 of this book is devoted to such unsteady-state, or transient, problems. In most heat transfer applications of Eq. (1.8), the potential and kinetic energy terms are negligible or zero in the R_{in}, R_{out}, and R_{stor} groups.

The R_{gen} term of Eq. (1.8) merits further discussion. When all energy transfers into or out of the control volume and all stored energy terms are correctly labelled, the generation term does not appear. However, it has been found that this term is useful in accounting for certain energy transfers and energy storage terms. For example, if an electric resistor of resistance R ohms with a current of I amperes flowing through it is inside a control volume, it is known that the effect of this resistor is to release or "generate" energy at the rate I^2R within the control volume. This is the electrical energy dissipated within the control volume, and it behaves as an apparent source of energy within the control volume. In reality, this I^2R loss is the difference between the rate at which electrical energy enters the control volume and the rate at which electrical energy leaves the control volume. It is preferable, however, to write it as a single term using the words "generated within the control volume" in Eq. (1.8), since the electrical energy dissipated is generally treated in terms of a single relation I^2R.

Another example of an energy term that might be treated as a generation term is in an exothermic chemical reaction, such as the burning of a hydrocarbon fuel. The chemical internal energy being liberated could be taken into account in the storage term or in the R_{in} term, depending upon how the fuel is supplied. However, it is often simpler to multiply the heat of combustion by the mass flow rate of fuel being burned and put the result under the R_{gen} term in Eq. (1.8). Hence, this can be treated either as an R_{in} or R_{gen} term, but naturally cannot appear both places in the same problem. The types of energy terms that are usually identified as R_{gen} terms also include the nuclear internal energy release rates which occur in the fission process, as well as in various radioactive decay reactions. The treatment of these as R_{gen} terms will be illustrated in a number of examples throughout the book.

In applying Eq. (1.8), the control volume used can be of any size or shape depending upon the problem and the information desired. Frequently, Eq. (1.8) is applied to a differential control volume and yields a partial or ordinary differential equation. It is also convenient, in some situations, to apply Eq. (1.8) to a control volume which is differential in extent in one direction and of finite extent in the other directions, in which case the result very often is an integral equation. Often Eq. (1.8) is applied to a control volume of finite extent in all directions, and the result might be an algebraic, integral, or ordinary differential equation.

It should be noted that a familiarity with the law of conservation of energy, Eq. (1.8), is quite useful for the solution of heat transfer problems. The correct application of Eq. (1.8) to all control volumes helps not only in the solution of numerical problems, but also in the derivation of general partial differential, integral, and algebraic equations governing temperature distributions within regions of interest. Also, the repeated use of Eq. (1.8), when appropriate, will improve both the student's understanding of the physics of heat transfer problems and his engineering judgment. As a result of the importance of this equation in heat transfer, a relatively large number of examples follow which integrate its use with the laws which govern the energy transfer rate for the various modes of heat transfer: Eq. (1.1) for simple conduction situations, Eq. (1.4) for convective heat transfer, and Eq. (1.7) for very simple radiation problems.

EXAMPLE 1.6

A small rocket nose cone of 1.4 m² outside surface area is to be maintained at a temperature of $-50°C$ in the vacuum of outer space when it is not struck by the rays of the sun. Determine the rate at which a device inside the nose cone must dissipate energy to accomplish this. Assume that none of the radiant energy from the nose cone surface ever returns and no other radiant energy is incident upon it. The emissivity of the nose cone material is 0.90.

Solution

Since information is given about the outside surface of the nose cone, but no detailed information about the inside, other than that there is a device dissipating energy, the entire nose cone may be treated as a control volume and the law of conservation of energy

$$R_{in} + R_{gen} = R_{out} + R_{stor} \qquad \text{[1.8]}$$

can be used to calculate the required rate.

The nose cone and the control surface as a dashed line enclosing the control volume are shown in Fig. 1.4. In Eq. (1.8), R_{in} is zero since the vacuum precludes any heat transfer by convection to or from the control surface. The R_{out} term is the total rate at which radiant energy is emitted from the surface, which is the emission rate per unit area W times the surface area A. The energy dissipated within the control volume is best treated as an apparent source and included as R_{gen}, although the device dissipating the energy may be doing so by chemical or nuclear reactions and might be correctly considered to be a storage term. Since we want the numerical value of the dissipation rate, it is best treated as an apparent source; hence,

$$R_{gen} = WA.$$

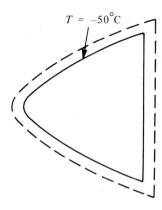

Figure 1.4 Control volume containing the rocket nose cone of Ex. 1.6.

But from Eq. (1.7), $W = \epsilon\sigma T^4$, so that $R_{gen} = \epsilon\sigma T^4 A$. The value $A = 1.4$ m², $\epsilon = 0.90$. From Eq. (1.6), $\sigma = 5.669 \times 10^{-8}$ W/m² °K⁴, and $T = 273 + (-50) = 223$°K, recalling that Kelvin degrees must be used in Eq. (1.7). Thus,

$$R_{gen} = (0.90)(5.669 \times 10^{-8})(223)^4(1.4) = 176.6 \text{ W}.$$

Hence the device inside the nose cone must dissipate energy at the rate of 177 W to maintain the outside surface at -50°C.

EXAMPLE 1.7

Figure 1.5 depicts a one-inch-thick wall of asbestos ($\rho = 36$ lbm/ft³) whose left side is being held at a temperature of 200°F by condensing steam. The temperature of the air on the right is 100°F, and the surface coefficient of heat transfer between the air on the right and the wall surface has been found, by using the methods advanced in Chapter 6, to be $h = 1.1$ Btu/hr ft² °F. If the wall is very long in the other coordinate directions and if steady-state conditions prevail with no generation, find the temperature T_w of the right-hand face.

Solution

Since the heat flow across the wall is one-dimensional, steady, and with no generation, Fourier's simplified law of conduction, Eq. (1.2), can be used by considering one square foot of wall perpendicular to the energy flow. The thermal conductivity will be found from Table A.2.b in Appendix A, where it is seen that the k for asbestos does have a variation with temperature. In Chapter 2, the proper way to handle the dependence of k upon T will be discussed. For the time being, an arithmetic average value of k will be assumed to be adequate.

The first estimate of k will be the value that would be appropriate if the entire slab, including the right face, were at 200°F—namely, $k = 0.11$. Eventually a value of k corrected for the fact that T_w is not 200°F will be used. Hence, using Eq. (1.2) yields

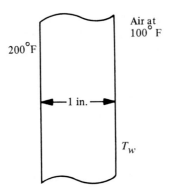

200°F

Air at 100°F

1 in.

T_w

Figure 1.5 The wall of Ex. 1.7.

$$q'' = \frac{k(T_1 - T_2)}{L} = \frac{0.11(200 - T_w)}{1/12}.$$

However, this equation contains two unknowns, T_w and q''; as a result, one additional equation is required. The energy that reaches the right face by conduction must be transferred into the fluid at the same rate by convection. This follows from the application of Eq. (1.8) to the interface taken as a control volume of zero volume (since the thickness of an interface is zero); hence, storage and generation terms do not appear at an interface even if there is storage and generation within the wall itself. At the interface, $R_{in} = R_{out}$. Using Eq. (1.4) for the convection heat transfer,

$$R_{out} = q'' = h(T_w - T_s) = 1.1(T_w - 100).$$

Equating the two expressions for q'',

$$\frac{0.11(200 - T_w)}{1/12} = 1.1(T_w - 100),$$

hence, $T_w = 155°F$.

Next, using the arithmetic average temperature of $(200 + 155)/2 = 177.5°F$, the value of $k = 0.10$ is found from Appendix A and inserted into the last equation in place of 0.11, yielding

$$T_w = 152°F.$$

This is close enough to the previously found value to justify the use of $k = 0.10$.

The beginning student in heat transfer is often tempted to say that the temperature of the right-hand face equals the undisturbed fluid temperature which, in Ex. 1.7, is 100°F. This can be seen to be incorrect, since by Newton's cooling law,

$$q'' = 1.1(T_w - 100).$$

The only way T_w can be close to 100 for a nonzero q'' is when h is extremely large. If h is infinitely large, then any heat flux q'' can be supported by essentially a zero temperature difference between the wall and the fluid. This is not the case in Ex. 1.7.

EXAMPLE 1.8

The spherical nuclear reactor of known outer radius R shown in Fig. 1.6 is releasing energy at a rate per unit volume equal to $q''' = B(R^2 - r^2)$ due to fission, where B is known and r is the general radial coordinate measured from the center of the sphere. Cooling fluid at known temperature T_s and known surface coefficient of heat transfer h is flowing over the outside of the sphere. If the temperature of the sphere does not vary with time, calculate the outside surface temperature in terms of known quantities.

Solution
The known quantities are T_s, h, B, and R, and the unknown surface temperature is T_w. The total rate at which energy is being released inside the sphere by the fission reactions must equal the rate at which energy leaves the sphere by convection to the fluid. Using Eq. (1.8), and taking the entire sphere as the control volume,

$$R_{in} + R_{gen} = R_{out} + R_{stor}. \qquad \text{[1.8]}$$

By Newton's cooling law, Eq. (1.4),

$$R_{out} = h(4\pi R^2)(T_w - T_s),$$

where $4\pi R^2$ is the outside surface area of the sphere. The total rate at which energy is generated within the sphere is obtained by multiplying the energy generation rate per unit volume q''' by a volume sufficiently small so that q''' is constant throughout the volume. Since q''' is a function of r only, the infinitesimally thin spherical shell located a distance r from the center is used. The total rate at which energy is generated within this differential volume is $q''' \, 4\pi r^2 \, dr = 4\pi B(R^2 r^2 - r^4) \, dr$, and the total rate at which energy is generated within the entire sphere is found by integration from $r = 0$ to $r = R$. Thus,

$$R_{gen} = \int_0^R 4\pi B(R^2 r^2 - r^4) \, dr = \frac{8}{15} \pi B R^5.$$

Substituting in Eq. (1.8),

$$0 + \frac{8}{15} \pi B R^5 = h 4\pi R^2 (T_w - T_s).$$

Solving for T_w,

$$T_w = T_s + \frac{2}{15} \frac{B}{h} R^3.$$

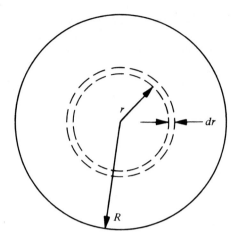

Figure 1.6 The spherical nuclear reactor of Ex. 1.8 showing the differential volume element.

Note that the use of the entire sphere as the control volume in this example provides the required outside surface temperature. However, details of the internal temperature distribution cannot be obtained using this approach.

Note the way Newton's cooling law has been used in conjunction with Eq. (1.8). In Ex. 1.8, the convection heat transfer rate is out of the control volume and the driving temperature difference in Newton's cooling law is chosen as $T_w - T_s$, rather than $T_s - T_w$, to ensure that the convection heat transfer is *out* of the control volume. If the convection heat transfer is chosen as R_{in} to the control volume, the driving potential difference is written with T_s greater than T_w, thus resulting in energy into the control volume. This leads to

$$h4\pi R^2(T_s - T_w) + \frac{8}{15} B\pi R^5 = 0 + 0,$$

which yields the same result for T_w.

EXAMPLE 1.9

A typical experiment for measuring the average surface coefficient of heat transfer in crossflow over a right circular cylinder is shown in Fig. 1.7. For a particular case, the tube is 0.15 m in diameter and 3 m long. Water at 93°C enters the tube at the rate of 0.15 kg/s and leaves at 82°C. The outside surface temperature of the tube is 71°C, and the fluid flowing over the outside surface is at 43°C. If steady-state conditions prevail, find the average surface coefficient of heat transfer between the outside surface of the tube and the fluid flowing over it in crossflow.

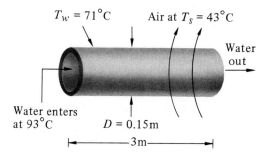

$T_W = 71°C$ Air at $T_S = 43°C$

Water out

Water enters at 93°C $D = 0.15m$

|←————3m————→|

Figure 1.7 The pipe of Ex. 1.9.

Solution

By Eq. (1.4), the heat transfer rate from the outside of the tube to the 43°C fluid is

$$q_c = hA\Delta T = h(\pi DL)(\Delta T) = h(\pi)(0.15)(3)(71 - 43)$$
$$= 39.58h. \tag{1.9}$$

Equation (1.9) has two unknowns, q_c and h. However, another equation relating q_c and h can be derived by applying the energy balance, Eq. (1.8), to a control volume whose control surface passes along the outside of the pipe, and hence includes both the pipe material and the water flowing inside the pipe. For this control volume, $R_{stor} = 0$ in Eq. (1.8) due to the steady-state condition. There are no energy generation terms, since there are no electric resistors within the control volume or any chemical or nuclear reactions going on; hence, $R_{gen} = 0$. For the total rate at which energy of all forms enters the control volume R_{in}, one has the energy transported in by the entering water.

Every kilogram of water which enters the pipe carries in internal energy, kinetic energy, gravitational potential energy, and flow work done by the upstream fluid to force it into the control volume. The sum of the internal energy per kilogram and the flow work per kilogram is defined as the enthalpy, here called i. The rate at which energy enters the control volume due to the incoming mass is given by the energy entering (per kilogram) multiplied by the mass flow rate $\dot{m}$. If V_1 represents the inlet velocity and z_1 the inlet elevation above a reference level, then the rate at which energy crosses the control surface as a result of mass entering is given by

$$\dot{m}(i_1 + \tfrac{1}{2}V_1^2 + z_1).$$

A similar expression holds at the exit of the pipe, where it is an R_{out} term. Conservation of mass requires that the mass flow rate at inlet and exit be equal; thus, part of the R_{out} term is $\dot{m}(i_2 + \tfrac{1}{2}V_2^2 + z_2)$. Another energy transfer rate out term is the convection from the outside pipe wall to the fluid flowing over the pipe, which is given by Eq. (1.9). Since $R_{in} = R_{out}$, we have that

$$\dot{m}(i_1 + \tfrac{1}{2}V_1^2 + z_1) = \dot{m}(i_2 + \tfrac{1}{2}V_2^2 + z_2) + 39.58h.$$

Neglecting elevation changes, $z_1 = z_2$, and treating water as incompressible, mass conservation $\dot{m} = \rho_1 A_1 V_1 = \rho_2 A_2 V_2$, where ρ is the mass density, yields $V_1 = V_2$. Therefore,

$$\dot{m}(i_1 - i_2) = 39.58h.$$

From thermodynamics, we have, approximately, that $i_1 - i_2 = c_p(T_1 - T_2)$, where c_p = 4180 J/kg °C from Table A.5.a in Appendix A. Hence, substituting,

$$.15(4180)(93 - 82) = 39.58h,$$
$$h = 174 \text{ W/m}^2 \text{ °C}.$$

One difficulty in this type of experimental setup is maintaining a constant wall temperature when the water flowing inside the tube varies in temperature from one end of the tube to the other. This causes the tube temperature to vary axially as well. This may be overcome by using a tube material of high enough thermal conductivity and with a large enough wall thickness to cause axial temperature variations to be small.

1.4 PROBLEMS

1.1   A gypsum board wall has dimensions 8 ft × 24 ft × 1/2 in. If one face is held at 70°F and the other face at 0°F, compute the heat transfer rate through this wall, where $k = 0.12$ Btu/hr ft °F.

1.2   A 0.3-m-diameter steam pipe, 3 m long, passes through a room. The room temperature is 20°C, and the surface temperature of the steam pipe is 40°C and is constant over the entire surface of the pipe. The average surface coefficient of heat transfer is 9 W/m² °C. Calculate the rate at which heat is transferred from the pipe to the room air, neglecting radiation effects.

1.3   A spherical storage tank of radius 10 ft has an outside surface temperature of 80°F. Air at 50°F is blowing over the spherical tank. The average surface coefficient of heat transfer is 5 Btu/hr ft² °F. Calculate the heat transfer rate from the spherical tank to the air.

1.4   A plane piece of insulation board is used to reduce the heat loss from a hot furnace wall into a cellar. One face of the board is at 82°C and the other face is at 15°C, and it is desired to keep the heat loss down to 125 W per square meter of insulation board. If the thermal conductivity of the board is 0.05 W/m °C, calculate the required thickness of the board.

1.5   The wall of a simple heat exchanger is plane and 0.25 in. thick. Conditions on one side of the wall fix the temperature on that side at 100°F. Find the surface area, perpendicular to the heat flow, which will yield a steady-state energy transfer rate across the wall of 20×10^5 Btu/hr, without the temperature on the other side of the wall exceeding 500°F. Assume that $k = 30$ Btu/hr ft °F.

1.6   Ice at 32°F is in a container whose outside surface area is 4 ft² and whose outside temperature is 32°F. The outside surface is surrounded by 75°F air, and the surface coefficient of heat transfer $h = 1.1$ Btu/hr ft² °F between the container surface and the air. If the latent heat of fusion of the ice is 144 Btu/lbm, calculate the rate in lbm/hr at which the ice is being changed into liquid water.

1.7   The side wall of an oven has dimensions 1 m × 1 m × 7.5 cm. It is primarily made of insulation with a thermal conductivity of 0.04 W/m °C. When its inside temperature is 150°C and outside surface temperature is 37°C, calculate the rate, in watts, at which the electric coils within the oven must dissipate electrical energy to make up for the loss through this wall.

1.8   A 0.6 cm steel plate with a thermal conductivity of 43.3 W/m °C receives a *net* heat input due to radiation at its upper face of 6300 W/m², and the surface coefficient of heat transfer between the upper face and the 150°C gas flowing over it is 60 W/m² °C. The upper face temperature is 93°C. If steady-state conditions prevail, determine the temperature at the lower face.

1.9   An electric resistance heater of 0.1 m² total outside surface area is surrounded by 21°C water, and the average surface coefficient of heat transfer is 60 W/m² °C. Electrical energy is dissipated within the heater at a rate of 300 W. If steady-state conditions prevail, calculate the average outside surface temperature of the heater.

1.10   Circle the single correct answer. Most metals are good heat conductors because of:
        a. Migration of neutrons from hot end to cold end.
        b. Energy transport due to molecular vibration.
        c. The presence of many free electrons.
        d. A unique bacterium which lives in the crystal structure.
        e. Lattice defects such as dislocations.

1.11   Circle the two correct statements that follow. Convection heat transfer is distinguished by:
        a. Continual emission of energy-carrying electromagnetic waves.
        b. Energy transport as a result of bulk fluid motion.
        c. Migration of valence electrons.
        d. A pure conduction process in the first few molecular layers of the fluid immediately adjacent to the solid surface.
        e. A small humming noise appearing to come from the fluid/solid interface.
        f. The absence of an intervening medium between two solids at different temperatures.

1.12 Referring to Ex. 1.1, suppose that the outside wall is in contact with air flowing over it, due to a fan, in such a way that the surface coefficient of heat transfer between the outside surface and the outside air is estimated to be $h = 20$ Btu/hr ft² °F. Neglecting any radiant loss, find the temperature of the outside air flowing over the wall.

1.13 For the steel member acting as a leg for the tank of hot chemicals above it, as shown in Fig. 1.8, the heat transfer rate from the steel leg during an average day is required. Rather than attempt to compute it completely analytically, the engineer in charge measured the steady-state temperature distribution down the leg with thermocouples and fitted his data to the curve

$$T = 300 - 100x + 25x^2,$$

where T is the temperature in °F at the point x in the steel leg. Gradients normal to the x direction within the steel were found to be negligible. (a) If $k_{steel} = 25$ Btu/hr ft °F and the cross-sectional area of the steel leg is 1 in.², compute the heat transfer rate from the tank through the leg. (b) Using the information given, determine the temperature of the material in the tank.

Figure 1.8

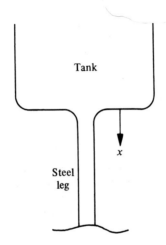

Tank

x

Steel
leg

1.14 Consider an earth satellite in the vacuum of outer space. Because of the instrumentation inside the satellite, it is desirable to maintain its outside surface at 400°R by dissipation of energy within the satellite. If the outside surface area is 3 ft² and its emissivity is 0.30, calculate the rate of energy dissipation needed when the satellite emits radiant energy but does not receive it. Steady-state conditions prevail.

1.15 Consider a braking system consisting of a brake shoe and drum. While continuously braking and moving at constant speed down a mountain, 9 W of energy is dissipated by the braking system, whose total outside area is 0.1 m². The surface coefficient of heat transfer between the drum and the air is 10 W/m² °C. The air is at 20°C. Calculate the average outside surface temperature, neglecting radiant effects. Assume that steady-state conditions prevail.

1.16 A wire of negligible thermal conductivity being drawn through a die at a constant velocity of 100 ft/hr is shown in Fig. 1.9. A force of 77.8 lb must be constantly applied to the wire. The surface of the die exposed to the 80°F surrounding air is 0.06 ft², and the surface coefficient of heat transfer between the die and the air is 1.5 Btu/hr ft² °F. Calculate the average steady-state surface temperature of the die. (Note that 778 ft lb = 1 Btu.)

Figure 1.9

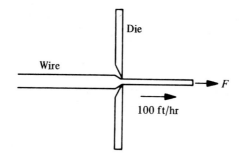

1.17 A solid cement wall of a building has a thermal conductivity k and a thickness Δx and is heated by convection on the inner side and cooled by convection on the outer side. Show that the heat flux through the wall can be expressed as

$$q'' = \frac{T_1 - T_2}{\dfrac{1}{h_1} + \dfrac{\Delta x}{k} + \dfrac{1}{h_2}},$$

where T_1 and T_2 are the temperatures of the fluids on each side of the wall and h_1 and h_2 are the corresponding surface coefficients of heat transfer.

1.18 Fluid is confined in a container whose total outside surface area is 0.05 m². A stirrer in the fluid has work supplied to it at the rate of 1.5 W to keep it rotating in the fluid. The container of fluid is surrounded by air at 15°C with a surface coefficient of heat transfer of 6 W/m² °C. If steady-state conditions prevail, find the average outside surface temperature of the container.

1.19 Consider a 0.3-m-thick layer of water confined between two plates, shown in Fig. 1.10. Compute the heat transfer rate per unit area between the upper and lower plates.

Figure 1.10

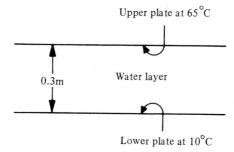

1.20   In order to calculate the cooling air requirements for a gas turbine blade, the total rate at which the main gas stream at temperature 1500°F transfers energy to the blade surface (which as shown in Fig. 1.11 can be idealized as a flat plate 1/3 ft deep into the paper and 1/2 ft long) must be determined. The surface temperature is constant at 1000°F. The surface coefficient of heat transfer varies with x as follows: $h = 60/x^{0.2}$. Compute the total convection heat transfer rate from the gas to the turbine blade.

Figure 1.11

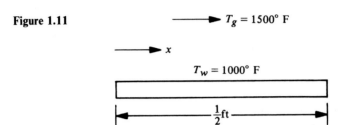

1.21   In an experiment to determine the surface coefficient of heat transfer h, water at 38°C was forced over the outside of a 2.5 cm diameter, right circular cylinder. If the cylinder surface temperature is 61°C and electrical energy is dissipated within the cylinder at the rate of 557 W to keep the 1.8-m-long cylinder at steady state, compute the average coefficient of heat transfer h over the outside surface of the cylinder.

1.22   Compute the rate at which electrical heaters must dissipate energy to keep copper at just under its melting point of approximately 2000°F if the copper is in a ladle with an exposed surface of 2 ft², in a vacuum chamber. The emissivity of the surface is 0.70. Assume that the copper emits, but does not receive, radiant energy.

1.23   Figure 1.12 is a sketch of a 75 W light bulb. The bulb is operating in the steady state in a room where the air is at 21°C and the free convection surface coefficient of heat transfer is estimated to be 8.5 W/m² °C. Calculate the highest temperature that can be reasonably anticipated for the surface of the bulb.

Figure 1.12

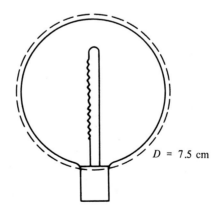

$D = 7.5$ cm

1.24 The outside surface of a gearbox has an area of 0.25 m² and must be kept at a temperature no higher than 65°C. To accomplish this, cooling fluid at 26°C is available for forcing over the outside surface of the gearbox. Since the surface coefficient of heat transfer h between the fluid and the gearbox depends upon the fluid velocity, h must be found so that the fluid velocity can be specified. Calculate the surface coefficient of heat transfer needed if the components in the gearbox are dissipating mechanical energy at the rate of 1000 W in the steady state.

1.25 An electrical resistance heater with a 750 W output has a total exposed surface area of 1 ft². It is designed to operate submerged in water under conditions leading to a surface coefficient of heat transfer $h = 50$ Btu/hr ft² °F. The heater operates under steady-state conditions in 100°F water. (a) Calculate the surface temperature of the heater under design conditions. (b) If the heater is mistakenly operated in 100°F air with a surface coefficient of heat transfer $h = 1.5$ Btu/hr ft² °F, determine the steady-state surface temperature. (c) Discuss the possible consequences of operating the heater in air.

1.26 A metal block whose outside surface area is 0.035 m² is sawed in half by a power hacksaw which dissipates mechanical energy at the rate of 600 W. The coolant is at 15°C and the surface coefficient of heat transfer between the coolant and the surface of the metal block is estimated to be 110 W/m² °C. Assuming that all the mechanical energy dissipated flows into the block rather than into the hacksaw blade, determine the average steady-state temperature of the outside of the block.

1.27 The wall of a gas turbine blade can be idealized as a flat plate 0.12 cm thick as shown in Fig. 1.13. Hot gases at 1000°C flow over the upper surface of the blade, and the surface coefficient of heat transfer is estimated to be 2800 W/m² °C between the upper surface and the gas. The lower surface is cooled by air bled off the compressor, and a surface coefficient of heat transfer equal to 1400 W/m² °C is estimated. Determine the temperature of the cooling air so that the blade temperature does not exceed 870°C. The blade material has a thermal conductivity of 17 W/m °C, and steady-state conditions have been reached.

Figure 1.13

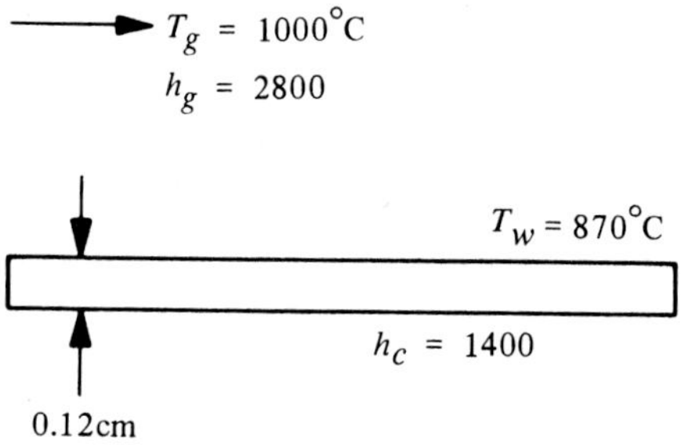

$T_g = 1000°C$

$h_g = 2800$

$T_w = 870°C$

$h_c = 1400$

0.12cm

1.28 The oven of an electric stove is maintained at the proper steady-state condition by dissipating electrical energy at the rate of 600 W. If the total outside surface area of the stove is 3 m² and is exposed to 21°C room air with an average surface coefficient of heat transfer between the outside stove surface and the room air of 11 W/m² °C, find (a) the average outside surface temperature of the stove and (b) the inside surface temperature if the wall thickness is 3.8 cm and the thermal conductivity is 0.07 W/m °C.

Introduction to Heat Transfer

1.29 Figure 1.14 shows a gas turbine vane having an outside surface area of 0.04 ft². Hot gases at 2100°F are flowing across the outside area. The outside surface coefficient of heat transfer is 400 Btu/hr ft² °F. Find the mass flow rate of cooling air needed for the outside surface temperature to be held at 1700°F if the cooling air enters the hollow vane at 1300°F and leaves at 1600°F. Steady-state conditions prevail and $c_p = 0.24$ Btu/lbm °F for air.

Figure 1.14

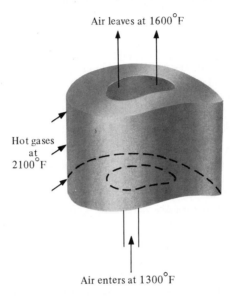

Air leaves at 1600°F

Hot gases at 2100°F

Air enters at 1300°F

1.30 Figure 1.15 shows a thin electric resistance heater which is dissipating electrical energy at the known rate of B_0 Btu/hr. The heater is sandwiched between two different materials. All thermal properties, such as k_a, k_b, etc., the total outside surface coefficients h_i and h_0, and the fluid temperatures T_0 and T_i are known. The sandwich is 1 ft deep into the paper and 1 ft high, and steady-state conditions prevail. Compute the heater temperature in terms of known quantities.

Figure 1.15

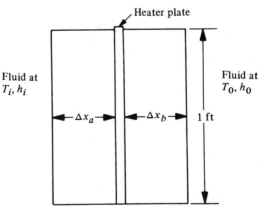

Heater plate

Fluid at T_i, h_i

Fluid at T_0, h_0

Δx_a Δx_b 1 ft

1.31   The wall of the low-temperature building shown in Fig. 1.16 is constructed as follows: The inside surface is 1/2-in. steel followed by 6 in. of glass wool insulation which is followed by 6 in. of building brick. The temperature of the air inside the room is to be maintained at $0°F$ while the temperature of the outside surroundings is $80°F$. The inside surface coefficient is 2.0 Btu/hr ft² °F while the total outside surface coefficient (i.e., h contains a contribution which approximates the net radiant loss) is estimated to be 3.5 Btu/hr ft² °F. Note that $k_{glass\ wool} = 0.023$ Btu/hr ft °F, $k_{brick} = 0.40$ Btu/hr ft °F, and $k_{steel} = 25$ Btu/hr ft °F. (a) Compute the steady heat transfer rate per unit area through the wall. (b) If the total wall surface is 2400 ft², calculate the total rate at which the refrigeration system must extract heat to keep the room at $0°F$ if only the heat leakage through the walls is considered.

Figure 1.16

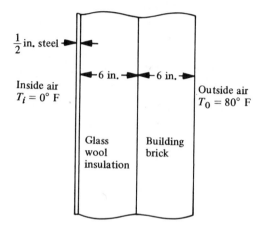

1.32   A circular pipe of 1 ft outside diameter and an outside surface temperature of $200°F$ is exposed to $300°F$ air flowing over the outside of the pipe, as shown in Fig. 1.7. The pipe is 20 ft long and the outside surface coefficient of heat transfer is estimated to be 15 Btu/hr ft² °F. Water is entering the pipe at the rate of 900 lbm/hr at $60°F$. If steady-state conditions prevail, find the exit temperature of the water.

1.33   Rework Prob. 1.23, assuming that the light bulb is operating in a vacuum and the bulb emits radiant energy but does not receive any. The emissivity of the bulb surface is 0.20.

1.34   A heat exchanger of 0.5 m² outside area and $-23°C$ outside temperature is in the vacuum of outer space. The emissivity of its outside surface is 0.88. Water in a closed circuit within the satellite enters the exchanger at $65°C$ and leaves at $4°C$. If steady-state conditions prevail, compute the mass flow rate of the water.

1.35 Figure 1.17 shows a plate which is perfectly insulated at $x = 0$ and exposed to a fluid at known temperature T_s and known surface coefficient of heat transfer h on the right-hand face. The plate is in the steady state, but has energy being generated at the rate per unit volume $q''' = \beta \cos (\pi x/2L)$. The quantities β, L, and the thermal conductivity k are known constants. Find the temperature at the right-hand face, i.e., at $x = L$, in terms of known quantities.

Figure 1.17

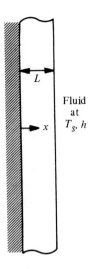

STEADY-STATE CONDUCTION

2.1 INTRODUCTION

The most fundamental information in a general heat transfer analysis is the temperature distribution within the region of interest. This knowledge allows one to determine, for example, the maximum temperature within the region, the temperature at critical locations such as interfaces, and the time it takes to reach a particular temperature of interest. Once the temperature distribution is known, one can also, by use of the appropriate heat transfer laws presented in Chapter 1, compute heat transfer rates.

Toward this end, the first topic considered in this chapter is the derivation of the general differential equation, which when solved leads to the temperature distribution within any region of interest when conduction is the mode of heat transfer. The commonly encountered boundary conditions on this equation are presented, and then attention is directed toward steady-state conduction, where the temperature is not changing with time at any space points. Solutions for the temperature distribution in some simple one-dimensional problems, where the temperature depends only upon one space coordinate, will be developed, and then the additional complicating factors of variable thermal conductivity and generation will be addressed.

Following this will be a discussion of temperature profiles which are quasi-one-dimensional. That is, the profiles, though actually two- or three-dimensional, can be adequately approximated as being one-dimensional. Included in this class of problems is the very important topic of fins which are often added to a surface to increase the heat transfer rate. Two- and three-dimensional conduction problems will be treated next. Use of the method of separation of variables, along with Fourier series expansions, leads to some exact analytical solutions for multi-dimensional steady-state conduction problems.

Lastly, in order to be able to attack the more complicated, and often more realistic, multidimensional problems, the powerful and very important numerical method termed the finite difference procedure is presented. In actual practice, this method almost always involves the use of a digital computer to solve the resulting large number of algebraic equations for the temperature at various points within the material.

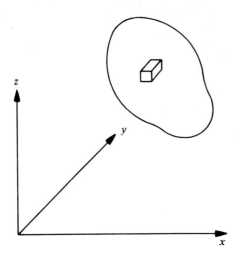

Figure 2.1 An arbitrarily shaped three-dimensional solid conduction region with imbedded differential control volume shown.

2.2 THE GENERAL CONDUCTION EQUATION

The general problem of conductive heat transfer within a solid involves the prediction of the temperature within the (solid) region of interest as a function of the space coordinates and time. That is, the problem is to find the temperature distribution $T = T(x, y, z, \tau)$, where the three space coordinates x, y, z locate any point of interest within the conduction region and τ is time.

To derive an equation which can be solved for the temperature distribution $T = T(x, y, z, \tau)$, the law of conservation of energy in the form

$$R_{in} + R_{gen} = R_{out} + R_{stor} \qquad [1.8]$$

is applied to a differential control volume fixed in space within a solid conducting region. Figure 2.1 shows a solid region with points located by a rectangular cartesian coordinate system, $x, y,$ and z. Also indicated is a differential control volume of volume $dxdydz$ located within the region. It is assumed that the solid is isotropic and rigid, where rigid implies that negligible work is done on the material by the action of mechanical forces. Figure 2.2 shows the differential control volume in more detail. Since all six faces of the control volume are in reality surrounded by other solid material, the only way in which energy can enter or leave the control volume is by conduction across the faces. In Fig. 2.2, the conduction heat transfer rates across the six faces are shown, with q_x indicating the rate at which energy is conducted into the control volume in the x direction at x, and q_{x+dx} indicating the rate at which energy is conducted out of the control volume in the x direction at $x + dx$. Similar meanings hold for the remaining four terms. Hence,

$$R_{in} = q_x + q_y + q_z, \quad R_{out} = q_{x+dx} + q_{y+dy} + q_{z+dz}.$$

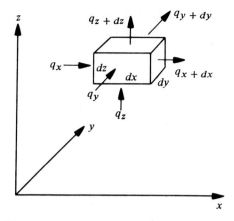

Figure 2.2 The differential control volume of Fig. 2.1 with conduction heat transfer rates shown.

However, q_{x+dx} can be put in terms of q_x by viewing q_x as a function for the conduction heat transfer rate in the x direction and expanding this function in a Taylor series about x to $x + dx$; thus,

$$q_{x+dx} = q_x + \frac{\partial q_x}{\partial x} dx + \frac{\partial^2 q_x}{\partial x^2} \frac{(dx)^2}{2!} + \frac{\partial^3 q_x}{\partial x^3} \frac{(dx)^3}{3!} + \cdots.$$

Since q_x is really a function of all three space coordinates, the Taylor series expansion should be about the point x, y, z and would contain partial derivatives of q_x with respect to y and z, as well as mixed partial derivatives. Since q_{x+dx} is at the same value of y and z as q_x, these terms go to zero, leaving the result shown. Now, since dx is infinitesimal, terms containing $(dx)^2$, $(dx)^3$, etc. may be neglected when compared with the second term on the right; thus,

$$q_{x+dx} = q_x + \frac{\partial q_x}{\partial x} dx.$$

Similarly,

$$q_{y+dy} = q_y + \frac{\partial q_y}{\partial y} dy, \quad q_{z+dz} = q_z + \frac{\partial q_z}{\partial z} dz.$$

Hence,

$$R_{\text{out}} = q_x + \frac{\partial q_x}{\partial x} dx + q_y + \frac{\partial q_y}{\partial y} dy + q_z + \frac{\partial q_z}{\partial z} dz.$$

To calculate the rate R_{gen} at which energy is generated within the control volume, the quantity q''' is defined as the generation rate per unit volume, with units of W/m^3 or Btu/hr ft^3. In general, the quantity q''' is specified as a function of space coordinates, time, and temperature. For example, in an exothermic chemical reaction q''' would be chemically determined and used by the engineer solving the heat transfer problem. In a problem involving the dissipation of electrical energy, a knowledge of the resistance characteristics of the material and the current flow would yield the value of q'''. Hence, in terms of q''', the generation term R_{gen} is the generation rate per unit volume q''' times the volume $dxdydz$ of the control volume,

$$R_{gen} = q''' \, dxdydz.$$

A solid stores energy by virtue of its molecular arrangement and motion, that is, as internal energy. Let u be the internal energy per kilogram or pound mass and ρ the density. Then the last term in Eq. (1.8), the time rate of change of stored energy within the control volume, will be

$$R_{stor} = \frac{\partial}{\partial \tau} [\rho(dxdydz)u].$$

The density ρ of a solid may be considered to be constant, so that

$$R_{stor} = \rho \frac{\partial u}{\partial \tau} dxdydz.$$

Substitution into Eq. (1.8) leads to

$$q_x + q_y + q_z + q'''dxdydz = q_x + \frac{\partial q_x}{\partial x} dx + q_y + \frac{\partial q_y}{\partial y} dy + q_z$$
$$+ \frac{\partial q_z}{\partial z} dz + \rho \frac{\partial u}{\partial \tau} dxdydz.$$

Cancelling terms,

$$q'''dxdydz = \frac{\partial q_x}{\partial x} dx + \frac{\partial q_y}{\partial y} dy + \frac{\partial q_z}{\partial z} dz + \rho \frac{\partial u}{\partial \tau} dxdydz. \qquad (2.1)$$

From thermodynamics, the caloric equation relates u to pressure and temperature for particular substances; for a rigid solid, since $c_p \approx c_v$,

$$u = u_0 + c_p(T - T_0),$$

where c_p is the constant pressure specific heat, which may be assumed constant for a solid, in J/kg °C or Btu/lbm °F, and u_0 is the internal energy per kilogram or pound mass at the reference temperature T_0. From this expression,

$$\frac{\partial u}{\partial \tau} = c_p \frac{\partial T}{\partial \tau} . \qquad (2.2)$$

Terms such as $\partial q_x/\partial x$ can be put in terms of the temperature by the exact form of Fourier's law of conduction,

$$q'' = -k \frac{\partial T}{\partial n},$$ [1.3]

where n is the direction of interest and q'' is the conduction heat transfer rate per unit area in that direction. [Note that expressions such as $u = u_0 + c_p(T - T_0)$ and $q'' = -k(\partial T/\partial n)$ are called constitutive relations or equations.] Since $q_x = dA_x q_x''$ where $dA_x = dydz$, and from Eq. (1.3), $q_x'' = -k(\partial T/\partial x)$,

$$q_x = -k \frac{\partial T}{\partial x} dydz.$$ (2.3a)

Similarly,

$$q_y = -k \frac{\partial T}{\partial y} dzdx,$$ (2.3b)

$$q_z = -k \frac{\partial T}{\partial z} dxdy.$$ (2.3c)

Differentiating q_x, q_y, q_z partially with respect to x, y, and z, respectively,

$$\frac{\partial q_x}{\partial x} = -\frac{\partial}{\partial x} \left(k \frac{\partial T}{\partial x} \right) dydz, \quad \frac{\partial q_y}{\partial y} = -\frac{\partial}{\partial y} \left(k \frac{\partial T}{\partial y} \right) dxdz,$$

$$\frac{\partial q_z}{\partial z} = -\frac{\partial}{\partial z} \left(k \frac{\partial T}{\partial z} \right) dydx.$$ (2.4)

Substituting Eqs. (2.4) and (2.2) into Eq. (2.1), and dividing through by the common volume element $dxdydz$,

$$\frac{\partial}{\partial x} \left(k \frac{\partial T}{\partial x} \right) + \frac{\partial}{\partial y} \left(k \frac{\partial T}{\partial y} \right) + \frac{\partial}{\partial z} \left(k \frac{\partial T}{\partial z} \right) + q''' = \rho c_p \frac{\partial T}{\partial \tau}.$$ (2.5)

Equation (2.5) is a partial differential equation whose solution within a given conduction region, when subjected to boundary conditions on the surface of the region and an initial temperature distribution throughout the region at time $\tau = 0$, gives the temperature as a function of the space coordinates and time for an isotropic, rigid solid body in a rectangular cartesian coordinate system.[1] It is difficult to solve Eq. (2.5) in its present

---

1. By starting with a control volume of finite size and arbitrary shape within the solid conduction region, and with the aid of Gauss' theorem and the more general Fourier law $q = -\mathbf{K}\nabla T$, a more general equation for a rigid anisotropic solid can be derived; i.e., $\nabla, (\mathbf{K} \nabla T) + q''' = \rho c_p(\partial T/\partial \tau)$. By properly expanding the first term, the equivalent of Eq. (2.5) in any general curvilinear coordinate system can be obtained.

form for exact analytic solutions of T. Hence available exact solutions include a variety of assumptions to simplify Eq. (2.5). A common case is to assume that the thermal conductivity k is independent of temperature and position; therefore, k can be brought outside the differential operators, and the equation divided by k. The group $k/\rho c_p$ is called the *thermal diffusivity* α and has a physical interpretation for transient problems (where temperature depends on space coordinates *and* time) in that it is a measure of the time required for the material to experience some temperature change; for example, as a result of a boundary condition change. Thus, if k is constant, Eq. (2.5) becomes

$$\frac{\partial^2 T}{\partial x^2} + \frac{\partial^2 T}{\partial y^2} + \frac{\partial^2 T}{\partial z^2} + \frac{q'''}{k} = \frac{1}{\alpha}\frac{\partial T}{\partial \tau}. \tag{2.6}$$

Since the sum of the first three terms on the left is the divergence of the gradient vector (i.e., the scalar Laplacian for the rectangular cartesian coordinate system), Eq. (2.6) can be written in vector notation as

$$\nabla^2 T + \frac{q'''}{k} = \frac{1}{\alpha}\frac{\partial T}{\partial \tau}. \tag{2.7}$$

Equation (2.7) can be expanded in other coordinate systems. (Alternatively, the equivalent of Eq. [2.6] or [2.5] in other coordinate systems could be obtained by performing a formal coordinate transformation between the rectangular system and the coordinate system of interest, or by going back to first principles and applying Eq. [1.8] to the appropriate differential control volume in the coordinate system of interest.) The equivalent of Eq. (2.6) in the circular cylindrical coordinate system is

$$\frac{\partial^2 T}{\partial r^2} + \frac{1}{r}\frac{\partial T}{\partial r} + \frac{1}{r^2}\frac{\partial^2 T}{\partial \theta^2} + \frac{\partial^2 T}{\partial z^2} + \frac{q'''}{k} = \frac{1}{\alpha}\frac{\partial T}{\partial \tau}, \tag{2.8}$$

where θ is the angle in the xy plane; and in the spherical coordinate system,

$$\frac{\partial^2 T}{\partial r^2} + \frac{2}{r}\frac{\partial T}{\partial r} + \frac{1}{r^2 \sin\theta}\frac{\partial}{\partial \theta}\left(\sin\theta\,\frac{\partial T}{\partial \theta}\right) + \frac{1}{r^2\sin^2\theta}\frac{\partial^2 T}{\partial \phi^2}$$
$$+ \frac{q'''}{k} = \frac{1}{\alpha}\frac{\partial T}{\partial \tau}, \tag{2.9}$$

where ϕ is the angle in the xy plane, and θ is the angle between r and the z axis. When steady-state conditions prevail, that is, $\partial T/\partial \tau = 0$, Eq. (2.6) becomes

$$\frac{\partial^2 T}{\partial x^2} + \frac{\partial^2 T}{\partial y^2} + \frac{\partial^2 T}{\partial z^2} + \frac{q'''}{k} = 0. \tag{2.10}$$

If the energy generation rate q''' per unit volume depends only upon the space coordinates, Eq. (2.10) is known as the *Poisson equation*. If there is no generation, that is, $q''' = 0$, the resulting equation is *Laplace's equation,*

$$\frac{\partial^2 T}{\partial x^2} + \frac{\partial^2 T}{\partial y^2} + \frac{\partial^2 T}{\partial z^2} = 0. \tag{2.11}$$

2.2a Side Conditions

Readers will recall, from their differential equations course, that associated with the solution of a partial differential equation such as Eq. (2.5) are a number of integration constants that must be determined in order to complete the solution for a particular problem. (More generally, depending upon the equation and solution technique, there may arise functions of integration to be determined.) The integration constants are determined by application of the side conditions which are usually referred to as boundary conditions when applied at a place in space on the body and as initial conditions when applied at an instant of time. The number of integration constants, and therefore the number of boundary and initial conditions needed, is dictated by the order of the highest derivatives in the partial differential equation. Thus, examination of Eq. (2.5) indicates that one initial condition is needed as well as two boundary conditions for each space coordinate.

Initial Condition

The one initial condition needed is the specification of the known temperature throughout the conduction region at some initial instant of time, usually called time $\tau = 0$.
    Hence, at $\tau = 0$, all x, y, z in the conduction region,

$$T = F(x, y, z) \tag{2.11a}$$

where F is a known function which in many cases may simply be a constant; namely, the known temperature of the entire conduction region just as a process begins. In other coordinate systems, the x, y, and z are replaced by the appropriate space coordinates for that system.

Boundary Conditions

Four kinds of boundary conditions are commonly encountered in engineering conduction heat transfer problems. These will be discussed by reference to the bounding surface at $x = x_0$ in Fig. 2.3.

Boundary Condition of the First Kind—Specified Temperature   Sometimes conditions are such that the temperature is known at the bounding surface. This may occur when a fluid bounding the surface does not present much resistance to convection heat transfer

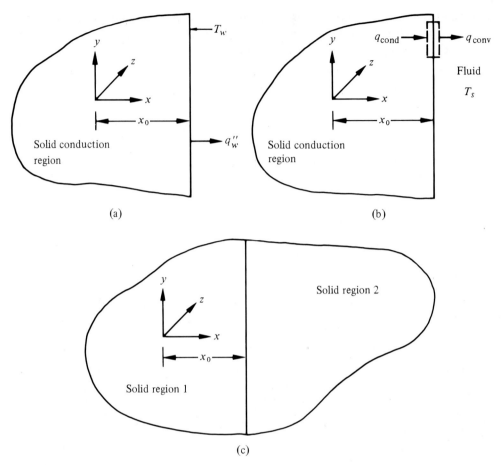

Figure 2.3 Possible boundary conditions on a solid region.

compared to the resistance to conduction within the solid in the x direction. Thus the boundary condition is then, at $x = x_0$, $\tau > 0$, all y and z on the boundary,

$$T = T_w = G(y, z, \tau) \tag{2.11b}$$

where G is a known function, often a constant.

Boundary Condition of the Second Kind—Specified Flux   When the energy being released within a conduction region, due to the electrical energy dissipation or to nuclear or chemical reactions, has no place to go except across the bounding surface at x_0, this fixes the value of the heat flux, the heat transfer rate per unit area, at x_0 because of the law of conservation of energy, Eq. (1.8). Since this known flux, q_w'' in Fig. 2.3a, arrives at

x_0 by conduction from within the solid region, it is related to the temperature gradient at x_0 by Fourier's law of conduction, Eq. (1.3). Hence the appropriate boundary condition becomes, at $x = x_0$, $\tau > 0$, all y and z on the boundary,

$$-k\frac{\partial T}{\partial x} = q''_w = F(y, z, \tau). \qquad (2.11c)$$

The value $F(y, z, \tau)$ is a known function which is most often independent of time τ and sometimes may simply be a constant. This type of boundary condition is often appropriate at the surface of an electric resistive heating element or a nuclear fuel rod. Note that when conditions are such that the flux q''_w is known, then the surface temperature T_w is *not* known at the beginning of the analysis and therefore cannot be used as a boundary condition. Similarly, when a boundary condition of the first kind occurs, in which T_w in Fig. 2.3a is known, then the flux q''_w would not be known at the start of the analysis.

Boundary Condition of the Third Kind—Convective-Type Boundary Condition   Perhaps the most common and realistic type of boundary condition encountered in engineering heat transfer analysis is the one that occurs at an interface between a solid conduction region and a bounding fluid, as shown in Fig. 2.3b, when there is a known surface coefficient of heat transfer h between the solid surface at x_0 and the fluid of undisturbed temperature T_s away from the boundary. The boundary condition for this circumstance can be found by applying the energy balance, Eq. (1.8), to a control volume of practically zero thickness, right at the solid-fluid interface as shown by the dashed control surface in Fig. 2.3b.

When there is no appreciable radiation from the surface, there is a conduction heat transfer rate into the control volume from within the solid region and a convection heat transfer rate out to the fluid. Equation (1.8) will have both R_{gen} and R_{stor} equal to zero because of the fact that there is no actual volume when the control volume is taken right along the interface. This is true even if generation and unsteady conditions exist within the solid region. Hence, setting $R_{in} = R_{out}$, replacing q_{cond} by Eq. (1.3) and q_{conv} by Newton's cooling law, Eq. (1.4), and cancelling out the areas yields the boundary condition at x_0, $\tau > 0$, all y and z on the boundary,

$$-k\frac{\partial T}{\partial x} = h(T - T_s). \qquad (2.11d)$$

Note that in Eq. (2.11d), neither T at x_0 nor $\partial T/\partial x$ at x_0 is known at the start except in terms of the integration constants whose values must be found from application of boundary conditions such as Eq. (2.11d).

Boundary Condition of the Fourth Kind—Conjugation Conditions   When two different solid materials, 1 and 2, meet and are joined intimately at an interface x_0, as shown in Fig. 2.3c, the temperature and the heat flux are continuous at the interface; that is, they have the same values, respectively, in materials 1 and 2. Hence, the appropriate conditions to be forced on the temperature distributions are as follows: at $x = x_0$, $\tau > 0$, all y and z on the interface,

$$T_1 = T_2, \quad -k_1 \frac{\partial T_1}{\partial x} = -k_2 \frac{\partial T_2}{\partial x}. \tag{2.11e}$$

The four different types of boundary conditions discussed in this section certainly do not cover every possible type of condition one may confront. In the event that one of these boundary conditions is not the proper one based upon what is known at the boundary—as for instance when there is significant radiation transfer at the bounding surface—then one can often derive the correct boundary condition by applying the energy balance, Eq. (1.8), to a zero volume control volume right at the surface, as was done in the development of Eq. (2.11d).

2.3 STEADY ONE-DIMENSIONAL HEAT TRANSFER WITHOUT HEAT GENERATION

In this section, we begin to study the solution of the relevant differential equation and the application of boundary conditions in order to arrive at the details of the temperature distribution within a solid material and to find the heat transfer rates of interest.

2.3a The Plane Slab

Consider the plane slab depicted in Fig. 2.4, in which there is no generation, steady-state conditions prevail, the thermal conductivity is constant at k, and the bounding temperatures T_1 and T_2 are known. The temperature distribution as a function of the relevant space coordinates, and the heat flux q'' are to be found.

The slab dimensions in the two remaining coordinate directions are taken to be very large when compared with the dimension L in the x direction; hence, the temperature distribution is one-dimensional. For steady-state problems with no generation and constant thermal conductivity, the governing partial differential equation for the temperature distribution is Eq. (2.11). Since the situation is one-dimensional, $\partial^2 T/\partial y^2 = \partial^2 T/\partial z^2 = 0$, leading to the ordinary differential equation

$$\frac{d^2 T}{dx^2} = 0.$$

Integrating twice,

$$T = c_1 x + c_2,$$

where c_1 and c_2 are constants of integration to be determined by the boundary conditions.

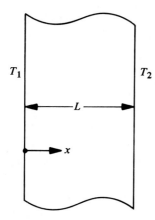

Figure 2.4 One-dimensional conduction through a plane slab.

The boundary conditions are the known temperatures,

$$\text{at } x = 0, \ T = T_1; \quad \text{at } x = L, \ T = T_2.$$

Inserting these conditions into the equation for the temperature distribution,

$$T_1 = 0 + c_2, \quad T_2 = c_1 L + c_2.$$

Solving for c_1 and c_2 and inserting these into the equation for T yields

$$T = (T_2 - T_1) \frac{x}{L} + T_1.$$

Thus, for this simple situation, the temperature linearly depends upon x. To obtain the heat flux, Eq. (1.3) is used replacing n with x. Hence,

$$q'' = -k \frac{\partial T}{\partial x},$$

and since

$$\frac{\partial T}{\partial x} = \frac{dT}{dx} = \frac{d}{dx} \left[(T_2 - T_1) \frac{x}{L} + T_1 \right] = \frac{T_2 - T_1}{L},$$

the heat flux is

$$q'' = k \frac{(T_1 - T_2)}{L}.$$

Note that for some given surface area A perpendicular to the x direction,

$$q = q''A = kA \frac{(T_1 - T_2)}{L}.$$

This result is Fourier's simplified law of conduction, Eq. (1.1). The assumptions implied by Eq. (1.1) can now be delineated. That is, the simplified form of Fourier's law of conduction can be used only when steady-state conditions prevail, no heat generation is present, the thermal conductivity is constant, one-dimensional heat flow prevails, and no area variation exists in the direction of the heat flow.

The overall procedure that was used to find the temperature distribution and heat transfer rate for the slab—starting with the partial differential equation that governs the temperature, solving it, and subjecting that solution to the known boundary conditions—is also usually the way to proceed in more involved and more general problems when the temperature field within a region is to be found.

2.3b The Circular Cylindrical Shell

Figure 2.5 shows an end view of a long circular cylindrical shell or pipe of known inner and outer radii r_1 and r_2, respectively. Steady-state conditions prevail with no generation, and the thermal conductivity k is constant. The outer surface is held at the temperature T_2, while a fluid at temperature T_0 flows inside the pipe (between $r = 0$ and $r = r_1$) with a known surface coefficient of heat transfer h_0 between the fluid and the inside pipe wall. Once again, our objective is to determine the temperature distribution and the heat transfer rate. In the analysis that follows, pay special attention to the way that the boundary condition at the inside surface, the convective-type condition, is handled. The author's experience in teaching heat transfer indicates that the implementation of the convective-type boundary condition is a particular source of difficulty for students.

The appropriate coordinate system is the cylindrical coordinate system, since the boundaries of the conduction region are circular cylinders. For steady state and no generation, Eq. (2.8) reduces to

$$\frac{\partial^2 T}{\partial r^2} + \frac{1}{r} \frac{\partial T}{\partial r} + \frac{1}{r^2} \frac{\partial^2 T}{\partial \theta^2} + \frac{\partial^2 T}{\partial z^2} = 0.$$

Since the shell is long, $\partial^2 T/\partial z^2 = 0$, and the symmetry of the physical system leads to the conclusion that $\partial^2 T/\partial \theta^2 = 0$. The above equation reduces to the ordinary differential equation

$$\frac{d^2 T}{dr^2} + \frac{1}{r} \frac{dT}{dr} = 0.$$

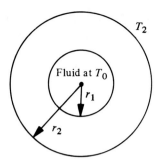

Figure 2.5 One-dimensional radial conduction through a circular cylindrical shell.

Since no term in T alone appears, the order of the equation can be reduced by the substitution $S = dT/dr$. Hence, $d^2T/dr^2 = dS/dr$, and the equation becomes

$$\frac{dS}{dr} + \frac{S}{r} = 0.$$

This linear, first-order, and homogeneous equation with variable coefficients can be solved by the method of the integrating factor. However, the formal method does not have to be applied, since multiplication by $r\, dr$ yields

$$r\, dS + S\, dr = 0 = d(Sr).$$

Integration leads to $Sr = c_1$. Since $S = dT/dr$, $dT/dr = c_1/r$, and integrating once more,

$$T = c_1 \ln r + c_2. \tag{2.12}$$

One boundary condition is that at $r = r_2$, $T = T_2$. This gives an equation relating c_1 and c_2:

$$T_2 = c_1 \ln r_2 + c_2. \tag{2.13}$$

The second boundary condition might be considered to be that at $r = r_1$, $T = T_1$. However, T_1 is not a given quantity, so application of this condition combined with Eqs. (2.12–13) would yield the temperature distribution with an unknown T_1 still in it. The fluid temperature T_0 and the surface coefficient h_0 are known. Note that T_1 does not equal T_0 unless h_0 is infinitely large, so that an energy transfer rate from the fluid to the inside surface can be accomplished by Newton's cooling law with a zero temperature difference. The boundary condition that should be used is the convective boundary condition at the inner surface. The rate at which energy is transferred by convection from the fluid to the surface at r_1 must equal the rate at which it is conducted into the solid at r_1. The general form required here is Eq. (2.11d), where the x_0 corresponds to r_1 in this analysis. Hence, at $r = r_1$,

$$h_0 2\pi r_1(1)(T_0 - T_1) = -k 2\pi r_1(1)(\partial T/\partial r)_{r_1},$$

where the left side is Newton's law of cooling, the right side is Fourier's exact law of conduction, and (1) refers to an axial length of cylinder. Note that the subscript r_1 on the derivative $\partial T/\partial r$ indicates that this derivative must be evaluated at the interface where $r = r_1$. Simplifying,

$$h_0(T_0 - T_1) = -k(\partial T/\partial r)_{r_1} . \tag{2.14}$$

For Eq. (2.14) to yield another relationship between c_1 and c_2, Eq. (2.12) must be used; that is, differentiating Eq. (2.12),

$$dT/dr = c_1/r.$$

Hence, at $r = r_1$,

$$(dT/dr)_{r_1} = c_1/r_1. \tag{2.15}$$

Now, T_1 is found in terms of c_1 and c_2 by substituting $r = r_1$ in Eq. (2.12),

$$T_1 = c_1 \ln r_1 + c_2. \tag{2.16}$$

Equation (2.14), after inserting Eqs. (2.15) and (2.16), becomes

$$h_0(T_0 - c_1 \ln r_1 - c_2) = -kc_1/r_1. \tag{2.17}$$

Equations (2.17) and (2.13) are solved simultaneously for c_1 and c_2 to yield

$$c_1 = -\frac{(T_0 - T_2)}{\ln(r_2/r_1) + (k/h_0 r_1)} \quad \text{and}$$

$$c_2 = T_2 - \frac{(T_0 - T_2) \ln r_2}{\ln(r_2/r_1) + (k/h_0 r_1)} .$$

Inserting these equations into Eq. (2.12), the temperature as a function of r for the conduction region between r_1 and r_2 is

$$T = -\frac{(T_0 - T_2)}{\ln(r_2/r_1) + (k/h_0 r_1)} \ln\left(\frac{r}{r_2}\right) + T_2. \tag{2.18}$$

The heat transfer rate per unit length of shell is obtained from Fourier's exact law of conduction

$$q' = -k2\pi r(1)\frac{dT}{dr}, \tag{2.19}$$

where q' is the heat transfer rate per unit length at any r. Differentiating Eq. (2.18) and combining with Eq. (2.19) gives

$$q' = \frac{2\pi k(T_0 - T_2)}{\ln(r_2/r_1) + (k/h_0 r_1)} . \tag{2.20}$$

It is of interest to note what happens as $h_0 \to \infty$. From Eq. (2.18), this implies that $T_1 \to T_0$ and, thus, the inside surface temperature would be known. The second term in the denominator of Eq. (2.20) becomes zero, leading to

$$q' = \frac{2\pi k(T_1 - T_2)}{\ln(r_2/r_1)} . \tag{2.21}$$

Equation (2.21) is the expression for q' across a circular cylindrical shell in the steady state with no generation, constant thermal conductivity, and one-dimensional radial heat flow. When T_1 and T_2 are not known, additional equations would be needed to solve for the unknowns. Equation (2.21) might be called *Fourier's simplified law of conduction for a purely radial system* and is the analogue of Eq. (1.1), which is for a planar one-dimensional system. Equation (2.21) is an important result which is often needed when, for example, one is analyzing heat transfer across pipes made of more than one material and across pipe walls which separate two fluids.

As was done in Chapter 1 for the planar geometry, Eq. (2.21) can be rearranged to a form which displays the thermal resistance to heat transfer, r_t, explicitly for the radial system as

$$q' = \frac{T_1 - T_2}{\ln(r_2/r_1)/2\pi k} = \frac{T_1 - T_2}{r_t} . \tag{2.21a}$$

2.3c The Spherical Shell

The last commonly encountered coordinate system in heat transfer problems is the one most suited to the spherical geometry. Hence, we will consider next the problem of the temperature profile and the heat transfer rate in a spherical shell of inside radius r_1 where the temperature is known to be T_1, and outside radius r_2 where the known value of the temperature is T_2. Steady-state conditions prevail, the thermal conductivity k is constant, and there is no generation. The geometry dictates that one start with the general conduction equation for spherical coordinates, Eq. (2.9).

The governing partial differential equation is

$$\frac{\partial^2 T}{\partial r^2} + \frac{2}{r}\frac{\partial T}{\partial r} + \frac{1}{r^2 \sin\theta}\frac{\partial}{\partial \theta}\left(\sin\theta \frac{\partial T}{\partial \theta}\right) + \frac{1}{r^2 \sin^2\theta}\frac{\partial^2 T}{\partial \phi^2}$$

$$+ q'''/k = \frac{1}{\alpha}\frac{\partial T}{\partial \tau} .$$

Since there is no generation and steady-state conditions prevail, $q''' = \partial T / \partial \tau = 0$. Also, since there are unbalanced thermal loadings (i.e., T_1 not equaling T_2) only in the radial direction, all terms, except those containing derivatives with respect to r, drop out. Hence,

$$\frac{d^2T}{dr^2} + \frac{2}{r}\frac{dT}{dr} = 0. \qquad (2.22)$$

As was the case for a similar equation in the circular cylindrical coordinate system, the transformation

$$S = dT/dr \qquad (2.23)$$

reduces Eq. (2.22) to the first-order equation

$$dS/dr + 2S/r = 0.$$

Separating the variables,

$$dS/S = -2dr/r.$$

Integrating,

$$\ln S = -2 \ln r + \ln C_1 = -\ln r^2 + \ln C_1.$$

Rearranging and taking antilogarithms,

$$Sr^2 = C_1. \qquad (2.24)$$

Substituting Eq. (2.23) into (2.24) and rearranging,

$$dT/dr = C_1/r^2.$$

Integrating,

$$T = -C_1/r + C_2.$$

With the boundary conditions at $r = r_1$, $T = T_1$, and at $r = r_2$, $T = T_2$, the temperature distribution as a function of radius r is

$$T = T_1 + \frac{(T_2 - T_1)(1/r_1 - 1/r)}{1/r_1 - 1/r_2}. \qquad (2.25)$$

The heat transfer rate at any radius r is given by Fourier's exact law of conduction as

$$q = -kA(dT/dr),$$

since $q'' = q/A$ and $n = r$ in Eq. (1.3). Differentiating Eq. (2.25) with respect to r,

$$\frac{dT}{dr} = \frac{(T_2 - T_1)(1/r^2)}{1/r_1 - 1/r_2} .$$

For the sphere at the general radius r, the area A for heat flow is

$$A = 4\pi r^2.$$

When we combine these last three equations, the heat transfer rate across the spherical shell is

$$q = \frac{4\pi k(T_1 - T_2)}{1/r_1 - 1/r_2} . \tag{2.26}$$

Equation (2.26) is a simplified form of Fourier's law of conduction for a spherical coordinate system and is analogous to Eq. (1.1) for a planar system in rectangular cartesian coordinates.

Equation (2.26) can also be rearranged in a form that is analogous to Ohm's law and which lends itself to interpretation in terms of a circuit element; i.e.,

$$q = \frac{T_1 - T_2}{(1/4\pi k)(1/r_1 - 1/r_2)} . \tag{2.27}$$

In this form, q is interpreted as the current flowing across the driving potential difference $T_1 - T_2$ through the thermal resistance r_t, where

$$r_t = \frac{1}{4\pi k}\left(\frac{1}{r_1} - \frac{1}{r_2}\right).$$

This last form is particularly convenient for composite spherical shell barriers.

2.3d Variable Thermal Conductivity

In this section, we will illustrate the overall approach to be taken in the analysis of steady-state one-dimensional problems with no generation when the material's thermal conductivity k is variable. We will do this by deriving a result for a pipe which has implications for other geometries as well.

Figure 2.6 The differential control volume.

Consider a circular pipe of inner radius r_1 where the temperature is T_1, and outer radius r_2 where the known temperature is T_2. The pipe's thermal conductivity k varies linearly with temperature according to the prescription

$$k = k_0 (1 + bT) \tag{2.28}$$

where k_0 is a known positive constant and b is a known constant. (The actual variation of k with T for many real materials can often be adequately represented as a linear variation over a significant temperature range around the expected temperatures within the material.) In order to find the temperature distribution through the pipe wall and the heat transfer rate per foot of pipe length, the governing differential equation is needed in the circular cylindrical coordinate system. Equation (2.5) is valid for variable k but is not in the form for easy application in the cylindrical coordinate system. Since Eq. (2.8) is limited to constant k, we instead will quickly derive the needed differential equation.

In circular cylindrical coordinates when the temperature depends only upon r, the appropriate control volume is the shell of inner radius r and thickness dr shown in Fig. 2.6. This control volume is imbedded within a solid conduction region. Application of the law of conservation of energy, Eq. (1.8), leads to $R_{\text{in}} + R_{\text{gen}} = R_{\text{out}}$, since $R_{\text{stor}} = 0$ by the steady-state condition imposed. Energy enters by conduction at r at a rate q_r, and leaves the control volume by conduction at a rate q_{r+dr}, which is put in terms of q_r by a Taylor series expansion of the function about r and across to $r + dr$. Hence,

$$R_{\text{out}} = q_{r+dr} = q_r + \frac{dq_r}{dr}\, dr.$$

(Note that, unlike the more general derivations at the beginning of this chapter, partial derivative notation is not required here, since there is only one independent variable, r.) To allow for increased generality, the generation term will be retained during the derivation of the equation.

Once again denoting q''' as the function for the energy generation rate per unit volume, $R_{\text{gen}} = q''' \, 2\pi r \, dr$. Hence, Eq. (1.8) becomes

$$q_r + q''' \, 2\pi r \, dr = q_r + \frac{dq_r}{dr}\, dr, \quad \text{or} \quad 2\pi r q''' = \frac{dq_r}{dr}. \tag{2.29}$$

But Fourier's exact law of conduction $q = -k\,A(dT/dr)$ gives $q_r = -k\,2\pi r(dT/dr)$. Thus,

$$\frac{dq_r}{dr} = -2\pi \frac{d}{dr}\left(rk\frac{dT}{dr}\right). \tag{2.30}$$

(Note that k cannot be brought outside of the differential operator because here k is a variable. Hence, it either depends upon r explicitly, as in the case of a nonhomogeneous material, or implicitly in the more usual situation where k is given as a function of temperature.) Combining Eqs. (2.29) and (2.30) and rearranging,

$$\frac{d}{dr}\left(rk\frac{dT}{dr}\right) + rq''' = 0. \tag{2.31}$$

Equation (2.31) holds for steady, radial heat flow in the circular cylindrical coordinate system when the thermal conductivity k is variable and when there is generation per unit volume, q''', within the material.

For the case at hand, where there is no generation, $q''' = 0$. With this, Eq. (2.31) reduces to the form which must be solved,

$$\frac{d}{dr}\left(rk\frac{dT}{dr}\right) = 0.$$

Integrating,

$$rk\frac{dT}{dr} = C_1 \quad \text{or} \quad k\frac{dT}{dr} = \frac{C_1}{r}.$$

Since $k = k_0(1 + bT)$,

$$\int k_0(1 + bT)dT = \int C_1 \frac{dr}{r} + C_2.$$

Integrating,

$$k_0(T + \tfrac{1}{2}bT^2) = C_1 \ln r + C_2.$$

Substituting the boundary conditions at $r = r_1$, $T = T_1$ and at $r = r_2$, $T = T_2$, leads to

$$T + \tfrac{1}{2}bT^2 = \frac{-[T_1 - T_2 + \tfrac{1}{2}b\,(T_1^2 - T_2^2)]\,\ln(r/r_2)}{\ln(r_2/r_1)} \tag{2.32}$$

$$+\ T_2 + \tfrac{1}{2}bT_2^2.$$

Thus, for any r between r_1 and r_2, Eq. (2.32) can be solved for the temperature at that position, or for a result which is explicit in T as a function of r, the quadratic formula can be applied to Eq. (2.32). To obtain the heat transfer rate per unit length, Fourier's exact law of conduction $q = -k\,2\pi r(dT/dr)$ is used. The derivative with respect to r of Eq. (2.32) is

$$(1 + bT)\frac{dT}{dr} = -\frac{1}{r}\frac{T_1 - T_2 + \frac{1}{2}b(T_1^2 - T_2^2)}{\ln(r_2/r_1)} \quad,$$

or

$$\frac{dT}{dr} = -\frac{1}{r(1 + bT)}\frac{T_1 - T_2 + \frac{1}{2}b\,(T_1^2 - T_2^2)}{\ln(r_2/r_1)} .$$

Multiplying by $-2\pi rk$, where $k = k_0(1 + bT)$,

$$q = 2\pi k_0 \frac{T_1 - T_2 + \frac{1}{2}b\,(T_1^2 - T_2^2)}{\ln(r_2/r_1)} .$$

Finally, factoring out $T_1 - T_2$,

$$q = 2\pi k_0[1 + \frac{1}{2}b\,(T_2 + T_1)]\frac{(T_1 - T_2)}{\ln(r_2/r_1)} . \qquad (2.33)$$

Note that if $b = 0$, the constant thermal conductivity case, the expression reduces to Eq. (2.21), as it should. In addition, the term $k_0[1 + \frac{1}{2}b\,(T_2 + T_1)]$ is simply the arithmetic average of the thermal conductivities at T_1 and T_2, the temperatures at the inside and outside of the shell. Thus Eq. (2.33) can serve as the basis for handling composite circular cylindrical shells if the thermal conductivity has the functional dependence upon temperature given by Eq. (2.28).

Hence, as long as the thermal conductivity varies *linearly* with temperature and as long as the process is steady, with no generation, and one-dimensional, the effect of variable thermal conductivity can be accounted for by the rather simple expedient of using the arithmetic average thermal conductivity in the constant conductivity result for the heat transfer rate, Eq. (2.21). It can also be proven that this procedure can be used for the planar and spherical problems if the previously mentioned conditions are satisfied. Note, however, by comparing Eq. (2.32) with Eq. (2.18), when $h_0 \rightarrow \infty$, that this simplicity of employing the arithmetic average value of k does *not* carry over to the actual temperature distribution.

EXAMPLE 2.1

A metal pistol barrel has the shape of a hollow right circular cylinder of inner radius r_1 and outer radius r_2, and its length is large compared to the outer radius. The barrel is mounted in a lathe undergoing a final machining operation at its inside radius, which dissipates mechanical energy at the rate W_0 per unit length of barrel while the outside of the barrel is held at the temperature T_0 by a spray of cooling fluid. If steady-state conditions prevail and if the temperature distribution in the barrel is primarily radial, derive an expression for the steady-state temperature distribution by beginning with Eq. (2.8). The thermal conductivity k is a known constant.

Solution
Since the temperature distribution is steady and radial with no generation (the mechanical work is being dissipated at the *surface* of the conduction region, not *within* it), Eq. (2.8) reduces to

$$\frac{d^2T}{dr^2} + \frac{1}{r}\frac{dT}{dr} = 0. \qquad \frac{1}{r}\frac{\partial}{\partial r}\left(r\frac{\partial T}{\partial r}\right) = 0 \qquad (2.34)$$

Equation (2.34) was solved in Sec. 2.3b with the result being,

$$T = C_1 \ln r + C_2. \qquad [2.12]$$

Two boundary conditions are needed to evaluate the integration constants C_1 and C_2. The boundary condition at the outer surface which is held at a known temperature by the coolant spray is that at $r = r_2$, $T = T_0$. Using this in Eq. (2.12),

$$C_2 = T_0 - C_1 \ln r_2. \qquad (2.35)$$

Combining Eqs. (2.35) and (2.12),

$$T = T_0 + C_1 \ln r - C_1 \ln r_2. \qquad (2.36)$$

At r_1, the mechanical energy dissipated by the machining goes into the barrel and along the machine tool. Most of it goes into the barrel, since it presumably is dissipated uniformly along the length of the barrel. This type of dissipation tends to set up radial temperature gradients rather than axial temperature gradients along the tool itself. Hence, the rate at which mechanical energy W_0 is dissipated per unit length at r_1 must equal the rate at which it is conducted away into the gun barrel at r_1. This rate is given by Fourier's exact law of conduction as

$$q = -kA\left.\frac{dT}{dr}\right|_{r_1} = -k2\pi r_1\left.\frac{dT}{dr}\right|_{r_1},$$

and since $q = W_0$,

$$W_0 = -k2\pi r_1\left.\frac{dT}{dr}\right|_{r_1}. \qquad (2.37)$$

Equation (2.37) is an example of a boundary condition of the second kind, the specified flux condition, once both sides of Eq. (2.37) are divided by $2\pi r_1$. Thus it represents a specific example of the general condition given by Eq. (2.11c).

From Eq. (2.36), upon differentiation with respect to r and evaluation at r_1,

$$\left.\frac{dT}{dr}\right|_{r_1} = \frac{C_1}{r_1}. \tag{2.38}$$

Combining Eqs. (2.37) and (2.38) and solving for C_1,

$$C_1 = -(W_0/2\pi k). \tag{2.39}$$

After combining Eqs. (2.36) and (2.39), the final result for the radial temperature distribution is

$$T = T_0 - (W_0/2\pi k)\ln(r/r_2).$$

In all likelihood, the quantity of most interest in this problem, and the reason for solving it in the first place, would be the maximum temperature of the pistol barrel during the machining process. Obviously, this will occur at the inside radius, r_1, where the lathe tool, possibly a reamer, is working against the barrel. Hence, setting $r = r_1$ and rearranging gives the maximum temperature as

$$T_{\max} = T_0 + (W_0/2\pi k)\ln(r_2/r_1).$$

EXAMPLE 2.2

Rework Ex. 2.1 by starting with Fourier's exact law of conduction at some general radial coordinate, assuming that the heat transfer rate per unit length is known and constant at the value W_0.

Solution
Fourier's exact law of conduction for the heat transfer rate per unit length at a general radial coordinate r is

$$q = -k2\pi r(dT/dr). \tag{2.40}$$

Since q is constant at the value W_0, Eq. (2.40) becomes

$$W_0 = -2\pi kr(dT/dr).$$

Since W_0 and k are constants, the variables may be separated to yield

$$-\frac{W_0}{2\pi k}\frac{dr}{r} = dT. \tag{2.41}$$

Integrating Eq. (2.41) between the general radial coordinate r where the temperature is the general one T, and the outside radius r_2 where the temperature is known to be T_0,

$$\int_T^{T_0} dT = -\frac{W_0}{2\pi k} \int_r^{r_2} \frac{dr}{r},$$

or

$$T = T_0 - \frac{W_0}{2\pi k} \ln\left(\frac{r}{r_2}\right). \tag{2.42}$$

Note that this seemingly simpler procedure yields the same result as Ex. 2.1, since q is not a function of r and its value is known.

The procedure illustrated in this example, namely, integrating Fourier's law of conduction directly, is actually more general than the conditions of this example show. It can be used even when the thermal conductivity is variable and when the area perpendicular to the heat flow direction varies, as long as the situation is *steady state, one dimensional,* and with *no generation* so that the heat transfer rate q does not vary with the space coordinate. This procedure will be discussed in more detail in the section on quasi-one-dimensional conduction later in the chapter.

EXAMPLE 2.3

The pistol barrel of Ex. 2.2 is made of steel, and $k = 35$ W/m °C with an inside diameter of 0.8 cm and an outside diameter of 2 cm. The machining operation causes a mechanical energy dissipation rate of $W_0 = 9000$ W per meter of barrel length. Find the highest temperature of the barrel during the machining operation if the coolant is at 15°C.

Solution
The maximum temperature of the barrel occurs at the inside radius r_1, so that from Eq. (2.42),

$$T_{\text{max}} = T_0 - \frac{W_0}{2\pi k} \ln\left(\frac{r_1}{r_2}\right).$$

Since $T_0 = 15$°C, $W_0 = 9000$ W/m, $k = 35$ W/m °C, and r_1/r_2 is in the same ratio as the diameters,

$$T_{\text{max}} = 60 - \frac{9000}{2\pi(35)} \ln\left(\frac{.8}{2}\right) = 97.5°C.$$

Actually, this calculated maximum temperature is some *average* inside surface temperature, since the mechanical energy dissipated in the machining operation is not distributed uniformly over the inside surface of the barrel, but instead is being dissipated at much smaller regions where the metal and the tool actually make contact.

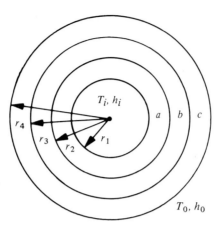

Figure 2.7 Radial conduction across a composite pipe.

2.3e Conduction Across Composite Barriers and Electric Analogy Interpretation

In heat transfer analysis and design work, one often has to contend with heat transfer across a number of different barriers, usually in series, but sometimes in parallel, between the regions exchanging energy. These barriers might be different solids or fluids or, most commonly, they might be composed of both solids and fluids. Important situations where this may occur include heat exchangers, pipelines, and the walls of buildings.

We begin by considering the important and frequently encountered case of heat transfer across the composite pipe shown in Fig. 2.7, where an end view of a pipe constructed of three different materials a, b, and c is depicted. The thermal conductivities k_a, k_b, and k_c are known, and the solid materials are in good contact at their interfaces with each other.

Material a could be a structural member, such as steel pipe, while materials b and c could be insulators used to reduce the heat transfer rate. A fluid at temperature T_i and surface coefficient of heat transfer h_i flows inside the pipe in the region $0 \leq r \leq r_1$ while another fluid at temperature T_0 with a surface coefficient h_0 flows over the outside of the composite pipe. If steady-state conditions prevail with no generation and radial heat flow only, we would like to derive an expression, in terms of known quantities, for the heat transfer rate per unit length from the inside fluid to the outside fluid, and eventually find any temperatures of interest as well.

The known quantities are T_i, T_0, h_i, h_0, k_a, k_b, k_c, r_1, r_2, r_3, and r_4. Note that the temperatures T_1, T_2, T_3, and T_4 at the corresponding radii are not known. Energy is transferred by convection from the inside fluid to the solid surface at r_1 in accordance with Newton's cooling law

$$q' = h_i 2\pi r_1 (T_i - T_1). \tag{2.43}$$

The heat transfer rate q' per unit length cannot be determined from this equation alone because T_1 is not known. However, this same q' must also be the rate at which energy is transferred by conduction across cylindrical shell a. This is made clear by the application of the law of conservation of energy, Eq. (1.8), taking shell a as the control volume and using the conditions of this problem: steady state, no heat generation, and one-dimensional heat flow. This reasoning is then extended to show that for this problem q' is independent of the local radius r. Using Eq. (2.21) on shell a,

$$q' = 2\pi k_a \frac{(T_1 - T_2)}{\ln(r_2/r_1)}. \tag{2.44}$$

Similarly, for shells b and c,

$$q' = 2\pi k_b \frac{(T_2 - T_3)}{\ln(r_3/r_2)}, \tag{2.45}$$

$$q' = 2\pi k_c \frac{(T_3 - T_4)}{\ln(r_4/r_3)}. \tag{2.46}$$

Using Newton's cooling law between the outside surface and the outside fluid,

$$q' = 2\pi r_4 h_0 (T_4 - T_0). \tag{2.47}$$

Equations (2.43–47) constitute five equations in the five unknowns q', T_1, T_2, T_3, and T_4. To solve for q', Eqs. (2.43–47) are solved for the temperature differences and then added to yield

$$q' = \frac{(T_i - T_0)}{\dfrac{1}{2\pi h_i r_1} + \dfrac{\ln(r_2/r_1)}{2\pi k_a} + \dfrac{\ln(r_3/r_2)}{2\pi k_b} + \dfrac{\ln(r_4/r_3)}{2\pi k_c} + \dfrac{1}{2\pi h_0 r_4}}. \tag{2.48}$$

Thus q' is in terms of known quantities. The form of Eq. (2.48) suggests an electrical analogue in which q' is the current flowing between two potentials T_i and T_0 separated by five thermal resistances in series whose values are the individual terms in the denominator of Eq. (2.48). This interpretation was discussed in connection with the simplest conduction problems in developing the concept of the internal thermal resistance to heat transfer, due to the finite thermal conductivity of material, in Eq. (1.1a), (2.21a), and (2.27). A simple rearrangement of Newton's cooling law, Eq. (1.4), leads to the concept of the external resistance to heat transfer due to a finite value of the surface coefficient of heat transfer h, shown next.

$$q = \frac{\Delta T}{1/hA} = \frac{\Delta T}{r_e} \tag{2.49}$$

Steady-State Conduction

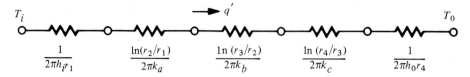

$$T_i \qquad\qquad \xrightarrow{\quad q' \quad} \qquad\qquad T_0$$

$$\underset{2\pi h_i r_1}{\dfrac{1}{}} \qquad \underset{2\pi k_a}{\dfrac{\ln(r_2/r_1)}{}} \qquad \underset{2\pi k_b}{\dfrac{\ln(r_3/r_2)}{}} \qquad \underset{2\pi k_c}{\dfrac{\ln(r_4/r_3)}{}} \qquad \underset{2\pi h_o r_4}{\dfrac{1}{}}$$

Figure 2.8 Electrical analogue of the heat transfer process of Fig. 2.7.

The electrical network *suggested* by the heat transfer result, Eq. (2.48), is shown in Fig. 2.8. The proof of the validity of the circuit shown in Fig. 2.8 is simply that the solution of the circuit for the "current," q', by standard electric network analysis gives the result, Eq. (2.48), that is *known* to be correct because it was derived by applying the proper relationships from heat transfer. Thus with experience one can put down the circuit directly, for a situation such as that shown in Fig. 2.7, and solve it for q' *without* going through the intermediate steps given by Eqs. (2.43–47). It should always be borne in mind, however, that this simple interpretation in terms of electric network elements can only be used for steady-state, zero generation, one-dimensional heat flow.

Once q' has been found from Eq. (2.48), any of the temperatures T_1, T_2, T_3, and T_4 can be found by properly combining Eqs. (2.43–47) or by applying electric network principles to the circuit given in Fig. 2.8. Furthermore, if the temperature at some r between, say, r_2 and r_3 is required, it can be found by solving the equation

$$\frac{d^2T}{dr^2} + \frac{1}{r}\frac{dT}{dr} = 0,$$

subject to the boundary conditions that at $r = r_2$, $T = T_2$, and at $r = r_3$, $T = T_3$. This yields

$$T = -(T_2 - T_3)\frac{\ln(r/r_3)}{\ln(r_3/r_2)} + T_3.$$

Note that this result could have been deduced directly by proper interpretation of Eq. (2.18).

Equation (2.48) can be generalized for any number of circular cylindrical shells. Letting n be the number of different shells and using j as an index, we have the following equation for the steady-state heat transfer rate per unit length with no generation, from an inside fluid at T_i and surface coefficient h_i to an outside fluid at temperature T_0 and surface coefficient h_0, across a composite barrier of n circular cylindrical shells, when the energy flow is purely radial:

$$q' = \frac{(T_i - T_0)}{\dfrac{1}{2\pi h_i r_1} + \displaystyle\sum_{j=1}^{n} \dfrac{\ln(r_{j+1}/r_j)}{2\pi k_j} + \dfrac{1}{2\pi h_0 r_{n+1}}}. \qquad\qquad (2.50)$$

The individual terms in the denominator of Eq. (2.50) are interpreted as being thermal resistances to heat transfer due either to the finite conductivities of the solids (the "internal" resistances) or to the finite values of the surface coefficients, h (the "external" resistances). Thus Eq. (2.50) can be rewritten in more compact form as follows:

$$q' = \frac{T_i - T_0}{\Sigma r_t}.$$ (2.51)

The denominator of Eq. (2.51), Σr_t, obviously represents the sum of all the thermal resistances to heat transfer between the points at which the temperature potentials, T_i and T_0, are known. Another compact form of Eq. (2.51) which is sometimes used is

$$q' = UA' (T_i - T_0).$$ (2.52)

Here A' is the area per unit of tube length and U is referred to as the *overall heat transmission coefficient* whose value would be found by equating Eq. (2.52) and (2.50) and solving for U in terms of h_i, k_j, r_{j+1}, etc. Thus the overall heat transmission coefficient U, like the sum of the thermal resistances, Σr_t, is a quantity defined, for convenience and compactness of form, in terms of the actual surface coefficients, thermal conductivities, areas, and lengths which occur in particular problems. This U concept is extensively used in heat exchanger analysis, as we will see in Chapter 8. It is also used by heating, air conditioning, and refrigeration engineers in their calculations of heating, cooling, and refrigeration loads.

Thus far in this section, the composite barriers to heat transfer have been in series. But parallel barriers occur as well. To develop the solution technique for parallel barriers, consider the situation shown in Fig. 2.9, where the slab of thickness Δx is made of a different material on the top than on the bottom and the slab of thickness Δx_c is a single material. The temperature T_1 is being held on the left face, whereas the known temperature T_3 is on the right-hand face. The thermal conductivities k_a, k_b, and k_c are all constant, A_a, A_b, and A_c are the areas seen in the direction of the thickness for the slabs, steady-state conditions prevail, and there is no generation. The problem is to determine the heat transfer rate across the slab of thickness Δx. Once this expression is found, the methods developed earlier will allow us to easily combine it with a suitable expression for slab c, thus giving the heat transfer rate across the entire composite slab.

Actually, a *two*-dimensional temperature distribution across the slabs might be expected as a result of $k_a \neq k_b$, despite the fact that T_1 and T_3 are uniform over the respective bounding surfaces. That is, if k_b is much less than k_a, then some of the energy entering A_b at the left face which is at T_1 finds it easier to reach the right face by flowing across the boundary separating a from b, even though this entails a longer path of travel to T_2 than across Δx. However, resistance to heat transfer also depends upon the k encountered, and the longer path length can be balanced by the much higher k in material

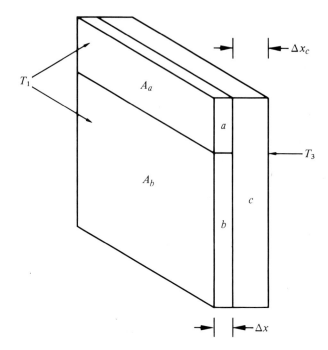

Figure 2.9 Parallel and series barriers to heat transfer.

a. In this problem, it will be assumed that k_a and k_b do not differ enough to cause a severe departure from a one-dimensional flow pattern in the direction Δx. In this case, Fourier's simplified law of conduction can be applied separately to materials a and b; thus,

$$q_a = k_a A_a \frac{(T_1 - T_2)}{\Delta x}, \quad q_b = k_b A_b \frac{(T_1 - T_2)}{\Delta x},$$

where the unknown temperature at the interface of the first slab with slab c has been called T_2.

Since the desired quantity is the total q across the slab, which is the sum of q_a and q_b,

$$q = q_a + q_b = \left(\frac{1}{\Delta x/k_a A_a} + \frac{1}{\Delta x/k_b A_b} \right)(T_1 - T_2). \tag{2.53}$$

Equation (2.53) can be interpreted in terms of an electric circuit, since the two thermal resistances $\Delta x/k_a A_a$ and $\Delta x/k_b A_b$ appear in the same way as two electric resistors in parallel would appear in Ohm's law. The circuit is shown in Fig. 2.10.

Now that the electric circuit which is analogous to the heat transfer process across the *parallel* barriers has been found, as shown in Fig. 2.10 which represents Eq. (2.53), the addition of the thermal resistance to heat transfer across slab c, which is in series with

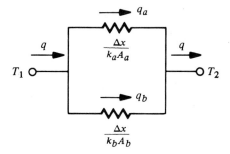

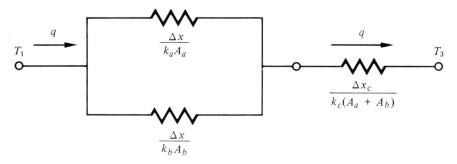

Figure 2.10 Electric circuit for the slab of thickness Δx.

Figure 2.11 The complete circuit for Fig. 2.9.

the element of Fig. 2.10, completes the circuit as is shown in Fig. 2.11. Application of electric network analysis to the circuit in Fig. 2.11 leads to the expression for q in terms of given quantities in the problem.

EXAMPLE 2.4

Steam at 120°C flows inside a pipe which has an inside radius of 5 cm and an outside radius of 5.25 cm. The surface coefficient of heat transfer between the steam and the inside pipe wall is 2800 W/m² °C. The outside of the 1% carbon steel pipe is exposed to atmospheric air at 26°C with a surface coefficient of heat transfer of 9 W/m² °C. Steady state prevails with no generation. Find the heat transfer rate per meter from the steam to the air across the pipe.

Solution
This is a special case of Eq. (2.50) with $n = 1$. Thus,

$$q' = \frac{T_i - T_0}{1/2\pi h_i r_1 + \ln(r_2/r_1)/2\pi k + 1/2\pi h_0 r_2}.$$

Substituting known quantities with $k = 43$ W/m °C from Table A.1.a,

$$q' = \frac{2\pi(120 - 26)}{\dfrac{1}{2800(.05)} + \dfrac{\ln(5.25/5)}{43} + \dfrac{1}{9(.0525)}}$$

$$= \frac{2\pi(94)}{0.0071 + 0.00113 + 2.12} = 278 \text{ W/m.}$$

An examination of the numerical values of the thermal resistances in the denominator of this equation leads to the conclusion that the last term, due to the outside surface coefficient of heat transfer, is predominant and is called the *controlling resistance* in Ex. 2.4. Thus, to increase the heat transfer rate per meter in Ex. 2.4 without changing the driving temperature difference, the first two resistances would be neglected, since even making them zero has a negligible effect upon q'. Rather, an attempt should be made to reduce the largest, controlling resistance, most likely by adding fins to the outside of the pipe. Fins will be discussed later in this chapter.

EXAMPLE 2.5

The wall of a guest house is constructed as shown in Fig. 2.12(a) with two different materials forming the inside wall to achieve a wainscotting effect. The temperature of the air in the room is 70°F and the surface coefficient of heat transfer between the room air and the wall is 1.1 Btu/hr ft² °F. The outside temperature is 20°F with an outside surface coefficient of heat transfer of 3 Btu/hr ft² °F. The wall is 7 ft high and 13 ft deep. The lower material extends three feet from the floor. The thicknesses and thermal conductivities are $\Delta x_a = 1/2$ in., $k_a = 0.09$, $\Delta x_b = 1/2$ in., $k_b = 0.12$, $\Delta x_c = 2$ in., $k_c = 0.023$, $\Delta x_d = 1$ in., and $k_d = 0.10$. Calculate the heat transfer rate across the wall in the steady state.

Solution

Using the ideas developed previously concerning analogous electric networks for combined conduction and convection through one-dimensional composite barriers, the circuit shown in Fig. 2.12(b) can be drawn.

Since $A_a = 3(13) = 39$ ft², $A_b = 4(13) = 52$ ft², and $A_d = 39 + 52 = 91$ ft², the resistances are

$$\frac{1}{h_i(A_a + A_b)} = \frac{1}{1.1(91)} = \frac{1}{100} = 0.01, \qquad \frac{\Delta x_a}{k_a A_a} = \frac{(\tfrac{1}{2}/12)}{0.09(39)} = 0.0119,$$

$$\frac{\Delta x_b}{k_b A_b} = \frac{(\tfrac{1}{2}/12)}{0.12(52)} = 0.0067, \qquad \frac{\Delta x_c}{k_c A_c} = \frac{(2/12)}{0.023(91)} = 0.0798,$$

$$\frac{\Delta x_d}{k_d A_d} = \frac{(1/12)}{0.10(91)} = 0.00915, \qquad \frac{1}{h_0 A_d} = \frac{1}{3(91)} = 0.00366.$$

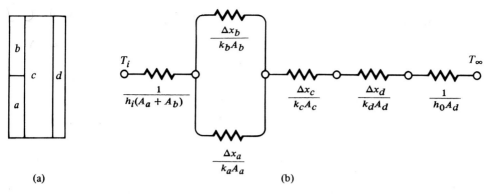

(a) (b)

Figure 2.12 (a) Wall of the guest house of Ex. 2.5. (b) Electrical analogue of heat transfer process through the wall.

The equivalent resistance of the two resistors in parallel is

$$\frac{0.0119(0.0067)}{0.0119 + 0.0067} = 0.00428.$$

This equivalent resistance is now in series with the first and the last three resistances, giving, for the equivalent resistance of the entire circuit,

$$0.01 + 0.00428 + 0.0798 + 0.00915 + 0.00366 = 0.1069.$$

Hence, the heat transfer rate through the wall, the current in the equivalent circuit, is

$$q = \frac{T_i - T_\infty}{R_{eq}} = \frac{70 - 20}{0.1069} = 468 \text{ Btu/hr.}$$

As mentioned previously when discussing composite parallel barriers to heat flow, the flow will not be one-dimensional if the thermal conductivities of the parallel barriers differ greatly. This is not the case in Ex. 2.5, since the wood wainscotting has $k = 0.09$, while the gypsum board above has $k = 0.12$.

2.3f Critical Insulation Thickness

Figure 2.13 shows two spherical shells in series, where the outer shell is surrounded by a fluid at temperature T_0 and the surface coefficient of heat transfer is h_0 between the outside shell and the fluid (in which case T_3 would not be known). We find the expression for the steady-state heat transfer rate for this situation by noticing that spherical shells a and b are in series; therefore, two resistors of the type given in Eq. (2.27) are needed.

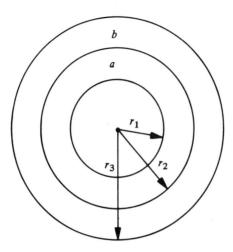

Figure 2.13 The spherical shells for the critical insulation thickness discussion.

In addition, shell b is in series with the resistance due to convection, given by Eq. (2.49). The resulting expression is,

$$q = \frac{4\pi(T_1 - T_0)}{(1/k_a)(1/r_1 - 1/r_2) + (1/k_b)(1/r_2 - 1/r_3) + 1/h_0 r_3^2}. \quad (2.54)$$

Thus, as the insulation thickness r_3 increases, the internal resistance of the insulation to heat transfer (the second resistance in the denominator) also increases and tends to reduce the heat flow (for a given driving potential difference). However, unlike a planar system, the increased insulation thickness also causes a change in another thermal resistance (the third term in the denominator): the resistance between the outside surface and the fluid. If h_0 does not change very much, an increase in insulation thickness *reduces* this thermal resistance, and hence one of the resistances increases while another decreases. Thus, whether the heat transfer rate q decreases or increases with increasing thickness of insulation depends upon which resistance has the greater change, and there is a possibility that the heat flow q might exhibit a maximum or minimum as r_3 is varied. Assuming that r_3 can be varied without changing anything in Eq. (2.54) except q, we will investigate the possible existence of some sort of critical insulation radius.

Since the governing equation for the heat flow in this problem is given by Eq. (2.54), we can determine whether q exhibits an extremum as the insulation outer radius r_3 is varied. By differentiating Eq. (2.54) with respect to r_3 and setting the result equal to zero, we can solve for any values of r_3 that cause the result to be zero.

$$\frac{dq}{dr_3} = \frac{-4\pi(T_1 - T_0)(1/k_b r_3^2 - 2/h_0 r_3^3)}{[(1/k_a)(1/r_1 - 1/r_2) + (1/k_b)(1/r_2 - 1/r_3) + 1/h_0 r_3^2]^2}$$

Now, $dq/dr_3 = 0$ when $1/k_b r_3^2 - 2/h_0 r_3^3 = 0$; that is, when

$$r_{3_C} = 2k_b/h_0. \tag{2.55}$$

To determine whether this one extremum is a maximum or a minimum, using the second derivative test yields

$$\left(\frac{d^2q}{dr_3^2}\right)_{r_{3_C}} < 0.$$

Hence, the critical insulation radius given by Eq. (2.55) represents a maximum in the heat transfer rate q. That is, $r_3 < r_{3_C}$ implies that if the insulation thickness is increased, the heat transfer rate will also increase until r_3 equals r_{3_C}. Further increases in insulation thickness cause reductions in the heat transfer rate. Hence, if insulation is put on a spherical tank, the outside radius of the insulation should be chosen greater than the value given by Eq. (2.55). In most practical situations this will probably not pose any real problem, since it would be unusual to have a value of h_0 too much smaller than about 5.5 W/m² °C or about 1 Btu/hr ft² °F, or an insulation with k_b greater than perhaps about 0.25 W/m °C or about 0.15 Btu/hr ft °F. Thus, these extreme values used in Eq. (2.55) yield a value for r_{3_C} of about 9.1 cm or 3.6 in. Most spherical tanks have a radius greater than this even without insulation; so that for the bulk of practical situations, an increase in the insulation thickness will decrease the heat transfer rate through the spherical shell.

By the same procedure, it can be shown that there is also a critical thickness of insulation for circular cylindrical pipes given by $r_C = k_b/h_0$. The same comments about its size apply as were just made for the spherical tank. However, this critical radius of insulation r_C for the circular cylindrical geometry also may have implications in the choice of thickness of electric insulation on thin wires used as electrical conductors. Here one might decide to try to have the outer radius of the electrical insulation be equal to r_C so as to minimize the wire's temperature for some maximum current-carrying capability.

2.4 STEADY ONE-DIMENSIONAL CONDUCTION WITH HEAT GENERATION

In many practical systems, one is confronted by heat generation distributed throughout a conduction region of interest. As has already been noted in an earlier section, the generation may be due to the dissipation of electrical energy in an electric resistive heating element or a current-carrying conductor. Or it may be due to chemical reactions, such as the combustion process, or to nuclear reactions such as occur in the fission process. Heat generation also occurs because of gamma ray absorption in nuclear shields, in the curing of concrete, and in ripening processes, as well as in biological decay processes.

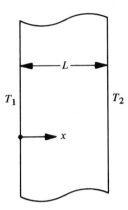

Figure 2.14 The slab with constant generation rate.

This section deals with the analytical prediction of the temperature distribution function in solid conduction regions when distributed generation is present within the material. The techniques presented can then be used on other problems involving generation where the boundary conditions and the generation rate per unit volume have different forms.

It is worth stressing once again that the electric analogy method of earlier sections *cannot* be used in problems with generation.

2.4a The Plane Slab

Figure 2.14 shows a plane slab of material which is L units thick and which has its two outside faces held at known temperatures. For the case where there is a *constant,* known generation rate per unit volume distributed uniformly throughout the slab, of strength q_0''' Btu/hr ft³ or W/m³, the temperature as a function of position is to be predicted when steady-state conditions prevail and the thermal conductivity is constant.

The appropriate governing partial differential equation for the temperature distribution is Eq. (2.10) with the terms involving y and z equal to zero due to the one-dimensionality of the situation. Eq. (2.10) reduces to the ordinary differential equation

$$d^2T/dx^2 + q_0'''/k = 0, \quad \text{or} \quad d^2T/dx^2 = -q_0'''/k.$$

Integrating twice with respect to x,

$$T = -\frac{q_0'''}{k}\frac{x^2}{2} + C_1 x + C_2. \tag{2.56}$$

Using the boundary conditions at $x = 0$, $T = T_1$, and at $x = L$, $T = T_2$, the temperature distribution as a function of x for steady, one-dimensional heat flow, with uniformly distributed constant generation q_0''', in a slab whose face temperatures are specified is

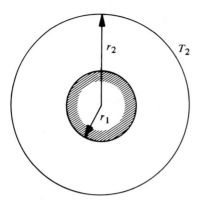

Figure 2.15 The hollow spherical shell.

$$T = \frac{q_0'''}{2k}\left(Lx - x^2\right) + \frac{T_2 - T_1}{L}x + T_1.\tag{2.57}$$

Note that as $q_0''' \to 0$, Eq. (2.57) reduces to the linear distribution of temperature within the slab found earlier when the only thermal loadings were at the boundaries. Equation (2.57) reflects the difference in the temperature distribution function caused by the additional thermal loading q_0''' throughout the material. Also, when $T_1 = T_2$, Eq. (2.57) reduces to a form which shows that the temperature distribution is symmetric about the midplane of the slab for this limiting case with generation.

2.4b Spherical Shell

A hollow spherical shell, which could, for instance, be an idealization of a small experimental nuclear reactor for a laboratory, is drawn in Fig. 2.15. The shell is perfectly insulated at its inside radius r_1, its outside surface r_2 is held at the known temperature T_2, and there is a known, uniformly distributed constant generation rate, q_0''', within the shell. We want to find the temperature as a function of the relevant space coordinates, and also the heat transfer rate across the outside surface when conditions are steady and k is constant.

The nature of the boundary conditions and the generation rate make the temperature distribution purely radial. Hence, for steady state, Eq. (2.9) is the appropriate governing partial differential equation in spherical coordinates; it reduces to

$$\frac{d^2T}{dr^2} + \frac{2}{r}\frac{dT}{dr} + \frac{q_0'''}{k} = 0.$$

To solve this equation, let $S = dT/dr$; then, rearranging,

$$\frac{dS}{dr} + \frac{2}{r}S = \frac{-q_0'''}{k}. \tag{2.58}$$

Equation (2.58) is a first-order, linear, nonhomogeneous, ordinary differential equation with variable coefficients and can be solved formally by the method of the integrating factor. However, as was shown earlier, this equation can be solved without proceeding formally with this method. Multiplying both sides by $r\, dr$,

$$r\, dS + 2S\, dr = -(q_0'''/k)r\, dr. \tag{2.59}$$

The left-hand side of Eq. (2.59) is not the perfect differential of Sr because of the factor 2. However, multiplication by r yields

$$r^2\, dS + 2rS\, dr = -(q_0'''/k)r^2\, dr,$$

and the left-hand side is $d(r^2 S)$. Hence,

$$d(r^2 S) = -(q_0'''/k)r^2\, dr,$$

and integrating once yields

$$r^2 S = -\frac{q_0'''}{k}\frac{r^3}{3} + C_1$$

or

$$S = \frac{dT}{dr} = -\frac{q_0'''}{3k}r + \frac{C_1}{r^2}. \tag{2.60}$$

Integrating Eq. (2.60) with respect to r,

$$T = -\frac{q_0''' r^2}{6k} - \frac{C_1}{r} + C_2. \tag{2.61}$$

One boundary condition is at $r = r_2$, $T = T_2$, and its use in Eq. (2.61) leads to one relation between the integration constants C_1 and C_2. The other boundary condition arises as a result of the surface at r_1 being *perfectly* insulated, that is, adiabatic. This means that no heat flows across the surface; mathematically, $q_{r=r_1} = 0$. But by Fourier's law of conduction, $q_{r_1} = -k\, A\, \dfrac{dT}{dr}\bigg|_{r_1}$. For $q_{r_1} = 0$, $\dfrac{dT}{dr}\bigg|_{r_1} = 0$; that is, the temperature derivative with respect to the coordinate perpendicular to the surface must be zero. The *adiabatic* surface is a special case of the boundary condition of the second kind, the specified flux, Eq. (2.11c) with $q''_w = 0$. Implementing this condition by differentiating Eq. (2.61) gives,

$$\left. \frac{dT}{dr} \right|_{r_1} = -\frac{q'''_0 r_1}{3k} + \frac{C_1}{r_1^2} = 0.$$

Hence, $C_1 = q'''_0 r_1^3/3k$. This result combined with the boundary condition at r_2 yields the temperature distribution

$$T = \frac{q'''_0}{6k} (r_2^2 - r^2) + \frac{q'''_0 r_1^3}{3k} \left(\frac{1}{r_2} - \frac{1}{r} \right) + T_2. \tag{2.62}$$

The heat transfer rate at the outer surface is given by Fourier's exact law of conduction as

$$q_{r_2} = -k(4\pi r_2^2) \left. \frac{dT}{dr} \right|_{r_2}. \tag{2.63}$$

From Eq. (2.62), dT/dr evaluated at $r = r_2$ yields

$$\left. \frac{dT}{dr} \right|_{r_2} = -\frac{q'''_0}{3k} \frac{(r_2^3 - r_1^3)}{r_2^2}. \tag{2.64}$$

Substituting Eq. (2.64) into Eq. (2.63) and rearranging yields the heat transfer rate

$$q_{r_2} = q'''_0 \frac{4}{3} \pi (r_2^3 - r_1^3). \tag{2.65}$$

If the only quantity desired is q_{r_2}, it can be found *without* first obtaining a detailed description of the internal temperature distribution (i.e., Eq. [2.62]). The entire spherical shell is taken as the control volume, and Eq. (1.8) is applied. In the steady state, Eq. (1.8) becomes $R_{in} + R_{gen} = R_{out}$.

Now, $R_{in} = 0$ because of the perfect insulation at $r = r_1$. The unknown heat transfer rate at the outer surface is R_{out}, previously called q_{r_2}. The total rate at which energy is generated within the control volume is R_{gen}. Since the energy generation rate q'''_0 per unit volume is independent of position within the volume, R_{gen} is simply the product of q'''_0 and the total volume of the spherical shell. The volume of the shell is $\frac{4}{3} \pi (r_2^3 - r_1^3)$. Equation (1.8) then gives $0 + q'''_0 \frac{4}{3} \pi (r_2^3 - r_1^3) = q_{r_2}$, which is the same result as the method used to give Eq. (2.65).

Also note that if a known temperature T_1 is specified at r_1 instead of perfect insulation, and it is desired to know only the heat transfer rate at the outside surface q_{r_2}, the temperature distribution must first be obtained and then Fourier's law of conduction used at $r = r_2$. Equation (1.8) would *not* yield the required information, since it only states that $q_{r_1} + q'''_0 \frac{4}{3} \pi (r_2^3 - r_1^3) = q_{r_2}$. Hence, in this type of problem, a detailed description of the temperature distribution would be needed.

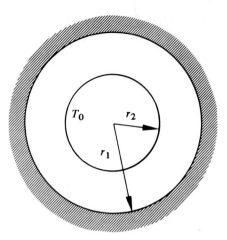

Figure 2.16 The circular cylindrical shell.

2.4c Circular Cylindrical Systems

Let us consider a hollow circular cylindrical shell which is perfectly insulated at its outside radius r_2 and is held at known temperature T_0 at its inside radius r_1 as shown in Fig. 2.16. The shell is in the steady state, has a constant k, and there is a known, constant uniform heat generation rate per unit volume q_0'''. This device might be an electric resistive heater with electrical energy being dissipated between r_1 and r_2 in order to add energy to a fluid flowing through the shell at values of the radial coordinate, r, less than r_1. We would like to predict the temperature distribution within the shell material.

For steady-state conditions with temperature gradients only in the radial direction, the general governing partial differential equation in the cylindrical coordinate system is Eq. (2.8), which reduces to

$$\frac{d^2T}{dr^2} + \frac{1}{r}\frac{dT}{dr} + \frac{q'''}{k} = 0.$$

The order of the equation is reduced by the transformation

$$S = dT/dr. \tag{2.66}$$

Substituting q_0''' for q''' and rearranging,

$$\frac{dS}{dr} + \frac{S}{r} = -\frac{q_0'''}{k}. \tag{2.67}$$

Multiplying through by $r\,dr$,

$$r\,dS + S\,dr = -(q_0'''/k)r\,dr.$$

Since the left-hand side is the derivative of the product of S and r,

$$d(Sr) = -(q_0'''/k)r \, dr. \tag{2.68}$$

(Alternatively, one could begin by classifying Eq. [2.67] as a first-order, ordinary, linear, nonhomogeneous differential equation with variable coefficients, a type which can be solved directly by the integrating factor method.)

Integrating both sides of Eq. (2.68),

$$Sr = -\frac{q_0'''}{k}\frac{r^2}{2} + C_1. \tag{2.69}$$

Substituting Eq. (2.66) into Eq. (2.69) and rearranging,

$$\frac{dT}{dr} = -\frac{q_0''' r}{2k} + \frac{C_1}{r}. \tag{2.70}$$

Integrating,

$$T = -\frac{q_0''' r^2}{4k} + C_1 \ln r + C_2. \tag{2.71}$$

The boundary conditions are at $r = r_1$, $T = T_0$, and at $r = r_2$, the conduction region is perfectly insulated, hence $q = 0$. By Fourier's law of conduction, the temperature derivative with respect to r must be zero at r_2, hence at $r = r_2$, $dT/dr = 0$. Using Eq. (2.70), the second boundary condition gives

$$0 = -\frac{q_0''' r_2}{2k} + \frac{C_1}{r_2},$$

or

$$C_1 = \frac{q_0''' \, r_2^2}{2k}.$$

With this value for C_1, Eq. (2.71) becomes

$$T = -q_0'''/2k \left(\tfrac{1}{2} r^2 - r_2^2 \ln r\right) + C_2. \tag{2.72}$$

The remaining boundary condition is now used, which leads to

$$T_0 = -\frac{q_0'''}{2k} \left(\tfrac{1}{2}r_1^2 - r_2^2 \ln r_1\right) + C_2 \quad \text{or}$$

$$C_2 = T_0 + \frac{q_0'''}{2k} \left(\tfrac{1}{2}r_1^2 - r_2^2 \ln r_1\right). \tag{2.73}$$

Steady-State Conduction

Combining Eqs. (2.73) and (2.72), the steady-state radial temperature distribution is

$$T = T_0 + \frac{q_0'''}{2k}\left[\frac{r_1^2 - r^2}{2} + r_2^2 \ln\left(\frac{r}{r_1}\right)\right]. \tag{2.74}$$

Note that in this problem, one can quite easily generalize the solution, Eq. (2.74), to the case where the inside surface temperature T_0 is not known, but instead the fluid temperature inside the pipe is known to be T_B, and the surface coefficient h between the fluid and the surface at r_1 is known. For this circumstance, an application of the law of conservation of energy, Eq. (1.8), to the entire shell between r_1 and r_2 taken as the control volume leads to the conclusion that the total energy generation rate must equal the energy transfer rate, by convection, from the inside surface, namely

$$q_0'''\pi\,(r_2^2 - r_1^2) = h(2\pi r_1)(T_0 - T_B). \tag{2.75}$$

Solving for T_0 from Eq. (2.75) and inserting this result into Eq. (2.74) yields the temperature distribution in this more general case. This can also be done for the spherical shell result, Eq. (2.62). It is important to keep in mind, however, that the approach which uses an overall energy balance such as Eq. (2.75) to find the surface temperature T_0, is limited to cases where the energy transfer rate across the surface can be found directly by application of Eq. (1.8) to the *entire* conduction region. If this cannot be done, and in general it cannot, then the boundary condition of the third kind, the convective-type condition, Eq. (2.11d), must be formally applied at the interface between the flowing fluid and the solid conduction region.

2.4d Variable Generation

Thus far the solutions developed for one-dimensional conduction problems with generation have all been for cases in which the generation rate per unit volume, q''', is a constant throughout the material. However, q''' sometimes varies in the space coordinates. Examples include the nonuniform distribution of chemical reactants, the attenuation of incoming gamma radiation by absorption, or the nonuniform distribution of neutron flux in a reactor, which causes a spatial variation of the nuclear internal energy release rate even if the fuel, such as U^{235}, is uniformly distributed. Generation may also vary from point to point within the conduction region due to a dependence of q''' upon temperature T which varies from point to point.

Consider the long bar of circular cross section shown in Fig. 2.17, in which the generation rate per unit volume is the following function of temperature:

$$q''' = b_0(1 + \beta T).$$

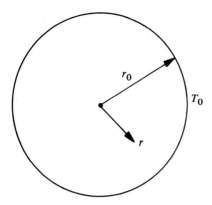

Figure 2.17 Cross section of the bar with variable q'''.

The values b_0 and β are known positive constants. The outside surface temperature is being held at the known value T_0, the thermal conductivity k is constant, and steady-state conditions prevail. Figure 2.17 might represent an electrical coil whose electrical energy dissipation rate depends upon temperature because the resistivity of the coil material is temperature dependent. The task is to predict the temperature distribution within the region shown in Fig. 2.17.

The governing partial differential equation in the circular cylindrical coordinate system is Eq. (2.8). Since steady-state conditions prevail and there is no axial conduction (i.e., z direction, perpendicular to the paper), and no circumferential temperature gradients (i.e., θ direction), Eq. (2.8) reduces to

$$\frac{d^2T}{dr^2} + \frac{1}{r}\frac{dT}{dr} + \frac{q'''}{k} = 0.$$

But $q''' = b_0(1 + \beta T)$; hence, defining $a_0 = b_0/k$,

$$\frac{d^2T}{dr^2} + \frac{1}{r}\frac{dT}{dr} + a_0(1 + \beta T) = 0. \qquad (2.76)$$

Equation (2.76) is simplified by making it homogeneous by the transformation $\phi = 1 + \beta T$ of the dependent variable. Then

$$\frac{dT}{dr} = \frac{1}{\beta}\frac{d\phi}{dr} \text{ and } \frac{d^2T}{dr^2} = \frac{1}{\beta}\frac{d^2\phi}{dr^2}.$$

After substitution and rearrangement, Eq. (2.76) becomes

$$\frac{d^2\phi}{dr^2} + \frac{1}{r}\frac{d\phi}{dr} + a_0\beta\phi = 0.$$

This equation is a second-order, linear, homogeneous ordinary differential equation with variable coefficients. Note that the substitution $S = d\phi/dr$ is not appropriate because of the linear term $a_0\beta\phi$ in ϕ. Multiplying every term by r^2,

$$r^2 \frac{d^2\phi}{dr^2} + r \frac{d\phi}{dr} + a_0\beta r^2\phi = 0. \qquad (2.77)$$

Equation (2.77) is Bessel's equation, which has been solved by infinite series methods. The various infinite series which satisfy it are called *Bessel functions*. (See [1] and [2] for values of Bessel functions and solutions to Bessel equations, respectively.) The solution is

$$\phi = AJ_0 \left(\sqrt{a_0\beta}\, r\right) + BY_0 \left(\sqrt{a_0\beta}\, r\right). \qquad (2.78)$$

The ordinary Bessel functions $J_0 \left(\sqrt{a_0\beta}\, r\right)$ and $Y_0 \left(\sqrt{a_0\beta}\, r\right)$ are of zero order and of the first and second kind, respectively. A *qualitative* indication of the behavior of these functions can be seen at various values of the argument n in Fig. 2.18. (The argument n here is identified as $n = \sqrt{a_0\beta}\, r$.) Values of selected Bessel functions for a moderate range of n are given in Appendix C.

Replacing ϕ in Eq. (2.78) by $1 + \beta T$ as defined and rearranging,

$$T = AJ_0 \left(\sqrt{a_0\beta}\, r\right) + BY_0 \left(\sqrt{a_0\beta}\, r\right) - 1/\beta, \qquad (2.79)$$

where $1/\beta$ in the first two terms on the right has been absorbed into the constants A and B. The boundary conditions determine the integration constants A and B of Eq. (2.79). One boundary condition is at $r = r_0$, $T = T_0$; this leads to

$$T_0 = AJ_0 \left(\sqrt{a_0\beta}\, r_0\right) + BY_0 \left(\sqrt{a_0\beta}\, r_0\right) - 1/\beta. \qquad (2.80)$$

The other boundary condition can be found from the radial symmetry of the temperature distribution about the origin which leads to $dT/dr = 0$ at $r = 0$. An equivalent way of stating this, in this particular problem, is to note that an infinite temperature is not expected anywhere in this conduction region. Figure 2.18 shows that $Y_0(0) = -\infty$, so that to keep the temperature bounded at $r = 0$, $B = 0$. Hence, from Eq. (2.80), using $B = 0$,

$$A = \frac{T_0 + 1/\beta}{J_0(\sqrt{a_0\beta}\, r_0)}. \qquad (2.81)$$

Combining Eqs. (2.81) and (2.79), the temperature distribution in the coil as a function of r is

$$T = (T_0 + 1/\beta) \frac{J_0(\sqrt{a_0\beta}\, r)}{J_0(\sqrt{a_0\beta}\, r_0)} - 1/\beta. \qquad (2.82)$$

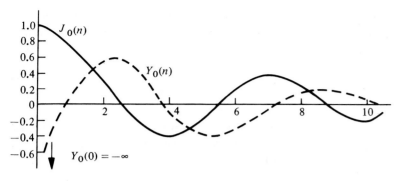

Figure 2.18 Graph of the ordinary Bessel functions of the first and second kind and of zero order.

The temperature distribution solution of Eq. (2.82) has an important and interesting implication. In order to discuss this aspect of the solution, consider the specific situation in which the bar of Fig. 2.17 is a coil with temperature-dependent resistivity leading to the temperature-dependent q''' used in the derivation of Eq. (2.82). Now, suppose that the numerical value of $\sqrt{a_0\beta}\, r_0$ is 2.405. From the tabulated values of $J_0(n)$ in Appendix C or Fig. 2.18, this corresponds to $J_0(2.405) = 0$. Hence, the temperature in Eq. (2.82) would be infinite for all r and the coil material would have melted or been destroyed before this upper limit on T could be reached. For a fixed r_0, the value of $\sqrt{a_0\beta}$ determines whether $\sqrt{a_0\beta}\, r_0 = 2.405$, and since a_0 depends upon the electric current carried by the coil, the condition $\sqrt{a_0\beta}\, r_0 = 2.405$ is used to estimate the maximum or "blow-out" current for the coil under consideration. The physical explanation of this effect is most easily seen by viewing the transient start-up of the coil from an initial temperature of T_0. As current passes through the coil and dissipates electrical energy, the result is an increase of the temperature throughout the coil and a conduction of some of the dissipated energy to the surface. But, as the temperature increases, the dissipation rate increases, since $q''' = b_0(1 + \beta T)$, so that the temperature further increases. This increases the conduction heat transfer rate out of the coil. Under ordinary circumstances the steady state is reached, where energy is conducted away at the same rate it is generated throughout the coil. But when $\sqrt{a_0\beta}\, r_0$ approaches 2.405, the temperatures needed to conduct it away at the required rate are above the melting point of the material, and destruction of the coil takes place.

EXAMPLE 2.6

The cylindrical shell in Fig. 2.16 is an electric resistive heater of inside diameter 1.3 cm and outside diameter 1.525 cm, with an electrical resistance per meter of 0.0274 ohms. The material's thermal conductivity is 17 W/m °C, and a fluid is holding the inside surface temperature at 37°C. Find the maximum allowable current if the temperature is not to exceed 50°C anywhere in the shell in the steady state.

Solution

Here the generation rate per unit volume is a constant, and Eq. (2.74) is the appropriate temperature distribution. By inspection of Fig. 2.16, it should be clear from physical considerations that the maximum temperature occurs at the insulated boundary, that is, at $r = r_2$. Alternately, this may be found more formally by setting the first derivative of Eq. (2.74) equal to zero and solving for the value of r at which this occurs. Thus, with $T_0 = 37°C$, $k = 17$, $r_1 = 1.3/2 = 0.65$ cm, $r_2 = 1.525/2 = 0.7625$ cm, and $T_{max} = 50°C$ at r_2, the only unknown in Eq. (2.74) is the generation rate per unit volume within the conductor, which will ultimately be related to the current passing through the conductor. Hence, at $r = r_2$, Eq. (2.74) becomes

$$50 = 37 + \frac{q_0'''}{2(17)} \left[\frac{0.0065^2 - 0.007625^2}{2} \right.$$

$$\left. + (0.007625)^2 \ln \left(\frac{0.007625}{0.0065} \right) \right].$$

Thus, $q_0''' = 3.31 \times 10^8$ W/m^3.

Since this energy generation rate per unit volume due to electrical dissipation is equal to the product of the current squared and the resistance per meter divided by the conductor volume per meter,

$$3.31 \times 10^8 = \frac{I^2(0.0274)}{(\pi/4)\,[.01525^2 - .013^2]}.$$

Solving for the current gives

$$I = 776.5 \text{ amps.}$$

2.5 STEADY QUASI-ONE-DIMENSIONAL CONDUCTION

Physical situations are often encountered in steady-state conduction heat transfer in which there are conduction heat transfer rates in all three coordinate directions and their relative magnitudes are such that none of these heat transfer rates can be neglected in solving for the temperature distribution within the conduction region. From a form of the general conduction equation such as Eq. (2.6), (2.7), or (2.8), temperature derivatives in all three directions would be expected, and hence a steady-state temperature distribution would depend upon all three space coordinates. However, in many of these situations, the temperature derivatives in two of the three directions, although not zero, are relatively small compared to temperature derivatives in the third direction. In such a case the average temperature over the two directions in which the derivatives are fairly small is used, and thus, the average temperature is a function only of the space variable in which the temperature derivatives are large. (Note that the temperature derivatives are not considered zero in two directions, since this would make the heat transfer rates zero in those directions, and as indicated above those heat transfer rates are *not* zero.) When these conditions prevail, that is, when the temperature derivatives are small enough in two directions

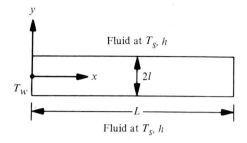

Figure 2.19 Solid region with conduction in both the x and y directions.

to allow the average temperature of the material over those directions (at a fixed value of the third space coordinate) to be close to the local temperature anywhere at the same fixed value of the third space coordinate, then the temperature is said to have been "lumped" or "averaged" over those directions. This *lumped or averaged temperature distribution* in the third space coordinate is *quasi-one-dimensional*. When the temperature distribution is quasi-one-dimensional, as in one-dimensional heat flow, an ordinary differential equation, rather than a partial differential equation, can be solved for the temperature. Examples of actual conduction regions in which the temperature distribution may often be quasi-one-dimensional are fins, devices which behave like fins such as thermometer stems and support struts, and wires which are carrying electric current.

It is necessary to determine whether the temperature distribution can be treated as quasi-one-dimensional with acceptable accuracy in a given situation. Unless the exact solutions for the actual two- or three-dimensional temperature distribution are known, a precise criterion cannot be given. However, following are rough guidelines which can be used in conjunction with engineering judgment, and in some cases actual experience, to determine whether the quasi-one-dimensional approximation will be accurate enough.

Consider a two-dimensional conduction region, as shown in Fig. 2.19, where there is conduction only in the x and y directions. Steady-state conditions prevail with no generation. At $x = 0$, the region is held at temperature T_w, while the fluid surrounding the rest of the region is at temperature T_s, and the surface coefficient of heat transfer between the fluid and the object is h.

Consider 1 ft or 1 m of the object in the z direction. Without any loss in generality, assume that $T_w > T_s$; hence, energy is flowing into the conduction region at $x = 0$ and then flowing to the fluid by having a distribution in x and y past $x = 0$. To estimate some of the various temperature differences, the simplified form of Fourier's law of conduction is used, although, strictly speaking, it is not applicable because of the two-dimensional heat flow. Hence, although the simplified form of Fourier's law would not be used to compute actual heat flow rates or temperatures in a situation such as this, it can be used to determine the parameters that are important in determining whether the temperature distribution is quasi-one-dimensional. If the temperature at $x = L$ is T_0, then the x conduction rate, very roughly, is

$$q_x \approx k2l(1)\,\frac{T_w - T_0}{L},$$

or

$$q_x \approx \frac{T_w - T_0}{r_x},$$

(2.83)

where

$$r_x = L/2kl.$$

(2.84)

(Note that r_x is a measure of the internal resistance of the material to heat transfer in the x direction.) Let the average surface temperature of the conduction region be T_2 at $y = l$ and $y = -l$ and some value of x between 0 and L. Also let the average centerline temperature be T_1 at $y = 0$ and the same x as for T_2. Then, roughly, the conduction heat transfer rate in the y direction is

$$q_y \approx k2L(1)\frac{T_1 - T_2}{l}$$

or

$$q_y \approx \frac{T_1 - T_2}{r_y},$$

(2.85)

where

$$r_y = \frac{l}{2kL}.$$

(2.86)

(Note that r_y is a measure of the internal resistance of the material to heat flow in the y direction.) Comparing r_y to r_x, note that in this situation $r_y < r_x$, since $l < L$, and that the ratio of these measures of thermal resistance is

$$r_y/r_x = (l/L)^2.$$

Hence, when $l < L$, $r_y << r_x$. From Eqs. (2.83) and (2.85) it is seen that, all other things being approximately equal, $r_y << r_x$ implies that $(T_1 - T_2) << (T_w - T_0)$, or that the temperature variation over the entire domain in y, from $y = 0$ to $y = l$, is much smaller than the temperature variation over the entire x domain, from $x = 0$ to $x = L$. Thus under these conditions an average temperature over y may be used, at any x, with relatively little error. The word *may* is used because this approximate qualitative analysis is not yet complete. For example, if h were extremely large, there would be no need for any difference in temperature between the surface and the fluid, and hence $T_2 = T_s$, and since $T_s \le T_0$, the inequality $T_1 - T_2 << T_w - T_0$ may not hold. Since all the energy conducted to the surface in the y direction must go by convection into the fluid at the rate prescribed by Newton's cooling law,

$$q_y \approx k2L\frac{T_1 - T_2}{l} = h2L(T_2 - T_s)$$

(2.87)

or

$$T_1 - T_2 = \frac{hl}{k}(T_2 - T_s). \tag{2.88}$$

From Eq. (2.88), it is seen that $T_1 - T_2$ will be small compared to $T_2 - T_s$; that is, T_2 will not be near T_s if the nondimensional group hl/k is small, say

$$hl/k << 1. \tag{2.89}$$

This group is called the *Biot number* and from Eq. (2.87) can be interpreted as the ratio of the internal resistance of the material to heat transfer in the y direction to the external resistance to heat transfer in the y direction due to the fluid.

This type of analysis holds for other shapes under the same type of thermal loading, as long as the shapes are not too irregular. Thus, when Eq. (2.89) is satisfied simultaneously with

$$r_y << r_x, \tag{2.90}$$

where r_x and r_y are defined by Eqs. (2.84) and (2.86), respectively, the temperature distribution can be considered quasi-one-dimensional in the x direction. This is also called *lumped* or *averaged* in the y direction, since a volume of material between x and $x + dx$ could be considered to have a single temperature over the entire y domain, and hence behave like a constant temperature "lump." Equations (2.89–90) can be generalized to three-dimensional conduction regions, and, by proper interpretation, even to other thermal loadings, such as insulated boundaries or distributed generation.

Once the criteria, Eqs. (2.89) and (2.90), are satisfied and one is reasonably satisfied that a quasi-one-dimensional analysis will be of acceptable accuracy, the next step is to obtain the governing differential equation for the quasi-one-dimensional temperature distribution function in x. One way of arriving at this governing equation is by an appropriate integration of the governing partial differential equation, such as Eq. (2.10). This will be discussed in an example at the end of this section. The more common technique for deriving the governing differential equation is to apply the energy balance, Eq. (1.8), to a control volume of the conduction region, which is differential in extent in the quasi-one-dimensional direction, x, and is finite in all other directions and extends to the outer boundaries of the region in these other directions.

2.5a The Generalized Rod with Convection at Its Tip

The quasi-one-dimensional approach will now be applied to the conduction region shown in Fig. 2.20, which is a generalized rod in the sense that $L >> l_y$ and $L >> l_z$. The constant cross sectional area perpendicular to the x direction is A, and the rod perimeter is P. The left face is held at the known temperature T_w, while the remainder of the region is immersed in a fluid whose undisturbed temperature is T_s with surface coefficient of heat transfer h between the fluid and the rod. Conditions are steady with no generation,

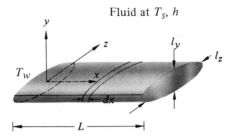

Fluid at T_S, h

Figure 2.20 Generalized rod.

and the rod's thermal conductivity is a constant. This generalized rod, when properly interpreted, could represent a fin, a support strap, or any one of a number of other devices. Thus we would like to predict the quasi-one-dimensional temperature distribution as a function of x and the heat transfer rate between the rod and the fluid.

First, in order to see whether or not the quasi-one-dimensional approximation is suitable for the generalized rod of Fig. 2.20, the criteria, represented by Eqs. (2.89) and (2.90), are extended to a three-dimensional region as follows:

$$hl/k << 1, \quad r_y << r_x, \quad r_z << r_x. \tag{2.91}$$

In Eq. (2.91), the l in hl/k would be interpreted as one-half of the largest transverse distance, that is, one-half of the largest y thickness or z thickness, whichever is greater, as long as there is heat transfer in both the y and z directions.

To compare the internal resistance of the solid to conduction in a given direction to that for the assumed quasi-one-dimensional direction, the x direction, a general expression for the *measure* of internal resistance in a general direction n is needed. This can be found by extending the development which culminated in Eq. (2.84) or by generalizing Eq. (1.1a).

Thus, a general measure of the internal resistance of a solid to heat transfer in a particular direction such as n can be expressed as

$$r_n = \overline{L}_n/k\overline{A}_n, \tag{2.92}$$

where $\overline{L}_n$ is the average path length for conduction heat flow in the n direction and $\overline{A}_n$ is the average area for conduction heat flow perpendicular to the n direction. (Note that an equation such as Eq. [2.92] is *not* a number for the internal resistance to heat transfer that would actually be used in a calculation of conduction heat transfer rates. It ordinarily is only a *qualitative* measure of the resistance, to be used in conjunction with Eq. [2.91] and the engineer's *judgment* to determine whether a quasi-one-dimensional analysis will be acceptable.)

Applying Eq. (2.92) to the region shown in Fig. 2.20 leads to

$$r_x = L/kA, \quad r_y = l_y/kL\overline{l}_z, \quad r_z = l_z/kL\overline{l}_y,$$

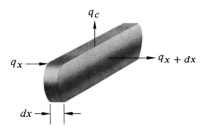

Figure 2.21 The differential control volume and energy transfer rates for the generalized rod.

where $\bar{l}_y$ and $\bar{l}_z$ are estimates of the average thickness of the region in the y and z directions, respectively. The use of these comes about as a result of estimating the average area perpendicular to the y direction and the z direction heat flows. Using, as an estimate, the area perpendicular to the x direction as being $A \simeq \bar{l}_z \bar{l}_y$ gives, after inserting this into r_x and using r_x, r_y, and r_z to form ratios,

$$\frac{r_y}{r_x} = \left(\frac{l_y \bar{l}_y}{L^2}\right), \quad \frac{r_z}{r_x} = \left(\frac{l_z \bar{l}_z}{L^2}\right).$$

For this long rod ($L \gg l_y$, $L \gg l_z$), it is obvious from these ratios that the second two inequalities of Eq. (2.91) are satisfied. Furthermore, it is assumed that the values of k and h are such as to ensure that the first inequality of Eq. (2.91) is also satisfied. Hence the generalized rod of Fig. 2.20 can be acceptably treated as quasi-one-dimensional in the x direction. Next, the governing differential equation for the temperature must be derived and solved.

Since the temperature distribution is quasi-one-dimensional, consider an energy balance on a control volume which is of differential extent in the x direction and of finite extent in the y and z directions, actually including all of the material in the y and z directions. This control volume and the associated energy transfer rates are shown in Fig. 2.21. In the steady state with no generation, Eq. (1.8) requires that

$$R_{\text{in}} = R_{\text{out}}, \quad \text{or} \quad q_x = q_{x+dx} + q_c,$$

or since, by a Taylor series expansion about x, $q_{x+dx} = q_x + \dfrac{dq_x}{dx}\, dx$,

$$0 = \frac{dq_x}{dx}\, dx + q_c.$$

But $q_c = hP\, dx\,(T - T_s)$, $q_x = -kA(dT/dx)$, and $dq_x/dx = -kA(d^2T/dx^2)$, where P is the given perimeter of the cross-sectional area A. Thus, after substitution and rearrangement,

$$\frac{d^2T}{dx^2} - \frac{hP}{kA}(T - T_s) = 0.$$

The boundary conditions are at $x = 0$, $T = T_w$, and at $x = L$, $-kA \, dT/dx = hA$ $(T - T_s)$. Defining $\theta = T - T_s$ and $N^2 = hP/kA$ and transforming the equation,

$$\frac{d^2\theta}{dx^2} - N^2 \, \theta = 0, \qquad (2.93)$$

with the boundary conditions

$$\text{at } x = 0, \quad \theta = \theta_0 = T_w - T_s, \qquad (2.94)$$

$$\text{at } x = L, \quad -k \, \frac{d\theta}{dx} = h\theta. \qquad (2.95)$$

Equation (2.93) is a linear, homogeneous, second-order, ordinary differential equation with constant coefficients which can be solved by the classical operator technique to yield

$$\theta = A_0 \, e^{-Nx} + B_0 \, e^{Nx}. \qquad (2.96)$$

If the x domain is finite, that is, if $L < \infty$, Eq. (2.96) can be put in terms of the hyperbolic sine and cosine as follows: if A and B are integration constants to be determined by the boundary conditions,

$$\theta = A \sinh Nx + B \cosh Nx. \qquad (2.97)$$

This can be proved by recalling that $\sinh z = (e^z - e^{-z})/2$ and $\cosh z = (e^z + e^{-z})/2$ and by making use of the fact that the sum of two or more valid solutions is also a valid solution to a linear homogeneous differential equation. Applying Eq. (2.94) to Eq. (2.97) gives $B = \theta_0$; hence, Eq. (2.97) becomes

$$\theta = A \sinh Nx + \theta_0 \cosh Nx. \qquad (2.98)$$

Using Eq. (2.98) in Eq. (2.95),

$$-k \, (AN \cosh NL + \theta_0 N \sinh NL) = h \, (A \sinh NL + \theta_0 \cosh NL).$$

Solving for A, the only unknown, after rearranging and making use of the fact that $\tanh NL = \sinh NL/\cosh NL$,

$$A = -\theta_0 \frac{h/kN + \tanh NL}{1 + (h/kN) \tanh NL}. \qquad (2.99)$$

Finally, combining Eqs. (2.99) and (2.98) and changing back to the original dependent variable T by using $\theta = T - T_s$ and $\theta_0 = T_w - T_s$, the temperature distribution as a function of x is

$$\frac{T - T_s}{T_w - T_s} = \cosh Nx - \left[\frac{h/kN + \tanh NL}{1 + (h/kN) \tanh NL}\right] \sinh Nx. \qquad (2.100)$$

where

$$N = + \sqrt{hP/kA}. \qquad (2.101)$$

To find the heat transfer rate, q_0, between the generalized rod and the fluid, it is noted that energy is transferred by convection from the rod's lateral surface and its right-hand end or tip. Thus, applying Newton's cooling law, Eq. (1.4), the energy transfer rate to the fluid between x and $x + dx$ is given by

$$hP(T - T_s)dx.$$

The total energy transfer rate to the fluid from the lateral surface is found by summing up all contributions of this type from $x = 0$ to $x = L$ in the definite integral,

$$\int_0^L hP(T - T_s)dx.$$

Adding the convective heat transfer rate from the right-hand tip to this result gives, as the heat transfer rate, the following:

$$q_0 = \int_0^L hP(T - T_s)dx + hA(T_{x=L} - T_s). \qquad (2.102)$$

Now, $T(x)$ and $T_{x=L}$ are given by Eq. (2.100); hence, combining Eqs. (2.100) and (2.102) and integrating yields the desired result for q_0. However, there is an alternate way of arriving at q_0 which is usually easier. Take the entire generalized rod of Fig. 2.20 as the control volume and apply Eq. (1.8). In this situation, $R_{in} = R_{out}$, and q_0 must also be the rate at which energy is *conducted* across the face at $x = 0$, since it is this energy which is then transferred to the fluid from the faces of the rod in contact with the fluid. Hence, by Fourier's law of conduction,

$$q_0 = -kA \left.\frac{dT}{dx}\right|_{x=0}.$$

Differentiating Eq. (2.100), evaluating the derivative at $x = 0$, and multiplying by $-kA$ finally yields

$$q_0 = kAN (T_w - T_s) \left[\frac{h/Nk + \tanh NL}{1 + (h/Nk) \tanh NL}\right]. \qquad (2.103)$$

2.5b The Generalized Rod with a Perfectly Insulated Tip

A condition that often occurs at the right end of the generalized rod in Fig. 2.20, the tip where $x = L$, is the adiabatic or perfectly insulated tip instead of convection to the fluid across the face at $x = L$. The adiabatic condition may result, at least to a good approximation, from the face at $x = L$ being the interface between the rod material and another, low thermal conductivity insulating material. It may also result from a thermal and geometric symmetry condition that insures that $\partial T/\partial x = 0$ at $x = L$. This would be the case if the rod of Fig. 2.20 was $2L$ in length with the face at $x = 2L$ also being held at the temperature T_w.

We follow the same procedure as led to Eqs. (2.100) and (2.103), except that instead of Eq. (2.95), we use the condition that at $x = L$, $\partial\theta/\partial x = 0$. This gives the following solutions:

$$\frac{T - T_s}{T_w - T_s} = \cosh Nx - \tanh NL \sinh Nx \qquad (2.104)$$

$$q_0 = kAN(T_w - T_s)\tanh NL \qquad (2.105)$$

where $N = \sqrt{hP/kA}$.

2.5c The Generalized Rod of Infinite Length

If L approaches infinity for the generalized rod of Fig. 2.20, the quasi-one-dimensional solution functions for temperature as a function of x and for the total heat transfer rate to the fluid take on a simple and easily used form. In reality, of course, a rod would be finite in length, but it might be long enough to behave as if it were infinitely long. If L is long, but finite, the energy entering the rod by conduction at $x = 0$ in Fig. 2.20 may be convected away to the fluid across the lateral rod surface *before* $x = L$ is reached. Such a rod then behaves as if it were indeed infinite.

Using as the rod tip boundary condition

$$\text{as } x \to \infty, \ T \to T_s, \text{ therefore } \theta \to 0,$$

instead of Eq. (2.95), in the development which led to Eqs. (2.100) and (2.103), yields the following results:

$$\frac{T - T_s}{T_w - T_s} = e^{-Nx} \qquad (2.106)$$

$$q_0 = kAN(T_w - T_s) \qquad (2.107)$$

where $N = \sqrt{hP/kA}$.

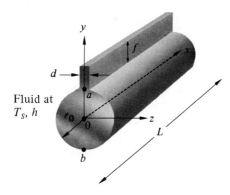

Figure 2.22 Solid region including a portion where the energy flow may not be quasi-one-dimensional.

By comparing the solutions, Eqs. (2.106) and (2.107), for the infinite length rod with their counterparts for finite length rods, either Eqs. (2.100) and (2.103) or Eqs. (2.104) and (2.105), we can see that they are the same if tanh $NL = 1.0$. If one accepts, as an approximation, a value of tanh $NL = 0.999$ instead of 1.0, then for $NL > 3.80$, an actual rod of finite length L can be treated as if it were infinitely long, and Eqs. (2.106) and (2.107) can be applied.

The quasi-one-dimensional solution functions for the generalized rod for the various tip conditions are important, and are extensively used because many different physical regions, such as fins, can be idealized well as such a generalized rod.

2.5d Qualitative Criteria for Quasi-One-Dimensional Analysis

From our discussion of conditions needed to insure that the quasi-one-dimensional approximation will be valid with acceptable accuracy, one might conclude that Eqs. (2.100) and (2.103) are valid for cross sections, at constant x in Fig. 2.20, of any shape as long as the conditions $hl/k << 1$, $r_y << r_x$, and $r_z << r_x$ hold. This is not true, since these criteria have to be tempered with judgment about the shape of the cross section. For example, consider the shape of the cross section shown in Fig. 2.22, where the x direction is perpendicular to the cross section. A measure of the internal resistance to heat transfer in the minus y direction from point 0 to point b on the interface with the fluid is, from Eq. (2.92),

$$r = \frac{r_0}{k\pi r_0 L/2}.$$

This would also be a measure of the internal resistance to conduction heat transfer between the points 0 and a in the solid. However, the energy arriving by conduction at b in the minus y direction is now transferred to the fluid, whereas the energy arriving at a from within the solid in the plus y direction still has another internal resistance to negotiate before it can be transferred to the fluid; namely, the thin piece of material jutting out at point a. From Eq. (2.92), a measure of the resistance to conduction heat transfer in the y direction through this material is given by

$$r^1 = f/kdL.$$

Note that if f is relatively large and/or d is relatively small, the sum $r^1 + r$ could be greater than r_x even though r by itself is much less than r_x. If this were true, the values of $\partial T/\partial y$ when $r_0 < y < (r_0 + f)$ would not be much smaller than $\partial T/\partial x$, and hence the temperature distribution could not be treated as quasi-one-dimensional, at least not in the thin piece of material jutting out of the bulk of the bar at position a. On the other hand, if the area of this thin piece df of the cross section is a very small fraction of the total cross-sectional area A, a quasi-one-dimensional analysis can still be made on the basis that $\partial T/\partial y << \partial T/\partial x$ throughout most of the cross section, as long as it is not deemed important to know the exact details of the temperature distribution within the region df.

As a result, a general statement can be made that the shape of a cross section that might be conducive to the use of the quasi-one-dimensional approximation should be such as to allow a fairly smooth curve for the perimeter. A fairly smooth curve is one in which the slope dy/dz or dz/dy of the perimeter curve in the yz plane is a gentle function of z or y, respectively. The yz plane is perpendicular to the x direction, the direction in which the temperature distribution is suspected to be quasi-one-dimensional. In the cross section shown in Fig. 2.22, the slope dy/dz of most of the perimeter curve varies gently with z, since most of the curve is the perimeter of a circle. However, just before point a, the slope is zero, then abruptly changes to ∞, and then just as abruptly changes to zero again, all at the same value of z. It then remains zero for a change in z equal to the dimension d, then becomes $-\infty$ and back to zero, again at the same z. Hence, in this local region of the perimeter curve of Fig. 2.22, the slope is not a gentle function of z and it is expected that, at least in this part of the cross section, the quasi-one-dimensional approximation might not be valid.

An alternate way of deciding whether the shape of a cross section will allow a quasi-one-dimensional analysis in a direction perpendicular to the cross section is to require that the *local* values of internal resistance to heat transfer in a particular direction also be much less than r_x, where x is the direction in which the temperature distribution is suspected to be quasi-one-dimensional. Figures 2.23–24 show examples of fairly smooth and fairly unsmooth perimeter curves of a cross section, respectively.

In Fig. 2.23(c), the local slope varies abruptly at the four sharp corners of the rectangle. The departure from being fairly smooth at any corner of the rectangle occupies zero distance along the perimeter and causes a very local change of the internal resistance of a small amount of material in the corner. This corner material is such a small fraction of the total cross-sectional area that it is not expected to significantly distort the temperature distribution over most of the cross section. As mentioned earlier, these "rules" are only broad guidelines to be combined with the engineer's judgment in arriving at a final decision about the suitability of a given conduction region for a quasi-one-dimensional analysis. Many of the ideas discussed in this section will be used in Chapter 3 when deciding whether the lumped parameter method of transient analysis can be used in a conduction region.

Chapter Two

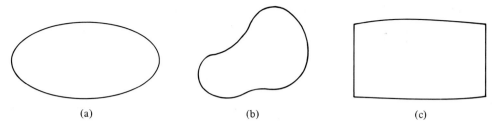

Figure 2.23 Fairly smooth curves.

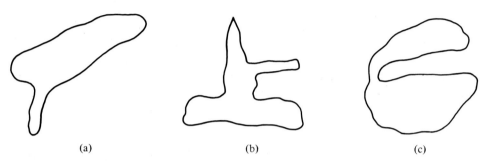

Figure 2.24 Fairly unsmooth curves.

EXAMPLE 2.7

Consider the two-dimensional conduction region, which is very long in the z direction, shown in Fig. 2.25. If the requirements, Eq. (2.91), for a quasi-one-dimensional analysis are satisfied, then a control volume dx by 2δ by 1 can be taken and the energy balance, Eq. (1.8), applied to yield the governing differential equation. But this has already been done for the generalized rod of Fig. 2.20. Therefore, with $A = 2\delta(1)$ and $P = 2$ units (note that the 2δ distance in y is *not* counted as part of the perimeter P because there is no heat transfer rate to the fluid across it in this very long region), Eq. (2.93), with N from Eq. (2.101), becomes the governing equation; therefore with $\theta = T - T_s$,

$$\frac{d^2\theta}{dx^2} - \frac{h}{k\delta}\theta = 0. \tag{2.108}$$

Now it has been stressed that in these quasi-one-dimensional problems the temperature derivatives are *not* zero in directions perpendicular to x, but small enough, compared to temperature derivatives with respect to x, so that at any x no significant variation in temperature exists over the entire cross section.

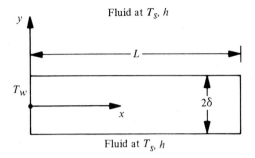

Figure 2.25 Diagram for Ex. 2.7.

Since the actual temperature depends on both x and y, denote it $\overline{T} = \overline{T}(x, y)$. Defining $\overline{\theta} = \overline{\theta}(x, y) = \overline{T} - T_s$, the true governing equation for this problem is given by Eq. (2.11) with the z direction absent. After changing the dependent variable to $\overline{\theta}$,

$$\frac{\partial^2 \overline{\theta}}{\partial x^2} + \frac{\partial^2 \overline{\theta}}{\partial y^2} = 0. \tag{2.109}$$

Instead of deriving Eq. (2.108) from an energy balance on an appropriate control volume, derive it by starting with Eq. (2.109).

Solution
From Eq. (2.109), it is clear that if the temperature derivative in the y direction is not only small compared to the one for the x direction, but is also set equal to zero, a completely incorrect result is obtained; namely, $d^2\overline{\theta}/dx^2 = 0$. Since the quasi-one-dimensional temperature $\theta(x)$ has also been called the lumped or average temperature at any x, Eq. (2.109) is averaged over y by multiplying by dy and integrating from $y = 0$ to $y = \delta$. (The symmetry exhibited by Fig. 2.25 allows the limits used.) Thus,

$$\int_0^\delta \frac{\partial^2 \overline{\theta}}{\partial x^2} \, dy + \int_0^\delta \frac{\partial^2 \overline{\theta}}{\partial y^2} \, dy = 0. \tag{2.110}$$

The first integral becomes

$$\frac{d^2}{dx^2} \int_0^\delta \overline{\theta} \, dy. \tag{2.111}$$

This is a consequence of applying the Leibnitz rule for differentiating an integral with respect to a variable which is not the integration variable [4]. The second integral becomes, after rearrangement and noting that the integration is performed at constant x,

$$\int_0^\delta \frac{\partial^2 \overline{\theta}}{\partial y^2} \, dy = \int_0^\delta \frac{\partial}{\partial y} \frac{\partial \overline{\theta}}{\partial y} \, dy = \int_0^\delta d\left(\frac{\partial \overline{\theta}}{\partial y}\right) = \frac{\partial \overline{\theta}}{\partial y}\bigg|_0^\delta.$$

Since $\partial \bar{\theta}/\partial y = 0$ at $y = 0$ because of the symmetry in the y direction, the second integral of Eq. (2.110) becomes

$$\frac{\partial \bar{\theta}}{\partial y}\bigg|_{y=\delta}.$$

However, when this equation is multiplied by $-k$, it becomes the heat transfer rate by conduction to the fluid/solid interface at $y = \delta$, and this must be equal to the rate at which it is convected into the fluid, which according to Newton's law of cooling is $h\,(\bar{T} - T_s)_\delta = h\,\bar{\theta}_\delta$. Hence,

$$-k\left(\frac{\partial \bar{\theta}}{\partial y}\right)_\delta = h\,\bar{\theta}_\delta,$$

$$\int_0^\delta \frac{\partial^2 \bar{\theta}}{\partial y^2}\, dy = -\frac{h}{k}\,\bar{\theta}_\delta. \tag{2.112}$$

Combining Eqs. (2.112), (2.111), and (2.110),

$$\frac{d^2}{dx^2}\int_0^\delta \bar{\theta}\, dy - \frac{h}{k}\,\bar{\theta}_\delta = 0. \tag{2.113}$$

Now, the integral on the left can be evaluated if the average $\bar{\theta}$ over y—call it θ—is defined by the equation

$$\theta\delta = \int_0^\delta \bar{\theta}\, dy. \tag{2.114}$$

With this, Eq. (2.113) becomes

$$\frac{d^2\theta}{dx^2} - \frac{h}{k\delta}\,\bar{\theta}_\delta = 0.$$

Up to this point, no approximations have been made. But the essence of the quasi-one-dimensional approximation is that in some situations the exact temperature $\bar{\theta}$ in Eq. (2.114) varies so weakly with y compared to x, that the average temperature θ over y is not very different from $\bar{\theta}$ at any y, including the surface coordinate $y = \delta$. Hence, it can now be assumed that $\theta \approx \bar{\theta}_{y = \delta}$, and this yields the ordinary differential equation

$$\frac{d^2\theta}{dx^2} - \frac{h}{k\delta}\,\theta = 0,$$

which is identical to Eq. (2.108).

Example 2.7 shows, from a different viewpoint, the implications of the quasi-one-dimensional approximation, and indicates that the governing equation for quasi-one-dimensional temperature distributions can also be arrived at by starting with the general partial differential equation and then proceeding in a formal mathematical fashion. This type of approach is ordinarily not preferred, since it is not quite as straightforward for complicated shapes or even for simple shapes in coordinate systems other than the rectangular cartesian. If, however, one wishes to employ the alternate approach in a coordinate system other than rectangular cartesian, the governing partial differential equation for the temperature distribution must be multiplied by a differential *volume* element in the coordinate system of interest and then integrated over the coordinates in which it is presumed that the temperature varies only slightly compared with the remaining coordinate.

EXAMPLE 2.8

A thermometric device is measuring the temperature of a liquid flowing in a 5-cm-diameter duct. The readout temperature of 71°C is the temperature at the tip of the device, as shown in Fig. 2.26. The device is joined intimately to the duct wall whose temperature is 54°C. The immersed length of the instrument is 2.5 cm, its diameter is 0.16 cm, and it has an equivalent thermal conductivity of 41 W/m °C. If the average surface coefficient of heat transfer between the thermometric device and the liquid is $h = 30$ W/m² °C, calculate the true temperature of the liquid.

Solution
The true temperature of the liquid is not the 71°C temperature recorded at the tip of the device, since the other end of the device is intimately joined to the duct wall. Hence, it is at the wall temperature of 54°C, implying that there is conduction through the device toward the duct wall, as well as convection along its surface. Since the device is presumably in the steady state, the energy transfer rate through the device at the wall must equal the energy transfer rate by convection from the liquid to the instrument, and to support a convective heat transfer rate, Newton's cooling law requires a nonzero temperature difference between the liquid and the surface. Hence, the temperature of the tip is not equal to that of the liquid. The difference between the tip temperature and liquid temperature is referred to as the *conduction error of the instrument*.

To find the actual liquid temperature, note that the physical situation is the same as that of the generalized rod with convection at its tip (Fig. 2.20) if x is measured from the duct wall. Thus, the temperature distribution is given by Eq. (2.100), assuming that the criteria for the use of the quasi-one-dimensional approximation are satisfied. These criteria are given by the first two inequalities of Eq. (2.91) with y interpreted as the radial coordinate from the thermometric device's center line. Hence,

$$\frac{hl}{k} = \frac{h(D/2)}{k} = \frac{30(0.0016/2)}{41} = 0.00059 << 1.$$

Using Eq. (2.92) for r_x and r_y, one has,

$$\frac{r_y}{r_x} = \frac{(\overline{L}_y/k\overline{A}_y)}{(\overline{L}_x/k\overline{A}_x)},$$

$$\overline{L}_y = D/2, \quad \overline{A}_y \simeq [\pi(D/2)L], \quad \overline{L}_x = L, \quad \overline{A}_x = \pi D^2/4.$$

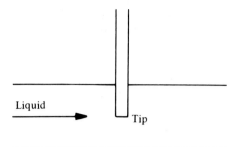

Figure 2.26 Liquid flowing over the thermometric device of Ex. 2.8.

With this,

$$\frac{r_y}{r_x} = \frac{D^2}{4L^2} = \frac{(0.0016)^2}{4(.0254)^2} = 0.00099 \ll 1.$$

Thus, the quasi-one-dimensional results will be of acceptable accuracy and Eq. (2.100) is applicable. From Eq. (2.101),

$$N = \sqrt{hP/kA} = \sqrt{\frac{30(\pi)(0.0016)}{41(\pi/4)(0.0016)^2}} = 42.77,$$

so $NL = 42.77(.0254) = 1.086$ and $h/Nk = 30/42.77(41) = 0.0171$. The unknown temperature of the liquid in which this thermometric device is submerged is T_s, while T_w is the duct wall temperature of 54°C. The tip temperature of 71°C occurs at $x = L = 0.0254$ m. Thus, substituting into Eq. (2.100),

$$\frac{71 - T_s}{54 - T_s} = \cosh 1.086 - \left(\frac{0.0171 + \tanh 1.086}{1 + 0.0171 \tanh 1.086}\right) \sinh 1.086.$$

Solving for the liquid temperature yields

$$T_s = 96.3°C.$$

The conduction error of this thermometric device is significant and must be properly accounted for when using the instrument.

2.5e Quasi-One-Dimensional Conduction When q Is Constant

Earlier in this chapter, reference was made to the possibility of using, in certain one-dimensional cases, an integrated form of Fourier's law of conduction, Eq. (1.3), to calculate the heat transfer rate without first explicitly solving for the temperature distribution function. This method, since it can also be used in the more general case of quasi-one-dimensional conduction, will be developed now.

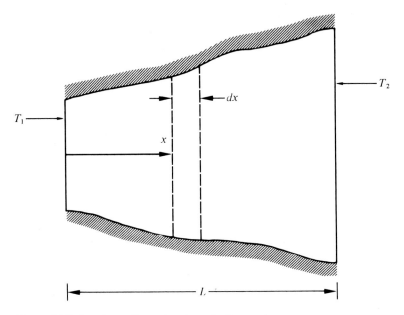

Figure 2.27 Quasi-one-dimensional conduction region.

Consider steady, quasi-one-dimensional heat flow, without generation, through the general conduction region shown in Fig. 2.27. The left and right faces are held at known temperatures T_1 and T_2, respectively. The cross-sectional area perpendicular to the x direction, across which energy is conducted, is a known function of x, $A(x)$. The region's thermal conductivity is a known function of the temperature, $k(T)$. We will derive an expression for the heat transfer rate, q, across the region and the temperature distribution within it.

Since the heat flow can be considered quasi-one-dimensional in the x direction, conditions are such that, at any x, the temperature variation over the cross section is very small, and hence a single value of temperature is assigned at every x. Make an energy balance on a control volume of material which is differential in extent in the x direction but extends over the entire cross section in the other directions. This control volume is shown in Fig. 2.27. Apply the law of conservation of energy to this control volume; that is,

$$R_{\text{in}} + R_{\text{gen}} = R_{\text{out}} + R_{\text{stor}}. \qquad [1.8]$$

Since conditions are steady state with no generation, R_{gen} and R_{stor} are zero. Energy enters the control volume by conduction across the left face of the control volume at the rate q_x, and energy leaves the control volume by conduction across the right face at the rate q_{x+dx}; hence Eq. (1.8) becomes

$$q_x = q_{x+dx}.$$

Again, use of a Taylor expansion leads to

$$\frac{dq_x}{dx} = 0. \tag{2.115}$$

Hence, the conduction heat transfer rate q in the x direction is a constant by integration of Eq. (2.115). This should be intuitively clear, since the energy entering the left-hand face of the entire conduction region cannot be added to or extracted from. Hence, the heat transfer rate must be invariant with x.

The constant q is also given by Fourier's exact law of conduction as

$$q = -k(T)A(x)\frac{dT}{dx}, \tag{2.116}$$

where the thermal conductivity k and local cross-sectional area A are general specified functions of T and x, respectively. Since q in Eq. (2.116) is not dependent upon x, the variables can be separated. Hence,

$$q\frac{dx}{A(x)} = -k(T)\,dT.$$

Now, both sides are integrated between the left- and right-hand faces of the conduction region to yield

$$q\int_0^L \frac{dx}{A(x)} = -\int_{T_1}^{T_2} k(T)\,dT, \tag{2.117}$$

since q, though unknown, is a constant. Finally, after rearrangement, the equation for q is found in terms of known quantities as

$$q = \frac{-\int_{T_1}^{T_2} k(T)\,dT}{\int_0^L \frac{dx}{A(x)}}. \tag{2.118}$$

Note that for a particular problem the actual functions $A(x)$ and $k(T)$ are known, the integrations in Eq. (2.118) can be performed, and the value of q can be determined.

Next, we seek the temperature distribution as a function of x. This can be found by first integrating Eq. (2.117) between different limits. Since T as a function of x is to be arrived at, it is required that the general temperature T at the general position x appear explicitly; hence, the left-hand side of Eq. (2.117) is integrated from 0 to x instead of L, and the right side from T_1 to T instead of T_2. Thus,

$$q = \frac{-\int_{T_1}^{T} k(T)\,dT}{\int_0^x \dfrac{dx}{A(x)}}.$$

However, it cannot yet be considered that this equation gives T as a function of x, because another quantity, q, is present. But q is given in terms of known quantities by Eq. (2.118). Hence, elimination of q between these two equations gives the temperature distribution as

$$\int_{T_1}^{T} k(T)\,dT = \left[\frac{\int_{T_1}^{T_2} k(T)\,dT}{\int_0^L \dfrac{dx}{A(x)}} \right] \int_0^x \frac{dx}{A(x)}. \tag{2.119}$$

When the actual functions $k(T)$ and $A(x)$ are given and the integrations performed, Eq. (2.119) is in a more useful form, giving $T = T(x)$.

Note that this method of approach, which finds both the heat transfer rate and the temperature distribution by use of Fourier's exact law of conduction alone, is predicated upon the local heat transfer rate being a *constant;* not known in advance, but nonetheless a constant. To satisfy this condition requires that the energy flow be *quasi-one-dimensional, or else truly one-dimensional, with no generation, with steady-state conditions, and with no energy transfer across the lateral surfaces of the region.* If one or more of these conditions is not satisfied, the method cannot be used. Instead one must begin by solving the appropriate differential equation for the temperature distribution, and then use Fourier's law of conduction to find the heat transfer rate, as was done in the earlier sections of the chapter.

2.5f Fins

Situations often arise in which energy is transferred from one fluid to another fluid across a solid barrier separating the fluids. Sometimes the surface coefficient of heat transfer between one side of the solid and the fluid is much lower than between the other side of the solid and the fluid, and, in addition, the solid itself presents very little resistance to heat transfer. Thus, in this situation the low surface coefficient of heat transfer represents the controlling resistance to heat transfer between the two fluids (see the discussion following Ex. 2.4), and if the heat transfer rate is to be increased, this controlling resistance must be decreased or else the temperature difference between the two fluids must be increased. However, outside factors may fix the temperature difference between the fluids.

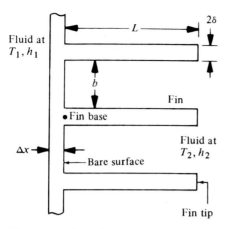

Figure 2.28 Straight fins of rectangular profile.

Even if it could be increased, the heat transfer rate is still limited by the large value of the controlling resistance on the side with the relatively low surface coefficient of heat transfer h. Since the form of this controlling resistance is $1/hA$ (see Eq. [2.49]), it may be decreased by increasing the surface coefficient h or the heat transfer surface area A. Usually, constraints such as pumping power requirements do not allow any significant increase in h. Furthermore, overall equipment size requirements may dictate the size of the area A. Generally what is done to decrease the controlling resistance is to increase the effective surface area—while keeping the projected area A the same—by adding surface area perpendicular to A. This is depicted in Fig. 2.28, where the original barrier between the fluids is of thickness Δx with a low surface coefficient between its right face and the fluid 2. Material of height 2δ and length L has been affixed perpendicular to the right surface to reduce the controlling resistance. This practice is called extending the surface, and usually these surface extensions are called fins, although they may be termed pins or spines, depending upon their shape.

The fins shown in Fig. 2.28 are classified as *straight* fins of *rectangular profile area*. Straight refers to the fact that the fin base is plane because it interfaces with a planar barrier between the two fluids, while rectangular profile area refers to the *shape* of the fin which, in this case, is rectangular with profile area $A_f = L(2\delta)$. Figure 2.29 shows straight fins of triangular profile area, whereas the inset of Fig. 2.30 portrays *circular* fins of rectangular profile area. Circular fins are also commonly referred to as *annular* or as *circumferential* and have a base which is curved such that they can conform to the surface of a circular cylinder, such as a pipe, when they are oriented perpendicular to the cylinder's long axis.

Since fins are added to a surface to increase the heat transfer rate, expressions for the heat transfer rate between a fin and the surrounding fluid are needed in fin analysis and design. However, fins perform their function best under the conditions $\delta << L$ in Fig. 2.28 and $h\delta/k << 1$, which give a quasi-one-dimensional temperature distribution within them in the direction of their length L. Hence, some of the needed expressions for a *single* fin of simple shape have already been derived in previous sections, and the methods of

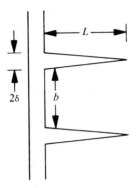

Figure 2.29 Straight fins of triangular profile.

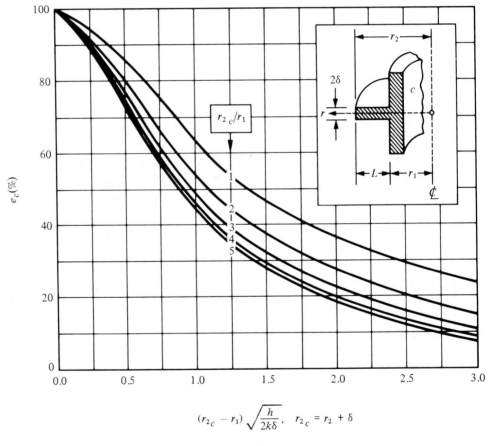

$$(r_{2_c} - r_1)\sqrt{\frac{h}{2k\delta}}, \quad r_{2_c} = r_2 + \delta$$

Figure 2.30 Efficiency of the circular rectangular fin.

these earlier sections can be applied to develop the expression for the heat transfer rate in the case of the more complicated fin shapes, such as the one of triangular profile area. Generally, a large number of fins are used on a surface, and the problem is to calculate the heat transfer rate between the two fluids when the fins are present.

The overall method of attack to determine the heat transfer rate when fins are being used will be developed by analyzing the case shown in Fig. 2.28, where straight fins of rectangular profile area, of thickness 2δ, and length L, are spaced a distance b apart on the surface. We would like to find an expression for the heat transfer rate per unit area, q'', in terms of known quantities when steady-state conditions prevail.

Let the *unknown* temperature of the bare surface, the surface on the right which is not covered by a fin, be T_0. Assume that the temperature at the base of the fin, where the fin is joined to the plate of thickness Δx, is also T_0. Consider a representative portion of the finned side, that is, a portion $b + 2\delta$ by 1 unit deep into the page, which is representative because it contains one fin and all the bare surface area to the next fin. Letting q_0 be the total rate at which a fin transfers energy to the fluid, the heat transfer rate to fluid 2 across this representative area is

$$q = q_0 + h_2 b (T_0 - T_2). \tag{2.120}$$

But q_0 is conducted across the base of the fin at $x = 0$; hence,

$$q_0 = -k \, 2\delta \frac{dT}{dx} \bigg|_{x=0}. \tag{2.121}$$

Thus, if $T = T(x)$ were known within the fin, Eq. (2.121) would yield q_0. If the fin dimensions are such as to allow a quasi-one-dimensional analysis, then the temperature as a function of x has already been found in Eq. (2.100) if T_0 is identified with T_w and T_2 with T_s. Hence, from Eq. (2.103),

$$q_0 = 2k\delta N (T_0 - T_2) \left[\frac{h_2/Nk + \tanh NL}{1 + (h_2/Nk) \tanh NL} \right], \tag{2.122}$$

where $N = \sqrt{h_2/k\delta}$. Now, q of Eq. (2.120) is also the heat transfer rate from fluid 1 to the plate across area $b + 2\delta$ and the conduction heat transfer rate through the plate; hence, two more expressions for q are

$$q = h_1 (b + 2\delta)(T_1 - T_p), \tag{2.123}$$

$$q = k(b + 2\delta) \frac{(T_p - T_0)}{\Delta x}, \tag{2.124}$$

where T_p is the unknown temperature of the left-hand side of the plate of thickness Δx. Rewriting Eq. (2.122),

$$q_0 = \psi (T_0 - T_2), \tag{2.125}$$

where

$$\psi = 2k\delta N \left[\frac{h_2/Nk + \tanh NL}{1 + (h_2/Nk) \tanh NL} \right]. \qquad \textbf{(2.126)}$$

Eliminating T_p and T_0 between Eqs. (2.120), (2.123), and (2.124) and dividing the result for q by $b + 2\delta$, the heat transfer rate q'' per unit area is

$$q'' = \frac{T_1 - T_2}{1/h_1 + \Delta x/k + (b + 2\delta)/(h_2 b + \psi)}. \qquad \textbf{(2.127)}$$

Thus, the heat transfer rate between the two fluids, including the effect of the fins shown in Fig. 2.28, is given in terms of quantities ordinarily available to the analyst by Eqs. (2.127) and (2.126).

The procedure by which Eq. (2.127) was developed is the one to be applied to other fin problems as well, in which the fins may be straight or circular, of any profile shape, and may even be located on both sides of the wall separating the fluids.

Criterion for Fin Addition

Figure 2.28 portrays a situation in which fins were added to the right surface in order to increase the heat transfer rate. One of these fins is isolated for further study in Fig. 2.31. However, does the addition of fins always act to increase the energy transfer rate, or is it possible that they may sometimes cause a decrease in the energy transfer rate? First this question will be addressed with some qualitative physical reasoning, and then a quantitative criterion which answers this question will be developed.

First, assume that the prime surface is at the temperature T_0 regardless of whether fins are attached, and also that h_2 is the same value for the fin to the fluid as for the prime surface to the fluid. Now, consider a portion of the prime surface 2δ high with no fin attached. The heat transfer rate between this portion and the fluid is given by

$$q = h_2 2\delta(T_0 - T_2) = \frac{T_0 - T_2}{1/2h_2\delta}.$$

Thus, an analogous electric circuit can be drawn for the case of no fins as shown in Fig. 2.32(a). If a fin is joined to this prime surface, as shown in Fig. 2.31, the energy flows from the surface at T_0 into the fin and encounters the internal resistance of the fin to conduction and, after arriving at the interface of the fin and fluid, flows into the fluid across a convective resistance. Let the average internal resistance of the fin to conduction heat transfer be R_f; then the analogous electric circuit is formed in Fig. 2.32(b). Let T_f be the average fin surface temperature. Thus, from Figs. 2.32 (a–b), the energy transfer rate without and with fins, respectively, depends upon the relative magnitude of the equivalent resistances which are shown. Calling the resistance with no fin R_a, and with the fin R_b,

$$R_a = \frac{1}{2h_2\delta}, \quad R_b = \frac{1}{2h_2(\delta + L)} + R_f.$$

Chapter Two

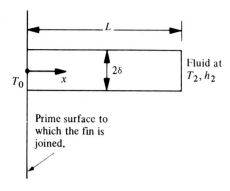

Figure 2.31 The fin of Fig. 2.28.

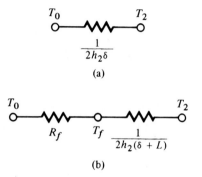

Figure 2.32 The analogous electric circuit of Fig. 2.31 for (a) a bare surface and (b) a fin attached.

Clearly, the convective resistance part of R_b is less than R_a, since $\delta + L > \delta$. However, it appears that situations could arise in which R_f is large enough to make $R_b > R_a$ and cause the fins to have an insulating effect rather than the beneficial effect of increasing the heat transfer rate. Consider, for example, the effect of adding relatively thick fins (large δ) constructed of asbestos. The low thermal conductivity of the asbestos could make R_f large enough so that $R_b > R_a$.

Since the preceding qualitative argument has demonstrated that fins possibly can have an effect opposite that which is normally intended, let us next develop a quantitative criterion that will tell us the conditions under which the addition of fins yields a *beneficial* effect. The expression for the heat transfer rate from the type of fin under consideration to the fluid is

$$q_0 = 2\delta k N(T_0 - T_2)\left[\frac{h_2/Nk + \tanh NL}{1 + (h_2/Nk)\tanh NL}\right],\qquad \textbf{[2.122]}$$

where $N = \sqrt{h_2/k\delta}$.

Now, if the coefficient of the bracket in Eq. (2.122) is the heat transfer rate from the surface with no fins, then the bracket would indicate the conditions under which the fins will give an increased heat transfer rate. Hence, factoring h_2 out of the numerator of the bracket and putting the factor kN inside the bracket yields

$$q_0 = 2\delta h_2(T_0 - T_2)\left[\frac{1 + (Nk/h_2)\tanh NL}{1 + (h_2/Nk)\tanh NL}\right]. \tag{2.128}$$

Now, $2\delta h_2(T_0 - T_2)$ is the heat transfer rate across the prime surface with no fins added. Calling it q_{nf}, using

$$\frac{Nk}{h_2} = \sqrt{\frac{k}{h_2\delta}}, \tag{2.129}$$

defining,

$$\frac{h_2\delta}{k} = N_{\text{Bi}}, \tag{2.130}$$

and inserting Eqs. (2.129) and (2.130) into Eq. (2.128) and rearranging,

$$\frac{q_0}{q_{nf}} = \left[\frac{1 + (1/\sqrt{N_{\text{Bi}}})\tanh NL}{1 + \sqrt{N_{\text{Bi}}}\tanh NL}\right]. \tag{2.131}$$

From Eq. (2.131) the addition of fins has a beneficial effect when $q_0 > q_{nf}$, and hence the term on the right side must be greater than 1. Solving the inequality

$$\frac{1 + (1/\sqrt{N_{\text{Bi}}})\tanh NL}{1 + \sqrt{N_{\text{Bi}}}\tanh NL} > 1$$

yields $N_{\text{Bi}} < 1$, or by Eq. (2.130),

$$\frac{h_2\delta}{k} < 1. \tag{2.132}$$

It is suggested in Kreith [3] that in a design situation involving fins, when all factors are taken into account, their use is ordinarily not justified unless

$$\frac{h_2\delta}{k} \leq 0.25. \tag{2.133}$$

This criterion could also be used, as a first approximation, in deciding when straight fins of some other profile shape would have a beneficial effect. Since in other shapes, δ is a function of x, engineering judgment would be used to estimate an average to use in Eq. (2.133).

Hence, unless the condition of Eq. (2.133) was met, one would not add fins to a surface. Normally, well designed fins are relatively thin—$\delta \ll L$—and have values of $h\delta/k$ far lower than the value of 0.25 given in the criterion.

EXAMPLE 2.9

For the physical situation shown in Fig. 2.28, fluid 1 is water at 93°C giving a value of $h_1 = 450$ W/m² °C and fluid 2 is air at 38°C being blown perpendicular to the page such that $h_2 = 23$ W/m² °C. The plate and fins are made of aluminum with $k = 208$ W/m °C, $\Delta x = 0.25$ cm, $\delta = 0.08$ cm, $L = 1.9$ cm, and $b = 1.25$ cm. Find the heat flux with and without the fins present.

Solution
Initially, a check is made to determine if the quasi-one-dimensional approximation is acceptable. If it is, the result, Eq. (2.127), can be used. For a quasi-one-dimensional analysis to be acceptable, from Eq. (2.91), $hl/k \ll 1$ and $r_y \ll r_x$. Since $l = \delta = 0.08$ cm,

$$hl/k = 23(0.0008)/208 = 0.000088.$$

This is not only much less than 1, but it also easily satisfies the beneficial effect criterion, Eq. (2.133). For this situation, $r_y = l/kL$ and $r_x = L/kl$, so that,

$$r_y/r_x = (l/L)^2 = (\delta/L)^2 = (0.08/1.9)^2 = 0.0018.$$

Thus, r_y is much less than r_x, and quasi-one-dimensional results will be of acceptable accuracy. Now, from Eq. (2.101) and as explained in Ex. 2.7,

$$N = \sqrt{h_2/k\delta} = \sqrt{23/208(0.0008)} = 11.76.$$

From Eq. (2.126),

$$\psi = 2(208)(0.0008)(11.76)$$
$$\times \frac{[23/11.76(208)] + \tanh [11.76(0.019)]}{1 + [23/11.76(208)] \tanh [11.76(0.019)]} = 0.8949.$$

Now, using Eq. (2.127),

$$q'' = \frac{93 - 38}{\dfrac{1}{450} + \dfrac{0.0025}{208} + \left[\dfrac{0.0125 + 2(0.0008)}{23(0.0125) + 0.8949}\right]} = 3885 \text{ W/m}^2.$$

The heat flux without fins is

$$q'' = \frac{T_1 - T_2}{1/h_1 + \Delta x/k + 1/h_2}.$$

Assuming that the presence of the fins does not significantly change the surface coefficient of heat transfer,

$$q''_{\text{no fins}} = \frac{93 - 38}{\dfrac{1}{450} + \dfrac{0.0025}{208} + \dfrac{1}{23}} = 1203 \text{ W/m}^2.$$

Certainly it is clear, from this last calculation, that the controlling resistance to heat transfer is the third one in the denominator, namely $1/h_2$, since it is about twenty times larger than the next largest resistance. This motivates the addition of the fins in the first place.

Fin Efficiency

Perhaps the most commonly used measure of fin performance, especially for comparing fins of different shapes, is *fin efficiency* η_f, defined as the actual rate at which a fin transfers energy to the fluid divided by the rate at which it would transfer energy if the entire fin surface were at the base temperature of the fin. Qualitatively, the efficiency can be thought of as being the ratio of the difference between the average fin temperature and fluid temperature to the difference between the base temperature and the fluid temperature.

Straight Rectangular Fin   Applying this definition to the very commonly used straight fin of rectangular profile area pictured in Fig. 2.31 gives, for a fin which is extremely deep perpendicular to the plane of the figure,

$$\eta_f = \frac{q_0}{2h(L + \delta)(T_0 - T_2)}.$$

The actual heat transfer rate from the fin to the fluid, q, is given by Eq. (2.125) for this case, and the denominator shown is the heat transfer rate that would occur if the entire fin were at the temperature of its base, T_0. Using Eqs. (2.125) and (2.126) gives

$$\eta_f = \frac{k \, \delta \, N}{h(L + \delta)} \left[\frac{h/Nk + \tanh NL}{1 + (h/Nk) \tanh NL} \right], \tag{2.134}$$

with

$$N = \sqrt{h/k\delta} . \tag{2.135}$$

Straight Triangular Fin   Another straight fin sometimes used is the one of triangular profile area shown in Fig. 2.29, where δ is the maximum half thickness, that of the fin base. Because this fin's thickness varies with distance x away from the base, it is more complicated to deal with analytically than is the fin of rectangular profile area. The details of the solution are available in Schneider [8]; only the final result for the efficiency will be given here.

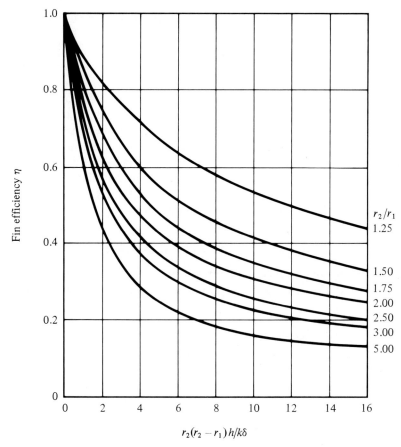

Figure 2.33 Efficiency of circular fins of triangular profile.

$$\eta_f = \frac{I_1(2NL)}{NL\, I_0(2NL)} \qquad\qquad (2.136)$$

$$N = \sqrt{h/k\delta} \qquad\qquad (2.137)$$

Circular Fins Figure 2.30 is a graphical representation of the solution for the common circular fin of rectangular profile area. The results in the figure are for the case in which the actual convective-type boundary condition at the fin tip is replaced by an approximately equivalent condition called the extended tip approximation. This approximation was introduced by Harper and Brown [6]. Schneider [8] indicates virtually no error due to this approximation as long as $h\delta/k \leqq 0.0625$, at least for the straight fin of rectangular profile area, and it is expected to be valid for the circular one as well.

The efficiency of circular fins of triangular profile area, from Smith and Sucec [7], is given in graphical form in Fig. 2.33, where δ refers to the maximum half thickness at the fin base.

These fin efficiency results can be used in a number of ways. For example, the actual fin heat transfer rate, q_0, can be conveniently calculated as the product of the efficiency, η_f, and the heat transfer rate which would occur if the entire fin surface were at the temperature of the base of the fin. In addition, the fin efficiency can be used to compare the performance of fins with different shapes. Theoretically, if straight fins of triangular profile were thought to be better than fins of rectangular profile as far as heat transfer rates are concerned, their efficiencies could be compared and a decision made about which to use. However, the comparison of fin efficiencies is only one factor to be considered in a design decision. Other significant factors are weight, ease of manufacture, cost, and space limitations. Ordinarily, no single factor can dictate the fin shape in a final design.

Additional details on fin analysis and design, including such considerations as optimum dimensions for a given profile and size, can be found in Schneider [8]. Extensive coverage of fins and their applications is available in Kern and Kraus [9].

2.6 STEADY TWO- AND THREE-DIMENSIONAL CONDUCTION

For steady, constant thermal conductivity, multidimensional conduction in a rectangular cartesian coordinate system, the temperature distribution $T = T(x, y, z)$ is governed by

$$\frac{\partial^2 T}{\partial x^2} + \frac{\partial^2 T}{\partial y^2} + \frac{\partial^2 T}{\partial z^2} + \frac{q'''}{k} = 0. \qquad [2.10]$$

This equation can sometimes be solved in an exact analytic fashion, although more complicated problems usually require an approximate solution by finite difference methods, which will be discussed in this section, and integral methods, which will not be discussed for conduction problems in this text. (Cf. [5] for integral methods.) This discussion of multidimensional conduction will begin with a number of problems for which so-called shape factors are available.

2.6a Shape Factors

In this section, we are concerned with steady-state two- and three-dimensional problems whose results can be put into the easy-to-use form of conduction shape factors. This form is simple for designers or analysts to use because the real work involved in finding the shape factors—the determination of the multidimensional temperature field in the solid—has already been carried out by engineers and researchers.

For certain multidimensional shapes with no generation, direct experimental methods, electrical analogue methods, and (for two-dimensional cases) conformal mapping and the graphical technique of flux plotting have yielded the temperature distribution between two isothermal surfaces of the solid when all other surfaces are adiabatic. These various techniques for finding the temperature profile will not be discussed here, but some of the results of these methods are presented in Table 2.1 in terms of the *conduction shape factor S*. The defining equation for this shape factor S is given as

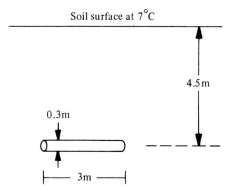

Figure 2.34 The buried cylinder of Ex. 2.10.

$$q = kS(T_1 - T_2), \tag{2.138}$$

where q is the heat transfer rate between the two isothermal surfaces whose temperatures are labelled as T_1 and T_2 in Table 2.1, k is the thermal conductivity of the solid material between the two isothermal surfaces, and S is the conduction shape factor given in the table. Steady-state conditions prevail, with no generation, k is constant, and additional restrictions, if any, are given in Table 2.1. Other configurations for which shape factors are available are given in Rohsenow and Hartnett [14].

EXAMPLE 2.10

Figure 2.34 shows a horizontal cylinder 0.3 m in diameter and 3 m long buried 4.5 m below the 7°C surface of the soil. The cylinder, which contains instruments for detecting radiation in the soil, has its surface held at 16°C. If the thermal conductivity of the soil is 0.6 W/m °C, find the total rate at which electrical energy must be dissipated within the cylinder to maintain steady-state conditions.

Solution
The total rate at which electrical energy must be dissipated to maintain the steady state is equal to the heat transfer rate at the outside surface. A check of Table 2.1 shows that the shape factor is available for this situation if end effects are neglected so that the cylinder of finite length is approximated by 3 m of an infinite cylinder.
Since $z = 4.5$ m and $D = 0.3$ m,

$$S = \frac{2\pi}{\cosh^{-1}(2z/D)} = \frac{2\pi}{\cosh^{-1}[2(4.5)/0.3]} = 1.532.$$

This is the shape factor per unit length, so Eq. (2.138) gives, after multiplying S by the length L,

$$q = kSL(T_1 - T_2) = 0.6(1.532)(3)(16 - 7) = 24.8 \text{ W}.$$

Table 2.1
Conduction Shape Factor S for Different Systems: $q = k S(T_1 - T_2)$

Physical System	Sketch	Shape Factor S	Limitations
a. Infinitely long cylinder of outside diameter D buried a distance z from an isothermal surface of a semi-infinite medium of thermal conductivity k.	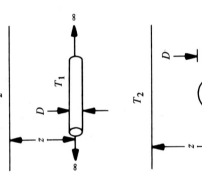	$\dfrac{2\pi}{\cosh^{-1}(2z/D)}$ (per unit of length)	
b. Isothermal sphere of diameter D buried a distance z from the isothermal surface of a semi-infinite medium of thermal conductivity k.		$\dfrac{2\pi D}{1 - D/4z}$	
c. Conduction heat transfer between two parallel isothermal cylinders buried in an infinite medium of thermal conductivity k.	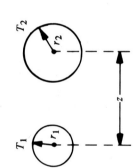	$\dfrac{2\pi}{\cosh^{-1}\left(\dfrac{z^2 - r_1^2 - r_2^2}{2r_1 r_2}\right)}$ (per unit of length)	

d. Conduction through two plane walls of equal thickness Δx and the edge along which they meet at right angles.

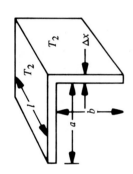

$$\frac{al}{\Delta x} + \frac{bl}{\Delta x} + 0.54l$$

(0.54l is the shape factor for the edge alone)

$b > \dfrac{\Delta x}{5}$

$a > \dfrac{\Delta x}{5}$

e. Conduction through a three-dimensional corner consisting of three plane walls of thickness Δx and thermal conductivity k meeting in a mutually perpendicular fashion.

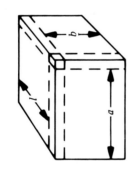

$0.15\,\Delta x$
(for the corner shown when one side of the three walls is at T_1 and the other side of the walls is at T_2)

$a > \dfrac{\Delta x}{5}$

$b > \dfrac{\Delta x}{5}$

$l > \dfrac{\Delta x}{5}$

Note that shape factors d and e can be used to compute the heat transfer from the inside of a hollow box, of inside dimensions a by b by l, to the outside surface when the wall thickness is Δx. For the hollow box, there are 8 three-dimensional corners, 12 two-dimensional edges, 4 each of length l, a, and b, and 6 walls, 2 of area a by b, 2 of area b by l, and 2 of area l by a, so that the expression for the heat transfer rate through the walls of the box becomes

$$q = k(T_1 - T_2)[(2)\frac{(ab + bl + al)}{\Delta x} + 1.2\Delta x + 0.54\,(4l + 4a + 4b)].$$

For boxes in which one or more of the inside edges is less than $\dfrac{\Delta x}{5}$, see [6]. This table was prepared principally from the information in [10], [12], and [13].

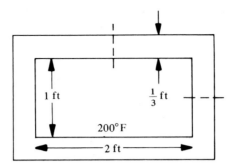

Figure 2.35 The cement duct of Ex. 2.11.

EXAMPLE 2.11

Figure 2.35 shows a cement duct whose inside surface temperature is 200°F, which is exposed to a 100°F fluid on the outside with an average surface coefficient of heat transfer of 3.0 Btu/hr ft² °F between the outside surface and the fluid. For cement, $k = 0.36$ Btu/hr ft °F. Calculate the steady-state heat transfer rate per foot of duct.

Solution

Table 2.1 does not have the configuration pictured. However, only one of the symmetrical quarters, such as the upper right one shown between dashed lines, must be considered. The corner for 1 ft of duct into the board is the same as (d) in Table 2.1, which is for conduction through two plane sections and the edge section of two walls of thermal conductivity k. (Note that our interior dimensions are greater than one-fifth the wall thickness.) Since $a = 1$ ft, $b = 1/2$ ft, $l = 1$ ft, and $\Delta x = 1/3$ ft,

$$S = \frac{al}{\Delta x} + \frac{bl}{\Delta x} + 0.54\, l = \frac{(1)\,(1)}{1/3} + \frac{\frac{1}{2}(1)}{1/3} + 0.54(1) = 5.04.$$

Since there are four of these sections, the heat transfer rate is

$$\begin{aligned} q = 4\, kS(T_1 - T_2) &= 4(0.36)(5.04)(200 - T_2) \\ &= 7.26(200 - T_2), \end{aligned} \tag{2.139}$$

where T_2 is unknown, but it is assumed that it is constant over the outside surface of the duct, since this assumption is needed to obtain the shape factor S. By Newton's law of cooling, $q = hA_2\,(T_2 - 100)$ and $A_2 = 2(1.667 + 2.667) = 8.67$ ft². Hence,

$$q = 26(T_2 - 100). \tag{2.140}$$

Combining Eqs. (2.139) and (2.140), the heat transfer rate is

$$q = 569 \text{ Btu/hr ft}.$$

Note that this problem utilizes the same concepts that are used to solve steady one-dimensional problems in Sec. 2.4, except that the shape factor was used to account for the two-dimensional effects.

2.7 EXACT ANALYTICAL SOLUTION OF MULTIDIMENSIONAL CONDUCTION PROBLEMS

Up to this point in the book, exact analytical solutions for the temperature distribution function have been developed only for the case of steady one-dimensional heat flow and for those actual multidimensional cases which can be adequately treated as being quasi-one-dimensional. Thus, so far, only ordinary differential equations have had to be solved. However, when the true two- or three-dimensional nature of a steady temperature distribution must be found, then by necessity one must deal with the solution of partial differential equations, such as Eqs. (2.10) and (2.11). The most common method of obtaining solutions to linear partial differential equations is the classical separation of variables technique, along with the aid of expansions in infinite series of orthogonal functions, such as the familiar Fourier series, to satisfy the nonhomogeneous boundary condition. The engineering student's mathematics background will include a good degree of familiarity with Fourier series expansions and with at least a short exposure to the method of separation of variables. In any event, the method of separation of variables will be presented in enough detail in this section to either refresh readers' memories or to allow them to learn the technique if they are seeing this type of material for the first time. Additional material for review is available in typical engineering mathematics texts such as Wylie [2].

The presentation of exact analytical solutions in this section on steady-state conduction will carry over, in large degree, to the exact solution of some very important unsteady one-dimensional problems in the next chapter.

It may be recalled that the method of separation of variables can be applied directly only to linear, homogeneous partial differential equations. Note that when there is generation q''', Eq. (2.10), while linear as long as q''' is linear, is not ordinarily homogeneous. This difficulty can often be overcome by transformations, which in effect shift the generation term to an ordinary differential equation and leave another homogeneous partial differential equation to be solved. Proper combination of the two solutions gives the required temperature distribution. (See [5] for further details when generation is present.) The following discussion is limited to the case of zero generation.

2.7a A Planar Case

Consider the steady-state, zero-generation, constant thermal conductivity, two-dimensional conduction region shown in Fig. 2.36 of dimension L by l. The thermal loadings at the boundaries are the known temperatures T_1 and T_2, and the bottom face is perfectly insulated. We would like to derive an exact solution for the temperature T as a function of x and y.

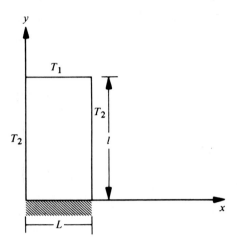

Figure 2.36 The two-dimensional conduction region.

The governing partial differential equation is Eq. (2.10) with the z conduction term and the generation rate per unit volume set equal to zero; that is,

$$\frac{\partial^2 T}{\partial x^2} + \frac{\partial^2 T}{\partial y^2} = 0. \tag{2.141}$$

The boundary conditions are ④

①   at $x = 0$ and $0 < y < l,$ $T = T_2,$

②   at $x = L$ and $0 < y < l,$ $T = T_2,$

and at $y = 0$ and $0 < x < L$, the surface is insulated perfectly so there can be no conduction normal to the surface, that is, no y conduction; hence $q_y'' = 0$ at $y = 0$. But by Fourier's exact law of conduction,

$$q_y'' = -k\,\frac{\partial T}{\partial y}.$$

Thus, $\partial T/\partial y = 0$ at $y = 0$ for $0 < x < L$, and, at $y = l$ and $0 < x < L, T = T_1$. Before solving Eq. (2.141) and applying the boundary conditions, note that the following transformation simplifies the boundary conditions:

$$\theta = T - T_2. \tag{2.142}$$

Note further that with θ as the new dependent variable, three zeroes on the boundary are obtained and the form of Eq. (2.141) remains the same. Solving for T from Eq. (2.142), $T = \theta + T_2$; hence,

$$\frac{\partial T}{\partial x} = \frac{\partial \theta}{\partial x} + 0, \quad \frac{\partial^2 T}{\partial x^2} = \frac{\partial^2 \theta}{\partial x^2}.$$

Similarly,

$$\frac{\partial^2 T}{\partial y^2} = \frac{\partial^2 \theta}{\partial y^2}.$$

The governing equation then becomes

$$\frac{\partial^2 \theta}{\partial x^2} + \frac{\partial^2 \theta}{\partial y^2} = 0, \qquad\qquad (2.143)$$

with the boundary conditions

at $x = 0$ and $0 < y < l$, $\theta = 0$,

at $x = L$ and $0 < y < l$, $\theta = 0$,

at $y = 0$ and $0 < x < L$, $\partial\theta/\partial y = 0$,

at $y = l$ and $0 < x < L$, $\theta = \theta_1 = T_1 - T_2$.

This is now a properly posed boundary value problem for the separation of variables technique, in that it consists of a linear homogeneous partial differential equation with all boundary conditions homogeneous except one. The separation of variables technique consists of assuming that the solution for $\theta = \theta(x, y)$ can be expressed as a product of a function of x(alone) times a function of y(alone). That is,

$$\theta(x, y) = R(x)Q(y), \qquad\qquad (2.144)$$

where $R(x)$ and $Q(y)$ must be determined by the insertion of Eq. (2.144) in the governing partial differential equation, Eq. (2.143). For convenience, write Eq. (2.144) as $\theta = RQ$, with the individual functional dependencies kept in mind from Eq. (2.144). Since the term $\frac{\partial^2 \theta}{\partial x^2} = \frac{\partial}{\partial x}\left(\frac{\partial \theta}{\partial x}\right)$ is needed, from Eq. (2.144),

$$\frac{\partial \theta}{\partial x} = \frac{dR}{dx} Q. \qquad\qquad (2.145)$$

Since Q does not depend on x, the partial derivative with respect to x of Eq. (2.145) yields

$$\frac{\partial^2 \theta}{\partial x^2} = \frac{d^2 R}{dx^2} Q. \qquad\qquad (2.146)$$

Similarly,

$$\frac{\partial^2 \theta}{\partial y^2} = R \frac{d^2 Q}{dy^2}. \qquad\qquad (2.147)$$

Steady-State Conduction

Inserting Eqs. (2.146) and (2.147) into Eq. (2.143) yields

$$\frac{d^2R}{dx^2} Q + R \frac{d^2Q}{dy^2} = 0. \tag{2.148}$$

Transposing the second term of Eq. (2.148) and dividing by the product RQ,

$$\frac{d^2R/dx^2}{R} = \frac{-d^2Q/dy^2}{Q}. \tag{2.149}$$

Since R depends only upon x and Q depends only upon y, the left-hand side of Eq. (2.149) depends only on x while the right-hand side depends only upon y. Yet the left-hand side must equal the right-hand side for all values of x and all values of y in the conduction region, even though x and y can be varied independently of one another. Thus, this equality can hold only if both sides of the equation are equal to the same constant denoted by $-\lambda^2$. The minus sign rather than the plus sign was selected for λ^2 because it allows all four boundary conditions to be eventually satisfied, whereas the plus sign does not. More formally, the algebraic sign is dictated by the need to have oscillatory functions in the x direction so that the two homogeneous boundary conditions can be satisfied in a nontrivial fashion. The value λ^2 is known as the *separation constant*. Setting the left-hand side of Eq. (2.149) equal to $-\lambda^2$ and then the right-hand side equal to $-\lambda^2$ gives the following ordinary differential equations, called the *separation equations:*

$$\frac{d^2R}{dx^2} + \lambda^2 R = 0, \tag{2.150}$$

$$\frac{d^2Q}{dy^2} - \lambda^2 Q = 0. \tag{2.151}$$

The solutions to Eqs. (2.150) and (2.151) are easily found using the classical operator technique, and since the domain is finite in the y direction, the solution for $Q(y)$ is in terms of the hyperbolic functions. Thus,

$$R = A_1 \sin \lambda x + A_2 \cos \lambda x, \quad Q = B_1 \sinh \lambda y + B_2 \cosh \lambda y.$$

Since $\theta = RQ$,

$$\theta = (A_1 \sin \lambda x + A_2 \cos \lambda x)(B_1 \sinh \lambda y + B_2 \cosh \lambda y). \tag{2.152}$$

The integration constants A_1, A_2, B_1, and B_2, and the value of λ will be determined by the application of the boundary conditions. The boundary condition at $x = 0$ and $0 < y < l$ is $\theta = 0$, which applied to Eq. (2.152) yields

$$0 = (0 + A_2)(B_1 \sinh \lambda y + B_2 \cosh \lambda y).$$

Note that this condition is not satisfied by the function of y because the condition must hold for *all* y in the conduction region. Hence, $A_2 = 0$, and calling $A_1 B_1 = C_1$ and $A_1 B_2 = C_2$, Eq. (2.152) is rewritten as

$$\theta = \sin \lambda x \, (C_1 \sinh \lambda y + C_2 \cosh \lambda y). \tag{2.153}$$

The second boundary condition requires that $\theta = 0$ at $x = L$ again for all y in the conduction region; so once again the function of y cannot satisfy the condition. (Note that the choice $C_1 = C_2 = 0$ is a trivial one, since it gives the solution that $\theta = 0$ for all x and y, which is incorrect here.) Hence, it must be required that

$$\sin \lambda L = 0. \tag{2.154}$$

This will be true if λL is some integral number of π radians; that is, if $\lambda L = n\pi$, where n is any positive integer. Thus,

$$\lambda = \frac{n\pi}{L}. \tag{2.155}$$

Equation (2.154) is called the *characteristic* equation or the *eigencondition*, the condition which determines the permissible values of the separation constants λ. The value λ is also referred to as a *characteristic value* or an *eigenvalue*.

Note that, from Eq. (2.154), $\lambda = 0$, a zero eigenvalue, also satisfies the equation. The zero eigenvalue must generally be treated separately, since it is unusual in the sense that it often changes the entire character of the separation equations and introduces new separation functions. This is most easily seen by examination of Eqs. (2.150) and (2.151) when $\lambda = 0$. In this case, these equations become the following:

$$\frac{d^2 R_0}{dx^2} = 0, \quad \text{and} \quad \frac{d^2 Q_0}{dy^2} = 0.$$

Solving these by integrating twice and then multiplying their solutions together yields the solution for θ—call it θ_0—corresponding to the zero eigenvalue.

$$\theta_0 = (A_0 + B_0 x)(C_0 + D_0 y) \tag{2.156}$$

Now, Eq. (2.156) has not yet been forced to satisfy the first two boundary conditions that have already been applied to Eq. (2.152). Hence, at $x = 0$, $0 < y < l$, $\theta_0 = 0$, and this condition applied to Eq. (2.156) leads to $A_0 = 0$. Thus, when the B_0 is absorbed in the C_0 and D_0, Eq. (2.156) becomes

$$\theta_0 = x(C_0 + D_0 y).$$

Next, applying the condition that at $x = L$, $0 < y < l$, $\theta_0 = 0$ gives,

$$\theta_0 = L(C_0 + D_0 y).$$

The only way this condition can be satisfied for *all* y in the interval from 0 to l is to make both C_0 and D_0 equal to zero, thus giving $\theta_0 = 0$. Hence, the existence of a zero eigenvalue, which introduces new separation functions, has no effect on the solution, since none of the new functions were needed to satisfy the boundary conditions. It should be stressed that there are many problems in which the zero eigenvalue yields new separation functions which *are* needed to solve the problem for the temperature distribution. Thus, whenever the zero eigenvalue appears, the new separation functions it introduces should be checked to see if they are needed in the final solution.

Next, the third boundary condition, that $\partial\theta/\partial y = 0$ at $y = 0$ for all x in the conduction region, will be applied to Eq. (2.153). For the condition that $\partial\theta/\partial y = 0$ at $y = 0$ for all x in the conduction region,

$$\frac{\partial\theta}{\partial y} = \sin \lambda x \, (\lambda C_1 \cosh \lambda y + \lambda C_2 \sinh \lambda y).$$

At $y = 0$,

$$\frac{\partial\theta}{\partial y} = 0 = \sin \lambda x \, (\lambda C_1 + 0).$$

Hence, if the condition is to hold for all x, $C_1 = 0$, and using this along with Eqs. (2.155) and (2.153), the solution becomes

$$\theta = C_2 \sin (n\pi x/L) \cosh (n\pi y/L). \tag{2.157}$$

The last condition to be satisfied requires that at $y = l$, for $0 < x < L$, $\theta = \theta_1$, a known constant. This condition cannot be satisfied by Eq. (2.157) as it stands. However, n can be any positive integer, so that it is expected that the eventual value of C_2 will depend on the value of n chosen. There are an infinite number of solutions of the type shown in Eq. (2.157) corresponding to values of n from 1 to ∞. A representative equation would be

$$\theta = C_n \cosh (n\pi y/L) \sin (n\pi x/L).$$

The principle of superposition for a linear homogeneous equation states that if there are a number of valid solutions, the sum of these is also a valid solution; hence, summing up all solutions of this type yields

$$\theta = \sum_{n=1}^{\infty} C_n \cosh (n\pi y/L) \sin (n\pi x/L). \tag{2.158}$$

Now at $y = l$ and $0 < x < L$, $\theta = \theta_1$; hence,

$$\theta_1 = \sum_{n=1}^{\infty} C_n \cosh (n\pi l/L) \sin (n\pi x/L). \tag{2.159}$$

The question is, in effect, whether C_n can be chosen in such a way that the infinite series of sines on the right-hand side of Eq. (2.159) converges to the left-hand side, θ_1, in the interval $0 < x < L$. Since the right-hand side of Eq. (2.159) is a Fourier sine series, the answer is yes, because a Fourier sine series can adequately represent any well behaved function in the interval $0 < x < L$; see [5]. By identifying the coefficients of a Fourier sine series with $C_n \cosh(n\pi l/L)$, C_n can be evaluated from the equation for the coefficients of such a series; that is,

$$A_n = \frac{2}{L} \int_0^L F(x) \sin\left(\frac{n\pi x}{L}\right) dx. \tag{2.160}$$

The A_n for a general function of x in the interval $0 < x < L$ appear in the function's Fourier sine series expansion as

$$F(x) = \sum_{n=1}^{\infty} A_n \sin\left(\frac{n\pi x}{L}\right). \tag{2.161}$$

Comparing Eqs. (2.160) and (2.161) to Eq. (2.159), it is found that here the function $F(x)$ is the constant θ_1, and the A_n are equal to

$$A_n = C_n \cosh\left(\frac{n\pi l}{L}\right) = \frac{2}{L} \int_0^L \theta_1 \sin\left(\frac{n\pi x}{L}\right) dx. \tag{2.162}$$

Integrating Eq. (2.162) and solving for C_n yields

$$C_n = 0 \quad \text{for even } n,$$

$$C_n = \frac{4\theta_1/n\pi}{\cosh(n\pi l/L)} \quad \text{for odd } n = 1, 3, 5, \ldots . \tag{2.163}$$

Substituting Eq. (2.163) into Eq. (2.158) and recalling that $\theta = T - T_2$ and $\theta_1 = T_1 - T_2$, the temperature distribution is

$$\frac{T - T_2}{T_1 - T_2} = \frac{4}{\pi} \sum_{n=1,3,5,\ldots}^{\infty} \frac{\cosh(n\pi y/L)\sin(n\pi x/L)}{n \cosh(n\pi l/L)}. \tag{2.164}$$

Now that the temperature profile, Eq. (2.164), has been found, values of the temperature at any space point x, y of interest in the region can be found. Also, the use of Eq. (2.164) in Fourier's exact law of conduction allows the calculation of the heat transfer rate across any surface of interest.

The rectangular plate situation whose solution has just been developed is one of the simplest multidimensional steady-state conduction problems. The more difficult ones, such as those involving the convective-type boundary condition, the boundary condition of the

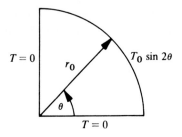

Figure 2.37 The one-quarter cylinder.

third kind, require additional background in the theory of sets of functions which exhibit orthogonality with respect to a weighting function within an interval. Arpaci [5] gives a fairly complete presentation of this material along with exact analytical solutions to many steady-state multidimensional problems.

2.7b A Circular Cylindrical Case

As a final application of exact solution methods for two-dimensional conduction problems, consider the cross section, shown in Fig. 2.37, of a long quarter cylinder whose left and lower faces are both held at a temperature excess of 0 while the curved lateral surface has a temperature excess T which varies with the angular coordinate θ as $T_0 \sin 2\theta$. There is no generation and conditions are steady. We wish to find the details of the temperature distribution's dependence upon the space coordinates r and θ.

The governing partial differential equation is given by Eq. (2.8) with the generation term, storage term, and z conduction term set equal to zero; that is,

$$\frac{\partial^2 T}{\partial r^2} + \frac{1}{r}\frac{\partial T}{\partial r} + \frac{1}{r^2}\frac{\partial^2 T}{\partial \theta^2} = 0. \qquad (2.165)$$

The boundary conditions are

$$
\begin{aligned}
&\text{at } r = 0 \text{ and } 0 < \theta < \tfrac{1}{2}\pi, \quad T \text{ should be finite,}\\
&\text{at } \theta = 0 \text{ and } 0 < r < r_0, \quad T = 0,\\
&\text{at } \theta = \tfrac{1}{2}\pi \text{ and } 0 < r < r_0, \quad T = 0,\\
&\text{at } r = r_0 \text{ and } 0 < \theta < \tfrac{1}{2}\pi, \quad T = T_0 \sin 2\theta.
\end{aligned}
$$

Equation (2.165), subject to the boundary conditions listed here, is solved by the method of separation of variables. Let

$$T(r, \theta) = Q(r)P(\theta), \qquad (2.166)$$

where $Q(r)$ and $P(\theta)$ are the separation functions to be found by insertion of Eq. (2.166) into Eq. (2.165). For convenience the notation $T = QP$ will be used. Partial differentiation of Eq. (2.166) with respect to r yields

$$\frac{\partial T}{\partial r} = P \frac{dQ}{dr} \tag{2.167}$$

since P is independent of r, and

$$\frac{\partial^2 T}{\partial r^2} = P \frac{d^2Q}{dr^2}. \tag{2.168}$$

The second derivative of T with respect to θ is found to be, noting that the function Q is independent of θ,

$$\frac{\partial^2 T}{\partial \theta^2} = Q \frac{d^2P}{d\theta^2}. \tag{2.169}$$

Inserting Eqs. (2.167), (2.168), and (2.169) into Eq. (2.165), multiplying through by r^2, and dividing by the product PQ gives

$$\frac{r^2(d^2Q/dr^2) + r(dQ/dr)}{Q} + \frac{d^2P/d\theta^2}{P} = 0,$$

or, by transposing,

$$\frac{r^2(d^2Q/dr^2) + r(dQ/dr)}{Q} = \frac{-d^2P/d\theta^2}{P}. \tag{2.170}$$

Since the left-hand side of Eq. (2.170) depends only on r and the right-hand side only on θ, and yet they must equal each other for all values of θ and r (which are independent of one another) in the conduction region, both sides must equal the same constant, the separation constant λ^2. Setting both sides of Eq. (2.170) individually equal to λ^2 and rearranging the results gives

$$\frac{d^2P}{d\theta^2} + \lambda^2 P = 0, \tag{2.171}$$

$$r^2 \frac{d^2Q}{dr^2} + r \frac{dQ}{dr} - \lambda^2 Q = 0. \tag{2.172}$$

The solution to Eq. (2.171) is

$$P = A_1 \sin \lambda\theta + A_2 \cos \lambda\theta. \tag{2.173}$$

Equation (2.172) is called *Euler's equation*, or the *equidimensional equation*, or sometimes *Cauchy's equation*, and its solution can be found in any standard engineering mathematics text, such as [2], as

$$Q = B_1 r^\lambda + B_2 r^{-\lambda}. \tag{2.174}$$

The product of Eqs. (2.173) and (2.174) is the general solution for the temperature distribution,

$$T = (B_1 r^\lambda + B_2 r^{-\lambda})(A_1 \sin \lambda\theta + A_2 \cos \lambda\theta). \tag{2.175}$$

The first boundary condition requires that T be finite at $r = 0$ for $0 < \theta < \frac{1}{2}\pi$. From Eq. (2.175), if $\lambda > 0$, the term $r^{-\lambda}$ becomes infinite at $r = 0$, and if $\lambda < 0$, r^λ would be infinite at the origin. The separation constants ordinarily are positive, so B_2 must be chosen equal to zero, and defining the products

$$B_1 A_1 = C_1, \quad B_1 A_2 = C_2,$$

Eq. (2.175) becomes, with the first boundary condition satisfied,

$$T = r^\lambda(C_1 \sin \lambda\theta + C_2 \cos \lambda\theta). \tag{2.176}$$

The second boundary condition requires that $T = 0$ at $\theta = 0$ for all r in $0 < r < r_0$. Hence, Eq. (2.176) becomes

$$0 = r^\lambda[C_1(0) + C_2].$$

Hence, $C_2 = 0$, and Eq. (2.176) now becomes with the first and second boundary conditions satisfied,

$$T = C_1 r^\lambda \sin \lambda\theta. \tag{2.177}$$

The third boundary condition requires that $T = 0$ at $\theta = \frac{1}{2}\pi$ for all values of r within the conduction region; hence Eq. (2.177) becomes

$$0 = C_1 r^\lambda \sin(\lambda \tfrac{1}{2}\pi). \tag{2.178}$$

Equation (2.178) can be satisfied in a nontrivial fashion for all values of r only if $\sin(\lambda\frac{1}{2}\pi)$ can be chosen equal to zero. Letting the separation constant λ be any positive, even integer such as $\lambda = n = 2, 4, 6, 8, \ldots$, allows $\sin \lambda\frac{1}{2}\pi$ to be zero. However, $\lambda = 0$ also satisfies Eq. (2.178), and thus once again the zero eigenvalue and its associated eigenfunctions must be investigated. When $\lambda = 0$, the solutions of the separation equations, Eqs. (2.171) and (2.172), when multiplied together give the form

$$(A_0 + B_0\theta)(C_0 + D_0 \ln r). \tag{2.179}$$

Next, the first three boundary conditions which have already been applied to Eq. (2.176), the solution for $\lambda \neq 0$, must be forced on Eq. (2.179). Application of the first one requires that $D_0 = 0$, the second dictates that $A_0 = 0$, and the third one results in $B_0 = 0$, thus giving a zero for Eq. (2.179) and no contribution due to the zero eigenvalue in this particular situation.

Hence, the solution with three of the four boundary conditions satisfied and with $\lambda = n = 2, 4, 6, 8, \ldots$ is given by,

$$T = C_1 r^n \sin n\theta, \quad n = 2, 4, 6, \ldots . \tag{2.180}$$

Usually, the next step is to sum all possible solutions of the type of Eq. (2.180) and attempt to satisfy the nonhomogeneous boundary condition by an expansion in an infinite series of orthogonal functions. However, before summing up all possible solutions, it is of interest to determine if one value of n can be chosen that will allow the last boundary condition to be satisfied. The fourth boundary condition demands that at $r = r_0$, $T = T_0 \sin 2\theta$; hence, at r_0, Eq. (2.180) becomes

$$T_0 \sin 2\theta = C_1 r_0^n \sin n\theta. \tag{2.181}$$

The two sides of Eq. (2.181) can be made identical by choosing $n = 2$ (which is a permissible eigenvalue since the third boundary condition required n to be an element of the set of positive even integers) and then solving for C_1 as

$$C_1 = T_0/r_0^2 . \tag{2.182}$$

Using $n = 2$ and Eq. (2.182) for C_1 in Eq. (2.180), the final solution for the temperature distribution as a function of r and θ is

---

$$T = T_0(r/r_0)^2 \sin 2\theta. \tag{2.183}$$

Note that, unlike the case of the rectangular plate previously treated, here an infinite series was not needed to satisfy the last, nonhomogeneous boundary condition, because the function on the boundary happened to be the same as one of the solution functions of the separation equation in θ. This is not a frequent occurrence, but it does happen at times.

EXAMPLE 2.12

Calculate the heat transfer rate across the surface at $r = r_0$ in Fig. 2.37 when $T_0 = 38°C$, $r_0 = 0.3$ m, $k = 35$ W/m °C, and the conduction region is 10 m long in the z direction.

Solution
From Eq. (2.183), the temperature distribution is

$$T = T_0(r/r_0)^2 \sin 2\theta = 38(r/0.3)^2 \sin 2\theta = 422 r^2 \sin 2\theta. \tag{2.184}$$

The heat transfer rate across a differential portion of the outside surface of area $10\, r_0\, d\theta$ is given by Fourier's exact law of conduction as

$$dq = -k\, 10r_0\, d\theta\, \frac{\partial T}{\partial r}\bigg|_{r_0}. \tag{2.185}$$

From Eq. (2.184),

$$\partial T / \partial r = 2(422)r \sin 2\theta.$$

Hence,

$$\frac{\partial T}{\partial r}\bigg|_{r_0} = 844\, r_0 \sin 2\theta. \tag{2.186}$$

Substituting Eq. (2.186) into Eq. (2.185) and using the known values of k and r_0 gives

$$dq = -26{,}586 \sin 2\theta\, d\theta. \tag{2.187}$$

The total heat transfer rate across the surface at $r = r_0$ is found by summing up all contributions of the type given by Eq. (2.187) as a definite integral from $\theta = 0$ to $\theta = \frac{1}{2}\pi$. As a result,

$$q = -26{,}586 \int_0^{\frac{1}{2}\pi} \sin 2\theta\, d\theta = -26{,}586\left(-\tfrac{1}{2} \cos 2\theta\right)\bigg|_0^{\frac{1}{2}\pi}$$
$$= -26{,}586 \text{ W.}$$

The negative sign indicates, in accordance with the convention of Chapter 1, that the heat flow is in the negative coordinate direction; that is, it is radially inward, which is expected by virtue of the values of the temperatures at the various faces.

This completes the formal discussion of exact solutions to multidimensional problems. Many other exact solutions, for shapes and boundary conditions different from those considered in this section, are available in books dealing with conduction heat transfer, such as the ones by Arpaci [5], Schneider [8], Ozisik [15], Myers [16], and the one which is still considered to be the "bible" of conduction, the treatise by Carslaw and Jaeger [17].

2.8 FINITE DIFFERENCE METHODS IN STEADY-STATE CONDUCTION

In a large number of practical problems involving steady-state heat transfer in a solid conduction region, an exact mathematical solution is precluded as a result of the complex shape of the conduction region, the type of boundary conditions, the energy generation rate per unit volume, variable thermal conductivity, or any combination of these. In such situations, the temperature distribution can frequently be determined by an approximate finite difference analysis. The finite difference equation approximates the governing partial differential equation at a finite number of points within the conduction region, called *nodes* or *nodal points, grid points,* or *lattice points,* by an algebraic finite difference equation at each point. Thus, if *n* nodal points are selected at which a solution for the approximate steady-state temperatures is desired, *n* simultaneous algebraic equations for the *n* unknown temperatures are solved.

Since the finite difference method ultimately requires the solution of *n* algebraic equations where *n* may often be a fairly large number, it is ideally performed with the use of high-speed digital computers and electronic calculators. A number of library programs for the solution of sets of algebraic equations are readily available. They require only that the numerical value of the various coefficients in the equation set of interest be specified by the engineer as input to the program. Hence the main task of the engineer, when using finite difference methods, is the setting up, or derivation, of the algebraic finite difference equations which are appropriate for the problem at hand. The ready availability of computers in industrial concerns, and even computer programming groups in the larger industries or the higher technology ones, has made the finite difference method an extremely valuable tool of analysis and design for the engineer, a tool which is often used to solve the complex conduction heat transfer problems that arise in industry.

2.8a Derivation of the Finite Difference Equations

Once *n* different points, the nodes, are selected within a conduction region at which a solution for the steady-state temperatures is desired, the next task of the engineer is to develop the proper algebraic finite difference equations for all *n* points at which the temperature is desired. There are three basic approaches that can be used to derive these algebraic equations. One approach is to take the governing partial differential equation such as, for constant thermal conductivity,

$$\frac{\partial^2 T}{\partial x^2} + \frac{\partial^2 T}{\partial y^2} + \frac{\partial^2 T}{\partial z^2} + \frac{q'''}{k} = 0. \qquad \textbf{[2.10]}$$

Since it must be satisfied everywhere in the conduction region, it must be satisfied at each node, say at node *j,* so that

$$\frac{\partial^2 T}{\partial x^2}\bigg|_j + \frac{\partial^2 T}{\partial y^2}\bigg|_j + \frac{\partial^2 T}{\partial z^2}\bigg|_j + \frac{q'''_j}{k} = 0. \qquad \textbf{(2.188)}$$

Now the various terms, such as $\dfrac{\partial^2 T}{\partial x^2}\bigg|_j$, are approximated by using the definition of the derivative, except that the spacing Δx between nodes in the *x* direction does not approach zero, but remains finite. This is a mathematical approach that can be used even by someone who does not understand the physics represented by the governing differential equation; that is, by a person who does not know that Eq. (2.10) is an energy balance, on a differential volume, which proclaims that the net conduction heat transfer rate in plus the generation must equal zero.

The second approach, the Taylor series method, is a more formal mathematical approach than the one just discussed, and provides to the finite difference analyst the algebraic finite difference equation as well as some important information about the equation. This approach consists of expanding the temperature at the jth node into a Taylor series about the node, evaluating the resulting series at other nodes that are clustered around node j, and then rearranging these series so that they can be solved for the individual partial derivatives of Eq. (2.188) in terms of nodal temperatures. When this is done, the concepts of *compatibility* (whether or not the finite difference equation properly simulates the partial differential equation) and of *truncation error* (a measure of how accurate the finite difference equation representation is) can be easily, and rather directly, studied. In addition, this method can also be used to study the considerably more complex topic of *convergence* (whether or not the solution of the finite difference equations approaches the solution to the partial differential equation) of the finite difference procedure. The Taylor series method and its application to the three topics just mentioned is illustrated nicely in the book by Smith [18].

The third approach to deriving algebraic finite difference equations, and the one which will be emphasized in this book, is a physical approach which employs energy balances. This approach usually appeals to engineers because it is rooted in an understanding of the physics of what is going on and because it is easier to apply than the Taylor series method. (This ease of application, relative to the Taylor series approach, is purchased at a price—namely, less information about certain characteristics of the finite difference equation. Ideally, in more advanced finite difference work, the individual strengths of both methods should be exploited by employing both of them in the same problem in a complementary fashion.)

In this physical approach, one identifies the small, but finite, volume of material associated with each nodal point and then makes an energy balance on this volume (that is, applies Eq. [1.8]) to derive the algebraic finite difference equation for each node. The phrase "material associated with each node" means that this material is essentially at the temperature of the node. The reader will recall that this procedure has already been used to derive the governing *partial* differential equation, Eq. (2.5), by an energy balance on a *differential* volume element.

General Interior Node

Consider the general three-dimensional conduction region shown in Fig. 2.38. For the purposes of a finite difference analysis, nodal points, or nodes, have been distributed throughout this region and over its surface in a regular pattern. Some representative nodes are shown as points in the figure. One has been labelled node 0, a general node *inside* the conduction region. We would like to derive the governing algebraic finite difference equation for this general interior node 0 in this three-dimensional region for steady-state conditions with generation, but with constant thermal conductivity. (Note that the variable thermal conductivity case is treated essentially the same way, but the resulting equation is more complicated algebraically. The important features of the method can be illustrated without this additional complexity, while at the same time it will be obvious how the variable thermal conductivity case would be considered.)

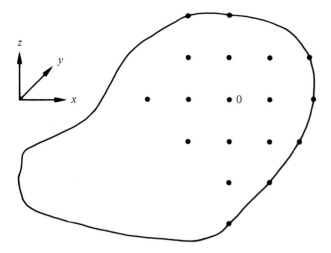

Figure 2.38 Nodes within and on a conduction region.

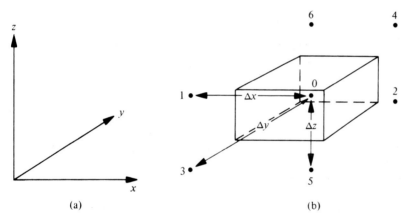

(a)   (b)

Figure 2.39 (a) Coordinate system and (b) general interior node in a three-dimensional conduction region.

Figure 2.39(b) shows an interior node, labelled node 0, in a three-dimensional conduction region surrounded by six other nodes in the three coordinate directions. The lattice spacing Δx in the x direction is the distance between any two nodes in the x direction, and the spacings in the y and z directions are called Δy and Δz, respectively. The rectangular parallelepiped shown in Fig. 2.39(b) with node 0 at its center is the material associated with node 0. The six nodes surrounding node 0 also have material associated with them, but are not outlined on the figure.

The spacing between nodes 1 and 0, and 0 and 2, is Δx, while between 3 and 0, and 0 and 4, it is Δy, and it is Δz between 5 and 0, and 0 and 6. Note that each of the six faces of the volume of material $\Delta x \Delta y \Delta z$ associated with node 0 lie halfway between node 0 and the adjacent node in the coordinate direction perpendicular to the face. Now, the law of conservation of energy, Eq. (1.8), applied to the control volume $\Delta x \Delta y \Delta z$ in the steady state gives

$$R_{in} + R_{gen} = R_{out}.$$

However, the only way energy can enter or leave a control volume which has all of its surfaces bounded by solid material is by conduction across the six faces of the control volume. Without loss of generality, it can be assumed that all of the conduction terms are into the node of interest; hence, Eq. (1.8) becomes

$$R_{in} + R_{gen} = 0.$$

Let q_{10} be the rate at which energy is conducted into the control volume from the material around node 1, and use a similar numbering scheme for the conduction heat transfer rates from the other five nodes surrounding node 0. Then,

$$R_{in} = q_{10} + q_{20} + q_{30} + q_{40} + q_{50} + q_{60}.$$

The total rate at which energy is generated within the control volume is given by the local generation rate per unit volume times the volume of material associated with node 0. Thus,

$$R_{gen} = q_0''' \, \Delta x \Delta y \Delta z.$$

The energy balance on node zero becomes

$$q_{10} + q_{20} + q_{30} + q_{40} + q_{50} + q_{60} + q_0''' \, \Delta x \Delta y \Delta z = 0.$$

Now, to relate q_{10}, for example, to the nodal temperatures, Fourier's simplified law of conduction is used. This is part of the finite difference approximation, since the conditions for the use of the simplified form of Fourier's law are not satisfied unless the finite spacings Δx, Δy, and Δz approach the infinitesimal quantities dx, dy, and dz. Fourier's simplified law of conduction for q_{10} gives

$$q_{10} = kA \frac{(T_1 - T_0)}{L}.$$

From Fig. 2.39(b), L is equal to Δx and the area A perpendicular to the heat flow is equal to $\Delta y \, \Delta z$; hence,

$$q_{10} = k \, \Delta y \, \Delta z \frac{(T_1 - T_0)}{\Delta x}.$$

For q_{20},

$$q_{20} = k \, \Delta y \, \Delta z \frac{(T_2 - T_0)}{\Delta x}.$$

Chapter Two

(Note that the temperature difference is written as $T_2 - T_0$, since it is assumed that q_{20} is *into* the control volume.)

The other heat transfer rates are, using the same type of reasoning as above,

$$q_{30} = k\,\Delta z\,\Delta x \frac{(T_3 - T_0)}{\Delta y}\,, \qquad q_{40} = k\,\Delta z\,\Delta x \frac{(T_4 - T_0)}{\Delta y}\,,$$

$$q_{50} = k\,\Delta y\,\Delta x \frac{(T_5 - T_0)}{\Delta z}\,, \qquad q_{60} = k\,\Delta y\,\Delta x \frac{(T_6 - T_0)}{\Delta z}\,.$$

After we insert these expressions for the heat transfer rates into Eq. (1.8) and divide each term by $k\,\Delta x\,\Delta y\,\Delta z$, the algebraic finite difference equation for the general interior node 0 is

$$\frac{T_1 + T_2 - 2T_0}{\Delta x^2} + \frac{T_3 + T_4 - 2T_0}{\Delta y^2} + \frac{T_5 + T_6 - 2T_0}{\Delta z^2}$$

$$+ \frac{q_0'''}{k} = 0. \tag{2.189}$$

Equation (2.189) is the finite difference equation for an interior node (one which has the volume of material $\Delta x\,\Delta y\,\Delta z$ associated with it) in a steady, three-dimensional, constant thermal conductivity conduction region, where the nodes are referenced to a rectangular cartesian coordinate system, as shown in Fig. 2.39(a).

There will be an equation of the same type as Eq. (2.189) for all interior nodes—that is, one for node 5, one for node 3, one for each node including the nodes that have not been shown, since the entire conduction region has nodes throughout spaced Δx apart in the x direction, Δy in the y direction, and Δz in the z direction. Since all these nodes are interior nodes, the finite difference equation is the same as Eq. (2.189) with just the subscripts changed. For instance, if the node labelled 40 is surrounded by nodes labelled 39 and 41 in the x direction, 17 and 18 in the y direction, and 178 and 180 in the z direction, the finite difference equation for node 40 becomes, by comparison with Eq. (2.189),

$$\frac{T_{39} + T_{41} - 2T_{40}}{\Delta x^2} + \frac{T_{17} + T_{18} - 2T_{40}}{\Delta y^2} + \frac{T_{178} + T_{180} - 2T_{40}}{\Delta z^2}$$

$$+ \frac{q_{40}'''}{k} = 0.$$

If the temperatures of the different portions of the outer surface of the entire conduction region are specified and the region's shape is such that all nodes are interior nodes of the type 0, as shown in Fig. 2.39, then an equation of the type of Eq. (2.189) applies at each of the n nodes. In the equations for the nodes that are exactly Δx or Δy or Δz from the outer surface, one or more of the temperatures appearing will be known, since at least one of the temperatures will apply to a point on the outer surface of the conduction region where the temperature has been specified. In this situation, the n linear algebraic equations are solved for the n unknown temperatures.

If the solid conduction region under consideration is two dimensional in the x, y plane, the general form Eq. (2.189) reduces to

$$\frac{T_1 + T_2 - 2T_0}{\Delta x^2} + \frac{T_3 + T_4 - 2T_0}{\Delta y^2} + \frac{q_0'''}{k} = 0, \qquad (2.190)$$

and for the one-dimensional case, Eq. (2.189) becomes

$$\frac{T_1 + T_2 - 2T_0}{\Delta x^2} + \frac{q_0'''}{k} = 0. \qquad (2.191)$$

Often one works with an equispaced lattice—that is, $\Delta y = \Delta z = \Delta x$. This may be done because of the expectation of temperature derivatives of about the same order of magnitude in all three directions, or because it may be easier to fit the lattice to the shape of the region's boundaries, or simply because the finite difference equations will have a much simpler form and are easier to work with. So, for equal lattice spacings, after multiplying each term by Δx^2, Eqs. (2.189) to (2.191) reduce for three-, two-, and one-dimensional problems, respectively, to

$$T_1 + T_2 + T_3 + T_4 + T_5 + T_6 - 6T_0 + \frac{q_0'''(\Delta x)^2}{k} = 0, \qquad (2.192)$$

$$T_1 + T_2 + T_3 + T_4 - 4T_0 + \frac{q_0'''(\Delta x)^2}{k} = 0, \qquad (2.193)$$

$$T_1 + T_2 - 2T_0 + \frac{q_0'''(\Delta x)^2}{k} = 0. \qquad (2.194)$$

When all the lattice spacings are equal, the general equation for an interior node in a three-dimensional conduction region becomes very simple to write for each of the interior nodes. From Fig. 2.39, it can be seen that Eq. (2.192) simply states that *for any interior node, the finite difference equation is found by summing the temperatures of the six surrounding nodes in the three coordinate directions, subtracting six times the temperature at the interior node of interest, adding the term $q_0'''(\Delta x)^2/k$ at this node, and setting the result equal to zero.*

Sometimes the shape of a conduction region makes a coordinate system other than the rectangular cartesian the most natural choice in a finite difference analysis. Consider the sphere shown in Fig. 2.40, whose generation rate per unit volume, q''', is such a complicated function of radial position that a finite difference analysis is needed to find the temperature distribution. Steady-state conditions prevail, k is constant, and the energy flow in the sphere is radial only. We would like to derive the appropriate finite difference equation at a general interior node 0. The nodes are spaced a distance apart Δr in the radial direction, and a few of the nodes closest to node 0 are labelled in Fig. 2.40.

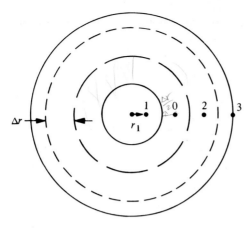

Figure 2.40 Nodes for the sphere.

Make an energy balance on the volume of material associated with node 0, that is, the spherical shell shown in Fig. 2.40 whose material is enclosed by the dashed surface and whose volume V_0 is

$$V_0 = \frac{4}{3}\pi[(r_0 + \frac{1}{2}\Delta r)^3 - (r_0 - \frac{1}{2}\Delta r)^3].$$

The law of conservation of energy is

$$R_{\text{in}} + R_{\text{gen}} = R_{\text{out}} + R_{\text{stor}},\qquad\qquad\qquad \textbf{[1.8]}$$

where

$$R_{\text{gen}} = q_0'''V_0 = q_0'''(\tfrac{4}{3})\pi[(r_0 + \tfrac{1}{2}\Delta r)^3 - (r_0 - \tfrac{1}{2}\Delta r)^3].\qquad \textbf{(2.195)}$$

Energy is conducted into the control volume from the material surrounding nodes 1 and 2. The conduction heat transfer rate from node 1 to node 0 is given by Eq. (2.26), Fourier's simplified law of conduction for the spherical coordinate system, as (after proper interpretation of the subscripts)

$$q_{10} = \frac{4\pi k(T_1 - T_0)}{1/r_1 - 1/r_0}.\qquad\qquad\qquad \textbf{(2.196)}$$

Similarly, the rate at which energy is conducted into shell 0 from the material around node 2 is given by

$$q_{20} = \frac{4\pi k(T_2 - T_0)}{1/r_0 - 1/r_2}.\qquad\qquad\qquad \textbf{(2.197)}$$

Steady-State Conduction

123

Inserting Eqs. (2.195–97) into Eq. (1.8), along with the fact that $R_{stor} = 0$, gives

$$\frac{4\pi k(T_1 - T_0)}{1/r_1 - 1/r_0} + \frac{4\pi k(T_2 - T_0)}{1/r_0 - 1/r_2}$$
$$+ q_0'''(\tfrac{4}{3})\pi \left[(r_0 + \tfrac{1}{2}\Delta r)^3 - (r_0 - \tfrac{1}{2}\Delta r)^3 \right] = 0.$$

After some rearrangement, the governing finite difference equation for interior node 0 becomes

$$\frac{T_1 - T_0}{1/r_1 - 1/r_0} + \frac{T_2 - T_0}{1/r_0 - 1/r_2}$$
$$+ \frac{q_0'''}{3k} \left[\left(r_0 + \frac{1}{2}\Delta r\right)^3 - \left(r_0 - \frac{1}{2}\Delta r\right)^3 \right] = 0. \qquad (2.198)$$

Figure 2.40 showed a very coarse lattice spacing Δr which led to a very small number of nodes within the region. In reality, to obtain highly accurate results, a far greater number of nodes may have to be used, such as 20 or 40 or more. Since most of these nodes will be interior nodes like node 0, Eq. (2.198) is written for each of them with simply a change to the appropriate subscripts, as was explained in the discussion which followed Eq. (2.189).

Noninterior Nodes

Now, in the general conduction problem some of the nodes are not interior nodes; that is, they will have a volume of material smaller than $\Delta x \, \Delta y \, \Delta z$ associated with them and will either be on the outer surface of the conduction region or have some part of their associated material on the outer surface, or perhaps be at an insulated surface or at the interface between two different solid materials. The finite difference equation for nodes such as these will differ from that for an interior node; surface nodes, for instance, can exchange energy by convection with a bounding fluid and also by radiation with other solid surfaces if the fluid is transparent to thermal radiation. However, for surface nodes or any noninterior node, the same approach is used to derive the governing finite difference equation as that used for interior nodes; namely, the amount of material associated with the node is identified and then this is taken as a control volume and the law of conservation of energy, Eq. (1.8), is applied.

For illustrative purposes, consider the derivation of the finite difference equation for some commonly encountered noninterior nodes. For simplicity, a two-dimensional region and an equispaced lattice $\Delta x = \Delta y$ is used. The equations for variable lattice spacing and three-dimensional regions are easily found once the basic procedure is well understood.

Surface Node with Convection   Consider the portion of a conduction region shown in Fig. 2.41. The temperatures of the surface nodes 1, 0, and 2 are not known. What is known is the undisturbed fluid temperature T_s, and the surface coefficient of heat transfer h

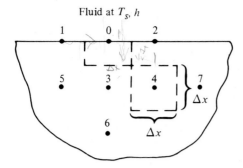

Figure 2.41 The surface node with convection.

between the fluid and the surface. Thus finite difference equations must be derived for these surface nodes, and attention will be focused on the surface node 0.

Clearly, unlike an interior node such as 4 which has the material Δx by Δx by 1 associated with it, node 0 has the material Δx by $\frac{1}{2} \Delta x$ by 1 associated with it as shown by dashed lines. Applying Eq. (1.8) to the control volume surrounding node 0, and assuming that all heat transfer rates are into the control volume,

$$R_{\text{in}} + R_{\text{gen}} = 0.$$

Energy enters the control volume by conduction from nodes 3, 1, and 2, and by *convection* from the fluid, since node 0 is at a fluid/solid interface. Assume that node 0 receives no net radiant gain from any surfaces beyond the fluid. Hence,

$$R_{\text{in}} = q_{30} + q_{10} + q_{20} + q_c,$$

where

$$q_{30} = k(\Delta x)(1) \frac{(T_3 - T_0)}{\Delta x}, \quad q_{20} = k \frac{\Delta x}{2} (1) \frac{(T_2 - T_0)}{\Delta x},$$

$$q_{10} = k \frac{\Delta x}{2} (1) \frac{(T_1 - T_0)}{\Delta x}.$$

Note that in the expressions for q_{10} and q_{20}, the area perpendicular to the heat flow is only $\frac{1}{2} \Delta x(1)$, rather than $(\Delta x)(1)$ as it is for the flow between 3 and 0. This is a consequence of the reduced amount of material associated with a noninterior node. By Newton's law of cooling,

$$q_c = h(\Delta x)(1)(T_s - T_0), \quad R_{\text{gen}} = q_0''' \frac{1}{2}(\Delta x)(\Delta x)(1) = q_0''' \frac{1}{2} \Delta x^2.$$

Note that the reduced size of the control volume also makes itself felt in the R_{gen} term, since R_{gen} is the *total* rate at which energy is generated within the control volume itself. After combining these results, the finite difference equation for node 0 becomes

$$T_3 + \tfrac{1}{2}T_2 + \tfrac{1}{2}T_1 + (h\Delta x T_s)/k - (2 + h\Delta x/k)\,T_0$$
$$+ \; q_0'''\Delta x^2/2k = 0. \tag{2.199}$$

Equation (2.199) can also be applied, with the proper changes in subscripts, to surface nodes 1 and 2, since they are in all respects similar to the surface node 0.

Node on a Perfectly Insulated Boundary   Next we derive the finite difference equation for the case in which node 0 is located on a perfectly insulated boundary as shown in Fig. 2.42. (The surface may be perfectly insulated, in effect, when it is a surface of symmetry of the conduction region. Symmetry here refers to both geometrical symmetry and thermal symmetry, symmetry with regard to thermal loadings.)

The material associated with node 0 is Δx by $\tfrac{1}{2}\Delta x$ by 1, as shown by the dashed line. Applying Eq. (1.8) to this control volume,

$$q_{10} + q_{20} + q_{30} + q_0'''(\Delta x)\,\tfrac{1}{2}\Delta x\,(1) = 0,$$

where there is no conduction from the left because of the perfect insulation. Again using Fourier's simplified law of conduction for the conduction heat transfer rates yields

$$q_{10} = k\,\frac{\Delta x}{2}\,(1)\,\frac{(T_1 - T_0)}{\Delta x}, \quad q_{20} = k\,\frac{\Delta x}{2}\,(1)\,\frac{(T_2 - T_0)}{\Delta x},$$

$$q_{30} = k\,\Delta x(1)\,\frac{(T_3 - T_0)}{\Delta x}.$$

Combining these four equations and rearranging, the finite difference equation for node 0 in Fig. 2.42 is

$$T_3 + \tfrac{1}{2}T_1 + \tfrac{1}{2}T_2 - 2T_0 + q_0'''(\Delta x)^2/2k = 0. \tag{2.200}$$

Actually, Eq. (2.200) could have been written down directly from Eq. (2.193) if one takes advantage of the symmetry by putting a fictitious node to the left of node 0 in Fig. 2.42 and calling this fictitious node's temperature T_3.

Surface Node on a Curved Boundary with Convection   All the noninterior nodes considered thus far were located on surfaces that were also the coordinate surfaces in the rectangular cartesian coordinate system. Suppose the conduction region has an outer surface shaped such that it is not a plane perpendicular to the *x, y,* or *z* axis shown in Fig. 2.39(a) and is not formed by the intersection of planes perpendicular to these same axes. In some cases, the shape of the conduction region might be such that its outer surface is formed by coordinate surfaces of some other common coordinate system, such as the circular cylindrical coordinate system or the spherical coordinate system. Then, an appropriately shaped finite control volume in the best suited coordinate system is considered,

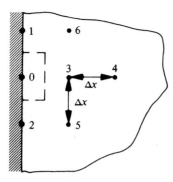

Figure 2.42 The node at the perfectly insulated boundary.

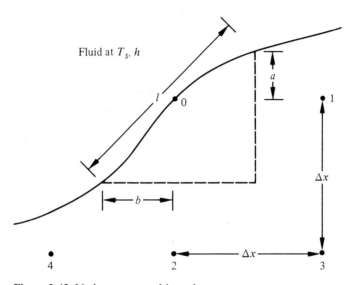

Figure 2.43 Node on a curved boundary.

and Eq. (1.8) is applied to obtain the analogue of Eqs. (2.189), (2.199), and (2.200) in the coordinate system being used. However, for a very irregularly shaped region, the outer surface would not coincide with the coordinate surfaces of the common coordinate systems, and therefore a rectangular cartesian coordinate system might still be used. Then the nodes on or near the actual outer surfaces are considered.

Perhaps the best way to handle a node at or near a curved boundary, such as node 0 in Fig. 2.43, is the Taylor series method referred to earlier. Another approach is to use a *variable* lattice spacing. In this approach, the smaller values of Δx that would be used near the boundary would obviate the need to deal with some fraction of the lattice spacing, such as the $2a/\Delta x$ and the $2b/\Delta x$ shown in Fig. 2.43 that must be dealt with when the lattice spacing is as large as that shown in the figure. Still another approach to the curved boundary problem is to approximate the actual curved boundary by *straight* lines, parallel to the x and y directions, that pass through nodes that fall closest to the actual boundary. If this were done in Fig. 2.43, the curved boundary would be replaced by a boundary that would look like a set of steps going up a hill. This approach necessitates the use of a very small lattice spacing Δx so that the approximated boundary of "steps" is practically coincident with the true, curved boundary. This method was used in the finite difference solution of some steady three-dimensional conduction problems in Warrington, Jr., et al. [19].

Here we will use the same approach we have previously, namely an energy balance, to take care of node 0 on the curved boundary of Fig. 2.43.

The finite difference equation for node 0 of Fig. 2.43 is determined by making an energy balance on the volume of material which is "naturally" associated with node 0. At an interior node, such as 2, the volume Δx by Δx by 1 is "naturally" associated with node 2. To associate the volume Δx by Δx by 1 with node 0 leads to running out of the solid and into the fluid. Hence, only the *solid material* that lies within Δx by Δx by 1 is included, as shown by the dashed lines in Fig. 2.43. (Some relevant lengths are labelled l and a in Fig. 2.43.)

Applying Eq. (1.8) to the control volume in dashed lines associated with node 0 gives

$$q_{20} + q_{10} + q_c + q_0''' V_0 = 0,$$

where V_0 is the volume of the control volume and is approximated by

$$V_0 = \Delta x^2/4 + a\Delta x/4 + b\Delta x/4, \tag{2.201}$$

since the upper and the left-hand portions of the volume have been approximated as triangular prisms. Now, the individual terms of the energy balance are as follows:

$$q_{20} = k\left(\frac{\Delta x}{2} + b\right)\frac{(T_2 - T_0)}{\Delta x}, \quad q_{10} = k\left(\frac{\Delta x}{2} + a\right)\frac{(T_1 - T_0)}{\Delta x}.$$

Newton's cooling law gives $q_c = hl(T_s - T_0)$. Thus the governing finite difference equation for node 0 of Fig. 2.43 becomes, after rearrangement,

$$\left(\frac{1}{2} + \frac{a}{\Delta x}\right)T_1 + \left(\frac{1}{2} + \frac{b}{\Delta x}\right)T_2 - \left(1 + \frac{a}{\Delta x} + \frac{b}{\Delta x} + \frac{hl}{k}\right)T_0$$
$$+ \frac{hlT_s}{k} + \frac{q_0''' V_0}{k} = 0. \tag{2.202}$$

A procedure similar to the one which led to Eq. (2.202) for node 0 on the curved boundary would be used to derive the appropriate algebraic finite difference equation for a node such as 4 in Fig. 2.43, which is located closer than $\Delta x/2$ to the curved boundary in one or more directions.

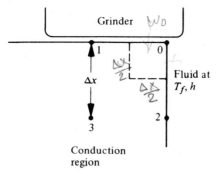

Figure 2.44 Diagram for Ex. 2.13.

EXAMPLE 2.13

Figure 2.44 shows part of a conduction region which is having its upper surface ground. Its right face is exposed to a fluid at T_f with a surface coefficient of heat transfer h between the surface and the fluid. For steady-state conditions, zero generation, and an equispaced lattice, derive the governing finite difference equation for node 0, assuming that the thermal conductivity of the conduction region is much greater than that of the grinding wheel and that the grinder dissipates mechanical energy at the upper surface at the rate per unit area W_0.

Solution
Apply Eq. (1.8) to the volume $\frac{1}{4}(\Delta x)^2$ of material associated with node 0. Energy is conducted in from around node 1 at the rate

$$q_{10} = k \tfrac{1}{2} \Delta x \, \frac{T_1 - T_0}{\Delta x}. \tag{2.203}$$

Energy is also conducted into the control volume surrounding node 0 from the material around node 2 at the rate

$$q_{20} = k \tfrac{1}{2} \Delta x \, \frac{T_2 - T_0}{\Delta x}. \tag{2.204}$$

The fluid at T_f causes a convective heat transfer to node 0 at the rate given by Newton's cooling law as

$$q_c = h \tfrac{1}{2} \Delta x (T_f - T_0). \tag{2.205}$$

Finally, since the thermal conductivity of the grinder is much smaller than that of the conduction region, the mechanical energy dissipated at the rate $W_0 \tfrac{1}{2} \Delta x$ at the interface between the grinder and one surface of node zero's material flows primarily into node 0. Hence, Eq. (1.8) becomes

$$q_{10} + q_{20} + q_c + W_0(\tfrac{1}{2} \Delta x) = 0. \tag{2.206}$$

Inserting Eqs. (2.203–5) into Eq. (2.206) gives

$$k \frac{1}{2} \Delta x \frac{T_1 - T_0}{\Delta x} + k \frac{1}{2} \Delta x \frac{T_2 - T_0}{\Delta x} + h \frac{1}{2} \Delta x (T_f - T_0)$$
$$+ \frac{1}{2} W_0 \Delta x = 0.$$

Dividing every term by $\frac{1}{2} k$, the finite difference equation at node 0 is

$$T_1 + T_2 - 2T_0 + (h\Delta x/k)(T_f - T_0) + W_0\Delta x/k = 0. \qquad \textbf{(2.207)}$$

EXAMPLE 2.14

Consider the conduction region shown in Fig. 2.45 which has a rectangular duct passing through it carrying a fluid at known temperature T_0 and with a known average surface coefficient of heat transfer between the fluid and the duct walls equal to h_0. The top face of the conduction region is exposed to another fluid at constant temperature T_∞ and constant surface coefficient of heat transfer h_∞. The right-hand face and the bottom face are held at known temperatures T_a and T_b, respectively. The upper half of the left-hand face is perfectly insulated, while the lower half is receiving a known constant heat flux q_0''. There is constant, uniformly distributed generation, due to the region carrying an electric current with known generation rate per unit volume q'''. The thermal conductivity is a constant k and steady-state conditions prevail. For an equispaced lattice, $\Delta x = \Delta y = l$, set up the network of nodes and derive the appropriate finite difference equation for each node.

Solution
Figure 2.45 is redrawn so that the nodes corresponding to $\Delta x = \Delta y = l$ are shown, with the nodes whose temperatures are not known, numbered in Fig. 2.46. For some of the noninterior nodes, the associated volumes of material are shown by dashed lines, since an energy balance may have to be made on these nodes to obtain the finite difference equations for them if the equations developed thus far are not applicable. Nodes 2, 9, 12, 13, 14, and 15 are ordinary interior nodes having the volume of material Δx by Δx so that the finite difference equations at these nodes are given by equations of the form of Eq. (2.193) with the subscripts properly interpreted. Thus, at node 2,

$$T_1 + T_9 + T_3 + T_a - 4T_2 + q''' \frac{(\Delta x)^2}{k} = 0; \qquad \textbf{(2.208)}$$

at node 9,

$$T_{10} + T_{12} + T_8 + T_2 - 4T_9 + q''' \frac{(\Delta x)^2}{k} = 0; \qquad \textbf{(2.209)}$$

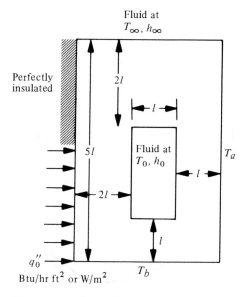

Figure 2.45 The conduction region of Ex. 2.14.

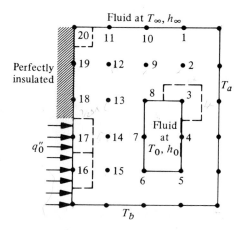

Figure 2.46 The finite difference network for Ex. 2.14.

at node 12,

$$T_{11} + T_{19} + T_{13} + T_9 - 4T_{12} + q''' \frac{(\Delta x)^2}{k} = 0; \qquad (2.210)$$

at node 13,

$$T_{12} + T_{18} + T_{14} + T_8 - 4T_{13} + q''' \frac{(\Delta x)^2}{k} = 0; \qquad (2.211)$$

at node 14,

$$T_{13} + T_{17} + T_{15} + T_7 - 4T_{14} + q''' \frac{(\Delta x)^2}{k} = 0; \qquad (2.212)$$

at node 15,

$$T_{14} + T_{16} + T_b + T_6 - 4T_{15} + q''' \frac{(\Delta x)^2}{k} = 0. \qquad (2.213)$$

Nodes 1, 10, 11, 4, and 7 are of the same general type, having a volume of material Δx by $\frac{1}{2} \Delta x$ and having a face of area Δx exposed to convection heat transfer. They, in fact, are the same type as node 0 of Fig. 2.41, for which the finite difference equation is Eq. (2.199). Applying Eq. (2.199) to these nodes with the subscripts properly interpreted yields, at node 1,

$$T_2 + \frac{1}{2}T_a + \frac{1}{2}T_{10} + \frac{h_\infty(\Delta x)T_\infty}{k} - \left[2 + \frac{h_\infty(\Delta x)}{k}\right]T_1$$

$$+ q''' \frac{(\Delta x)^2}{2k} = 0, \qquad (2.214)$$

at node 10,

$$T_9 + \frac{1}{2}T_1 + \frac{1}{2}T_{11} + \frac{h_\infty(\Delta x)T_\infty}{k} - \left[2 + \frac{h_\infty(\Delta x)}{k}\right]T_{10}$$

$$+ q''' \frac{(\Delta x)^2}{2k} = 0, \qquad (2.215)$$

at node 11,

$$T_{12} + \frac{1}{2}T_{20} + \frac{1}{2}T_{10} + \frac{h_\infty(\Delta x)T_\infty}{k} - \left[2 + \frac{h_\infty(\Delta x)}{k}\right]T_{11}$$

$$+ q''' \frac{(\Delta x)^2}{2k} = 0, \qquad (2.216)$$

at node 4,

$$T_a + \frac{1}{2}T_3 + \frac{1}{2}T_5 + \frac{h_0(\Delta x)T_0}{k} - \left[2 + \frac{h_0(\Delta x)}{k}\right]T_4$$

$$+ q''' \frac{(\Delta x)^2}{2k} = 0, \qquad (2.217)$$

and at node 7,

$$T_{14} + \frac{1}{2}T_8 + \frac{1}{2}T_6 + \frac{h_0(\Delta x)T_0}{k} - \left[2 + \frac{h_0(\Delta x)}{k}\right]T_7$$

$$+ q''' \frac{(\Delta x)^2}{2k} = 0. \qquad (2.218)$$

Nodes 18 and 19 are of the same general type, since they are on an insulated surface and have a volume of material Δx by $\frac{1}{2}\Delta x$ associated with them. This type has been considered, and Eq. (2.200) applies to node 0 in Fig. 2.42. Using Eq. (2.200), with the subscripts interpreted properly, yields at node 19,

$$T_{12} + \frac{1}{2}T_{20} + \frac{1}{2}T_{18} - 2T_{19} + q''' \frac{(\Delta x)^2}{2k} = 0, \qquad (2.219)$$

and at node 18,

$$T_{13} + \frac{1}{2}T_{19} + \frac{1}{2}T_{17} - 2T_{18} + q''' \frac{(\Delta x)^2}{2k} = 0. \qquad (2.220)$$

Since nodes like 20, 17, and 3 were not considered in the previous section, the relevant finite difference equations are derived by making an energy balance using Eq. (1.8) on the material associated with these nodes. First, for node 20 with its associated material $\frac{1}{2}\Delta x$ by $\frac{1}{2}\Delta x$, assuming all energy transfer rates are into the control volume of volume $\frac{1}{4}\Delta x^2$,

$$h_\infty \left(\frac{1}{2}\Delta x\right)(T_\infty - T_{20}) + k\left(\frac{1}{2}\Delta x\right)\frac{T_{19} - T_{20}}{\Delta x}$$

$$+ k\left(\frac{1}{2}\Delta x\right)\frac{T_{11} - T_{20}}{\Delta x} + q''' \frac{1}{4}(\Delta x)^2 = 0,$$

or rearranging, the finite difference equation at node 20 is

$$\frac{h_\infty \Delta x T_\infty}{2k} + \frac{1}{2}T_{19} + \frac{1}{2}T_{11} - \left(1 + \frac{h_\infty \Delta x}{2k}\right)T_{20}$$

$$+ \frac{q'''(\Delta x)^2}{4k} = 0. \qquad (2.221)$$

Nodes 17 and 16 are similar, since both have the volume of material Δx by $\frac{1}{2}\Delta x$ associated with them and are being subject to a known flux on a face Δx long. Applying Eq. (1.8) to node 17,

$$k(\Delta x)\frac{T_{14} - T_{17}}{\Delta x} + k\left(\frac{1}{2}\Delta x\right)\frac{T_{18} - T_{17}}{\Delta x} + k\left(\frac{1}{2}\Delta x\right)\frac{T_{16} - T_{17}}{\Delta x}$$

$$+ q_0'' \Delta x + q''' \frac{1}{2}(\Delta x)^2 = 0,$$

or rearranging, the finite difference equation appropriate to node 17 is obtained, and then by interpreting the subscripts properly, the equation appropriate to node 16 is obtained. As a result, at node 17,

$$T_{14} + \frac{1}{2}T_{18} + \frac{1}{2}T_{16} + \frac{q_0'' \Delta x}{k} - 2T_{17} + \frac{q'''(\Delta x)^2}{2k} = 0, \qquad (2.222)$$

and at node 16,

$$T_{15} + \frac{1}{2}T_{17} + \frac{1}{2}T_b + \frac{q_0''\Delta x}{k} - 2T_{16} + \frac{q'''(\Delta x)^2}{2k} = 0. \qquad (2.223)$$

Finally, the four remaining nodes, 3, 5, 6, and 8, are of the same general type; that is, inside corner nodes with a volume of material $\frac{3}{4}(\Delta x)^2$ associated with them and convecting to a fluid over a total area of Δx by 1 square unit. Applying Eq. (1.8) to the control volume of material associated with node 3 gives

$$k(\Delta x)\frac{T_2 - T_3}{\Delta x} + k(\Delta x)\frac{T_a - T_3}{\Delta x} + k\left(\frac{1}{2}\Delta x\right)\frac{T_4 - T_3}{\Delta x}$$

$$+ k\left(\frac{1}{2}\Delta x\right)\frac{T_8 - T_3}{\Delta x} + h_0(\Delta x)(T_0 - T_3) + q'''\frac{3}{4}(\Delta x)^2 = 0,$$

or at node 3,

$$T_2 + T_a + \frac{1}{2}T_4 + \frac{1}{2}T_8 + \frac{h_0(\Delta x)T_0}{k} - \left[3 + \frac{h_0(\Delta x)}{k}\right]T_3$$

$$+ \frac{q'''3(\Delta x)^2}{4k} = 0. \qquad (2.224)$$

Reinterpreting the subscripts of Eq. (2.224) for nodes 5, 6, and 8, respectively,

$$T_a + T_b + \frac{1}{2}T_4 + \frac{1}{2}T_6 + \frac{h_0(\Delta x)T_0}{k} - \left[3 + \frac{h_0(\Delta x)}{k}\right]T_5$$

$$+ \frac{q'''3(\Delta x)^2}{4k} = 0, \qquad (2.225)$$

$$T_b + T_{15} + \frac{1}{2}T_5 + \frac{1}{2}T_7 + \frac{h_0(\Delta x)T_0}{k} - \left[3 + \frac{h_0(\Delta x)}{k}\right]T_6$$

$$+ \frac{q'''3(\Delta x)^2}{4k} = 0, \qquad (2.226)$$

$$T_9 + T_{13} + \frac{1}{2}T_3 + \frac{1}{2}T_7 + \frac{h_0(\Delta x)T_0}{k} - \left[3 + \frac{h_0(\Delta x)}{k}\right]T_8$$

$$+ \frac{q'''3(\Delta x)^2}{4k} = 0. \qquad (2.227)$$

Thus, Eqs. (2.208–27) are the twenty finite difference equations for the nodes shown in Fig. 2.46 at which the temperatures are to be found.

This problem illustrates the procedure one would use to handle some of the wide variety of noninterior nodes that one may be confronted with in actual practice. This applies both to those for which appropriate finite difference equations were already available in an earlier section, and to those for which energy balances on the associated material have to be made to derive the finite difference equation. Notice that in this problem, the governing partial differential equation, Eq. (2.10) with the term containing z set equal to zero, has been replaced by a set of twenty simultaneous linear *algebraic* finite difference equations.

The next task of the engineer responsible for this problem is to insert the known numerical values of l, k, h_∞, h_0, T_a, T_b, etc., and then to solve the system of algebraic equations for the various nodal temperatures. Algebraic finite difference equations are solved using either hand techniques or a computer. The usual way to complete a finite difference solution is to use a computer; hand techniques are rarely used except as a check on the integrity of a computer program or as a crude calculation using a small number of nodes for rough estimates or "ballpark" figures.

2.8b Refinement of Nodal Spacings

It is evident from the derivation of the finite difference equations, such as Eq. (2.189), that an *approximation* is involved because of the *finite* spacings, Δx, Δy, Δz used between nodes. For example, heat transfer rates in the energy balances, leading to the algebraic finite difference equations, use approximate expressions such as $k\Delta y \Delta z (T_2 - T_0)/\Delta x$ instead of the exact expression given by Fourier's law of conduction, $-k\,dy\,dz\,\partial T/\partial x$. From a comparison of these expressions, one would certainly feel that as Δx, Δy, and Δz were made smaller and smaller, the various finite difference expressions and equations should approach the exact relationships. This intuitive observation, or hope, is verified and put on a firm theoretical foundation by the appropriate Taylor series expansions. These clearly show that, under the proper conditions, as Δx, Δy, and Δz approach zero, the solution of the algebraic finite difference equations goes to the solution of the partial differential equation (see Smith [18]).

As a consequence of this, refinement of nodal spacings Δx, Δy, and Δz is *required* in finite difference solutions, since the finite difference solution is usually treated as if it were the exact solution to the problem. Thus in a problem such as Ex. 2.14, where twenty finite difference equations were derived, the finite difference solution is *not* complete when these twenty equations are solved by the methods of the next section. One does not know in advance how accurate the finite difference results are for the one nodal spacing, $\Delta x = \Delta y = l$, which has been used so far. Hence, *refinement* of the nodal, or lattice, spacing *must be done*. One might reduce Δx by a factor of 2 to $l/2$, solve the greater number of finite difference equations, and compare the solutions at nodes common to both the original lattice spacing l and the finer spacing, $l/2$, to see if they agree. If agreement is close enough, the refinement of the nodal spacing can be stopped. If it is not close enough, the spacings can be cut in half again to $l/4$, the new, greater number of finite difference equations solved, and solutions compared again. This is continued until there is no longer any dependence of the finite difference solution on the nodal spacings.

Thus whenever one wants the finite difference solution to a problem to be the true solution, not just an approximation, the problem must be solved a number of times with smaller and smaller nodal spacings until the solution becomes independent of the nodal spacing used. Doing this entails a lot of work and perhaps even quite a large amount of computer time, so for most of the problems in this book we will employ only a single, fairly coarse, nodal spacing and will be content to treat the solution as an approximate result.

2.8c Solution of the Finite Difference Equations

After the required finite difference equations have been obtained for the nodes of the particular lattice structure chosen for a conduction region (that is, equations of the form of Eq. [2.189] for all interior nodes and of the forms shown in Eqs. [2.199–202] for the noninterior nodes), the result is a set of n linear algebraic equations in the n unknown nodal temperatures. Methods for solving sets of simultaneous linear algebraic equations include:

1. The systematic elimination of variables until all unknowns except one are eliminated, solving for the single remaining variable and then obtaining the other variables.
2. The use of Cramer's rule, which involves computation of the value of determinants.
3. The use of matrix algebra and the reduction of the solution to one of inverting a coefficient matrix.
4. The use of various iterative procedures, including relaxation analysis.

The elimination methods, matrix methods, and most of the iterative procedures (but not relaxation analysis) are extremely well suited to the capabilities of the digital computer. However, unless the number n of equations to be solved is very small, say less than four, the only method that is feasible to use in hand calculations is relaxation analysis.

As pointed out in the beginning of the discussion of finite difference methods, almost all finite difference solutions are carried out with the aid of the ubiquitous computer. However, hand calculations have a place as well in making rough preliminary estimates using a very small number of nodes, and in serving as check cases for computer programs and the input to these programs. Hence, hand calculations will be discussed next, followed by computer calculations.

Hand Solutions by Relaxation Analysis

A *relaxation analysis* is ordinarily used when solving simultaneous equations by hand when $n \geq 4$. Consider the following set of n simultaneous equations for the n unknown temperatures T_i, $i = 1, \ldots, n$, at the n nodal points of a given lattice structure:

$$a_1 T_1 + a_2 T_2 + \ldots + a_n T_n + a_0 q_1''' = 0,$$
$$b_1 T_1 + b_2 T_2 + \ldots + b_n T_n + b_0 q_2''' = 0,$$

$$\cdot$$
$$\cdot$$
$$\cdot$$

$$\tag{2.228}$$

$$c_1 T_1 + c_2 T_2 + \ldots + c_n T_n + c_0 q_n''' = 0,$$

where a_i, b_i, and c_i, $i = 1, 2, \ldots, n$, are the known coefficients of T_i.

The first step in relaxation analysis is to set all the equations equal to *residuals* R_1, $R_2, \ldots, R_n$ at the individual nodes, rather than to zero; that is,

$$a_1 T_1 + a_2 T_2 + \ldots + a_n T_n + a_0 q_1''' = R_1,$$
$$b_1 T_1 + b_2 T_2 + \ldots + b_n T_n + b_0 q_2''' = R_2,$$

$$\vdots$$

(2.229)

$$c_1 T_1 + c_2 T_2 + \ldots + c_n T_n + c_0 q_n''' = R_n.$$

(Note that when the temperatures T_i which satisfy Eq. [2.228] are inserted into Eq. [2.229], all the residuals $R_1, R_2, \ldots, R_n$ in Eq. [2.229] are identically zero.)

The next step is to make an educated guess of all the unknown temperatures to be used as a first approximation. Using a superscript to indicate the first and subsequent approximations to the true solutions, let the first approximation be the temperatures T_1^1, T_2^1, $\ldots, T_n^1$. Insert these into Eq. (2.229) and calculate the residuals. If all the residuals equal zero, then the initial guess is the correct solution to the problem. Ordinarily, of course, all the residuals on the right of Eq. (2.229) are nonzero after the initial guess. The procedure now becomes one of changing the estimated temperatures in a systematic way so that *all* of the residuals approach zero and the correct solution is then obtained. In order to *relax* these residuals toward zero, the equation whose residual is largest in *absolute value* is considered. Thus, if R_2 is the largest residual after inserting the temperatures $T_1^1, T_2^1, \ldots, T_n^1$, and is equal to, say, -100, a new T_2, call it T_2^2, is chosen in such a way that when it is inserted into the equation for R_2, it reduces R_2 to near zero. Then all the residuals must be recomputed, since a change in T_2 changes the residuals in some of the other equations as well. After the residuals have been recomputed, they are scanned again for the largest in absolute value, the temperature at that nodal point is changed accordingly, and that residual reduced, and all the other residuals recomputed. This procedure is repeated until all the residuals are so close to zero that small fractional changes in the temperatures would be needed to reduce them any further. *The final step and check is to insert the last-used values of temperatures into Eq. (2.228) to be sure that the right-hand sides are all indeed close to zero.* (The relaxation analysis procedure outlined here is the basic operation and is readily carried out. However, after the repeated use of relaxation analyses, shortcuts may be developed that allow the reduction of the residuals to zero more quickly. The three most common shortcuts are *overrelaxation,* *block relaxation,* and *group relaxation* and are discussed and illustrated in [3] and [5].)

One feature of the relaxation method is its ability to overcome arithmetic errors, so long as the last set of steps is error-free. Thus, if an error is made in the middle of the calculation procedure it is reflected in the distribution of the residuals, since this distribution would be different from the one that would occur without the error. However, since the aim of the method is to reduce the residuals to zero in any event, the error will eventually be corrected if the last set of steps is performed error-free.

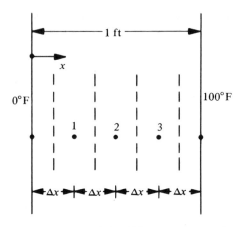

Figure 2.47 The finite difference network for Ex. 2.15.

EXAMPLE 2.15

Figure 2.47 shows a 1-ft-thick plane slab whose left-hand face is held at 0°F and right-hand face at 100°F. The energy generation rate per unit volume is given by $q''' = 6000x$ Btu/hr ft³, which is a linear function of x. The thermal conductivity is 10 Btu/hr ft °F. If steady-state conditions prevail, find the temperature distribution by the approximate finite difference method.

Solution

This problem can be solved in an exact analytic fashion very easily, but it is being used to illustrate the finite difference technique. The first step is to set up a network of nodes at which the temperature is to be determined. Since this problem is one-dimensional, the nodes are placed a distance Δx apart across the 1-ft dimension in the x direction. The greater the number of nodes (i.e., the smaller the value of Δx), the greater the accuracy of the result. To illustrate the procedure, a crude lattice spacing of $\Delta x = 1/4$ ft will be used. This leads to the five nodes shown in Fig. 2.47, but only three are numbered, since the first and last are on boundaries where the temperatures are already known. The material associated with each node at which the temperature is to be found is shown with dashed lines. All nodes are interior with the material of volume Δx by 1 ft by 1 ft associated with them. For an interior node of a one-dimensional equispaced lattice, Eq. (2.194) is the appropriate finite difference equation. That is,

$$T_1 + T_2 - 2T_0 + q_0'''(\Delta x)^2/k = 0, \qquad \textbf{[2.194]}$$

where the subscripts apply to Fig. 2.39, and must be reinterpreted for Fig. 2.47. The subscript 0 in Eq. (2.194) refers to the node of interest, while the subscript 1 refers to the node on the left of the node of interest and the subscript 2 refers to the node on the right of the node of interest. Thus, for node 1 in Fig. 2.47, the finite difference equation becomes

$$0 + T_2 - 2T_1 + q_1'''(\Delta x)^2/k = 0. \tag{2.230}$$

Now, q_1''' is the local generation rate per unit volume at node 1. The generation rate per unit volume as a function of x is $q''' = 6000x$. At node 1, $x = 1/4$ ft; thus, $q_1''' = (1/4)(6000) = 1500$. Hence, Eq. (2.230) becomes, after inserting Δx and k,

$$0 + T_2 - 2T_1 + 9.4 = 0. \tag{2.231}$$

Applying Eq. (2.194) to node 2 gives

$$T_1 + T_3 - 2T_2 + q_2''' (\Delta x)^2/k = 0,$$

or, since $q_2''' = (1/2) 6000$,

$$T_1 + T_3 - 2T_2 + 18.8 = 0. \tag{2.232}$$

Finally, applying Eq. (2.194) to node 3 and using $q_3''' = (3/4)6000$ gives

$$100 + T_2 - 2T_3 + 28.2 = 0. \tag{2.233}$$

Thus the finite difference equations, Eqs. (2.231–33), must be solved for the unknown nodal temperatures T_1, T_2, T_3. The first step is to rewrite the equations, setting them equal to the residual at the respective nodes; hence,

$$T_2 - 2T_1 + 9.4 = R_1, \tag{2.233a}$$

$$T_1 + T_3 - 2T_2 + 18.8 = R_2, \tag{2.233b}$$

$$T_2 - 2T_3 + 128.2 = R_3. \tag{2.233c}$$

The next step is to make educated guesses for T_1, T_2, and T_3. For instance, a guess of $T_1 = -40°F$ does not seem logical, since the boundary temperatures are $0°F$ and $100°F$ and the generation rate tends to cause higher temperatures at the interior points than would exist without generation. Also, the finite difference equations in the forms of Eqs. (2.231–33) permit an estimation of how much the generation rate influences the temperature at the point. Hence, an educated estimate of the temperatures would appear to be $T_1^1 = 30°F$, $T_2^1 = 70°F$, and $T_3^1 = 110°F$. Now, inserting these temperatures into Eqs. (2.233a–c) and computing the residuals yields

$$R_1 = 70 - 2(30) + 9.4 = +19.4,$$

$$R_2 = 30 + 110 - 2(70) + 18.8 = 18.8, \tag{2.234}$$

$$R_3 = 70 - 2(110) + 128.2 = -21.8.$$

The largest residual in absolute value is $R_3 = -21.8$; thus, the estimate of T_3 is revised to reduce this residual to practically zero. It can be seen from Eq. (2.233c) that if the estimated T_3 is reduced, then the residual will move toward zero. In particular, if ΔT_3 is the change in T_3 to reduce the residual to zero, and ΔR is the residual after the change minus the residual before the change, then

$$-2\Delta T_3 = \Delta R, \quad \text{or} \quad -2(T_3^2 - T_3^1) = 0 - (-21.8).$$

Thus, $T_3^2 \approx 99°$ and this gives $R_3 \approx 0$. However, note that just because $R_3 = 0$, this value of T_3 is not necessarily one of the solutions to the entire equation set. *All* residuals must be zero before the values of T_1, T_2, and T_3 can be considered the correct solutions.

Using $T_3^2 = 99°$, the other residuals as well as R_3 are

$$R_1 = +19.4, \quad R_2 = 30 + 99 - 2(70) + 18.8 = +8,$$
$$R_3 = 0. \tag{2.235}$$

(Note that in this particular step in this example, the change in T_3 also had a beneficial effect on the residual at the other node, node 2, influenced by T_3. This is not a general occurrence, however; most often the change at node 3 would increase the residuals at the neighboring points. The decrease experienced here results from the opposite algebraic signs of R_3 and R_2 in the first step.)

The residuals of Eq. (2.235) show that the largest residual in absolute value is $R_1 = 19.4$; hence, the estimate of T_1 is changed. From Eq. (2.233a), T_1 must be increased to reduce the positive R_1 to zero; or $-2(T_1^2 - T_1^1) = 0 - 19.4$, or $T_1^2 = 30 + 10$, rounding off $\frac{1}{2}(19.4)$. Thus, $T_1^2 = 40$. From this, the residuals are

$$R_1 = -0.6, \quad R_2 = 40 + 99 - 2(70) + 18.8 \approx +18,$$
$$R_3 = 0. \tag{2.236}$$

This time, notice that reducing the residual at node 1 has an adverse effect on the residual at node 2, but examination of Eqs. (2.234–36) indicates that, on the average, the residuals are being relaxed toward zero.

From Eq. (2.236), the largest residual in absolute value is the residual at node 2. Hence, increase the estimate of T_2, $-2(T_2^2 - T_2^1) = 0 - 18$, or $T_2^2 = 79$. Thus, the residuals become

$$R_1 = 79 - 2(40) + 9.4 = +8.4, \quad R_2 = 0,$$
$$R_3 = 79 - 2(99) + 128.2 = +9.2.$$

Now, considering the estimate $-2(T_3^3 - T_3^2) = 0 - 9.2$, or $T_3^3 = 99 + 5 = 104$ leads to

$$R_1 = +8.4, \quad R_2 = 40 + 104 - 2(79) + 18.8 = +4.8,$$
$$R_3 = -0.8.$$

Continuing this process of relaxing the residuals toward zero for nine additional steps finally yields,

$$R_1 = +1.4, \quad R_2 = +0.8, \quad R_3 = 0.$$

At this step the residuals have been relaxed to the point where fractional changes of a degree would be needed to get them closer to zero; thus, the analysis is stopped here using the last values for T_1, T_2, and T_3,

$$T_1 = 47°F, \quad T_2 = 86°F, \quad T_3 = 107°F.$$

However, the relaxation analysis is not complete until the final check is made by inserting these values into the original Eqs. (2.231–33). Substituting,

$$86 - 2(47) + 9.4 = +1.4,$$

$$47 + 107 - 2(86) + 18.8 = +0.8,$$

$$100 + 86 - 2(107) + 28.2 = +0.2.$$

The right-hand sides can be considered to be sufficiently close to zero. Using the methods of Sec. 2.4, the exact solution for the temperature distribution is

$$T = 200x - 100x^3.$$

A comparison of the finite difference solution with the exact solution and the percent error is shown in the following chart.

	Exact	Finite Difference, $\Delta x = 1/4$ ft	Error
x (ft)	T (°F)	T (°F)	%
1/4	48.44	47	−3.0
1/2	88.5	86	−2.8
3/4	107.8	107	−0.74

Note that in Ex. 2.15 high accuracy is achieved with a relatively crude lattice spacing. However, no general statement can be made about the accuracy achieved with a given lattice spacing in any particular problem other than that, in steady-state conduction problems, the accuracy increases as the lattice spacing decreases if all calculations are made with negligible round-off errors. This was one of the points stressed in the section that dealt with the refinement of the nodal spacings, Sec. 2.8b.

One technique for arriving at what usually is very close to the exact solution is to solve the problem by finite differences for three different lattice spacings. Then the temperatures at any space point are plotted as a function of the size of the lattice spacing. This curve is extrapolated to zero lattice spacing. The solution to the problem is the temperature extrapolated to when the spacing is equal to zero. This is called *Richardson's technique.*

In Ex. 2.15, the use of relaxation analysis to solve the three simultaneous equations at the nodal points involved quite a bit of labor compared to either Cramer's rule or systematic elimination of variables. However, when the number of equations to be solved is more than three, relaxation analysis is preferred for hand calculations. Also in Ex. 2.15, one of the simplest modifications of the procedure used, that of *overrelaxation,* would have shortened the time and effort considerably. Overrelaxation consists simply of reducing the largest residual not just to zero but past zero. Hence, if R_1 is positive, T_1 is changed by an amount that would make R_1 slightly negative rather than zero. Or if R_1 is initially negative, a change in T_1 is made that would cause R_1 to be slightly positive. This procedure, on the average, gives a better overall distribution of residuals. How far to overrelax must be determined by experience in every problem. However, a rule of thumb is to overrelax by one-third to one-half of the original residual (see Arpaci [5]).

Computer Solution of Finite Difference Equations

As we have seen in earlier sections, the finite difference method applied to steady-state conduction problems leads to a set of n simultaneous algebraic equations that must be solved for the n unknown nodal temperatures, $T_1, T_2, \ldots, T_n$. (Finite difference equations for *unsteady* conduction problems also lead to such a set of simultaneous equations if implicit finite difference schemes are used, as will be discussed in the next chapter.) If attention is focused on linear algebraic equations, the task is to solve the simultaneous equation set which follows, in which all the a_{ij} and b_i have known numerical values and the unknowns are $T_1, T_2, \ldots, T_n$; that is, all the T_j.

$$
\begin{aligned}
a_{11}T_1 + a_{12}T_2 + a_{13}T_3 + \ldots + a_{1n}T_n &= b_1 \\
a_{21}T_1 + a_{22}T_2 + a_{23}T_3 + \ldots + a_{2n}T_n &= b_2 \\
a_{31}T_1 + a_{32}T_2 + a_{33}T_3 + \ldots + a_{3n}T_n &= b_3 \\
&\;\;\vdots \\
a_{n1}T_1 + a_{n2}T_2 + a_{n3}T_3 + \ldots + a_{nn}T_n &= b_n
\end{aligned}
$$

(2.237)

By using the rules of matrix multiplication, Eq. (2.237) can be cast into matrix form in which a square matrix **A,** the coefficient matrix, multiplies or maps a single column row matrix **t,** or temperature vector, to give a single column row matrix **b,** or *B* vector. These various matrices are shown next.

$$A = \begin{bmatrix} a_{11} & a_{12} & a_{13} & \cdots & a_{1n} \\ a_{21} & a_{22} & a_{23} & \cdots & a_{2n} \\ a_{31} & a_{32} & a_{33} & \cdots & a_{3n} \\ \cdot & & & & \cdot \\ \cdot & & & & \cdot \\ \cdot & & & & \cdot \\ a_{n1} & a_{n2} & a_{n3} & \cdots & a_{nn} \end{bmatrix}$$

(2.238)

$$t = \begin{bmatrix} T_1 \\ T_2 \\ T_3 \\ \cdot \\ \cdot \\ \cdot \\ T_n \end{bmatrix} \qquad b = \begin{bmatrix} b_1 \\ b_2 \\ b_3 \\ \cdot \\ \cdot \\ \cdot \\ b_n \end{bmatrix}$$

(2.239)

Thus, in matrix form, Eq. (2.237) becomes

$$A \, t = b.$$

(2.240)

Generally in finite difference analysis, and certainly for the type of finite difference equations we will be dealing with in this book, many of the coefficients a_{ij} have zero values, thus giving a sparse coefficient matrix A in Eq. (2.238). Although theoretically A could contain as many as n^2 nonzero elements, it often contains only $5n$ nonzero elements for two-dimensional problems and $7n$ for three-dimensional problems. Equation sets such as Eq. (2.237) or the matrix form, Eq. (2.240), are solved by the digital computer using both direct and iterative methods.

Matrix Inversion   With the rules for finding the inverse of a coefficient matrix A (see Wylie [2]), computer programs have been written to perform this operation and give the inverse A^{-1}. Pre-multiplication by this inverse in Eq. (2.240) yields

$$t = A^{-1}b.$$

(2.241)

The rules covering multiplication of matrices are then used in the computer program to perform the indicated operation on the right-hand side of Eq. (2.241), thus yielding, as output from the program, the elements of the temperature vector t, namely, T_1, $T_2, \ldots, T_n$.

Gaussian Elimination   In the elimination methods for solving the equation set Eq. (2.237), one first adds some proper multiple of the first equation in the set to the second equation in the set; this is done to eliminate the term containing T_1, a term to the left of the main diagonal, from the resulting equation, which now becomes the new second equation. Then the computer program uses suitable multiples of the first two equations of the set to eliminate the terms containing T_1 and T_2 in the third equation of the set. Similarly for the remaining equations of the set, the preceding equations are combined with them in such a way as to eliminate all terms to the left of the main diagonal. When this is complete, the computer then has in storage the following equation set, where the various c's and d's are related to the a's and b's (for example, $c_{22} = a_{22} - a_{12}[a_{21}/a_{11}]$):

$$
\begin{aligned}
a_{11}T_1 + a_{12}T_2 + a_{13}T_3 + \cdots + a_{1n}T_n &= b_1 \\
c_{22}T_2 + c_{23}T_3 + \cdots + c_{2n}T_n &= d_2 \\
c_{33}T_3 + \cdots + c_{3n}T_n &= d_3 \\
&\vdots \\
c_{n-1n-1}T_{n-1} + c_{n-1n}T_n &= d_{n-1} \\
c_{nn}T_n &= d_n
\end{aligned}
\tag{2.242}
$$

The equation set (2.242) is now in upper, or right, triangular form. The program now solves for T_n, as $T_n = d_n/c_{nn}$, from the last equation of the set (2.242) and uses this value of T_n, which is now known, in the second to last equation to solve for T_{n-1}. Then this process of back substitution is continued, thus yielding as the program output the temperature vector **t,** that is, the numerical values of the nodal temperatures $T_1, T_2, \ldots, T_n$.

The elimination procedure described here is particularly simple to program in the special case in which the original equation set, Eq. (2.237), is tri-diagonal; that is, when the only nonzero values of the a_{ij} are on the two closest parallel diagonals, one just above and one just below the main diagonal. In this case, a fairly simple recursion formula can be developed for the c_{ij} and d_i and this is shown in Ozisik [15].

Iterative Methods   The iterative approach to the solution of equation set (2.237) employs an initial educated guess, or estimate, for the values of $T_1, T_2, \ldots, T_n$, which is inserted into Eq. (2.237) in order to generate a new estimate of these values. Then this new estimate, which is, in some sense, closer to the true solution than the original estimate, is inserted into the equation set to yield another new estimate, and this iteration process is continued until two successive iterations yield values of the nodal temperatures that are deemed close enough to one another. The factor which distinguishes one iterative method from another is the procedure used to generate the new estimate of all the unknowns. The one iterative method which will be described here, the *Gauss-Seidel* method, is used often and is easily programmed for the digital computer.

Step one in the Gauss-Seidel iterative technique is to rewrite equation set (2.237) so that the unknown temperature at the node, for which the equation was derived, appears on a side by itself with a coefficient of $+1$. Thus, the first equation of set (2.237) would be solved for T_1 in terms of the other unknown temperatures, the second equation would be solved for T_2, and so on, yielding the following result:

$$
\begin{aligned}
T_1 &= b_1 - (d_{12}T_2 + d_{13}T_3 + \cdots &&+ d_{1n}T_n) \\
T_2 &= b_2 - (d_{21}T_1 + d_{23}T_3 + \cdots &&+ d_{2n}T_n) \\
T_3 &= b_3 - (d_{31}T_1 + d_{32}T_2 + d_{34}T_4 + \cdots &&+ d_{3n}T_n) \\
&\quad\vdots \\
T_n &= b_n - (d_{n1}T_1 + d_{n2}T_2 + \cdots &&+ d_{nn-1}T_{n-1})
\end{aligned}
\tag{2.243}
$$

The iteration procedure is begun by making an initial estimate of the nodal temperatures. Next this original estimate of the temperatures $T_2, T_3, \ldots, T_n$ is inserted into the right side of the *first* equation in equation set (2.243) and the *new* estimate of T_1 is calculated on the left. This *new* estimate of T_1 and the original estimates of $T_3, T_4, \ldots, T_n$ are inserted on the right of the second equation in equation set (2.243) and the new estimate of T_2 is calculated on the left. This process is continued through the remaining equations in the set, always using the *new* estimates of the temperatures when they are available from a previous calculation. After the last equation in the set is solved for the new estimate of T_n, one now has new estimates for all the unknowns T_1, $T_2, \ldots, T_n$. These new estimates are now used in the right side of the first equation of the set and a newer estimate of T_1 is calculated. This is then used, with the most recent estimates of $T_3, T_4, \ldots, T_n$, in the second equation of the set to solve for the newer estimate of T_2. This iteration process continues until nodal temperatures are close enough to one another in two successive iteration cycles. Thus, if k, used as a superscript, denotes the previous iteration and $k + 1$ denotes the iteration just completed, then the computer program would contain a test of the following form:

$$Is \left| T_i^{k+1} - T_i^k \right| \leq \epsilon$$

for each node i, $i = 1, 2, 3, \ldots, n$, where ϵ is a small enough number so that it is deemed that the temperature at node i is no longer changing its value significantly from iteration to iteration.

Additional information on iterative methods of solution, including the addition of successive overrelaxation to speed up the convergence of the Gauss-Seidel method, is available in Smith [18] and in Ames [20]. This concludes the brief description of three of the primary techniques used in conjunction with the digital computer to solve simultaneous sets of linear algebraic equations.

Computer Programs

Standard computer programs for solving sets of algebraic equations using one of the previously described methods, or some variant thereof, are available at university, industrial, government, or research and development laboratory computing centers. These library program subroutines often have titles such as SIMQ, GAUJOR, and INVER. The input required for such subroutines is the number of simultaneous equations to be solved, the elements of the coefficient matrix **A** of Eq. (2.238) which appear as shown in Eq. (2.237) in the original equation set to be solved, and the elements of the B matrix, **b**, shown in Eqs. (2.239) and (2.237). The output will be the temperature vector **t** whose elements are the nodal temperatures $T_1, T_2, \ldots, T_n$, which one is trying to find. Often a number of different subroutines for solving equation sets will be available. The choice of the one that is "best" is determined by the number of equations to be solved, as well as the characteristics of the coefficient matrix **A,** such as whether or not it is sparsely populated. This is usually the case for finite difference methods using a large number of nodes.

In some cases the library programs may have been written to take advantage of the strengths and avoid the weaknesses of the particular computer, and generally they have been optimized with regard to computer storage and run time. In light of this, and also in light of their availability, it would seem best to use these subroutines rather than write one's own. Perhaps in cases where only a small number of equations are to be solved, readers might prefer to use the programs they have developed in a computer programming course, since storage and computer run time will not be large.

Finally, in the event that a computer program must be written by the reader to solve equation set (2.237), a typical beginning reference is Hornbeck [21], where the methods are discussed, flow charts are given, and some illustrative examples are presented.

EXAMPLE 2.16

Consider the two-dimensional conduction region in Fig. 2.48, where the temperature is specified on three sides as shown while the fourth side is in contact with a fluid at a temperature of 100°C, which is flowing from left to right. This flow causes the local surface coefficient of heat transfer to decrease about linearly with distance along the top from a value of 200 W/m² °C at the left corner to 20 W/m² °C at the right corner. The thermal conductivity of the material is $k = 30$ W/m °C, and there is no generation. Use finite differences with equispaced nodes 0.1 m apart to get an estimate of the steady-state temperature distribution as well as the heat transfer rate across the upper face to the fluid.

Solution
A fairly crude lattice spacing is chosen to get an estimate of this two-dimensional temperature distribution, namely, $\Delta x = \Delta y = 0.1$ m. This spacing implies the nodes numbered 1 through 4 in Fig. 2.48. Nodes 3 and 4 are interior nodes with the generation rate per unit volume equal to zero; hence, Eq. (2.193), with the subscripts interpreted properly, applies to nodes 3 and 4. In Eq. (2.193), the temperatures surrounding the node of interest are summed in the two coordinate directions, and then four multiplied by the temperature at the node of interest is subtracted. Thus, for node 3,

$$T_1 + 300 + 400 + T_4 - 4T_3 = 0.$$

For node 4,

$$T_2 + 200 + 400 + T_3 - 4T_4 = 0.$$

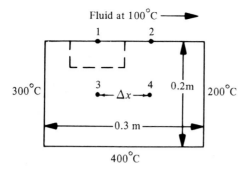

Figure 2.48 The finite difference network for Ex. 2.16.

Nodes 1 and 2 are surface nodes with an associated amount of material $\frac{1}{2} \Delta x$ by Δx by 0.1 m as shown by the dashed line for node 1 in Fig. 2.48. Since there is convection heat transfer to those nodes, values for the local surface coefficient of heat transfer in the neighborhood of nodes 1 and 2 are needed. Since h varies linearly from 200 on the left to 20 on the right, $h_1 = 140$ W/m² °C and $h_2 = 80$ W/m² °C.

An energy balance yielding a finite difference equation for nodes of the type of 1 and 2 has already been made in Eq. (2.199). Thus, for nodes 1 and 2, respectively,

$$T_3 + \frac{T_2}{2} + \frac{300}{2} + \frac{h_1(\Delta x)100}{k} - \left[2 + \frac{h_1(\Delta x)}{k}\right]T_1 = 0.$$

$$T_4 + \frac{200}{2} + \frac{T_1}{2} + \frac{h_2(\Delta x)100}{k} - \left[2 + \frac{h_2(\Delta x)}{k}\right]T_2 = 0.$$

Since $h_1\Delta x/k = 140(0.1)/30 = 0.467$ and $h_2\Delta x/k = 80(0.1)/30 = 0.267$, the equation set to be solved becomes, after rearrangement of the previous four equations,

$$
\begin{aligned}
-2.467T_1 + 0.5T_2 \quad + T_3 \qquad\qquad &= -196.7 \\
0.5T_1 - 2.267T_2 \qquad\quad + T_4 &= -126.7 \\
T_1 \qquad\qquad\quad - 4T_3 + T_4 &= -700 \\
T_2 + T_3 \; - 4T_4 &= -600
\end{aligned}
\tag{2.244}
$$

Writing this equation set in matrix form gives the following:

$$
\begin{bmatrix}
-2.467 & 0.5 & 1.0 & 0 \\
0.5 & -2.267 & 0 & 1.0 \\
1 & 0 & -4 & 1 \\
0 & 1 & 1 & -4
\end{bmatrix}
\begin{bmatrix}
T_1 \\ T_2 \\ T_3 \\ T_4
\end{bmatrix}
=
\begin{bmatrix}
-196.7 \\ -126.7 \\ -700 \\ -600
\end{bmatrix}
\tag{2.245}
$$

Inserting the elements of the square coefficient matrix of Eq. (2.245) into any standard library computer program, such as SIMQ, yields the solution for the temperature vector **t** as,

$$\mathbf{t} = \begin{bmatrix} 253.85 \\ 238.57 \\ 310.27 \\ 287.21 \end{bmatrix}$$

Thus the individual nodal temperatures are, after rounding off, $T_1 = 253.9°C$, $T_2 = 238.6°C$, $T_3 = 310.3°C$, $T_4 = 287.2°C$.

At this point it is recommended, as an overall check on the integrity of one's work, that these temperatures be inserted back into the *original* equation set, Eq. (2.244), to insure that the equations are satisfied by the nodal temperature values. Most computer programs for solving equation sets have internal checks (such as calculation of the residuals of the equations solved or multiplication of **A** by its inverse to see if the identity matrix is formed) to verify that the equation set which has been *input* to the program has been solved correctly. Often, especially when the number of nodes is larger than the number used here, mistakes are made during the input of the coefficient matrix **A**, Eq. (2.238), such as typing the wrong value of a coefficient or typing it in the wrong location. The internal checks in a standard program will not be able to detect such an error. Naturally one can compare the coefficient matrix **A** and the *B* matrix **b** of Eq. (2.240), which are usually printed out in the standard computer programs, to the ones that should have been input. However, it has been the author's experience that the most foolproof overall check is to see if the solutions satisfy the original equation set.

In the case of a larger number of nodes, a check of this type definitely involves some significant time and effort, but this has to be weighed against the importance of wanting to know the correct temperature distribution in the first place. Actually, with the aid of a hand or desk calculator, checks of this nature can be made in a reasonable period of time.

Thus, inserting the temperatures we have found into the left-hand sides of the equations in the set, Eq. (2.244), yields for the right-hand side the following: -196.69, -126.70, -700.02, and -600.00. Comparing these to the actual right-hand side of Eq. (2.244), the vector **b**, indicates that the equations are balanced adequately.

Next, to compute the heat transfer rate across the upper surface of Fig. 2.48, Newton's cooling law is used. The surface area associated with node 1 is Δx by 1 m deep perpendicular to the figure, and is similar for node 2. The corner nodes, at 300°C and 200°C, have the upper surface area $\Delta x/2$ by 1 m associated with each of them and have surface coefficients of heat transfer h of 200 W/m^2 °C and 20 W/m^2 °C, respectively, from the original given information. Thus, since $\Delta x = 0.1$ m, $T_1 = 253.9°C$, and $T_2 = 238.6°C$,

$$q = 200(0.05)(1)(300 - 100) + 140(0.1)(1)(253.9 - 100)$$
$$+ 80(0.1)(1)(238.6 - 100) + 20(0.05)(1)(200 - 100),$$

$$q = 5363.4 \text{ W.}$$

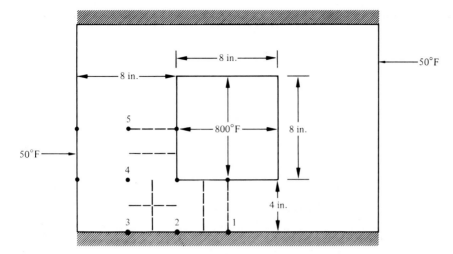

Figure 2.49 The proposed chimney of Ex. 2.17.

In this problem, the energy transfer rate across the surface of interest was computed using Newton's cooling law, because the fluid temperature and surface coefficients were known. The calculation of energy transfer rate across a surface whose temperature is specified will be illustrated in the next example.

As was stressed in an earlier section, the answers for the nodal temperatures calculated by the finite difference method are approximate unless the nodal spacing is continually made smaller until the nodal temperatures no longer exhibit any dependence upon the spacing. Hence, if it were crucial to know accurately the temperatures within the region in Fig. 2.48, the next step would be to cut the nodal spacing in half (or reduce it by some factor) to 0.05 m and solve the problem, and continue to do this until nodal temperatures no longer change as the spacing is reduced.

We have used the finite difference method in Ex. 2.16 mainly because of the variable surface coefficient of heat transfer across the upper surface. The shape of the region alone would allow an exact analytical solution by the methods discussed in an earlier section, but the variation of h renders the problem intractable for an exact analytical solution.

EXAMPLE 2.17

Consider the proposed chimney design shown in Fig. 2.49 in which the brick is in good contact with the wood of the house on two sides, thus causing those sides to be adiabatic. In certain operating modes of a stove, the inside surface of the chimney will be at 800°F while the uninsulated outside surfaces will be about 50°F. Steady-state conditions prevail, and one of the things we would like to find out is whether or not this design could cause the wood to be above the 700°F ignition temperature.

a. By using a coarse lattice spacing, $\Delta x = \Delta y = 4$ in., and the finite difference method, predict the value of the nodal temperatures.

b. Use the computer to find the nodal temperatures for the refined lattice where $\Delta x = \Delta y = 2$ in.

c. If the thermal conductivity of the brick is $k = 0.6$ Btu/hr ft °F, find the heat transfer rate per foot across the 800°F surface of the chimney for (a) and (b).

Solution

Because of thermal and geometric symmetry, only one-quarter of the region must be dealt with, and the lower left corner will be used. Using the lattice spacing of 4 inches from (a), the nodes at which the temperature is to be found are numbered 1 through 5 on Fig. 2.49. Remember that because of the symmetry, the upper surface of this quarter of the region where node 5 is located, and the right-hand surface of this quarter where the nodes are node 1 and another at 800°F, are *adiabatic* surfaces.

Node 4 is an interior node, so the finite difference equation is given by Eq. (2.193) when the subscripts are properly interpreted. Nodes 2, 3, and 5 are nodes on adiabatic surfaces, so Eq. (2.200) is appropriate; otherwise use Eq. (2.193) and invoke the symmetry condition. For example, at node 5, the temperature at the node not shown just above 5 is the same as at the node just below 5, T_4. This, when used in Eq. (2.193), gives the same result as Eq. (2.200). Node 1 has an adiabatic or symmetry condition with respect to both directions. Thus one can use Eq. (2.193), with the node (not shown) to the right of 1 being at T_2, the same as that to the left of 1, and with the fictitious node below 1 being at 800°F, as is the node directly above 1. Alternately, one may derive the equation for node 1 by an energy balance on the material associated with 1, namely a volume $\Delta x^2/4$ whose bottom and right-hand side are adiabatic.

After rearrangement, we find that the equation set to be solved in (a) is as follows:

$$
\begin{aligned}
-2T_1 + T_2 &= -800 \\
T_1 - 4T_2 + T_3 &= -1600 \\
T_2 - 4T_3 + 2T_4 &= -50 \\
T_3 - 4T_4 + T_5 &= -850 \\
2T_4 - 4T_5 &= -850
\end{aligned}
\tag{2.246}
$$

Writing equation set (2.246) in matrix form, according to Eq. (2.240), gives,

$$
\begin{bmatrix}
-2 & 1 & 0 & 0 & 0 \\
1 & -4 & 1 & 0 & 0 \\
0 & 1 & -4 & 2 & 0 \\
0 & 0 & 1 & -4 & 1 \\
0 & 0 & 0 & 2 & -4
\end{bmatrix}
\begin{bmatrix}
T_1 \\ T_2 \\ T_3 \\ T_4 \\ T_5
\end{bmatrix}
=
\begin{bmatrix}
-800 \\ -1600 \\ -50 \\ -850 \\ -850
\end{bmatrix}
\tag{2.247}
$$

Using the elements of the coefficient matrix on the left of this equation and of the matrix **b** on the right as input to a standard computer program yields the solution vector **t** in terms of the following nodal temperatures, after rounding: $T_1 = 741.6$°F, $T_2 = 683.1$°F, $T_3 = 390.9$°F, $T_4 = 415.3$°F, $T_5 = 420.1$°F. These temperatures were then inserted back into the original equations, Eq. (2.246), and it was verified that they did satisfy these equations. Thus we have solved part a.

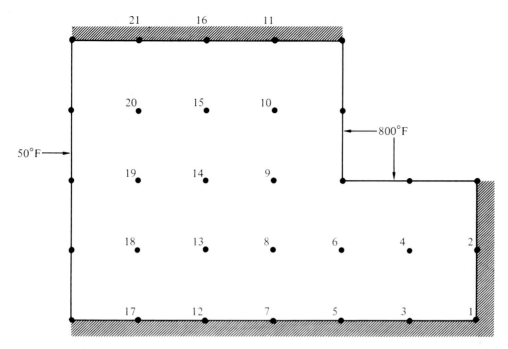

Figure 2.50 Quarter of chimney of Ex. 2.17.

Now in part b, we refine the nodal spacing by cutting it in half to $\Delta x = \Delta y = 2$ in. This gives rise to the twenty-one nodes shown and numbered in Fig. 2.50, which displays only the symmetrical lower quarter section.

Proceeding now as in part a, the twenty-one needed finite difference equations are derived for the numbered nodes. These equations, when put into matrix form, are shown in Fig. 2.51. As a shorthand in the matrix $\mathbf{A}$ of Eq. (2.248) in Fig. 2.51, an empty space is understood to mean that the element of that space is a zero element.

Introducing the elements of the coefficient matrix $\mathbf{A}$ and of the matrix $\mathbf{b}$ of Fig. 2.51 into a standard computer program gives the following nodal temperatures, all in Fahrenheit degrees:

$$T_1 = 735.8, \ T_2 = 753.7, \ T_3 = 717.8, \ T_4 = 739.5,$$
$$T_5 = 656.5, \ T_6 = 686.6, \ T_7 = 535.1, \ T_8 = 550.2,$$
$$T_9 = 589.2, \ T_{10} = 601.8, \ T_{11} = 604.8, \ T_{12} = 383.5,$$
$$T_{13} = 390.0, \ T_{14} = 404.7, \ T_{15} = 413.1, \ T_{16} = 415.7,$$
$$T_{17} = 219.1, \ T_{18} = 221.4, \ T_{19} = 226.6, \ T_{20} = 230.3,$$
$$T_{21} = 231.6.$$

These nodal temperatures were checked by inserting them into the original twenty-one equations, and it was verified that they do satisfy these equations and hence are the solution.

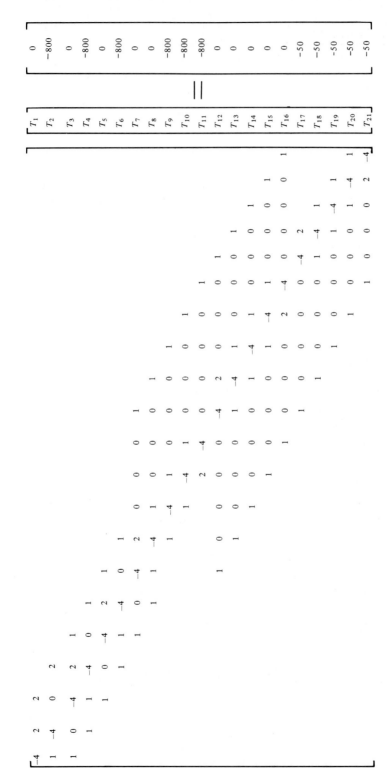

Figure 2.51 Equation 2.248

Notice that nodes 1, 2, 3, 4, and 5 of the *larger* 4-inch nodal spacing of part a correspond to the nodes numbered 1, 5, 12, 14, and 16, respectively, in the *smaller,* refined nodal spacing of two inches in part b. Comparing the nodal temperatures at these points in (b) to those at the same five space points in (a) gives one an indication of how close the nodal temperatures are to the true temperatures, the ones given by a solution as nodal spacing gets very small. It is seen that the largest change in nodal temperature occurs at node 5 of Fig. 2.50, where the temperature changed by 27° out of about 680° when the nodal spacing was cut in half. Hence, depending upon how close to the true solution one feels the finite difference solution must be, one either may or may not reduce the nodal spacing again and solve the problem over again.

With the nodal temperatures now known, the heat transfer rate across the inside surface of the chimney can be calculated. To do this for the quarter section of part a, Fourier's simplified law of conduction, Eq. (1.1) (which it will be recalled was also used in the derivation of the finite difference equations) will be applied to the four heat flow channels between the 800°F nodes and the nearest numbered nodes, shown as dashed lines in Fig. 2.49. Thus,

$$
\begin{aligned}
q &= k(\Delta x/2)(1)\frac{(800 - T_1)}{\Delta x} + k(\Delta x)(1)\frac{(800 - T_2)}{\Delta x} \\
&\quad + k(\Delta x)(1)\frac{(800 - T_4)}{\Delta x} + k(\Delta x/2)(1)\frac{(800 - T_5)}{\Delta x} \\
&= 0.6 \left[\frac{800 - 741.6}{2} + 800 - 683.1 + 800 - 415.3 \right. \\
&\quad \left. + \frac{800 - 420.1}{2} \right] = 432.5 \text{ Btu/hr.}
\end{aligned}
$$

Thus, for the entire chimney in part a, one has

$$q_{\text{total}} = 4(432.5) = 1730 \text{ Btu/hr.} \tag{a}$$

In like fashion, in part b, for the quarter section shown in Fig. 2.50, the heat flow channels will be basically the same as in (a) except that there will be two additional full heat flow channels corresponding to the conduction between the 800°F surface and the additional nodes 4 and 10. After cancelling out the Δx, which appears both in the area perpendicular to heat flow and in the path length L for heat flow in every heat flow term, the expression for q becomes

$$
\begin{aligned}
q = k &\left[\frac{800 - T_2}{2} + 800 - T_4 + 800 - T_6 + 800 - T_9 + 800 \right. \\
&\left. - T_{10} + \frac{800 - T_{11}}{2} \right].
\end{aligned}
$$

Inserting the nodal temperatures from (b) and the value of k and multiplying the result by four yields

$$q_{\text{total}} = 1689 \text{ Btu/hr.} \tag{b}$$

As is evident by looking at the predicted nodal temperatures in (b) for the nodes at the locations where the chimney is in good contact with the wood, particularly nodes 1 and 3, this is not a good chimney design, since the wood at nodes 1 and 3 would be above the ignition temperature.

Finite difference methods had to be used to solve Ex. 2.17 because of the complicated geometric shape of the conduction region depicted in Fig. 2.49.

2.9 PROBLEMS

2.1   A steel pipe is 15 m long with an inside diameter of 10 cm and a wall thickness of 0.6 cm. Its inside surface is held at 93°C, while its outside surface has a temperature of 54°C. (a) Compute the steady-state heat transfer rate. (b) Compare the result of (a) with that which would be obtained if the pipe were treated as a plane wall 15 m by 0.6 cm by $2\pi(5)$ cm.

2.2   The steel pipe described in Prob. 2.1 has a 5-cm-thick layer of asbestos insulation installed. The outer surface of the asbestos is now 54°C, while the inner surface of the pipe is still held at 93°C. Compute (a) the steady-state heat transfer rate and (b) the temperature at the asbestos-steel interface. (c) Draw the analogous electric circuit for this thermal situation.

2.3   A spherical shell has an inside temperature T_0 and a known outside surface temperature T_1. Steady-state conditions prevail with no generation and constant thermal conductivity. The inside and outside radii are r_0 and r_1, respectively. Starting with Fourier's exact law of conduction and noting that the conduction heat transfer rate is independent of the local radius r, find an expression for the heat transfer rate across the shell q, in terms of known quantities.

2.4   Water at 80°F flows through a composite pipe whose inside radius is 2 in. and whose steel inside wall is 1/2 in. thick. There is also a 1-in. layer of 85% magnesia on the steel. The outside air is at 50°F with an outside surface coefficient of heat transfer of 3.0 Btu/hr ft² °F. If the inside surface coefficient of heat transfer is 40 Btu/hr ft² °F, calculate the steady-state heat transfer loss per foot of pipe.

2.5   A 12-m-long pipe has an inside radius of 7.5 cm, an outside radius of 8.25 cm, and is made of steel. Its inside surface is held at 150°C, while its outside surface is at 40°C. There is no generation and steady-state conditions hold. Find the rate at which energy is being transferred across the pipe wall.

2.6   You are given the equation

$$T = 50 \, x^3 \, y^2 \, e^{-z}$$

as a solution to a three-dimensional steady-state conduction problem with no generation. Determine whether this is a valid solution, and carry out the necessary steps to prove whether this is a true solution to a steady-state, zero generation problem.

2.7 Figure 2.52 shows a composite slab with a vacuum to the left and a liquid at known temperature T_∞ and known surface coefficient of heat transfer h_∞ on the right-hand side. The thicknesses and thermal conductivities Δx_b, Δx_a, k_a, and k_b are also known. The left-hand face is absorbing radiant energy at the known net rate of q_0'' Btu/hr ft^2 or W/m^2 due to an infrared source which is not shown. Find the temperature of the left-hand face in terms of known quantities. Steady-state conditions prevail.

Figure 2.52

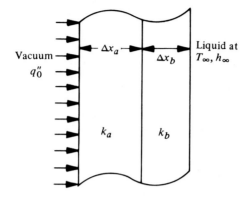

2.8 For the steady-state, zero generation, one-dimensional case shown in Fig. 2.53, sketch a solution for finding the heat transfer rate, from the fluid on the left to the fluid on the right, by use of the appropriate electrical circuit.

Figure 2.53

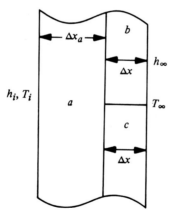

2.9 To reduce the heat loss from the steam pipe of Ex. 2.4, a 5-cm layer of 85% magnesia insulation ($k = 0.06$ W/m °C) is added. If the outside and inside surface coefficients of heat transfer remain the same, calculate the percentage reduction in the heat loss with the insulation as compared to the bare pipe of Ex. 2.4.

2.10 Derive the differential equation for the temperature distribution in steady, one-dimensional radial heat flow, with variable thermal conductivity, in a circular cylindrical coordinate system by making an energy balance on the appropriate differential control volume.

2.11 Figure 2.54 shows a spherical liquid air storage tank of 2 ft inside radius. The outside radius of the structural metal a is 2.1 ft and $k_a = 25$ Btu/hr ft °F. Material b is insulation of 1 ft thickness and $k_b = 0.045$ Btu/hr ft °F. The temperature of the inside surface is $-220°$F, while the outside surface is at 80°F. Calculate the steady-state heat transfer rate through the tank to the liquid air contained inside.

Figure 2.54

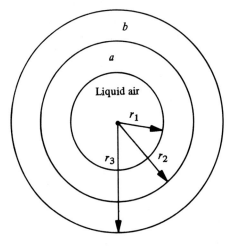

2.12 Figure 2.55 shows a hollow sphere of inside radius 0.3 m and outside radius 1 m. The surface temperature of the sphere is 38°C at the inside and 10°C at the outside. The thermal conductivity of the sphere material is 17 W/m °C. If steady-state conditions prevail with no generation, calculate the rate at which heat is transferred from the sphere.

Figure 2.55

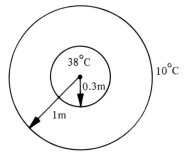

2.13 A plate of thickness L is held at temperature T_1 on one face and T_2 on the other face. The plate's thermal conductivity varies with temperature in the manner $k = k_0 + BT^2$, where T is in °F and B and k_0 are known constants. (a) Derive an expression for the steady-state heat transfer rate across the plate. (b) If $L = 1/2$ ft, $T_1 = 150°$F, $T_2 = 50°$F, $k_0 = 50$, and $B = 8 \times 10^{-4}$, compute the heat transfer rate per square foot of surface.

2.14 Consider the conduction region shown in Fig. 2.56. Steady-state conditions prevail and there is no generation. The region is perfectly insulated along $r = r_0$ and $r = r_1$, the thermal conductivity k is constant, and the conductive region is very long perpendicular to the paper. (a) The governing partial differential equation is

$$\frac{\partial^2 T}{\partial r^2} + \frac{1}{r}\frac{\partial T}{\partial r} + \frac{1}{r^2}\frac{\partial^2 T}{\partial \theta^2} + \frac{\partial^2 T}{\partial z^2} + \frac{q'''}{k} = \frac{1}{\alpha}\frac{\partial T}{\partial \tau}.$$

Reduce this equation to the form appropriate to this problem and solve for the temperature's distribution in space. Note that T_1 and T_2 are known temperatures. The conduction region lies between r_0 and r_1. (b) Find an expression for the heat transfer rate across the face $\theta = 0$.

Figure 2.56

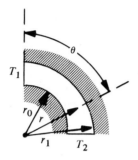

2.15 Derive Eq. (2.8) by making an energy balance on a differential control volume in the circular cylindrical coordinate system.

2.16 Derive the governing differential equation for purely radial heat flow in a spherical shell with no generation and for steady-state conditions, but where the thermal conductivity is a function of the local radius r.

2.17 A board is being sanded with a belt sander which is doing work on the board at the known rate W_0 per unit area of surface. The other side of the board is exposed to air at known temperature T_∞ and surface coefficient h_∞. If steady-state conditions prevail, the board's thickness is L, and thermal conductivity is k, (a) derive an expression for the temperature distribution as a function of position within the board and (b) find an expression for the temperature of the face being sanded.

2.18 A fluid at temperature T_0 is flowing in a composite two-material pipe where r_1 is the inside radius of the first material, r_2 is the inside radius of the second, which is insulating material, and r_3 is the outside radius of the insulating material. The steady-state heat transfer rate per foot from the fluid at T_0 to a fluid at the outside at temperature T_∞ is

$$q = \frac{2\pi(T_0 - T_\infty)}{\dfrac{1}{h_0 r_1} + \dfrac{\ln(r_2/r_1)}{k_a} + \dfrac{\ln(r_3/r_2)}{k_b} + \dfrac{1}{h_\infty r_3}},$$

where h_0 and h_∞ are the inside and outside surface coefficients, respectively, and k_a and k_b are the thermal conductivities of the first material and the outer insulating material. Determine if there are any extrema of the function q if the outer radius r_3 of the insulation can be varied while keeping the quantities, other than q, constant. In particular, determine if there is some optimum thickness of the insulation or some thickness for which the insulation does not perform its function properly. Explain the results.

Steady-State Conduction

2.19 Figure 2.57 shows a slab of thickness L in which there is a uniformly distributed source of constant strength q'''. The left-hand face, at $x = 0$, is insulated, while the right-hand face is bounded by a fluid at temperature T_0. The surface coefficient is constant at h. Find the temperature distribution as a function of x.

Figure 2.57

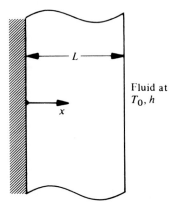

2.20 The infinite slab shown in Fig. 2.58 is an idealization of a nuclear reactor core. As a result of fission, the generation rate per unit volume is the following function of the space coordinate x: $q''' = q_0''' x^2$, where q_0''' is a known constant. The right-hand face of the core is perfectly insulated, while the left-hand face is held at temperature T_0 by the reactor coolant. Predict the steady-state temperature distribution within the slab as a function of x. The thermal conductivity is a known constant.

Figure 2.58

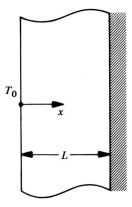

2.21 If in Prob. 2.20 the thermal conductivity is 8.5 W/m °C, the coolant temperature is 370°C, and $q''' = 100,000 \, x^2$ W/m³, find the highest temperature in the reactor core when $L = 0.3$ m.

2.22 An infinite slab, such as that shown in Fig. 2.14, has both faces held at 100°F by a coolant and is to be a conductor of electrical current in the direction perpendicular to the plane of the figure. The electrical energy dissipated within the slab by this current flow depends upon the slab thickness L, because the material resistance to current flow in the indicated direction is inversely proportional to L. It has been calculated that, with L in feet, the electrical energy dissipation rate per unit volume will be $(36)10^5/L$ Btu/hr ft^3. If the maximum slab temperature is not to exceed 600°F, calculate the largest value of slab thickness L that can be used. The thermal conductivity is 120 Btu/hr ft °F and steady-state conditions prevail.

2.23 Referring to Fig. 2.14, suppose that the slab is a 1-m-thick poured concrete wall with both sides held at a temperature of 20°C. Because of the curing of the concrete, chemical internal energy is released at a rate per unit volume of 100 W/m^3. The thermal conductivity is $k = 1.5$ W/m °C and the temperature does not vary with time. Find the maximum temperature of the concrete.

2.24 In Prob. 2.23, assume that the thickness L of the wall is not known. Find the thickest wall that can be poured without causing a temperature gradient greater than 100°C per meter anywhere in the wall.

2.25 Measurement of temperature within a certain solid conduction region in the steady state yields the temperature distribution

$$T = 200 + (100 - x^2)(\sin 5y) z^3.$$

If the thermal conductivity of the material is constant at $k = 10$ Btu/hr ft °F, find the energy generation rate per unit volume as a function of the space coordinates.

2.26 Consider a solid circular cylindrical bar shown in Fig. 2.59 with outer radius r_0 and constant thermal conductivity k. The energy release rate per unit volume is $q''' = br^2$, where b is a known constant and r is the general radial coordinate measured from the bar's center. The surrounding fluid is at known temperature T_∞, the surface coefficient of heat transfer is known to be h_∞, and steady-state conditions prevail. The governing general equation is

$$\frac{\partial^2 T}{\partial r^2} + \frac{1}{r}\frac{\partial T}{\partial r} + \frac{1}{r^2}\frac{\partial^2 T}{\partial \theta^2} + \frac{\partial^2 T}{\partial z^2} + \frac{q'''}{k} = \frac{1}{\alpha}\frac{\partial T}{\partial \tau}.$$

In terms of known quantities, find (a) the temperature distribution within the bar as a function of the general radial coordinate r and (b) the heat transfer rate at the outer surface, $r = r_0$.

Figure 2.59

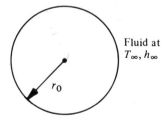

Fluid at T_∞, h_∞

r_0

2.27 An exothermic chemical reaction takes place in a 0.3-m-thick wall of thermal conductivity 20 W/m °C. The chemical internal energy release rate per unit volume is 10,000 W/m^3. Both faces of the wall are held at 38°C. Compute the maximum steady-state temperature of the wall.

2.28 The slab shown in Fig. 2.60 is releasing energy within itself because of internal nuclear reactions. The generation rate due to this is approximately constant at q_0''' Btu/hr ft³ or W/m³. The left-hand face is separated by a vacuum from a high-temperature source which delivers energy at the net rate of q_0'' Btu/hr ft² or W/m² to the left-hand face by radiation. The slab is bounded by a fluid on the right whose undisturbed temperature is T_∞, and the surface coefficient of heat transfer is known to be h_∞. Find (a) the temperature distribution as a function of x within the slab, and (b) the rate at which energy is crossing the right-hand face at $x = L$.

Figure 2.60

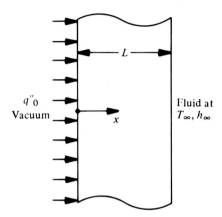

2.29 Consider a solid sphere of radius R and thermal conductivity k whose surface temperature is maintained at T_0 while electrical energy is being dissipated at the constant known rate per unit volume q_0''' within the sphere. Derive an expression for the steady-state temperature distribution as a function of the general radial coordinate r.

2.30 One side of a space station can be approximated by the infinite slab L ft or m thick shown in Fig. 2.61. There is constant distributed internal generation at the rate of q''' Btu/hr ft³ or W/m³ as a result of using the side as an electrical conductor from the station's nuclear power source. The left-hand side of the slab is insulated as shown and the right-hand side is exposed to the vacuum of space. Space behaves approximately as a black body at 0°R or °K. The slab surface also behaves like a black body. If steady-state conditions prevail, derive an expression for the temperature as a function of x.

Figure 2.61

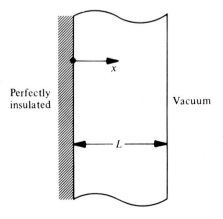

2.31 By solving the relevant differential equation, show that the solution for the temperature within a solid rod of radius R, in the steady state, with a constant uniform generation rate per unit volume q_0''', and with a constant surface temperature T_0 is given by

$$T = T_0 + \frac{q_0''' R^2}{4k}\left(1 - \frac{r^2}{R^2}\right).$$

2.32 A copper wire 0.32 cm in diameter has an electrical resistance per meter of 22×10^{-4} ohms. Its outside surface is held at 37°C by oil flowing over it. Compute the maximum wire temperature if the current flow is 200 amperes.
(*Hint:* Use the result of Prob. 2.31.)

2.33 A copper wire of radius r_1 is covered by electrical insulation of outer radius r_2. The voltage drop per foot or meter of wire is V_0 volts while passing a current of I amperes. The outer surface of the insulation is exposed to a fluid at temperature T_∞ and the surface coefficient of heat transfer between the insulator and the fluid is h_∞. Find the steady-state radial temperature distribution within the insulator.

2.34 For the situation of Prob. 2.33, $r_1 = 0.16$ cm, the resistance per meter is 20×10^{-4} ohms, and the outer radius of the insulation is 0.32 cm and its thermal conductivity is 0.16 W/m °C. If the fluid temperature is 37°C and the outside surface coefficient of heat transfer is 11 W/m² °C, calculate the allowable current if the maximum temperature of the wire is not to exceed 94°C.

2.35 A steel strap is serving as a support strut for the steam pipe shown in Fig. 2.62. The strap is welded to the pipe and bolted to the ceiling. The outside temperature of the steam pipe is 220°F. The juncture between the support strut and the ceiling is adiabatic. The strut is 2 ft high, 5 in. wide and 1/8 in. thick. Assume that $k_{steel} = 26$ Btu/hr ft °F, the total outside surface coefficient is 3.0 Btu/hr ft² °F, and the surrounding air is at 90°F. Calculate the rate at which heat is lost to the surrounding air by the support strut.

Figure 2.62

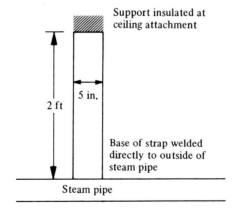

Support insulated at ceiling attachment

5 in.

2 ft

Base of strap welded directly to outside of steam pipe

Steam pipe

2.36 A 1-in.-diameter rod, which is used to rearrange objects in a heat treating oven, projects out of the oven side into an 80°F room and a surface coefficient of heat transfer equal to 1 Btu/hr ft² °F exists between the room air and the rod. In order to avoid burns, the point on the rod in the room farthest from the oven must have a temperature no higher than 130°F. If the point on the rod where it projects out of the oven is at 800°F, calculate the minimum distance the rod should project out of the furnace. The thermal conductivity of the rod material is 10 Btu/hr ft °F.

2.37 Figure 2.63 shows a 5-cm-diameter rod, 0.9 m long, which is having its lower face ground smooth. The remainder of the rod is exposed to the 32°C room air and a surface coefficient of heat transfer equal to 7 W/m² °C exists between the rod surface and the room air. If the grinder dissipates mechanical energy at the rate of 35 W and if the grinding material has a much lower thermal conductivity than the value of 40 W/m °C of the rod material, find the temperature of the rod at the point where the grinding is taking place.

Figure 2.63

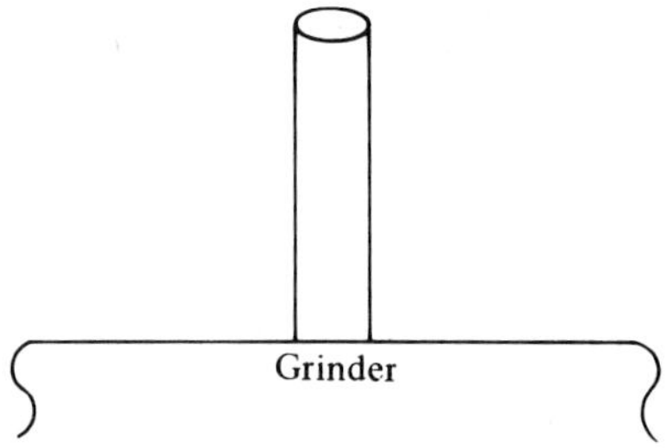

Grinder

2.38 A 3.2-cm-diameter bar has one of its ends in a heat treating oven at 427°C. The remaining 9 m of the bar are exposed to the room air at 26°C with an average surface coefficient of heat transfer between the room air and the bar of 24 W/m² °C. The thermal conductivity of the bar material is 34 W/m °C, and steady-state conditions prevail. (a) Compute the heat transfer rate to the room from this bar. (b) To see if a warning sign should be put near the end of the bar farthest away from the furnace, calculate the lowest temperature in the bar and specify whether a sign warning of high temperature is needed.

2.39 A handle for a saucepan is to be designed such that it is 12 in. long and 3/4 in. in diameter. One end of the handle will be subjected to a temperature of 212°F during certain cooking situations. An average surface coefficient of heat transfer of 1.3 Btu/hr ft² °F is expected between the handle and the kitchen air at 75°F. Someone cooking is likely to grasp the last 4 in. of the handle, hence the temperature in this region should not exceed 100°F. (a) Calculate the thermal conductivity of the handle material needed to accomplish this. (b) Based on the value of the thermal conductivity found in (a), can any other problems be anticipated, such as structural ones? If so, explain briefly how these problems might be overcome.

2.40 Figure 2.64 shows a solid blade from one of the last turbine stages in a jet engine. The blade has a height of 3 in., a cross-sectional area A of 0.25 in.², and a perimeter P of 2 in. Gases at 1000°F flow over the entire surface of the blade except the bottom surface, that is, the root where it is attached to the turbine disk. The average surface coefficient of heat transfer between the gases and the blade surface is about 81 Btu/hr ft² °F, and the thermal conductivity of the blade material is 100 Btu/hr ft °F. Assuming that a quasi-one-dimensional analysis is acceptable, estimate the temperature the blade root must be kept at if the tip temperature is not to exceed 900°F.

Figure 2.64

Root

2.41 Consider the trapezoidal-shaped conduction region shown in Fig. 2.65, in which there is no heat transfer in the z direction. The top and bottom are perfectly insulated, while the left-hand side is at temperature T_1 and the right-hand side is at temperature T_2. The thermal conductivity k is constant, the dimensions of the region are shown, and there is no generation. Find the steady-state temperature distribution and the heat transfer rate in the conduction region between the faces at T_1 and T_2.

Figure 2.65

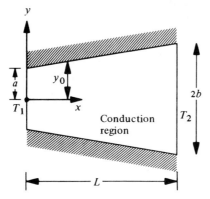

2.42 In the preceding figure, $a = 0.3$ m, $b = 0.6$ m, $L = 3$ m, $k = 80$ W/m °C, $T_1 = 150°C$, and $T_2 = 100°C$. (a) Compute the heat transfer rate across the conduction region per meter of z direction. (b) Compare the answer of (a) with the result that Fourier's simplified law of conduction would yield if the arithmetic average area were used.

Steady-State Conduction   163

2.43 Derive the expression for the steady-state temperature distribution and the heat transfer rate to the fluid when the generalized rod of Fig. 2.20 is infinitely long in the x direction.

2.44 Consider the generalized rod of Fig. 2.20 when the face at $x = L$ is adiabatic, and derive expressions for the steady-state temperature profile and heat transfer rate to the surrounding fluid.

2.45 A support strut of length L separates two tanks, 1 and 2, which have different temperatures T_1 and T_2, respectively. The support strut is joined to tank 1 at $x = 0$ and to tank 2 at $x = L$, while the remainder of the strut is exposed to a fluid environment at temperature T_∞ and with an average surface coefficient of heat transfer h_∞ between the fluid and the strut. Solve for the quasi-one-dimensional steady-state temperature distribution, and show that it is

$$\theta = \left(\frac{\theta_2 - \theta_1 \cosh NL}{\sinh NL} \right) \sinh Nx + \theta_1 \cosh Nx,$$

where $\theta_1 = T_1 - T_\infty$, $\theta_2 = T_2 - T_\infty$, $N^2 = h_\infty P / kA$, $P = $ perimeter.

2.46 An aluminum slab $1/2$ in. thick separates $150°F$ water from $70°F$ air. The water-side surface coefficient of heat transfer is 50 Btu/hr ft² °F, while the air-side surface coefficient is 5 Btu/hr ft² °F. Determine (a) the heat transfer rate per unit area, and (b) the heat transfer rate per unit area if aluminum straight fins of rectangular profile, 0.10 in. high and 2 in. long spaced 1 in. apart on centers, are added on the air side.

2.47 It is proposed that the heat transfer rate between a surface and a fluid be increased by adding straight fins of rectangular profile to the surface. The fins are to be made of steel ($k = 43$ W/m °C), 0.64 cm high, and 13 cm from tip to base. If the average surface coefficient of heat transfer between fin and fluid is estimated to be 4000 W/m² °C, would you advise the installation of these fins and why?

2.48 Consider the thin ring of thickness t and radius R shown in Fig. 2.66, which is partially immersed in fluid 1 and partially immersed in fluid 2. The fluid temperatures are T_1 and T_2, while the average surface coefficients of heat transfer between the respective fluids and the ring surface are h_1 and h_2. Using a quasi-one-dimensional approach, find the steady-state circumferential temperature distribution in the ring.

Figure 2.66

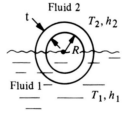

2.49 The bar shown in Fig. 2.67 is being used as a simple heat exchanger between fluid 1 at $250°F$ and fluid 2 at $150°F$. The bar ($k = 120$ Btu/hr ft °F) is perfectly insulated at both ends and there is a thin membrane at the bar center to prevent mixing of the fluids. The average surface coefficient of heat transfer between fluids 1 and 2 and the bar's surface are 2 and 10 Btu/hr ft² °F, respectively. Find the heat transfer rate between fluids 1 and 2 through the bar.

Figure 2.67

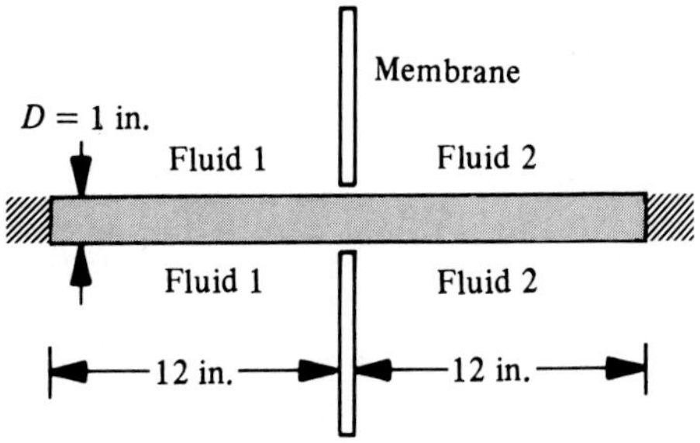

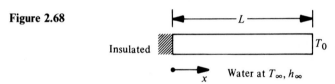

2.50 Consider the rod of perimeter P and cross-sectional area A shown in Fig. 2.68. Energy is released at the known uniform rate q_0''' Btu/hr ft³ or W/m³ as a result of electrical dissipation. The left-hand end is insulated while the right-hand end is held at T_0. The outer surface is immersed in water at temperature T_∞ and known surface coefficient h_∞. The temperature gradients are small in directions other than the x direction, and steady-state conditions prevail. (a) Derive the governing differential equation and put down the boundary conditions. (b) Solve the equation for the temperature distribution in the rod.

Figure 2.68

2.51 Figure 2.69 shows a circular fin of inside radius r_i and outside radius r_0 of constant thickness δ. Since $\delta << r_0 - r_i$, and there is no generation, and steady-state conditions prevail with a known temperature T_1 at r_i, the temperature can be lumped in the direction of the thickness and T can be treated as a function only of the general radial coordinate r. Derive the governing ordinary differential equation for T by making an energy balance on the ring-shaped control volume of thickness δ and width dr. Except at r_i the entire fin is exposed to fluid at T_e with surface coefficient of heat transfer h_e. Note that it is only necessary to derive the equation, not solve it.

Figure 2.69

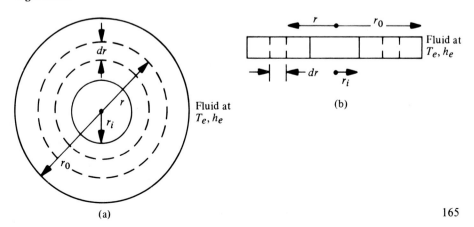

(a)

2.52 For the straight fin of triangular profile shown in Fig. 2.70, derive the governing differential equation for the temperature distribution by assuming that it is quasi-one-dimensional and by making an energy balance on the control volume shown in dashed lines. Also assume that $dl \approx dx$, and show that the resulting equation is given by

$$x \frac{d^2T}{dx^2} + \frac{dT}{dx} - \frac{hL}{k\delta} (T - T_\infty) = 0.$$

Figure 2.70

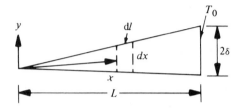

Fluid at T_∞, h_∞

2.53 Figure 2.71 shows a conduction region shaped like a truncated cone which is perfectly insulated on its lateral surfaces. The circular faces of radii r_0 and r_b are held at known temperatures T_1 and T_2, respectively. The outer radius of this cone at any x is

$$r = r_0 + (r_b - r_0) \, x/b.$$

The thermal conductivity is constant at the value k. (a) If the temperature distribution can be assumed to be quasi-one-dimensional, find an expression for the heat transfer rate across this truncated cone. (b) Using the result of part a, find the shape factor S for this conduction region. (c) For $T_1 = 500°F$, $T_2 = 100°F$, $k = 220$ Btu/hr ft °F, $r_0 = 2$ in., $r_b = 4$ in., and $b = 8$ in., compute the heat transfer rate across the cone.

Figure 2.71

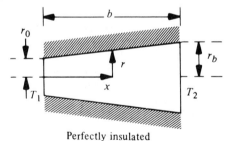

Perfectly insulated

2.54 In Prob. 2.53, suppose that there is also electrical energy being dissipated throughout the conduction region at the known rate per unit volume of q_0''' Btu/hr ft³ or W/m³. Derive an expression for the temperature as a function of x.

2.55 In a simple heat exchanger consisting of a 0.64-cm-thick aluminum flat plate separating two fluids, water at 82°C flows along one side, giving an average surface coefficient of heat transfer of 480 W/m² °C on that side, while nitrogen at 15°C flows over the opposite side, giving a surface coefficient of heat transfer of 36 W/m² °C between that side and the nitrogen. Straight aluminum fins are to be added to the nitrogen side to increase the heat transfer rate. The fins are 0.13 cm high and, because of a space problem, can only be 2.5 cm long from base to tip. (a) Calculate the fin spacings if a flux of 15,700 W/m² is desired across the plate. (b) Determine the fin spacings if a flux of 95,000 W/m² is desired across the plate. Also determine if this is possible and explain your conclusions. (c) Determine the maximum flux possible across the plate if the fin spacing on the nitrogen side is the only allowable change.

2.56 Water at 150°F flows over the outside surface of a 2.40-in. outside diameter copper tube, giving rise to a surface coefficient of heat transfer of 200 Btu/hr ft² °F between the water and the tube. Air at 70°F flows inside the tube of inside diameter 2 in. The air-side surface coefficient is 4.5 Btu/hr ft² °F. To increase the heat transfer rate, four straight copper fins of rectangular profile and 1/16 in. thick are added to the inside of the tube in a symmetrical fashion as shown in Fig. 2.72. Calculate the heat transfer rate per foot of pipe (a) with and (b) without the internal fins.

Figure 2.72

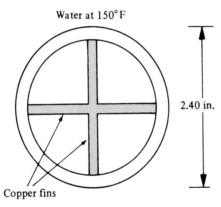

Water at 150°F

2.40 in.

Copper fins

2.57 Referring to Prob. 2.51, the figure showed a circular fin of rectangular profile, since the edge view between r_i and r_0 is a rectangle. For a circular fin of triangular profile, having thickness δ at r_i and 0 at r_0, derive the governing ordinary differential equation for the temperature as a function of the general radial coordinate r.

2.58 The valve shown in Fig. 2.73 is undergoing a final grinding operation on its upper and lower faces. The valve itself is circular, of outer radius r_0, and thickness t. The valve and its stem are held stationary in a jig while the grinders work on the valve. The jig holding the valve is such that at r_0, the face is perfectly insulated. The grinding wheels dissipate mechanical energy at the rate W_0 Btu/hr per square foot or W/m² of surface being ground, and essentially all of this energy goes into the valve, because its thermal conductivity is much higher than that of the grinding wheels. The valve stem, which is not touched by the upper grinding wheel, is exposed to air at known temperature T_∞ and with an average surface coefficient of heat transfer of h_∞ between the stem and the air. Find the maximum temperature of the valve stem in terms of known quantities during this machining operation, assuming that D is much smaller than r_0.

Figure 2.73

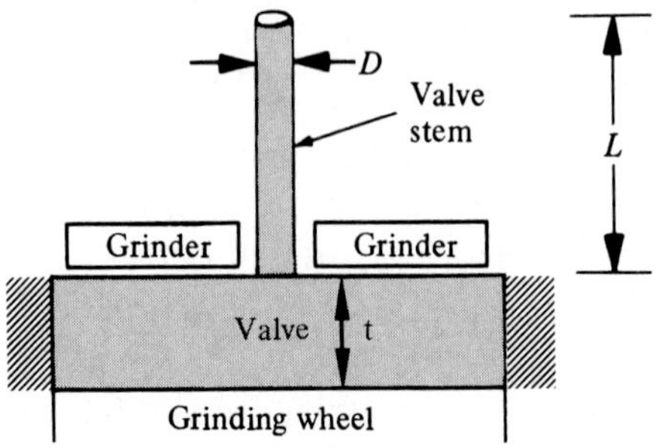

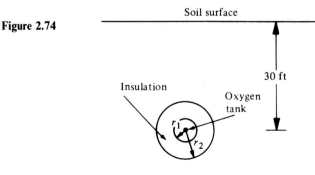

2.59 A 0.3-m-diameter sphere containing instrumentation has been buried to a depth of 1.5 m in sand ($k = 0.37$ W/m °C). In order to maintain the instruments properly, the sphere's surface is to be kept at a constant temperature of 32°C by an electric resistance heater. If the ground surface temperature is 18°C, find the rate in watts at which electrical energy must be dissipated within the sphere.

2.60 A steam pipe 50 ft long and 6 in. in diameter runs between two buildings. The pipe surface is at 300°F and is buried in soil ($k = 0.40$ Btu/hr ft °F) to a depth of 3 ft. If the soil surface is at 35°F, calculate the heat transfer loss from the pipe.

2.61 A long electrical transmission cable of 1.25 cm outside diameter is buried in soil whose thermal conductivity is 0.7 W/m °C and whose surface temperature is 21°C. If the electrical resistance per meter of the transmission line is 0.013 ohms, and the current is 100 amps, find the surface temperature of the transmission cable when the cable is at a depth of 0.61 m.

2.62 A small cubical electric laboratory furnace has inside dimensions 1/2 ft by 1/2 ft by 1/2 ft, a wall thickness of 4 in., and the wall material has a thermal conductivity of 0.4 Btu/hr ft °F. The furnace contains a 600 W electric resistance heater. The average surface coefficient of heat transfer between the outside of the furnace and the 75°F room air is expected to be 1.2 Btu/hr ft² °F. Find the steady-state temperature of both the (a) inside and (b) outside of the furnace wall.

2.63 Figure 2.74 shows a spherical liquid oxygen tank with 2-ft-thick insulation ($k_a = 0.03$ Btu/hr ft °F), buried 30 ft below the surface of the soil ($k = 0.30$ Btu/hr ft °F). The inside radius of the insulation is 4 ft and the inside surface temperature of the insulation is $T_0 = -300$°F. If the soil surface temperature is 50°F, calculate the heat transfer rate to the liquid oxygen.

Figure 2.74

Chapter Two

2.64 The fastener shown in Fig. 2.75 is to be used in many different places in a new product and will be subjected to a different set of temperatures T_1 and T_2 at each different location in the product. Furthermore, the material the fastener is constructed of may change because of a desired decorative effect on the outside of the product. In order to compute steady-state heat transfer rates across the fastener, an experiment is performed to find its shape factor. The experimental data is $T_1 = 58°C$, $T_2 = 31°C$, $k = 43$ W/m °C, and the measured heat transfer rate is 4.4 W. If all surfaces of the fastener except those at T_1 and T_2 are perfectly insulated, compute the experimental shape factor.

Figure 2.75

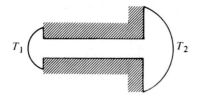

2.65 A long 0.15-m-diameter steam pipe has an outside temperature of 82°C and is buried in soil whose average thermal conductivity is $k = 1.0$ W/m °C. The pipe is 0.6 m below the soil surface, which is at 10°C. Compute the heat transfer rate per meter from the steam pipe.

2.66 Determine how deep the steam pipe of Prob. 2.65 would have to be below the ground to reduce its heat transfer rate per meter to 94 W/m.

2.67 If the heat transfer rate from the steam pipe of Prob. 2.65 is to be reduced to 95 W/m by adding insulation while keeping it at the same depth of 0.6 m, calculate the required outside diameter of insulation when its thermal conductivity is $k_a = 0.05$ W/m °C. The temperature at the inside of the insulation is 82°C.

2.68 Water at 5°C is pumped from a well through a 10-m-long pipe which is 2 m below the surface of the soil, has an outside diameter of 2.5 cm, and an outside surface temperature of about 5°C. The soil surface is at 27°C and its thermal conductivity is 1 W/m °C. Find (a) the heat transfer rate to the pipe, and (b) the lowest mass flow rate of water which will allow the water to leave the pipe with no more than a three-degree temperature increase.

2.69 A 6-in.-diameter steam line, which has an outside surface temperature of 250°F, runs parallel to a 2-in.-diameter cold water pipe at 40°F. These pipes are buried in soil ($k = 0.45$ Btu/hr ft °F) at the same depth from the surface. In order to prevent the cold water from heating up appreciably, it is desirable to limit the heat transfer rate from the hot water pipe to the cold water to a maximum of 200 Btu/hr per foot of pipe. Find the minimum spacing of the pipes needed to accomplish this.

2.70 A hot water pipe of diameter D and length L passes between two buildings through soil of thermal conductivity k and surface temperature T_0, at a depth of z units below the surface. The hot water enters the pipe at temperature T_1 with a mass flow rate and specific heat of $\dot{m}$ and c_p, respectively. The pipe wall is so thin that the outside surface of the pipe is at the local temperature of the water, which decreases continually along the pipe as a result of the heat loss, since $T_1 > T_0$. Explain in detail how to calculate the water temperature as a function of position along the pipe, from inlet to outlet, utilizing the given information and the table of shape factors, and find this temperature distribution.

2.71   Figure 2.76 shows a two-dimensional conduction region in the steady state with no generation. The strip is infinitely long in the x direction. Solve the appropriate partial differential
equation by separation of variables to obtain the temperature distribution

$$T(x, y) = T_0 + T_1 \sin (\pi y/l) \, e^{-\pi x/l}.$$

Figure 2.76

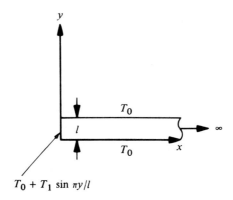

$T_0 + T_1 \sin \pi y/l$

2.72   (a) Referring to Prob. 2.71, use the temperature distribution to obtain an expression for the
total rate at which energy is conducted across the face at $x = 0$. (b) Using the result of (a),
write down directly an expression for the heat transfer rate across the face at $y = 0$.

2.73   Suppose that for Prob. 2.71, $T_0 = 38°C$, $T_1 = 260°C$, $l = 0.3$ m, and $k = 26$ W/m °C.
Find (a) the maximum temperature of the strip and (b) the temperature at the space point
$x = 0.6$ m, $y = 0.15$ m.

2.74   Figure 2.77 shows a two-dimensional conduction region extending to $+\infty$ in the y direction.
Its bottom face at $y = 0$ is exposed to a fluid whose known temperature is $T_\infty \sin (4\pi x/L)$,
and there is a constant surface coefficient of heat transfer h between this fluid and the bottom
face. Find an exact analytical solution for the steady-state temperature distribution as a function of x and y.

Figure 2.77

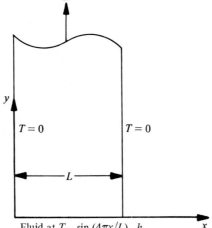

Fluid at $T_\infty \sin (4\pi x/L)$, h

2.75 Explain how the exact analytical solution which was developed for Fig. 2.36 can be used almost directly to express the temperature as a known function of x and y for the conduction region shown in Fig. 2.78.

Figure 2.78

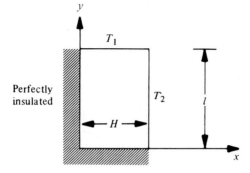

2.76 Find an exact analytical solution for the temperature distribution as a function of x and y for the steady two-dimensional conduction region shown in Fig. 2.79.

Figure 2.79

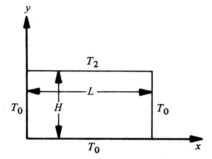

2.77 Find an exact analytical solution for the steady-state temperature distribution in a cubical conduction region with no generation if five of the six faces are perfectly insulated, and the sixth face, which is at $x = 0$, has the known temperature T_0. The cube side length is L.

2.78 Figure 2.80 shows a cross section of a long one-eighth-cylinder, perfectly insulated on one face, held at temperature 0 on the bottom face, and at the varying temperature $T_0 \sin 2\theta$ on the surface at $r = r_0$. If steady-state conditions prevail with no generation, find an exact analytical solution for the temperature distribution within the conduction region.

Figure 2.80

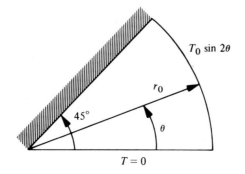

2.79 Figure 2.81 shows a right circular cylinder of constant thermal conductivity k and zero generation rate per unit volume. All surfaces are at 0 except for the face at $z = L$, which has the temperature $T = T_0 J_0(2.405\ r/R)$, which is a function of the radial coordinate r, where T_0 is a known positive constant and $J_0(2.405\ r/R)$ is the ordinary Bessel function of the first kind and order zero. Find an exact analytic expression for the steady-state temperature distribution within the cylinder.

Figure 2.81

$$T = T_0 J_0(2.405\tfrac{r}{R})$$

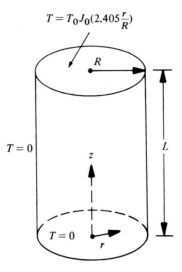

2.80 The conduction region shown in Fig. 2.82 has constant thermal conductivity k and zero generation. The temperature is known to be the value T_1 on the upper and lower faces, and the heat flux is given on the left-hand face as $q'' = q_0 \sin(\pi y/l)$. The region extends infinitely far in the plus-x direction. Find an exact analytical solution for the steady-state temperature as a function of the space coordinates x and y.

Figure 2.82

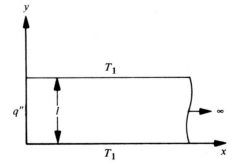

2.81 Find an expression for the total heat transfer rate across the lower surface, at $y = 0$, between $x = 0$ and $x = L$, $0 < L < \infty$, for the conduction region of Prob. 2.80.

2.82 Figure 2.83 shows a support bar of radius r_0 and length L which has its ends held at the constant temperature T_0, while its curved surface is at temperature T_f as a result of fluid flowing over it with a very high surface coefficient of heat transfer. Find the steady-state temperature distribution if there is no generation and if the thermal conductivity is constant.

Figure 2.83

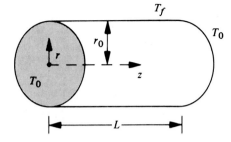

T_f

T_0

r_0

z

T_0

L

2.83 Show that the solution for the steady-state temperature distribution within the plate pictured in Fig. 2.84 is the sum of two other solutions, one being the same plate with $T = 0$ along $y = 0$ and T_2 at $x = 0$, and the other being the same plate with $T = 0$ along $x = 0$ and T_1 along $y = 0$. This is most easily done by assuming that the solutions to the other two problems are known, and then showing that their sum satisfies the governing partial differential equation and the boundary conditions for the figure.

Figure 2.84

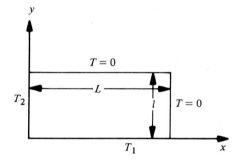

y

$T = 0$

L

T_2

l

$T = 0$

T_1

x

2.84 Consider the steady-state two-dimensional conduction region shown in Fig. 2.85. Find an exact analytical solution for the temperature distribution by solving the conduction equation in circular cylindrical coordinates, Eq. (2.8), subject to the boundary conditions shown.

Figure 2.85

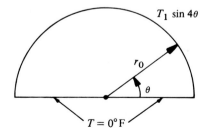

$T_1 \sin 4\theta$

r_0

θ

$T = 0°F$

2.85 For the conduction region shown in Fig. 2.86, find an exact analytical solution for the steady-state temperature as a function of x, y, and z. Note that *two* separation constants will have to be used to obtain the three separation equations.

Figure 2.86

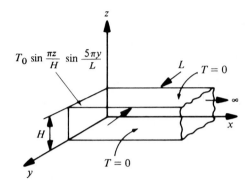

2.86 Consider a 0.3-m-thick plane slab, of thermal conductivity 85 W/m °C, with one surface held at 260°C and the other surface held at 38°C. Steady-state conditions prevail with no generation. (a) Using three or more interior nodes, solve for the temperature distribution by finite difference methods. (b) Compare the answers to part a with the exact solution (linear variation of temperature from 260°C to 38°C), and explain the agreement.

2.87 Figure 2.87 shows a lattice point, numbered 0, at the interface of two different materials a and b. Their thermal conductivities are k_a and k_b, respectively. Derive the finite difference equation for node 0.

Figure 2.87

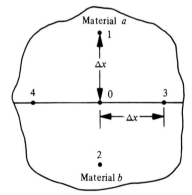

2.88 In the conduction region between r_i and R shown in Fig. 2.88, the temperature is a function only of r, so that it seems reasonable to space the nodes a distance Δr apart, as shown. Identify the volume of material associated with node 0, and then derive the governing finite difference equation for node 0.

Figure 2.88

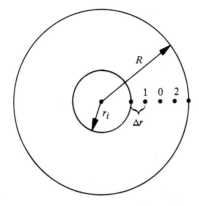

2.89 Figure 2.89 shows a two-dimensional conduction region which is in the steady state and has no generation. The temperatures and dimensions are as shown. (a) Using a square lattice network, $\Delta x = \Delta y = 0.075$ m, and taking advantage of the obvious line of symmetry, use finite difference techniques to obtain the values of the temperatures at the grid points. (b) If the thermal conductivity of the material is 85 W/m °C, calculate the heat transfer rate across the 90°C surface.

Figure 2.89

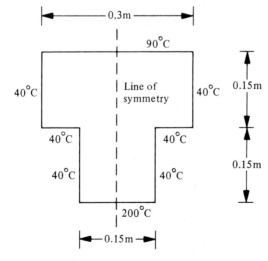

2.90 Figure 2.90 shows a two-dimensional conduction region with steady-state conditions prevailing and no generation. Using a square grid network with $\Delta x = \Delta y = 0.08$ m and $k = 44$ W/m °C, and taking advantage of the line of symmetry, find the temperature at the various nodal points by numerical methods.

Figure 2.90

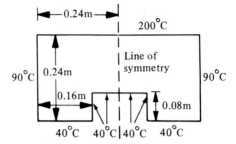

2.91 The surface node 0 shown in Fig. 2.91 is exchanging energy by convection with the fluid and also losing radiant energy at a rate prescribed by the Stefan-Boltzmann law (see Chapter 1). If it does not gain any radiant energy and if there is a constant generation rate per unit volume within the solid, derive the governing finite difference equation at node 0.

Figure 2.91   Fluid at T_∞, h_∞

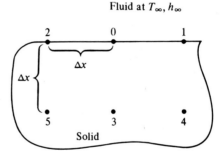

2.92 The corner node 0 shown in Fig. 2.92 is exposed to a fluid at known temperature T_∞ on two sides, but because of the orientation of the solid, the surface coefficients of heat transfer differ on the top and side as shown. Derive the finite difference equation for node 0.

Figure 2.92   Fluid at
T_∞, h_a

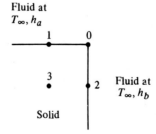

2.93   Consider an interior node in a three-dimensional conduction region with generation throughout the region. The region's shape is such that the bounding surfaces are also coordinate
surfaces in the circular cylindrical coordinate system where a point in space is located by the
coordinates r, θ, and z. For this node, derive the general finite difference equation using the
circular cylindrical coordinate system.

2.94   Use an equispaced lattice and find the steady-state temperature distribution in the two-
dimensional triangular conduction region shown in Fig. 2.93 if $k = 77$ W/m °C.

Figure 2.93

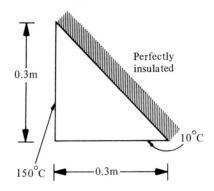

0.3m

Perfectly
insulated

$10°$C

$150°$C |←——0.3m——→|

2.95   Determine what other shape, with what boundary conditions, Prob. 2.94 is the proper solution
to.
(*Hint:* Utilize geometric and thermal symmetries.)

2.96   Derive the finite difference equation for an interior node in a two-dimensional conduction
region which has a temperature-dependent thermal conductivity. Determine (a) in what important respect the resulting equation differs from one for constant thermal conductivity, and
(b) if a set of equations of this type is amenable to solution by Cramer's rule or by relaxation
analysis. Explain your answers.

2.97   Figure 2.94 shows a cross section of a master duct in a heating system. Hot water at 90°C
flows in the center feeder duct and causes the sides to be at 90°C, while cooler return water
is flowing in the four smaller ducts, which causes the surfaces of these four ducts to be at
32°C. The insulating material separating the five ducts has a thermal conductivity of 0.078
W/m °C, and the outside surface of this material is exposed to air at 15°C with an average
surface coefficient of heat transfer of 10 W/m² °C between the air and the outside surface.
(a) If $l = 5$ cm, use an equispaced lattice of 5 cm to find, by numerical methods, the steady-
state temperature distribution within the insulating material. Take advantage of all the sym-

metry of the situation. (b) Estimate the heat transfer rate to or from one of the small return ducts. (c) Given the freedom to use the center duct for the cooler water and the four smaller ducts for the flow of the hot water, would you make this change? Explain your reply.

Figure 2.94

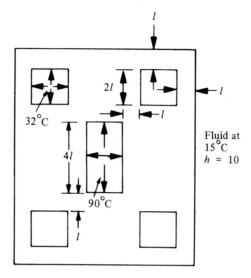

2.98   The two-dimensional conduction region shown in Fig. 2.95 has a thermal conductivity of 10 Btu/hr ft °F. The left face is held at 300°F, while the remaining three faces are exposed to a 100°F fluid with a surface coefficient of heat transfer of 120 Btu/hr ft² °F. (a) Take a 1-in. lattice spacing and solve for the temperature distribution by finite difference methods. Note the symmetry of the problem. (b) Compare the answers to part a to those that would be obtained using a quasi-one-dimensional analysis.

Figure 2.95

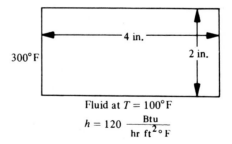

Fluid at $T = 100°F$

$$h = 120 \ \frac{Btu}{hr \ ft^2 \ °F}$$

2.99   Suppose that the sphere shown in Fig. 2.40 is surrounded by a fluid at known temperature T_f, and there is a known average surface coefficient of heat transfer h between the outside surface and the fluid. There is generation and the heat flow is radial and steady. Derive the governing finite difference equation for node 3.

2.100   Figure 2.96 shows a cross section of a long structural member in a wall whose thermal conductivity is so low, compared to that of the structural member, that the structural member is effectively insulated. The fluid at the bottom surface of the member is at 1000°F, and a surface coefficient of heat transfer equal to 30 Btu/hr ft² °F exists between the member and the bottom side. The corresponding quantities for the fluid on the top side are 100°F and $h = 100$ Btu/hr ft² °F. Using a 1/2-ft equispaced lattice, find the steady-state temperature distribution using finite difference methods. Assume that $k = 30$ Btu/hr ft °F.

Figure 2.96

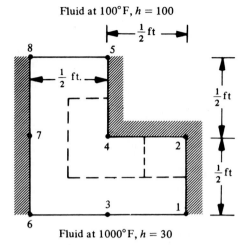

Fluid at 100°F, $h = 100$

Fluid at 1000°F, $h = 30$

2.101 The surface of the 6-in. diameter steam pipe shown in Fig. 2.97 is at 300°F, while the material surrounding it has a thermal conductivity of 1.0 Btu/hr ft °F. Find (a) the steady-state temperature distribution in the material using numerical methods, and (b) the heat transfer rate from the steam pipe and the shape factor S.

Figure 2.97

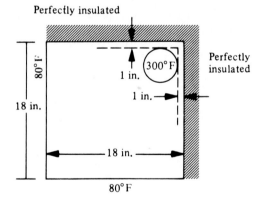

2.102 A very long concrete column of cross-sectional dimensions 0.6 m by 1.2 m has been poured and is setting. Assuming that the setting of the concrete results in a uniformly distributed generation rate per unit volume of 1000 W/m³, that the thermal conductivity of the concrete is 1.56 W/m °C, and that the outside of the column is maintained at 10°C, find the temperature distribution and the maximum temperature within the concrete using numerical methods. Use an equispaced lattice of 0.15 m and take advantage of symmetry.

2.103   Depicted in Fig. 2.98 is a steel support for a tank in a foundry. The hot tank bears against the 1000°F surfaces. Steady-state conditions prevail with no generation, and the support is deep enough into the paper so that conduction can be neglected in that direction. Using numerical methods and a crude equispaced lattice $\Delta x = \Delta y = 1$ in. and utilizing the obvious line of symmetry, find the temperature at the various nodal points. (a) Use the computer and a solution technique of your choice for the finite difference equations. (b) For a finer lattice spacing $\Delta x = \Delta y = 1/2$ in., use the computer to find the temperature at the various nodal points. (c) If $k = 25$ Btu/hr ft °F, estimate the heat transfer rate across the 500°F face, for one foot of depth into the paper, for both the 1-in. lattice spacing and the 1/2-in. spacing.

Figure 2.98

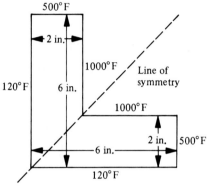

2.104   In Fig. 2.99, a long structural member is idealized as being perfectly insulated at its lower surface with temperatures as shown on its other surfaces due to its contact with other bodies. Steady-state conditions prevail and there is no generation. The thermal conductivity is essentially constant at 25 W/m °C. (a) Use the finite difference method and the coarse lattice spacing $\Delta x = \Delta y = 0.075$ m and find the temperature at the various nodal points. (b) Use the computer to find the nodal temperatures for the finer grid spacing $\Delta x = \Delta y = 0.0375$ m. (c) For part b, calculate the heat transfer rate across the 200°C face for a one-meter length of member.

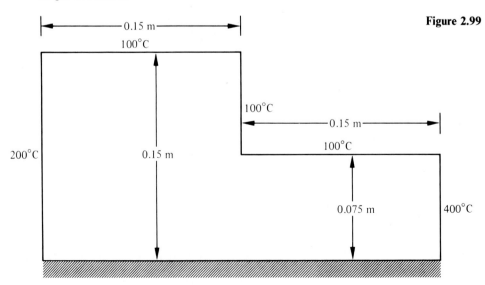

Figure 2.99

2.105 Shown in Fig. 2.100 is a solid conduction region with surface temperatures as shown. The thermal conductivity is known and constant at the value of 25 Btu/hr ft °F. There is no generation, steady-state conditions prevail, and the region can be considered to be very long in the direction perpendicular to the plane of the paper. Using finite difference methods and a crude equispaced lattice in which $\Delta x = \Delta y = 3$ in. and taking advantage of the obvious line of symmetry, find the nodal temperatures. (a) With the computer, utilize any appropriate solution procedure for the finite difference equations. (b) Repeat for the finer lattice spacing $\Delta x = \Delta y = 1.5$ in. by use of the computer. (c) Using the solution of part b, calculate the heat transfer rate across the upper face per foot of member into the paper.

Figure 2.100

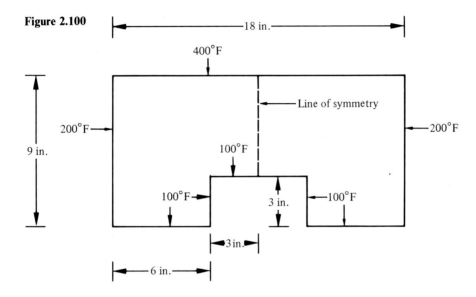

2.106 Figure 2.101(a) shows a cross section of a turbine blade for a jet engine with a rectangular hole carrying 1200°F cooling air which keeps the hole surfaces at 1200°F. The blade is made of Udimet 700, which has an average thermal conductivity of about 18 Btu/hr ft °F in the temperature range the blade will experience. Hot gases at an effective temperature of 1800°F flow around the blade, giving rise to a local surface coefficient of heat transfer h_0 at the stagnation point of 900 Btu/hr ft² °F. The ratio of the local surface coefficient of heat transfer anywhere on the top of the blade (suction surface) to h_0 is given in (b). To

Figure 2.101

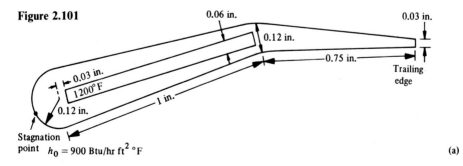

(a)

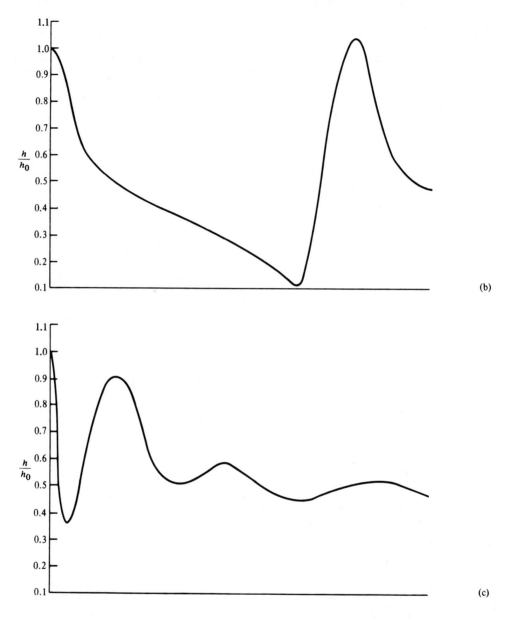

(b)

(c)

obtain the local h anywhere on the top surface, simply run a vertical line from the point of interest to the curve in (b). Referring to (c), use the same procedure to obtain the local surface coefficient on the bottom of the blade (pressure surface). The leading edge of the blade is simply the arc of a circle with the radius given in the figure. Using numerical methods, find the steady-state temperature distribution within the turbine blade. Use any network of nodes which seems to be most appropriate. In particular, once out of the leading edge region, a network which tends to follow the curvature of the blade is suggested, thus giving nodes right on the surface all the way to the trailing edge.

REFERENCES

1. Jahnke, E., and F. Emde. *Tables of Functions with Formulae and Curves.* 4th ed. New York: Dover, 1945.
2. Wylie, C. R., and L. C. Barrett. *Advanced Engineering Mathematics.* 5th ed. New York: McGraw-Hill, 1982.
3. Kreith, F. *Principles of Heat Transfer.* 2d ed. Scranton, Pa.: International Textbook Co., 1965.
4. Hildebrand, F. B. *Advanced Calculus for Applications.* Englewood Cliffs, N.J.: Prentice-Hall, 1963.
5. Arpaci, V. S. *Conduction Heat Transfer.* Reading, Mass.: Addison-Wesley, 1966.
6. Harper, D. R., 3d, and W. B. Brown. "Mathematical Equations for Heat Conduction in the Fins of Air Cooled Engines." *NACA Report,* No. 158 (1922).
7. Smith, P. J., and J. Sucec. "Efficiency of Circular Fins of Triangular Profile." *Journal of Heat Transfer,* Vol. 91 (1969), pp. 181–82.
8. Schneider, P. J. *Conduction Heat Transfer.* Reading, Mass.: Addison-Wesley, 1955.
9. Kern, D. Q., and A. D. Kraus. *Extended Surface Heat Transfer.* New York: McGraw-Hill, 1972.
10. Andrews, R. V. "Solving Conductive Heat Transfer Problems with Electrical-Analogue Shape Factors." *Chemical Engineering Progress,* Vol. 51, No. 2 (1955), pp. 67–71.
11. Smith, J. C., J. E. Lind, and D. S. Lermond. "Shape Factors for Conductive Heat Flow." *American Institute of Chemical Engineering Journal,* Vol. 4, No. 3 (September 1958), pp. 330–31.
12. Rudenberg, V. R. "Die Ausbreitung der Luft-und Erdfelder um Hochspannungs leitungen besonders bei Erd-und Kurzschlüssen." *Elektrotechnische Zeitschrift,* No. 36 (September 1925), pp. 1342–46.
13. Langmuir, I., E. Q. Adams, and F. S. Meikle. "Flow of Heat Through Furnace Walls: The Shape Factor." *Transactions of the American Electrochemical Society,* Vol. 24 (1913), pp. 53–84.
14. Rohsenow, W. M., and J. P. Hartnett, eds. *Handbook of Heat Transfer.* New York: McGraw-Hill, 1973.
15. Ozisik, M. N. *Boundary Value Problems of Heat Conduction.* Scranton, Penn.: International Textbook Co., 1968.
16. Myers, G. E. *Analytical Methods in Conduction Heat Transfer.* New York: McGraw-Hill, 1971.
17. Carslaw, H. S., and J. C. Jaeger. *Conduction of Heat in Solids.* London: Oxford University Press, 1959.
18. Smith, G. D. *Numerical Solution of Partial Differential Equations.* New York: Oxford University Press, 1965.
19. Warrington, R. O., Jr., R. E. Powe, and R. L. Musselman. "Steady Conduction in Three Dimensional Shells." *Journal of Heat Transfer,* Vol. 104 (1982), pp. 393–94.
20. Ames, W. F. *Numerical Methods for Partial Differential Equations.* 2d ed. New York: Academic Press, 1977.
21. Hornbeck, R. W. *Numerical Methods.* New York: Quantum Publishers, 1975.
22. Holman, J. P. *Heat Transfer.* 5th ed. New York: McGraw-Hill, 1981.
23. Lienhard, J. H. *A Heat Transfer Textbook.* Englewood Cliffs, N.J.: Prentice-Hall, 1981.
24. Incropera, F. P., and D. P. DeWitt. *Fundamentals of Heat Transfer.* New York: Wiley, 1981.
25. Karlekar, B. V., and R. M. Desmond. *Heat Transfer.* 2d ed. St. Paul, Minn.: West, 1982.
26. Sucec, J. *Heat Transfer.* New York: Simon and Schuster, 1975.
27. Jakob, M. *Heat Transfer.* Vol. 1. New York: Wiley, 1959.
28. Ingersoll, L. R., O. I. Zobel, and A. C. Ingersoll. *Heat Conduction.* Madison, Wisc.: The University of Wisconsin Press, 1954.
29. Boelter, L. M. K., V. H. Cherry, H. A. Johnson, and R. C. Martinelli. *Heat Transfer Notes.* New York: McGraw-Hill, 1965.
30. Grober, H., S. Erk, and U. Grigull. *Fundamentals of Heat Transfer.* 3d ed. New York: McGraw-Hill, 1961.
31. Ozisik, M. N. *Heat Conduction.* New York: Wiley, 1980.

UNSTEADY-STATE CONDUCTION

3.1 INTRODUCTION

When the temperature at any space point within a conduction region is changing with time, the temperature distribution is termed *unsteady;* that is, an *unsteady-state condition* prevails. The general unsteady-state conduction problem requires the determination of the temperature distribution in a solid conduction region as a function of space coordinates and time τ.

Two types of unsteady-state conduction are:

1. *Regular periodic conduction.* The temperature at the various space points, although changing with time, changes according to a definite pattern, namely, a cyclic variation in time which is repeated. An example of regular periodic unsteadiness is the temperature distribution within the cylinder wall of a reciprocating automobile engine operating at constant speed over a period of time. Measurement of the temperature as a function of time at some point within the cylinder wall yields a definite pattern of T vs τ that is repeated each time the piston cylinder arrangement completes the sequence of thermodynamic processes that is termed a cycle. Thus, at the beginning of the intake stroke, the temperature at a point in the cylinder wall may have the value T_0. The temperature then increases during the compression stroke, and increases further during the combustion process. It then decreases during the power and exhaust strokes until it again reaches the value T_0, at which time the thermal cycle starts again.

2. *Transient conduction.* The temperature variation at a space point, as time goes on, is not cyclic. Transient problems are by far the larger and more important class of unsteady-state problems and include, in the most general situation, the regular periodic problems as degenerate cases as time gets very large. For this reason, attention will be focused here on transient conduction problems. Examples of actual situations in which knowledge of transient temperature

distributions is important include jet engines and rocket engines during start-up and shut-down, heat-treating operations, start-up and shut-down of heat exchangers and of nuclear reactors, and the design of safes which must afford fire protection to their contents for some specified period of time.

In this discussion of transient temperature distributions, some exact analytical solutions to the governing partial differential equation are presented first. Here the results for several important one-dimensional transients can be presented in chart form. Then, certain two- and three-dimensional transients are treated by the appropriate combination of the elementary one-dimensional transients already determined. Next, the *lumped parameter method* of transient analysis is presented, in which, under certain conditions, the temperature variation in the space coordinates can be neglected and attention focused on the average temperature of the conduction region as a function of time alone. The lumped parameter method is extremely useful to the engineer, when it can be used, because of the relative ease of application. Finally, in order to solve the most general transient problems, the approximate and very powerful finite difference method will be extended to transient problems.

3.2 EXACT ANALYTICAL SOLUTIONS AND CHARTS

Shown here is the governing partial differential equation for the temperature distribution as a function of space and time within a conduction region consisting of an isotropic, rigid solid with constant thermal conductivity, for the rectangular cartesian coordinate system.

$$\frac{\partial^2 T}{\partial x^2} + \frac{\partial^2 T}{\partial y^2} + \frac{\partial^2 T}{\partial z^2} + \frac{q'''}{k} = \frac{1}{\alpha}\frac{\partial T}{\partial \tau}. \qquad \textbf{[2.6]}$$

In many problems, the separation of variables method will lead to an exact analytical solution of Eq. (2.6), subject to certain boundary conditions and an initial condition, just as it did for some steady-state problems in Chapter 2. However, the exact analytical solutions to transient problems are more involved than for "equivalent" steady-state problems, because of the presence of the additional independent variable, time τ.

In spite of this additional complicating feature, exact analytical solutions of Eq. (2.6) can be found for a number of elementary one-dimensional transient problems that are of great importance from the standpoint of practical applications of their solutions. In addition to being important in their own right, these one-dimensional transient solutions form the basis for certain two- and three-dimensional transient conduction solutions. The overall approach to be used in attempting an exact analytical solution of Eq. (2.6) will be demonstrated by solving the infinite slab problem in detail.

3.2a Transient in the Infinite Slab

Figure 3.1 shows a slab of thickness $2L$ in the x direction which is large enough in the y and z directions to have essentially zero temperature derivatives in those directions. The slab is initially at a constant known temperature T_0 throughout when suddenly, at time

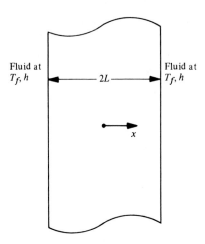

Figure 3.1 Slab undergoing the transient conduction process.

$\tau = 0$, it is immersed in a fluid whose temperature is, and remains, T_f, with a constant average surface coefficient of heat transfer h between both faces of the slab and the fluid. If there is no heat generation and if the thermal conductivity of the slab is assumed constant, we would like to solve for the temperature profile within the slab as a function of the relevant space coordinates and of time.

The governing partial differential equation is given by Eq. (2.6) with the terms containing y and z derivatives, along with the generation term, set equal to zero; hence, the temperature will depend only upon x and τ. Thus,

$$\frac{\partial^2 T}{\partial x^2} = \frac{1}{\alpha} \frac{\partial T}{\partial \tau} .$$ (3.1)

Figure 3.1 shows that the temperature distribution at all times is symmetric about the midplane of the slab, and, hence, $x = 0$ is situated at the center of the slab. As a result, the temperature at $x = 0$ must be either a maximum or a minimum; that is, the slope $\partial T / \partial x = 0$ at $x = 0$. Hence, by Fourier's exact law of conduction

$$q''_x = -k \frac{\partial T}{\partial x} ,$$

we can see that the heat transfer rate across $x = 0$ is zero, and the midplane at $x = 0$ behaves as a perfect insulator, an adiabatic surface. The result is that the solution to this problem also applies to the slab shown in Fig. 3.2, which is perfectly insulated at one face, initially at known temperature T_0, and then exposed on one face to fluid at the constant temperature T_f (with constant surface coefficient of heat transfer h). Thus the boundary condition at $x = 0$ is

$$\frac{\partial T}{\partial x} = 0 \quad \text{for } \tau > 0.$$ (3.2)

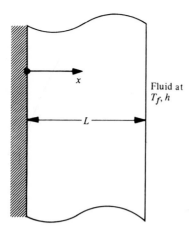

Figure 3.2 The solution for Fig. 3.1 also applies to this slab.

Since no energy can be stored at the solid/fluid interface at $x = L$, the law of conservation of energy requires that the rate at which energy is conducted to the interface from within the solid at $x = L$ is equal to the rate at which energy is convected into the fluid. This equality must hold at every instant of time; hence, the boundary condition at $x = L$ is

$$-k \left. \frac{\partial T}{\partial x} \right|_{x=L} = h \, (T_{x=L} - T_f) \quad \text{for } \tau > 0. \tag{3.3}$$

Inspection of Eq. (3.1) shows the presence of a first derivative with respect to time, thus indicating that something must be specified about the temperature at some value of time. It is clear that temperatures which occur at future times cannot influence present or past events, although past events may influence the present and future. Hence, an *initial condition* is ordinarily specified, such as the temperature as a function of the space coordinates at time $\tau = 0$. Here, the initial condition is that at $\tau = 0$,

$$T = T_0 \quad \text{for } 0 < x < L. \tag{3.4}$$

To solve the problem posed by Eq. (3.1) subject to the boundary conditions Eqs. (3.2) and (3.3) and the initial condition Eq. (3.4), a change of the dependent variable is made to

$$\theta = T - T_f. \tag{3.5}$$

The reason for this choice of new dependent variable is that it allows us to work with a properly posed boundary and initial value problem for Eq. (3.1). That is, one that can be solved by separation of variables and leads to orthogonal functions in the space coordinates. From Eq. (3.5), $T = \theta + T_f$; hence, $\partial^2 T/\partial x^2 = \partial^2\theta/\partial x^2$ and $\partial T/\partial \tau = \partial\theta/\partial \tau$, so that Eq. (3.1) becomes

$$\frac{\partial^2 \theta}{\partial x^2} = \frac{1}{\alpha} \frac{\partial \theta}{\partial \tau}. \tag{3.6}$$

The boundary condition of Eq. (3.2) becomes that at $x = 0$,

$$\frac{\partial \theta}{\partial x} = 0 \quad \text{for } \tau > 0. \tag{3.7}$$

The boundary condition of Eq. (3.3) becomes that at $x = L$,

$$-k \left. \frac{\partial \theta}{\partial x} \right|_{x=L} = h\theta_{x=L} \quad \text{for } \tau > 0. \tag{3.8}$$

The initial condition of Eq. (3.4) becomes that at $\tau = 0$,

$$\theta = T_0 - T_f = \theta_0 \quad \text{for } 0 < x < L. \tag{3.9}$$

Using the method of separation of variables, assume that

$$\theta = R(x)\, G(\tau), \tag{3.10}$$

where $R(x)$ is a function of x only, $G(\tau)$ is a function of time τ only, and both R and G must be determined. Substituting Eq. (3.10) into Eq. (3.6) and rearranging,

$$\frac{d^2R/dx^2}{R} = \frac{dG/d\tau}{\alpha G} = -\lambda^2.$$

The individual separation equations are then

$$\frac{d^2R}{dx^2} + \lambda^2 R = 0, \tag{3.11}$$

$$\frac{dG}{d\tau} + \lambda^2 \alpha G = 0. \tag{3.12}$$

(The choice of algebraic sign for the separation constant is rationalized by noting that if $+\lambda^2$ were used, the solution to Eq. [3.12] would give $G(\tau)$ as a function which becomes infinite as $\tau \rightarrow \infty$ and, hence, by Eq. [3.10], so would θ. Since θ is bounded, $-\lambda^2$ is used. Although this argument is quite correct, the principal reason for using $-\lambda^2$ is to ensure that the solution of Eq. [3.11] gives sines and cosines, since Eq. [3.11], combined with Eqs. [3.7] and [3.8], constitutes a characteristic value problem which requires characteristic functions for a nontrivial solution.)

Solving Eqs. (3.11) and (3.12) by the classical operator technique yields, after combining the solutions according to Eq. (3.10),

$$\theta = (A \sin \lambda x + B \cos \lambda x)e^{-\alpha\lambda^2\tau},$$

where A and B are integration constants. Now

$$\frac{\partial \theta}{\partial x} = (\lambda A \cos \lambda x - \lambda B \sin \lambda x)e^{-\alpha\lambda^2\tau} \tag{3.13}$$

and Eq. (3.7) requires that this be zero at $x = 0$ for all τ. Hence,

$$0 = (\lambda A - 0)e^{-\alpha\lambda^2\tau}.$$

Thus, A must be chosen equal to zero and the solution becomes

$$\theta = B \cos \lambda x \, e^{-\alpha\lambda^2\tau}. \tag{3.14}$$

Now, for Eq. (3.8), $(\partial\theta/\partial x)_{x=L}$ and $\theta_{x=L}$ are required. From Eq. (3.13),

$$\frac{\partial \theta}{\partial x}\bigg|_{x=L} = -\lambda B \sin \lambda L \, e^{-\alpha\lambda^2\tau},$$

and from Eq. (3.14),

$$\theta_{x=L} = B \cos \lambda L \, e^{-\alpha\lambda^2\tau}.$$

Thus, Eq. (3.8) becomes

$$+k \lambda B \sin \lambda L \, e^{-\alpha\lambda^2\tau} = h B \cos \lambda L \, e^{-\alpha\lambda^2\tau},$$

or after cancelling common terms, multiplying both sides by L, and rearranging,

$$\lambda L \sin \lambda L = (hL/k) \cos \lambda L.$$

This is the transcendental equation that determines the infinite number of eigenvalues or characteristic values of λ. Any particular value is designated λ_n, hence,

$$\lambda_n L \sin \lambda_n L = (hL/k)\cos \lambda_n L. \tag{3.15}$$

Inspection of Eq. (3.15) yields the conclusion that $\lambda = 0$ will *not* satisfy the equation. Hence the separation functions associated with the zero separation constant need not be investigated here.

Next, Eq. (3.15) must be solved for the permissible values of the separation constants λ_n. The nondimensional group hL/k is called the *Biot number* and is denoted by N_{Bi}. Equation (3.15) is solved for the values $\lambda_n L$ by first rearranging and then solving the resulting equation graphically. Rearrangement gives

$$\cot \phi_n = \phi_n/N_{\text{Bi}}, \tag{3.16}$$

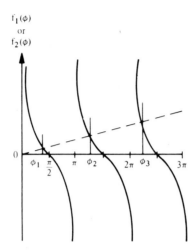

Figure 3.3 Graphical solution for the transcendental Eq. (3.16).

where $\phi_n \equiv \lambda_n L$. Now, the left-hand side of Eq. (3.16) is a function of the parameter ϕ. Letting $f_1(\phi) = \cot \phi$, this function is plotted as the solid lines in Fig. 3.3. The right-hand side of Eq. (3.16) is the function of ϕ defined by $f_2(\phi) = \phi/N_{Bi}$ and is plotted as the dashed curve in Fig. 3.3 for the value of N_{Bi} in this problem. Equation (3.16) has for its solutions the particular values of ϕ, called ϕ_n, which occur when $f_1(\phi) = f_2(\phi)$. This occurs when the curves intersect in a common plane; hence, the intersection points in Fig. 3.3 define the solutions $\phi_n = \lambda_n L$ of Eq. (3.16). The first three intersection points ϕ_1, ϕ_2, and ϕ_3 are shown. Actually, Eq. (3.16) can be solved more accurately by using an iterative method, such as Newton's method, where the original estimate of the roots of Eq. (3.16) might come from a very rough graphical solution as described here.

Since Eq. (3.16) is a commonly encountered transcendental equation, its first five roots, at least, are known and tabulated as functions of the *Biot number* and are available [8]. Hence the λ_n that satisfy Eq. (3.15) have been determined, and Eq. (3.14) becomes

$$\theta = B \cos \lambda_n x \, e^{-\alpha \lambda_n^2 \tau} . \tag{3.17}$$

However, the initial condition Eq. (3.9) cannot be satisfied by Eq. (3.17) in its present form. But since the partial differential equation is linear and Eq. (3.17) represents a valid solution to it for any value of λ_n which satisfies Eq. (3.15), the principle of linear superposition is used to sum up all possible solutions of the form of Eq. (3.17) to obtain

$$\theta = \sum_{n=1}^{\infty} B_n \cos \lambda_n x \, e^{-\alpha \lambda_n^2 \tau} . \tag{3.18}$$

Now the last condition that Eq. (3.18) must satisfy is that at $\tau = 0$,

$$\theta = \theta_0 = T_0 - T_f \text{ for } 0 < x < L.$$

At $\tau = 0$ with $\theta = \theta_0$, Eq. (3.18) becomes

$$\theta_0 = \sum_{n=1}^{\infty} B_n \cos \lambda_n x, \tag{3.19}$$

and thus B_n must be chosen in such a way that the right-hand side of Eq. (3.19) is an infinite series which converges to θ_0 in the interval $0 < x < L$. If λ_n were equal to $n\pi/L$, this would be a Fourier cosine series expansion whose coefficients are given by

$$B_n = \frac{\int_0^L f(x) \cos (n\pi x/L) \, dx}{\int_0^L \cos^2 (n\pi x/L) \, dx} \tag{3.20}$$

for the expansion of any well-behaved function $f(x)$ in the interval $0 < x < L$ [2]. However, here λ_n are *not* equal to $n\pi/L$. A study of orthogonal functions which come from a characteristic value problem involving a linear, homogeneous, second-order differential equation subject to homogeneous boundary conditions (the Sturm-Liouville system) leads to the conclusion that any well-behaved function $f(x)$ can be expanded into an infinite series of orthogonal functions with the coefficients of such a series given by a "formula" of a type similar to Eq. (3.20) [5, p. 185]. A study of this type leads directly to the relation

$$B_n = \frac{\int_0^L \theta_0 \cos \lambda_n x \, dx}{\int_0^L \cos^2 \lambda_n x \, dx} \tag{3.21}$$

for the unknown coefficients B_n. Carrying out the integration leads to

$$B_n = \frac{2\theta_0 \sin \lambda_n L}{\lambda_n L + \sin \lambda_n L \cos \lambda_n L}. \tag{3.22}$$

Note that another way of arriving at an equation for B_n is to proceed from Eq. (3.19) using the same techniques described in [2] for finding the coefficients of a Fourier cosine series. Recall that Eq. (3.20) resulted from the fact that the integral

$$\int_0^L \cos \frac{m\pi x}{L} \cos \frac{n\pi x}{L} \, dx$$

is zero when $m \neq n$ and nonzero when $m = n$. Thus, beginning with Eq. (3.19) and following the same steps which were used for the development of Eq. (3.20), the integral

$$\int_0^L \cos \lambda_n x \cos \lambda_m x \, dx \tag{3.23}$$

must ultimately be determined. Since λ_n and λ_m are given by Eq. (3.16) as

$$\cot \lambda_n L = \frac{\lambda_n L}{N_{Bi}}, \quad \cot \lambda_m L = \frac{\lambda_m L}{N_{Bi}}, \tag{3.24}$$

integration of Eq. (3.23) combined with Eq. (3.24) yields the result that the integral in Eq. (3.23) is zero unless $\lambda_n = \lambda_m$. Following the rest of the procedure for an ordinary Fourier cosine series yields Eq. (3.21). It may be felt that this second approach is preferable because it builds upon knowledge of finding the coefficients for a Fourier series expansion. However, in most problems the integral $\int_0^L \phi_n(x)\,\phi_m(x)\,dx$ will not be zero when $n \neq m$ unless a weighting function $w(x)$ is introduced; then, the new integral $\int_0^L w(x)\,\phi_n(x)\,\phi_m(x)\,dx = 0$ when $m \neq n$. The systematic study of the general Sturm-Liouville system, which was first mentioned in arriving at Eq. (3.21), shows what $w(x)$ must be in terms of the functions of x in the differential equation for which $\phi(x)$ are a solution. As a result, although the second method is a logical extension of the technique used for Fourier series, it has a very serious drawback when a weighting function $w(x)$ is needed to make the integral zero when $m \neq n$.[1]

Since Eq. (3.22) determines B_n such that the initial condition of Eq. (3.9) is satisfied, combining Eqs. (3.22) and (3.18) completes the exact analytical solution for the temperature distribution. That is,

$$\frac{\theta}{\theta_0} = \frac{T - T_f}{T_0 - T_f} = \sum_{n=1}^{\infty} \frac{2 \sin \lambda_n L \cos \lambda_n x}{\lambda_n L + \sin \lambda_n L \cos \lambda_n L} e^{-\lambda_n^2 \alpha \tau}, \tag{3.25}$$

where the characteristic values, or eigenvalues λ_n, are given as the solution of transcendental Eq. (3.16), rewritten as

$$\cot \lambda_n L = \frac{\lambda_n L}{hL/k}. \tag{3.26}$$

Summarizing, Eqs. (3.25) and (3.26) are the exact analytical solution for the temperature as a function of position x and time τ in an infinite slab of thickness $2L$ (shown in Fig. 3.1) which at time $\tau = 0$ is at the known temperature T_0 throughout, when suddenly both

---

1. When the integral $\int_0^L \phi_m(x)\,\phi_n(x)\,dx = 0$ if $m \neq n$, it is said that the set of functions $\{\phi(x)\}$, of which $\phi_m(x)$ and $\phi_n(x)$ are elements, are simply orthogonal in the interval $0 < x < L$; whereas, if a weighting function is needed to insure that $\int_0^L w(x)\,\phi_m(x)\,\phi_n(x)\,dx = 0$ when $n \neq m$, the set of functions $\{\phi\}$ is said to be orthogonal with respect to a weighting function $w(x)$ on the interval $0 < x < L$. Ordinarily, the set of real-valued functions of a real variable $\{\phi(x)\}$ will be elements of a linear vector space, and the integrals are an inner product or dot product function for the vectors $\phi_m(x)$ and $\phi_n(x)$ in that space. Thus, when the integral $\int_0^L \phi_m(x)\,\phi_n(x)\,dx = 0$ [when the dot product of the vectors $\phi_m(x)$ and $\phi_n(x)$ are zero], it is easily seen why the functions would be called orthogonal. This concept of functions being vectors in "function" space is especially useful in understanding why some of the so-called "approximate integral methods," to be discussed in Chapter 5, are applicable and are convergent.

faces are exposed to a constant temperature fluid T_f, with a constant surface coefficient of heat transfer h existing between the fluid and the faces of the infinite slab.

From Eq. (3.26), it is seen that the product $\lambda_n L$ is nondimensional. This fact is used to rearrange the right-hand side of Eq. (3.25) so that only nondimensional groups appear. This is done by multiplying the argument of $\cos \lambda_n x$ by L/L and the argument of the exponential by L^2/L^2, yielding

$$\frac{\theta}{\theta_0} = \frac{T - T_f}{T_0 - T_f} = \sum_{n=1}^{\infty} \frac{2 \sin \lambda_n L \cos [\lambda_n L (x/L)]}{\lambda_n L + \sin \lambda_n L \cos \lambda_n L} e^{-\lambda_n^2 L^2(\alpha\tau/L^2)}. \qquad (3.27)$$

The nondimensional temperature excess ratio θ/θ_0 depends upon the nondimensional groups $\lambda_n L$, x/L, and $\alpha\tau/L^2$; but, by Eq. (3.26), $\lambda_n L$ depends upon hL/k, so that in implicit functional form we have

$$\frac{\theta}{\theta_0} = F\left(\frac{hL}{k}, \frac{x}{L}, \frac{\alpha\tau}{L^2}\right),$$

where the function F is defined by Eqs. (3.27) and (3.26). The group $hL/k = N_{Bi}$ has been defined as the Biot number N_{Bi}, while the group $\alpha\tau/L^2$ (the nondimensional time parameter) is called the *Fourier number* and is denoted by N_{Fo}. Hence,

$$N_{Bi} = hL/k \qquad (3.27a)$$

$$N_{Fo} = \alpha\tau/L^2. \qquad (3.27b)$$

Utilizing these definitions, Eq. (3.27) has the implicit representation

$$\frac{\theta}{\theta_0} = \frac{T - T_f}{T_0 - T_f} = F(N_{Bi}, x/L, N_{Fo}). \qquad (3.28)$$

3.2b Charts for One-Dimensional Transient in the Slab

In working a numerical example using Eqs. (3.27) and (3.26), a considerable amount of effort is involved in evaluating the various terms of the infinite series. Most of this work has already been accomplished, however, by Heisler [9] and others. Thus in the chart of Fig. 3.4, which will be called *Heisler's main chart,* the nondimensional temperature excess ratio at $x/L = 0$, the center plane of the infinite slab, is plotted against the Fourier number, at various values of the reciprocal Biot number. If the temperature is needed at any other value of x/L, Eqs. (3.28) and (3.27) show that θ/θ_0 would have to be recalculated if x/L is different from zero, and a new chart similar to the chart of Fig. 3.4 would have to be constructed. It would appear that a different chart for every different x/L of interest would have to be constructed. The charts at six different values of x/L are given in [3]. However, careful study of the behavior of the solution function in Eq.

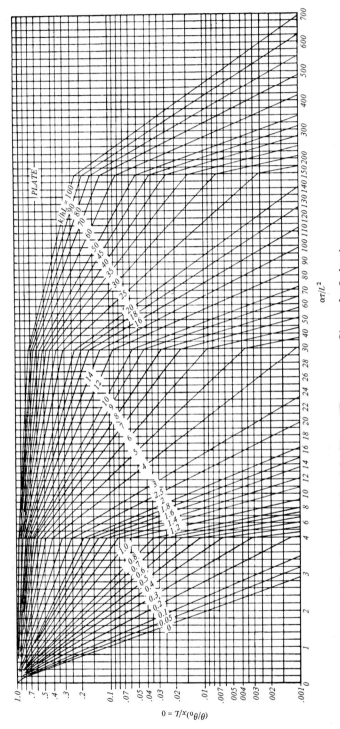

Figure 3.4 Heisler's main chart for the infinite slab. (From "Temperature Charts for Induction and Constant Temperature Heating" by M. P. Heisler. *Transactions of the A.S.M.E.*, Vol. 69 [1947], pp. 227–36.)

(3.27) shows that for a wide range of the Fourier number, the ratio of θ at any x/L of interest to θ at $x/L = 0$ is essentially independent of time, and therefore depends only upon x/L and the Biot number; that is,

$$\frac{\theta_{x/L}}{\theta_{x/L=0}} = f\left(\frac{x}{L}, N_{Bi}\right). \tag{3.29}$$

Equation (3.29) results from the fact that, at large enough nondimensional time, all of the terms in the infinite series representation of the solution function, Eq. (3.25) or Eq. (3.27), are negligible except for the first term. This is true because $\lambda_1 < \lambda_2 < \cdots < \lambda_n < \cdots$, which causes all of the exponential terms, the e terms, to be practically zero compared to the one for which $n = 1$ when $\alpha\tau/L^2$ is large enough. Then, writing Eq. (3.27) for $x/L = 0$ and for any general x/L, and dividing one by the other results in Eq. (3.29).

Equation (3.29) implies a type of profile similarity as time goes on, since it states, in effect, that the θ profile when divided by θ at $x/L = 0$ becomes independent of N_{Fo} and dependent only upon two variables, x/L and N_{Bi}. Note that this does not mean that θ at any x/L is independent of time; only that it is the same function of time as is θ at the centerline $x/L = 0$. The relation implied by Eq. (3.29) is plotted as Fig. 3.5, called the *auxiliary chart*. By using Fig. 3.5 in combination with the main chart, Fig. 3.4, the temperature at any point in the slab can be found at almost any instant of time. Note that the phrase "at *almost* any instant of time" is used, because there are times when Eq. (3.29) does not and cannot hold. For instance, if the Biot number is 1.0 and we are interested in the surface temperature $x/L = 1.0$, Fig. 3.5 yields the information that

$$\frac{\theta_{x/L=1}}{\theta_{x/L=0}} = \frac{T_s - T_f}{T_{x/L=0} - T_f} = 0.65,$$

where T_s is the surface temperature. However, there are times when this ratio is obviously not 0.65, namely at $\tau = 0$ when $T_s = T_{x/L=0} = T_0$. Hence, at very short times, the solutions using the charts can be grossly incorrect if the auxiliary chart is employed. When this happens, the charts in [3] or [5] can be utilized, or Eq. (3.27) can be used directly, or the solution presented in the next section for the *semi*-infinite body can be used. Problems for which the auxiliary chart is not valid do not seem to occur often.

For the case in which the thermal conductivity of the slab depends linearly on temperature, Figs. 3.4 and 3.5 can be used in conjunction with correction factors given by Sucec and Hegde [10].

3.2c Charts for Transient in the Infinite Cylinder and the Sphere

Exact analytical solutions and main and auxiliary charts are also available for two other geometries commonly encountered in engineering situations. These are the infinitely long cylinder of radius r_0 and the sphere of radius r_0, when these bodies are initially at temperature T_0 throughout and are then suddenly plunged into, or exposed to, a constant

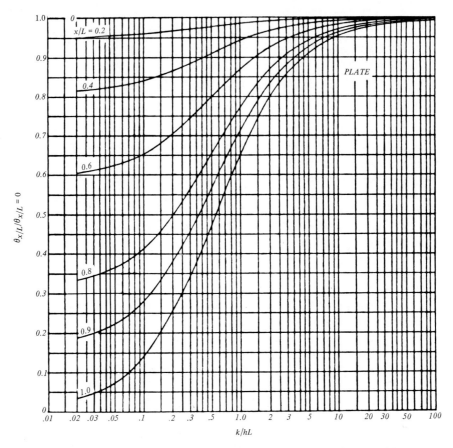

Figure 3.5 Heisler's auxiliary chart for the infinite slab. (From Heisler, 1947.)

temperature fluid at T_f with an average surface coefficient of heat transfer which is constant at the value h. Details of these exact solutions, by separation of variables and expansions in infinite series of orthogonal functions, are given in [5]. Here we present these solutions only in the easy-to-use chart form analogous to Figs. 3.4 and 3.5, which are appropriate to the infinite slab.

Once again there is a main chart, which connects the temperature excess ratio at the body center to the nondimensional time, and an auxiliary chart which, except at very short times, connects the temperature excess ratio at any nondimensional position to the temperature excess ratio at the body center at the same instant of time. Figures 3.6 and 3.7 are for the infinitely long cylinder of outer radius r_0, while Figs. 3.8 and 3.9 are the charts to be used on a sphere of outer radius r_0.

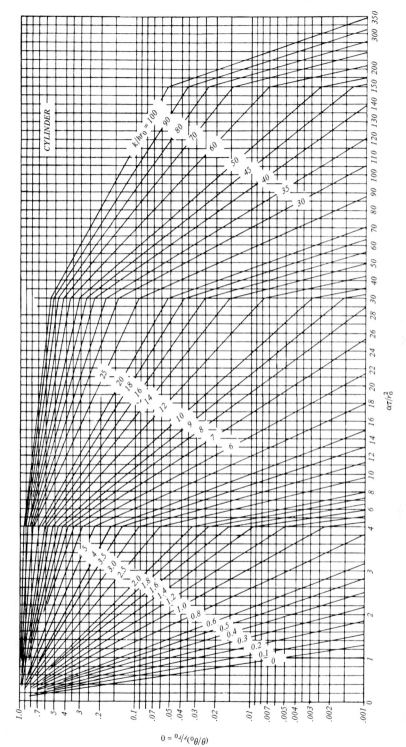

Figure 3.6 Heisler's main chart for the infinite cylinder. (From Heisler, 1947.)

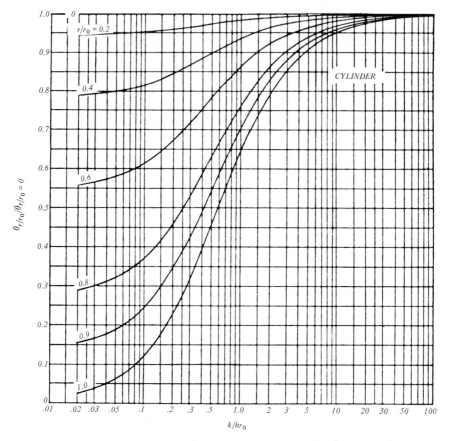

Figure 3.7 Heisler's auxiliary chart for the infinite cylinder. (From Heisler, 1947.)

EXAMPLE 3.1

A large-diameter rocket engine nozzle has a wall thickness of $1/3$ in., and the wall material properties are

$$\rho = 538 \text{ lbm/ft}^3, \quad k = 15 \text{ Btu/hr ft }°F, \quad c_p = 0.130 \text{ Btu/lbm }°F.$$

In a static thrust test, the walls, which are initially at $80°F$, are subjected to hot combustion gases at $3200°F$ with a surface coefficient of heat transfer of 360 Btu/hr ft² °F between the gases and the wall (Fig. 3.10). If the maximum wall temperature that can be tolerated is $2000°F$, find the time the rocket motor can be allowed to run if the outside of the wall is assumed to be perfectly insulated.

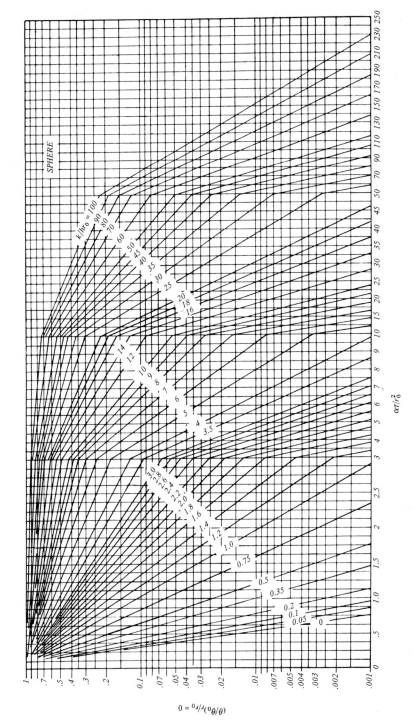

Figure 3.8 Heisler's main chart for the sphere of radius r_0. (From Heisler, 1947.)

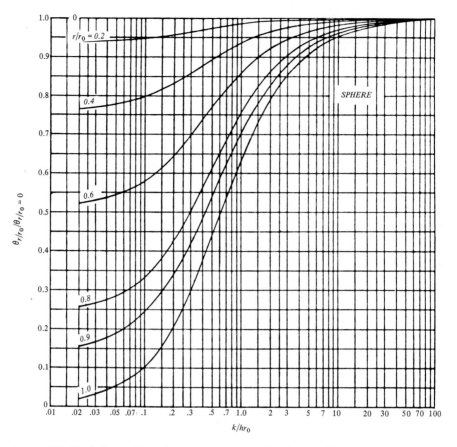

Figure 3.9 Heisler's auxiliary chart for the sphere of radius r_0. (From Heisler, 1947.)

Solution

Since the rocket nozzle is of large diameter and the wall thickness is relatively thin, the effects of wall curvature can be idealized as an infinite slab insulated on one face and subject to the hot fluid on the other face, as shown in Fig. 3.2. Earlier it was noted that the solution for the temperature as a function of x and τ in an infinite slab of thickness $2L$, initially at temperature T_0, whose two faces are then exposed to a fluid at temperature T_f and with a surface coefficient of heat transfer h, could also be applied to a slab of thickness L where one face is insulated and the other conditions are the same as for the $2L$-thick slab. So for this problem, Heisler's charts, Figs. 3.4–5, can be used when x is measured with its origin at the insulated face, as shown in Fig. 3.2. Hence, when $T_0 = 80°\text{F}$ and $T_f = 3200°\text{F}$,

$$\theta_0 = T_0 - T_f = 80 - 3200 = -3120°.$$

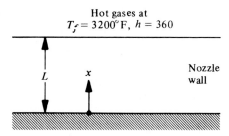

Figure 3.10 Wall of the rocket motor of Ex. 3.1.

At the time to be determined, the maximum temperature is to be 2000°F, and it occurs at $x/L = 1.0$ on the surface interacting directly with the rocket gases. Thus, $\theta_{x/L=1.0}$ = 2000 − 3200 = −1200. To find the time needed for the surface at $x = L$ to reach 2000°F, $\theta_{x/L=1.0}$ is related to θ at the center $\theta_{x/L=0}$ by Fig. 3.5. Then Fig. 3.4 must be used, since it relates θ at the center $x/L = 0$ to time; hence, the value of time being sought can be determined. Thus, since $k/hL = 15/360\ [\frac{1}{3(12)}] = 1.5$, from Fig. 3.5 at k/hL = 1.50 and $x/L = 1.0$,

$$\frac{\theta_{x/L=1.0}}{\theta_{x/L=0}} = 0.75, \quad \text{or} \quad \theta_{x/L=0} = \frac{\theta_{x/L=1.0}}{0.75} = \frac{-1200}{0.75} = -1600.$$

Hence,

$$\frac{\theta_{x/L=0}}{\theta_0} = \frac{-1600}{-3120} = 0.512.$$

This occurs when the surface temperature is 2000°F. From Fig. 3.4, for the infinite plate at $(\theta/\theta_0)_{x/L=0} = 0.512$ and $k/hL = 1.5$,

$$\frac{\alpha\tau}{L^2} = 1.43, \quad \alpha = \frac{k}{\rho c_p} = \frac{15}{538(0.130)} = 0.214 \text{ ft}^2/\text{hr}.$$

Hence,

$$\tau = \frac{1.43}{0.214}\left[\frac{1}{3(12)}\right]^2 = \frac{1}{194} \text{ hr} = 18.6 \text{ sec}.$$

EXAMPLE 3.2

A roast of beef can be idealized as a 0.2 m diameter sphere. The roast is initially at 15°C when it is placed into a 165°C oven, with an average surface coefficient of heat transfer h between the beef and the oven air of 17 W/m² °C. The cooking time for the roast is about 5 hrs. Find the center temperature of the roast when it is done. The thermophysical properties of the beef are given as $\rho = 1025$ kg/m³, $k = 0.675$ W/m °C, and $c_p = 4186.6$ J/kg °C.

Solution

This physical situation satisfies all the requirements for the validity of Heisler's charts, Figs. 3.8 and 3.9, for the sphere. The Fourier number and the quantity k/hr_0, unity divided by the Biot number, are needed to use Fig. 3.8 for the ratio of the center temperature excess to the initial temperature excess. The initial temperature excess θ_0 is $T_0 - T_f$ = 15 − 165 = −150°. Since $r_0 = 0.2/2 = 0.1$ m,

$$k/hr_0 = 0.675/17(0.1) = 0.397.$$

Also, since the thermal diffusivity α is $\alpha = k/\rho c_p$,

$$\alpha = 0.675/1025(4186.6) = 1.6 \times 10^{-7} \text{ m}^2/\text{s}.$$

Thus, the Fourier number, N_{Fo}, is found as, at $\tau = 5(3600) = 18{,}000$ s,

$$N_{Fo} = \alpha\tau/r_0^2 = 1.6 \times 10^{-7}(18{,}000)/(0.1)^2 = 0.288.$$

Using these values on Fig. 3.8 leads to

$$\frac{\theta_{r/r_0=0}}{\theta_0} = 0.4,$$

or since $\theta_0 = -150$, one has that

$$\theta_{r/r_0=0} = T_{r/r_0=0} - T_f = T_{r/r_0=0} - 165 = 0.4(-150).$$

Hence,

$$T_{r/r_0=0} = 105°\text{C}.$$

EXAMPLE 3.3

It is planned to heat long shafts of various diameters to 1200°F to prepare them for a hot machining operation needed along the shaft centerline. The shafts are carried from the heat treatment area to the machine shop by being suspended from a carrier moving on an overhead rail. This motion causes an average surface coefficient of heat transfer of 7 Btu/hr ft² °F between the shafts and the surrounding 80°F air, and the trip from heat treatment to the machine shop requires 17.2 min (0.287 hr). If the center temperature is not to drop below 864°F before the shafts reach the machine shop, find the diameter of the smallest shaft that can be processed. Assume that $k = 10$ Btu/hr ft °F and $\alpha = 0.20$ ft²/hr.

Solution

The diameter of a shaft whose center temperature is 864°F after 0.287 hr and whose initial temperature was 1200°F is the smallest diameter allowable, since a smaller diameter shaft would cool more rapidly. Heisler's charts for the infinite cylinder, Figs. 3.6–7, are used. Since $T_0 = 1200°F$ and $T_f = 80°F$,

$$\theta_0 = 1200 - 80 = 1120°.$$

After 0.287 hours, the center temperature excess will be

$$\theta_{r/r_0=0} = 864 - 80 = 784°.$$

Thus, the nondimensional temperature excess ratio at the cylinder center is

$$\frac{\theta}{\theta_0}\bigg|_{r/r_0=0} = \frac{784}{1120} = 0.70.$$

The Fourier number is

$$N_{\text{Fo}} = \frac{\alpha\tau}{r_0^2} = \frac{0.2(0.287)}{r_0^2} = \frac{0.0574}{r_0^2}$$

and the Biot number is

$$\frac{k}{hr_0} = \frac{10}{7r_0} = \frac{1.43}{r_0}.$$

The chart of Fig. 3.6 at $(\theta/\theta_0)_{r/r_0=0} = 0.70$ indicates that since both the Fourier number and the Biot number contain the unknown radius r_0, the solution must be found by trial and error. A value for r_0 is selected to be used in the reciprocal of the Biot number, and the chart of Fig. 3.6 is used with the actual temperature excess and the reciprocal Biot number which the guess of r_0 implies. Then we solve for r_0 from the Fourier number, and if it equals the assumed value, it is the solution. If it does not, either the value calculated from N_{Fo} is used as a revised estimated in k/hr_0, or another r_0 is estimated and the process is continued until a value of r_0 is found that gives a Fourier number and reciprocal Biot number which correspond to $(\theta/\theta_0)_{r/r_0=0} = 0.70$ in Fig. 3.6.

As a first estimate, select $(r_0)_1 = 0.5$ ft; then, $k/h(r_0)_1 = 1.43/(r_0)_1 = \frac{1.43}{0.5} = 2.86$. In Fig. 3.6 at $k/h(r_0)_1 = 2.86$ and $(\theta/\theta_0)_{r/r_0=0} = 0.70$,

$$N_{\text{Fo}_1} = 0.70 = \frac{0.0574}{(r_0^2)_2}.$$

Hence, $(r_0)_2 = 0.286$ ft, which is not equal to the first estimate of 0.5 ft, so the calculation is repeated using $(r_0)_2$ as a revised estimate of r_0. That is, $k/h(r_0)_2 = \frac{1.43}{0.286} = 5.0$. In Fig. 3.6 at $k/h(r_0)_2 = 5.0$ and $(\theta/\theta_0)_{r/r_0=0} = 0.70$,

$$N_{Fo_2} = 1.20 = \frac{0.0574}{(r_0^2)_3}.$$

Hence, $(r_0)_3 = 0.218$ ft, which again does not equal the estimate used to calculate $(r_0)_3$; hence, $(r_0)_3$ is employed as a revised estimate of the correct r_0. That is $k/h(r_0)_3 = 1.43/0.218 = 6.57$. From Fig. 3.6 at $k/h(r_0)_3 = 6.57$ and $(\theta/\theta_0)_{r/r_0=0} = 0.70$,

$$N_{Fo_3} \simeq 1.50 = \frac{0.0574}{(r_0^2)_4}.$$

Hence, $(r_0)_4 = 0.195$ ft. This is the new estimate of r_0. Using the procedure developed above, subsequent computations proceed as follows:

Since $1.43/(r_0)_4 = 7.34$,

$$N_{Fo_4} = 1.70 = \frac{0.0574}{(r_0^2)_5},$$

and hence, $(r_0)_5 = 0.183$ ft.

Since $1.43/(r_0)_5 = 7.83$,

$$N_{Fo_5} = 1.80 = \frac{0.0574}{(r_0^2)_6},$$

and hence, $(r_0)_6 = 0.178$ ft.

Since $1.43/(r_0)_6 = 8.0$,

$$N_{Fo_6} \simeq 1.80 = \frac{0.0574}{(r_0^2)_7},$$

and hence, $(r_0)_7 = 0.178$ ft.

Thus, the smallest diameter shaft which can be processed in the manner described by the problem statement is

$$D = 2r_0 = 2(0.178) = 0.354 \text{ ft.}$$

Note that the systematic procedure used to obtain the final answer converged rather slowly. This procedure might have been faster if, after noting that the procedure was indicating a smaller radius than the initial estimate of 0.5 ft, we made another estimate, such as 0.1 ft, and then employed the systematic procedure again.

These examples not only illustrate the proper way to use Heisler's charts, but also display a few of the many situations in which the charts can be used to find a solution. They verify our earlier statement concerning the importance of these charts in transient conduction problems.

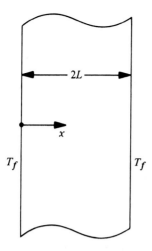

Figure 3.11 The infinite slab of Ex. 3.4.

EXAMPLE 3.4

Figure 3.11 shows an infinite slab in which steady conditions prevail, since the electrical energy dissipated within the slab is conducted out at the same rate through the two faces, giving rise to the temperature distribution

$$T = T_f + (T_m - T_f)\sin(\pi x/2L),$$

where T_f and T_m are known. If the source of the electrical energy is turned off at time $\tau = 0$, and the faces are kept at the known temperature T_f, determine the temperature as a function of space coordinates and time within the slab. The material properties are known constants.

Solution
The governing partial differential equation for the temperature is Eq. (2.6) with no y- and z-conduction terms and no generation rate per unit volume q'''. This term is set equal to zero, because turning off the electric power stops the dissipation of electrical energy and creates the transient situation. Hence, the equation to be solved is

$$\frac{\partial^2 T}{\partial x^2} = \frac{1}{\alpha}\frac{\partial T}{\partial \tau}, \qquad (3.30)$$

with the following boundary and initial conditions:

$$
\begin{aligned}
&\text{at } x = 0 \text{ and } \tau > 0, &&T = T_f; \\
&\text{at } x = 2L \text{ and } \tau > 0, &&T = T_f; \\
&\text{at } \tau = 0, &&T = T_f + (T_m - T_f)\sin(\pi x/2L).
\end{aligned}
\qquad (3.31)
$$

Unsteady-State Conduction

To solve Eq. (3.30) by separation of variables, let

$$\theta = T - T_f. \tag{3.32}$$

Using this transformation and noting that $\partial\theta/\partial\tau = \partial T/\partial\tau$ and that $\partial^2\theta/\partial x^2 = \partial^2 T/\partial x^2$ since T_f is a constant, Eqs. (3.30) and (3.31) become

$$\frac{\partial^2\theta}{\partial x^2} = \frac{1}{\alpha}\frac{\partial\theta}{\partial\tau}, \tag{3.33}$$

$$
\begin{array}{lll}
\text{at } x = 0 \text{ and } \tau > 0, & \theta = 0, & \\
\text{at } x = 2L \text{ and } \tau > 0, & \theta = 0, & \\
\text{at } \tau = 0, & \theta = (T_m - T_f)\sin(\pi x/2L). &
\end{array} \tag{3.34}
$$

Now, using separation of variables, let $\theta = R(x)\,G(\tau)$; substituting this into Eq. (3.33), rearranging, and setting both sides equal to the separation constant gives

$$\frac{d^2R/dx^2}{R} = \frac{dG/d\tau}{\alpha G} = -\lambda^2,$$

leading to the separation equations,

$$\frac{d^2R}{dx^2} + \lambda^2 R = 0, \quad \frac{dG}{d\tau} + \alpha\lambda^2 G = 0.$$

Solving these ordinary differential equations for $G(\tau)$ and $R(x)$ and forming their product yields

$$\theta = (A\sin\lambda x + B\cos\lambda x)e^{-\alpha\lambda^2\tau}. \tag{3.35}$$

Applying the first boundary condition from Eq. (3.34), at $x = 0$ and $\tau > 0$, $\theta = 0$, gives

$$0 = (0 + B)e^{-\alpha\lambda^2\tau}.$$

Since this must hold for all τ, B must be chosen equal to zero. Hence, Eq. (3.35) becomes

$$\theta = A\sin\lambda x\, e^{-\alpha\lambda^2\tau}. \tag{3.36}$$

The next boundary condition, at $x = 2L$ and $\tau > 0$, $\theta = 0$, yields

$$0 = A\sin\lambda 2L\, e^{-\alpha\lambda^2\tau}.$$

The only nontrivial solution to this equation that holds for all τ is

$$\sin\lambda 2L = 0.$$

That is, the argument of the sine function is chosen to be $n\pi$, where n is a positive integer. Hence, $\lambda 2L = n\pi$ or $\lambda = n\pi/2L$. With the two boundary conditions satisfied, Eq. (3.36) becomes

$$\theta = A \sin(n\pi x/2L)e^{-\alpha(n\pi/2L)^2\tau}. \qquad (3.37)$$

Finally, the initial condition must be satisfied. In a general situation, the initial condition could not be satisfied until all possible solutions of the type of Eq. (3.37) were summed up from $n = 0$ to $n = \infty$, and then an infinite series expansion of the initial condition in terms of the sine functions in this case would allow the determination of the various A_n. However, notice that the initial condition in this problem is of the same general form as Eq. (3.37), when τ is set equal to zero, so that the initial condition for this problem can be satisfied without a summation. (Situations such as this also appeared in certain two-dimensional, steady conduction problems in Chapter 2.)

Now, at $\tau = 0$, $\theta = (T_m - T_0)\sin(\pi x/2L)$. Hence, setting this equal to the right-hand side of Eq. (3.37) when $\tau = 0$,

$$(T_m - T_0)\sin(\pi x/2L) = A \sin(n\pi x/2L).$$

This is an equality for all values of x in the conduction region when $n = 1$ and $A = T_m - T_0$. Using these values in Eq. (3.37) and transforming back to the original dependent variable T yields as the solution,

$$T = T_f + (T_m - T_f)\sin(\pi x/2L)e^{-(\pi/2L)^2\alpha\tau}. \qquad (3.38)$$

From a physical standpoint, the slab must have the fluid temperature T_f at all x as the time gets very large. By letting $\tau \to \infty$ in Eq. (3.38), the mathematical form of the solution predicts this result for the eventual slab temperature.

This problem illustrates further the techniques one must use to find exact analytical solutions of unsteady-state conduction problems. It complements the previous derivation which led to Eq. (3.25). Notice also the similarities between the exact solution methods being used in this chapter and the ones used for multidimensional conduction problems in Chapter 2. In both chapters, separation of variables is used to solve the partial differential equations. Expansions in infinite series of orthogonal functions are then employed to satisfy a boundary condition in Chapter 2, but they are used to handle the initial condition in this chapter.

3.2d Transients in the Semi-Infinite Slab

Figure 3.12 illustrates the semi-infinite slab or semi-infinite body. This is a region that extends to infinity in both the positive and negative y directions (vertical direction in the figure) and the positive and negative z directions (direction perpendicular to the plane of the figure), has a face at $x = 0$, and extends to infinity in the positive x direction.

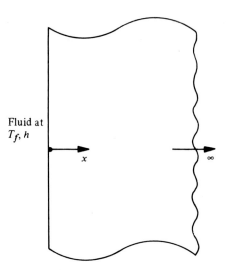

Fluid at
T_f, h

x

∞

Figure 3.12 The semi-infinite slab.

An example of a physical region that is approximated well as a semi-infinite slab is the ground, with its surface being the face of the slab and x being measured toward the earth's center. In addition, a slab of *finite* thickness $2L$, such as that shown in Fig. 3.1, may often behave as if it were semi-infinite from the viewpoint of its temperature distribution. This occurs when the temperature at the center of Fig. 3.1 has not yet changed appreciably from its original value due to the sudden change of the fluid temperature. Hence, as mentioned earlier, during the brief period when Heisler's auxiliary chart, Fig. 3.5, is not yet applicable, the solution for the semi-infinite body can be used for the body in Fig. 3.1.

Thus, because of its practical importance, we seek the solution for the temperature distribution within the semi-infinite slab of Fig. 3.12. The material is initially at a known constant temperature T_0 throughout, when suddenly the face at $x = 0$ is exposed to a fluid whose temperature T_f is constant, and a constant surface coefficient of heat transfer h exists between the fluid and the surface. Defining a temperature excess $\theta = T - T_f$ as in the case of the slab in Fig. 3.1, the governing partial differential equation, Eq. (2.6), for the temperature distribution $T(x, \tau)$ within the slab reduces to

$$\frac{\partial^2 \theta}{\partial x^2} = \frac{1}{\alpha} \frac{\partial \theta}{\partial \tau}$$

with the following boundary and initial conditions:

at $x = 0$ and $\tau > 0$, $-h\theta = -k \dfrac{\partial \theta}{\partial x}$;

as $x \to \infty$ and $\tau > 0$, θ remains finite;

at $\tau = 0$ and $0 \le x < \infty$, $\theta = \theta_0 = T_0 - T_f$.

It might appear that this problem can be solved by separation of variables, as was the example of the infinite slab. However, the infinite extent of the domain in the x direction will not allow the condition $\theta = \theta_0$ at $\tau = 0$ for $0 \leq x < \infty$ to be satisfied with the type of series expansion of orthogonal functions already discussed. Hence, if separation of variables is used, the Fourier integral must be used to accomplish the expansion of θ_0 in the *infinite* half interval $0 \leq x \leq \infty$. See [5] and [8] for the details of the solution as well as the development of the Fourier integral. A simpler way to arrive at the solution to this problem is to use the Laplace transform to transform time τ and yield an ordinary differential equation to be solved in the transformed plane. Once the boundary conditions are applied to this solution in the transformed plane, a standard table of transforms inverts the result back to the x, τ variables. For details of this procedure, see [5] and [8]. By either method, the result for the nondimensional temperature excess ratio in the semi-infinite slab which is initially at temperature T_0 and is then exposed to a fluid at constant temperature T_f with surface coefficient of heat transfer h is

$$
\frac{\theta}{\theta_0} = \frac{T - T_f}{T_0 - T_f}
$$

$$
= \mathrm{erf}\left(\frac{x}{2\sqrt{\alpha\tau}}\right) + \exp\left(\frac{hx}{k} + \frac{h^2\alpha\tau}{k^2}\right) \tag{3.39}
$$

$$
\times \left[1 - \mathrm{erf}\left(\frac{x}{2\sqrt{\alpha\tau}} + \frac{h\sqrt{\alpha\tau}}{k}\right)\right],
$$

where the error function erf is defined by

$$
\mathrm{erf}\left(\frac{x}{2\sqrt{\alpha\tau}}\right) = \frac{2}{\sqrt{\pi}} \int_0^{x/2\sqrt{\alpha\tau}} e^{-z^2} \, dz. \tag{3.40}
$$

The quantity z is a dummy variable. A plot of the error function versus its argument is shown in Fig. 3.13, and selected values of the error function are presented in Appendix C.

If the surface temperature of the semi-infinite slab is suddenly changed to T_f, when the entire semi-infinite slab was initially at T_0 throughout, rather than suddenly being exposed to a fluid at T_f with a finite surface coefficient of heat transfer h, the transient temperature distribution can be formed by letting $h \to \infty$ in Eq. (3.39) as

$$
\frac{T - T_f}{T_0 - T_f} = \mathrm{erf}\left(\frac{x}{2\sqrt{\alpha\tau}}\right). \tag{3.41}
$$

Unsteady-State Conduction

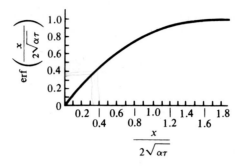

Figure 3.13 Graph of the error function.

EXAMPLE 3.5

The concrete walls of a jet engine test cell are extremely thick and are initially at the constant temperature of 20°C. The combination of exhaust gases from a turbojet engine and cooling water sprayed into the gases suddenly raises the surface temperature of the wall to 315°C. Determine how long it will take a point 7.5 cm from the surface to reach 205°C. For the concrete wall, $\alpha = 4.39 \times 10^{-7}$ m²/s.

Solution

Since the actual wall thickness is not given, while it is mentioned that the wall is very thick, it is assumed that it behaves as if it were a semi-infinite wall initially at 20°C throughout when its face is suddenly raised to 315°C. The solution for this situation is given by Eq. (3.41).

The temperature that the face is suddenly brought to is $T_f = 315°C$; the initial temperature of the material is $T_0 = 20°C$. Hence,

$$\frac{T - T_f}{T_0 - T_f} = \text{erf}\left(\frac{x}{2\sqrt{\alpha\tau}}\right).$$

At $x = 0.075$ m, and $T = 205°C$,

$$\frac{205 - 315}{20 - 315} = 0.373 = \text{erf}\frac{x}{2\sqrt{\alpha\tau}}.$$

From Fig. 3.13, or from Appendix C, the argument of the error function corresponding to this value of 0.373 is

$$\frac{x}{2\sqrt{\alpha\tau}} = 0.344 = \frac{0.075}{2\sqrt{4.39 \times 10^{-7}\tau}}.$$

Hence,

$$\tau = 27,069 \text{ s} = 7.52 \text{ hr.}$$

EXAMPLE 3.6

Frozen food containers stacked together are stored at $-20°C$ in a storage bin. The properties of the frozen food are approximately $k = 0.52$ W/m °C and $\alpha = 1.3 \times 10^{-7}$ m²/s. Suddenly a compressor fails, causing the refrigeration system to cease functioning and when the trouble is investigated, the top surface of the frozen food in the bin is exposed to 20°C air, with a surface coefficient of heat transfer of 6.8 W/m² °C between the air and the surface of the frozen food. Indications are that 12.5 hr will be required to complete the repairs to the refrigeration system. Assuming that the frozen food in the bin behaves as a semi-infinite slab, decide whether the food must be transferred to another bin to prevent spoilage.

Solution
If the surface of the frozen food does not rise to 0°C within the repair time of 12.5 hr, the food can remain in the bin. The temperature distribution within the food, as a function of depth below the surface and time, is given by Eq. (3.39). Since the surface temperature is of interest, set $x = 0$. Hence since erf (0) = 0, Eq. (3.39) becomes

$$\frac{T - T_f}{T_0 - T_f} = \exp\left(\frac{h^2 \alpha \tau}{k^2}\right)\left[1 - \operatorname{erf}\frac{(h\sqrt{\alpha\tau})}{k}\right].$$

Since $T_f = 20°C$ and $T_0 = -20°C$,

$$\frac{T - 20}{-20 - 20} = \exp\left[\frac{(6.8)^2(1.3 \times 10^{-7})(12.5 \times 3600)}{(0.52)^2}\right]$$

$$\times \left\{1 - \operatorname{erf}\left[\frac{6.8\sqrt{1.3 \times 10^{-7}(12.5 \times 3600)}}{0.52}\right]\right\}.$$

Thus,

$$T = 3°C.$$

This calculation indicates that at a time less than 12.5 hours, the surface of the frozen food will have reached 0°C and melting will have begun. Thus, the food should be moved to another bin.

3.2e Two- and Three-Dimensional Transients

As has been seen so far, the one-dimensional transient solution functions, as embodied in Heisler's charts and in the results for the semi-infinite body, are important because they cover a wide range of physical applications. However, sometimes an actual two- or three-dimensional body cannot be idealized adequately as having only a one-dimensional temperature distribution within it. Therefore, the details of the multidimensional transient temperature distribution must be predicted. Fortunately, a large number of important two- and three-dimensional bodies have their multidimensional temperature distributions determined by a proper combination of the already known solutions to the one-dimensional transients considered previously. The proof of the validity of such combinations of one-dimensional solutions will now be shown for a *finite* length right circular cylinder. From this analysis, we will then see what other multidimensional shapes can be handled in the same fashion. We will then show how additional shapes can be generated by taking advantage of internal surfaces of thermal and geometric symmetry. Finally, some examples will be solved to illustrate the solution details in this extremely important class of transient conduction problems.

Figure 3.14 shows a right circular cylinder of finite length $2L$ and of outer radius r_0, which is initially at the known constant temperature T_0 throughout and then at time $\tau = 0$ suddenly has all of its surface exposed to a fluid at known constant temperature T_f with a known constant surface coefficient of heat transfer between surface and fluid. An exact analytical solution for the temperature as a function of the relevant space coordinates and time is required. Taking advantage of the geometric and thermal symmetry, the origin of the coordinate system is located at the geometric center of the conduction region, and the temperature T will be a function of x, r, and τ. The governing partial differential equation is Eq. (2.8) with the θ terms set equal to zero, since T does not depend upon θ, and $q''' = 0$. Also, x is used here instead of z. Hence,

$$\frac{\partial^2 T}{\partial r^2} + \frac{1}{r}\frac{\partial T}{\partial r} + \frac{\partial^2 T}{\partial x^2} = \frac{1}{\alpha}\frac{\partial T}{\partial \tau}. \tag{3.42}$$

The boundary and initial conditions are shown in equation set (3.43).

$$\text{at } \tau > 0, x = 0, \text{ and } 0 < r < r_0, \qquad \frac{\partial T}{\partial x} = 0 \text{ (by symmetry)};$$

$$\text{at } \tau > 0, x = L, \text{ and } 0 < r < r_0, \qquad -k\frac{\partial T}{\partial x} = h(T - T_f);$$

$$\text{at } \tau > 0, r = 0, \text{ and } 0 < x < L, \qquad \frac{\partial T}{\partial r} = 0 \text{ (by symmetry)}; \quad \textbf{(3.43)}$$

$$\text{at } \tau > 0, r = r_0, \text{ and } 0 < x < L, \qquad -k\frac{\partial T}{\partial r} = h(T - T_f);$$

$$\text{at } \tau = 0, 0 < x < L, \text{ and } 0 < r < r_0, T = T_0.$$

This problem, after defining a new dependent variable $\theta = T - T_f$, could be solved using the method of separation of variables. However, the special shape of this body combined with the boundary conditions suggests that the solution can be expressed in terms of transient solutions already obtained. In particular, by "special shape" of the finite length cylinder is meant that it can be generated by the common points in the mutually perpendicular intersection of an infinite slab of thickness $2L$ and an infinite cylinder of outer radius r_0, as shown in Fig. 3.15. The shaded portion represents the common points of the intersection of the infinite cylinder and the infinite slab, and this intersection is the finite cylinder of Fig. 3.14. To see that the solution to this problem, the finite cylinder, is related to the

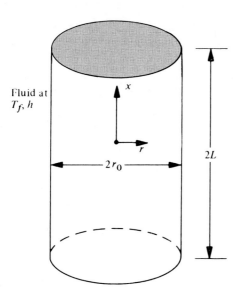

Figure 3.14 Right circular cylinder of finite length.

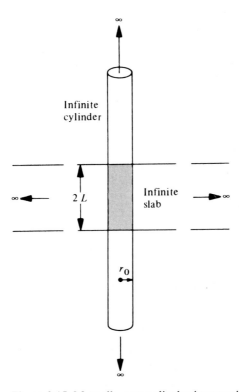

Figure 3.15 Mutually perpendicular intersection of an infinite slab of thickness $2L$ and an infinite cylinder of radius r_0 generating a finite length right circular cylinder.

solutions for the infinite slab and the infinite cylinder, change the dependent variable in Eqs. (3.42) and (3.43) to the nondimensional temperature excess ratio ϕ, where

$$\phi = \frac{T - T_f}{T_0 - T_f}.$$

Since T_0 and T_f are constants, Eq. (3.42) becomes, in terms of ϕ,

$$\frac{\partial^2 \phi}{\partial r^2} + \frac{1}{r}\frac{\partial \phi}{\partial r} + \frac{\partial^2 \phi}{\partial x^2} = \frac{1}{\alpha}\frac{\partial \phi}{\partial \tau}, \tag{3.44}$$

and the boundary conditions of Eq. (3.43) become equation set (3.45)

$$\text{at } \tau > 0,\ x = 0,\ \text{and } 0 < r < r_0, \qquad \frac{\partial \phi}{\partial x} = 0;$$

$$\text{at } \tau > 0,\ x = L,\ \text{and } 0 < r < r_0, \qquad -k\frac{\partial \phi}{\partial x} = h\phi;$$

$$\text{at } \tau > 0,\ r = 0,\ \text{and } 0 < x < L, \qquad \frac{\partial \phi}{\partial r} = 0; \tag{3.45}$$

$$\text{at } \tau > 0,\ r = r_0,\ \text{and } 0 < x < L, \qquad -k\frac{\partial \phi}{\partial r} = h\phi;$$

$$\text{at } \tau = 0,\ 0 < r < r_0,\ \text{and } 0 < x < L, \qquad \phi = 1.$$

For the infinite cylinder, a new dependent variable is defined as

$$\phi_1 = \left(\frac{T - T_f}{T_0 - T_f}\right)_{\text{inf. cyl.}}$$

In terms of ϕ_1, the partial differential equation for an infinite cylinder of radius r_0 initially at temperature T_0 throughout, then suddenly exposed to a fluid at T_f and h is

$$\frac{\partial^2 \phi_1}{\partial r^2} + \frac{1}{r}\frac{\partial \phi_1}{\partial r} = \frac{1}{\alpha}\frac{\partial \phi_1}{\partial \tau}, \tag{3.46}$$

with boundary conditions:

$$\text{at } \tau > 0 \text{ and } r = 0, \qquad \frac{\partial \phi_1}{\partial r} = 0;$$

$$\text{at } \tau > 0 \text{ and } r = r_0, \qquad -k\frac{\partial \phi_1}{\partial r} = h\phi_1;$$

$$\text{at } \tau = 0 \text{ and } 0 < r < r_0, \qquad \phi_1 = 1.$$

For the infinite slab, a new variable is defined as

$$\phi_2 = \left(\frac{T - T_f}{T_0 - T_f}\right)_{\text{inf. slab}}$$

The governing equation is

$$\frac{\partial^2 \phi_2}{\partial x^2} = \frac{1}{\alpha} \frac{\partial \phi_2}{\partial \tau}, \tag{3.47}$$

with boundary conditions:

$$\text{at } \tau > 0 \text{ and } x = 0, \qquad \frac{\partial \phi_2}{\partial x} = 0;$$

$$\text{at } \tau > 0 \text{ and } x = L, \qquad -k \frac{\partial \phi_2}{\partial x} = h\phi_2;$$

$$\text{at } \tau = 0 \text{ and } 0 < x < L, \qquad \phi_2 = 1.$$

It might be expected that the addition of ϕ_1 and ϕ_2 is equal to ϕ. Although this satisfies the differential equation, Eq. (3.44), it does not satisfy the boundary and initial conditions in Eq. (3.45). In particular, at $\tau = 0$, $\phi = 2$ instead of 1. Of greater consequence is the fact that the convective boundary conditions would not be satisfied. Hence, since the addition of ϕ_1 and ϕ_2 does not yield ϕ, the product might be tried, particularly since product-type solutions are used in the separation of variables technique.

If $\phi = \phi_1 \phi_2$, then $\phi_1 \phi_2$ must satisfy Eq. (3.44), the partial differential equation for ϕ, as well as all the conditions given by Eq. (3.45). Substituting $\phi = \phi_1 \phi_2$ into Eq. (3.44) yields

$$\frac{\partial^2}{\partial r^2}(\phi_1 \phi_2) + \frac{1}{r}\frac{\partial}{\partial r}(\phi_1 \phi_2) + \frac{\partial^2}{\partial x^2}(\phi_1 \phi_2) = \frac{1}{\alpha}\frac{\partial}{\partial \tau}(\phi_1 \phi_2).$$

Recall that $\phi_1 = \phi_1(r, \tau)$ and does not depend on x, while ϕ_2 depends on x and τ, but not on r. This is a consequence of ϕ_1 being the solution to the infinite cylinder alone, and ϕ_2 the solution to the infinite slab alone. Taking the indicated partial derivatives, noting that

$$\frac{\partial}{\partial r}(\phi_1 \phi_2) = \phi_2 \frac{\partial \phi_1}{\partial r},$$

and using this same procedure on the other terms, yields

$$\phi_2 \frac{\partial^2 \phi_1}{\partial r^2} + \frac{\phi_2}{r}\frac{\partial \phi_1}{\partial r} + \phi_1 \frac{\partial^2 \phi_2}{\partial x^2} = \frac{\phi_1}{\alpha}\frac{\partial \phi_2}{\partial \tau} + \frac{\phi_2}{\alpha}\frac{\partial \phi_1}{\partial \tau}.$$

Rearranging,

$$\phi_2 \left(\frac{\partial^2 \phi_1}{\partial r^2} + \frac{1}{r}\frac{\partial \phi_1}{\partial r} - \frac{1}{\alpha}\frac{\partial \phi_1}{\partial \tau} \right) = -\phi_1 \left(\frac{\partial^2 \phi_2}{\partial x^2} - \frac{1}{\alpha}\frac{\partial \phi_2}{\partial \tau} \right). \tag{3.48}$$

However, Eq. (3.48) is identically satisfied because the terms in parentheses both equal zero as a result of Eqs. (3.46) and (3.47); hence, $\phi = \phi_1 \phi_2$ satisfies the partial differential equation, Eq. (3.44), for $\phi(x, r, \tau)$.

The first boundary condition in Eq. (3.45) is

$$\text{at } \tau > 0, \, x = 0, \text{ and } 0 < r < r_0, \quad \frac{\partial}{\partial x} (\phi_1 \phi_2) = 0.$$

However,

$$\frac{\partial}{\partial x} (\phi_1 \phi_2) = \phi_1 \frac{\partial \phi_2}{\partial x} \text{ and } \frac{\partial \phi_2}{\partial x} = 0 \text{ at } x = 0$$

by virtue of the boundary conditions following Eq. (3.47). Also at $\tau > 0$, $x = L$, and $0 < r < r_0$,

$$-k \frac{\partial}{\partial x} (\phi_1 \phi_2) = h \phi_1 \phi_2.$$

Since ϕ_1 is independent of x,

$$-k \frac{\partial \phi_2}{\partial x} = h \, \phi_2,$$

which is correct by the boundary conditions following Eq. (3.47). In a similar fashion, it can be shown that letting $\phi = \phi_1 \phi_2$ also satisfies the last three conditions of Eq. (3.45).

Since $\phi = \phi_1 \phi_2$ satisfies both the governing partial differential equation for ϕ and all the boundary conditions on ϕ, it is the solution for $\phi(x, r, \tau)$; using the definitions of ϕ, ϕ_1, and ϕ_2,

$$\left(\frac{T - T_f}{T_0 - T_f} \right)_{\text{fin. cyl.}} = \left(\frac{T - T_f}{T_0 - T_f} \right)_{\text{inf. cyl.}} \left(\frac{T - T_f}{T_0 - T_f} \right)_{\text{inf. slab}} \tag{3.49}$$

Thus, the charts of Figs. 3.4–7 serve as the solution to the two-dimensional transient conduction region shown in Fig. 3.14. The temperature T anywhere within the finite cylinder is obtained by solving for ϕ_1 at the point of interest, using Figs. 3.6–7, then solving for ϕ_2 at the same point using Figs. 3.4–5, and then multiplying ϕ_1 by ϕ_2. Finally, T is calculated from Eq. (3.49) as

$$\phi = \left(\frac{T - T_f}{T_0 - T_f} \right)_{\text{fin. cyl.}} = \phi_1 \phi_2.$$

Note that ϕ_1, ϕ_2, and ϕ must be used, since only in terms of these nondimensional temperature excess ratios is the product form correct. Examples of product forms which are *not correct* are $\theta = \theta_1 \theta_2$ and $T = T_1 T_2$. Also note that when the product solution $\phi = \phi_1 \phi_2$ is used, T is *not* calculated from ϕ_2, that is, from

$$\phi_2 = \left(\frac{T - T_f}{T_0 - T_f} \right)_{\text{inf. slab}},$$

since the T in this expression is the temperature for the infinite slab.

There are other combinations of the known solutions to one-dimensional transients which yield solutions to certain two- and three-dimensional transients. For example, if $\phi_2(L)$ is the nondimensional temperature excess ratio for an infinite slab of thickness $2L$ and if $\phi_2(a)$ and $\phi_2(b)$ are the temperature excess ratios for infinite slabs of thickness $2a$ and $2b$, respectively, then the product $\phi_2(L)\phi_2(a)\phi_2(b)$ of these three excess ratios is the nondimensional temperature excess ratio for the right rectangular parallelepiped, formed by the mutually perpendicular intersection of the three separate infinite slabs, when the origin of the rectangular cartesian coordinate system is at the geometric center of the body whose dimensions are $2L$ by $2a$ by $2b$. Similarly, the product $\phi_2(L)\phi_2(a)$ is the temperature excess ratio for a long rectangular bar of cross section $2L$ by $2a$.

Let the nondimensional temperature excess ratio of Eq. (3.39), which is the solution to a semi-infinite slab, be ϕ_3. Then the product $\phi_3 \, \phi_2(L)\phi_2(b)$ is the solution to a semi-infinite rectangular bar, and since ϕ_1 is the solution to the infinite cylinder, $\phi_1\phi_3$ is the solution to a semi-infinite cylinder.

Thus, for certain two- and three-dimensional conduction regions, which are initially at a constant temperature T_0 throughout, and are suddenly, at time $\tau = 0$, exposed to a fluid at constant temperature T_f with a constant surface coefficient of heat transfer h between the fluid and the surface of the conduction region, the solution for the transient temperature distribution is given by the product of the one-dimensional transient solutions treated earlier. The preceding statement concerning the constant h can be modified, for in the analysis of the finite cylinder which led to the product form, all that is needed is a constant surface coefficient, say h_1, on the curved surface of the cylinder (the part associated with an infinite cylinder) and a different constant surface coefficient h_2 on the plane circular ends (which are associated with the infinite slab). *Thus, the definitions of the ϕ's which permit solutions to certain multidimensional transients to be expressed in terms of known solutions to some one-dimensional transients are:*

$$\phi = \left(\frac{T - T_f}{T_0 - T_f}\right)_{\text{general conduction region}}, \qquad \begin{array}{l}\text{given by the product of}\\ \text{certain of the } \phi_1, \phi_2, \phi_3;\end{array}$$

$$\phi_1 = \left(\frac{T - T_f}{T_0 - T_f}\right)_{\text{infinite cylinder of radius } r_0}, \qquad \begin{array}{l}\phi_1 \text{ given by Figs.}\\ 3.6\text{--}7;\end{array}$$

$$\phi_2(L) = \left(\frac{T - T_f}{T_0 - T_f}\right)_{\text{infinite slab of thickness } 2L}, \qquad \begin{array}{l}\phi_2 \text{ given by Figs.}\\ 3.4\text{--}5 \text{ or by Eqs.}\\ (3.25\text{--}26);\end{array} \qquad (3.50)$$

$$\phi_3 = \left(\frac{T - T_f}{T_0 - T_f}\right)_{\text{semi-infinite slab}}, \qquad \phi_3 \text{ given by Eq. (3.39)}.$$

No generation term q''' is allowed by the individual solutions for ϕ_1, ϕ_2, and ϕ_3.

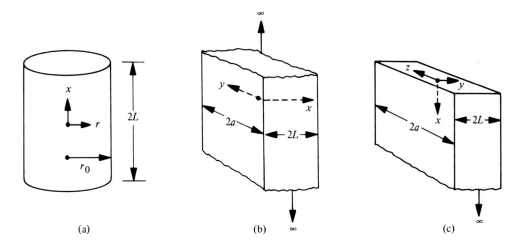

Figure 3.16 (a) Finite cylinder, $\phi = \phi_1\phi_2\,(L)$, where the origin is at geometric center.
(b) Infinite rectangular bar, $\phi = \phi_2\,(L)\phi_2(a)$, where the origin is at geometric center.
(c) Semi-infinite rectangular bar, $\phi = \phi_2\,(L)\phi_2\,(a)\phi_3$, where the origin is at the center of the top face.

Permissible Multidimensional Bodies

Figure 3.16 shows the conduction regions for which the product-type solution is valid; it also shows the proper location of the origin of the coordinate system. The systems shown in Fig. 3.16 are by no means the only systems that can be handled by the type of product solution discussed herein. An infinite number of systems can be devised for which the product solution is valid by taking advantage of the planes of symmetry (since these are adiabatic surfaces) for the conduction regions shown in Fig. 3.16. In Fig. 3.16(d), there is symmetry about all midplanes, so that the solution in Fig. 3.16(d) is also valid for the body shown in Fig. 3.17, which is a right rectangular parallelepiped whose bottom face is perfectly insulated. Thus, taking advantage of the planes of symmetry, an infinite number of new problems can be generated which can be solved using the same relations given in Fig. 3.16. A representative sample of these new conduction regions is shown in Fig. 3.18. The key to creating these new bodies is simply to look for planes of symmetry within the shapes shown in Fig. 3.16, and then cut away any part of those shapes which is isolated, in effect, from the rest of the body by adiabatic surfaces.

Exact analytical solutions to multidimensional transient conduction problems, other than the solutions discussed so far in this section, are beyond the scope of an introductory course. When certain conditions are satisfied, it is possible to treat these more complicated problems, within an acceptable degree of approximation, by the lumped parameter method we will discuss in Sec. 3.3. Multidimensional transient conduction problems will also be solved by the powerful finite difference methods advanced in Sec. 3.4.

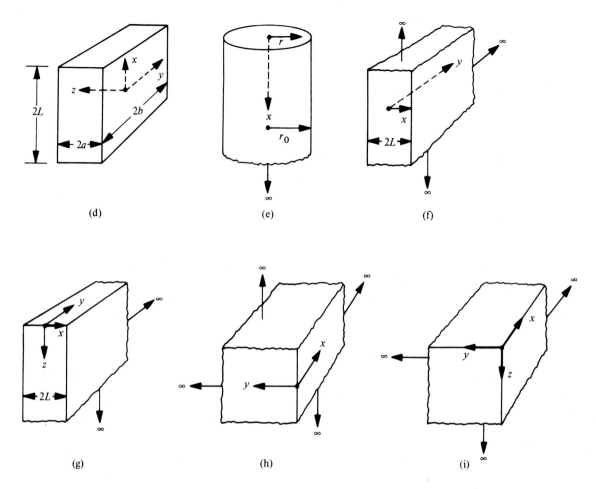

Figure 3.16 *(Continued)* (d) Right rectangular parallelepiped, $\phi = \phi_2(L)\phi_2(a)\;\phi_2(b)$, where the origin is at geometric center.

(e) Semi-infinite cylinder, $\phi = \phi_1\phi_3$, where the origin is at the center of the top face.

(f) Semi-infinite plate, $\phi = \phi_2(L)\phi_3$, where the origin is centered on the exposed face of width $2L$.

(g) Quarter-infinite plate, $\phi = \phi_2(L)\phi_3(y)\phi_3(z)$, where the origin is centered on the upper corner.

(h) Quarter-infinite body, $\phi = \phi_3(x)\phi_3(y)$, where the origin is on the corner.

(i) Eighth-infinite body, $\phi = \phi_3(x)\phi_3(y)\phi_3(z)$, where the origin is on the corner.

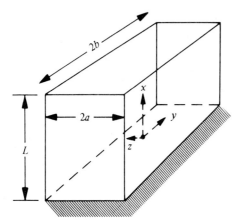

Figure 3.17 The solution for Fig. 3.16(d) applies to this right rectangular parallelepiped, which has an insulated bottom face.

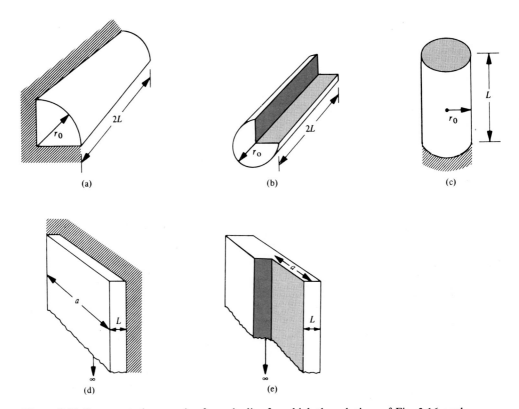

Figure 3.18 Representative sample of new bodies for which the solutions of Fig. 3.16 apply.

EXAMPLE 3.7

To illustrate the use of the charts for one-dimensional problems and the product solutions for multidimensional problems, consider a very long brass bar of radius 2 in. which is initially at 1000°F when it is suddenly put into a 100°F fluid to cool. The average surface coefficient of heat transfer between the bar's surface and the fluid is 36 Btu/hr ft² °F, while the thermal conductivity and thermal diffusivity of the brass are 60 Btu/hr ft °F and 1.14 ft²/hr, respectively. (a) Find the center temperature and surface temperature of the bar after 5.32 min. (b) Suppose that the bar, instead of being very long, is only 8 in. long. Would you expect the temperature at the geometric center to be greater or less than the center temperature of part a? Explain. (c) For the conditions of part b, find the temperature at both the geometric center and the center of the curved outside surface.

Solution

a. The brass bar is first idealized as an infinite cylinder of outer radius $r_0 = 2$ in. $= 1/6$ ft, initially at constant temperature $T_0 = 1000°F$ throughout; then, it is exposed to a fluid at constant temperature $T_f = 100°F$ and with a constant surface coefficient of heat transfer $h = 36$ Btu/hr ft² °F. The conditions described satisfy the restrictions on the use of Heisler's charts for the infinite cylinder, Figs. 3.6–7. Figure 3.6 gives the non-dimensional temperature excess ratio at the center as a function of a Fourier number and one over the Biot number; hence, they are computed to be

$$\frac{1}{N_{Bi}} = \frac{k}{hr_0} = \frac{60}{36(\frac{1}{6})} = 10,$$

$$N_{Fo} = \frac{\alpha\tau}{r_0^2} = \frac{1.14(\frac{1}{11.3})}{\frac{1}{36}} = 3.64,$$

since $\tau = 5.32$ min $= \frac{1}{11.3}$ hr. From Fig. 3.6 at $1/N_{Bi} = 10$ and $N_{Fo} = 3.64$,

$$\left(\frac{\theta}{\theta_0}\right)_{r/r_0=0} = 0.50 \quad \text{or} \quad \left(\frac{T - T_f}{T_0 - T_f}\right)_{r/r_0=0} = 0.50;$$

hence, $T_{r/r_0=0} = 0.50 (1000 - 100) + 100 = 550°F$.

Thus, the center temperature is 550°F. For the surface temperature, the auxiliary chart of Fig. 3.7 for the infinite cylinder is used, since it relates the temperature excess at any point in the cylinder to the temperature excess at the center at any time, except for very short times.

From Fig. 3.7 at $1/N_{Bi} = 10$, and at the surface coordinate $r/r_0 = 1$,

$$\frac{\theta_{r/r_0=1}}{\theta_{r/r_0=0}} = 0.95, \quad \theta_{r/r_0=0} = T_{r/r_0=0} - T_f = 550 - 100 = 450,$$

$$\theta_{r/r_0=1} = T_{r/r_0=1} - T_f = 0.95(450) = 427;$$

Unsteady-State Conduction

hence, $T_{r/r_0=1} = 427 + 100 = 527°\text{F}$. Thus, the center temperature after 5.32 minutes of cooling is 550°F, while the surface temperature of the bar is 527°F.

b. If the bar is only 8 in. long, the temperature of the geometric center of the bar after 5.32 min is expected to be lower than that computed in (a). In (a), energy could flow out of the bar to the fluid only across the curved outside surface, while here energy can flow across the two plane circular ends, as well as the curved surface. Hence, the bar will lose energy at a greater rate, leading to a faster temperature drop in the same time than for the conditions of (a). Generally, for a given volume of material to be cooled (or heated), it is reasonable to expect the cooling rate to increase as greater surface area for heat transfer becomes available for the given volume. Hence, if all other factors are equal, the conduction region with the greatest surface area for heat transfer to volume of material ratio will respond fastest, in terms of time, when it is being cooled (or heated).

c. To find the center temperature and the temperature at the center of the curved surface, the product solution is used. The situation is shown in Fig. 3.19, where $r_0 = 1/6$ ft and $L = 4$ in. $= 1/3$ ft.

First, the temperature at the geometric center, which is at the coordinate $r/r_0 = 0$ within the infinite cylinder and the coordinate $x/L = 0$ within the infinite slab is determined. The point of interest in the two-dimensional body under consideration is thus located in the two one-dimensional bodies whose perpendicular intersection generates the body of interest shown in Fig. 3.19. Hence,

$$\phi_{r/r_0=0,\,x/L=0} = \phi_{1\,r/r_0=0}\,\phi_{2\,x/L=0}(L).$$

From (a),

$$\phi_{1\,r/r_0=0} = \left(\frac{\theta}{\theta_0}\right)_{r/r_0=0,\,\text{inf. cyl.}} = 0.50.$$

The chart of Fig. 3.4 for the center temperature excess ratio for the infinite slab by itself is used to determine $\phi_{2\,x/L=0}(L)$. Thus, for the infinite slab,

$$\frac{1}{N_{\text{Bi}_L}} = \frac{k}{hL} = \frac{60}{36(1/3)} = 5.0, \quad N_{\text{Fo}_L} = \frac{\alpha\tau_0}{L^2} = \frac{1.14(\frac{1}{11.3})}{1/9} = 0.91.$$

Using these values on Fig. 3.4,

$$\phi_{2\,x/L=0}(L) = \left(\frac{\theta}{\theta_0}\right)_{x/L=0,\,\text{inf. slab}} = 0.86.$$

The product yields

$$\left(\frac{T - T_f}{T_0 - T_f}\right)_{r/r_0=0,\,x/L=0} = \phi_{r/r_0=0,\,x/L=0} = (0.50)(0.86) = 0.43.$$

Hence, the center temperature $T = 100 + 0.43(900) = 487°\text{F}$ of the finite cylinder is less than in part a, as was correctly reasoned in part b.

Chapter Three

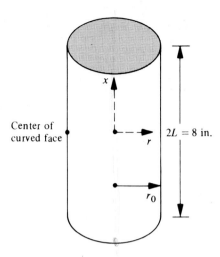

Center of
curved face

$2L = 8$ in.

r_0

Figure 3.19 The short bar of Ex. 3.7 (b-c).

A point at the center of the curved face in Fig. 3.19 has the coordinates $x/L = 0$ and $r/r_0 = 1.0$. Hence, $\phi_{r/r_0=1.0,\, x/L=0} = \phi_{1_{r/r_0=1.0}} \phi_{2_{x/L=0}}(L)$. But from (b), $\phi_{2_{x/L=0}} (L)$ $= 0.86$, and from (a), $\phi_{1_{r/r_0=1}} = 0.95(0.50) = 0.475$. Note that $\phi_{1_{r/r_0=1}}$ can also be calculated directly from the temperatures found in (a) as

$$\phi_{1_{r/r_0=1}} = \left(\frac{T - T_f}{T_0 - T_f}\right)_{r/r_0=1} = \frac{527 - 100}{1000 - 100} = 0.475.$$

Thus,

$$\left(\frac{T - T_f}{T_0 - T_f}\right)_{r/r_0=1.0,\, x/L=0} = 0.86(0.475) = 0.409,$$

and the temperature of the curved surface is $T = 100 + 0.409(900) = 468°F$.

Thus the temperatures at the geometric center and center of the curved surface are $487°F$ and $468°F$, respectively.

EXAMPLE 3.8

A cut of meat is packaged so that its shape is that of a rectangular parallelepiped 10 cm by 10 cm by 20 cm. Initially, it is in a freezer at $-21°C$ throughout and then is exposed to $19°C$ air. A surface coefficient of heat transfer equal to $8.5 \ W/m^2 \ °C$ exists between the air and the meat surface. Treating the meat as if it has the properties of ice, calculate the time the meat can be left out without any part of it thawing. For ice, $k = 2.14$ $W/m \ °C$ and $\alpha = 1.25 \times 10^{-6} \ m^2/s$.

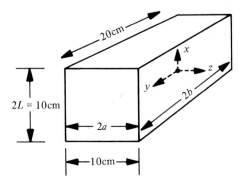

Figure 3.20 Shape of the meat package of Ex. 3.8.

Solution

The conduction region is pictured in Fig. 3.20, and the problem involves a three-dimensional transient. But it satisfies the conditions needed to use the product of known solutions to one-dimensional transients, and in fact is the body shown in Fig. 3.16(d). Hence, the solution will be formed as the product of solutions for three infinite planes, two of thickness 10 cm and one of thickness 20 cm, and can be represented by the expression

$$\phi = \phi_2(L)\,\phi_2(a)\,\phi_2(b). \qquad (3.51)$$

The point of interest in the actual three-dimensional body is the point at which thawing would first occur. This would be a corner, since the material at one of the corners in Fig. 3.20 has a greater surface area for heat transfer to volume ratio than the same volume of material located elsewhere; that is, the corners respond most quickly in a temperature transient.

The initial temperature is $T_0 = -21\,°C$, while the fluid temperature is $T_f = 19\,°C$. Thus, $\theta_0 = T_0 - T_f = -21 - 19 = -40°$, and the actual temperature excess at a corner when thawing begins is

$$\theta_c = \theta_{x/L=1,\, y/b=1,\, z/a=1} = 0 - 19 = -19.$$

Hence,

$$\phi = \frac{\theta_c}{\theta_0} = \frac{-19}{-40} = 0.475.$$

Since $a = L = 5$ cm, Eq. (3.51) becomes, after inserting the known value for ϕ,

$$0.475 = \phi^2_{2_{x/L=1.0}}(L)\,\phi_{2_{y/b=1.0}}(b). \qquad (3.52)$$

But the chart of Fig. 3.5 relates $\phi_{2_{y/b=1.0}}(b)$ to $\phi_{2_{y/b=0}}(b)$ and also relates $\phi_{2_{x/L=1.0}}(L)$ to $\phi_{2_{x/L=0}}(L)$. So, treating the portion of the body that has thickness $2b = 20$ cm as an infinite slab by itself, the quantity k/hb is formed as

$$k/hb = 2.14/8.5(0.1) = 2.52.$$

Now, from Fig. 3.5 at $y/b = 1.0$,

$$0.83 = \frac{\theta_{y/b=1.0}}{\theta_{y/b=0}} = \frac{\phi_{y/b=1.0}(b)}{\phi_{y/b=0}(b)}.$$

Hence,

$$\phi_{y/b=1.0}(b) = 0.83\ \phi_{y/b=0}(b), \qquad (3.53)$$

where $\phi_{y/b=0}(b)$ is the nondimensional temperature excess ratio at $y/b = 0$. That is,

$$\phi_{y/b=0}(b) = \left(\frac{T - T_f}{T_0 - T_f}\right)_{\text{inf. slab at } y/b=0}$$

For the other two equal width (10 cm) infinite slabs,

$$k/hL = 2.14/8.5(0.05) = 5.04,$$

and again from Fig. 3.5,

$$\frac{\theta_{x/L=1}}{\theta_{x/L=0}} = 0.91 = \frac{\phi_{x/L=1.0}(L)}{\phi_{x/L=0}(L)}.$$

As a result,

$$\phi_{x/L=1.0}(L) = 0.91\ \phi_{x/L=0}(L), \qquad (3.54)$$

where

$$\phi_{x/L=0}(L) = \left(\frac{T - T_f}{T_0 - T_f}\right)_{\text{inf. slab at } x/L=0}$$

Inserting Eqs. (3.53) and (3.54) into Eq. (3.52) gives

$$
\begin{aligned}
0.475 &= (0.91)^2\ \phi_{x/L=0}^2(L)(0.83)\ \phi_{y/b=0}(b) \quad \text{or} \\
0.692 &= \phi_{x/L=0}^2(L)\ \phi_{y/b=0}(b).
\end{aligned}
\qquad (3.55)
$$

Recall that $\phi_{x/L=0}(L)$ is the ordinate of the chart of Fig. 3.4 for an infinite slab $2L$ by itself, and $\phi_{y/b=0}(b)$ is the ordinate of the same chart for an infinite slab $2b$ by itself. For the 10-cm slabs,

$$k/hL = 5.04, \quad N_{\text{Fo}_L} = \alpha\tau/L^2 = 1.25 \times 10^{-6}\ \tau/(0.05)^2 = 0.0005\tau,$$

and for the 20-cm slab,

$$k/hb = 2.52, \quad N_{\text{Fo}_b} = \alpha\tau/b^2 = 1.25 \times 10^{-6}\ \tau/(0.1)^2 = 0.000125\tau.$$

The unknown parameter τ appears in the two Fourier numbers for slabs $2L$ and $2b$. Since $\phi_{x/L=0}(L)$ and $\phi_{y/b=0}(b)$ depend upon these Fourier numbers, which in turn both contain the unknown quantity, the solution must be found by trial and error. Note that there are, in effect, three equations in three unknowns. One is Eq. (3.55) containing the two unknowns $\phi_{x/L=0}(L)$ and $\phi_{y/b=0}(b)$. Another equation is the chart of Fig. 3.4 for slab $2L$, since it, in effect, gives a relation between $\phi_{x/L=0}(L)$ and τ. The last equation is the chart of Fig. 3.4 for slab $2b$, since it gives a relation between $\phi_{y/b=0}(b)$ and time τ. Thus, there are three equations in the three unknowns $\phi_{x/L=0}(L)$, $\phi_{y/b=0}(b)$, and τ. A trial-and-error procedure is needed, because the two equations represented by Fig. 3.4 cannot be written out in simple, explicit, mathematical forms.

To solve Eq. (3.55), first assume that $\phi_{y/b=0}(b) = 1$. Certainly $\phi_{y/b=0}(b)$ is expected to be much greater than $\phi_{x/L=0}(L)$, since $b > L$, and hence slab $2L$ responds more quickly to transient thermal loading than does the larger slab $2b$. With $\phi_{y/b=0}(b) = 1$, Eq. (3.55) gives $0.692 = \phi_{x/L=0}^2(L)$ or $\phi_{x/L=0}(L) = 0.833$. From Fig. 3.4, at $\phi_{x/L=0}(L) = 0.833$ and $k/hL = 5.04$,

$$N_{Fo_1} = 1.20 = 0.0005\tau_1.$$

Hence, $\tau_1 = \frac{1.2}{0.0005} = 2400$ s. The subscript 1 refers to the first approximation, since this time of 2400 s was found by assuming that $\phi_{y/b=0}(b) = 1$. Now this assumption has to be checked by using τ_1 to get $\phi_{y/b=0}(b)_2$, and if it is unity, τ_1 is the correct time; if not, use the new value $\phi_{y/b=0}(b)_2$ in Eq. (3.55), find a new $\phi_{x/L=0}(L)$ and a new time τ_2, and continue this procedure until the times no longer change. Hence,

$$N_{Fo_b} = 0.000125\tau_1 = 0.000125(2400) = 0.30, \quad k/hb = 2.52.$$

Using these values in Fig. 3.4 gives $\phi_{y/b=0}(b)_2 = 0.95$; using this in Eq. (3.55) leads to $0.692 = \phi_{x/L=0}^2(L)_2(0.95)$. Hence, $\phi_{x/L=0}(L)_2 = 0.852$. From Fig. 3.4 at 0.852 and $k/hL = 5.04$, $N_{Fo_L} = 1.1 = 0.0005\tau_2$, yielding $\tau_2 = 1.1/0.0005 = 2200$ s. To check this revised time, calculate

$$N_{Fo_b} = 0.000125\tau_2 = 0.000125(2200) = 0.275,$$

and again use Fig. 3.4 at 0.275 and $k/hb = 2.52$ to obtain $\phi_{y/b=0}(b)_3 = 0.96$. Since the graph cannot be read this closely, the value of $\tau = 2200$ s is acceptable as the final answer; that is,

$$\tau = 0.612 \text{ hr} = 36.7 \text{ min.}$$

In slightly more than one-half hour, the calculation indicates that some parts of the meat will begin the thawing process. Naturally, once thawing begins, the charts no longer hold, because a phase transformation is occurring which makes the material behave as if it were absorbing thermal energy in the regions where thawing is taking place. The solution to the partial differential equation which Heisler's charts are based upon does not take into account this type of generation.

As this problem shows, when the temperature within or on a multidimensional conduction region is specified and one must find the time needed to get to that temperature, an iterative procedure may be needed when the problem's solution is given by the product of known one-dimensional transients. The iterative procedure described attempts to exploit the possibility that one of the elementary one-dimensional bodies in the product solution may, because of a smaller characteristic thickness, respond more rapidly than the others. Thus, a good first approximation to the unknown Fourier number is found by setting the values of ϕ equal to unity for the thicker one-dimensional bodies involved in the product solution.

3.3 LUMPED PARAMETER METHOD OF TRANSIENT ANALYSIS

In Sec. 3.2, exact analytical solutions to transient problems were treated; that is, for certain shapes of conduction regions, the temperature $T(x, y, z, \tau)$ was found as a function of both the space coordinates and time.

However, in a great many transient problems, conditions are such that the dependence of the temperature on the space coordinates is so weak that the local temperature anywhere within the conduction region is very close to the average temperature of the region. When this is true, only the average temperature as a function of time must be solved for, and this involves an ordinary differential equation rather than the partial differential equations of Sec. 3.2. Thus, at any instant of time τ, if the temperature anywhere throughout the control volume is close enough to the average temperature of all the material in the control volume at the same time τ, the material of the entire control volume can be treated as a single "lump" at the average temperature; hence the name *lumped parameter method*.

In many respects the lumped parameter approximation is similar to the quasi-one-dimensional approximation discussed in detail in Chapter 2, in that temperature derivatives in the space coordinates, although definitely not zero, are small enough to justify assigning a single value of temperature to all points in the conduction region at any instant of time. Thus the lumped parameter method might be termed a quasi-zero-dimensional approximation, since an average temperature over all space coordinates is used which varies only with the single variable time.

What is meant by the local temperature throughout a region being close enough to the average at any instant of time to justify the use of a single temperature for the entire conduction region? Consider a sphere of radius R which is initially (at time $\tau = 0$) at 500°F throughout. Suppose this sphere is plunged into a cooling bath of oil and the following temperatures are measured:

Time, hr	Center Temperature, °F	Surface Temperature, °F
0	500	500
0.2	460	454
0.6	400	390
2.0	100	96

Clearly, at any instant of time, the temperature of the points in the sphere between the center and the surface will have a temperature between the center temperature and the surface temperature. Therefore, the average temperature of the entire sphere will lie between the center temperature and the surface temperature. For simplicity, it is assumed here that the average temperature at any time is about halfway between the center and surface temperatures. Calling the average temperature T_m, the following chart can be constructed:

Time, hr	T_m, °F
0	500
0.2	457
0.6	395
2.0	98

From these two charts note that at 0.6 hr, for example, the average temperature is only 5°F from both the surface and the center temperatures; hence, if it is said that every point in the solid at this instant is at the average temperature (basic lumped parameter approximation), the maximum error in the local temperature would be only 1.28%. Thus, in a problem such as this, it can be said with acceptable accuracy for engineering calculations that at any instant of time the local temperature is close enough to the average temperature to justify the prediction only of the average temperature as a function of time.

Once we decide that the lumped parameter method will give results of acceptable engineering accuracy, the next step is to apply the energy balance, Eq. (1.8), to the entire region and derive the *ordinary* differential equation whose solution gives the region's average, or lumped, temperature as a function of time.

3.3a Criterion for Validity of the Lumped Parameter Method

The problem to be considered is that if the lumped parameter method of analysis is to be used intelligently as a tool of the practicing engineer, does a criterion exist for deciding when the approximation will yield results of acceptable accuracy? Referring to the sphere discussed previously, under what conditions will the surface temperature be sufficiently close to the center temperature to permit the use of the lumped parameter method? Consider a three-dimensional conduction region of fairly regular shape, as shown in Fig. 3.21, which is initially at the known temperature T_0 throughout. Suddenly at time $\tau = 0$ it is immersed in a fluid of constant temperature T_f, with a constant surface coefficient of heat transfer h existing between the fluid and the surface of the conduction region. Denote the temperature at the center of the region as T_c (which varies with time) and the temperature of the surface as T_s (this temperature will not vary much from point to point along the surface if the region is of fairly regular shape, and it is also a function of time τ). If the material of the conduction region is cooling, then temperature derivatives within the material in the various space directions cause energy to be conducted to the surface, where it is convected into the surrounding fluid. As this occurs, the stored energy of the material

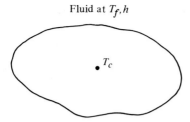

Fluid at T_f, h

$\cdot T_c$

Figure 3.21 General conduction region used to develop the lumped parameter method criterion.

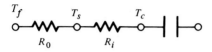

$T_f \qquad T_s \qquad T_c$

$R_0 \qquad\quad R_i$

Figure 3.22 Circuit used to deduce the lumped parameter method criterion.

is decreasing. This description of what is occurring can be replaced by the simple electrical analogue shown in Fig. 3.22, wherein what is actually happening is idealized by showing a capacitor, which represents the stored internal energy of the system, discharging this energy through a series of resistances. First, the energy must flow through the internal resistance R_i of the material itself, between the center temperature T_c and the surface temperature T_s at this instant of time, and then the energy must flow from the surface to the surrounding fluid across the resistance R_0 due to the outside surface coefficient of heat transfer. It is required that the conditions that must be satisfied be determined so that at any instant of time, including this one, T_s and T_c are sufficiently close to each other, and therefore are both close to the average temperature T. The circuit shown in Fig. 3.22 can support large current flow (i.e., heat flows) with only a small difference in the potentials T_c and T_s if R_0 is much greater than R_i. In this case, almost all of the total temperature drop from T_c to T_f takes place across R_0, and hence T_s will be close to T_c. Thus, it is required that $R_i << R_0$, or

$$\frac{R_i}{R_0} << 1. \tag{3.56}$$

Thus *if the ratio of the internal resistance of the material to conduction heat transfer to the external resistance of the adjacent fluid to convection heat transfer is small, then the lumped parameter approximation is acceptable.* Now, methods of calculating R_i and R_0 are required. Equation (2.92), properly interpreted, gives

$$R_i = \overline{L}/k\overline{A}, \quad R_0 = 1/hA_s,$$

where $\overline{L}$ is some average path length for conduction from the center, or the extreme interior point in the region, to the surface of the region across which energy exchange with the fluid occurs. The value $\overline{A}$ is some average area across which energy flows on the way

Unsteady-State Conduction

to the surface whose area is A_s, and k is the thermal conductivity of the region. Hence, forming the resistance ratio gives

$$\frac{R_i}{R_0} = \frac{h}{k}\left(\frac{A_s\overline{L}}{\overline{A}}\right). \tag{3.57}$$

The term in parentheses, say S', has the dimensions of length. If Eq. (3.57) is being used for a slab of half-thickness L, then $S' = L$.

How does one compute this critical or characteristic dimension S' for other shapes which are more complicated than the slab? One way, of course, is to make an estimate of the quantity in parentheses in Eq. (3.57), $A_s\overline{L}/\overline{A}$, and use this as S'. Looking again at the simple slab of half-thickness L, one sees that L is also $L_{\max}$, *the maximum linear distance between the center and the surface in a direction in which there is nonzero heat transfer.* Thus this can be used as S' in a more general conduction region. Finally, again for the slab, the numerator of the parenthesis in Eq. (3.57) is V_0, *the volume of material having its temperature changed during the transient.* The denominator is also, for the slab, A_s, *the surface area for convection heat transfer.*

Combining Eqs. (3.56) and (3.57), the criterion takes the form

$$hS'/k \ll 1. \tag{3.58}$$

Now, Eq. (3.58) requires that $hS'/k \ll 1$, but how much less than 1 has not yet been determined.

This question can be answered by reference to the auxiliary charts of Figs. 3.5, 3.7, and 3.9, wherein the ratio of the surface temperature excess to the center temperature excess is given at any instant of time for the infinite slab, infinite cylinder, and the sphere, as a function of the Biot number for these geometries. Note, for example, in Fig. 3.5 that at $k/hL = 10$ or $hL/k = 0.10$, $(T_s - T_f)/(T_c - T_f) = 0.95$, which could be taken as the criterion under which the accuracy of the lumped parameter method would be acceptable.

Using this as a criterion, the lumped parameter method of transient analysis can be used with an error of 5% or less, an error induced by assuming that at any instant of time the temperature throughout the conduction region is essentially uniform. Hence, for bodies whose shape is not too irregular, this result is extended or generalized by using S' in place of L. This gives the following quantitative criterion for validity of the lumped parameter method:

$$\frac{hS'}{k} \le 0.10. \tag{3.59}$$

The value S' can be estimated from one of the following expressions:

$$S' = A_s\overline{L}/\overline{A} \tag{3.60}$$

$$S' = L_{\max} \tag{3.61}$$

$$S' = V_0/A_s \tag{3.62}$$

Normally, the value of S' should be calculated from Eq. (3.61) or Eq. (3.60); these generally give the most conservative, and the safest, estimate of S' to be used in the criterion Eq. (3.59), according to the author's calculations. Equation (3.62) should probably be used for S' only in the case of an extremely complicated shape in which it is difficult to make estimates of S' via Eqs. (3.60) or (3.61). As a matter of fact, if one calculates S' for a sphere using Eq. (3.62), and with this value of S' finds that the inequality, Eq. (3.59), is barely satisfied, it turns out that the lumped parameter method does *not* give the 5% accuracy previously referred to. Hence the value of S' should be found from Eqs. (3.60) or (3.61) when possible. If Eq. (3.62) is used instead, one would want hS'/k to be considerably less than 0.10.

If the conduction region is of highly irregular shape, it might have local regions in which the internal resistance to conduction is not much smaller than the external resistance of the fluid; therefore, at least in these regions, the temperature would vary significantly from point to point at any instant of time. Whether the lumped parameter method can still be used depends on whether or not these regions constitute a very large fraction of the total, and also on whether a very accurate value of temperature is needed in these regions. (There is a strong similarity between the use of the lumped parameter method in irregularly shaped regions and the use of the quasi-one-dimensional analysis in irregularly shaped cross sections—i.e., cross sections that are not very smooth. Thus, the discussion in Chapter 2 on the use of the quasi-one-dimensional approximation on not-so-smooth cross sections can also help the engineer decide when the lumped parameter method of transient analysis can be applied to irregular shapes.)

The lumped parameter method of transient analysis is valuable and powerful because the single temperature associated with the entire conduction region at any instant of time is a function only of time, and hence satisfies an ordinary differential equation rather than a partial differential equation. Since generally an ordinary differential equation is easier to solve than a partial differential equation, more complicated boundary conditions and generation terms can be handled in an exact analytical fashion. This, along with the fact that complicated shapes can be dealt with easily and that the solution for $T = T(\tau)$ is often a function which is easy to use in calculation of actual numbers, is what makes the lumped parameter method such a powerful tool for attacking transient conduction problems. Following is the formal procedure for using this method:

1. If the conduction region is not too irregular in shape, determine if Eq. (3.59) is satisfied. If it is, proceed to step 2. If it is not, the temperature's dependence upon the space coordinates has to be taken into account explicitly, either by the methods of Sec. 3.2 or by the finite difference procedure to be discussed in Sec. 3.4.

2. Take the entire conduction region as the control volume, use Eq. (1.8) on the control volume, and identify the various energy transfer, generation, and storage terms. This leads to an ordinary differential equation for the temperature T as a function of time τ.

3. Solve the ordinary differential equation for $T = T(\tau)$ and apply the initial condition.

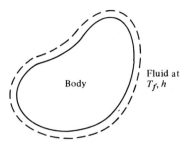

Figure 3.23 Arbitrarily shaped body.

Arbitrarily Shaped Body with Convection

Consider an arbitrarily shaped body, as shown in Fig. 3.23, which is initially at a constant temperature T_0 throughout, and then suddenly has its outside surface exposed to a fluid at known constant temperature T_f with constant surface coefficient of heat transfer h. Let us suppose that the Biot number check, Eq. (3.59), has been made and that the criterion is satisfied. With this done, the next step in a lumped parameter analysis is to take the entire conduction region as a control volume and apply the energy balance, Eq. (1.8). The control surface is depicted as a dashed line in Fig. 3.23. Since there is no generation, Eq. (1.8) becomes

$$R_{\text{in}} = R_{\text{out}} + R_{\text{stor}}. \tag{3.63}$$

At any instant of time, there will be convection heat transfer from the surface of the body to the fluid at a rate given by Newton's cooling law; that is,

$$R_{\text{out}} = hA_s(T - T_f), \tag{3.64}$$

where T is the temperature of the body at this instant of time.

Now, an expression is needed for R_{stor}, the time rate of change of stored energy within the entire conduction region. At any instant of time, the total stored energy is the product of the mass of material in the conduction region and the internal energy per unit mass. (Since all of the material is at the same temperature, then all of it has the same internal energy per unit mass at any one instant of time.) The total internal energy at this instant is then

$$\rho V_0 c_p T,$$

where ρ is the mass density and V_0 is the total volume of the conduction region. Hence, R_{stor} is the time derivative of the internal energy; that is,

$$R_{\text{stor}} = \frac{d}{d\tau}(\rho V_0 c_p T) = \rho c_p V_0 \frac{dT}{d\tau}. \tag{3.65}$$

Combining Eqs. (3.65), (3.64), and (3.63),

$$0 = hA_s(T - T_f) + \rho c_p V_0 \frac{dT}{d\tau}. \tag{3.66}$$

Defining $\phi = hA_s/\rho c_p V_0$ and the temperature excess $\theta = T - T_f$, $d\theta/d\tau = dT/d\tau$, and using the definitions in Eq. (3.66) leads to

$$\frac{d\theta}{d\tau} + \phi\theta = 0.$$

Thus, a first-order, linear, homogeneous, ordinary differential equation with constant coefficients is obtained to solve for θ as a function of time. Using the classical operator method or separating the variables and integrating yields

$$\theta = Be^{-\phi\tau}, \tag{3.67}$$

where B is an integration constant. Now, at $\tau = 0$, $T = T_0$; hence $\theta = \theta_0 = T_0 - T_f$. Applying this condition to Eq. (3.67) leads to $B = \theta_0$, and Eq. (3.67) becomes

$$\frac{\theta}{\theta_0} = \frac{T - T_f}{T_0 - T_f} = e^{-\phi\tau}. \tag{3.68}$$

Note how simple in form the solution to this type of transient becomes when Eq. (3.59), the criterion, is satisfied, thereby permitting the use of the lumped parameter method. (By comparing Eq. [3.68] with Eq. [3.25], the exact analytical solution to a similar problem, an idea of the relative complexities of the two methods is obtained.)

Arbitrarily Shaped Body with Convection and Constant Generation

An object, such as an electric resistance heater or an iron, is initially at the known constant temperature T_0 throughout. Suddenly, energy is dissipated within the body at the total rate Q_0, while at the same time part of the body's outside surface area A_s is exposed to a fluid at the constant temperature T_f with a constant surface coefficient of heat transfer h between the fluid and the body. Assuming that the Biot number criterion, Eq. (3.59), is satisfied (if this is not the case, a multidimensional transient analysis, probably using finite difference methods, would have to be made), we would like to derive an expression for the lumped temperature as a function of time for this constant generation case. The volume of the body comprising the conduction region is V_0, and the body's properties are known. The physical situation is sketched in Fig. 3.23.

Take the entire body as the control volume, shown enclosed by the dashed control surface in Fig. 3.23, and apply the law of conservation of energy,

$$R_{in} + R_{gen} = R_{out} + R_{stor}. \tag{1.8}$$

Now R_{gen}, the total rate at which energy is being generated or released within the control volume itself, is given as Q_0. The convection heat transfer rate at this instant of time will be considered as an energy transfer rate out of the control volume; hence, $R_{out} = hA_s (T - T_f)$, where T is the instantaneous temperature of all the material in the control volume, and R_{stor} is $mc_p\, dT/d\tau = \rho V_0 c_p\, dT/d\tau$. Thus, from Eq. (1.8),

$$Q_0 = hA_s (T - T_f) + \rho c_p V_0\, dT/d\tau. \qquad (3.69)$$

Defining $\phi = hA_s/\rho c_p V_0$ and introducing the temperature excess, so that $\theta = T - T_f$ and $d\theta/d\tau = dT/d\tau$, Eq. (3.69) becomes

$$\frac{d\theta}{d\tau} + \phi\theta = \frac{Q_0}{\rho c_p V_0}. \qquad (3.70)$$

Equation (3.70) is a linear, first-order, nonhomogeneous, ordinary differential equation with constant coefficients, and as such can be solved by adding the complementary and particular solutions. However, a simple transformation allows the nonhomogeneity $Q_0/\rho c_p V_0$ to be effectively removed. First, Eq. (3.70) is rewritten as

$$\frac{d\theta}{d\tau} + \phi\theta - \frac{Q_0}{\rho c_p V_0} = 0, \qquad (3.71)$$

and ψ is defined as

$$\psi = \phi\theta - \frac{Q_0}{\rho c_p V_0}, \qquad (3.72)$$

so that

$$\frac{d\psi}{d\tau} = \phi\,\frac{d\theta}{d\tau}. \qquad (3.73)$$

Substituting Eqs. (3.72) and (3.73) into Eq. (3.71) and rearranging,

$$\frac{d\psi}{d\tau} + \phi\psi = 0. \qquad (3.74)$$

The solution to Eq. (3.74) is

$$\psi = Be^{-\phi\tau},$$

or, since $\psi = \phi\theta - Q_0/\rho c_p V_0$ and $\phi = hA_s/\rho c_p V_0$,

$$\theta = \frac{Q_0}{hA_s} + Be^{-\phi\tau}, \qquad (3.75)$$

where a ϕ has been absorbed into the integration constant B. The initial condition is at $\tau = 0$, $T = T_0$. Therefore,

$$\theta = T_0 - T_f = \theta_0.$$

Using this in Eq. (3.75) gives

$$\theta_0 - \frac{Q_0}{hA_s} = B.$$

Hence,

$$\theta = \frac{Q_0}{hA_s} + \theta_0\, e^{-\phi\tau} - \frac{Q_0}{hA_s}\, e^{-\phi\tau}$$

or

$$T - T_f = \frac{Q_0}{hA_s}\, (1 - e^{-\phi\tau}) + (T_0 - T_f)e^{-\phi\tau}. \tag{3.76}$$

$$\phi = hA_s/\rho c_p V_0 \tag{3.77}$$

Note that at $\tau = 0$, $T = T_0$ as it must, and as $\tau \to \infty$, T approaches $T_f + Q_0/hA_s$, which is exactly the result which a steady-state energy balance would yield after the transient dies out.

Equation (3.76) clearly shows that in this case the temperature variation with time is due to two separate thermal loadings; for even if $Q_0 = 0$, the exposure of the object to a fluid at a different temperature T_f would cause a temperature variation with time as shown by the second term of Eq. (3.76). Even if the object originally had the same temperature as the fluid T_f, the start of the generation Q_0 would cause the temperature variation with time shown by the first term of Eq. (3.76). More explicitly, the two different thermal loadings are the fluid at the temperature T_f different from T_0, and the total generation rate Q_0.

Arbitrarily Shaped Body with Radiation to Low-Temperature Surroundings

Consider a body which is originally at the constant temperature T_0 and is then exposed, in a vacuum, to very low-temperature black surroundings. The shape of the body is such that none of its own emitted radiant energy strikes itself. The mass of the body is m, specific heat is c_p, total hemispherical emissivity of its surface is ϵ_s, and the outside surface area is A_s. For the case in which the ratio, R_i/R_0, of internal to external resistance to heat transfer is very small (note that this ratio is *not* given by Eq. [3.57], which is only for convection from the surface), we would like to predict the body temperature as a function of time τ. This body that we refer to could be an earth satellite in the vacuum of space, initially held at constant temperature by an internal power source, when suddenly the power source fails, initiating the transient.

Since it is assumed that a lumped parameter analysis will be sufficiently accurate, the entire body is taken as the control volume and the law of conservation of energy, Eq. (1.8), is applied. There is no generation, so $R_{gen} = 0$. Also, there is no convection in the vacuum: no radiant energy *absorbed* because the surroundings are at a very low temperature, approaching $0°K$ or $0°R$ in the limit, compared to the body temperature; no energy returns from the black surroundings because of their being a perfect absorber; and furthermore, the body, due to its shape, does not absorb any of its own emission.

Hence, $R_{in} = 0$ and R_{out} is the rate at which the body emits radiant energy at any instant of time. This is given by Eq. (1.7), the Stefan-Boltzmann law modified by the total hemispherical emissivity ϵ_s of the body. Hence, $R_{out} = \epsilon_s A_s \sigma T^4$, where T in $°K$ or $°R$ is the body temperature at any instant of time. The storage term is $R_{stor} = mc_p(dT/d\tau)$ since, in the lumped parameter method, all of the material at any instant of time is very near the average temperature of the body. Hence, Eq. (1.8) becomes

$$0 = \epsilon_s A_s \sigma T^4 + mc_p \frac{dT}{d\tau}.$$

Rearranging and defining,

$$B = \frac{\epsilon_s A_s \sigma}{mc_p}, \tag{3.78}$$

which leads to

$$\frac{dT}{d\tau} + BT^4 = 0,$$

which is a first-order, nonlinear, ordinary differential equation which can be solved by separating the variables. Hence,

$$dT/T^4 = -B \, d\tau.$$

Since at $\tau = 0$, $T = T_0$ (the initial temperature), and at the general time τ, the temperature is T,

$$\int_{T_0}^{T} \frac{dT}{T^4} = -B \int_{0}^{\tau} d\tau.$$

Integrating,

$$\frac{-1}{3T^3}\bigg|_{T_0}^{T} = -B\tau, \quad \text{or} \quad \frac{1}{T^3} - \frac{1}{T_0^3} = 3 B\tau.$$

Rearranging,

$$T = \left(\frac{1}{1/T_0^3 + 3 B\tau} \right)^{1/3}. \tag{3.79}$$

The value B is given by Eq. (3.78).

Equation (3.79) is the solution for the temperature versus time in a body initially at the known temperature T_0 throughout, which then loses radiant energy to a $0°K$ or $°R$ surrounding, when none of its own emitted radiant energy is absorbed, and when the conditions are such that the lumped parameter method can be used. This result is useful not only for a body in space, but also for the initial part of a transient in a body which is so much hotter than its surroundings that any radiant energy received from the surroundings is negligible and, in addition, the surface coefficient of heat transfer between the body and the air is low enough for the convection heat transfer rate to be neglected. A situation such as this could arise for a short period of time just after a metal bar has been taken out of an extremely hot heat-treating oven and is being allowed to cool in air. In such a circumstance, it may be said that the initial cooling period is *radiation-controlled,* or *radiation-dominated.*

Body with Convection to a Fluid Whose Temperature Varies Harmonically with Time

Here we consider an arbitrarily shaped body at initial temperature T_0 which is then brought into contact with a fluid whose temperature varies with time as

$$T_f = T_m + T_1\cos \omega\tau, \qquad (3.80)$$

and the surface coefficient of heat transfer h is a known constant. The average fluid temperature, T_m, the angular frequency ω, and T_1 are constants, and the body's properties are known. Possible physical situations that may satisfy these conditions include the thin steel wall of a reciprocating-type internal combustion engine as well as some regenerative-type heat exchangers.

Assuming that conditions are satisfied, according to Eq. (3.59), for the validity of the lumped parameter method, we now apply the energy balance, Eq. (1.8), to the entire body as the control volume in order to predict the temperature's dependence upon time.

$$hA_s (T_f - T) + 0 = \rho V_0 c_p \frac{dT}{d\tau}. \qquad (3.81)$$

Defining $\phi = hA_s/\rho c_p V_0$, and rearranging, Eq. (3.81) becomes

$$\frac{dT}{d\tau} + \phi T = \phi T_f. \qquad (3.82)$$

(Note that the transformation $\theta = T - T_f$ is not particularly useful in this problem, because T_f depends upon time by Eq. [3.80], and hence $dT/d\tau$ would have to be replaced not just by $d\theta/d\tau$, but by $d\theta/d\tau + dT_f/d\tau$. Thus, the transformation does not simplify Eq. [3.82].) Inserting Eq. (3.80) into Eq. (3.82) gives

$$\frac{dT}{d\tau} + \phi (T - T_m) = \phi T_1\cos \omega\tau. \qquad (3.83)$$

Now, defining a new dependent variable θ as

$$\theta = T - T_m,\tag{3.84}$$

so that $d\theta/d\tau = dT/d\tau$, Eq. (3.83) becomes

$$\frac{d\theta}{d\tau} + \phi\theta = \phi T_1 \cos \omega\tau.\tag{3.85}$$

Equation (3.85) is a linear, first-order, ordinary differential equation with constant coefficients, which can be solved by the method of the integrating factor or by summing the complementary and particular solutions. The latter method is used here. The complementary solution θ_c is

$$\theta_c = Be^{-\phi\tau}.$$

The particular solution θ_p when operated on by the left side of Eq. (3.85) gives the right side.

The method of *undetermined coefficients* can be used to find the particular solution of Eq. (3.85) if the right-hand side is composed of a function or functions of time, which upon repeated differentiation with respect to time either repeat themselves or go to zero. If this condition is satisfied, the particular solution is a coefficient times the function on the right plus another coefficient times its first derivative with respect to time, plus another coefficient times its second derivative, and so on, until the differentiation either repeats the original function on the right or goes to zero. The resulting sum is inserted into the left side of Eq. (3.85), and the undetermined coefficients are determined by equating the coefficients of like functions of time on both sides of the equation.

In Eq. (3.85), the function of time on the right is $\cos \omega\tau$; hence, θ_p is constructed as

$$\theta_p = a_1\cos \omega\tau + a_2\frac{d}{d\tau}(\cos \omega\tau) + a_3\frac{d^2}{d\tau^2}(\cos \omega\tau) + \cdots.\tag{3.86}$$

However, since

$$\frac{d}{d\tau}(\cos \omega\tau) = -\omega \sin \omega\tau,$$

and the higher-order derivatives, which multiply a_3, a_4, etc., are either sines or cosines, and therefore no new functions of time are being generated, Eq. (3.86) becomes the following:

$$\theta_p = a_1\cos \omega\tau + a_2\sin \omega\tau.$$

Inserting this into Eq. (3.85) yields

$$-\omega a_1\sin \omega\tau + \omega a_2\cos \omega\tau + \phi a_1\cos \omega\tau + \phi a_2\sin \omega\tau = \phi T_1\cos \omega\tau,$$

or

$$(\phi a_2 - \omega a_1)\sin \omega\tau + (\phi a_1 + \omega a_2)\cos \omega\tau = \phi T_1\cos \omega\tau.$$

For this equation to hold for all values of time τ, the coefficient of $\sin \omega\tau$ on the left must equal the coefficient of $\sin \omega\tau$ on the right, which is zero. Similarly, the coefficient of $\cos \omega\tau$ on the left must equal the coefficient of $\cos \omega\tau$ on the right. This leads to

$$\phi a_2 - \omega a_1 = 0, \quad \phi a_1 + \omega a_2 = \phi T_1.$$

Solving these equations for a_1 and a_2 yields

$$a_1 = \frac{\phi^2 T_1}{\phi^2 + \omega^2}, \tag{3.87}$$

$$a_2 = \frac{\omega\phi T_1}{\phi^2 + \omega^2}. \tag{3.88}$$

Thus, the solution for θ is

$$\theta = Be^{-\phi\tau} + a_1\cos \omega\tau + a_2\sin \omega\tau. \tag{3.89}$$

Now, at $\tau = 0$, $\theta = T_0 - T_m$; hence, from Eq. (3.89),

$$B = (T_0 - T_m) - a_1. \tag{3.90}$$

Inserting Eqs. (3.87), (3.88), and (3.90) into Eq. (3.89) and using Eq. (3.84) with $\phi = hA_s/\rho c_p V_0$ gives as the cylinder wall temperature,

$$T = T_m + \left(T_0 - T_m - \frac{\phi^2 T_1}{\phi^2 + \omega^2}\right)e^{-\phi\tau}$$
$$+ \frac{\phi T_1}{\phi^2 + \omega^2}(\phi\cos \omega\tau + \omega\sin \omega\tau). \tag{3.91}$$

This particular unsteady temperature distribution is interesting because both types of unsteadiness discussed at the beginning of the chapter are present. While the $e^{-\phi\tau}$ term has nonzero values, the distribution is transient; at long times, as $\tau \to \infty$ and $e^{-\phi\tau} \to 0$, only the terms T_m and the last term in parentheses remain in Eq. (3.91). This result is called a regular periodic unsteady state, because the body temperature executes a cycle over and over again. Also, the term containing $e^{-\phi\tau}$ represents the body's attempt to change from the initial temperature T_0 and to follow the fluid temperature variation given by Eq. (3.80). However, as the last term in Eq. (3.91) indicates, the body cannot follow the fluid temperature variation exactly. A study of Eq. (3.91) or the physics of the situation will show that, because of the finite surface coefficient of heat transfer h, the body temperature exhibits a lag in time, and a diminished amplitude, relative to the fluid temperature variation.

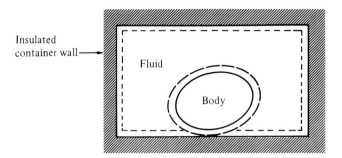

Figure 3.24 Body and fluid combination.

Body with Convection to a Finite Amount of Fluid

In Fig. 3.24, a body of initial temperature T_0 is thrust into a perfectly insulated chamber which contains a fluid at known original temperature T_{0_f} with known surface coefficient of heat transfer h between the body and the fluid. The mass of the fluid in the container is too small for its temperature to remain sensibly constant while exchanging energy with the body. Hence *both* the body temperature and the fluid temperature will vary with time. The volume, surface area, mass, and specific heat of the body and the fluid are given, respectively, by V_s, A_s, m_s, c_{ps}, and V_f, A_s, m_f, and c_{pf}. When the lumped parameter method criterion, Eq. (3.59), is satisfied for *both* the body and the fluid, the body temperature T as a function of time must be found.

Since both the body and the fluid temperatures are unknown as a function of time, the energy balance, Eq. (1.8), must be applied to both control volumes shown in Fig. 3.24, the one for the body and the other for the fluid. Letting T be the instantaneous body temperature and T_f the instantaneous fluid temperature, the application of Eq. (1.8) to the body gives

$$0 + 0 = hA_s(T - T_f) + m_s c_{ps} \frac{dT}{d\tau}. \tag{3.92}$$

Letting $\phi_s = hA_s/m_s c_{ps}$, Eq. (3.92) becomes

$$\frac{dT}{d\tau} + \phi_s T = \phi_s T_f. \tag{3.93}$$

However, at this point Eq. (3.93) cannot be solved for $T = T(\tau)$ because it contains the fluid temperature which also depends upon time in an unknown fashion. To obtain $T_f = T_f(\tau)$, the energy balance, Eq. (1.8), is applied to the fluid alone as a control volume, leading to

$$hA_s(T - T_f) + 0 = 0 + m_f c_{pf} \frac{dT_f}{d\tau}. \tag{3.94}$$

Chapter Three

Letting $\phi_f = hA_s/m_f c_{pf}$, Eq. (3.94) becomes

$$\frac{dT_f}{d\tau} + \phi_f T_f = \phi_f T. \tag{3.95}$$

Thus, Eqs. (3.93) and (3.95) comprise a set of mutually coupled, linear, first-order, ordinary differential equations with constant coefficients for the body and fluid temperature as functions of time. (Note that *mutually coupled* means that neither dependent variable T nor T_f can be solved for only from its own differential equation; the two are coupled because both dependent variables appear in both equations.)

Before solving these equations, note that if a large amount of fluid is present so that T_f barely changes at all from its initial value of T_{0_f} during the transient, then with a constant T_f, Eq. (3.93) could be solved directly for $T = T(\tau)$, as it has been previously.

Rearranging Eqs. (3.93) and (3.95) in operator form yields

$$(D + \phi_s)\, T - \phi_s\, T_f = 0, \tag{3.96}$$

$$-\phi_f\, T + (D + \phi_f)\, T_f = 0. \tag{3.97}$$

The differential equation for the sphere temperature T alone is arrived at by using Cramer's rule on Eqs. (3.96) and (3.97) as if they were algebraic equations; hence,

$$\begin{vmatrix} (D + \phi_s) & -\phi_s \\ -\phi_f & (D + \phi_f) \end{vmatrix} T = \begin{vmatrix} 0 & -\phi_s \\ 0 & (D + \phi_f) \end{vmatrix} = 0.$$

Expanding the left-hand side gives

$$(D^2 + \phi_f D + \phi_s D + \phi_f \phi_s - \phi_f \phi_s)\, T = 0, \quad \text{or}$$
$$[D^2 + (\phi_f + \phi_s)D]\, T = 0. \tag{3.98}$$

Solving Eq. (3.98) by the classical operator technique yields

$$T = a_1 + a_2\, e^{-\phi\tau}, \tag{3.99}$$

where

$$\phi = \phi_f + \phi_s.$$

Now, at $\tau = 0$, $T = T_0$; hence, from Eq. (3.99), $a_1 + a_2 = T_0$. Another condition is needed at $\tau = 0$. This can be arrived at by noting that Eq. (3.93) at $\tau = 0$ gives $dT/d\tau$ at $\tau = 0$. That is,

$$\left(\frac{dT}{d\tau}\right)_0 = \phi_s\, (T_{0_f} - T_0). \tag{3.100}$$

Unsteady-State Conduction   241

(Note that another relation between the a_1 and a_2 which is the equivalent of Eq. [3.100] can also be arrived at in a formal mathematical way by the establishment of compatibility conditions for the system of equations.) From Eq. (3.99),

$$\frac{dT}{d\tau} = -\phi \, a_2 \, e^{-\phi\tau}.$$

Evaluating at $\tau = 0$ and combining it with Eq. (3.100),

$$a_2 = -\frac{\phi_s}{\phi} (T_{0_f} - T_0), \tag{3.101}$$

and since $a_1 + a_2 = T_0$,

$$a_1 = T_0 + \frac{\phi_s}{\phi} (T_{0_f} - T_0). \tag{3.102}$$

Hence, combining Eqs. (3.99), (3.101), and (3.102), the body temperature is

$$T = T_0 + \frac{\phi_s}{\phi} (T_{0_f} - T_0) (1 - e^{-\phi\tau}), \tag{3.103}$$

where

$$\phi_s = hA_s/m_s c_{ps}$$
$$\phi = hA_s/(m_s c_{ps} + m_f c_{pf}).$$

Now that the body temperature is known and is given by Eq. (3.103), the fluid temperature T_f as a function of time can be solved for by inserting Eq. (3.103) into Eq. (3.93), giving the following result:

$$T_f = T_0 + \frac{\phi_s}{\phi} (T_{0_f} - T_0) (1 - e^{-\phi\tau}) + (T_{0_f} - T_0) e^{-\phi\tau}. \tag{3.104}$$

These solution functions, Eqs. (3.103) and (3.104), can be applied to heat-treating operations in which the solid body is quenched in a fairly small amount of cooling fluid.

In this section, the lumped parameter method has been applied to a number of different, fairly general, cases which display its simplicity, power, and importance. In particular, the importance of the first two solutions, Eqs. (3.68) and (3.76), is attested to by the large number of applications these solutions cover. The power of the method is perhaps best displayed by the last solutions, Eqs. (3.91), (3.103), and (3.104), where exact analytical solutions, accounting for *both* the time and space dependency, would be difficult exercises for a simple shape and impossible to find for the more complicated shapes. As far as simplicity is concerned, the strength of the lumped parameter method is the *relative* simplicity with which it can be applied, and solutions found, to transient problems complicated by geometric shape, generation, and time varying fluid temperatures. This simplicity is demonstrated by all of the solutions found by this method.

The preceding discussion indicates that the fastest, simplest way of handling transient conduction problems is by the lumped parameter method, as long as the criteria, such as Eq. (3.59), which allow acceptable engineering accuracy are satisfied. Thus, the first method of attack on a transient conduction problem is to see whether or not the lumped parameter method can be used.

This section on the lumped parameter method will conclude with a number of examples.

EXAMPLE 3.9

A long 2-in.-radius brass bar is initially at $1000°$F when it is suddenly put into a $100°$F fluid to cool. The surface coefficient of heat transfer between the surface of the bar and the fluid is 36 Btu/hr ft^2 $°$F, while the conductivity of the brass is 60 Btu/hr ft $°$F and its thermal diffusivity is 1.14 ft^2/hr. Find the bar's temperature after 5.32 min have passed.

Solution
First determine if the lumped parameter method can be used by employing the criterion, Eq. (3.59). The value S' will be found, as was emphasized earlier, from Eq. (3.61), where L_{max}, the maximum linear dimension from the center to the surface where convection occurs, is two inches in this problem. (Note that the "long" direction referred to in the problem statement is not to be used as L_{max}, since there is presumably no significant heat transfer in that direction.)

Thus,

$$hS'/k = 36(1/6)/60 = 0.10,$$

and the criterion, Eq. (3.59), is just satisfied; hence the lumped parameter method can be used.

The problem statement satisfies all the conditions for which Eq. (3.68) is valid. Here, $T_0 = 1000°$F, $T_f = 100°$F, h $= 36$, $A_s/V_0 = 12$, and $\alpha = k/\rho c_p$; hence, $\rho c_p = k/\alpha = \dfrac{60}{1.14} = 52.6$. Since $\tau = \dfrac{5.32}{60} = \dfrac{1}{11.3}$ hr and $\phi = \dfrac{hA_s}{\rho c_p V_0} = \dfrac{36(12)}{52.6} = 8.21$, Eq. (3.68) yields

$$\frac{T - 100}{1000 - 100} = e^{-8.21(1/11.3)}$$

Thus, $T = 535°$F.

A check on this answer already exists from Ex. 3.7, which is the same problem using exact solutions. The results of Ex. 3.7 were $T_c = 550°$F and $T_s = 527°$F after 5.32 min for the center and surface temperatures, respectively. The lumped parameter method of calculating the temperature at any time yields a result of 535°F, which is in close agreement with the "exact" values. Because of the simplicity of this method, the first step in any transient problem should be to use Eq. (3.59) to determine if the lumped parameter method will yield results of acceptable accuracy.

EXAMPLE 3.10

Consider a small water immersion heater whose total volume is 1.6×10^{-5} m³, and whose total outside surface area is 0.0032 m². The heater is made of a metal with $\rho = 8940$ kg/m³, $k = 260$ W/m °C, and $c_p = 419$ J/kg °C. The heater dissipates electrical energy at the rate of 40 W. The heater was designed to be used in water, where the high surface coefficients would keep the heater temperature within safe limits. If the heater is initially at 21°C, the temperature of the room air, and is inadvertently plugged in while in air, compute the time it takes for it to reach the melting temperature of 540°C. The surface coefficient of heat transfer between the room air and the heater is assumed constant at 11.3 W/m² °C.

Solution
Here the particular geometric shape of the heater was not specified. Hence, essentially by default, the lumped parameter method must be used, since all other methods require the details of the shape to be known. Nevertheless, the first step is still to check and see whether or not the criterion, Eq. (3.59), is satisfied. Based on the information given in the problem statement, there is no choice but to use Eq. (3.62) to compute S'. Thus,

$$S' = V_0/A_s = 1.6 \times 10^{-5}/0.0032 = 0.005 \text{ m},$$

$$hS'/k = 11.3(0.005)/260 = 0.00022.$$

This value is so much smaller than the criterion cutoff of 0.10 that the lumped parameter method should give results of very good accuracy even if the shape is fairly complex. The solution for the conditions of this problem has already been found in Eq. (3.76). Here,

$$T_0 = T_f = 21°C, \quad Q_0 = 40 \text{ W},$$

$$\phi = hA_s/\rho c_p V_0 = 11.3(0.0032)/8940(419)(1.6 \times 10^{-5})$$
$$= 0.000603,$$

and the temperature at the unknown time of interest is the melting value, $T = 540°C$. Substitution into Eq. (3.76) gives,

$$540 - 21 = \frac{40}{11.3(0.0032)} (1 - e^{-0.000603\tau}),$$

$$\tau = 1050 \text{ s} = 0.292 \text{ hr} = 17.52 \text{ min}.$$

Hence, if the heater is unplugged or is immersed in water in less than 17.5 minutes, melting will not occur.

EXAMPLE 3.11

An electrical coil dissipates electrical energy at a rate which is a function of temperature; that is,

$$\dot{Q} = \dot{Q}_0 (1 + BT), \tag{3.105}$$

where $\dot{Q}_0$ and B are known constants, and T is the local coil temperature. Initially, the coil of surface area A_s and volume V_0 is at the temperature T_f of the surrounding fluid, when it is suddenly turned on and a known surface coefficient of heat transfer h exists between the coil surface and the fluid. (a) Assuming that the conditions are such that the lumped parameter method can be used, derive an analytical expression for the coil temperature as a function of time τ. (b) Investigate any conditions under which the coil could not attain a steady state after being turned on and left on.

Solution
Take the entire coil as the control volume and apply the law of conservation of energy, Eq. (1.8), to this control volume, where $R_{\text{gen}} = \dot{Q}_0(1 + BT)$ is given by Eq. (3.105). Since all of the coil material is assumed to be near the average temperature T at any instant of time, the storage term is

$$R_{\text{stor}} = \rho c_p V_0 \frac{dT}{d\tau}.$$

Energy leaves the control volume across the outside surface by convection to the surrounding fluid at the rate prescribed by Newton's cooling law; that is,

$$R_{\text{out}} = hA_s(T - T_f).$$

Hence, Eq. (1.8) becomes

$$0 + \dot{Q}_0(1 + BT) = hA_s(T - T_f) + \rho c_p V_0 \frac{dT}{d\tau}.$$

Dividing every term by $\rho c_p V_0$, defining $\phi = hA_s/\rho c_p V_0$, and rearranging gives

$$\frac{dT}{d\tau} + \left(\phi - \frac{B\dot{Q}_0}{\rho c_p V_0} \right) T = \frac{\dot{Q}_0}{\rho c_p V_0} + \phi T_f. \tag{3.106}$$

Defining $\psi = \phi - B\dot{Q}_0/\rho c_p V_0$ and $E = \dot{Q}_0/\rho c_p V_0 + \phi T_f$, Eq. (3.106) becomes

$$\frac{dT}{d\tau} + \psi T = E. \tag{3.107}$$

This is a first-order, linear, ordinary, nonhomogeneous differential equation with constant coefficients and is solved by the method of summing the complementary and particular solutions. The particular solution, by the method of undetermined coefficients, is taken to be $T_p = C_1$, a constant to be determined. Inserting this into Eq. (3.107) yields

$$0 + \psi C_1 = E, \quad \text{or} \quad C_1 = E/\psi.$$

Thus,

$$T_p = E/\psi. \tag{3.108}$$

The complementary solution T_c of Eq. (3.107) satisfies the complementary equation

$$\frac{dT_c}{d\tau} + \psi T_c = 0. \tag{3.109}$$

By the operator method, the solution to Eq. (3.109) is

$$T_c = A_1 e^{-\psi\tau}. \tag{3.110}$$

Hence the solution to Eq. (3.107) is the sum of the solutions, Eq. (3.108) and (3.110); that is,

$$T = A_1 e^{-\psi\tau} + E/\psi. \tag{3.111}$$

The single integration constant A_1 is determined by application of the initial condition, i.e., at $\tau = 0$, $T = T_f$. Hence, Eq. (3.111) becomes

$$T_f = A_1 + E/\psi, \quad \text{or} \quad A_1 = T_f - E/\psi.$$

a. Thus, the final solution for the instantaneous coil temperature as a function of time τ is

$$T = (T_f - E/\psi) e^{-\psi\tau} + E/\psi, \tag{3.112}$$

where

$$E = \frac{\dot{Q}_0}{\rho c_p V_0} + \phi T_f, \quad \phi = \frac{hA_s}{\rho c_p V_0}, \quad \psi = \phi - \frac{B \dot{Q}_0}{\rho c_p V_0}.$$

b. If a steady-state temperature could not be reached, it would be because the function of time in the temperature distribution function never vanishes or becomes constant. The temperature distribution is Eq. (3.112). If $\psi > 0$, then as $\tau \to \infty$, $e^{-\psi\tau} \to 0$ and some steady-state temperature is reached. If, however, $\psi < 0$, then $e^{-\psi\tau}$ always increases with time and no steady state is possible, and the coil will melt. Since, from part a,

$$\psi = \phi - \frac{B \dot{Q}_0}{\rho c_p V_0} = \frac{hA_s}{\rho c_p V_0} - \frac{B \dot{Q}_0}{\rho c_p V_0},$$

$\psi < 0$ corresponds to

$$\frac{hA_s}{\rho c_p V_0} - \frac{B \dot{Q}_0}{\rho c_p V_0} < 0, \quad \text{or} \quad hA_s < B \dot{Q}_0. \tag{3.113}$$

Thus, Eq. (3.113) is the condition under which a steady-state temperature would not be attained. Physically, this would occur as a result of the temperature-dependent generation rate always being greater than the rate at which energy can be convected away from the coil surface at any temperature T; hence, T tends to become higher in an attempt to balance off the generation rate, but this causes the generation rate $\dot{Q}_0(1 + BT)$ to go even higher, thus making a balance of the two, and a steady state, impossible.

3.4 FINITE DIFFERENCE METHODS IN UNSTEADY-STATE CONDUCTION

Chapter 2 showed that very often the shape of the conduction region, the boundary conditions, the energy generation rate per unit volume, variable thermal conductivity, or any combination of these causes the determining of the steady-state temperature distribution to become mathematically intractable. In these cases, finite difference techniques are generally employed. These difficulties also occur in transient situations along with an additional complicating factor; that is, the initial distribution of temperature throughout the conduction region at time $\tau = 0$, which is called the initial condition. Hence, in transient problems where the lumped parameter method cannot be used and where one or more of the factors just mentioned precludes the possibility of an exact analytical solution, the powerful finite difference method is very often used.

The transient finite difference equation at any node of interest is arrived at in a manner similar to that described in Chapter 2 for the steady-state finite difference equation; that is, by making an energy balance on the volume of material associated with each node, leading to a set of algebraic equations for the n nodal temperatures that are to be solved at a finite number of times. At this point, the similarity between the steady-state finite difference equations and the transient finite difference equations often ends because of differences in the way the equation set is solved and the appearance of a phenomenon called *stability* of the equation set.

The finite difference method for transient conduction problems can lead to large numbers of algebraic equations to be solved. Once again, as was also the case for steady-state finite difference methods, the digital computer is an invaluable ally. Thus, the remarks made at the beginning of Sec. 2.8 concerning the computer and the ready availability of computer programs for solving systems of equations apply in this section as well. For one type of transient finite difference equation set, termed the *implicit* type, the equation system has the same overall form as did the steady-state difference equations of Chapter 2; therefore, the same standard library computer programs can be used. The other type of transient finite difference equation is the *explicit* type. In most respects, this is simpler to deal with than the implicit type; the equations in the system are uncoupled and can be solved one at a time. For this type of finite difference equation set, very simple and short computer programs are easily written to solve the equations.

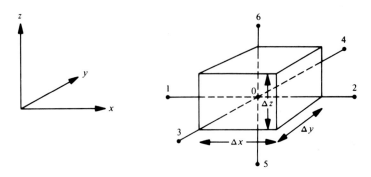

Figure 3.25 Coordinate system and a general interior node in a three-dimensional conduction region.

In a fashion analogous to the steady-state finite difference equations, the transient finite difference equations are *algebraic* equations which approximate the governing partial differential equation, such as Eq. (2.6), at a finite number of space points, the nodal or lattice or mesh points, at a finite number of different times.

3.4a Derivation of the Transient Finite Difference Equations

The three basic approaches for deriving the transient finite difference equations are the same as for the steady-state difference equations. These approaches have already been discussed, in some detail, in Sec. 2.8. Our attention will focus only upon the approach that employs the energy balance, Eq. (1.8). One new decision we must make during the derivation of the transient finite difference equations concerns the *time* τ at which the nodal temperatures, in the conduction heat transfer terms, should be evaluated. When these temperatures are evaluated at the *present time* τ, the finite difference equations will be of the *explicit* type; if these nodal temperatures are evaluated at the *future time*, $\tau + \Delta\tau$, the resulting equations will be of the *implicit* type. Evaluation at times *between* τ and $\tau + \Delta\tau$ also leads to implicit equations. Because of their simplicity of solution, most of the derivations and solutions presented here will be for explicit difference equations.

General Interior Node

The first step is to derive the governing transient finite difference equation at an interior node of interest, such as node 0 in Fig. 3.25, which is also Fig. 2.39. Node 0 is surrounded by six other nodes in the three coordinate directions and, for simplicity, constant properties are assumed. The analysis can be easily modified for the variable property case. In the transient finite difference equations, when the temperature appears without a superscript, such as T_0, it is interpreted as the temperature at the node 0 at the *present time* τ. If the temperature symbol appears with a superscript, such as $T_0^{\Delta\tau}$, it is interpreted as the temperature at node 0 at the *future time* $\tau + \Delta\tau$, that is, the present time plus one time increment. Since the temperatures throughout the conduction region at an initial time, usually called zero time, are known (i.e., the initial condition is known), the problem is to compute the temperatures at time $0 + \Delta\tau$, then $0 + 2\Delta\tau$, and so on, until either steady state or the time of interest is reached.

Hence, apply

$$R_{\text{in}} + R_{\text{gen}} = R_{\text{out}} + R_{\text{stor}} \qquad\qquad [1.8]$$

to the control volume shown surrounding node zero in Fig. 3.25. The R_{in} and R_{gen} terms will have the same form as for the steady-state finite difference equation discussed in Chapter 2 *if the temperatures of the nodes at the present time τ are used.* Hence,

$$
\begin{aligned}
R_{\text{in}} &= k\Delta y \Delta z \frac{(T_1 - T_0)}{\Delta x} + k\Delta y \Delta z \frac{(T_2 - T_0)}{\Delta x} \\
&\quad + k\Delta z \Delta x \frac{(T_3 - T_0)}{\Delta y} + k\Delta z \Delta x \frac{(T_4 - T_0)}{\Delta y} \\
&\quad + k\Delta x \Delta y \frac{(T_5 - T_0)}{\Delta z} + k\Delta x \Delta y \frac{(T_6 - T_0)}{\Delta z},
\end{aligned}
\qquad (3.114)
$$

$$R_{\text{gen}} = q_0''' \,\Delta x \Delta y \Delta z. \qquad\qquad (3.115)$$

The time rate of change of the stored internal energy R_{stor} can be written for the finite time increment $\Delta\tau$ as the internal energy of all the material in the control volume at time $\tau + \Delta\tau$ minus the internal energy of all the material in the control volume at time τ divided by the time increment $\Delta\tau$; that is,

$$
\begin{aligned}
R_{\text{stor}} &= \frac{\rho c_p \Delta x \Delta y \Delta z\, T_0^{\Delta\tau} - \rho c_p \Delta x \Delta y \Delta z\, T_0}{\Delta\tau} \\
&= \rho c_p \Delta x \Delta y \Delta z \frac{(T_0^{\Delta\tau} - T_0)}{\Delta\tau}.
\end{aligned}
\qquad (3.116)
$$

Note that as Δx, Δy, Δz, and $\Delta\tau$ approach zero, R_{stor} in Eq. (3.116) goes to the correct storage term for the partial differential equation $\rho c_p\, dxdydz\, \partial T/\partial\tau$. Hence, Eq. (3.116) can be considered to be approximating the storage term in the general partial differential equation, Eq. (2.5). Combining Eqs. (3.114), (3.115), (3.116), and (1.8), dividing every term by $k\Delta x \Delta y \Delta z$, and recognizing that $\alpha = k/\rho c_p$ yields the general transient finite difference equation

$$
\begin{aligned}
\frac{T_1 + T_2 - 2T_0}{(\Delta x)^2} &+ \frac{T_3 + T_4 - 2T_0}{(\Delta y)^2} + \frac{(T_5 + T_6 - 2T_0)}{(\Delta z)^2} + \frac{q_0'''}{k} \\
&= \frac{1}{\alpha} \frac{(T_0^{\Delta\tau} - T_0)}{\Delta\tau}.
\end{aligned}
\qquad (3.117)
$$

This is valid for any interior node in a three-dimensional conduction region described by a rectangular cartesian coordinate system. The reduction of Eq. (3.117) to a form suitable for a one- or two-dimensional problem is accomplished by simply dropping the terms on the left that pertain to directions in which there is no temperature variation. An equation like Eq. (3.117) is written for every interior node in the conduction region by a suitable change in the subscripts, as was discussed for its steady-state counterpart in Chapter 2.

Often, either for the sake of simplicity or because temperature derivatives of about the same severity are expected in all space directions, an equispaced lattice, $\Delta y = \Delta z = \Delta x$, is used. When this is done, and after defining

$$M = (\Delta x)^2/\alpha\Delta\tau, \tag{3.118}$$

and rearranging, Eq. (3.117) takes the following form at a general interior node when all nodal spacings are equal:

$$T_0^{\Delta\tau} = \frac{1}{M}(T_1 + T_2 + T_3 + T_4 + T_5 + T_6) + \frac{q_0'''(\Delta x)^2}{kM}$$
$$+ \left(1 - \frac{6}{M}\right)T_0. \tag{3.119}$$

For an interior node of a two-dimensional equispaced lattice,

$$T_0^{\Delta\tau} = \frac{1}{M}(T_1 + T_2 + T_3 + T_4) + \frac{q_0'''(\Delta x)^2}{kM}$$
$$+ \left(1 - \frac{4}{M}\right)T_0, \tag{3.120}$$

and for an interior node of a one-dimensional lattice,

$$T_0^{\Delta\tau} = \frac{1}{M}(T_1 + T_2) + \frac{q_0'''(\Delta x)^2}{kM} + \left(1 - \frac{2}{M}\right)T_0. \tag{3.121}$$

By inspection of Eqs. (3.117) through (3.121), observe that in each equation only *one* nodal temperature at the future time appears, $T_0^{\Delta\tau}$, the temperature at the node under consideration at the new or future time. This structure of the equation results from the decision to evaluate all nodal temperatures in the conduction terms of Eq. (3.114) at the *present* time. This type of transient finite difference equation in which the nodal temperature at the future time can be isolated on a side by itself, with the only other temperatures appearing being those evaluated at the present time, is an *explicit finite difference equation*. As will be seen shortly, when the equations are of the explicit type, each equation of the set can be solved, one at a time, independently of the others while advancing the solution in time by one time increment.

If the nodal temperatures in the conduction terms of Eq. (3.114) are evaluated at the future time in the derivation, $\tau + \Delta\tau$, then it can be shown that the transient finite difference equations are given by the following forms:

For the general interior node with unequal nodal spacings,

$$\frac{T_1^{\Delta\tau} + T_2^{\Delta\tau} - 2T_0^{\Delta\tau}}{(\Delta x)^2} + \frac{T_3^{\Delta\tau} + T_4^{\Delta\tau} - 2T_0^{\Delta\tau}}{(\Delta y)^2}$$
$$+ \frac{T_5^{\Delta\tau} + T_6^{\Delta\tau} - 2T_0^{\Delta\tau}}{\Delta z^2} + \frac{q_0'''}{k} = \frac{1}{\alpha} \frac{(T_0^{\Delta\tau} - T_0)}{\Delta\tau}. \tag{3.122}$$

For an interior node of a three-dimensional equispaced lattice,

$$T_1^{\Delta\tau} + T_2^{\Delta\tau} + T_3^{\Delta\tau} + T_4^{\Delta\tau} + T_5^{\Delta\tau} + T_6^{\Delta\tau} + \frac{q_0'''(\Delta x)^2}{k}$$
$$+ MT_0 = (6 + M)T_0^{\Delta\tau}. \tag{3.123}$$

For an interior node of a two-dimensional equispaced lattice,

$$T_1^{\Delta\tau} + T_2^{\Delta\tau} + T_3^{\Delta\tau} + T_4^{\Delta\tau} + \frac{q_0'''(\Delta x)^2}{k} + MT_0$$
$$= (4 + M)T_0^{\Delta\tau}. \tag{3.124}$$

For an interior node of a one-dimensional lattice,

$$T_1^{\Delta\tau} + T_2^{\Delta\tau} + \frac{q_0'''(\Delta x)^2}{k} + MT_0 = (2 + M)T_0^{\Delta\tau}. \tag{3.125}$$

Looking at Eqs. (3.122) through (3.125), one notes that, unlike the explicit-type equations, such as Eq. (3.119), more than one nodal temperature at the future time appears in each equation. Thus $T_0^{\Delta\tau}$ cannot be solved directly in terms of nodal temperatures at the present time. This type of equation is an *implicit finite difference equation,* and all such equations must be solved simultaneously, not one at a time, in order to advance the solution in time.

EXAMPLE 3.12

Derive the transient explicit finite difference equation for a general interior node in a sphere when there is no generation and when all nodes are equally spaced a distance Δr apart, as shown in Fig. 3.26.

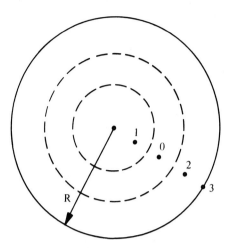

Figure 3.26 The interior node in the sphere of Ex. 3.12.

Solution

Because of the radial symmetry, Eq. (1.8) is applied to the control volume of material in the spherical shell bounded by the dashed surfaces shown in Fig. 3.26. This volume is the material associated with node 0. Here,

$$R_{\text{stor}} = \rho c_p V_0 \frac{T_0^{\Delta\tau} - T_0}{\Delta\tau},$$

where the volume of material associated with node 0 is

$$V_0 = \frac{4\pi}{3}\left[\left(r_0 + \frac{\Delta r}{2}\right)^3 - \left(r_0 - \frac{\Delta r}{2}\right)^3\right].$$

Then R_{stor} becomes

$$R_{\text{stor}} = \frac{4\pi}{3}\rho c_p \left[\left(r_0 + \frac{\Delta r}{2}\right)^3 - \left(r_0 - \frac{\Delta r}{2}\right)^3\right]\frac{T_0^{\Delta\tau} - T_0}{\Delta\tau}. \tag{3.126}$$

Energy is conducted into the control volume from the material associated with nodes 1 and 2 at rates given by the expression derived in Chapter 2 for one-dimensional conduction across a spherical shell. Since an explicit-type equation was asked for, all temperatures in the conduction terms are evaluated at the present time. Thus the conduction heat transfer rates are given by

$$q_{10} = \frac{4\pi k(T_1 - T_0)}{1/r_1 - 1/r_0}, \tag{3.127}$$

$$q_{20} = \frac{4\pi k(T_2 - T_0)}{1/r_0 - 1/r_2}. \tag{3.128}$$

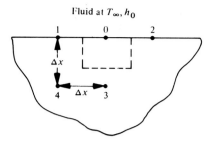

Figure 3.27 The surface node with convection.

Inserting Eqs. (3.127), (3.128), and (3.126) into Eq. (1.8), dividing by $4\pi k$, and noting that $\alpha = k/\rho c_p$, the transient finite difference equation for this type of interior node is

$$\frac{T_1 - T_0}{1/r_1 - 1/r_0} + \frac{T_2 - T_0}{1/r_0 - 1/r_2}$$
$$= \left[\frac{(r_0 + \Delta r/2)^3 - (r_0 - \Delta r/2)^3}{3\alpha\Delta\tau}\right](T_0^{\Delta\tau} - T_0). \qquad (3.129)$$

Noninterior Nodes

For noninterior nodes, which usually have a different volume of material associated with them and might be located on exterior surfaces where they may interact with a fluid by convection and with other solids by radiation, the finite difference equation can be found by the basic rule of making an energy balance on the volume of material associated with the node.

Surface Node with Convection   One of the most important and frequently occurring noninterior nodes is one such as node 0 in Fig. 3.27. This node is situated on a surface of a two-dimensional conduction region with the surface material exchanging energy by convection with the surrounding fluid. Let us develop an explicit finite difference equation for this surface node when an equispaced lattice is being used with all nodes spaced Δx apart in both directions.

The volume $\Delta x^2/2$ of material associated with node 0 is shown by dashed lines in the figure. To give an explicit equation, all temperatures in the conduction terms of the energy balance will be used at the present time. So, applying Eq. (1.8) to this control volume gives

$$k\Delta x\,\frac{(T_3 - T_0)}{\Delta x} + k\frac{\Delta x}{2}\frac{(T_1 - T_0)}{\Delta x} + k\frac{\Delta x}{2}\frac{(T_2 - T_0)}{\Delta x}$$
$$+ h_0\Delta x(T_\infty - T_0) + q_0'''\frac{(\Delta x)^2}{2} = \rho c_p\frac{(\Delta x)^2}{2}\frac{(T_0^{\Delta\tau} - T_0)}{\Delta\tau}. \qquad (3.130)$$

Defining

$$N = \frac{h_0\Delta x}{k}, \qquad (3.131)$$

Unsteady-State Conduction   253

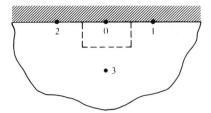

Figure 3.28 Node on adiabatic surface.

and rearranging Eq. (3.130) using Eqs. (3.118) and (3.131) gives the transient finite difference equation

$$T_0^{\Delta\tau} = \frac{1}{M}(2T_3 + T_1 + T_2 + 2NT_\infty) + \frac{q_0'''(\Delta x)^2}{Mk}$$
$$+ \left[1 - \frac{(4 + 2N)}{M}\right] T_0 \qquad\qquad \textbf{(3.132)}$$

for the surface node shown in Fig. 3.27.

Node on a Perfectly Insulated Boundary　To derive the transient explicit finite difference equation for a node on a perfectly insulated boundary or, equivalently, on a surface of thermal and geometric symmetry, we apply the energy balance, Eq. (1.8), to the volume of material Δx by $\Delta x/2$ by 1 shown in Fig. 3.28. Equal nodal spacings are being used in both directions.
This gives

$$k\frac{(\Delta x)}{2}\frac{(T_2 - T_0)}{\Delta x} + k\frac{(\Delta x)}{2}\frac{(T_1 - T_0)}{\Delta x} + k\,\Delta x\,\frac{(T_3 - T_0)}{\Delta x}$$
$$+ q_0'''\frac{(\Delta x)^2}{2} = \rho c_p \frac{(\Delta x)^2}{2}\frac{(T_0^{\Delta\tau} - T_0)}{\Delta\tau}.$$

Rearranging and using the definition of M, Eq. (3.118), gives the following explicit difference equation for the node on an adiabatic surface of an equispaced lattice:

$$T_0^{\Delta\tau} = \frac{1}{M}\left[T_1 + T_2 + 2T_3 + q_0'''\frac{(\Delta x)^2}{k}\right] + \left(1 - \frac{4}{M}\right)T_0. \qquad \textbf{(3.133)}$$

Node at an Inside Corner with Convection　Figure 3.29 shows a node 0 on an inside corner interacting by convection with a fluid at known temperature T_f and with known surface coefficient of heat transfer h_f. The explicit finite difference equation will be derived for this node both for the case of unequal nodal spacings, Δx and Δy, and for equal nodal spacings, $\Delta y = \Delta x$.

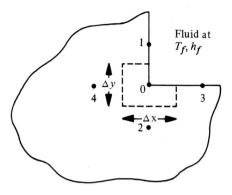

Figure 3.29 The inside corner node.

The volume of material associated with node 0 is seen to be $\frac{3}{4}\Delta x\Delta y$, and Eq. (1.8) applied to this as the control volume yields

$$
k\,\Delta y\,\frac{(T_4 - T_0)}{\Delta x} + k\,\frac{\Delta y}{2}\,\frac{(T_3 - T_0)}{\Delta x} + k\Delta x\,\frac{(T_2 - T_0)}{\Delta y}
$$
$$
+\,k\,\frac{\Delta x}{2}\,\frac{(T_1 - T_0)}{\Delta y} + h_f\left(\frac{\Delta y}{2} + \frac{\Delta x}{2}\right)(T_f - T_0)
$$
$$
+\,\frac{3}{4}\,q_0'''\Delta x\Delta y = \rho c_p\,\frac{3}{4}\Delta x\Delta y\,\frac{(T_0^{\Delta\tau} - T_0)}{\Delta\tau}\ .
$$

Dividing by $k\Delta x\Delta y$ leads to

$$
\frac{(T_4 - T_0)}{\Delta x^2} + \frac{(T_3 - T_0)}{2\Delta x^2} + \frac{(T_2 - T_0)}{\Delta y^2} + \frac{(T_1 - T_0)}{2\Delta y^2}
$$
$$
+\,\frac{h_f}{2k}\,\frac{(\Delta y + \Delta x)}{\Delta x\Delta y}\,(T_f - T_0)
$$
$$
+\,\frac{3}{4}\,\frac{q_0'''}{k} = \frac{3}{4}\,\frac{\rho c_p}{k\Delta\tau}\,(T_0^{\Delta\tau} - T_0). \tag{3.134}
$$

For the case in which the nodal spacings are equal, set $\Delta y = \Delta x$ in Eq. (3.134) and rearrange, using the definition of M given in Eq. (3.118), to yield

$$
T_0^{\Delta\tau} = \frac{4}{3M}\left[T_4 + T_2 + \frac{T_3}{2} + \frac{T_1}{2} + N_f T_f\right] + \frac{q_0'''\Delta x^2}{kM}
$$
$$
+\left[1 - \frac{4}{3}\left(\frac{3 + N_f}{M}\right)\right]T_0. \tag{3.135}
$$

$$
N_f = h_f\,\Delta x/k \tag{3.136}
$$

Unsteady-State Conduction

255

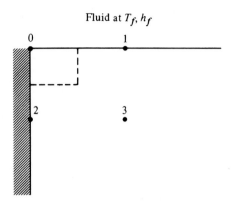

Figure 3.30 The corner node of Ex. 3.13.

When transient finite difference equations are needed for surface nodes with conditions different than the ones just treated—conditions such as a three-dimensional unequal nodal spacing, variable thermal conductivity of the solid, or implicit rather than explicit equations—they can be derived using the same basic procedure of an energy balance applied to the small but finite volume of material associated with the node.

EXAMPLE 3.13

For equal nodal spacings Δx in both directions, derive the *implicit* transient finite difference equation for node 0 in Fig. 3.30.

Solution

First the material associated with the node of interest, node 0, is determined. Since all nodes are spaced a distance Δx apart in both directions, the halfway point between adjacent nodes determines the extent of material associated with the nodes. Thus the two dashed lines shown in Fig. 3.30 act as the boundary between material associated with node 0 and node 2 (the horizontal dashed line) and material associated with node 0 and node 1 (the vertical dashed line). Hence the material associated with this corner node 0 is $\Delta x/2$ by $\Delta x/2$, or $\Delta x^2/4$. Next the energy balance, Eq. (1.8), is applied to this control volume. We find that there is conduction in, from nodes 2 and 1 to node 0, and convection in, from the fluid to the portion of the upper surface associated with node 0. Since an implicit equation was asked for, the nodal temperatures in the conduction and the convection terms are evaluated at the future time $\tau + \Delta\tau$. Hence, Eq. (1.8) becomes

$$\frac{k\Delta x}{2}\frac{(T_2^{\Delta\tau} - T_0^{\Delta\tau})}{\Delta x} + \frac{k\Delta x}{2}\frac{(T_1^{\Delta\tau} - T_0^{\Delta\tau})}{\Delta x} + h_f\frac{\Delta x}{2}(T_f - T_0^{\Delta\tau})$$
$$+ q_0'''\frac{\Delta x^2}{4} = \rho c_p\frac{\Delta x^2}{4}\frac{(T_0^{\Delta\tau} - T_0)}{\Delta\tau}.$$

Defining N_f as in Eq. (3.136), and with M given by Eq. (3.118), this equation can be rearranged to yield the following:

$$2(T_2^{\Delta\tau} + T_1^{\Delta\tau} + N_f T_f) + \frac{q_0'''(\Delta x)^2}{k} + MT_0 = (4 + M + 2N_f)T_0^{\Delta\tau}.$$

3.4b Solution Methods for the Transient Finite Difference Equations and Refinement of Increment Sizes

In a transient conduction problem being solved by finite difference methods, the total number of nodes n, some of them interior nodes and some of them noninterior, gives rise to n *algebraic* finite difference equations to be solved. The appropriate solution method for these n equations depends upon whether the equations are explicit or implicit.

Explicit Finite Difference Equations

Suppose all of the finite difference equations are explicit, such as Eq. (3.120) for all interior nodes, and equations such as Eqs. (3.132), (3.133), (3.135), or others yet to be derived for all noninterior nodes. Then, since all of the nodal temperatures are known at time $\tau = 0$ because this is the initial, *known*, condition, these known nodal temperatures at $\tau = 0$ are viewed as being the *present* temperatures. They are inserted into the right-hand sides of Eqs. (3.120), (3.132), (3.133), and so on, to calculate the temperature at each nodal point after one time increment $\Delta\tau$; that is, to calculate the *future* temperature at every node. Note that to do this for these explicit finite difference equations, we do not solve a set of n *simultaneous* equations. Instead, we solve each equation individually for the temperature at each node.

For example, suppose a problem exists in which the generation rate per unit volume is zero, and at time zero, the temperatures at the various nodal points are $T_1 = 500°C$, $T_0 = T_2 = T_3 = T_4 = T_5 = T_6 = T_7 = T_8 = 100°C$, and M is chosen to be 10 for this two-dimensional conduction region. Calculate the temperature at node 0, an interior node, after one time increment $1\Delta\tau$ has passed. The relevant transient finite difference equation is Eq. (3.120) with $q_0''' = 0$. Hence,

$$T_0^{\Delta\tau} = \frac{1}{M}(T_1 + T_2 + T_3 + T_4) + \left(1 - \frac{4}{M}\right)T_0.$$

Substituting the given values,

$$T_0^{\Delta\tau} = \frac{1}{10}(500 + 100 + 100 + 100) + (1 - \tfrac{4}{10})\,100 = 140°C.$$

Hence, each of the equations in the set is solved one at a time for the nodal temperature after one $\Delta\tau$ has passed. After all the nodal temperatures at $1\Delta\tau$ have been computed, these temperatures are used in the right-hand sides of, for example, Eqs. (3.120), (3.132), and (3.133) to determine the temperature on the left side at each node after two time increments $2\Delta\tau$ have passed. This procedure is repeated until the steady state or the time of interest is reached.

As is evident from the discussion, the explicit type of transient finite difference equation is easy to solve either by hand (when a fairly coarse nodal spacing and a fairly large time increment, $\Delta\tau$, will suffice for one's purposes) or by use of the digital computer. The computer is usually used.

Implicit Finite Difference Equations

When all of the finite difference equations are implicit, such as Eq. (3.124) for interior nodes and equations like the answer to Ex. 3.13 for noninterior nodes, *more* than one unknown (more than one nodal temperature at the future time $\tau + \Delta\tau$) appears in each equation. Hence, one has a set of mutually coupled algebraic equations that must be solved *simultaneously*. Thus, one must first insert all the known initial temperatures at $\tau = 0$ as the temperatures at the present time, then solve the equations *simultaneously* to get all the nodal temperatures at the future time, which is one time increment away from the initial time $\tau = 0$. Now these new nodal temperatures at $\tau = 1\Delta\tau$ are viewed as the present temperatures and are inserted into the equations, which are then solved *simultaneously again* to give the nodal temperatures at $\tau = 2\Delta\tau$. This process of solving the equation set simultaneously is continued until the steady state or the time of interest is reached. The standard library computer programs or subroutines for solving equation sets, discussed in Sec. 2.8, can be used here as well.

Since the solution of the implicit equations obviously entails much more work than does the solution of the explicit equations, one may ask why implicit equations would ever be used in an analysis. The answer lies with the favorable *stability* characteristics the implicit equations may possess, which we will discuss shortly.

Refinement of Increment Sizes

The solution of the transient finite difference equations, like that of the steady-state difference equations previously discussed, represents an *approximate* solution to the governing partial differential equation, unless one has continually refined the increment sizes in space and time so that the solution no longer depends upon the increment sizes Δx, Δy, Δz, and $\Delta\tau$ that were used. Hence, as was also the case for the steady-state problems, the transient finite difference solution can only be considered the exact solution when the refinement of increment sizes has been carried out.

3.4c Stability of the Finite Difference Equations

From the discussion thus far, it appears that complicated transient problems are solved by the finite difference method by selecting a lattice spacing Δx, which determines the number and location of all the nodes and yields a set of equations of the type of Eq. (3.120) (for two-dimensional conduction regions) for the interior nodes and other equations of the form of Eqs. (3.132) and (3.133) for the noninterior nodes if explicit finite difference equations are employed. Similarly, equations such as Eq. (3.125) arise when an implicit scheme is used.

Once a finite time interval $\Delta\tau$ is chosen, M can be calculated from Eq. (3.118), and the equations can then be solved for each nodal temperature at the first time level. These temperatures are then used to arrive at those for the second time level, and so on. However, M, which by Eq. (3.118) is

$$M = \frac{(\Delta x)^2}{\alpha\Delta\tau},$$ [3.118]

cannot have its value assigned in a completely arbitrary manner. The problem that can arise by using an inadmissible value of M is shown by the following example. Suppose a two-dimensional conduction region exists with no generation. At a certain time τ, the temperatures at an interior node 0 and the four surrounding nodes are $T_0 = 200°F$, $T_1 = 180°F$, $T_2 = 150°F$, $T_3 = 130°F$, $T_4 = 120°F$, and the temperature at node 0 at time $\tau + \Delta\tau$ must be determined using a value of $M = 2$. The transient finite difference equation is Eq. (3.120) with the generation rate per unit volume set equal to zero. That is

$$T_0^{\Delta\tau} = \frac{1}{M}(T_1 + T_2 + T_3 + T_4) + \left(1 - \frac{4}{M}\right)T_0.$$

Substituting the known temperatures at the present time τ and the value of M,

$$T_0 = \tfrac{1}{2}(180 + 150 + 130 + 120) + (1 - \tfrac{4}{2})200 = 90°F.$$

Thus, the temperature at node 0 is $90°F$. But this is not physically possible, since it is lower than any of the temperatures of the surrounding nodes. Equation (3.120) is the result of an energy balance on the material surrounding node 0, but for node 0 to drop in temperature from $200°F$, energy must be conducted away to lower-temperature nodes at a greater rate than it is conducted in from higher-temperature nodes. Initially, this is not a problem, since all four of the surrounding nodes are at a lower temperature than T_0. However, after T_0 drops past $120°F$, something goes wrong, since node 0 is presumably losing energy, and to accomplish this, the energy must be conducted to *higher-temperature* surrounding nodes. This is a clear violation of the Clausius statement of the second law of thermodynamics.

If the calculation of the transient temperature distribution, with a value of M which violated the second law, was nonetheless carried out and the results at the various nodes plotted against the different times 0, $1\Delta\tau$, $2\Delta\tau$, etc., the nodal temperature would oscillate with increasing amplitude as time went on. If an exact solution to the same problem is available, it will not show this oscillatory nature. When the solution to a set of transient finite difference equations exhibits this oscillatory behavior, it is said to be *unstable*. If it does not exhibit this oscillatory behavior, it is termed *stable*. Clearly, stable transient finite difference equations are required. The condition that they be stable implies a restriction on the possible values of M. This restriction is referred to as the *stability criterion*.

3.4d Stability Criterion

Mathematically, a set of finite difference equations is unstable if a numerical error, due perhaps to round-off of a number or to an arithmetic error, introduced at some point in time, amplifies as time goes on even though no further errors of any type are made.[2] If the numerical error introduced remains the same or (preferably) decays as time goes on, the finite difference equation set is said to be *mathematically stable*. One way to develop the condition under which this is true—the stability criterion—is to use the matrix method of stability analysis.

All the known temperatures at the present time are viewed as a row matrix which, when operated on by a square coefficient matrix and added to a constant row matrix, is mapped into a row matrix whose elements are the temperatures of the nodes at the new time. In this way, the behavior of the error vector (matrix) as time goes on can be studied, and the condition for stability turns out to be that *the modulus of the largest eigenvalue of the coefficient matrix should not exceed unity* (see Smith [1]).

A less powerful method of stability analysis, but one which is easier to use and to understand physically, is the one used by Arpaci [5]. The essence of Arpaci's method, in words, can be stated as follows:

The temperature at any nodal point cannot be allowed to jump beyond its ultimate steady-state value as the finite difference solution is advanced by one time increment, $\Delta\tau$.

One way this statement can be used to develop the quantitative form of the stability criterion is to write down the finite difference equation for node 0 in the standard form, that is, with $T_0^{\Delta\tau}$ on a side by itself with a coefficient of $+1$. Next, by setting both $T_0^{\Delta\tau}$ and T_0 equal to the unknown steady-state value T_0^{ss}, write down the steady-state version of the same equation. Finally, by subtracting these two equations and rearranging, one will find the stability condition.

Next, let us use these statements to develop the stability criterion for Eq. (3.121). This equation is already in the standard form referred to previously, since the temperature at the future time for the node 0, for which the equation was derived, is on a side by itself with a coefficient of unity.

$$T_0^{\Delta\tau} = \frac{1}{M}(T_1 + T_2) + \frac{q_0'''(\Delta x)^2}{k} + \left(1 - \frac{2}{M}\right)T_0 \qquad \textbf{[3.121]}$$

Next, the steady-state version of this equation (steady at node 0) is written as follows, where T_0^{ss} refers to the unknown steady-state temperature at node 0:

$$T_0^{ss} = \frac{1}{M}(T_1 + T_2) + q_0'''\frac{(\Delta x)^2}{k} + \left(1 - \frac{2}{M}\right)T_0^{ss}. \qquad (3.137)$$

---

2. More generally, no component of the *initial value vector* is allowed to amplify as time goes on. The initial value vector is the temperature vector **t** of Eq. (2.239) at $\tau = 0$ when its elements are the initial nodal temperatures. This rule assures stability of the equation set and automatically includes the more easily understood numerical errors referred to previously.

Subtracting the transient version of the finite difference equation, Eq. (3.121), from its steady-state counterpart, Eq. (3.137), gives

$$T_0^{ss} - T_0^{\Delta\tau} = \left(1 - \frac{2}{M}\right)(T_0^{ss} - T_0).$$

Now, one divides both sides of the equation by $T_0^{ss} - T_0$, yielding

$$\frac{T_0^{ss} - T_0^{\Delta\tau}}{T_0^{ss} - T_0} = 1 - \frac{2}{M} \qquad (3.138)$$

From Eq. (3.138), it is now seen that $T_0^{\Delta\tau}$ will not be able to jump beyond the ultimate, unknown steady-state value T_0^{ss} in any time increment $\Delta\tau$ as long as the ratio on the left is greater than or equal to zero; that is, as long as the ratio is positive. That this is the case can be easily seen by considering the two possibilities for the temperature at node 0 at the present time, $T_0 < T_0^{ss}$ or $T_0 > T_0^{ss}$. Thus the stability requirement for Eq. (3.121) becomes

$$1 - \frac{2}{M} \geq 0, \qquad (3.139)$$

or, after solving this inequality for M,

$$M \geq 2. \qquad (3.140)$$

We can see, from the inequality Eq. (3.139), that what Arpaci's stability analysis finally comes down to, for Eq. (3.121), is the requirement that the coefficient of the nodal temperature T_0 at the present time be greater than or equal to zero. That is, it must not be negative *when Eq. (3.121) is in standard form* with $T_0^{\Delta\tau}$ on a side by itself with a coefficient of unity. This statement can be shown to be true for a general finite difference equation, not just Eq. (3.121), when it is arranged in standard form. Hence, for present purposes, the stability criterion used is that a negative coefficient of T_0, when the equation is arranged in standard form, is what causes instability in some cases. This may show up as a second law violation like the one previously discussed.

Formally stated, the stability criterion used here is that when the transient finite difference equation is arranged so that the future temperature of the node of interest is on a side by itself with a unity coefficient (the standard form of the equation), then the coefficient of the temperature of the same node at the present time should not be negative.

Applying this criterion directly to the explicit finite difference Eqs. (3.119), (3.120), and (3.121), gives the following results, when q_0''' does not depend upon T_0:

For a three-dimensional, equispaced lattice, interior node, Eq. (3.119) requires that

$$M \geq 6. \qquad (3.141)$$

For a two-dimensional, equispaced lattice, interior node, Eq. (3.120) requires that

$$M \geq 4. \tag{3.142}$$

For a one-dimensional, equispaced lattice, interior node, Eq. (3.121) requires that

$$M \geq 2. \tag{3.143}$$

The preceding stability statement must be true for all nodes, not just interior nodes for which Eqs. (3.141–43) apply.

Suppose one of the nodes of interest in a two-dimensional conduction region with an equispaced lattice is at a surface and convects to a fluid. The governing transient finite difference equation for this node from Eq. (3.132) is

$$T_0^{\Delta \tau} = \frac{1}{M}(T_1 + T_2 + 2T_3 + 2NT_\infty) + q_0''' \frac{(\Delta x)^2}{Mk} + \left(1 - \frac{4 + 2N}{M}\right)T_0.$$

This equation is already in the proper form to apply the stability criterion, since $T_0^{\Delta \tau}$ is on one side with a unity coefficient. The coefficient of T_0 at the present time on the other side of the equation should not be negative; hence,

$$1 - \frac{4 + 2N}{M} \geq 0.$$

Solving this inequality for M, the stability criterion for this type of node becomes

$$M \geq 4 + 2N. \tag{3.144}$$

Thus, in a problem that has interior nodes and surface nodes of the type just described, the interior nodes would require that $M \geq 4$; but the surface nodes require an even *larger* value of M, that is $M \geq 4 + 2N$, since $N = h\Delta x / k$ is inherently positive. A value of M must be used that simultaneously satisfies both inequalities. Hence, M would have to be determined by

$$M \geq 4 + 2N$$

in a problem such as this.

In a general problem which contains interior nodes as well as a number of noninterior nodes of various types, the inequality that M must satisfy is found by scanning the equations and using the value of M that will not allow negative coefficients in any of the equations.

So far, the stability criterion has been applied only to explicit finite difference equations. Let us now consider its application to an implicit finite difference equation by considering Eq. (3.125) for the case in which the generation does not depend upon temperature. First, Eq. (3.125) is cast into standard form by dividing both sides by $M + 2$. This gives

$$T_0^{\Delta\tau} = \frac{T_1^{\Delta\tau} + T_2^{\Delta\tau} + q_0'''(\Delta x)^2/k}{M + 2} + \left(\frac{M}{M + 2}\right) T_0. \tag{3.145}$$

For stability, the coefficient of T_0 must not be negative; it must be greater than or equal to zero. Hence,

$$\frac{M}{M + 2} \geq 0. \tag{3.146}$$

However, by inspection of the definition of M, Eq. (3.118), we see that M is inherently positive. It follows that Eq. (3.146) is satisfied regardless of the value of M. Since there is no limit on the size of M for which the implicit difference equation, Eq. (3.145), is stable, the equation is termed *unconditionally stable*. The implications of this, and of the conditional stability criteria given in Eqs. (3.141–44), will be discussed next.

3.4e Implication of the Stability Criterion

Stability is necessary if the solution to the transient finite difference equations is to be physically reasonable. In transient conduction problems, generally a lattice spacing Δx is first chosen which is usually dictated by some compromise between accuracy, which increases with decreasing Δx, and time and effort expended, which also increase with decreasing Δx. After Δx is chosen, $\Delta \tau$ cannot be selected arbitrarily but rather must be chosen so that the inequality in M is satisfied. For example, if a two-dimensional conduction region has only interior nodes and an equispaced lattice, M must satisfy Eq. (3.142); that is, $M \geq 4$. But, since $M = (\Delta x)^2/\alpha\Delta\tau$,

$$\frac{(\Delta x)^2}{\alpha\Delta\tau} \geq 4,$$

or

$$\Delta\tau \leq \frac{(\Delta x)^2}{4\alpha}.$$

Hence once a lattice spacing is chosen, this, along with the thermal diffusivity α and the stability criterion, fixes the largest size time increment that can be used. The time increment should be large, because this reduces the number of times that the equation set must be solved to obtain the nodal temperatures at the various times. On the other hand, a smaller time increment $\Delta\tau$ increases the accuracy of the solution. Hence, a compromise must again be made, this time subject to the limitation on the size of $\Delta\tau$ imposed by the stability criterion.

The preceding discussion dealt with an explicit finite difference equation which was conditionally stable, with the condition being Eq. (3.142). For the implicit finite difference Eq. (3.145), which is unconditionally stable, the time and space increments $\Delta\tau$ and Δx can be chosen arbitrarily and the equation will be stable. Hence, for this unconditionally stable equation, one could use large time increments $\Delta\tau$ to advance the solution in time as long as this time increment $\Delta\tau$ was still small enough to allow the finite difference solution to give accurate results, results very close to the exact solution of the problem. (Recall the previous discussion of the need to refine the increments in both space and time. It is precisely this point which sometimes may cause the best choice of finite difference equations for a problem to be implicit ones rather than the much simpler-to-solve explicit ones.)

Thus, in deciding which of these two types of finite difference equations will require less effort to solve a specific problem, we must consider the following competing factors: Should one use a fairly large time step $\Delta\tau$ allowed in the implicit equations, which then have to be solved as a simultaneous set at each time increment? Or, should one use explicit equations, whose conditional stability criteria might demand the use of a very small $\Delta\tau$, and then solve these simple, uncoupled equations one at a time at each time increment for the fairly large number of time steps needed to reach the time of interest? In practice, this decision is most often made by comparing computer run times. Generally, if the computer run times are not prohibitive for the explicit finite difference equations, their utter simplicity of structure, which allows solution of each equation one at a time, makes them the usual choice. As a consequence of this, the explicit type of equation will be stressed here.

It was shown that a surface node has a more restrictive stability criterion than does an interior node. This is generally true, and recalling the derivation of the transient finite difference equation for the surface node shown in Fig. 3.27, we find that the additional restriction is due not only to the convection as embodied in the quantity $N = h\Delta x/k$, but also is due to the fact that the amount of material associated with these nodes is smaller than that for an interior node. (This is where the factor of 2 comes from in the expression $M \geq 4 + 2N$.) Hence, a problem involving, say, 50 interior nodes and, perhaps, 4 or 5 surface nodes will force the use of a smaller value of $\Delta\tau$ than we might prefer. Three recourses are available if we want to use the largest $\Delta\tau$ which the interior nodes alone would allow.

1. The storage term at the surface nodes can be neglected so that there is no stability criterion associated with them. For example, for the surface node pictured in Fig. 3.27, Eq. (3.130) is valid; that is,

$$
k\Delta x \frac{(T_3 - T_0)}{\Delta x} + k \frac{\Delta x}{2} \frac{(T_1 - T_0)}{\Delta x} + k \frac{\Delta x}{2} \frac{(T_2 - T_0)}{\Delta x}
$$
$$
+ h_0\Delta x(T_\infty - T_0) + q_0''' \frac{(\Delta x)^2}{2} = \rho c_p \frac{(\Delta x)^2}{2} \frac{(T_0^{\Delta\tau} - T_0)}{\Delta\tau}. \qquad \text{[3.130]}
$$

If the storage term at this node is neglected, the right-hand side of the equation is zero, and the equation for the temperature of node zero becomes

$$T_3 - T_0 + \frac{T_1}{2} - \frac{T_0}{2} + \frac{T_2}{2} - \frac{T_0}{2} + \frac{h_0 \Delta x}{k} T_\infty$$
$$- \frac{h_0 \Delta x}{k} T_0 + \frac{q_0'''}{k} \frac{(\Delta x)^2}{2} = 0,$$

or after rearrangement,

$$T_0 = \frac{T_3 + T_1/2 + T_2/2 + (h_0 \Delta x/k) T_\infty + (q_0'''/k)(\Delta x)^2/2}{(2 + h_0 \Delta x/k)}. \quad \textbf{(3.147)}$$

Note that neglecting the storage term at these surface nodes reduces the accuracy of the entire approximation, especially in the vicinity of the surface nodes whose energy storage terms were neglected. Even though there is no longer any stability restriction associated with the surface node when its finite difference equation is Eq. (3.147)—and therefore the stability restriction would be the less severe one for an interior node (or for some other less severe surface node)—the use of Eq. (3.147) necessitates a smaller nodal spacing Δx in order to achieve the same accuracy of prediction as would occur if the storage term had not been neglected.

2. Implicit finite difference equations that exhibit unconditional stability can be used at all nodes, both interior and surface nodes. Then the size of $\Delta \tau$ will be determined only by the required accuracy of the solution and not by any separate stability requirement at the boundary. This course of action, however, requires us to solve a simultaneous set of equations at every $\Delta \tau$ rather than being able to solve the equations independently, one at a time.

3. One can use implicit finite difference equations on the boundary only. In this case, if the implicit equations used are unconditionally stable, the stability criterion reverts back to the less severe one at the interior nodes where explicit difference equations are being used. When using this technique in a one-dimensional problem, the implicit difference equation at the surface node can be combined with the explicit difference equation for the interior node nearest the surface. The result is a new surface node equation which, though explicit, is subject at most to the less severe stability criterion at an internal node. If on the other hand the problem is two- or three-dimensional, then the implicit equations being used on the boundary have to be solved as a simultaneous set at each time increment. However, since explicit algorithms are being used at all the interior nodes, and since, for the typical fairly small nodal spacings that might be used in practice, the surface nodes are far fewer in number than the interior nodes, the number of equations that must be solved as a simultaneous set is considerably smaller than option 2 would require.

EXAMPLE 3.14

Derive the stability criterion for Eq. (3.134).

Solution
Equation (3.134) is

$$\frac{T_4 - T_0}{\Delta x^2} + \frac{T_3 - T_0}{2\Delta x^2} + \frac{T_2 - T_0}{\Delta y^2} + \frac{T_1 - T_0}{2\Delta y^2}$$
$$+ \frac{h_f}{2k} \frac{\Delta y + \Delta x}{\Delta x \Delta y} (T_f - T_0) + \frac{3}{4} \frac{q_0'''}{k} = \frac{3}{4} \rho \frac{c_p}{k\Delta\tau} (T_0^{\Delta\tau} - T_0).$$

This equation must be rearranged with $T_0^{\Delta\tau}$ on one side and all the terms containing T_0 collected on the other side. Expanding,

$$\frac{T_4}{\Delta x^2} + \frac{T_3}{2\Delta x^2} + \frac{T_2}{\Delta y^2} + \frac{T_1}{2\Delta y^2} + \left(\frac{h_f}{2k} \frac{\Delta y + \Delta x}{\Delta x \Delta y} \right) T_f + \frac{3}{4} \frac{q_0'''}{k}$$
$$- T_0 \left(\frac{1}{\Delta x^2} + \frac{1}{2\Delta x^2} + \frac{1}{\Delta y^2} + \frac{1}{2\Delta y^2} + \frac{h_f}{2k} \frac{\Delta y + \Delta x}{\Delta x \Delta y} \right)$$
$$= \frac{3}{4} \rho \frac{c_p}{k\Delta\tau} (T_0^{\Delta\tau} - T_0).$$

Dividing through by $\frac{3}{4} \rho(c_p/k\Delta\tau)$ gives

$$T_0^{\Delta\tau} = \frac{4\Delta\tau k}{3\rho c_p} \left(\frac{T_4}{\Delta x^2} + \frac{T_3}{2\Delta x^2} + \frac{T_2}{\Delta y^2} + \frac{T_1}{2\Delta y^2} + \frac{h_f}{2k} \frac{\Delta y + \Delta x}{\Delta x \Delta y} T_f \right.$$
$$\left. + \frac{3}{4} \frac{q_0'''}{k} \right) + \left[1 - \frac{4\Delta\tau k}{3\rho c_p} \left(\frac{3}{2\Delta x^2} + \frac{3}{2\Delta y^2} + \frac{h_f}{2k} \frac{\Delta y + \Delta x}{\Delta x \Delta y} \right) \right] T_0.$$

For stability, the coefficient of T_0 must be greater than or equal to zero; hence, for stability,

$$1 - \frac{4\Delta\tau k}{3\rho c_p} \left(\frac{3}{2\Delta x^2} + \frac{3}{2\Delta y^2} + \frac{h_f}{2k} \frac{\Delta x + \Delta y}{\Delta x \Delta y} \right) \geq 0.$$

Thus, for a given material, once the lattice spacings Δx and Δy are chosen, this inequality can be solved for the largest time increment that can be used, with the stability of Eq. (3.134) insured.

3.5 TRANSIENT FINITE DIFFERENCE SOLUTIONS

Earlier we discussed the derivation of transient finite difference equations, the condition needed for stability of the equation set, and an overview of the solution procedures for these equations both for the explicit and implicit type of equation. In this section, all of this previous material will be integrated in the complete solution of transient conduction problems by finite differences, from increment size selection to calculation of the numerical values of the nodal temperatures at various times.

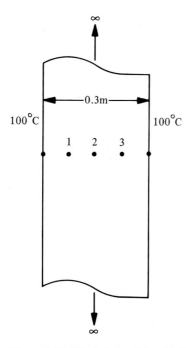

Figure 3.31 The infinite slab with finite difference mesh.

To illustrate the procedure and to learn something about the accuracy one might expect for fairly large increment sizes, we will begin with a simple one-dimensional transient problem, with known boundary temperatures. This will be followed by more difficult transient problems which begin to display the power and importance of the transient finite difference method as well as the need to use the computer for almost all problems, except those in which very large increment sizes are sufficient.

Consider the infinite slab of thickness 0.3 meters which is shown in Fig. 3.31. Initially, because of an electrical current passing through the slab, the slab is in the steady state with a temperature distribution, in °C, given by $T = 100 + 400 \sin[\pi x/0.3]$. However, at time $\tau = 0$, the electric current is shut off, thus causing a transient in the temperature, while the slab faces continue to be held at 100°C by a coolant. The thermal diffusivity of the material is constant at the value $\alpha = 2.84 \times 10^{-5}$ m²/s. We would like to find the temperature within the slab as a function of position and time by use of finite difference methods.

To begin, let us choose a fairly coarse nodal spacing, with $\Delta x = 0.075$ m, to give the nodes shown in Fig. 3.31. These are all interior nodes in a one-dimensional conduction region. Choosing to use explicit finite difference equations, it is seen that Eq. (3.121) applies. That is,

$$T_0^{\Delta\tau} = \frac{1}{M}(T_1 + T_2) + q_0''' \frac{(\Delta x)^2}{kM} + \left(1 - \frac{2}{M}\right)T_0, \qquad \textbf{[3.121]}$$

where 0 represents the node of interest, 1 and 2 are the nodes on either side of node 0 in the x direction, and q_0''' is the generation rate per unit volume of the material associated with node 0. Here the generation is zero, since it is the turning off of the electrical energy that causes the transient.

Finite difference equations are not needed for the surface nodes because their temperature is being held at 100°C. This constitutes the x direction boundary conditions.

Writing Eq. (3.121) for nodes 1, 2, and 3 by properly interpreting the subscripts yields the transient finite difference equations

$$T_1^{\Delta\tau} = \frac{1}{M}(100 + T_2) + \left(1 - \frac{2}{M}\right)T_1, \tag{3.148}$$

$$T_2^{\Delta\tau} = \frac{1}{M}(T_1 + T_3) + \left(1 - \frac{2}{M}\right)T_2, \tag{3.149}$$

$$T_3^{\Delta\tau} = \frac{1}{M}(T_2 + 100) + \left(1 - \frac{2}{M}\right)T_3. \tag{3.150}$$

Because of the symmetry of the problem, $T_1 = T_3$ at every instant of time; hence only Eqs. (3.148) and (3.149) need be solved after replacing T_3 in Eq. (3.149) by T_1. (Note that symmetry encompasses both geometric symmetry and symmetry of the thermal loadings. The conduction region is geometrically symmetrical about the midplane of the slab. Since both faces of the slab are kept at 100°C, the boundary conditions, one of the thermal loadings, are symmetrical. In addition, the initial temperature distribution $T = 100 + 400 \sin[\pi x/0.3]$, which is the other thermal loading, is also symmetrical about the midplane of the slab.) Rewriting Eqs. (3.148–49) using the symmetry,

$$T_1^{\Delta\tau} = \frac{1}{M}(100 + T_2) + \left(1 - \frac{2}{M}\right)T_1, \tag{3.151}$$

$$T_2^{\Delta\tau} = \frac{1}{M}(T_1 + T_1) + \left(1 - \frac{2}{M}\right)T_2. \tag{3.152}$$

The stability criterion is given in Eq. (3.143) as $M \geq 2$. If $M = 3$ is chosen, using Eq. (3.118) the size of the time increment can be calculated. That is,

$$\Delta\tau = (\Delta x)^2/\alpha M = (0.075)^2/(2.84 \times 10^{-5})(3) = 66 \text{ s} = 1.1 \text{ min.}$$

Hence, using the value of $M = 3$, the temperatures at the various nodes will be calculated every 1.1 minutes from $\tau = 0$.

Note that, for stability satisfaction, any value of M greater than or equal to 2 could have been chosen. For larger values of M the time step $\Delta\tau$ will be smaller, and this will give a more accurate solution at the expense of having to solve the equations at a large number of time steps. The value of $M = 3$ that was chosen simply represents a compromise between these two opposing effects. Sometimes, however, the decision is made to perform the calculations right on the stability limit, $M = 2$, in order to work with even simpler equations, because the last term on the right side drops out.

In every transient problem, the initial distribution of the temperature in the space coordinates must be known to supply the starting values of the nodal temperatures in the finite difference equations. The initial temperature profile was given as

$$T = 100 + 400 \sin[\pi x/0.3].$$

Substituting $x = 0.075$ m and then $x = 0.15$ m leads to T_1 and T_2 at $\tau = 0$. Thus, at $\tau = 0$,

$$T_1 = 100 + 400 \sin(\pi/4) = 383°C,$$
$$T_2 = 100 + 400 \sin(\pi/2) = 500°C.$$

Using $M = 3$ in Eqs. (3.151) and (3.152),

$$T_1^{\Delta\tau} = \frac{1}{3}(100 + T_2) + \frac{T_1}{3}, \qquad\qquad (3.153)$$

$$T_2^{\Delta\tau} = \frac{2}{3} T_1 + \frac{T_2}{3}. \qquad\qquad (3.154)$$

Now, to determine the temperatures at $\tau = 0 + 1\Delta\tau = 1\Delta\tau$, simply insert the values of the nodal temperatures at $\tau = 0$, the present time, into the right-hand sides of Eqs. (3.153) and (3.154). Then calculate the value of the temperatures of the nodes at the future time, $\tau = 1\Delta\tau$. Hence, at $\tau = 1\Delta\tau$,

$$T_1^{\Delta\tau} = \frac{1}{3}(100 + 500) + \frac{383}{3} = 328°C,$$

$$T_2^{\Delta\tau} = \frac{2}{3}(383) + \frac{500}{3} = 423°C.$$

Next, to advance the solution in time to $\tau = 2\Delta\tau$, these computed values of nodal temperatures at $\tau = 1\Delta\tau$ are viewed as the present temperatures and inserted into the right-hand sides of Eqs. (3.153) and (3.154). Thus, at $2\Delta\tau$,

$$T_1^{\Delta\tau} = \frac{1}{3}(100 + 423) + \frac{328}{3} = 283°C,$$

$$T_2^{\Delta\tau} = \frac{2}{3}(328) + \frac{423}{3} = 360°C.$$

These nodal temperatures at $\tau = 2\Delta\tau$, 283°C at node 1 and 360°C at node 2, are now the present temperatures to be inserted into the right sides of Eqs. (3.153) and (3.154) to give the nodal temperatures at the next new future time, $\tau = 3\Delta\tau$:

$$T_1^{\Delta\tau} = \frac{1}{3}(100 + 360) + \frac{283}{3} = 247°C,$$

$$T_2^{\Delta\tau} = \frac{2}{3}(283) + \frac{360}{3} = 309°C.$$

It is important to keep in mind that the equations being used, Eqs. (3.153) and (3.154), always require the use of the *present* nodal values on the right side. Hence, one would never mix present values with future values that may have been just calculated from an earlier equation. For instance, in the last numerical calculation of $T_2^{\Delta\tau}$ at $\tau = 3\Delta\tau$, 283°C was correctly used as T_1, *not* the 247°C just computed. Mixing present and future values of the nodal temperatures on the right-hand sides of these equations not only leads to incorrect numerical values, but may also cause the equation's solution to be unstable.

The solution procedure for these explicit equations which was just illustrated for the first three time steps is now continued in a similar fashion for $\tau = 4\Delta\tau$, $5\Delta\tau$, etc. Carrying this out to $\tau = 18\Delta\tau$ leads to

$$T_1^{\Delta\tau} = 105°C, \quad T_2^{\Delta\tau} = 108°C \quad \text{at} \quad \tau = 18\Delta\tau.$$

Thus we can see that at $18\Delta\tau = 18(1.1) = 19.8$ minutes, the nodal temperatures are within 5°C and 8°C of their asymptotic final value of 100°C, and the calculations indicate that they are approaching 100°C very slowly.

At this point in a transient finite difference solution, the next step would be refinement of the increment sizes. One would decrease the size of Δx and $\Delta\tau$ and re-solve the problem, comparing nodal temperatures found using the smaller increment sizes with those previously found using the original increment sizes. This process of decreasing the increment size, re-solving, and comparing results with the previous increment sizes would continue until, finally, the nodal temperatures no longer have changed significantly from the previous increment size used. Since this problem was done principally to demonstrate the overall procedure, we will not refine the increment sizes. Instead, we will compare the solution to an available exact solution to try to get some very rough qualitative idea of what to expect in the way of accuracy when using fairly coarse increment sizes.

The exact analytical solution to this problem was worked out in detail in Ex. 3.4. After substituting the numerical values appropriate to the present problem, we have

$$T(x,\tau) = 100 + 400\, e^{-0.00311\tau} \sin[\pi x/0.3].$$

The results of this finite difference solution are presented in Table 3.1 along with the exact solution values at selected times and the percent error between the finite difference solution and the exact solution. For this particular problem, excellent agreement prevails, even for the relatively crude lattice spacing and time increment used. As in the case of

Table 3.1
Finite Difference Solution and Comparison with Exact Results

		Node 1			Node 2		
Number of Time Increments	Time (min)	Finite Difference (°C)	Exact (°C)	Percent Error	Finite Difference (°C)	Exact (°C)	Percent Error
0	0	383	383		500	500	
1	1.1	328			423		
2	2.2	283			360		
3	3.3	247	252	2	309	315	1.9
4	4.4	218			267		
5	5.5	195			234		
6	6.6	176	182	3.3	208	216	3.7
7	7.7	162			186		
8	8.8	149			169		
9	9.9	140			156		
10	11.0	132	135	2.2	145	150	3.3
11	12.1	126			136		
12	13.2	121			129		
13	14.3	116			124		
14	15.4	113	116	2.6	118	122	3.3
15	16.5	110			115		
16	17.6	108			112		
17	18.7	107			109		
18	19.8	105	107	1.87	108	110	1.82

steady-state finite differences, no general statement can be made about the magnitude of the error, since it depends upon the problem. Agreement would not have been as good here, for example, if the surfaces had suddenly been raised to 1000°C when the generation was turned off.

In the transient finite difference problem just worked, the relatively small number of space nodes and small number of time increments needed to come close to the steady state made hand calculation feasible. However, when increment size refinement takes place, the number of calculations required makes the use of the computer mandatory, and one would write a simple program to solve these explicit difference equations. The basic steps in such a program would simply be to (1) read in the nodal temperatures at $\tau = 0$ in a dimensioned array, say $TP(J)$, where TP is the present temperature at the node J; (2) compute the future temperatures $TF(J)$, by use of a DO loop on J, employing Eq. (3.149) properly subscripted; and (3) set $TP(J) = TF(J)$ and return to step 2. Naturally, nodal temperatures would be printed out at selected times, and the program would be ended at either some preselected time of interest or when one satisfies some criterion one has chosen for closeness to the steady state. This straightforwardness of the solution procedure is the reason for earlier reference to the "utter simplicity" of the structure of explicit finite difference equations.

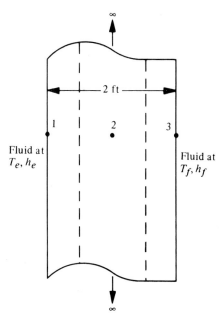

Figure 3.32 The slab and nodal network of Ex. 3.15.

EXAMPLE 3.15

Figure 3.32 shows a 2-ft-thick infinite slab which is initially at 50°F throughout when suddenly the left-hand face is exposed to a fluid at temperature $T_e = 100$°F, while the right-hand face is exposed to a fluid at a temperature of $T_f = 0$°F. The surface coefficient of heat transfer on the left face is $h_e = 20$ Btu/hr ft² °F, and on the right face it is $h_f = 5$ Btu/hr ft² °F. The properties of the slab are $k = 10$ Btu/hr ft °F and $\alpha = 1$ ft²/hr. Determine the time required for the right-hand face to drop to 37°F. (Note that this problem cannot be solved using Heisler's charts, because the fluid temperatures are different on both sides and also because of the different surface heat transfer coefficients on the right and left sides.)

Solution

A fairly crude grid spacing $\Delta x = 1$ ft is chosen, and the nodes appropriate to this spacing are shown in Fig. 3.32. Next, the transient finite difference equation for each of the three nodes is written. The center node 2 is an interior node in a one-dimensional situation, and the equation for it is Eq. (3.121). When the subscripts in Eq. (3.121) are properly interpreted, for node 2,

$$T_2^{\Delta\tau} = \frac{1}{M}(T_1 + T_3) + \left(1 - \frac{2}{M}\right)T_2, \qquad \textbf{(3.155)}$$

where $M = (\Delta x)^2/\alpha\Delta\tau$ from Eq. (3.118).

To arrive at the finite difference equations for nodes 1 and 3, an energy balance is made on the material of volume $(\Delta x/2) \times 1 \times 1$, associated with each of those nodes. For node 1, using Eq. (1.8),

$$k(1)^2 \frac{(T_2 - T_1)}{\Delta x} + h_e(1)^2 (T_e - T_1) = \rho \frac{\Delta x}{2} (1)^2 c_p \frac{(T_1^{\Delta\tau} - T_1)}{\Delta \tau};$$

dividing by $k/\Delta x$ and using Eq. (3.118),

$$T_2 - T_1 + \frac{h_e \Delta x}{k} T_e - \frac{h_e \Delta x}{k} T_1 = \frac{M}{2} (T_1^{\Delta\tau} - T_1).$$

Letting $N_e = h_e \Delta x/k$ and then solving for $T_1^{\Delta\tau}$,

$$T_1^{\Delta\tau} = \frac{2}{M} T_2 + \frac{2N_e}{M} T_e + \left[1 - \frac{(2 + 2N_e)}{M}\right] T_1. \tag{3.156}$$

Node 3 is similar to node 1, so that Eq. (3.156) can be used directly after reinterpreting its subscripts and letting $N_f = h_f \Delta x/k$ replace N_e. Thus,

$$T_3^{\Delta\tau} = \frac{2}{M} T_2 + \frac{2N_f}{M} T_f + \left[1 - \frac{(2 + 2N_f)}{M}\right] T_3. \tag{3.157}$$

To insure stability of the set of transient finite difference equations, Eqs. (3.155–57), the following three inequalities must be satisfied:

$$1 - \frac{2}{M} \geq 0, \quad \text{or} \quad M \geq 2; \tag{3.158}$$

$$1 - \frac{(2 + 2N_e)}{M} \geq 0, \quad \text{or} \quad M \geq 2 + 2N_e; \tag{3.159}$$

$$1 - \frac{(2 + 2N_f)}{M} \geq 0, \quad \text{or} \quad M \geq 2 + 2N_f. \tag{3.160}$$

Since

$$N_e = \frac{h_e \Delta x}{k} = \frac{20(1)}{10} = 2, \quad N_f = \frac{h_f \Delta x}{k} = \frac{5(1)}{10} = \frac{1}{2},$$

Eqs. (3.159) and (3.160) become, respectively,

$$M \geq 2 + 2(2) = 6, \quad M \geq 2 + 2(\tfrac{1}{2}) = 3.$$

The value of M chosen must satisfy all three inequalities, Eqs. (3.158–60); for this to be true, $M \geq 6$. Choosing $M = 6$ gives the largest possible time increment $\Delta\tau$ (which is also the crudest possible "spacing" in time) and from Eq. (3.118), $6 = (1)^2/(1)\Delta\tau$ and $\Delta\tau = 1/6$ hr $= 10$ min. Using $M = 6$ in the transient finite difference equations, Eqs. (3.155–57), along with $T_e = 100$, $T_f = 0$, $N_e = 2$, and $N_f = 1/2$ gives

$$T_1^{\Delta\tau} = \frac{T_2}{3} + 66.7, \quad T_2^{\Delta\tau} = \frac{T_1 + T_3}{6} + \frac{2}{3} T_2,$$

$$T_3^{\Delta\tau} = \frac{T_2}{3} + \frac{T_3}{2}. \tag{3.161}$$

To get the temperatures at nodes 1, 2, and 3 after $1\Delta\tau$ has passed, the initial temperatures are inserted into the right-hand sides of the set of Eq. (3.161). The initial temperatures are all 50°F. Thus, at time $1\Delta\tau$,

$$T_1^{\Delta\tau} = \frac{50}{3} + 66.7 = 83°\text{F}, \quad T_2^{\Delta\tau} = \frac{100}{6} + \frac{2}{3}(50) = 50°\text{F},$$

$$T_3^{\Delta\tau} = \frac{50}{3} + \frac{50}{2} = 42°\text{F}. \tag{3.162}$$

Now use results of Eq. (3.162) in the right-hand side of Eq. (3.161) to obtain at time $2\Delta\tau$,

$$T_1^{\Delta\tau} = \frac{50}{3} + 66.7 = 83°\text{F}, \quad T_2^{\Delta\tau} = \frac{83 + 42}{6} + \frac{2}{3}(50) = 54°\text{F},$$

$$T_3^{\Delta\tau} = \frac{50}{3} + \frac{42}{2} = 38°\text{F}.$$

The procedure is repeated until T_3 reaches 37°F. Hence, at time $3\Delta\tau$,

$$T_1^{\Delta\tau} = \frac{54}{3} + 66.7 = 85°\text{F}, \quad T_2^{\Delta\tau} = \frac{83 + 38}{6} + \frac{2}{3}(54) = 56°\text{F},$$

$$T_3^{\Delta\tau} = \frac{54}{3} + \frac{38}{2} = 37°\text{F}.$$

Hence, after a time equal to $3\Delta\tau$, the right-hand face of the slab in Fig. 3.32 has dropped to about 37°F. As a result, the required time is

$$3\Delta\tau = 3(1/6) = 1/2 \text{ hr} = 30 \text{ min}.$$

This value of the time is, of course, approximate because of the relatively large increment sizes, Δx and $\Delta\tau$, used. To achieve increased accuracy, the increment sizes would have to be decreased and the entire problem solved over as has been explained. Note that node 1's temperature remained the same between $1\Delta\tau$ and $2\Delta\tau$. This is a consequence of the relatively coarse lattice spacing Δx and also, in part, of using $M = 6$ rather than some value greater than 6.

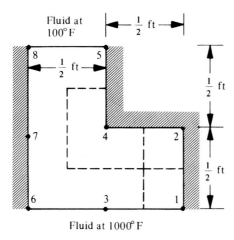

Figure 3.33 The cross section of the structural member and nodes of Ex. 3.16.

This problem demonstrates the frequent need to apply the energy balance to noninterior nodes in order to derive their finite difference equations when the appropriate equations are not already available. Also, specifically shown by Eqs. (3.158–60) is the scanning of the stability criteria for *all* nodes so that selection of the most severe stability condition, the one that satisfies the stability requirement at *every* node, can be made.

EXAMPLE 3.16

The structural member shown in cross section in Fig. 3.33 is joined on the faces shown to such a low thermal conductivity material that those faces are effectively insulated. Initially, the member and both the fluid bounding the top face and the fluid bounding the lower face are at 100°F. Suddenly, the lower face is exposed to 1000°F fluid with a surface coefficient of heat transfer of 30 Btu/hr ft² °F between the fluid and the lower face. The fluid bounding the upper face remains at 100°F and the surface coefficient of heat transfer is 100 Btu/hr ft² °F between the upper face and the fluid. Using nodes spaced one-half foot apart, determine the transient temperature distribution for the first few time increments and estimate the time required for any point on the upper face to reach 125°F. Here, $k = 30$ Btu/hr ft °F and $\alpha = 0.6$ ft²/hr.

Solution

The nodal spacing of 0.5 ft in both directions gives rise to the nodes numbered in Fig. 3.33. The next step is to put down the governing transient finite difference equations for the eight nodes. We choose to deal with the simple explicit type of finite difference equation. Nodes 1, 6, 8, and 5 are all of the same general type. However, since this type has not yet been treated, the finite difference equation will be derived by use of an energy balance on the volume of material associated with node 1, namely, $\Delta x^2/4$ as shown by the dashed lines in Fig. 3.33. Applying Eq. (1.8) yields, with the fluid temperature designated as T_f,

$$k \frac{\Delta x}{2} \frac{(T_3 - T_1)}{\Delta x} + k \frac{\Delta x}{2} \frac{(T_2 - T_1)}{\Delta x} + h_1 \frac{\Delta x}{2} (T_f - T_1)$$
$$= \rho c_p \frac{\Delta x^2}{4} \frac{(T_1^{\Delta \tau} - T_1)}{\Delta \tau}.$$

Using the definition of M from Eq. (3.118) and rearranging, one has

$$T_1^{\Delta \tau} = \frac{2}{M} \left[T_2 + T_3 + \frac{h_1 \Delta x T_f}{k} \right] + \left[1 - \frac{2}{M} \left(2 + \frac{h_1 \Delta x}{k} \right) \right] T_1. \quad \textbf{(3.163)}$$

Now, since $T_f = 1000$ and $h_1 \Delta x/k = 30(0.5)/30 = 0.5$, Eq. (3.163), with these numerical values appropriate to node 1, becomes

$$T_1^{\Delta \tau} = \frac{2}{M} (T_2 + T_3 + 500) + \left(1 - \frac{5}{M} \right) T_1. \quad \textbf{(3.164)}$$

By proper reinterpretation of the subscripts in Eq. (3.163), the difference equations for nodes 6, 8, and 5 are written down next. Thus, for node 6, Eq. (3.163) becomes,

$$T_6^{\Delta \tau} = \frac{2}{M} (T_7 + T_3 + 500) + \left(1 - \frac{5}{M} \right) T_6. \quad \textbf{(3.165)}$$

For node 8,

$$T_8^{\Delta \tau} = \frac{2}{M} \left[T_7 + T_5 + h_2 \frac{\Delta x}{k} (100) \right] + \left[1 - \frac{2}{M} \left(2 + h_2 \frac{\Delta x}{k} \right) \right] T_8,$$

and since $h_2 \Delta x/k = \frac{100}{30} \left(\frac{1}{2} \right) = 1.67$,

$$T_8^{\Delta \tau} = \frac{2}{M} (T_7 + T_5 + 167) + \left(1 - \frac{7.34}{M} \right) T_8. \quad \textbf{(3.166)}$$

Using Eq. (3.163) for node 5,

$$T_5^{\Delta \tau} = \frac{2}{M} (T_4 + T_8 + 167) + \left(1 - \frac{7.34}{M} \right) T_5. \quad \textbf{(3.167)}$$

Node 3 is governed by the same finite difference equation as node 0 in Fig. 3.27; hence, with $N = h_1 \Delta x/k = 1/2$, $T_\infty = 1000°F$, $q_0'' = 0$, and identifying in Eq. (3.132) 0 with 3, 1 with 6, 2 with 1, and 3 with 4,

$$T_3^{\Delta \tau} = \frac{1}{M} (2T_4 + T_6 + T_1 + 1000) + \left(1 - \frac{5}{M} \right) T_3. \quad \textbf{(3.168)}$$

Node 7 is of the same type as node 0 of Fig. 3.28; hence, with proper interpretation of the subscripts, in Eq. (3.133),

$$T_7^{\Delta\tau} = \frac{1}{M}(T_8 + T_6 + 2T_4) + \left(1 - \frac{4}{M}\right)T_7. \tag{3.169}$$

Finite difference equations for noninterior nodes 2 and 4 have not been previously developed; therefore, they are now derived by making energy balances on the control volume of material associated with each node. These control volumes are shown by the dashed lines in Fig. 3.33. The volume of material associated with node 2 is $(\Delta x/2)(\Delta x/2) = (\Delta x)^2/4$. Using Eq. (1.8) on this control volume,

$$k\frac{\Delta x}{2}\frac{(T_4 - T_2)}{\Delta x} + k\frac{\Delta x}{2}\frac{(T_1 - T_2)}{\Delta x} = \rho c_p \frac{(\Delta x)^2}{4}\frac{(T_2^{\Delta\tau} - T_2)}{\Delta\tau},$$

or

$$T_4 + T_1 - 2T_2 = \frac{1}{2}\frac{(\Delta x)^2}{\alpha\Delta\tau}(T_2^{\Delta\tau} - T_2).$$

Since $M = \Delta x^2/\alpha\Delta\tau$, solving for $T_2^{\Delta\tau}$ yields

$$T_2^{\Delta\tau} = \frac{1}{M}(2T_4 + 2T_1) + \left(1 - \frac{4}{M}\right)T_2. \tag{3.170}$$

Also using Eq. (1.8) on the volume of material $\frac{3}{4}\Delta x^2$ associated with node 4 gives

$$k\Delta x\frac{(T_7 - T_4)}{\Delta x} + k\Delta x\frac{(T_3 - T_4)}{\Delta x} + k\frac{\Delta x}{2}\frac{(T_5 - T_4)}{\Delta x}$$
$$+ k\frac{\Delta x}{2}\frac{(T_2 - T_4)}{\Delta x} = \frac{3}{4}(\Delta x)^2 \rho c_p \frac{(T_4^{\Delta\tau} - T_4)}{\Delta\tau},$$

or

$$T_7 + T_3 + T_2/2 + T_5/2 - 3T_4 = \frac{3}{4}M(T_4^{\Delta\tau} - T_4).$$

Rearrangement yields the transient finite difference equation for node 4 as

$$T_4^{\Delta\tau} = \frac{4}{3M}\left(T_7 + T_3 + \frac{T_5}{2} + \frac{T_2}{2}\right) + \left(1 - \frac{4}{M}\right)T_4. \tag{3.171}$$

Thus, Eqs. (3.164–71) are the governing finite difference equations for the eight nodes of this problem. The next step is to select a value of M which will insure stability. This means that the coefficient of the temperature at a node, before the time increment $\Delta\tau$ has passed, should not be negative. This is satisfied for Eqs. (3.164–65) if $M \geq 5$. For Eqs. (3.166–67), $M \geq 7.34$, for Eq. (3.168), $M \geq 5$, and for Eqs. (3.169–71), $M \geq 4$. The only value of M that satisfies all of these conditions is $M \geq 7.34$; hence, the stability criterion for this problem is

$$M \geq 7.34.$$

Since $M = (\Delta x)^2 / \alpha \Delta \tau$, the size of the time increment can be determined from

$$M = \frac{(\frac{1}{2})^2}{0.6 \Delta \tau} = 7.34, \quad \text{or} \quad \Delta \tau = 0.0568 \text{ hr} = 3.4 \text{ min.}$$

Using $M = 7.34$ in Eqs. (3.164–71) and reordering the equations leads to

$$T_1^{\Delta \tau} = 0.273 \, (T_2 + T_3 + 500) + 0.317 \, T_1,$$

$$T_3^{\Delta \tau} = 0.1363 \, (2T_4 + T_6 + T_1 + 1000) + 0.317 \, T_3,$$

$$T_6^{\Delta \tau} = 0.273 \, (T_7 + T_3 + 500) + 0.317 \, T_6,$$

$$T_7^{\Delta \tau} = 0.1363 \, (T_8 + T_6 + 2T_4) + 0.455 \, T_7,$$

$$T_4^{\Delta \tau} = 0.182 \left(T_7 + T_3 + \frac{T_5}{2} + \frac{T_2}{2} \right) + 0.455 \, T_4. \tag{3.172}$$

$$T_2^{\Delta \tau} = 0.273 \, (T_4 + T_1) + 0.455 \, T_2,$$

$$T_5^{\Delta \tau} = 0.273 \, (T_4 + T_8 + 167),$$

$$T_8^{\Delta \tau} = 0.273 \, (T_7 + T_5 + 167).$$

Next a short computer program, along the lines of the discussion just preceding Ex. 3.15, was written for these explicit equations, and the results are given in Table 3.2.

Notice from the evolution of the nodal temperatures as given by the computer solution that a node on the upper face, node 5, has reached 126°F, or one degree past the 125°F required by the problem statement. This has happened after the passage of five time increments, so the approximate time required is 17 minutes. If it is critically important to

Table 3.2
Computer Solution for Temperatures in °F at Node Number

Number of Time Increments	Time (minutes)	1	2	3	4	5	6	7	8
0	0	100	100	100	100	100	100	100	100
1	3.4	224	100	223	100	100	224	100	100
2	6.8	295	134	295	122	100	295	117	100
3	10.2	348	175	344	151	106	343	140	105
4	13.6	388	218	381	182	115	377	166	113
5	17.0	423	255	411	213	126	405	191	122

find a highly accurate value of the time needed for some point on the upper surface to reach 125°F, then increment size refinement must be carried out and the problem resolved until the predicted temperatures no longer are influenced by the size of Δx and $\Delta \tau$ being used.

Example 3.16 is a multidimensional transient conduction problem in which the complexity of the geometric shape renders the problem intractable from an exact analytic solution standpoint; it therefore requires the use of finite difference methods.

3.5a A Nonlinear Problem

A class of problems in transient conduction for which one almost always must employ finite difference techniques occurs when the governing partial differential equation, or boundary conditions, or both, contain nonlinear terms in the dependent variable. This may occur when the material's thermal conductivity depends significantly on temperature in the temperature range anticipated for a particular circumstance. As mentioned in Chapter 1, the thermal conductivity of many materials can be adequately represented, in fairly wide temperature ranges, as depending linearly upon the temperature. (Note that even though k depends linearly upon T, the governing partial differential equation is nonlinear; this is evident by looking at Eq. [2.5].) To provide a vehicle for discussing the application of finite difference methods to nonlinear problems, we will deal with a situation in which the material's thermal conductivity varies linearly with temperature.

Consider the slab shown in Fig. 3.34. This is a 2% tungsten steel slab, of half thickness $L = 1.75$ cm, initially at temperature $T_i = 760°C$ throughout. Suddenly, the slab is plunged into a coolant at temperature $T_f = 27°C$ with surface coefficient of heat transfer $h = 17,500$ W/m² °C between the coolant and both faces of the slab. We must find the time it takes for the center temperature to drop to 200°C, and also find the surface temperature at that time. As a result of the thermal conductivity of the tungsten steel varying by a factor of almost two in the expected temperature range, Heisler's charts will not be used. Instead, finite difference methods must be employed. A plot of k versus T, using the data in Appendix A, indicates practically a linear variation which can be written as

$$k = k_f[1 + b\,(T - T_f)]. \tag{3.173}$$

Often a problem such as this one, in which specific numerical values are given and the actual numerical values of temperatures and times are sought, may have to be solved again for different numerical values of T_f, T_0, h, different materials, and so on. When this is the case, it is to one's advantage to set up the problem for the more general situation, using nondimensional groups, and then to solve it for the particular set of conditions of present interest. This is the procedure we will follow.

The nodal spacing and numbering scheme are shown in Fig. 3.34 with all nodes spaced a distance Δx apart across only one-half of the slab, because of the thermal and geometric symmetry. We choose to deal with explicit finite difference equations, and now they must be written for each node in the lattice. However, previously derived equations for an interior node in a one-dimensional region, such as Eq. (3.121), cannot be used, because they

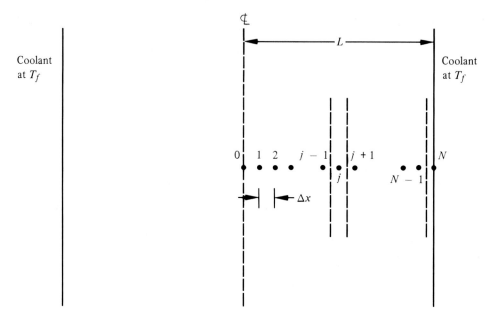

Figure 3.34 Infinite slab with nodal system.

were found for the case of constant thermal conductivity k. Hence we must make an energy balance on a general interior node, node j in Fig. 3.34, to obtain the difference equations appropriate to a variable thermal conductivity condition. The volume of material associated with the jth node is Δx by one unit by one unit, as shown in the figure by the dashed surfaces around node j. The following equation results when we apply Eq. (1.8) to the control volume enclosing node j, and when we take into account the temperature dependence of the thermal conductivity by use of the arithmetic average conductivity between adjacent nodes when calculating the conduction heat transfer rate between those nodes:

$$
\left(\frac{k_{j-1} + k_j}{2}\right) \frac{(T_{j-1} - T_j)}{\Delta x} + \left(\frac{k_{j+1} + k_j}{2}\right) \frac{(T_{j+1} - T_j)}{\Delta x}
$$
$$
= \rho c_p \Delta x \, \frac{(T_j^{\Delta \tau} - T_j)}{\Delta \tau}. \tag{3.174}
$$

In Eq. (3.174), the subscript denotes the node at which that quantity is to be evaluated. The thermal conductivity values will be evaluated at the *present* temperature of each node; that is, k_j is evaluated at T_j. Dividing every term in Eq. (3.174) by k_f, and defining,

$$
M_f = \Delta x^2 / \alpha_f \Delta \tau, \tag{3.175}
$$
$$
\alpha_f = k_f / \rho c_p, \tag{3.176}
$$

$$K_{j-1} = \frac{k_{j-1} + k_j}{2k_f},$$ (3.177)

$$K_{j+1} = \frac{k_{j+1} + k_j}{2k_f}.$$ (3.178)

Equation (3.174) can be rewritten as

$$K_{j-1}(T_{j-1} - T_j) + K_{j+1}(T_{j+1} - T_j) = M_f(T_j^{\Delta\tau} - T_j).$$ (3.179)

Equation (3.179) applies at all interior nodes, including node 0 when the symmetry condition is invoked; that is, for $j = 0$ to $j = N - 1$. Now we need the equation for the surface node N. We note, however, that if we derive an explicit finite difference equation at node N, the associated stability criterion could be much more severe than at any of the interior nodes. (See the discussion of Eq. [3.144] for the constant thermal conductivity counterpart of this node.) This could then force us to use such a small $\Delta\tau$ that the solution of the explicit equations becomes prohibitively expensive in computer time. To alleviate this problem, we will use the third option mentioned in the discussion that follows Eq. (3.147); we will use an implicit equation *only* for the surface node. Hence, Eq. (1.8) applied to the control volume associated with node N in Fig. 3.34 gives

$$\left(\frac{k_{N-1} + k_N}{2}\right)\frac{(T_{N-1}^{\Delta\tau} - T_N^{\Delta\tau})}{\Delta x} + h(T_f - T_N^{\Delta\tau}) = \rho c_p \frac{\Delta x}{2}$$
$$\times \frac{(T_N^{\Delta\tau} - T_N)}{\Delta\tau}.$$ (3.180)

Using the definitions of K and M_f given by Eqs. (3.175) and (3.177), and defining,

$$N_f = h\Delta x/k_f,$$ (3.181)

Eq. (3.180) becomes

$$K_{N-1}(T_{N-1}^{\Delta\tau} - T_N^{\Delta\tau}) + N_f(T_f - T_N^{\Delta\tau}) = \frac{M_f}{2}(T_N^{\Delta\tau} - T_N).$$ (3.182)

Next we put Eqs. (3.173), (3.174) and (3.182) into nondimensional form by introducing the temperature excess ratio

$$\phi_j = \frac{T_j - T_f}{T_i - T_f}.$$ (3.183)

With this, Eq. (3.173) evaluated at node j becomes

$$k_j = k_f(1 + \gamma\phi_j)$$ (3.184)

with

$$\gamma = b/(T_i - T_f). \tag{3.185}$$

Using Eq. (3.183) in Eq. (3.179) gives

$$K_{j-1}(\phi_{j-1} - \phi_j) + K_{j+1}(\phi_{j+1} - \phi_j) = M_f(\phi_j^{\Delta\tau} - \phi_j). \tag{3.186}$$

As mentioned earlier, Eq. (3.179), and therefore also Eq. (3.186), applies to node 0 where $j = 0$, $j + 1 = 1$, and $j - 1$ is also 1 because of the symmetry condition at the midplane. Next we use Eq. (3.183) to put Eq. (3.182) into dimensionless form and then put all equations into the standard form with the nodal temperature excess ratio on a side by itself with a multiplier of unity. This gives the following finite difference equations for the nodes:

$$\phi_0^{\Delta\tau} = \frac{2K_1}{M_f} \phi_1 + \left(1 - \frac{2K_1}{M_f}\right) \phi_0$$

$$.$$
$$.$$
$$.$$

$$\phi_j^{\Delta\tau} = \frac{1}{M_f} (K_{j-1}\phi_{j-1} + K_{j+1}\phi_{j+1}) + \left[1 - \frac{(K_{j-1} + K_{j+1})}{M_f}\right] \phi_j \tag{3.187}$$

$$.$$
$$.$$
$$.$$

$$\phi_N^{\Delta\tau} = \frac{2K_{N-1} \phi_{N-1}^{\Delta\tau}}{M_f + 2K_{N-1} + 2N_f} + \left(\frac{M_f}{M_f + 2K_{N-1} + 2N_f}\right) \phi_N$$

At this point, since the equations are in standard form, the stability of the equations is checked by applying the stability condition, which requires that there be no negative coefficients of the temperature at the node at the present time. Although the stability criterion was, strictly speaking, derived for the linear situation, we assume that it is approximately valid when applied locally in time to nonlinear equations. Experience seems to bear out this assertion (see Richtmeyer and Morton [11]). Thus, at an interior node, the stability criterion requires that

$$M_f \geq K_{j-1} + K_{j+1}. \tag{3.188}$$

The last equation of equation set (3.187) is for the surface node N, and is seen to be unconditionally stable (by our criterion) because the coefficient of ϕ_N is greater than zero if $K_{N-1} > 0$. If on the other hand $K_{N-1} < 0$, then one can show that as long as the stability condition at interior nodes, Eq. (3.188), is satisfied, stability is also insured at the surface node.

Examining Eq. (3.188), one sees that the permissible value of M_f also depends upon the time, since both values of K, by Eqs. (3.184) and (3.177–78), depend upon the instantaneous nodal temperatures. Note that if $\gamma > 0$ in Eq. (3.184), then the most severe condition, as far as stability is concerned, occurs when the values of ϕ are highest: that is, at $\tau = 0$ when all ϕ values are 1.0. So the following conditions insure stability at *all* time steps $\Delta\tau$, if $\gamma > 0$:

$$M_f \geq 2 + 2\gamma \quad \text{for} \quad \gamma > 0. \tag{3.189}$$

If $\gamma < 0$, then condition (3.188) is most severe when ϕ at each node has the lowest possible value, which by Eq. (3.183) is seen to be zero. Hence, the sufficient condition for stability at all time steps when $\gamma < 0$ is

$$M_f \geq 2 \quad \text{for} \quad \gamma < 0. \tag{3.190}$$

Look at the group of difference equations, Eq. (3.187), and note that all of these equations except the last one are explicit and therefore can be solved one at a time. In fact, as long as one solves the second to last equation for $\phi_{N-1}^{\Delta\tau}$ *before* solving the last equation, this computed value of $\phi_{N-1}^{\Delta\tau}$ can be inserted into the last equation and then the equation solved directly for $\phi_N^{\Delta\tau}$.

A computer program is now written to solve these explicit equations, Eq. (3.187), subject to the stability limitation, Eq. (3.189) or (3.190), with the input being the values of Δx, N, h, γ, and α_f so that the nondimensional parameters N_f and M_f can be calculated. Additional input would be values of T_i and T_f so that the actual temperatures, if desired, and not only the temperature excess ratio ϕ, become the output at different times τ. Once this is done, we can return to the original numerical problem that was posed earlier for this 2% tungsten steel slab.

The known numerical values are as follows:

$$T_i = 760°C, \quad T_f = 27°C, \quad h = 17,500 \text{ W/m}^2 \text{ °C},$$
$$L = 0.0175 \text{ m}.$$

The center temperature that the slab is to be cooled to is $T = 200°C$. Hence the value of ϕ at the center at this unknown cooling time is, from Eq. (3.183),

$$\phi_0 = \frac{200 - 27}{760 - 27} = 0.236.$$

Fitting Eq. (3.184) to the experimental thermal conductivity data given in Appendix A yields

$$k = 61.25(1 - 0.5\phi).$$

Thus, $\gamma = -0.5$. Also, $\alpha_f = 1.76 \times 10^{-5} \text{ m}^2/\text{s}$.

Table 3.3
Computer Results for the Slab of Fig. 3.34

Number of Time Increments	Time in Seconds	Computer Results for ϕ	
		Center Node O	Surface Node N
0	0	1.000	1.000
98	0.872	0.9999	0.3528
196	1.743	0.9980	0.2739
587	5.22	0.8518	0.1729
978	8.7	0.6325	0.1319
1369	12.18	0.4585	0.1011
1760	15.66	0.3301	0.0759
2150	19.14	0.2365	0.0560
2542	22.62	0.1690	0.0409

The refinement of nodal spacings Δx was carried out by solving the problem for part of the transient using 21 nodes, then 31 nodes, and finally 41 nodes. From this it was ascertained that 31 nodes allow a solution essentially independent of increment size. Hence the solution was then obtained, for the total duration of the transient, using 31 nodes; this corresponds to a nodal spacing $\Delta x = 0.0005833$ m. Since M_f was held constant as Δx was decreased during the increment refinement process, this meant that the time increment $\Delta\tau$ was simultaneously being refined (see Eq. [3.175]). Using 31 nodes, Eq. (3.181) gives

$$N_f = 17,500(0.0005833)/61.25 = 0.167.$$

For stability, the value of M_f chosen was in accord with the inequality Eq. (3.190) to be slightly above the limit in that equation, since $\gamma = -0.5$. Thus $M_f = 2.174$. This gives the size of the time increment as $\Delta\tau = 0.0089$ s.

The results of the computer solution to Eq. (3.187) for these numerical values are shown in Table 3.3 for the slab center and its surface at selected times during the cooling transient.

The computer results indicate that the required ϕ value at the center is reached after 19.1 seconds. At this same time the ϕ value at the surface is 0.0560, corresponding to a surface temperature of 68°C.

Note that in this problem, an accurate finite difference solution required thirty-one nodes and over two thousand time increments; hence the need for a computer solution of the finite difference equations.

3.6 PROBLEMS

3.1 At a given instant of time, the temperature distribution within a two-dimensional solid is $T = 5y^2 \cos x$. At this instant of time, show whether the temperature at the point $x = 0$, $y = 1$ is increasing or decreasing. Assume no generation.

3.2 For the conditions of Ex. 3.4, find an expression for the energy transferred, per square foot of surface, across the face at $x = 0$ between the times τ_0 and τ_1. The thermal conductivity of the material is k.

3.3 The temperature distribution within a very long cylinder at a particular instant of time is known to be $T = 500 + 100 \cos 3r + 200\, r^2 + 50\, r^3$ °C with the general radial coordinate r in meters. If there is no generation and the thermal diffusivity $\alpha = 2.6 \times 10^{-5}$ m²/s, find the rate at which the cylinder center temperature is changing with time at this instant of time.

3.4 Using Eq. (3.25), derive an expression for the heat transfer rate at any time τ at the face $x = L$.

3.5 Rework the slab problem of Sec. 3.2 for the simpler case in which the slab faces are held at the known temperature T_f, and show by exact analytical methods that the solution is

$$\frac{T - T_f}{T_0 - T_f} = 2 \sum_{n=0}^{\infty} \frac{(-1)^n}{\lambda_n L} e^{-\alpha \lambda_n^2 \tau} \cos(\lambda_n x),$$

where $\lambda_n = \pi(2n + 1)/2L$.

3.6 An infinite cylinder of radius R initially has the following temperature distribution as a function of the general radial coordinate r:

$$T = T_0 J_0[2.405\, (r/R)] + T_f,$$

where T_0 and T_f are known constants and J_0 is an ordinary Bessel function of the first kind and of zero order. Suddenly, the source which is causing the steady-state temperature distribution is shut off, and the surface is held at the constant temperature T_f. Show, by exact analytical solution, that the temperature distribution as a function of r and time τ is

$$\theta = T - T_f = T_0 J_0[2.405(r/R)]\, e^{-\alpha \frac{(2.405)^2}{R^2} \tau}.$$

[*Hint:* Note that $J_0(2.405) = 0$.]

3.7 A solid sphere of radius R is initially at a temperature T_0 throughout when, suddenly, its surface is held at known temperature T_f. Using exact analytical methods, show that the solution for the temperature as a function of r and time τ is

$$\frac{T - T_f}{T_0 - T_f} = 2 \sum_{n=1}^{\infty} (-1)^{n+1} e^{-\alpha \frac{n^2 \pi^2}{R^2} \tau} \frac{\sin(n\pi r/R)}{n\pi(r/R)}.$$

[*Hint:* The transformation $\psi(r, \tau) = r\theta(r, \tau)$, where $\theta(r, \tau) = T - T_f$, of the dependent variable T in the governing partial differential equation for the sphere transient yields a partial differential equation in ψ which is exactly like the one for rectangular cartesian coordinates:

$$\frac{\partial^2\psi}{\partial r^2} = \frac{1}{\alpha}\frac{\partial\psi}{\partial\tau}.$$

Then solve for $\psi(r, \tau)$.]

3.8 Because of the dissipation of electrical energy within the slab shown in Fig. 3.35, the initial steady-state temperature distribution is $T = T_f + T_m \sin(\pi x/L)$. Suddenly, the electrical power is shut off and simultaneously both faces of the slab at $x = 0$ and $x = L$ are perfectly insulated. (a) Using exact analytical methods, find the slab temperature as a function of x and τ. (b) Find the final steady-state temperature of the slab $(\tau \rightarrow \infty)$ in two different ways.

Figure 3.35

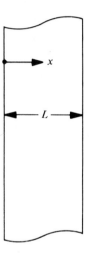

3.9 The relation for the temperature as a function of distance and time in an infinite plate is $T = 100 + 100 \, e^{-0.5\tau} \sin x$. Given that $k = 50$ W/m °C, and that the slab is 1 m wide, compute the time necessary for 2×10^6 J of energy to leave the slab across an area 1 m² on the right-hand face $(x = L)$. The value τ is in hours.

3.10 A long cylindrical bar of radius 1/2 ft is initially at 80°F throughout. It is then placed into a heat-treating oven containing 1000°F gases, and an average surface coefficient of heat transfer of 50 Btu/hr ft² °F has been calculated. Find the time required for the surface to reach 800°F. Assume that $k = 25$ Btu/hr ft °F and $\alpha = 0.578$ ft²/hr.

3.11 Long 0.15-m-diameter shafts come out of an oven at 815°C throughout and are cooled by quenching in a large bath of 38°C coolant. If the surface coefficient of heat transfer between the shaft surface and the coolant is 170 W/m² °C, calculate the time it takes for the shaft center to reach 115°C. Assume that $k = 17.3$ W/m °C and $\alpha = 5.16 \times 10^{-6}$ m²/s.

3.12 For the conditions of Prob. 3.11, (a) calculate the surface temperature of the shaft when its center temperature is 115°C, and (b) estimate the temperature gradient at the outside surface at the same instant of time.

3.13   A 1-in.-thick slab of wood initially at 80°F is accidentally left behind in a large heat-treating oven, where it is subjected to 1500°F gases with an average surface coefficient of heat transfer of 5 Btu/hr ft² °F between the wood surface and the air. If the ignition temperature of the wood is 700°F, determine the time between initial exposure and possible ignition. Assume that $k = 0.30$ Btu/hr ft °F, $\rho = 50$ lbm/ft³, and $c_p = 0.60$ Btu/lbm °F.

3.14   A large 2-in.-thick slab of brick-like material is to be used as the floor of a large high-temperature oven. The bottom side of this slab is essentially insulated. When the oven is turned on, the upper face of the slab is exposed to 3000°F gases and the surface coefficient of heat transfer h is 6 Btu/hr ft² °F. Since the oven is not to be used until all surfaces are at least 2800°F, calculate the time required for the floor to meet this condition. The initial temperature of the slab is 70°F, $\alpha = 0.02$ ft²/hr, and $k = 2.0$ Btu/hr ft °F.

3.15   Large 5-cm-thick sheets of a ceramic material have been stored outside in the winter cold, giving them a uniform temperature of -29°C. They are brought inside a building where the air temperature is 21°C and a surface coefficient of heat transfer of 8.5 W/m² °C exists between the sheets and the air. The ceramic is too brittle to work with until its temperature reaches 10°C. Find how much time must pass before the sheets can be worked. Assume that $k = 19$ W/m °C, $\rho = 3200$ kg/m³, and $c_p = 837$ J/kg °C.

3.16   In the experimental determination of the average surface coefficient of heat transfer between a 2-in.-thick plate and a fluid which will flow over its surface, the plate is initially at 70°F throughout and is then exposed to a 300°F fluid. Measurement indicates a plate center temperature of 254°F after one-half hour has elapsed. If $k = 0.58$ Btu/hr ft °F and $\alpha = 0.02$ ft²/hr, calculate the surface coefficient of heat transfer between the plate and the fluid.

3.17   An orange is idealized as a 5-cm-radius sphere which is initially at 7°C with the properties $k = 0.52$ W/m °C and $\alpha = 1.29 \times 10^{-7}$ m²/s. A severe weather system causes the air surrounding the orange to drop rather abruptly to -5°C. If frost forms on the outside of the orange at 0°C and if the surface coefficient of heat transfer between the orange surface and the air is 5.7 W/m² °C, calculate the time which must elapse before frost occurs.

3.18   A hemisphere of glass which is eventually to be made part of a telescope lens system has a radius of 0.15 m and is initially at 27°C. For stress relief prior to grinding, the hemisphere is placed, flat-side down, on an insulating board and put into a 425°C oven with a surface coefficient of heat transfer of 9.1 W/m² °C between the curved surface of the hemisphere and the gases. Find the maximum temperature after 10 hr have passed. Assume that $k = 0.78$ W/m °C, $\rho = 2723$ kg/m³, and $c_p = 837$ J/kg °C.

3.19   If the glass hemisphere of Prob. 3.18 is taken out of the oven after 10 hr and allowed to cool in 27°C room air with a known surface coefficient of heat transfer between the hemisphere and the air, explain what procedure might be used to estimate the time for the maximum glass temperature to reach 38°C. In particular, explain how to handle the nonuniform initial temperature of this problem while still making use of Heisler's charts. A numerical result is not desired; only a description of a possible procedure.

3.20   For the purpose of experimentally determining the surface coefficient of heat transfer between a sphere and water in a certain forced convection situation, a 4-in.-diameter sphere, initially at 70°F, is plunged into a 200°F water flow. After 3.34 min have passed, a surface temperature of 183°F is measured. Find the surface coefficient of heat transfer between the sphere and the water. Assume that $k = 20$ Btu/hr ft °F and $\alpha = 0.40$ ft²/hr.

Unsteady-State Conduction   287

3.21 Figure 3.36 shows a long cylindrical billet passing through an oven which is 20 ft long. The billet is 1 ft in diameter and its center must be raised to 1400°F. The billet is at 400°F before it enters the furnace, whose gases are at 2700°F, where the surface coefficient of heat transfer between the gases and the billet is $h = 18$ Btu/hr ft² °F. Find the maximum speed at which a continuous billet can travel through the furnace. List all the assumptions required for the analysis and then, in light of the answer arrived at, comment on how valid the assumptions appear to be. Assume that $\alpha = 0.578$ ft²/hr and $k = 30$ Btu/hr ft °F.

Figure 3.36

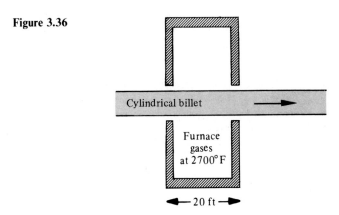

3.22 A 1-in. slab of teflon, initially at 60°F, is to be melted by suspending it above a drip tray in a 1000°F oven. If the surface coefficient of heat transfer between the teflon surface and the furnace gases is 2 Btu/hr ft² °F, find the time required for the surface to reach the melting temperature of 620°F. Assume that $k = 0.14$ Btu/hr ft °F and $\alpha = 0.0041$ ft²/hr.

3.23 Long 4-in.-diameter driveshafts, initially at 80°F, are to be heat-treated in a furnace where a surface coefficient of heat transfer of 20 Btu/hr ft² °F is expected between the shaft surface and the gases. The properties of the driveshafts are $k = 10$ Btu/hr ft °F, and $\alpha = 0.14$ ft²/hr. Determine the temperature of the furnace gases if the driveshaft center temperature reaches 1000°F in 48 min.

3.24 A concrete blast pad in a jet and rocket engine test cell is initially at 26°C when, due to the start of a test, it is suddenly exposed to 815°C gases with an average surface coefficient of heat transfer of 2270 W/m² °C existing between the gases and the concrete. If the concrete is thick enough to be treated as a semi-infinite slab, calculate (a) the surface temperature and (b) the temperature 5 cm below the surface after 10 seconds have passed. Assume $\alpha = 4.39 \times 10^{-7}$ m²/s and $k = 1.73$ W/m °C.

3.25 Show that Eq. (3.39) degenerates to Eq. (3.41) in the limit as $h \to \infty$.

3.26 The ground near a house is to be idealized as a semi-infinite body initially at 50°F. The air above the ground, due to an extreme cold wave, drops quickly to 10°F and remains there. High winds cause a surface coefficient of heat transfer of 5 Btu/hr ft² °F between the air and the surface of the ground. Calculate the time for the soil 6 ft beneath the surface (where a water pipe is buried) to reach 32°F. Assume that $\alpha = 0.015$ ft²/hr and $k = 0.25$ Btu/hr ft °F.

3.27 An oil line running between two buildings is buried at a depth of 0.6 m in the ground. An unusually severe winter has caused the ground to be at a temperature of 4°C. However, the oil is so viscous at this temperature that the correct flow rate cannot be maintained. Hence, it is proposed to warm the line to 10°C by spraying 60°C water along the ground surface, causing the surface temperature to rise abruptly to 60°C. Estimate the time required for the ground at the oil level to reach 10°C. Assume that $\alpha = 3.87 \times 10^{-7}$ m²/s and $k = 0.43$ W/m °C.

3.28 Figure 3.37 shows an experimental rig as set up for measuring the surface coefficient of heat transfer in normal impingement heating or cooling by a clustered array of nozzles. In this type of cooling (or heating), a jet of fluid is sprayed on the surface to be heated or cooled in a direction perpendicular (or normal) to the surface. To measure the surface coefficient of heat transfer, the nozzles cause the fluid to impinge on a flat surface, originally at 80°F, which is large enough to be considered a semi-infinite solid, and then a surface temperature is measured and the time recorded to allow calculation of the value of h from Eq. (3.39). For the condition in which the impinging fluid is at 300°F and the measured surface temperature is 212°F after 40.5 sec, calculate the surface coefficient of heat transfer. Assume that $\alpha = 0.10$ ft²/hr and $k = 10$ Btu/hr ft °F.

Figure 3.37

Fluid at T_f impinging on top surface

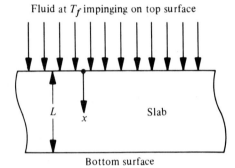

Bottom surface

3.29 A firewall is to be constructed out of bricks, such that a temperature of 1100°C impressed on one side of the wall hardly causes any temperature change at all on the other side of the wall for 5 hr. Determine the thickness of the wall. Note that $\alpha = 5.42 \times 10^{-7}$ m²/s.

3.30 In Ex. 3.5, assume that there is a finite surface coefficient of heat transfer $h = 11$ W/m² °C between the 315°C gas mixture and the wall of the test cell. Find the temperature 7.5 cm from the surface after 7.52 hr have passed, and compare it to the 205°C temperature at that point in Ex. 3.5. Assume that $k = 1.7$ W/m °C.

3.31 In Prob. 3.28, the slab is actually of finite thickness L and only behaves as if it were infinitely thick until the thermal loading caused by the impinging jets on the upper surface, as shown in Fig. 3.37, gives rise to a temperature change at the bottom surface. The slab is initially at 26°C. For 150°C fluid, the worst possible condition for the slab (in terms of the thickness needed to make the slab behave as if it were semi-infinite) is when the surface coefficient is extremely large, leading to the top surface having the temperature of the fluid, 150°C. In some experimental runs, 60 seconds will be needed before the temperatures are recorded. Estimate the minimum slab thickness L needed for the slab to behave as if it were semi-infinite if the condition at the bottom face is $\dfrac{T - T_f}{T_0 - T_f} \geq 0.95$.

3.32 When an infinite slab of thickness $2L$, as shown in Fig. 3.38, is initially at temperature T_0 and is suddenly exposed on both faces to a fluid at temperature T_f and surface coefficient of heat transfer h, the solution for the temperature can be found from Heisler's charts of Figs. 3.4–5, except at very short times for the reasons discussed earlier. Describe a method for finding the temperature distribution within the slab which might be expected to be valid at very short times.

Figure 3.38

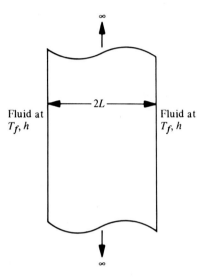

Fluid at T_f, h $2L$ Fluid at T_f, h

3.33 A block of wood has dimensions of 1 ft × 1 ft × 1 ft and is initially at 50°F throughout. It is then placed in an 850°F atmosphere with a surface coefficient of heat transfer of 1.0 Btu/hr ft² °F. Determine how long it will take before the wood first reaches the ignition temperature of 800°F somewhere. Assume that $k = 0.30$ Btu/hr ft °F, $c_p = 0.60$ Btu/lbm °F, and $\rho = 50$ lbm/ft³.

3.34 A circular cylindrical bar is 0.6 m in diameter and 0.6 m high and is initially at 38°C throughout. The properties of the bar material are $k = 47$ W/m °C and $\alpha = 1.47 \times 10^{-5}$ m²/s. The bar is put into a heat-treating oven where the air is at 425°C and a constant surface coefficient of heat transfer of 153 W/m² °C exists between the bar surface and the air. Calculate the center temperature of the bar after 1.73 hr have passed.

3.35 Consider a very long square beam 0.3 m by 0.3 m which is initially at 260°C when it is exposed to a 38°C environment with an average surface coefficient of heat transfer of 568 W/m² °C. The thermal diffusivity is $\alpha = 2.58 \times 10^{-5}$ m²/s and the thermal conductivity is 260 W/m °C. Find the temperature at the center after 1 hr has elapsed.

3.36 A long rectangular bar with a rectangular cross section of 2 in. by 4 in. is to be stress-relieved in a 1100°F heat-treating oven. The blowing of the hot gases over the bar gives rise to a surface coefficient of heat transfer of 20 Btu/hr ft² °F. The bar, which is initially at 70°F, is placed on the oven floor resting on one of its 4-in. faces, as shown in Fig. 3.39. This causes the face of the bar in contact with the oven floor to be effectively insulated. Calculate the lowest temperature in the bar after one hour has passed. The properties of the bar are $k = 20$ Btu/hr ft °F and $\alpha = 0.40$ ft²/hr.

Figure 3.39

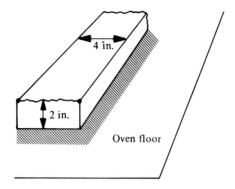

4 in.

2 in.

Oven floor

3.37 It is desired to anneal a cylinder of plate glass 1 ft in diameter by 8 in. high to relieve the stresses in the disk preparatory to optical grinding. The disk is initially at 80°F and is placed in an oven at 800°F. Determine the maximum temperature in the disk after 10 hr have passed. The surface coefficient of heat transfer is $h = 1.6$ Btu/hr ft² °F and the glass properties are $k = 0.45$ Btu/hr ft °F, $\rho = 170$ lbm/ft³ and $c = 0.2$ Btu/lbm °F. Use Heisler's charts in the solution.

3.38 A company is manufacturing custom-made bricks. These bricks are 4-in. cubes, and their properties are $\alpha = 0.02$ ft²/hr and $k = 2.0$ Btu/hr ft °F. Initially the brick is at 80°F throughout when it is thrust into an oven containing gases at 3000°F. If the average surface coefficient of heat transfer h is about constant at 6 Btu/hr ft² °F, calculate the time required for the maximum brick temperature to be 2800°F.

3.39 Rework Prob. 3.21 for the case in which the billet described is just entering the furnace, as shown in Fig. 3.40, and compute the maximum speed of the billet that allows the center temperature to be 1400°F during the start-up of the continuous process described in Prob. 3.21. List assumptions and comment on their validity.

Figure 3.40

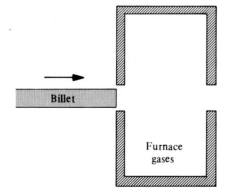

Billet

Furnace gases

3.40　A cube of ice 0.3 m × 0.3 m × 0.3 m is taken from a freezer where its temperature was −29°C and put into an atmosphere where the temperature is 4.5°C. The surface coefficient of heat transfer is 5.67 W/m² °C. The properties of ice are $k = 2.18$ W/m °C and $\alpha = 10^{-6}$ m²/s. Determine (a) where the ice will first reach its melting temperature of 0°C, and (b) how long it will take for the point found in (a) to reach the melting temperature.

3.41　Figure 3.41 shows a turbine disk blank in the shape of a right circular cylinder of radius 9 in. and thickness 4 in. which must undergo a heat treatment before final machining. The blank is initially at 80°F throughout and is then put into a 1000°F oven with an average surface coefficient of heat transfer equal to 16 Btu/hr ft² °F between the oven gases and the blank surface. Calculate the center temperature after 1.13 hr. Assume that $k = 12$ Btu/hr ft °F and $\alpha = 0.25$ ft²/hr.

Figure 3.41

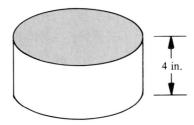

4 in.

3.42　One end of the long 2.5-cm-radius shaft shown in Fig. 3.42, after being at 425°C, is to be quenched in 40°C coolant. A surface coefficient of heat transfer equal to 341 W/m² °C is expected between the shaft surface and the coolant. If $k = 26$ W/m °C and $\alpha = 8.6 \times 10^{-6}$ m²/s, calculate the center temperature of the end face after 1.25 min have elapsed.

Figure 3.42

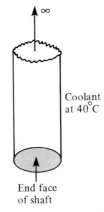

∞

Coolant at 40°C

End face of shaft

3.43　A brick box-shaped building with extremely thick walls serves to protect occupants who observe and operate instrumentation for controlled fires and explosions in the vicinity of the building. In one experiment, a fire outside is to burn for 0.45 hr and cause the outside of the building to be exposed to 1400°C gases with a surface coefficient of heat transfer equal to 57 W/m² °C between the gases and the surface of the building. The hottest point on the outside of the building occurs at a top corner, as shown in Fig. 3.43. The walls and ceiling are initially at 32°C throughout and are thick enough to behave as semi-infinite solids. If k = 1.73 W/m °C and α = 5.7 × 10^{-7} m²/s, find the corner temperature at the end of the fire.

Figure 3.43

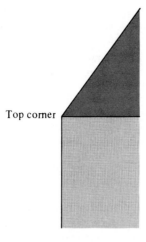

Top corner

3.44 The right circular cylindrical column shown in Fig. 3.44 is 4 ft long by 4 ft in diameter and is formed from two different materials. One-quarter of the cylinder is the metal structural member, while the remaining three-fourths is a decorative high temperature plastic. Initially the entire cylinder is at 500°F when it is placed standing on a circular end on an insulating pad to cool in a 100°F fluid. The heat transfer coefficient h is 10 Btu/hr ft² °F. If the thermal conductivity of the plastic is much smaller than that of the metal portion, find the highest metal temperature after 5.2 hr. For the metal, α = 0.578 ft²/hr and k = 20 Btu/hr ft °F.

Figure 3.44

Plastic

Metal

3.45 Explain why a body generated by the intersection of an infinite cylinder of radius R with a sphere of radius R_s, where $R < R_s$ and the axis of the cylinder passed through the sphere center, is not one which is amenable to a solution of the type which is a product of those for the sphere and the cylinder separately.

3.46 A long carbon steel plate 2.5 cm by 15 cm has been cold-worked and must be heat-treated. It is required that the plate center remain at 538°C for 1 hr when the plate, initially at 38°C, is placed into a 713°C oven with a surface coefficient of heat transfer estimated to be 142 W/m² °C. Find the total time that must pass before the plate can be taken out of the oven. Assume that $k = 43$ W/m °C and $\alpha = 1.29 \times 10^{-5}$ m²/s.

3.47 A spherical thermocouple bead used to measure some fluid temperature T_f is to be designed for transient responses such that, when the thermocouple is originally at the temperature T_0 and then is exposed to the fluid, the temperature indicated by the thermocouple within 1 second causes a nondimensional temperature excess ratio of $0.02 = (T - T_f)/(T_0 - T_f)$. The expected surface coefficient of heat transfer between the bead surface and the fluid is 60 W/m² °C, $\rho = 8010$ kg/m³, $c_p = 418.7$ J/kg °C, and $k = 52$ W/m °C. Find the largest permissible radius of the thermocouple bead.

3.48 A 1/8-in.-diameter aluminum wire has undergone a deep drawing operation, and before proceeding with additional drawing, it must be held at 175°F for 1 hr. If its original temperature is 70°F and it is placed in a 300°F oven with a surface coefficient h of 3 Btu/hr ft² °F, find the time when the oven temperature can be reduced to 175°F.

3.49 A chef wishes to sample some of the soup he is making. However, the soup is at 93°C, so the chef scoops up 0.00091 kg of soup into a wooden spoon and blows on it, causing 24°C air to pass over the liquid surface and giving rise to a surface coefficient of heat transfer of 28 W/m² °C. If the surface area of the soup is 0.00084 m², the energy loss from the soup through the wood neglected, and the soup's properties (for heat transfer calculations, but hopefully not as far as taste is concerned) are those of water, find the time he must continue blowing if he desires to cool the soup to 38°C before tasting.

3.50 Figure 3.45 shows two objects of the same material and dimensions exposed to the same environment and the same surface coefficient of heat transfer, but partially insulated in different ways. Would the lumped parameter method give better results for the case of Fig. 3.45(a) or (b)? No heat flows perpendicular to the paper.

Figure 3.45

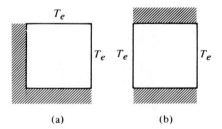

(a)   (b)

3.51 Rework Ex. 3.7 using the lumped parameter method of transient analysis.

3.52 Rework Prob. 3.15 by the lumped parameter method.

Chapter Three

3.53 A 0.91-kg clothes iron has a 100 W heating element. The surface area is 0.037 m², the air temperature is 21°C, and the surface coefficient of heat transfer is 14.2 W/m² °C. Calculate the time required for the iron to reach 104°C after being plugged in. Assume $k = 208$ W/m °C, $\rho = 2720$ kg/m³, and $c = 879$ J/kg °C.

3.54 A long thin wire is used as a sensor in a fire alarm system. When the wire melts and breaks, the alarm is triggered. The melting temperature of the wire is 950°F and its properties are $k = 120$ Btu/hr ft °F, $\rho = 450$ lbm/ft³, and $c_p = 0.10$ Btu/lbm °R. If the wire, which is initially at 70°F, is suddenly exposed to 1200°F gases with a constant surface coefficient of heat transfer of 2.0 Btu/hr ft² °F, determine what the wire diameter should be in order for the alarm system to go off in 80 sec.

3.55 A transient technique is to be used to arrive at the surface coefficient of heat transfer at a particular point on a gas turbine blade. A cube of copper, $\frac{1}{4}$ in. on a side, has been buried flush with the surface of the blade; that is, only one of the six sides of the cube will be directly exposed to the hot gas. The other five sides are effectively insulated by the very low conductivity turbine blade material. The copper is initially at 100°F when the blades are exposed to a 1000°F gas stream which causes the temperature of the copper to change to 450°F in 3.7 sec. Estimate the local surface coefficient of heat transfer at this point on the blade. For copper, $k = 220$ Btu/hr ft °F, $\rho = 558$ lbm/ft³, and $c_p = 0.092$ Btu/lbm °F.

3.56 A block of metal of outside surface area 0.033 m², volume 0.00042 m³, and thermal conductivity 52 W/m °C is initially at 16°C. Suddenly, a power hacksaw begins cutting the metal in half and causes a mechanical energy dissipation rate of 600 W, while simultaneously the outside surface of the metal is cooled by a 16°C coolant with an average surface coefficient of heat transfer of 110 W/m² °C. Find the metal temperature (a) after 15 min have elapsed and (b) after a long enough period of time has passed, while the hacksaw is cutting, so that steady-state conditions prevail.

3.57 A large 1.25-cm aluminum plate suddenly has 150°C air blown over the top of the plate, giving rise to a surface coefficient h of 70 W/m² °C, while the bottom of the plate convects to 21°C air with a surface coefficient of heat transfer of 6 W/m² °C. The initial plate temperature is 21°C. (a) Determine the time it will take for the aluminum to reach 115°C, and justify your assumptions. (b) Also determine if the plate can reach 146°C and explain.

3.58 A small sphere has a very small pool of liquid at its center. The sphere is used as a sensing device in a controlled temperature room. If the liquid at the center reaches its freezing point of -20°F, the change in electrical resistance triggers a circuit to warm up the room. Treat the entire sphere as if it had the properties $k = 224$ Btu/hr ft °F, $\rho = 550$ lbm/ft³, and $c_p = 0.10$ Btu/lbm °R. If the sphere was initially at 10°F and the environmental temperature suddenly drops to -40°F as a result of a defect in the refrigeration equipment, predict the radius of the sphere needed to trigger the circuit after 50 sec have passed. Assume that $h = 10$ Btu/hr ft² °F.

3.59 A brass blank in the shape of a coin 5 cm in diameter and 1.25 cm high is to be brought up to 600°C before a hot-forming operation. If the blank is originally at 50°C and is placed into a 760°C oven with a surface coefficient of heat transfer of 20 W/m² °C, determine how much time the brass must spend in the oven. Assume that $k = 111$ W/m °C and $\alpha = 3.4 \times 10^{-5}$ m²/s.

3.60 By the time the hacksaw has cut through the metal in Prob. 3.56, it is at the temperature in part b of that problem. When the cutting process is completed, the hacksaw shuts off and the metal piece falls apart into two equal size pieces, each of which has an exposed surface area of 0.022 m² for the 16°C cooling fluid. If the surface coefficient of heat transfer remains at 110 W/m² °C, calculate the time for the metal to cool to 37°C.

3.61 A spherical satellite of radius $R = 0.3$ m and mass equal to 270 kg is in orbit and is being maintained at a constant temperature by its 300 W power supply. If the power supply fails, calculate the temperature of the satellite after 8 hr using the lumped parameter method of analysis. The properties of the satellite material are $c_p = 628$ J/kg °C and $\epsilon_s = 0.70$.

3.62 A solid body at some initial temperature T_0 is suddenly placed in a room where the air temperature is T_∞ and the walls of the room are very large. The heat transfer coefficient is h and the surface of the solid may be considered to be a black body. Assuming that the temperature of the solid is uniform at any instant, derive the differential equation for the variation of temperature with time considering both convection and radiation.
(*Hint:* Assume the solid receives no radiation, but only emits it.)

3.63 A can of beer is originally at 60°F and is put into a refrigerator where the air is at 36°F. Determine how long the beer will take to reach 40°F. Assume beer has the properties of water and that the surface coefficient is about 1 Btu/hr ft² °F. The beer can is 5 in. high and 2 in. in diameter.

3.64 By using the ideas which led to the establishment of the criterion given in Eq. (3.58), develop a criterion for use of the lumped parameter method of transient analysis when an object simultaneously exchanges energy by radiation with another object and convection with a fluid.

3.65 An electric heater whose total area is 2 ft² and whose volume is 1/12 ft³ is dissipating electrical energy at the rate of 5000 watts = 17,000 Btu/hr in the steady state while suspended in a room where the air is at 80°F and the average surface coefficient of heat transfer is 30 Btu/hr ft² °F. Suddenly, the heater falls (but remains plugged in) on to a rug on the floor which effectively insulates half the heater. For the heater, $k = 25$ Btu/hr ft °F, $\rho = 500$ lbm/ft³, and $c_p = 0.1$ Btu/lbm °R. (a) Calculate the initial steady-state temperature of the heater. (b) If the ignition point of the rug is 500°F, calculate the time needed for it to reach the ignition temperature.

3.66 Wooden rods L cm long and D cm in diameter are to be ignited by being exposed to hot gases. If it is desired that ignition occur more quickly (assuming that the initial temperature, the gas temperature, and the heat transfer coefficient do not change), explain qualitatively the reasoning used in making one of the following choices: (1) Use 1-cm shorter rods or (2) use 1-cm smaller diameter rods. Note that only an explanation of the reasoning used in making such a choice is required, not the actual choice itself.

3.67 Long wooden dowels of diameter 0.15 cm are at 15°C when they are exposed to 540°C gases with a surface coefficient of heat transfer of 12 W/m² °C existing between the gases and the dowel. If the ignition temperature of the wood is 315°C, find the exposure time before possible ignition. Assume that $\alpha = 2.58 \times 10^{-7}$ m²/s and $k = 0.43$ W/m °C.

3.68 A metal blank in the shape of a short right circular cylinder is initially at the temperature T_0 in a heat-treating oven when it is taken out and suspended in a cooling fluid in the position shown in Fig. 3.46. The orientation of its surfaces gives rise to three different surface coefficients of heat transfer: h_1 on the bottom face, h_2 on the curved lateral face, and h_3 on the upper face between the object and the fluid at temperature T_f. If the temperature gradients at any instant of time, within the blank, are relatively small, find an expression for the blank's temperature as a function of time.

Figure 3.46

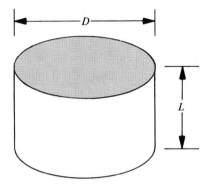

3.69 The small electronic component shown in Fig. 3.47 is originally at the temperature T_0 = 70°F, when suddenly a transient discharge from a capacitor bank causes an electrical energy dissipation rate that varies with time within the component according to the relation $\dot{Q} = \dot{Q}_0 e^{-\beta\tau}$, where $\dot{Q}_0$ = 25 Btu/hr and β = 30 (hr)$^{-1}$. The surrounding air is at 70°F, h = 1.0 Btu/hr ft^2 °F, the surface area of the component is 0.04 ft^2, the volume of the component is 0.02 ft^3, the mass of the component is 0.05 lbm, c_p = 0.20 Btu/lbm °F, and k = 10 Btu/hr ft °F. (a) Derive a relation for the component temperature as a function of time and known parameters such as T_0, $\dot{Q}_0$, etc. (b) Find the component temperature after 5 min have passed.

Figure 3.47

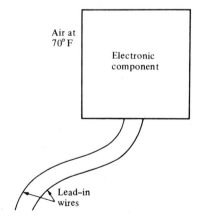

Air at 70° F

Electronic component

Lead–in wires

3.70 The electrical input to a 3-lb aluminum household iron depends upon the temperature of the iron in the following manner: Electrical input $= 200 - (T - 70)/10$ Btu/hr, when the temperature is in °F. The surface area is 0.5 ft^2 and the surface coefficient of heat transfer is about 2.0 Btu/hr ft^2 °F. The properties of aluminum are $\rho = 170$ lbm/ft^3, $c_p = 0.21$ Btu/lbm °F, and $k = 120$ Btu/hr ft °F. The air temperature is 70°F. (a) Derive a relationship for the temperature of the iron as a function of time after the iron is plugged in. (b) Find the time needed to raise the iron's temperature to 220°F. (c) To reduce the time required in part b, the iron could be placed on an asbestos pad which essentially insulates it. If this were done, how long would it take to reach 220°F? (d) Determine the final equilibrium temperature of the iron.

3.71 In Eq. (3.91), show that as $h \rightarrow \infty$, the cylinder wall temperature becomes equal to the gas temperature T_f at any instant of time.

3.72 Using the results of Eq. (3.91), show that the cylinder wall temperature lags the gas temperature and also find the amplitude of the wall temperature and compare it to the amplitude of the gas temperature. What conclusions can be drawn about the effect of the magnitude of the angular frequency ω on the amplitude of the temperature of the cylinder wall?

3.73 A metal component is undergoing a critical final machining process which is causing a release of heat energy within the metal at an initial rate of $\dot{Q}$ Btu/hr. Because of the importance of keeping the component's temperature low, a cooling stream of air at constant temperature T_m is directed on the metal. This leads to a low constant metal temperature of T_i. The Biot number is less than 0.1 and quantities such as ρ, c_p, V_0 and A_s are known. Suddenly, a partial breakdown in the equipment keeping the air cool occurs, causing the air temperature to vary such that $T_e = T_m + (T_0 - T_m) e^{-\tau} \sin \tau$. A feedback control system senses the breakdown and starts shutting down the machining process so that the generation rate becomes $\dot{Q} e^{-\tau/2}$. (a) Derive an expression for the component temperature as a function of time. (b) Calculate the final steady-state temperature of the component in terms of known quantities.

3.74 A steel cube 1 in. on a side is originally at the constant temperature T_0, when suddenly it is exposed to a flowing fluid whose temperature varies with time in the following manner: $T_f = T_2 + T_3 (\frac{1}{2} - \tau)$ for $0 \leq \tau \leq 1/2$ hr, and $T_f = T_2$ for $\tau \geq 1/2$ hr. The surface coefficient of heat transfer is 10 Btu/hr ft^2 °F. (a) Derive expressions for the cube temperature at all values of time $0 < \tau < \infty$. (b) If $T_0 = 60$°F, $T_2 = 400$°F, and $T_3 = 200$°F, find the maximum temperature reached by the cube.

3.75 A soldering iron has an outside surface area of 0.011 m^2, a mass of 0.62 kg, and a 50 W heating element with $\rho = 8000$ kg/m^3 and $c_p = 418.7$ J/kg °C. The soldering iron is initially at 26°C and is then covered with insulation and plugged in. Assuming that the lumped parameter method is valid, estimate the time it takes to reach 430°C.

3.76 A small piece of metal with known properties ρ, k, and c_p is initially at a uniform temperature T_2. Suddenly, at $\tau = 0$, there is constant generation established in the material at a rate $\dot{Q}$ and the fluid surrounding the metal has the following temperature variation with time: $T_e = T_i + T_0 \tau e^{-\tau}$, where h is known and constant. The surface area for heat transfer is A_s and the volume is V_0. (a) Derive an expression for the metal temperature as a function of time if it is known that the Biot number is less than 0.10. (b) Determine the final steady-state temperature of the metal.

3.77 An object of surface area A_s and volume V_0 is initially at the temperature T_0 when it is exposed to a fluid whose temperature is $T_f = T_1 + T_2 \sin \omega\tau + T_3 \cos \omega\tau$ and a constant surface coefficient of heat transfer h prevails between object and fluid. If conditions are such as to allow use of the lumped parameter method of transient analysis, find the object's temperature as a function of time τ and explain how this solution could be used to treat the case of a fluid temperature which is periodic, but is otherwise an arbitrary function of time.

3.78 A bearing has an outside surface area of 0.09 m² exposed to the air and a total volume of 0.0012 m³. The thermal conductivity is $k = 45$ W/m °C and the surface coefficient of heat transfer between the exposed bearing surface and the air is $h = 300$ W/m² °C. The air temperature is constant at T_∞, and the other bearing material properties such as α, ρ, and c_p are known. Initially, steady-state conditions prevail with work being done on the bearing by the rotating shaft in the amount of A_0 W. Then, because of a partial lubrication failure, the work rate on the bearing varies with time in the manner $(A_0 + B_0 e^{-b\tau})$ W where A_0, B_0, and b are known constants and τ is time. (a) Derive an expression for the bearing temperature as a function of time and known quantities. (b) If A_0, B_0, and b are positive quantities, make a qualitative sketch of the bearing temperature as a function of time. (c) Decide whether the failure in the lubrication system is being remedied.

3.79 Show by derivation that the transient finite difference equation for the situation described in Prob. 2.87 is

$$
k_b(T_2 - T_0) + k_a(T_1 - T_0) + \tfrac{1}{2}(k_a + k_b)(T_4 - T_0)
$$
$$
+ \tfrac{1}{2}(k_a + k_b)(T_3 - T_0) + q_0'''(\Delta x)^2 = \frac{(\Delta x)^2(\rho_a c_{p_a} + \rho_b c_{p_b})}{2\Delta\tau}(T_0^{\Delta\tau} - T_0).
$$

3.80 For the corner node described in Prob. 2.92, show that the transient finite difference equation is

$$
T_0^{\Delta\tau} = \frac{2}{M}(T_1 + T_2) + \frac{2}{M}(N_a + N_b)\,T_\infty + \left[1 - \frac{4 + 2(N_b + N_a)}{M}\right]T_0,
$$

where $N_b = h_b\Delta x/k$ and $N_a = h_a\Delta x/k$.

3.81 Derive Eq. (3.120) by performing an energy balance on an interior node in a two-dimensional conduction region when all nodes are equally spaced.

3.82 Show that the transient finite difference equation for the conditions of Prob. 2.93 is

$$
k r_{10}\,\Delta\theta\,\Delta z\,\frac{T_1 - T_0}{\Delta r} + k(r_{10} + \Delta r)\,\Delta\theta\,\Delta z\,\frac{T_2 - T_0}{\Delta r}
$$
$$
+ k\,\Delta r\,\Delta z\,\frac{T_3 - T_0}{(r_{10} + \tfrac{1}{2}\Delta r)\,\Delta\theta} + k\,\Delta r\,\Delta z\,\frac{T_4 - T_0}{(r_{10} + \tfrac{1}{2}\Delta r)\,\Delta\theta}
$$
$$
+ k(r_{10} + \tfrac{1}{2}\Delta r)\,\Delta\theta\,\Delta r\,\frac{T_5 - T_0}{\Delta z} + k(r_{10} + \tfrac{1}{2}\Delta r)\,\Delta\theta\,\Delta r\,\frac{T_6 - T_0}{\Delta z}
$$
$$
+ q_0'''(r_{10} + \tfrac{1}{2}\Delta r)\,\Delta\theta\,\Delta r\,\Delta z = \rho c_p(r_{10} + \tfrac{1}{2}\Delta r)\,\Delta\theta\,\Delta r\,\Delta z\,\frac{T_0^{\Delta\tau} - T_0}{\Delta\tau}.
$$

3.83 Show that the transient finite difference equation for an interior node for an equispaced lattice in the circular cylindrical coordinate system when the temperature depends only upon radius r and time is

$$T_1 - T_0 + \left(1 - \frac{\Delta r}{r_{10}}\right)(T_2 - T_0) + \frac{q_0'''}{k}\left(1 + \frac{\Delta r}{2r_{10}}\right)\Delta r^2$$

$$= \left(1 + \frac{\Delta r}{2r_{10}}\right)\frac{\Delta r^2}{\alpha\Delta\tau}(T_0^{\Delta\tau} - T_0),$$

where r_{10} is the radius to a point midway between nodes 1 and 0.

3.84 Figure 3.48 shows node 0 on a surface which is receiving mechanical energy at the known rate W_0 Btu/hr ft² or W/m² because of a grinding operation being performed on the surface. Find the transient finite difference equation for this node. Lattice spacings are all equal to Δx and there is no generation within the material.

Figure 3.48

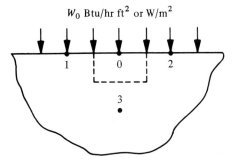

W_0 Btu/hr ft² or W/m²

3.85 Derive the transient finite difference equation for the noninterior node shown in Fig. 3.49 when the temperature distribution depends only upon radius and time and an equispaced lattice in the circular cylindrical coordinate system is used.

Figure 3.49

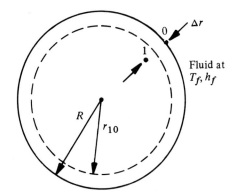

Fluid at T_f, h_f

3.86 Show that the result of Prob. 3.83 reduces to a form exactly like Eq. (3.121) when the lattice spacing is such that $\Delta r \ll r_{10}$. Even for very small spacings Δr, is there any position in a conduction region at which the inequality $\Delta r \ll r_{10}$ would not be expected to hold?

3.87 Show by derivation that the transient finite difference equation for node 0 in Prob. 2.91 is

$$T_0^{\Delta\tau} = \frac{2}{M}\left[T_3 + \tfrac{1}{2}T_2 + \tfrac{1}{2}T_1 + N_\infty \, T_\infty + q_0''' \, \frac{(\Delta x)^2}{2k}\right]$$
$$+ \left\{1 - \frac{[4 + 2N_\infty + (2\sigma\Delta x/k)T_0^3]}{M}\right\} T_0.$$

3.88 For a one-dimensional transient with no generation in a rectangular cartesian coordinate system, show that the transient finite difference equation, when the thermal conductivity is temperature dependent, can be written as

$$T_0^{\Delta\tau} = \frac{\Delta\tau}{\rho c_p (\Delta x)^2}\left[\tfrac{1}{2}(k_1 + k_0)T_1 + \tfrac{1}{2}(k_2 + k_0)T_2\right]$$
$$+ \left[1 - \frac{\Delta\tau}{\rho c_p (\Delta x)^2}(\tfrac{1}{2}k_1 + \tfrac{1}{2}k_2 + k_0)\right] T_0.$$

3.89 Show that the result in Prob. 3.88 reduces to Eq. (3.121) (with $q_0''' = 0$) when the thermal conductivity does not vary with temperature.

3.90 Show that the stability criterion for Prob. 3.79 is

$$1 - \frac{4(k_a + k_b) \, \Delta\tau}{(\Delta x)^2 \, (\rho_a c_{p_a} + \rho_b c_{p_b})} \geq 0.$$

3.91 For the finite difference equation of Prob. 3.80, show that

$$M \geq 4 + 2(N_b + N_a)$$

is required for stability.

3.92 Show that the stability criterion for the transient finite difference equation of Prob. 3.83 is

$$1 - \frac{(2 + \Delta r/r_{10})\alpha\Delta\tau}{(1 + \Delta r/2r_{10}) \, (\Delta r)^2} \geq 0.$$

3.93 Show that for the finite difference equation given in Prob. 3.87, the stability criterion is (assuming that the statement of the criterion holds for nonlinear boundary conditions)

$$M \geq 4 + 2N_\infty + 2\sigma(\Delta x/k) \, T_0^3 \, .$$

Discuss the major differences between this criterion and the ones developed earlier.

3.94 Show that the stability criterion for the equation given in Prob. 3.88 is

$$\rho \, \frac{c_p \, (\Delta x)^2}{\Delta\tau} \geq \tfrac{1}{2}k_1 + \tfrac{1}{2}k_2 + k_0.$$

3.95 Show that to insure stability of Eq. (3.117),

$$1 - 2\alpha\Delta\tau \left[\frac{1}{(\Delta x)^2} + \frac{1}{(\Delta y)^2} + \frac{1}{(\Delta z)^2}\right] \geq 0$$

must be satisfied.

3.96 Find the stability criterion for the finite difference equation of Prob. 3.84.

Unsteady-State Conduction

3.97 Derive the governing finite difference equation and the stability criterion for the noninterior node 3 which occurs on the surface of the sphere in Fig. 3.26. The sphere surface is surrounded by a fluid at known temperature T_f and with a known surface coefficient of heat transfer h between the sphere surface and the fluid.

3.98 If in Eq. (3.121) the generation rate per unit volume around node 0 is given as the linear function of the temperature at node 0, $q_0''' = a_0 + b_0 T_0$, where a_0 and b_0 are known constants, find the new stability criterion.

3.99 Show that the result of Prob. 3.95 reduces to Eq. (3.141) when the lattice spacings are equal.

3.100 Figure 3.50 shows a metal tank support initially at $120°F$ throughout. Suddenly, a tank containing molten metal is placed on the support, causing the surface temperatures to change to those shown in Fig. 3.50 and to remain constant at those values. Using an equispaced lattice of $\Delta x = \Delta y = 1$ in. and $M = 4$, find (a) the size of the time increments and (b) the temperatures at the nodal points at the various times to steady state. Assume $\alpha = 0.6$ ft²/hr.

Figure 3.50

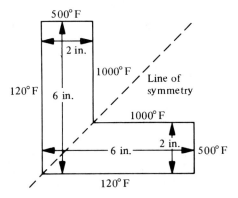

3.101 Consider a plane slab initially at $38°C$ which suddenly has one face raised to $260°C$ while the other face is held at $38°C$. The slab is 0.3 m thick with $k = 87$ W/m °C and $\alpha = 1.29 \times 10^{-5}$ m²/s. Using three or more interior nodes, find the temperature at the various nodes as a function of time until the steady state is approached.

3.102 Figure 3.51 shows a thick fin which is initially at the temperature of the fluid flowing over it; that is, $40°C$. The average surface coefficient of heat transfer between the fluid and the top, bottom, and right-hand surfaces of the fin is 700 W/m² °C. Suddenly, because of the start-up of the heat exchanger to which the left face of the fin is attached, the left face reaches $150°C$. Using a 2.5-cm equispaced lattice, solve for the transient temperature distribution by finite difference methods. Assume $k = 17$ W/m °C and $\alpha = 10^{-5}$ m²/s.

Figure 3.51

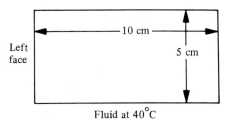

Fluid at $40°C$

3.103 Figure 3.52 shows a two-dimensional conduction region with no generation. Originally, every point within the region is at 100°F, when, suddenly, at $\tau = 0$, the upper edge is raised to 400°F and the sides to 200°F and held there. The bottom of the region continues to be held at 100°F. With a grid spacing $\Delta x = \Delta y = 3$ in. and $M = 5$, use numerical techniques to arrive at the node temperatures as a function of time. Continue until the steady state has practically been reached. The thermal diffusivity α of the material is $1/8$ ft²/hr. Make a table showing the nodal temperatures at the various times.

Figure 3.52

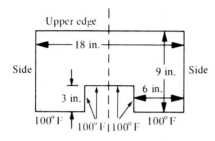

3.104 A 3-m-long, 1.25-cm-diameter steel rod connects two surfaces. One surface is at 200°C and the other is at 40°C, and the connections between the two surfaces and the ends of the rod are intimate ones. The curved surface of the steel rod is essentially perfectly insulated. Suddenly, a current is passed through the bar, giving rise to an energy generation rate per unit volume of 2700 W/m³ within the rod as a result of the electrical dissipation. Taking $k = 35$ W/m °C and $\alpha = 1.55 \times 10^{-5}$ m²/s and using nine interior nodes, find the temperature distribution for the first four time increments.

3.105 Figure 3.53 shows a two-dimensional conduction region which is initially at 100°F throughout, when suddenly, at $\tau = 0$, the lowest face is held at 400°F, the top face at 200°F, and all others at 100°F. The thermal diffusivity is $\alpha = 1/8$ ft²/hr. Using a square grid spacing of $\Delta x = \Delta y = 3$ in., (a) compute the size of the time increment when using $M = 5$, and (b) find the temperatures at the grid points as a function of time up to where the steady state has practically been reached.

Figure 3.53

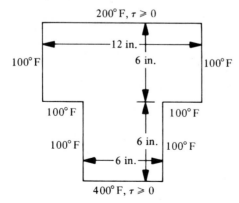

Unsteady-State Conduction 303

3.106 The triangular conduction region in Fig. 3.54 is initially at 10°C throughout when suddenly the left face is held at 150°C while the bottom face is held at 10°C. The face corresponding to the hypotenuse is perfectly insulated. Using an equispaced lattice, find the transient temperature distribution for the first five time increments. Assume $\alpha = 1.29 \times 10^{-5}$ m²/s.

Figure 3.54

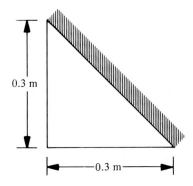

0.3 m

0.3 m

3.107 A square column of dimensions 2 ft by 2 ft is formed by pouring 60°F concrete into a form. The column's cross section is shown in Fig. 3.55. After pouring, the outside of the column is held at 60°F with a spray of cooling water, and the setting of the concrete results in a uniformly distributed source $q''' = 14$ Btu/hr ft³. The thermal conductivity of the concrete is $k = 0.7$ Btu/hr ft °F, and the thermal diffusivity is $\alpha = 0.025$ ft²/hr. Find the transient two-dimensional temperature distribution by finite difference methods to see if there will be any difficulty with high interior temperatures or with thermal stresses resulting from the temperature distribution.

Figure 3.55

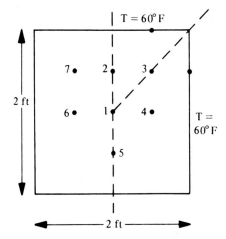

T = 60° F

2 ft

T = 60° F

2 ft

3.108 Refer to the physical situation described in Prob. 2.97. Initially, the insulating material is at 15°C everywhere, when suddenly the heating system is turned on, causing the surface of the large duct to abruptly reach 90°C while the surfaces of the small ducts reach 32°C. The surrounding fluid remains at 15°C while the surface coefficient h_0 of heat transfer between the 15°C fluid and the duct surface is 10 W/m² °C. Given that $l = 5$ cm, $k = 0.078$ W/m °C, and $\alpha = 5.16 \times 10^{-7}$ m²/s, and using an equispaced 5-cm lattice, find the transient temperature distribution.

3.109 Refer to the physical situation in Prob. 2.101 where, initially, all the material surrounding the steam pipe is at 80°F. Suddenly, the steam is turned on, causing the pipe surface to come to 300°F. If $\alpha = 0.03$ ft²/hr, solve for the transient temperature distribution using finite differences.

3.110 Refer to the turbine blade of Prob. 2.106. Initially the blade is at 100°F when the start-up of the jet engine causes the boundary conditions to be those given in Prob. 2.106. If $\alpha = 0.36$ ft²/hr, solve for the transient temperature distribution.

References

1. Smith, G. D. *Numerical Solution of Partial Differential Equations.* New York: Oxford University Press, 1965.
2. Wylie, C. R. *Advanced Engineering Mathematics.* 2d ed. New York: McGraw-Hill, 1960.
3. Kreith, F. *Principles of Heat Transfer.* 2d ed. Scranton, Pa.: International Textbook Company, 1965.
4. Obert, E. F. *Concepts of Thermodynamics.* New York: McGraw-Hill, 1960.
5. Arpaci, V. S. *Conduction Heat Transfer.* Reading, Mass.: Addison-Wesley, 1966.
6. Gebhart, B. *Heat Transfer.* 2d ed. New York: McGraw-Hill, 1971.
7. Holman, J. P. *Heat Transfer.* 2d ed. New York: McGraw-Hill, 1968.
8. Schneider, P. J. *Conduction Heat Transfer.* Reading, Mass.: Addison-Wesley, 1955.
9. Heisler, M. P. "Temperature Charts for Induction and Constant Temperature Heating," *Transactions of the A.S.M.E.,* Vol. 69 (1947), pp. 227–36.
10. Sucec, J., and S. Hegde. "Transient Conduction in a Slab with Temperature Dependent Thermal Conductivity." *Journal of Heat Transfer,* Vol. 100, No. 1 (1978), pp. 172–74.
11. Richtmeyer, R. D., and K. W. Morton. *Difference Methods for Initial Value Problems.* New York: Interscience Publishers, 1967.
12. Incropera, F. P., and D. P. DeWitt. *Fundamentals of Heat Transfer.* New York: Wiley, 1981.
13. Karlekar, B. V., and R. M. Desmond. *Heat Transfer.* 2d ed. St. Paul, Minn.: West, 1982.
14. Sucec, J. *Heat Transfer.* New York: Simon and Schuster, 1975.
15. Myers, G. E. *Analytical Methods in Conduction Heat Transfer.* New York: McGraw-Hill, 1971.

RADIATION HEAT TRANSFER

4.1 INTRODUCTION

The mode of heat transfer termed *radiation* implies any of the various energy transfer processes which arise as a result of the emission of energy-carrying electromagnetic waves. These waves are emitted by materials because of changes in the energy levels of atoms and molecules. As an example, consider the core of a nuclear reactor in the case where a fuel or a moderator nucleus absorbs a wandering neutron in a nonproductive capture, that is, one which does not result in a fission. The resultant compound nucleus will be unstable and may return to a more stable state by the emission of a gamma photon— gamma radiation—with a wavelength of perhaps about 10^{-6} micrometers (1 micrometer $= 1 \ \mu m = 10^{-6}$ m). This is one way in which radiation can occur. Emission of radiant energy at a much different wavelength can arise due to braking radiation, the deceleration of a high-speed electron in the Coulomb force field near a nucleus. In radiation heat transfer, however, all attention is focused upon *thermal radiation,* that radiant energy emitted by matter simply as a consequence of its temperature being above absolute zero. Thermal radiation occurs primarily between the wavelengths of 0.1 μm and about 100 μm. One part of thermal radiation is the visible wavelength range, from about 0.3 μm to 0.7 μm. The portion of the thermal spectrum of wavelengths below 0.3 μm is called the ultraviolet range, while the portion above about 0.7 μm is termed the infrared range.

As indicated here and in Chapter 1, thermal radiation is based on a material's temperature level. In Chapter 1 we noted that the rate at which a surface emits thermal radiation depends upon something close to the fourth power of the absolute temperature. Hence at high temperature levels, as might be found in a furnace, a jet engine combustor, or an electric resistive heating element, one would expect radiation to be an important mode of heat transfer which might occur simultaneously with convection, conduction, or with both. In a vacuum, convection is, of course, absent and radiation becomes the only possible mode of heat transfer between unconnected surfaces. Actually, the surface coefficient of heat transfer h that usually occurs between a solid surface and an adjacent gas, under *natural* convection circumstances, is generally so low that the radiation contribution to the heat transfer rate from the surface is significant even at low temperature levels and at low differences in temperature between different solid surfaces. The heating and air conditioning engineer has to consider solar radiation when sizing temperature-controlling equipment for a building. It is for these and other reasons that we study radiation heat transfer.

We noted in Chapter 1 that the radiation calculation of primary importance in engineering radiation heat transfer is the calculation of the *net radiant loss from a surface;* that is, the difference between the rate at which a surface *emits* radiant energy and the rate at which it *absorbs* radiant energy from all sources. It is this net radiant loss which must be balanced by other types of energy transfer terms, such as convection, if the radiating body is to be in the steady state. And it is the difference between this net radiant loss and all other energy transfer and generation terms which causes the body temperature to vary with time in a transient. Thus the primary focus of our study in this chapter will be on the calculation of the net radiant loss. First to be discussed will be the terminology, definitions, and concepts of thermal radiation which we need to understand in order to calculate the net radiant loss. We will discuss the distribution in wavelength and in space of emitted radiant energy as well as what can occur when radiant energy from one body strikes the surface of another body. This knowledge will then be used to develop a number of techniques to calculate the net radiant loss from any surface of interest.

4.2 CONCEPTS OF THERMAL RADIATION

The total emissive power W is the total rate at which a surface, at absolute temperature T, emits radiant energy per unit area of that surface. It is one of the quantities which must be found in order to eventually calculate the net radiant loss. The units of W are W/m^2 or $Btu/hr\ ft^2$.

Experiment shows that, in general, W is a complicated function of temperature, type of material, and surface condition. However, as was discussed in Chapter 1, there is a class of ideal surfaces, called *black surfaces,* which absorb all incident radiant energy (and hence would appear black to the eye as long as they are not at too high a temperature level) and emit radiant energy at the highest possible rate per unit area at any given temperature level. The total rate, W_b, at which a black surface emits radiant energy per unit area at the absolute temperature T was found experimentally by Stefan and later shown theoretically by Boltzmann, who used thermodynamic concepts applied to a photon gas (see Jakob [1]). It is given by the Stefan-Boltzmann law as,

$$W_b = \sigma T^4. \tag{4.1}$$

In Eq. (4.1), T is the temperature in Kelvin degrees or Rankine degrees, while σ is the Stefan-Boltzmann constant with the value

$$\sigma = 5.669 \times 10^{-8}\ W/m^2\ {}^\circ K^4$$

or

$$\sigma = 0.1714 \times 10^{-8} \text{ Btu/hr ft}^2 \text{ °R}^4.$$

No perfectly black surfaces exist in nature, though some surfaces are very close to being black. Actual surfaces emit radiant energy at a total rate per unit area which is less than that of a black surface at the same temperature. Thus, the total emissive power W of a nonblack surface is given as follows.

$$W = \epsilon \sigma T^4. \tag{4.2}$$

The variable ϵ is called the total hemispherical emissivity, and its value depends upon temperature, surface condition, and type of material. A selection of experimentally measured values of ϵ is given in Appendix B.

Both the total emissive powers, Eq. (4.1) for black surfaces and Eq. (4.2) for general nonblack surfaces, and the total hemispherical emissivity ϵ apply to all of the radiant energy emitted from a surface; that is, radiant energy in every possible wavelength and direction in space. *Total* means including every possible wavelength, while *hemispherical* refers to radiation emitted in all possible directions. We will discuss the distribution of emitted radiant energy both in terms of wavelength and in terms of direction in space.

4.2a Wavelength Distribution of Emitted Radiant Energy

Black Body

It has already been mentioned that the total emissive power of a black body given by the Stefan-Boltzmann law, Eq. (4.1), represents the total emission over all possible wavelengths. Planck, using quantum theory, derived the distribution law for the *monochromatic emissive power of a black surface*. This is the rate $W_{b\lambda}$ at which a black surface emits radiant energy per unit area and per unit wavelength around the wavelength λ. *Planck's distribution law* is

$$W_{b\lambda} = \frac{C_1 \lambda^{-5}}{e^{C_2/\lambda T} - 1}. \tag{4.3}$$

Here $W_{b\lambda}$ has units of W/m^2 μm or Btu/hr ft^2 μm, λ is the wavelength of the emitted radiant energy in micrometers (μm), T is the surface temperature in °K or °R, $C_1 = 3.7420 \times 10^8$ W(μm)^4/m^2 or $C_1 = 1.187 \times 10^8$ Btu(μm)^4/hr ft^2, and $C_2 = 1.4388 \times 10^4$ μm °K or $C_2 = 2.5897 \times 10^4$ μm °R.

In Fig. 4.1, the monochromatic black body emissive power $W_{b\lambda}$ is plotted against wavelength λ at the two temperatures 2000°R and 3000°R. As can be seen, at a given temperature, the wavelength for which $W_{b\lambda}$ is a maximum shifts to lower and lower wavelengths as the temperature of the surface increases. The wavelength at which $W_{b\lambda}$ is a maximum

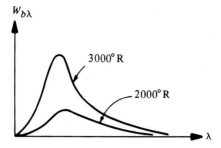

Figure 4.1 The monochromatic emissive power of a black body as a function of wavelength and temperature.

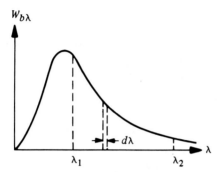

Figure 4.2 Plot of $W_{b\lambda}$ vs. λ indicating the determination of the emissive power of a black body between two given wavelengths.

at any given temperature is denoted by λ_m, and it can be shown that λ_m and T obey *Wien's displacement law:*

$$\lambda_m T = 2897.7 \; \mu m \; °K$$

$$\lambda_m T = 5215.6 \; \mu m \; °R \tag{4.4}$$

An important application of Planck's distribution law, Eq. (4.3), determines what fraction of the total radiant energy emitted by a black surface at temperature T is emitted between two particular wavelengths λ_1 and λ_2. To determine this fraction, first compute the rate at which a black surface emits radiant energy between λ_1 and λ_2, and then divide this result by W_b, that is, by σT^4. Figure 4.2 shows a sketch of $W_{b\lambda}$ vs. λ at a given temperature, as well as the wavelength interval of interest between λ_1 and λ_2. Consider an interval of wavelength $d\lambda$ around λ as shown in Fig. 4.2. The rate at which the black surface emits radiant energy per unit area between λ and $\lambda + d\lambda$ is the *monochromatic emissive power* $W_{b\lambda}$ at λ times the wavelength interval $d\lambda$:

$$W_{b\lambda} \, d\lambda. \tag{4.5}$$

Radiation Heat Transfer

To find the rate at which radiant energy is emitted in the interval λ_1 to λ_2, add all contributions of the fundamental type of Eq. (4.5) according to the definite integral $\int_{\lambda_1}^{\lambda_2} W_{b\lambda} \, d\lambda$. Hence

$$W_{b_{\lambda_1 \to \lambda_2}} = \int_{\lambda_1}^{\lambda_2} W_{b\lambda} \, d\lambda, \qquad (4.6)$$

where $W_{b_{\lambda_1 \to \lambda_2}}$ is the rate per unit area of black body emission between the wavelengths λ_1 and λ_2. Hence, the fraction emitted between these wavelengths is

$$\frac{W_{b_{\lambda_1 \to \lambda_2}}}{W_b} = \frac{\int_{\lambda_1}^{\lambda_2} W_{b\lambda} \, d\lambda}{\sigma T^4}. \qquad (4.7)$$

This fraction can be rewritten as

$$\frac{W_{b_{\lambda_1 \to \lambda_2}}}{W_b} = \frac{\int_0^{\lambda_2} W_{b\lambda} \, d\lambda}{\sigma T^4} - \frac{\int_0^{\lambda_1} W_{b\lambda} \, d\lambda}{\sigma T^4}. \qquad (4.8)$$

The fraction $\int_0^{\lambda} W_{b\lambda} \, d\lambda / \sigma T^4$ has been tabulated by Dunkle [2] as a function of the product λT and is given in Table 4.1. This table, used in conjunction with Eq. (4.8), permits quick calculation of the rate at which radiant energy is emitted per unit area by a black surface between any two wavelengths of interest λ_1 and λ_2, while at the temperature T.

Nonblack Surfaces

The *monochromatic emissive power W_λ of a nonblack surface* is usually written as

$$W_\lambda = \epsilon_\lambda \, W_{b\lambda}, \qquad (4.9)$$

where ϵ_λ is the *monochromatic hemispheric emissivity* and depends upon the wavelength, temperature, material, and surface conditions; its value lies between zero and unity.

The total emissive power of a nonblack surface can be found by summing up all contributions of the type $\epsilon_\lambda \, W_{b\lambda} \, d\lambda$ in a definite integral from $\lambda = 0$ to $\lambda = \infty$ yielding

$$W = \int_0^\infty \epsilon_\lambda \, W_{b\lambda} \, d\lambda. \qquad (4.10)$$

Earlier, however, the expression used for the emissive power of a nonblack surface was given as

$$W = \epsilon \sigma T^4. \qquad [4.2]$$

Table 4.1
The Fraction of Emitted Radiant Energy of a Black Surface Between the Wavelengths 0 and λ as a Function of λT

λT		$W_{b_{0\to\lambda}}/W_b$	λT		$W_{b_{0\to\lambda}}/W_b$	λT		$W_{b_{0\to\lambda}}/W_b$	λT		$W_{b_{0\to\lambda}}/W_b$
$\mu m\,°K$	$\mu m\,°R$	W_b	$\mu m\,°K$	$\mu m\,°R$	W_b	$\mu m\,°K$	$\mu m\,°R$	W_b	$\mu m\,°K$	$\mu m\,°R$	W_b
555.6	1000	0	3222.2	5800	0.3230	5888.9	10,600	0.7282	9,444.4	17,000	0.9017
666.7	1200	0	3333.3	6000	0.3474	6000	10,800	0.7378	10,000	18,000	0.9142
777.8	1400	0	3444.4	6200	0.3712	6111.1	11,000	0.7474	10,555.6	19,000	0.9247
888.9	1600	0.0001	3555.6	6400	0.3945	6222.2	11,200	0.7559	11,111.1	20,000	0.9335
1000	1800	0.0003	3666.7	6600	0.4171	6333.3	11,400	0.7643	11,666.7	21,000	0.9411
1111.1	2000	0.0009	3777.8	6800	0.4391	6444.4	11,600	0.7724	12,222.2	22,000	0.9475
1222.2	2200	0.0025	3888.9	7000	0.4604	6555.6	11,800	0.7802	12,777.8	23,000	0.9531
1333.3	2400	0.0053	4000	7200	0.4809	6666.7	12,000	0.7876	13,333.3	24,000	0.9589
1444.4	2600	0.0098	4111.1	7400	0.5007	6777.8	12,200	0.7947	13,888.9	25,000	0.9621
1555.6	2800	0.0164	4222.2	7600	0.5199	6888.9	12,400	0.8015	14,444.4	26,000	0.9657
1666.7	3000	0.0254	4333.3	7800	0.5381	7000	12,600	0.8081	15,000	27,000	0.9689
1777.8	3200	0.0368	4444.4	8000	0.5558	7111.1	12,800	0.8144	15,555.6	28,000	0.9718
1888.9	3400	0.0506	4555.6	8200	0.5727	7222.2	13,000	0.8204	16,111.1	29,000	0.9742
2000	3600	0.0667	4666.7	8400	0.5890	7333.3	13,200	0.8262	16,666.7	30,000	0.9765
2111.1	3800	0.0850	4777.8	8600	0.6045	7444.4	13,400	0.8317	22,222.2	40,000	0.9881
2222.2	4000	0.1051	4888.9	8800	0.6195	7555.6	13,600	0.8370	27,777.8	50,000	0.9941
2333.3	4200	0.1267	5000	9000	0.6337	7666.7	13,800	0.8421	33,333.3	60,000	0.9963
2444.4	4400	0.1496	5111.1	9200	0.6474	7777.8	14,000	0.8470	38,888.9	70,000	0.9981
2555.6	4600	0.1734	5222.2	9400	0.6606	7888.9	14,200	0.8517	44,444.4	80,000	0.9987
2666.7	4800	0.1979	5333.3	9600	0.6731	8000	14,400	0.8563	50,000	90,000	0.9990
2777.8	5000	0.2229	5444.4	9800	0.6851	8111.1	14,600	0.8606	55,555.6	100,000	0.9992
2888.9	5200	0.2481	5555.6	10,000	0.6966	8222.2	14,800	0.8648	∞	∞	1.0000
3000	5400	0.2733	5666.7	10,200	0.7076	8333.3	15,000	0.8688			
3111.1	5600	0.2983	5777.8	10,400	0.7181	8888.9	16,000	0.8868			

This table is adapted from Dunkle, R. V., "Thermal Radiation Tables and Applications," *Transactions of the A.S.M.E.*, Vol. 76 (1954), p. 549.

Equating Eqs. (4.2) and (4.10) yields, for the total hemispheric emissivity of a nonblack surface in terms of the more fundamental monochromatic hemispheric emissivity, the relation

$$\epsilon = \frac{\int_0^\infty \epsilon_\lambda W_{b\lambda} \, d\lambda}{\sigma T^4}.$$

(4.11)

For surfaces that behave practically as perfect absorbers, that is, black surfaces, ϵ_λ and ϵ are both unity.

In order to use Eqs. (4.10) and (4.11) for nonblack surfaces, the monochromatic hemispherical emissivity ϵ_λ would have to be known as a function of wavelength λ, and this would most likely come from experimental measurements. The value $W_{b\lambda}$ is given by Planck's distribution law, Eq. (4.3), and this, along with ϵ_λ, allows the required integrations in Eqs. (4.10) or (4.11) to be made. These integrals might have to be evaluated graphically, or numerically on the computer, depending upon the functional form of the dependence of ϵ_λ upon wavelength λ.

Next, as was previously done for the case of a black surface, an expression for the fraction of a nonblack surface's emitted radiant energy between any two wavelengths of interest, λ_1 and λ_2, will be derived.

The procedure is that which led to Eq. (4.8), except that the nonblackness of the surface is taken into account by use of the monochromatic hemispheric emissivity ϵ_λ. The rate at which radiant energy is emitted between the wavelengths λ and $\lambda + d\lambda$ is given by the product of the rate W_λ, at which the body emits radiant energy per unit area per unit wavelength around λ and the small wavelength interval $d\lambda$; that is, $W_\lambda \, d\lambda$. But, from Eq. (4.9), $W_\lambda = \epsilon_\lambda W_{b\lambda}$, so that the emission rate per unit area of the nonblack body in the wavelength interval λ to $\lambda + d\lambda$ is $\epsilon_\lambda W_{b\lambda} \, d\lambda$. Adding up all such contributions in the definite integral from λ_1 to λ_2, the rate at which the body emits radiant energy between λ_1 and λ_2 is

$$W_{\lambda_1 \rightarrow \lambda_2} = \int_{\lambda_1}^{\lambda_2} \epsilon_\lambda W_{b\lambda} \, d\lambda.$$

The total rate at which the nonblack body emits radiant energy over all wavelengths is given by Eq. (4.2) as $W = \epsilon \sigma T^4$. Hence, the desired fraction is $\int_{\lambda_1}^{\lambda_2} \epsilon_\lambda W_{b\lambda} \, d\lambda / \epsilon \sigma T^4$, or rearranging the numerator, the generalization of Eq. (4.8) for a nonblack surface is

$$\frac{W_{\lambda_1 \rightarrow \lambda_2}}{W} = \frac{\int_0^{\lambda_2} \epsilon_\lambda W_{b\lambda} \, d\lambda}{\epsilon \sigma T^4} - \frac{\int_0^{\lambda_1} \epsilon_\lambda W_{b\lambda} \, d\lambda}{\epsilon \sigma T^4},$$

(4.12)

where the ratio $W_{\lambda_1 \rightarrow \lambda_2} / W$ is the fractional rate at which a nonblack surface emits radiant energy per unit area between the wavelengths λ_1 and λ_2. As a check to see if Eq. (4.12) properly reduces to Eq. (4.8) when the surface is black, insert $\epsilon = \epsilon_\lambda = 1$. In this case, the two equations are the same.

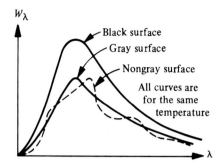

Figure 4.3 The monochromatic emissive power of various surfaces at the same temperature.

Gray Surfaces   For some nonblack surfaces, ϵ_λ is a relatively complex function of the wavelength λ. For these cases, the relations of the previous section, Eqs. (4.10), (4.11), and (4.12), cannot be simplified in general. However, other real nonblack surfaces behave more simply. Sometimes ϵ_λ is such a weak function of the wavelength λ, in the range of wavelengths in which almost all of the emission occurs, that it can be considered to be independent of wavelength. Therefore, by study of Eq. (4.11), we see that the monochromatic hemispherical emissivity ϵ_λ and the total hemispherical emissivity ϵ are equal to one another. When this is true, the surface is called a gray surface or gray body. Hence, *for a gray surface,*

$$\epsilon = \epsilon_\lambda \neq f(\lambda). \tag{4.13}$$

A great many engineering radiation calculations are made assuming that all surfaces interacting radiantly are gray. This greatly simplifies the calculation procedure, and is very often the only way a problem can be solved due to a lack of information relating ϵ_λ to λ for many materials and surface conditions.

Since, for the gray surface, ϵ_λ is independent of λ and equals ϵ, the monochromatic emissive power of a gray surface at every wavelength λ is equal to the fraction ϵ of the black surface emissive power at every wavelength for a given temperature. This is shown qualitatively in Fig. 4.3, where for comparison purposes the monochromatic emissive power of a nongray surface at the same temperature is also shown.

As a result of the fact that, on a monochromatic basis, a gray surface behaves *qualitatively* like a black surface, a number of previously presented relations for black surfaces are also valid for gray surfaces. Wien's displacement law, Eq. (4.4), is seen to be correct, without modification, for gray surfaces. Examination of Eq. (4.12) for the general nonblack surface reveals that when the surface is gray, so that from Eq. (4.13), $\epsilon_\lambda = \epsilon$, the various emissivities cancel out and Eq. (4.12) reduces to Eq. (4.8), the black body result. Hence, the fraction of a gray surface's radiant emission between wavelengths λ_1 and λ_2, at temperature T, is the same as for a black body and is given by Eq. (4.8), whose integrals are tabulated in Table 4.1.

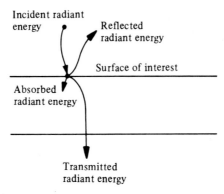

Figure 4.4 Example of what can happen to radiation incident upon a surface.

If real surfaces can be idealized with acceptable accuracy as being gray rather than the more general nonblack condition where ϵ_λ depends on λ, this saves us effort when calculating the *net radiant loss,* since no explicit attention need be paid to the wavelength distribution of the radiant energy.

4.2b Radiant Energy Incident Upon a Surface

To this point, only the emission of radiant energy from a surface has been discussed in detail; however, the emitting surface is generally also being struck (irradiated) by radiant energy from many sources. Fig. 4.4 shows a surface which is being irradiated at the total rate G per unit area called the *irradiation;* that is, *the rate at which radiant energy from all sources is incident on the surface of interest per unit area of surface.* In general, some of the radiant energy incident upon the surface is absorbed by the material, some is transmitted through the material, and some is reflected from the surface of the material, as shown in Fig. 4.4. The *absorptivity* α is defined as the fraction of the incident radiant energy absorbed by the material, the *transmissivity* τ as the fraction of the incident radiant energy transmitted through the material, and the *reflectivity* ρ as the fraction of the incident radiant energy reflected from the material. Since energy is conserved and the incident irradiation G is either absorbed, transmitted, and/or reflected, these three fractions must sum to unity;

$$\alpha + \tau + \rho = 1. \tag{4.14}$$

Most solids absorb almost all of the incident radiant energy that is not reflected within a depth of between about 1 μm and 0.25 cm or $1/10$ in. from the surface, and hence transmit almost none. As a result, most solids are considered to be opaque—$\tau = 0$—and Eq. (4.14) reduces to

$$\alpha + \rho = 1. \tag{4.15}$$

Equation (4.15) also holds well for most liquid bodies.

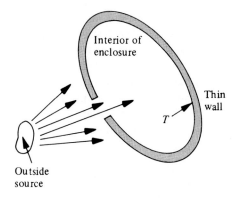

Figure 4.5 Radiant energy entering an enclosure from an outside source.

If the radiant energy which is reflected from the surface shown in Fig. 4.4 leaves at the same angle to the surface as the incident radiation made with the surface, the reflection is called *specular* or *regular*. This type of reflection ordinarily would only occur on a very smooth, highly polished surface. If the reflected radiant energy leaves in all directions with no preferential direction, the reflection is termed *diffuse,* and this is the type more likely to occur on a rough surface. In actuality, radiant energy reflected off a surface is a combination of specular and diffuse reflections. However, it has been found that as far as calculations of the net radiant loss from industrially rough surfaces are concerned—surfaces which have been machined, painted, sandblasted, or plated—it is adequate to treat reflections from these surfaces as being diffuse (see Schornhorst and Viskanta [3]).

Equation (4.14) holds for all surfaces including black surfaces, and since by definition such surfaces absorb all incident radiant energy, it follows that $\alpha = 1$, $\rho = 0$ and $\tau = 0$ for black surfaces.

The properties α, τ, and ρ defined here are total values of absorptivity, transmissivity, and reflectivity, respectively; that is, they represent integrated values over all wavelengths of the incident irradiation. In analogy with the monochromatic hemispherical emissivity ϵ_λ, there are also monochromatic values of α, τ, and ρ. However, for the most part all calculations of the net radiant loss in this text will be for the case in which all surfaces can be idealized as being gray. As a result, monochromatic values of these properties will not be needed here. Information concerning these monochromatic, or spectral, values is available in radiation textbooks, such as Sparrow and Cess [4].

4.2c The Laboratory Black Body

It was noted earlier that no truly black surfaces exist in nature; however, a laboratory black surface can be constructed in a fairly simple fashion. To prove this, consider the thin-walled enclosure shown in Fig. 4.5 whose interior surface is made of a nonblack opaque material of total hemispheric emissivity ϵ and absorptivity α, neither of which are unity. Furthermore, this interior surface is held at the constant temperature T. Now, a very small hole is made in the side of the enclosure and is irradiated with an outside source, as shown in Fig. 4.5. The radiation from this outside source that enters the enclosure will be partially absorbed when it hits the enclosure wall and partially reflected. But the reflected portion, for the most part, will impinge on other parts of the enclosure and almost none will leave through the hole, since the hole area is an extremely small fraction of the total inside area. This process of successive absorption and reflection will continue until practically all of the incident radiant energy from the outside source is absorbed by the enclosure. Hence, the small hole in the side of the enclosure is acting like a black body in that it is absorbing all incident radiant energy. However, for this hole to be truly behaving as a black surface, it must appear to be emitting radiant energy (per unit area) at a rate prescribed by the Stefan-Boltzmann law; that is, σT^4. To decide whether or not it is doing so, insert a plug of black material at the enclosure temperature T into the hole of area A_0. Since the black plug and the nonblack interior surfaces of the enclosure are both at the same temperature T, the second law of thermodynamics prohibits any net energy transfer between the plug and the enclosure surfaces; hence, the rate at which the black plug emits radiant energy must equal the rate at which it absorbs radiant energy. It emits at the total rate $A_0 \sigma T^4$ and, since none of its own radiant energy returns, the rate at which radiant energy from all of the enclosure surface (which is comprised of both radiation emitted by the enclosure surface and reflections of the enclosure's emitted radiant energy) strikes the black plug must also equal $A_0 \sigma T^4$ to keep the plug in thermal equilibrium. If the plug is now removed from the hole, a great change is not anticipated in the radiant energy from the enclosure which strikes the hole, so that enclosure radiation streams out of the hole at the rate $A_0 \sigma T^4$, which is the black surface rate. Hence, a small hole in the side of a large, thin-walled enclosure behaves like a black surface at the temperature of the enclosure material regardless of the emissivity of the enclosure material. As a result, laboratory black bodies can be created at almost any temperature without great difficulty.

The previous development of the laboratory black body can be extended to prove that any small body, such as a disk, regardless of where it is located inside a constant-temperature enclosure, receives, in effect, black body radiation from the enclosure, even though the enclosure walls are not black.

The situation is shown in Fig. 4.6. The procedure is similar to that used when a black plug was inserted into a small hole in the enclosure wall. Taking the black disk as a control volume and applying the energy balance, Eq. (1.8), noting that steady-state conditions must prevail and there is no generation, yields

$$R_{in} = R_{out}. \tag{4.16}$$

The total rate at which energy leaves the control volume, the black disk, is given by its radiative emissive power W_b times the area, or

$$R_{out} = A_0 \sigma T^4. \tag{4.17}$$

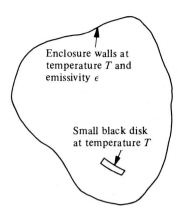

Enclosure walls at
temperature T and
emissivity ϵ

Small black disk
at temperature T

Figure 4.6 The small black disk in the enclosure.

The total rate at which radiant energy strikes the disk is R_{in}, since all that strikes is absorbed by virtue of the disk being black. This radiant energy striking the disk could conceivably arise from two different sources: (1) the disk's own emission and (2) the emission from the cavity that strikes the disk. However, by the argument previously used in connection with proving that a small hole in the side of a large thin-walled enclosure behaves as if it were black, it can be reasoned that practically none of the disk's radiant energy comes back; for, even though ϵ of the enclosure is less than one, the disk's emitted radiant energy will be absorbed by the enclosure walls after many reflections. Thus, combining Eqs. (4.16–17) gives $R_{in} = A_0\,\sigma T^4$; that is, the enclosure irradiates the disk with black body radiation that, at any time, is the sum of direct emission from the enclosure surface and reflected radiant energy from the enclosure surface. Hence, any small volume in this enclosure receives black body radiation, since the result did not depend upon a particular spatial location of the disk, and certainly, if the small disk were removed, no change would be expected in the radiant energy streaming toward the volume that had been occupied by the disk. Although this development was for a body, the disk, whose shape is such that its own emitted radiant energy cannot strike the body directly, the final conclusion can be shown to apply to *any* shape body, as long as its surface area is small compared to the surface area of the surrounding enclosure walls.

The laboratory black body, once constructed, can be used to calibrate certain types of radiation and temperature measuring equipment, such as the optical pyrometer (refer to Jakob [1] for details). In addition, an understanding of what constitutes a laboratory black body is often important in calculating net radiant losses later in this chapter. If one can recognize that a given configuration is actually behaving as a laboratory black body, the calculations required are greatly simplified. An example of this might be an opening on a furnace wall due to a door being left open. If the wall is thin enough, and the opening small enough relative to the total inside wall area, the opening will behave as if it were black and at the temperature of the interior walls of the furnace. Hence, in this case, details of the furnace interior do not have to be known, since to an observer outside the furnace the opening appears to act as if it were a black solid surface of known area at the furnace temperature.

4.2d Kirchhoff's Law

An equation which will be extremely useful in the calculation of net radiant loss later in this chapter will be developed next. This equation relates the total hemispherical emissivity ϵ to the total hemispherical absorptivity α, under certain conditions.

Consider the enclosure shown in Fig. 4.6. Now, instead of the small black disk shown, we propose to put a small *gray* disk, of area A_0, emissivity ϵ, and temperature T equal to that of the enclosure walls, at the same general location inside the enclosure. Next, the small gray disk is taken as a control volume and the law of conservation of energy, Eq. (1.8), is applied to it. The enclosure is either evacuated or filled with a gas at the same temperature T as that of the small gray body and the enclosure. In either event, the convection is zero, and hence only radiation terms will appear in the energy balance. Since both the small gray body and the enclosure walls have the same temperature, the small body must remain in thermal equilibrium during the radiation exchange process. This small gray body emits radiant energy at the rate given by Eq. (4.2) multiplied by the surface area, namely, $A_0 \epsilon \sigma T^4$. Because of its shape, none of this emission strikes the small body directly, but instead strikes the enclosure and none ever gets back to the small body because of the behavior of such an enclosure, as was detailed in Sec. 4.2c. Again from Sec. 4.2c, the enclosure's behavior is such that black body radiation is delivered to the small gray disk at the rate of $A_0 \sigma T^4$. However, only the fraction α of this is absorbed by the gray disk, with the remainder reflected back to the enclosure where, because of the behavior of the large enclosure, it is in effect completely absorbed. Hence, for the gray disk, the balance, required by Eq. (1.8), between the emission rate and the absorption rate is given by

$$A_0 \epsilon \sigma T^4 = \alpha(A_0 \sigma T^4).$$

Cancelling common factors gives

$$\epsilon = \alpha. \tag{4.18}$$

This result is called *Kirchhoff's law* and states that the emissivity ϵ of a gray surface at temperature T is equal to its absorptivity α for *black body* radiation from a source at the *same* temperature T. Hence, the equality in Eq. (4.18) embodies two major restrictions at this point: that the incident radiation be from a black body, and that the black body have the same temperature as the body it irradiates. However, if a body is truly gray, these restrictions are not needed, and Eq. (4.18) is valid for a gray surface regardless of the nature of the incident radiant energy and regardless of the temperatures of the bodies supplying this incident radiant energy. This is explained in Sparrow and Cess [4], where they also discuss the special conditions under which Eq. (4.18) applies to more general nonblack surfaces.

EXAMPLE 4.1

Evidence indicates that the sun behaves approximately like a black body at 5555°K, whereas the tungsten filament of a 100 W light bulb may behave as a gray surface with $\epsilon = 0.30$ at a temperature of 2777°K. The range of visible wavelengths of thermal radiation is between about 0.3 μm and 0.7 μm. Calculate the fraction of the emitted radiant energy in the visible range for both the sun and the light bulb filament, and compute the wavelength, λ_m, at which their monochromatic emissive power is a maximum at their given temperatures.

Solution

Consider the sun first. Since it is a black body, Eq. (4.8) in conjunction with Table 4.1 can be used to determine the fraction of its emission in the desired wavelength interval. The individual fractions on the right of Eq. (4.8) are given as a function of the product, λT, in Table 4.1. Hence,

$$\lambda_1 T = 0.3(5555) = 1666.5,$$
$$\lambda_2 T = 0.7(5555) = 3888.5.$$

From Table 4.1, at $\lambda_1 T = 1666.5$,

$$W_{b_{0-\lambda_1}}/W_b = 0.0254,$$

and at $\lambda_2 T = 3888.5$,

$$W_{b_{0-\lambda_2}}/W_b = 0.4604.$$

Subtracting these two fractions, as called for by Eq. (4.8), yields $0.4604 - 0.0254 = 0.4350$. Hence, 43.5% of the sun's emitted radiant energy lies in the visible range of wavelengths. To find the wavelength at which emission is a maximum for this black body, Wien's displacement law, Eq. (4.4), is used. With $T = 5555$°K, Eq. (4.4) becomes λ_m (5555) = 2897.7. So, $\lambda_m = 0.522$ μm. Hence, for the sun, the maximum radiant emission is squarely in the center of the visible range of wavelengths.

Now, the tungsten filament is gray with $\epsilon = 0.30$, but as explained in the section on gray surfaces, both Eqs. (4.8) and (4.4) are also valid for gray bodies. Hence, at temperature $T = 2777$°K, one has

$$\lambda_1 T = 0.3(2777) = 833.1,$$
$$\lambda_2 T = 0.7(2777) = 1944.$$

Using these values in Table 4.1 and interpolating yields a fraction of 0.0 when $\lambda_1 T = 833.1$ and gives a fraction of 0.0586 when $\lambda_2 T = 1944$. Subtraction of these values gives the fraction emitted in the desired wavelength range as 0.0586.

Thus, only 5.86% of the filament's emitted radiant energy is in the visible range. Therefore only about 6 W from the 100 W light bulb is performing the intended function of the bulb, namely illumination. The other 94 W simply tend to heat the room. (Actually the entire 100 W should be taken into account when a heating or air conditioning system is being designed for a building if the light stays on most of the day, as might be the case in an office.)

To find the wavelength at which maximum emission occurs at this temperature for the gray filament, Eq. (4.4) is used. So, with $T = 2777°$K, Eq. (4.4) becomes $\lambda_m (2777) = 2897.7$. Therefore, $\lambda_m = 1.04$ μm. Thus, for the filament, λ_m lies outside the range of visible wavelengths, and serves as a rough qualitative indication that one would not expect a very large fraction of the emission of such a surface to fall within the visible range.

EXAMPLE 4.2

A certain nonblack surface at 2000°R can be characterized as having an emissivity of 0.6 between the wavelengths 1 μm and 4 μm and an emissivity of 0.4 between 4 μm and 10 μm. The emissivity is essentially zero outside the wavelength range 1 to 10 μm. Find the total emissive power of the body and the value of the total hemispheric emissivity.

Solution

The total emissive power of any nonblack body is given in Eq. (4.10) as $W = \int_0^\infty \epsilon_\lambda\, W_{b\lambda}\, d\lambda$. For this body, ϵ_λ has a nonzero value only between $\lambda_1 = 1$ μm and $\lambda_3 = 10$ μm. Thus, the expression for the total emissive power is $W = \int_{\lambda_1}^{\lambda_3} \epsilon_\lambda\, W_{b\lambda}\, d\lambda$, which can be rewritten as

$$W = \int_{\lambda_1}^{\lambda_2} \epsilon_\lambda\, W_{b\lambda}\, d\lambda + \int_{\lambda_2}^{\lambda_3} \epsilon_\lambda\, W_{b\lambda}\, d\lambda,$$

where $\lambda_1 = 1$ μm, $\lambda_2 = 4$ μm and $\lambda_3 = 10$ μm. However, between 1 and 4 μm, ϵ_λ is given as 0.6 and between 4 and 10 μm, ϵ_λ is given as 0.4. Hence,

$$W = 0.6 \int_{\lambda_1}^{\lambda_2} W_{b\lambda}\, d\lambda + 0.4 \int_{\lambda_2}^{\lambda_3} W_{b\lambda}\, d\lambda. \tag{4.19}$$

The two integrals in this equation can be evaluated by using Eq. (4.8) and Table 4.1 for each integral. For the first integral, $\lambda_1 T = (1)(2000) = 2000$ and $\lambda_2 T = (4)(2000) = 8000$. Using Table 4.1,

$$\frac{\int_{\lambda_1}^{\lambda_2} W_{b\lambda}\, d\lambda}{\sigma T^4} = \frac{\int_0^{\lambda_2} W_{b\lambda}\, d\lambda}{\sigma T^4} - \frac{\int_0^{\lambda_1} W_{b\lambda}\, d\lambda}{\sigma T^4}$$
$$= 0.5558 - 0.0009 = 0.5549,$$

or

$$\int_{\lambda_1}^{\lambda_2} W_{b\lambda}\, d\lambda = 0.5549\, \sigma T^4. \tag{4.20}$$

Similarly, for the second integral of Eq. (4.19), $\lambda_2 T = 4(2000) = 8000$ and $\lambda_3 T = 10(2000) = 20,000$. From Table 4.1,

$$\frac{\int_{\lambda_2}^{\lambda_3} W_{b\lambda}\, d\lambda}{\sigma T^4} = 0.9335 - 0.5558 = 0.3777,$$

or

$$\int_{\lambda_2}^{\lambda_3} W_{b\lambda}\, d\lambda = 0.3777\, \sigma T^4. \tag{4.21}$$

Substituting Eqs. (4.20–21) into Eq. (4.19) gives

$$\begin{aligned}
W &= 0.6(0.5549)\, \sigma T^4 + 0.4(0.3777)\, \sigma T^4 = 0.484\, \sigma T^4 \\
&= 0.484(0.1714)10^{-8}(2000)^4 = \epsilon \sigma T^4 \\
&= 13,300 \text{ Btu/hr ft}^2.
\end{aligned}$$

Thus, the total hemispheric emissivity of this nonblack body at 2000°R is $\epsilon = 0.484$.

Example 4.2 illustrates the procedure which can be used to calculate the total hemispheric emissivity of a nonblack surface when the monochromatic hemispheric emissivity is specified. Even if ϵ_λ is given as a continuous function of λ, the wavelength range can be divided into a finite number of intervals in which the average ϵ_λ is specified, and the procedure developed here used over the total number of intervals. This could lead to a considerably greater number of intervals than the two which were used in Ex. 4.2.

EXAMPLE 4.3

Figure 4.7 shows two large furnaces which share a thin separating wall. The interior surface temperature of furnace 1 is 1100°K and its emissivity is 0.85, while the corresponding quantities for furnace 2 have the values 800°K and 0.77. If there is a hole of area 0.045 m² in the wall between the furnaces, calculate the net rate at which radiant energy crosses this opening.

Figure 4.7 The furnaces of Ex. 4.3.

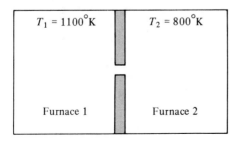

Solution

The net rate at which radiant energy crosses the opening between the furnaces is the difference between the rates at which radiant energy from furnace 1 and radiant energy from furnace 2 cross the opening. To calculate the rate at which radiant energy from furnace 1 crosses the opening, the opening is considered to be a black surface at T_1, since it is a small hole in the side of a large thin-walled enclosure at temperature T_1. The same reasoning is used for the opening in relation to furnace 2. Hence, the net radiant transfer across the opening between the furnaces is

$$
q_{\text{net}} = A_0 \sigma T_1^4 - A_0 \sigma T_2^4 = (0.045)(5.669 \times 10^{-8}) \, [1100^4 - 800^4]
$$
$$
= 2690 \text{ W.}
$$

Thus, radiant energy is being transferred at a *net* rate of 2690 W from the hotter furnace to the cooler one across the opening in their common wall.

4.2e Intensity of Radiation

In addition to the emissive power, a description of the distribution of the emitted radiant energy above the surface of interest is also needed to make calculations of the net radiant loss from any surface of interest. For example, Fig. 4.8 shows two surfaces which exchange radiant energy. To calculate the net radiant loss from surface 1, something about the emission from surface 2 must be stated; in particular, how much of its emission, if any, will strike surface 1. If all the emitted radiant energy leaves a surface as shown in Fig. 4.8, then emission from surface 1 can strike surface 2, but the emission from surface 2 cannot strike surface 1 directly; or, if both surfaces emitted all their radiant energy perpendicular to the surface, they would not even be interacting as far as their direct emission is concerned. Since in computing the net radiant loss from surface 1, the rate at which surface 1 absorbs radiant energy from all sources, including surface 2, must be known, it becomes clear that we must know the distribution of emitted radiant energy in space above a surface.

In discussing the distribution in space of a surface's emitted radiant energy, it is useful to introduce and define the intensity of radiation. Figure 4.9 shows a differential area dA_1 emitting radiant energy in the half-space above it with the emitted radiant energy, in general, leaving in all directions. Hence, it passes through the imaginary hemisphere shown constructed above dA_1. The *intensity I* of radiation emitted by dA_1 is defined as the rate dq at which radiant energy is emitted by dA_1 in a particular direction per unit solid angle and per unit area of the projection of dA_1 perpendicular to the direction r. It is mathematically defined as

$$
I = \frac{dq}{d\omega \, dA_1 \cos \phi_1}.
\tag{4.22}
$$

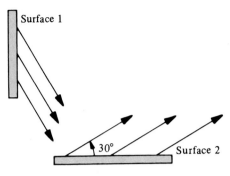

Figure 4.8 Surfaces which emit radiant energy in a hypothetical single direction.

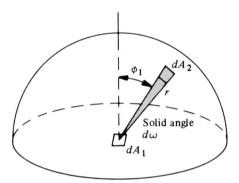

Figure 4.9 Hemispherical space above an emitting differential area.

The solid angle $d\omega$ can be expressed, using the notation of Fig. 4.9 and noting that dA_2 is perpendicular to r, as $d\omega = dA_2/r^2$, and $dA_1 \cos \phi_1$ is the projection of dA_1 perpendicular to the direction r. In general, the intensity I varies from point to point in space. Hence, it can be a function both of ϕ_1 and an angle θ which is measured in the plane of dA_1 as a rotation about the normal to dA_1. A qualitative way of thinking about the intensity I is to say that it is equivalent to brightness; hence, if dA_1 is at a high enough constant temperature, the eye will detect differences in intensity as varying shades of brightness as it moves to different points in space.

Black and Gray Surfaces

For black and gray surfaces, it can be proven and experimentally verified that the intensity I of the emitted radiant energy is a constant. Hence, if dA_1 is gray at a high enough constant temperature, it appears to have the same brightness regardless of the position of the eye in the half-space above dA_1. Looking directly down on dA_1 along its normal, a viewer sees all of dA_1 at some level of brightness. Looking at dA_1 along a line which made the angle ϕ_1 with the normal to dA_1, as shown in Fig. 4.10, presents a view of only the projected area $dA_1 \cos \phi_1$ at the same level of brightness; that is, the same intensity as when viewed along the normal to dA_1.

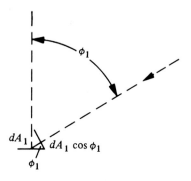

Figure 4.10 Details of the view observed when looking through a general angle to the normal of dA_1.

Surfaces which emit in such a way that the intensity defined by Eq. (4.22) is a constant are called *diffuse emitters* or, simply, *diffuse*. It may be recalled that earlier, when discussing *reflected* radiant energy, the term "diffuse" was used to characterize energy that was reflected in all directions with no preferred direction. Diffuse reflection is formally defined here as occurring when the intensity of the reflected radiant energy is a constant in the space above the reflecting surface.

Nonblack, Nongray Surfaces

Nonblack, nongray surfaces have intensities that vary primarily with the angle ϕ between the normal to the surface and the direction of interest. A measure of the intensity as a function of ϕ is the *directional emissivity* ϵ_ϕ. Its qualitative behavior is shown in Fig. 4.11 for black and gray surfaces as well as for highly polished metals and electrical insulators. For black and gray surfaces, the directional emissivity is a constant; that is, it does not vary with ϕ, and it equals the total hemispheric emissivity. The electrical insulators, as a class, have a fairly uniform intensity except at very large values of ϕ. Hence the gray approximation, and therefore the assumption of diffuse emission, would probably not be too bad for this class of materials. The highly polished metals, as a class, have a relatively low intensity at moderate values of ϕ which increases rapidly at the larger values of ϕ, finally decreasing to zero at the very large ϕ near 90°. (Note that this decrease is not shown in Fig. 4.11.)

Discussion here will be restricted to diffuse emitters, gray and black surfaces, where the intensity I of emitted radiant energy is a constant above a surface at temperature T. Further discussion of the directional characteristics of emitted radiant energy, as well as reflected and absorbed radiant energy, can be found in specialized texts on radiation heat transfer such as Sparrow and Cess [4], and also [6] through [9].

Relation Between Intensity and Emissive Power   As is evident by the previously given word definition of intensity and also by its mathematical equivalent, Eq. (4.22), the concept of intensity is important because the intensity, I, once its numerical value is known, tells one where the radiant energy emitted by a differential area is going, and therefore

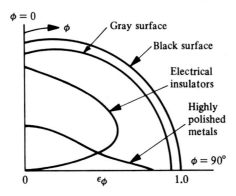

Figure 4.11 Directional emissivity as a function of angle, from the normal, for various surfaces.

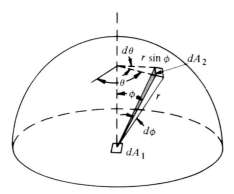

Figure 4.12 The solid angle and hemisphere used in relating intensity to emissive power.

what it must strike in its passage, and also how much radiant energy is going there. Since the intensity of emitted radiant energy is caused, at its root, by emission by the surface, we would like to relate the local intensity I to the emission characteristics of the surface, specifically, the total emissive power W. To do this, consider the differential emitting area dA_1 shown in Fig. 4.12. The radiant energy emitted from dA_1 must all pass through the imaginary hemisphere shown, which is of radius r where r is taken so that no other surfaces intercept any of the emission from dA_1 until distances greater than r are reached. By computing the rate at which radiant energy, emitted by dA_1, streams across the surface of the hemisphere and then equating that to the total emission rate, $W dA_1$, we can find an expression connecting I and W.

The rate at which radiant energy emitted by dA_1 passes through dA_2 on the hemisphere is given, because of the general definition of intensity I, as dq in Eq. (4.22), as long as the solid angle $d\omega$ is the one shown in Fig. 4.12, subtended by dA_2. Hence,

$$dq = I \cos \phi \, dA_1 \, d\omega.$$

But $d\omega = dA_2/r^2$, since dA_2 is perpendicular to r as is required by the definition of a solid angle, and dA_2 itself is $dA_2 = r \sin\phi\, d\theta\, r\, d\phi = r^2 \sin\phi\, d\phi\, d\theta$. Hence,

$$dq = I \sin\phi \cos\phi\, d\phi\, d\theta\, dA_1. \tag{4.23}$$

The total rate at which radiant energy emitted by dA_1 passes through the entire surface of the hemisphere is obtained by adding all contributions of the type shown in Eq. (4.23) in the definite integral

$$q = \int_0^{2\pi} d\theta \int_0^{\frac{1}{2}\pi} I \sin\phi \cos\phi\, d\phi\, dA_1. \tag{4.24}$$

But q is also equal to the emissive power of dA_1 times the area dA_1, that is,

$$q = W\, dA_1. \tag{4.25}$$

Upon equating Eqs. (4.25) and (4.24), one has a general relationship between the intensity of radiation I and the total emissive power W regardless of whether or not the emitted radiation is diffuse.

Now for a diffuse emitter, which is all we will concern ourselves with from this point on, the intensity I is a constant. Therefore, I and dA_1 are independent of both ϕ and θ, and can be factored out of the integral operators in Eq. (4.24). Doing this and integrating with respect to θ yields,

$$q = I\, dA_1\, 2\pi \int_0^{\frac{1}{2}\pi} \sin\phi \cos\phi\, d\phi.$$

Now, $\frac{1}{2}\sin 2\phi = \sin\phi \cos\phi$; hence, the integral becomes

$$-\left.\frac{\cos 2\phi}{4}\right|_0^{\frac{1}{2}\pi} = -\frac{1}{4}(\cos\pi - \cos 0) = -\frac{1}{4}(-1 - 1) = +\frac{1}{2},$$

and thus,

$$q = \pi I\, dA_1.$$

Eliminating q between this equation and Eq. (4.25) gives

$$I = W/\pi. \tag{4.26}$$

Equation (4.26) is the relation connecting the intensity I to the total emissive power W for a *diffuse emitter* (where I is a constant in the half space above the emitter). This expression will be used in the next section to develop an expression for the angle factor, the final concept needed before one can make calculations of the net radiant loss in circumstances other than the very simplest ones.

4.2f Angle Factors

The previous sections in this chapter concerning the total emission rate of radiant energy, its distribution in the space above the emitting surface, and its distribution in the thermal range of wavelengths, have laid the foundation needed to develop the last relation required before calculation of net radiant losses can be made in earnest. This last relation is the angle factor, which basically tells us how much of the emission from one surface directly strikes some other surface. Remember, however, that this type of information can also be found by use of the intensity of emitted radiant energy; in fact, the derivation of the general relationship for the angle factor does involve the use of the local intensity I.

The angle factor F_{12} from a constant-temperature surface 1 to surface 2 is defined as the fraction of the radiant energy emitted by surface 1 which strikes surface 2 directly. Two points in the definition of the angle factor should be noted: (1) As the definition states, the radiant energy from surface 1 only has to strike surface 2, not be absorbed completely. In general, it would be partially absorbed and partially reflected. (2) The fraction of the emission from surface 1 defined by the angle factor must be the part that strikes directly. A much greater fraction of the radiant energy from surface 1 may strike surface 2 indirectly by being reflected from other surfaces. Also, the definition of the angle factor allows a value of F_{11}, the fraction of emission from surface 1 which strikes surface 1 directly. When this happens, $F_{11} \neq 0$ and it is said that surface 1 "sees" itself. More generally, whenever an angle factor such as F_{12} or F_{73} is nonzero, it is said that the surfaces involved "see" each other. The angle factor is also referred to by various other names such as *view factor, configuration factor, geometrical factor,* or *shape factor.*

Angle Factor Expressions

We would like to translate our verbal definition for the angle factor into mathematical expressions in terms of the quantities it depends upon. Hence, in order of increasing complexity, expressions will be developed for the angle factor between two differential areas, one differential area and a finite area, and finally, the most general case, a finite area and another finite area. All of these expressions will be for diffuse emitters only.

Differential Area to Differential Area Shown in Fig. 4.13, on two finite surfaces 1 and 2, are two differential surfaces of area dA_1 and dA_2 respectively. The value r is the length of the line connecting dA_1 and dA_2, ϕ_1 is the angle between r and the normal to dA_1 and ϕ_2 is the angle between r and the normal to dA_2. The task is to derive an expression for $F_{dA_1 \rightarrow dA_2}$.

By definition, the required angle factor is the fraction of radiant energy emitted from surface dA_1 which directly strikes dA_2. This fraction, then, is simply the ratio of the rate at which emitted radiant energy from dA_1 directly strikes dA_2 to the total rate at which dA_1 emits radiant energy. From the concept of intensity, it is seen that all of the emitted radiant energy from dA_1 which passes through the solid angle $d\omega$, seen from dA_2 along r, directly strikes dA_2. Hence, after rearranging Eq. (4.22),

$$dq_{dA_1 \rightarrow dA_2} = I \, dA_1 \cos \phi_1 \, d\omega,$$

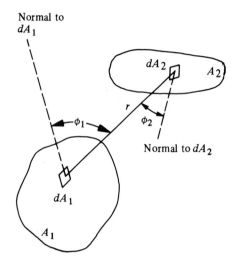

Figure 4.13 The areas and angles used in the derivation of angle factor expressions.

where the solid angle $d\omega$ is the area perpendicular to r at dA_2 divided by r^2, or $d\omega = dA_n/r^2$. But, dA_n, the area perpendicular to r at dA_2, is the normal projection of dA_2 along r, which is $dA_2 \cos \phi_2$. Hence,

$$dq_{dA_1-dA_2} = I \frac{\cos \phi_1 \cos \phi_2}{r^2} dA_1 \, dA_2.$$

From Eq. (4.26) for a diffuse emitter, the relation between the intensity and the emissive power of the body is $I = W/\pi$. As a result,

$$dq_{dA_1-dA_2} = W \frac{\cos \phi_1 \cos \phi_2}{\pi r^2} dA_1 \, dA_2 \qquad (4.27)$$

is the rate at which radiant energy from dA_1 directly strikes dA_2. The total rate at which dA_1 emits radiant energy is its emissive power W times its area dA_1, or $W_1 dA_1$. Dividing by this term, the required angle factor between the differential areas dA_1 and dA_2 is

$$F_{dA_1-dA_2} = \frac{\cos \phi_1 \cos \phi_2}{\pi r^2} dA_2. \qquad (4.28)$$

One use of Eq. (4.28) is as a building block for eventually obtaining the angle factor F_{12}. Another use for Eq. (4.28) is in finding the angle factor between two relatively small finite areas spaced a large distance apart. For example, consider the two finite areas shown in Fig. 4.14 with the line r joining the centers of these plane disks. Because of their mutual spatial relationship, it is seen that as the line connecting the disks has its endpoints moved

328   Chapter Four

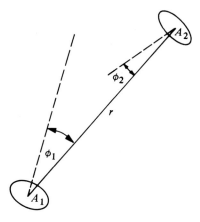

Figure 4.14 Two finite plane areas which behave as differential areas if r is large enough.

to portions of the disks not at the center, the length r of the line, and the angles ϕ_1 and ϕ_2 will hardly change, so that the two finite area plane disks are behaving, since they are spaced so far apart relative to their size, as differential areas. Hence, Eq. (4.28) can be used in this case.

Differential Area to Finite Area   Finding an expression for the angle factor between differential area dA_1 and finite area A_2 in Fig. 4.13 involves an extension of the analysis which led to Eq. (4.28). Again, the required angle factor $F_{dA_{1-2}}$ is, by definition, the fraction of the radiant energy emitted by dA_1 which strikes finite surface 2 directly. An expression for the rate at which emission from dA_1 strikes dA_2 directly has already been found as Eq. (4.27). Summing all contributions of this type by letting dA_2 sweep over surface 2 by means of a definite integral, the result is the rate at which radiant energy emitted by dA_1 directly strikes finite surface 2. Hence,

$$dq_{dA_{1-2}} = W \, dA_1 \int_{A_2} \frac{\cos \phi_1 \cos \phi_2}{\pi r^2} \, dA_2.$$

Dividing this by the total rate at which surface dA_1 emits radiant energy, that is, $W_1 \, dA_1$, the required angle factor is

$$F_{dA_{1-2}} = \int_{A_2} \frac{\cos \phi_1 \cos \phi_2}{\pi r^2} \, dA_2. \tag{4.29}$$

Remarks about the uses of this expression, Eq. (4.29), for the angle factor between a *differential* area 1 and a *finite* area 2 parallel those made just after Eq. (4.28) for the corresponding expression between two differential areas.

Finite Area to Finite Area   This is the most general case, involving diffuse emission. We must find the angle factor between a finite *isothermal* surface 1 and another finite surface 2 as depicted in Fig. 4.13. The required result, F_{12}, is the ratio of the rate at which surface 1 emits radiant energy which directly strikes surface 2 to the rate at which surface 1 emits radiant energy. From the derivation which led to the previous result, Eq. (4.29), the rate at which dA_1 emits radiant energy which strikes finite surface 2 is

$$dq_{dA_{1-2}} = W_1 \, dA_1 \int_{A_2} \frac{\cos \phi_1 \cos \phi_2}{\pi r^2} \, dA_2. \tag{4.30}$$

Hence, the total rate at which radiant energy emitted by finite surface 1 directly strikes finite surface 2 is obtained by summing up all contributions of the type of Eq. (4.30) in the definite integral

$$q_{1-2} = W_1 \int_{A_1} \int_{A_2} \frac{\cos \phi_1 \cos \phi_2}{\pi r^2} \, dA_2 \, dA_1.$$

(Note that since surface 1 is at constant temperature, W_1 is not a function of dA_1.) Dividing q_{1-2} by the total rate at which finite surface 1 emits radiant energy, $W_1 A_1$, the angle factor from finite surface 1 to finite surface 2 is

$$F_{12} = \frac{1}{A_1} \int_{A_1} \int_{A_2} \frac{\cos \phi_1 \cos \phi_2}{\pi r^2} \, dA_2 \, dA_1. \tag{4.31}$$

The angle factor from a general diffuse surface i at constant temperature to any surface j of interest can be written directly from Eq. (4.31) as

$$F_{ij} = \frac{1}{A_i} \int_{A_i} \int_{A_j} \frac{\cos \phi_i \cos \phi_j}{\pi r^2} \, dA_j \, dA_i. \tag{4.32}$$

By examining the various equations for the angle factor, Eqs. (4.28–29) and (4.31–32), note that the angle factors are purely geometric quantities that do not depend upon the temperatures or the emissivities of the surfaces involved. *The angle factor depends only upon the spatial relationship of one surface to another, as long as these surfaces are diffuse emitters and are at constant temperature.*

Equation (4.32) is the angle factor from surface i to surface j. For F_{ji}, interchange the subscripts of Eq. (4.32) to yield

$$F_{ji} = \frac{1}{A_j} \int_{A_j} \int_{A_i} \frac{\cos \phi_j \cos \phi_i}{\pi r^2} \, dA_i \, dA_j. \tag{4.33}$$

Comparing Eqs. (4.32) and (4.33), note that the double integrals differ only in the order of integration; hence, both yield the same result. Multiplying Eq. (4.32) by A_i and Eq. (4.33) by A_j and equating the double integrals leads to

$$A_i F_{ij} = A_j F_{ji}. \tag{4.34}$$

This result is called the *reciprocity theorem* and is extremely valuable when computing angle factors, since one of the two angle factors F_{ij} or F_{ji} is often much easier to evaluate than the other; hence, the easier one is found first and the more difficult one is determined from Eq. (4.34).

Enclosure Relationship A general surface i usually interacts radiantly with a number of other surfaces. Some of surface i's radiant emission may strike itself directly as well as striking directly other surfaces which, with surface i, form an enclosure of n surfaces. By taking advantage of the fact that the emission from i must strike directly the surfaces in the enclosure, one can develop an expression that will aid in finding angle factors.

The fraction F_{i1} of the emitted radiant energy of surface i directly strikes surface 1, F_{i2} directly strikes 2, F_{ii} directly strikes i, and so on, with F_{in} directly striking n. Since all of the emitted radiant energy from surface i must directly strike something in the enclosure, these fractions must sum to unity:

$$F_{i1} + F_{i2} + F_{i3} + \cdots + F_{ii} + \cdots + F_{ij} + \cdots + F_{in} = 1,$$

or

$$\sum_{j=1}^{n} F_{ij} = 1. \tag{4.35}$$

In an enclosure of n surfaces, there are n relations of the type of Eq. (4.35), one for each value of i from $i = 1$ to $i = n$. These relations, like the reciprocity theorem, are of great aid in reducing the work necessary to find the angle factors required in an enclosure. It should be noted that requiring that an enclosure exist is not really a restriction, since all radiation problems involving surfaces are really enclosure problems. This is so because the calculation of the net radiant loss from any surface of interest requires the identification of every surface from which the surface of interest can receive radiant energy. In particular, all of the surfaces that can "see" the surface of interest must completely enclose the surface of interest. Some of these "surfaces" enclosing the surface of interest may not be solid material, but may be openings which will be assigned a temperature and emissivity that corresponds to the way they radiantly behave. This point will be considered in more detail in Sec. 4.3.

4.2g Calculation of Numerical Values of Angle Factors

When calculating the net radiant loss from a surface, the actual numerical values of the angle factors involved are needed; indeed, the determination of the values of the angle factors is often the most difficult part of the problem of calculating the net radiant loss.

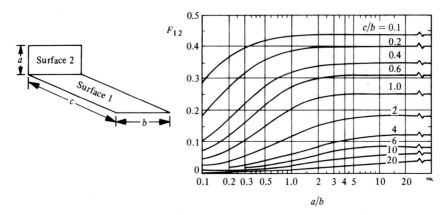

Figure 4.15 The angle factor between two rectangles meeting in a perpendicular fashion along a common edge. (From "Radiant Interchange Configuration Factors" by D. C. Hamilton and W. R. Morgan, *NACA TN 2836*, 1952.)

The following list contains four general procedures that can be used to obtain numerical values of the angle factor (note that procedures 2 and 4 will be used most frequently herein, and procedure 1 will be used sparingly).

1. Perform whatever integrations are needed in Eqs. (4.28–29) and (4.31), the general mathematical expressions for the angle factor, to obtain explicit equations from which the actual numerical values can be calculated. This procedure is limited by the difficulty in performing the required integrations. The method of contour integration described, for example, in [4] is often useful in this regard.

2. Numerical values of the angle factors can be found from formulas and charts that were prepared by researchers in the past who, in most cases, performed the integrations discussed in procedure 1. Some of these charts are reproduced here in Figs. 4.15–17. More extensive compilations of charts and formulas can be found in [4], [6], [7], [8], [10], and [24].

3. Various experimental and graphical techniques, as well as a mechanical device, have been devised to arrive at numerical values of the angle factors. These techniques are described in [7] and [11].

4. This procedure, some portions of which are sometimes referred to as *flux algebra* or *shorthand methods,* employs the definition of the angle factor, common sense, the reciprocity theorem, and the relationship among the angle factors in an enclosure. Very extensive and rigorous coverage of flux algebra is given in [4], [7], [8], and especially [9].

Angle Factors by Integration

Here, the mechanics of using equations such as (4.29) and (4.32) will be demonstrated for a couple of different surface configurations.

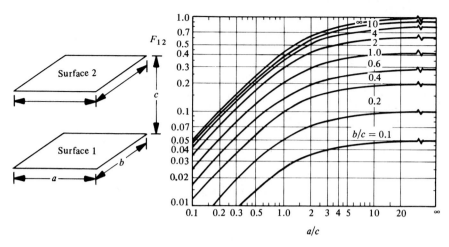

Figure 4.16 The angle factor between two rectangles of equal size. (From Hamilton and Morgan, 1952.)

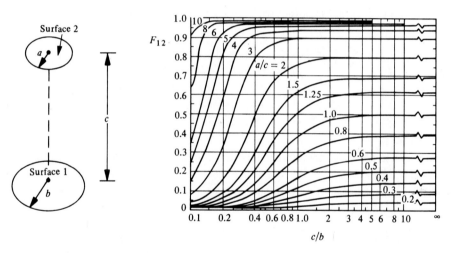

Figure 4.17 The angle factor between two parallel circular disks centered on each other. (From Hamilton and Morgan, 1952.)

Differential Area to Portion of Hemisphere　Consider Fig. 4.18, which shows a small plane area A_1 interacting radiantly with surface 2, a portion of a hemisphere located directly above 1 as shown. The radius R and the angle β defining the limit of the hemispheric portion, surface 2, are both given. We must find the angle factor F_{12} in terms of known quantities. Since surface 1 is small compared to surface 2 and is plane, it will be treated as a differential area, and the expression that will be used for the angle factor is Eq. (4.29).

From Fig. 4.18 we can see that $r = R$ is a constant, $\phi_2 = 0°$ always, since dA_2 is on the surface of a sphere and therefore its normal will always coincide with the direction

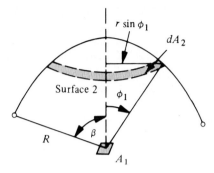

Figure 4.18 A differential area interacting radiantly with a portion of the hemisphere.

r, and dA_2 can be taken to be a ring-shaped surface area element, as shown, since all along this ring ϕ_1, ϕ_2, and r do not vary. Thus,

$$dA_2 = 2\pi R \sin \phi_1 \, R \, d\phi_1.$$

Substituting this into Eq. (4.29), along with $\cos \phi_2 = 1$, yields

$$F_{12} = \int_0^\beta 2 \sin \phi_1 \cos \phi_1 \, d\phi_1 = \int_0^\beta \sin 2\phi_1 \, d\phi_1 = \frac{-\cos 2\phi_1}{2} \bigg|_0^\beta.$$

Hence,

$$F_{12} = -\frac{1}{2} (\cos 2\beta - \cos 0) = \frac{1}{2} (1 - \cos 2\beta).$$

After making use of a trigonometric identity, the final expression for the angle factor becomes

---

$$F_{12} = \sin^2 \beta. \qquad (4.36)$$

Differential Disk to Finite Disk A small plane disk is interacting radiantly with a much larger disk which is directly above and parallel to the small disk, as is shown in Fig. 4.19. Since disk 1 is so much smaller than disk 2, it will be idealized as being differential in extent with surface area dA_1, and the application of Eq. (4.29) is again appropriate. Here dA_2 is taken as the ring-shaped differential area of magnitude $2\pi x \, dx$, which is shown in Fig. 4.19. (Note that x is used as a surface coordinate on disk 2 because r was used as the distance between dA_1 and dA_2.) Also, $\phi_1 = \phi_2$, since the disks are parallel, and $r^2 = x^2 + D^2$; thus $\cos \phi_1 = D/\sqrt{x^2 + D^2} = \cos \phi_2$. Substitution into Eq. (4.29) gives

$$F_{12} = \int_0^{r_0} \frac{2D^2 x \, dx}{(x^2 + D^2)^2} = 2D^2 \int_0^{r_0} \frac{x \, dx}{(x^2 + D^2)^2}.$$

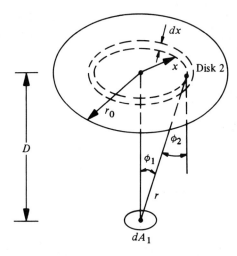

Figure 4.19 The parallel disks.

Let $u = x^2 + D^2$, from which $du = 2x\,dx$. As a result, the integral just presented becomes

$$\int \frac{2x\,dx}{(x^2 + D^2)^2} = \int \frac{du}{u^2} = -\frac{1}{u} = -\frac{1}{x^2 + D^2}.$$

Hence,

$$F_{12} = -\frac{D^2}{x^2 + D^2}\bigg|_0^{r_0} = -D^2\left(\frac{1}{r_0^2 + D^2} - \frac{1}{D^2}\right),$$

and the angle factor is

$$F_{12} = \frac{r_0^2}{r_0^2 + D^2}. \tag{4.37}$$

As a check on the result, Eq. (4.37), several limiting cases in which the angle factor's value can be seen directly will be examined. If the spacing D between the plates becomes very small, it is expected that all of the emission of the smaller disk will directly strike the larger disk, hence $F_{12} \rightarrow 1$. As $D \rightarrow 0$ in Eq. (4.37), this limit of 1 is also approached. Also, if r_0 is very large relative to a finite D, it is expected that $F_{12} \rightarrow 1$; indeed, in Eq. (4.37), when $r_0 \rightarrow \infty$, L'Hospital's rule shows that the limit is one. Finally, as D becomes very large at a finite value of r_0, most of the emission from surface 1 should miss disk 2, hence $F_{12} \rightarrow 0$. Allowing $D \rightarrow \infty$ in Eq. (4.37) gives the result that $F_{12} \rightarrow 0$, which confirms the physical reasoning.

These two angle factor results, Eqs. (4.36) and (4.37), were arrived at by use of the first method proposed at the beginning of this section, namely performing the required

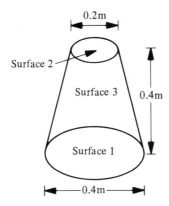

Figure 4.20 A three-surface enclosure.

integrations. Next, we will use some of the other methods for finding values of angle factors for the configuration of surfaces shown in Fig. 4.20. We would like to find the angle factors F_{12}, F_{21}, and F_{13}. Since surface 3 does not affect the way surface 1 sees surface 2, F_{12} can be found directly from the chart for two parallel disks, Fig. 4.17. Relating the figure on the chart to the situation given in Fig. 4.20 leads to $c = 0.4$ m, $b = 0.4/2 = 0.2$ m, $a = 0.2/2 = 0.1$ m, $c/b = 0.4/0.2 = 2$, and $a/c = 0.1/0.4 = 0.25$. At these values on Fig. 4.17, $F_{12} \simeq 0.04$. To obtain F_{21}, one simply uses the reciprocity theorem, Eq. (4.34), which gives, between surfaces 1 and 2,

$$A_1 F_{12} = A_2 F_{21},$$

or

$$\frac{\pi}{4} (0.4)^2 (0.04) = \frac{\pi}{4} (0.2)^2 F_{21}$$

Hence, $F_{21} = 0.16$.

Finally, to find F_{13} in Fig. 4.20, the special relationship among the angle factors of an enclosure, Eq. (4.35), can be used. Here $n = 3$, and since F_{13} is to be found, $i = 1$. Expanding Eq. (4.35) yields

$$F_{11} + F_{12} + F_{13} = 1.$$

However, $F_{11} = 0$ because surface 1's shape is such that none of its own diffuse emission can strike it directly. Thus since F_{12} is already known to be 0.04, we have

$$0 + 0.04 + F_{13} = 1,$$

or $F_{13} = 0.96$. Hence the required angle factor values are $F_{12} = 0.04$, $F_{21} = 0.16$, and $F_{13} = 0.96$.

To find these angle factor values, both method 2, the use of available charts, and method 4, shorthand methods discussed at the beginning of this section, were employed.

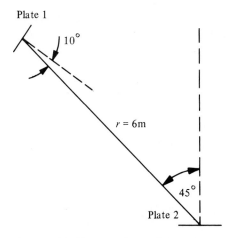

Plate 1

10°

r = 6m

45°

Plate 2

Figure 4.21 The two plates of Ex. 4.4.

EXAMPLE 4.4

Figure 4.21 shows an edge view of two square plates. Plate 1 has an area of 0.05 m², while plate 2 has an area of 0.1 m². Find the numerical values of the angle factors F_{12} and F_{21}. Also, if the opening shown in Fig. 4.21 is treated as a third "surface," surface 3, find F_{13}.

Solution
Since both surfaces are plane and relatively small compared to the distance separating them, they approximate differential areas. Hence,

$$F_{12} = \frac{\cos \phi_1 \cos \phi_2}{\pi r^2} \, dA_2 \qquad\qquad [4.28]$$

can be used. Here $dA_2 = 0.1$ m², $dA_1 = 0.05$ m², $r = 6$ m, $\phi_1 = 10°$, and $\phi_2 = 45°$. Hence, $\cos \phi_1 = 0.985$ and $\cos \phi_2 = 0.707$. Substituting into Eq. (4.28) yields

$$F_{12} = \frac{0.985(0.707)(0.1)}{\pi(6)^2} = 0.000616.$$

To find F_{21}, either Eq. (4.28) with the subscripts interchanged or the reciprocity theorem, Eq. (4.34), can be used. The latter approach will be taken here so that Eq. (4.34) becomes

$$A_1 F_{12} = A_2 F_{21}, \quad \text{or,} \quad 0.05(0.000616) = 0.1 \, F_{21},$$

yielding

$$F_{21} = 0.000308.$$

In order to calculate the value of F_{13}, note that the opening, which we will call surface 3 and which is not labelled explicitly in Fig. 4.21, together with surfaces 1 and 2 forms an enclosure. Hence the relationship among the angle factors in an enclosure, Eq. (4.35), can be used here. Using $n = 3$ and $i = 1$ in Eq. (4.35) gives

$$F_{11} + F_{12} + F_{13} = 1. \tag{4.38}$$

However, $F_{11} = 0$, since surface 1 is plane and therefore does not see itself, and F_{12} was found earlier to be 0.000616. Inserting these values into Eq. (4.38) gives

$$F_{13} = 0.9994.$$

This problem illustrates how the analytical expression for the angle factor between two *differential* areas, Eq. (4.28), can also be used to advantage on two areas of finite extent, as long as these areas can be idealized well as differential areas. This was the case here, because first of all *both* finite areas are *plane* (all differential areas are plane), and secondly if the line r has its ends moved to different points on plates 1 and 2, the angles of 10° and 45° and the length of r of 6 m hardly change at all. (This, of course, is also the situation for true differential areas.) In addition, this example also demonstrates the ease with which *some* of the angle factors can be found by use of the reciprocity theorem and the enclosure relationship, as long as some other angle factors have been found first by use of the mathematical definition, the charts, or experiment. Here, F_{12} was first found by employing the mathematical definition, and then use of reciprocity and the enclosure relation led to F_{21} and F_{13}.

EXAMPLE 4.5

Figure 4.22 shows three surfaces with surfaces 2 and 3 being in the same plane, and that plane is perpendicular to surface 1 at the common edge. All dimensions are known. Show how the angle factor F_{13} can be found, utilizing charts and equations already developed.

Solution
From the chart of Fig. 4.15, the angle factor from surface 1 to the combined surface 2 plus 3 can be found as $F_{1(2 + 3)}$. Also, the angle factor from surface 1 to surface 2 only can be found from the same chart as F_{12}. The difference between these two angle factors must be the fraction of the radiant energy emitted by surface 1 which directly strikes surface 3, since the fraction that strikes 2 + 3, $F_{1(2 + 3)}$, minus the fraction which strikes 2 alone, F_{12}, must strike three directly. Thus, $F_{13} = F_{1(2 + 3)} - F_{12}$, where both of the angle factors on the right-hand side can be found from Fig. 4.15. Equation (4.35) cannot be used here, because surfaces 1, 2, and 3 by themselves do not form an enclosure.

An important aspect of the solution to Ex. 4.5 is the use of "common sense," or perhaps more properly, "physical reasoning," which was mentioned in method 4 at the beginning of this section on angle factors. The physical reasoning referred to here is simply the recognition that the difference between the radiant energy emitted by 1 which strikes

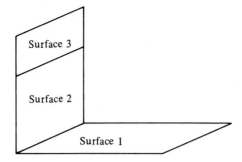

Figure 4.22 The surfaces of Ex. 4.5.

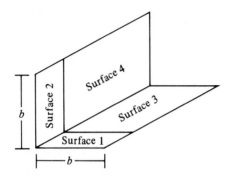

Figure 4.23 The surfaces of Ex. 4.6.

surface $(2 + 3)$ and that which strikes surface 2 is equal to that which strikes surface 3. This realization is an important tool when using the shorthand methods, or flux algebra, described in method 4.

EXAMPLE 4.6

Using relationships already developed, set up a system of algebraic equations from which the angle factor F_{14} in Fig. 4.23 can be found.

Solution
The fraction of the radiant energy emitted by surface 1 which strikes the combined surface $(2 + 4)$ is the angle factor $F_{1(2+4)}$, and this equals the sum of the fractions which strike surfaces 2 and 4 separately. Thus,

$$F_{12} + F_{14} = F_{1(2+4)}. \tag{4.39}$$

Similarly, an expression like Eq. (4.39) can be written for surface 3:

$$F_{32} + F_{34} = F_{3(2+4)}. \tag{4.40}$$

Since the radiant emission from surface $(1 + 3)$ which directly strikes surface $(2 + 4)$ comes from the individual surfaces 1 and 3 which comprise surface $(1 + 3)$, a relationship can be found among $F_{(1+3)(2+4)}$, $F_{1(2+4)}$, and $F_{3(2+4)}$. The radiant energy emitted by $(1 + 3)$ which directly strikes $(2 + 4)$ is the fraction which strikes, $F_{(1+3)(2+4)}$, times the surface area A_{1+3} times the emissive power per unit area $W_{(1+3)}$, or $F_{(1+3)(2+4)} A_{(1+3)} W_{(1+3)}$. Similarly, the rates at which radiant energy (not just the fraction) emitted by surfaces 1 and 3, individually, strike surface $(2 + 4)$ are, respectively, $F_{1(2+4)} A_1 W_1$ and $F_{3(2+4)} A_3 W_3$. Since the radiant energy emitted by surface $(1 + 3)$ that directly strikes surface $(2 + 4)$ must have come from the individual surfaces 1 and 3,

$$F_{(1+3)(2+4)} A_{(1+3)} W_{(1+3)} = F_{1(2+4)} A_1 W_1 + F_{3(2+4)} A_3 W_3. \tag{4.41}$$

[Note that Eq. (4.41) is the definition of $W_{(1+3)}$.] In general, W_1, W_3, and W_{1+3} are not equal; however, as has already been noted, the angle factors depend only on relative spatial relationships of surfaces to one another and not on the emissive power. Also, the areas do not depend on the emissive powers; hence, in Eq. (4.41), the angle factors are the same regardless of W_1 and W_3. For the purpose of determining the angle factors, letting $W_1 = W_3 = W_{1+3}$, Eq. (4.41) leads to

$$F_{(1+3)(2+4)} A_{(1+3)} = F_{1(2+4)} A_1 + F_{3(2+4)} A_3. \tag{4.42}$$

Equations (4.39–40) and (4.42) are three equations in the four unknowns $F_{14}, F_{32}, F_{1(2+4)}$, and $F_{3(2+4)}$. [Note that angle factors F_{12}, F_{34}, and $F_{(1+3)(2+4)}$ can be found from the chart of Fig. 4.15, and hence were not included among the four unknown angle factors.] Referring to Fig. 4.23, note that the relative spatial relationship between surfaces 1 and 4 is exactly the same as between surfaces 2 and 3; hence,

$$F_{14} = F_{23}. \tag{4.43}$$

But, by the reciprocity theorem,

$$A_2 F_{23} = A_3 F_{32}. \tag{4.44}$$

Combining Eqs. (4.43–44),

$$A_2 F_{14} = A_3 F_{32}. \tag{4.45}$$

Equations (4.39–40), (4.42), and (4.45) can be solved simultaneously for $F_{14}, F_{32}, F_{1(2+4)}$, and $F_{3(2+4)}$.

One of the new ideas introduced by Ex. 4.6 is that of writing down an expression for the rate at which emitted radiant energy from a surface such as $(1 + 3)$ strikes another surface such as $(2 + 4)$ and equating this to the rate at which emitted radiant energy from the surfaces which comprise $(1 + 3)$, such as surfaces 1 and 3, individually strike surface $(2 + 4)$. This led to Eq. (4.41), which is valid for all values of W_1 and W_3. But

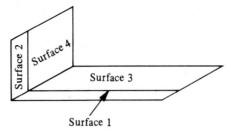

Figure 4.24 General surface arrangement for the use of the special reciprocity theorem derived in Ex. 4.6.

the angle factors do not depend on the values of the emissive powers W_1 and W_3. Hence, Eq. (4.41) must hold when $W_1 = W_3 = W_{1+3}$, which gives another relation among the angle factors. A step such as the one just described can be used in even more complicated problems and is one of the most valuable steps in "flux algebra."

Equation (4.45) has a form roughly similar to the reciprocity theorem; it is not, however, the reciprocity theorem. From Fig. 4.23, note that $A_2 = A_1$, so that Eq. (4.45) can be rearranged to yield

$$A_1 F_{14} = A_3 F_{32}. \tag{4.46}$$

This is a degenerate form of what is called the *special reciprocity theorem,* or the *law of corresponding corners.* For example, in Fig. 4.24, where the two basic rectangles are not even the same size, it can be shown that Eq. (4.46) still holds, even though, quite obviously, $F_{14} \neq F_{23}$. It can be proved that Eq. (4.46) is still valid by setting up the expressions for both $A_1 F_{14}$ and $A_3 F_{32}$ from Eq. (4.31) and then noting the symmetry in the integrand and in the limits on the double integrals. A catalogue of the many cases for which the special reciprocity theorem holds can be found in [7] and [8].

4.3 CALCULATION OF THE NET RADIANT LOSS

We have now developed thermal radiation concepts to the point at which we can make the calculation which is normally of most interest in engineering radiation heat transfer: the net radiant loss from any surface of interest. The *net radiant loss* is the difference between the rate at which a surface *emits* radiant energy and the rate at which it *absorbs* radiant energy from all sources. Sometimes the net radiant loss from a surface is known in advance, as might be the case for an electric resistive heater in a vacuum, or a surface that is perfectly insulated and in a vacuum, thus leading to a net radiant loss of zero. When we know the net radiant loss from a surface, the engineering radiation heat transfer problem becomes that of finding the *surface temperature*. Since the procedures developed to calculate the net radiant loss also can (at least for black and gray surfaces) be used to calculate the temperature when given the net radiant loss, most of the subsequent development will refer explicitly to net radiant loss calculations.

Three different methods for net radiant loss calculations will be presented: (1) the *reflection method,* (2) the *electric network method,* and (3) the *absorption factor method.*

The basic rationale for the use of three different methods is that the full spectrum of radiation heat transfer problems contains some that would be most appropriately handled by method 1, some by method 2, and some by method 3. By "most appropriate" or "best" method, we mean the one which yields the net radiant loss with the smallest investment of time and effort on the part of the engineer doing the analysis.

Presented next is a classification scheme which gives the author's recommendation for the particular method to be used, and which highlights any special features of the methods.

4.3a Method Classification Scheme

1. *Reflection Method.* This is the method to be applied to the *simplest* radiation problems. In particular, this method should be used when the surfaces interacting radiantly are such that there are a finite number of reflections to be considered. Hence, this method would be employed if all surfaces are black, since then there are no reflections at all. This method is also important because it is the one most intimately connected to the physics of the radiant transfer process between surfaces, and thus gives the analyst a feel for what is going on physically.

2. *Electric Network Method.* This would be applied to the radiation problems which have an intermediate degree of difficulty. Specifically, this method would be applied when there are infinitudes of reflections to consider, due to a general gray surface seeing itself or seeing another general gray surface, and when the number of surfaces involved is less than about five. A nice aspect of this method is that the standard electric network reduction techniques, which can sometimes permit the collapse of a complicated circuit to a simpler one, may be applied directly to many radiation heat transfer problems.

3. *Absorption Factor Method.* This method handles the most difficult radiation problems. Thus, when one has infinitudes of reflections and more than about four surfaces, this is usually the most appropriate method. Since the result of the method is a system of linear algebraic equations for the unknown absorption factors, the same library computer programs that were used in the finite difference solutions to conduction problems in Chapter 2 can also be used here. An important feature of this method is the relative ease with which it can accommodate the calculation of the net radiant loss for a fixed geometrical configuration of surfaces when the temperatures of the surfaces are changed, perhaps a number of different times.

When confronted by a radiation heat transfer problem involving the net radiant loss, one should first use the Method Classification Scheme to determine the most suitable method. Then apply that method to arrive at the unknown quantity in the problem.

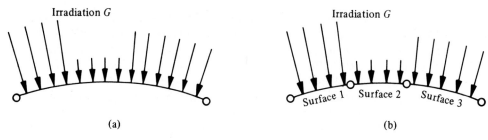

Figure 4.25 (a) Nonuniform irradiation on a single surface. (b) Surface broken up into three uniformly irradiated surfaces. The length of the arrow is proportional to the value of G.

4.3b Restrictions on Methods

Regardless of the method to be used, a number of restrictions and idealizations are normally employed when justified, both in a first course in heat transfer and in radiation problems in industry. These restrictions are listed and discussed below, and will be assumed valid throughout the chapter unless information to the contrary is provided.

1. The surfaces involved are either black or gray.
2. Each surface, at any instant of time, has a single value of temperature (i.e., temperature does not vary on the surface) and has a single value of the total hemispheric emissivity ϵ.
3. Kirchhoff's law, $\epsilon = \alpha$, where α is the total absorptivity, is valid for each surface. This is rigorously true for a gray body, as is discussed in [4].
4. Reflected radiant energy is diffuse, just as the emission from gray bodies is diffuse.
5. The surfaces involved are opaque; hence, from Eq. (4.14), $\alpha + \rho + \tau = 1$, and, since $\tau = 0$, $\alpha + \rho = 1$. But, since $\epsilon = \alpha$, the reflectivity can also be related to ϵ by the equation $\rho = 1 - \epsilon$.
6. All surfaces are uniformly irradiated; that is, the total rate at which radiant energy strikes any one surface per unit area of surface, previously defined as the irradiation G, does not vary with position on the surface. Theoretically, this condition can only be satisfied by choosing differential areas as the surfaces, which leads to an integral equation or equations to be solved for the net radiant loss. In practice, if it is suspected that the irradiation G on a geometric surface varies greatly with position, as in Fig. 4.25(a), the surface is divided into a number of other surfaces each of which has essentially a constant irradiation. This technique is shown in Fig. 4.25(b).
7. A consequence of restrictions 2 through 6 is that the angle factors apply to the reflected radiant energy. Originally, the angle factor was defined in terms of the fraction of the diffuse emitted radiant energy from one surface that directly strikes another surface. However, for a uniformly irradiated surface, radiant energy from all sources strikes at the rate GA, where A is the surface area.

Also, $\rho G A$ is reflected from the surface in a diffuse fashion. Hence, the reflected radiation $\rho G A$ behaves like the total emitted radiant energy rate WA to which the angle factor applies; hence, it must also apply to reflected radiant energy $\rho G A$ since, once the reflected energy leaves the surface, it behaves (diffusely) just like emitted radiant energy. (Note that this statement does not apply unless the surface is *uniformly* irradiated. If it is not, G depends upon position, and this is analogous to W depending upon position on the surface. By noting the steps that led to the general mathematical relationship for the angle factor, Eq. [4.31], W, and hence G, cannot depend upon position on the surface.)

8. Any gases between the opaque surfaces are to be considered completely transparent to thermal radiation. That is, they do not take part in the radiative transfer process; they neither emit nor absorb thermal radiation. This is approximately true in a large number of situations, and this restriction will be discussed further in this chapter.

4.3c Reflection Method

The essence of the *reflection method* of calculating net radiant loss is to follow the emission and subsequent reflections from each surface in order to determine how much of these emissions and reflections are absorbed by the surface of interest. The calculation of the net radiant loss when all surfaces are black is treated first, since this is the simplest case involving no reflections at all, as a consequence of the zero reflectivity for a black surface.

Black Surfaces

First the methodology of the reflection method for black surfaces will be illustrated by applying it in detail to a very simple two-black-body problem. Since there can be, by definition, no reflections from black surfaces, the Method Classification Scheme dictates the use of the reflection method on such problems.

Figure 4.26 shows an enclosure consisting of a black disk, surface 1, with a radius of 0.3 m, which is being held at $T_1 = 560°$K, and a black hemisphere, surface 2, whose temperature is $T_2 = 725°$K. We propose to find an expression for the net radiant loss—call it q_2—from surface 2 and to calculate its numerical value, as well as the value of the net radiant loss from surface 1, q_1.

The net radiant loss q_2 from surface 2 is the difference between the rate at which surface 2 emits radiant energy and the rate at which it absorbs radiant energy from all sources. It emits radiant energy at the rate $W_2 A_2$. It absorbs radiant energy from itself and surface 1. The angle factor F_{22} of $W_2 A_2$ directly strikes surface 2, and it is all absorbed, since surface 2 is black and by definition has an absorptivity of unity. The remainder of $W_2 A_2$ strikes surface 1 and can never be reflected back to surface 2, because surface 1 is black and will not reflect any portion of $W_2 A_2$ which directly strikes it. Hence, $F_{22} W_2 A_2$ of surface 2's own emission is absorbed by surface 2. All of the emission from surface 1, $W_1 A_1$, directly strikes surface 2 because $F_{12} = 1$; hence, all that strikes it is absorbed, since surface 2 is black. As a result, surface 2 absorbs radiant energy at the

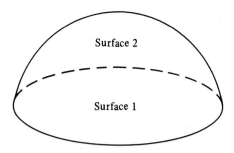

Figure 4.26 The capped hemisphere.

rate $W_1 A_1$ from surface 1. Since q_2 is the difference between the rate at which surface 2 emits radiant energy and the rate at which surface 2 absorbs radiant energy from all sources,

$$q_2 = W_2 A_2 - F_{22} W_2 A_2 - W_1 A_1. \tag{4.47}$$

However, $W_2 = \sigma T_2^4$ and $W_1 = \sigma T_1^4$. Thus, $q_2 = (1 - F_{22}) A_2 \, \sigma T_2^4 - A_1 \, \sigma T_1^4$.

To arrive at the actual numerical value of the net radiant loss q_2, the angle factor F_{22} must be found.

By inspection of Fig. 4.26, we see that surface 1 cannot see itself and can only see surface 2. Hence, all of surface 1's emitted radiant energy must strike surface 2 directly; therefore, $F_{12} = 1$. By the reciprocity theorem, Eq. (4.34), one has

$$A_1 \, F_{12} = A_2 \, F_{21} \quad \text{or} \quad \pi(0.3)^2(1) = 2\pi(0.3)^2 \, F_{21},$$

so $F_{21} = 0.5$.

Next using the relationships among the angle factors in an enclosure, Eq. (4.35), with $n = 2$ and $i = 2$,

$$F_{21} + F_{22} = 1.$$

Hence, $F_{22} = 0.5$.

Inserting the known quantities into Eq. (4.47),

$$q_2 = (1 - 0.5)2\pi(0.3)^2 5.669 \times 10^{-8}(725)^4 \\ - \pi(0.3)^2 5.669 \times 10^{-8}(560)^4,$$

$$q_2 = 2852 \text{ W}.$$

To obtain the net radiant loss from surface 1, note that since only two surfaces are present, the net radiant loss from surface 2 must be the net radiant gain of surface 1; that is, the energy cannot escape elsewhere; hence, $q_2 = -q_1$ or $q_1 = -2852$ W, with the minus sign indicating that the loss is actually a gain; i.e., energy from all sources is being absorbed by surface 1 at a greater rate than surface 1 emits radiant energy.

Since surface 2 has a net radiant loss, if steady-state conditions are to prevail, something must be supplying surface 2 with energy at exactly the same rate of 2852 W. This might be electric resistive heaters buried in material 2 or a hot fluid flowing over the outside of surface 2 and causing energy to be convected to it at the rate of 2852 W.

We can see that the net radiant loss is the calculation of greatest interest when the surface temperature is specified, because it is this net radiant loss that must be balanced by other energy transfer terms if the surface is to remain in the steady state. Also, in a transient situation it is the net radiant loss, along with other types of energy transfer terms, that cause the temperature of the material to vary with time.

Now the reflection method of analysis will be applied to a fairly general situation of an enclosure consisting of n black surfaces all at different known temperatures. All the areas and angle factors are considered to be already found and known. We would like to develop a general analytical expression for the net radiant loss from some surface of interest, surface i.

The net radiant loss from the ith surface is the rate at which surface i emits radiant energy minus the rate at which it absorbs radiant energy from all sources. Since all surfaces in the enclosure are black, i absorbs radiant energy from every surface at the rate of the product of the angle factor from each surface to the ith surface times the area and emissive power of each surface; i.e.,

$$q_i = W_i A_i - F_{1i} A_1 W_1 - F_{2i} A_2 W_2 - \cdots - F_{ii} A_i W_i - \cdots$$
$$- F_{ji} A_j W_j - \cdots - F_{ni} A_n W_n.$$

This can be put into a more compact form by using the reciprocity theorem $A_i F_{ij} = A_j F_{ji}$ on every term except the first and the term containing F_{ii}. Thus,

$$q_i = A_i (W_i - F_{i1} W_1 - F_{i2} W_2 - \cdots - F_{ii} W_i - \cdots - F_{ij} W_j$$
$$- \cdots - F_{in} W_n),$$

which can be written

$$q_i = A_i \left(W_i - \sum_{j=1}^{n} F_{ij} W_j \right), \tag{4.48}$$

$$W_j = \sigma T_j^4. \tag{4.49}$$

This general expression, Eq. (4.48), for the net radiant loss from a black surface i in an enclosure of n black surfaces degenerates, of course, to Eq. (4.47) when conditions are such that $n = 2$, $i = 2$, and $F_{12} = 1$, as can easily be verified.

Surfaces in Radiant Balance

Thus far, only surfaces whose temperature was specified have been considered and the net radiant loss from one or more of these surfaces was required. In many engineering radiation heat transfer problems, one or more surfaces are present for which the net radiant loss is zero, and this surface is said to be in *radiant balance*. It is also called a *reradiating* or *adiabatic surface* from the radiation viewpoint. Examples of such a surface could be the interior walls of a room, in some instances even the ceiling of a room, some furnace walls, certain panel sections on a spacecraft, or, in general, any object which is in the steady state but has no way of gaining or losing energy except by radiation. In that case, its net radiant loss must be zero if it is to remain in the steady state. For instance, consider a satellite in outer space whose internal power supply has failed. If the entire satellite is taken as the control volume, the only way energy can cross the control surface is by radiation. Yet, if steady-state conditions prevail, the energy balance, Eq. (1.8), requires that the rate at which the surface emits radiant energy be exactly equal to the rate at which it absorbs radiant energy from all sources, or, that its net radiant loss be zero. Hence, the satellite described would be in radiant balance if a steady state is achieved.

Throughout the remainder of this chapter, surfaces in radiant balance figure prominently, and ordinarily the temperature of a surface in radiant balance is not given. Since it is in radiant balance its temperature is determined by its radiant interaction with the surrounding surfaces, and this temperature can be found from the statement that the net radiant loss is zero.

EXAMPLE 4.7

In Fig. 4.27 electric resistance heaters are buried uniformly throughout the hemispheric surface 2 and are dissipating electrical energy at the rate of 500 W with the outside of the hemisphere perfectly insulated. Surface 1 is at 600°R while surface 3 is an opening into a large room where the room surfaces are at 500°R. If surfaces 1 and 2 are black and the hemisphere radius is 1 ft, find the temperature of surface 2.

Figure 4.27 The hemisphere of Ex. 4.7.

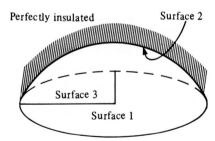

Solution

Since the opening, surface 3, opens into a room which presumably has large dimensions relative to the 1-ft radius of the hemisphere, surface 3 can be treated as a small opening in the side of a large enclosure (the room), and hence behaves as a black surface at 500°R, the temperature of the interior of the enclosure.

To find T_2, the expression for the net radiant loss from surface 2 is written. Since the net radiant loss from surface 2 is the rate at which surface 2 emits radiant energy minus the rate at which it absorbs radiant energy from all sources, and since all surfaces are black,

$$q_2 = W_2 A_2 - F_{22} W_2 A_2 - F_{12} W_1 A_1 - F_{32} W_3 A_3.$$

Note that $F_{12} = F_{32} = 1$ and, from previous work, $F_{22} = \frac{1}{2}$; hence,

$$\begin{aligned}
q_2 &= \tfrac{1}{2} W_2 A_2 - W_1 A_1 - W_3 A_3 \\
&= \tfrac{1}{2}[(0.1714)10^{-8} T_2^4][2\pi(1)^2] - (0.1714)10^{-8}(600)^4 \tfrac{3}{4}(\pi)(1)^2 \\
&\quad - (0.1714)10^{-8}(500)^4 \tfrac{1}{4}\pi(1)^2 = (0.538)10^{-8} T_2^4 - 438. \quad \textbf{(4.50)}
\end{aligned}$$

There are still the two unknowns q_2 and T_2 in Eq. (4.50). Using Eq. (1.8) on the hemispherical surface 2, for steady state, $R_{in} + R_{gen} = R_{out}$. Since the outside surface is perfectly insulated, the only energy transfer rate is the net radiant loss, while the total rate at which electrical energy is dissipated within the control volume is 1703 Btu/hr = 500 W; hence,

$$q_2 = 1703.$$

Combining this with Eq. (4.50) yields

$$1703 = (0.538)10^{-8} T_2^4 - 438,$$

$$3980 = 10^{-8} T_2^4 = \frac{T_2^4}{10^8} = \left(\frac{T_2}{10^2}\right)^4 = \left(\frac{T_2}{100}\right)^4.$$

(Note that the 10^{-8} term is usually handled as shown on the right of the equation, since if T_2 is known, the division by 100 usually gives a number, to be raised to the fourth power, which is convenient to work with.) Thus, $T_2 = 794°R$.

One of the things which Ex. 4.7 illustrates is the procedure to be applied in the case where one of the black surfaces has a specified net radiant loss and the surface temperature is desired. Although in this case the net radiant loss from surface 2 is not zero but 500 W, and therefore surface 2 is *not* in radiant balance, basically the same procedure would be used if 2 *were* in radiant balance. To illustrate, reconsider Ex. 4.7 with the only change being that surface 2 is in radiant balance. The procedure up to and including Eq. (4.50) would be exactly the same. However, if 2 is in radiant balance, then $q_2 = 0$. Putting this into Eq. (4.50) and solving for T_2 leads to

$$T_2 = 534.2°R.$$

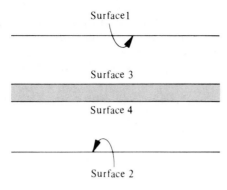

Figure 4.28 Black shield.

Black Radiation Shield

Sometimes the problem in radiation heat transfer is to *reduce* the net radiant loss from a surface. This is also sometimes the case in conduction, convection, and cases where two or even all three modes of heat transfer are acting simultaneously. One way to do this is to interpose other bodies between this surface and the other surfaces with which it exchanges radiant energy. These other bodies are referred to as *radiation shields*.

Figure 4.28 shows two infinite black planes, surfaces 1 and 2, of known temperatures T_1 and T_2, respectively. In order to reduce the net radiant loss from surface 1, a radiation shield consisting of an infinite thin black plate made of a good conductor is put between surfaces 1 and 2 as shown in Fig. 4.28. The upper surface of this shield is called surface 3 and the lower is surface 4.

We would like to derive an expression, in terms of known quantities, for the net radiant loss from surface 1 when the shield is present.

First we write down the expression for q_1, on a per unit area basis, using the reflection method result, Eq. (4.48), with $i = 1$, $n = 2$, and $j = 1$ and 3. Or we can derive the expression for q_1 by applying first principles directly to the situation. Either method results in

$$q_1 = W_1 - W_3. \tag{4.51}$$

Using the Stefan-Boltzmann law gives, for Eq. (4.51),

$$q_1 = \sigma (T_1^4 - T_3^4). \tag{4.52}$$

However, the shield temperature T_3 is still unknown and must be found.

Taking the thin plate as the control volume and using Eq. (1.8), noting that steady-state conditions prevail with no generation, yields $R_{in} = R_{out}$, or

$$0 = q_3 + q_4, \tag{4.53}$$

where q_3 is the net radiant loss from the upper side of the plate while q_4 is the net radiant loss from the lower side of the plate. Since q_3 equals the rate at which surface 3 emits radiant energy minus the rate at which it absorbs radiant energy from all sources, $q_3 = W_3 A_3 - F_{13} W_1 A_1$. By inspection, $F_{13} = 1$ and dealing with 1 ft² or 1 m² of surface $A_3 = A_1 = 1$, and

$$q_3 = W_3 - W_1. \tag{4.54}$$

By the same procedure, the net radiant loss from the lower surface 4 is

$$q_4 = W_4 - W_2. \tag{4.55}$$

Substituting Eqs. (4.54–55) into Eq. (4.53) yields $W_3 + W_4 = W_1 + W_2$ or, using $W_1 = \sigma T_1^4$, etc., and dividing out σ,

$$T_3^4 + T_4^4 = T_1^4 + T_2^4.$$

This equation still contains the two unknowns T_3 and T_4. However, the plate is thin and a good conductor; hence, the internal resistance to heat transfer by conduction through the plate is so small that $T_3 = T_4$ and hence,

$$T_3^4 = \frac{T_1^4 + T_2^4}{2}. \tag{4.56}$$

Combining Eqs. (4.52) and (4.56) yields the expression for the net radiant loss from surface 1, in the presence of the black shield of Fig. 4.28, as

$$q_1 = \frac{\sigma(T_1^4 - T_2^4)}{2}. \tag{4.57}$$

Examination of Eq. (4.57) indicates that the black shield has reduced, by one-half, the net radiant loss from that which would prevail if the shield were not being used. Later in the chapter it will be seen that far greater reductions are possible with a gray shield for which ϵ_3 and ϵ_4 might be far below unity.

Gray Surfaces

The reflection method will now be applied to situations in which at least one of the surfaces is gray. This generally leads to a large number of reflections and re-reflections to be accounted for in the calculation of the net radiant loss. Since the rate at which the surface of interest absorbs radiant energy from all sources must be determined, each source is considered one at a time, the absorption rate by the surface of interest from each individual source of emission is calculated, and the results are added together to obtain the total absorption rate by the surface of interest.

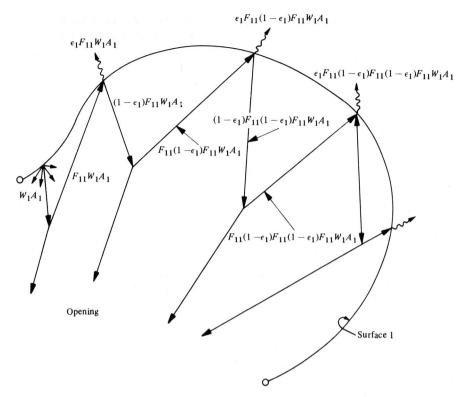

$\epsilon_1 F_{11} W_1 A_1$

$\epsilon_1 F_{11} (1 - \epsilon_1) F_{11} W_1 A_1$

$\epsilon_1 F_{11} (1 - \epsilon_1) F_{11} (1 - \epsilon_1) F_{11} W_1 A_1$

$(1 - \epsilon_1) F_{11} W_1 A_1$

$(1 - \epsilon_1) F_{11} (1 - \epsilon_1) F_{11} W_1 A_1$

$F_{11} (1 - \epsilon_1) F_{11} W_1 A_1$

$F_{11} W_1 A_1$

$W_1 A_1$

$F_{11} (1 - \epsilon_1) F_{11} (1 - \epsilon_1) F_{11} W_1 A_1$

Opening

Surface 1

Figure 4.29 Details of the emissions, reflections, and absorptions from the single gray surface.

Recall from the Method Classification Scheme that the reflection method was recommended only when there was a *finite* number of reflections to be considered. For the moment, we will disregard that recommendation and instead apply the method to a case where there is an infinitude of reflections. This will be done both to carefully illustrate the bookkeeping needed in this method, even for cases where the number of reflections is finite, and also to show the reader why this method is very difficult to implement in complicated radiation problems involving infinitudes of reflections. After this first derivation is complete, the recommendation of the Method Classification Scheme will be adhered to and the reflection method used *only* when the number of reflections is a reasonable finite number.

Consider the simplest possible situation involving an infinitude of reflections, namely, a *gray* surface 1 which can see itself and can also see an opening which is behaving as if it were black and at 0° absolute. Recalling the discussion of the laboratory black body, the opening would behave in this way if the surroundings into which it opens are at a very low temperature compared to the known temperature T_1 of the gray surface. The physical situation being considered is sketched in Fig. 4.29. The emissivity, area, and all angle factors are known, and an analytical expression for the net radiant loss, q_1, from gray surface 1 is to be found.

The net radiant loss q_1 is the rate at which surface 1 emits radiant energy, W_1A_1, minus the rate at which it absorbs radiant energy from all sources. The only source from which it can absorb radiant energy here is itself. Hence, the procedure is to follow the emission from surface 1, determine how much of this is directly absorbed and then how much is absorbed as each reflection hits surface 1, finally summing up all these partial absorption rates to obtain the total absorption rate. Begin by noting that the emission rate W_1A_1 leaves the surface diffusely, in all directions; that is, from normal to the surface to parallel to the surface, and this occurs at every point on surface 1. However, for ease of visualization, the total emission is shown in Fig. 4.29 coming from one point on the surface. Furthermore, for clarity, the diffuse emission and reflections are characterized by a single arrow showing the strength of the radiation. Absorption rates are shown as wavy arrows in surface 1 with their strengths written alongside.

Now, a portion of the emission rate W_1A_1 goes through the opening, and because of the nature of the opening, never returns; hence, that portion is of no concern. The only concern is the portion of the radiant energy that has a chance of being absorbed at surface 1, since only this can contribute to the net radiant loss from surface 1. The only part of the original emission that can possibly be absorbed by surface 1 is the part that directly strikes surface 1; this part, by definition, is the angle factor F_{11}. Hence, $F_{11}W_1A_1$ is shown heading toward and striking surface 1. Part of this is absorbed and part reflected. By definition of the absorptivity α_1 and the reflectivity ρ_1, the absorbed part is $\alpha_1F_{11}W_1A_1$ and the reflected part is $\rho_1F_{11}W_1A_1$. However, by the restrictions discussed earlier in this section, Kirchhoff's law gives $\alpha_1 = \epsilon_1$, and the fact that surface 1 is opaque, along with Kirchhoff's law (see restriction 5), gives $\rho_1 = 1 - \epsilon_1$. With these changes the absorption rate on the first direct strike is

$$\epsilon_1 F_{11} W_1 A_1, \tag{4.58}$$

while the portion reflected is $(1 - \epsilon_1)F_{11}W_1A_1$, as shown in Fig. 4.29. Part of this reflection leaves the opening and that part can never be absorbed by surface 1 (hence, it is not considered) and part of this reflection strikes surface 1. Remember that reflections, like emissions, are assumed to be diffuse and, since they are, the same angle factor F_{11} which applied to the emission also applies to the reflection. As a result, $F_{11}(1 - \epsilon_1)F_{11}W_1A_1$ is the portion of the first reflection which heads back to and strikes the surface as shown. Now, ϵ_1 (which is equal to the absorptivity α_1, by Kirchhoff's law) is the fraction of the incident reflection absorbed, while $\rho_1 = 1 - \epsilon_1$ is the fraction of the incident reflection which is again reflected. Thus the absorption rate for the first reflection is

$$\epsilon_1 F_{11}(1 - \epsilon_1)F_{11}W_1A_1. \tag{4.59}$$

Again part of the reflected radiant energy leaving in a diffuse fashion strikes the surface; that is, the angle factor F_{11} times the strength of the leaving reflections. Again, ϵ_1 of it is absorbed, while $1 - \epsilon_1$ of it is reflected. The rate at which radiant energy is absorbed from this second reflection is

$$\epsilon_1 F_{11}(1 - \epsilon_1)F_{11}(1 - \epsilon_1)F_{11}W_1A_1. \tag{4.60}$$

This process continues and, since the net radiant loss q_1 equals $W_1 A_1$ minus the rate at which it absorbs radiant energy from all sources, the sum of Eqs. (4.58–60), as well as additional terms of the same type, must be subtracted from $W_1 A_1$. Thus,

$$q_1 = W_1 A_1 - [\epsilon_1 F_{11} W_1 A_1 + \epsilon_1 F_{11} (1 - \epsilon_1) F_{11} W_1 A_1 \\ + \epsilon_1 F_{11} (1 - \epsilon_1) F_{11} (1 - \epsilon_1) F_{11} W_1 A_1 + \cdots]. \tag{4.61}$$

Rearranging yields

$$q_1 = W_1 A_1 \{ 1 - \epsilon_1 F_{11} [1 + F_{11} (1 - \epsilon_1) \\ + F_{11}^2 (1 - \epsilon_1)^2 + \cdots]\}. \tag{4.62}$$

We can see from Eq. (4.62) that the form of the additional terms for the absorption of successive reflections is $F_{11}^3 (1 - \epsilon_1)^3$, $F_{11}^4 (1 - \epsilon_1)^4$, etc., for an infinite number of terms. This infinite series can be put into a compact, closed form by using the expansion

$$1/(1 - x) = 1 + x + x^2 + x^3 + x^4 + x^5 + \cdots,$$

for $0 \le x < 1$; i.e., x is a positive fraction less than 1. In Eq. (4.62), the infinite series in the square bracket suggests that x should be identified as $F_{11} (1 - \epsilon_1)$. A check of the product $F_{11} (1 - \epsilon_1)$ yields the conclusion that it is a positive fraction less than 1, since $0 < F_{11} < 1$ and $0 \le (1 - \epsilon_1) < 1$. Hence,

$$1 + F_{11} (1 - \epsilon_1) + F_{11}^2 (1 - \epsilon_1)^2 + F_{11}^3 (1 - \epsilon_1)^3 + \cdots \\ = \frac{1}{1 - F_{11} (1 - \epsilon_1)} . \tag{4.63}$$

Since surface 1 is gray, its emissive power W_1 is equal to $\epsilon_1 \, \sigma T_1^4$. Using this, and substituting Eq. (4.63) into Eq. (4.62), the expression for the net radiant loss from this cavity, surface 1, is

$$q_1 = A_1 \epsilon_1 \sigma T_1^4 \left[1 - \frac{\epsilon_1 F_{11}}{1 - F_{11} (1 - \epsilon_1)} \right]. \tag{4.64}$$

The same procedures which led to Eq. (4.64) by the reflection method are also the ones to use when dealing with the considerably simpler problems this method is normally used for, those in which there is a reasonable *finite* number of reflections. These simpler problems obviate the need for most of the procedures beyond Eq. (4.60), because one is dealing with only a finite number of reflections. However, even in these simpler problems, careful attention must be paid to keeping proper track of all relevant emissions, reflections, and absorptions which determine the net radiant loss from the surface under consideration.

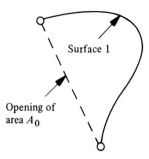

Figure 4.30 The gray cavity used in determining the effective emissivity of the opening.

Effective Emissivity of an Opening

One use of Eq. (4.64) is in the determination of the *effective emissivity, ϵ_0*, of the opening of area A_0 of a gray cavity, surface 1. Figure 4.29 is shown again in Fig. 4.30 in a less cluttered form for defining and finding ϵ_0.

Since the net radiant loss from gray surface 1 in Fig. 4.30 must pass out through the cavity opening 0, it can also be viewed as the net radiant loss for the opening itself. The effective emissivity of the opening, ϵ_0, is defined as the emissivity the opening would have to have if it were a solid surface of area A_0, at the cavity surface temperature T_1, with its net radiant loss being that of the actual cavity. The net radiant loss from the cavity, in this situation, is given by Eq. (4.64). Since this passes through the opening, the apparent emissivity of the opening is defined by

$$q_1 = A_0 \epsilon_0 \sigma T_1^4.$$

Equating this and Eq. (4.64), the apparent emissivity of the opening of the cavity is

$$\epsilon_0 = \frac{A_1}{A_0} \epsilon_1 \left[1 - \frac{\epsilon_1 F_{11}}{1 - F_{11}(1 - \epsilon_1)} \right]. \tag{4.65}$$

Rearranging Eq. (4.65) leads to

$$\epsilon_0 = \frac{\epsilon_1}{1 - F_{11}(1 - \epsilon_1)}. \tag{4.66}$$

This equation is especially useful in deciding whether or not a hole in the side of a large enclosure is the appropriate size (relative to the interior surfaces of the enclosure) for it to act like a black body at the enclosure temperature. Our earlier physical arguments concerning the requirements to construct a laboratory black body centered on an opening which was small compared to the inside surface area of the enclosure. When this is the case, it means that $F_{11} \approx 1$ is very, very close to unity. When this is true, Eq. (4.66) shows that ϵ_0 approaches unity without regard to the value of the emissivity ϵ_1 of the enclosure's interior walls.

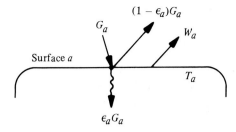

Figure 4.31 A gray surface in radiant balance showing incident, absorbed, reflected, and emitted radiation.

Gray Surfaces in Radiant Balance

Surfaces in radiant balance, by definition, have a net radiant loss of zero, and their temperature is determined by their radiant interaction with the surrounding surfaces. *An important characteristic of a gray surface in radiant balance is that it behaves, regardless of its actual emissivity, exactly like a black surface in radiant balance.* This fact vastly simplifies calculations of the net radiant loss by the reflection method when some of the gray surfaces present are in radiant balance. First, however, it will be shown that a gray surface in radiant balance can be treated as a perfect reflector, regardless of its actual emissivity.

Figure 4.31 shows a gray surface a in radiant balance with emissivity ϵ_a and some unknown temperature T_a. Considering 1 m² or 1 ft² of area, radiation from all sources strikes surface a at the rate G_a, the irradiation of the surface. Now, $\epsilon_a G_a$ is absorbed as shown in Fig. 4.31, while $(1 - \epsilon_a)G_a$ is reflected from the surface. However, the surface is also emitting radiant energy at the rate per unit area W_a, also shown in Fig. 4.31. Since surface a is in radiant balance, its net radiant loss q_a is zero, which means that it is emitting radiant energy at exactly the same rate at which it is absorbing radiant energy from all sources. Hence, $q_a = 0 = W_a - \epsilon_a G_a$, or

$$W_a = \epsilon_a G_a. \tag{4.67}$$

The total rate at which radiant energy leaves the surface, both by emission and reflection, is $W_a + (1 - \epsilon_a)G_a$; however, from Eq. (4.67), $W_a = \epsilon_a G_a$, so that the total rate at which radiant energy leaves the surface is $\epsilon_a G_a + (1 - \epsilon_a)G_a = G_a$. Hence, radiant energy leaves the surface at precisely the same rate per unit area G_a as radiant energy strikes the surface from all sources. Thus, to an observer off the surface who sees only G_a strike and leave the surface, the surface in radiant balance appears to behave as a perfect reflector, i.e., $\epsilon = 0$, regardless of its actual emissivity. This fact will also be used in a method of calculating net radiant losses to be subsequently discussed. (Note that the correct net radiant loss from all other surfaces with which the surface in radiant balance interacts can be found by treating the surface in radiant balance as a perfect reflector, rather than using its actual emissivity.)

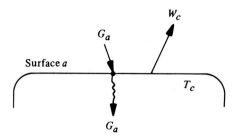

Figure 4.32 A black surface in radiant balance.

We will now prove that a gray surface in radiant balance behaves exactly like a black surface in radiant balance *at the same temperature*. (Note that the last part of the statement is very important, since it permits the easy calculation of the temperature of the surface in radiant balance, whereas the treatment of the surface in radiant balance as a perfect reflector does not allow determination of its temperature.)

Figure 4.32 shows a black surface of exactly the same size and shape as the gray surface pictured in Fig. 4.31, which is in radiant balance at temperature T_c and which receives an irradiation G_a. By the definition of a black surface, all of the irradiation G_a is absorbed and none reflected. However, as shown in Fig. 4.32, the black surface emits radiant energy at the rate W_c, and since the black surface is in radiant balance, the net radiant loss must be zero; that is, $0 = W_c - G_a$, or $W_c = G_a$. Hence, exactly as the gray surface in radiant balance, the other surfaces see radiant energy at the rate G_a strike and leave the surface. The black surface behaves exactly like the gray surface in radiant balance in that it has exactly the same effect on other surfaces (not shown in the Fig. 4.32) as does the gray surface in radiant balance.

The temperature of the black surface in radiant balance can be compared with the temperature T_a of the actual gray surface in radiant balance using Eq. (4.67). Since $W_a = \epsilon_a \sigma T_a^4$, substitution into Eq. (4.67) yields

$$T_a^4 = G_a/\sigma. \tag{4.68}$$

It has been shown that $G_a = W_c$; however, since the surface is black, that is, $W_c = \sigma T_c^4$,

$$T_c^4 = G_a/\sigma. \tag{4.69}$$

A comparison of Eqs. (4.68) and (4.69) shows that $T_c = T_a$, thereby completing the proof. Hence, any gray surface with emissivity ϵ *which is in radiant balance* can be treated as a black surface in radiant balance at the same temperature, and the correct results will be obtained for the net radiant losses of all surfaces which interact with the surface in radiant balance. The correct temperature of the surface in radiant balance can also be obtained, if this is desired. This method of treating surfaces in radiant balance is the preferred one when using the reflection method of analysis, since it greatly limits the number of reflections that must be kept track of in the accounting procedure.

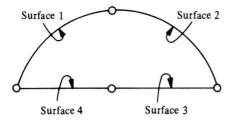

Figure 4.33 The enclosure of Ex. 4.8.

EXAMPLE 4.8

Surface 1 in Fig. 4.33 is gray with emissivity ϵ_1 and is in radiant balance, surface 2 is black at temperature T_2, while surfaces 3 and 4 are gray with temperatures T_3 and T_4 and emissivities ϵ_3 and ϵ_4, respectively. Also surfaces 3 and 4 do not see themselves or each other. All angle factors and areas are known. Derive an expression for the temperature of the surface in radiant balance.

Solution

It has been shown that a gray surface in radiant balance behaves like a black surface in radiant balance at the same temperature, as far as its temperature and its effect on the surroundings is concerned. Hence, surface 1 will be treated as black in radiant balance. With surface 2 actually black, surface 1 effectively black, and surfaces 3 and 4 gray, but not able to see themselves or each other, a quick mental check indicates that there will be a finite number of reflections to be considered here. Hence, our Method Classification Scheme recommends the reflection method for this problem.

The net radiant loss q_1 from surface 1 is zero and equal to the rate at which surface 1 emits radiant energy $W_1 A_1$ minus the rate at which surface 1 absorbs radiant energy from all sources. It absorbs some of its own emission both on the direct strike and by reflection from gray surfaces 3 and 4. Thus, surface 1 emits at the rate $W_1 A_1$, with $F_{11} W_1 A_1$ striking surface 1 and being absorbed and $F_{14} W_1 A_1$ striking surface 4 where $(1 - \epsilon_4) F_{14} W_1 A_1$ is reflected and F_{41} of this returns to surface 1 and is absorbed there. Hence, the rate at which surface 1 absorbs its own emission which is reflected from surface 4 is $F_{41}(1 - \epsilon_4) F_{14} W_1 A_1$. The rest of the emission from surface 1 which is reflected from surface 4 goes to surface 2, which is black; as a result, none can return to surface 1. Using the same reasoning, the rate at which surface 1 absorbs radiant energy which it emitted and which was reflected off gray surface 3 is $F_{31}(1 - \epsilon_3) F_{13} W_1 A_1$. Hence, the total rate at which surface 1 absorbs its own emission is the sum of that absorbed on the direct strike, because surface 1 sees itself, and that absorbed indirectly by reflection from gray surfaces 3 and 4. This total rate is

$$W_1 A_1 [F_{11} + F_{41}(1 - \epsilon_4) F_{14} + F_{31}(1 - \epsilon_3) F_{13}]. \tag{4.70}$$

Surface 1 also absorbs radiant energy on the direct strike as a result of emission from surfaces 3 and 4, and that emission from surfaces 3 and 4 which does not strike surface

1 strikes surface 2 and cannot return to surface 1, since surface 2 is black. Hence, the rate at which surface 1 absorbs radiant energy from surfaces 3 and 4 is

$$F_{31}W_3A_3 + F_{41}W_4A_4. \tag{4.71}$$

Finally, surface 1 absorbs radiant energy emitted by black surface 2, both on the direct strike from surface 2 and by reflection of some of the emission from surface 2 off gray surfaces 3 and 4. The reasoning used to calculate the total rate at which surface 1 absorbs radiant energy emitted by surface 2 exactly parallels the development leading to Eq. (4.70); thus, the result can be written directly as

$$W_2A_2[F_{21} + F_{31}(1 - \epsilon_3)F_{23} + F_{41}(1 - \epsilon_4)F_{24}]. \tag{4.72}$$

As a result, the expression for the net radiant loss of surface 1 is W_1A_1 minus the sum of Eqs. (4.70–72), or

$$\begin{aligned} q_1 = 0 = W_1A_1 &- W_1A_1[F_{11} + F_{41}(1 - \epsilon_4)F_{14} + F_{31}(1 - \epsilon_3)F_{13}] \\ &- F_{41}W_4A_4 - F_{31}W_3A_3 \\ &- W_2A_2[F_{21} + F_{31}(1 - \epsilon_3)F_{23} + F_{41}(1 - \epsilon_4)F_{24}]. \end{aligned}$$

Rearranging, and noting that since surface 1 is treated as black in radiant balance, $W_1 = \sigma T_1^4$,

$$A_1\sigma T_1^4 = \frac{F_{41}W_4A_4 + F_{31}W_3A_3 + W_2A_2[F_{21} + F_{31}(1 - \epsilon_3)F_{23} + F_{41}(1 - \epsilon_4)F_{24}]}{1 - [F_{11} + F_{41}(1 - \epsilon_4)F_{14} + F_{31}(1 - \epsilon_3)F_{13}]} \tag{4.73}$$

where $W_4 = \epsilon_4\sigma T_4^4$, $W_3 = \epsilon_3\sigma T_3^4$, and $W_2 = \sigma T_2^4$.

Thus, Eq. (4.73) gives the temperature of the surface in radiant balance in terms of known quantities in the example.

Note that in Ex. 4.8, the expression for the total emissive power W_4 of a surface such as 4 in terms of its temperature T_4 is $\epsilon_4\sigma T_4^4$. Often there is a tendency to believe that the grayness of the surfaces most certainly has been accounted for, and to incorrectly write σT_4^4 for W_4. However, the emission must be reduced by the grayness of the surface, so that W_4 is correctly written as $\epsilon_4\sigma T_4^4$.

Another common error made in a problem such as Ex. 4.8 occurs when performing the analysis to treat the gray surface 1, which is in radiant balance, as a black body in radiant balance throughout the entire problem, and then at the last step to incorrectly replace W_1 by $\epsilon_1\sigma T_1^4$, where ϵ_1 is the actual emissivity of the gray surface in radiant balance. This is incorrect, since the analysis would be inconsistent; one cannot treat the gray surface in radiant balance as being black in radiant balance for part of the analysis and then suddenly use its actual emissivity in relations that were derived on the basis of its behaving as if it were black.

Example 4.8 represents the type of radiation heat transfer problem for which, according to our Method Classification Scheme, the reflection method of analysis is the most appropriate tool, because there is a reasonable, *finite,* number of reflections of radiant

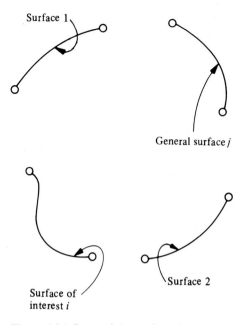

Surface 1

General surface j

Surface of
interest i

Surface 2

Figure 4.34 Some of the surfaces in a general enclosure of n gray surfaces.

energy. The more involved radiation problems, which contain infinitudes of reflections, make it extremely difficult to keep proper account of all needed reflections and absorptions by use of the reflection method. Hence, the next two sections will develop the electric network method and the absorption factor method for these problems.

4.3d Electric Network Method

Earlier remarks and developments have made it clear that for radiant exchange calculations more complicated than those explored so far, such as two gray surfaces which can see each other as well as themselves, the reflection method of calculating net radiant losses becomes very cumbersome. Furthermore, in these more complicated situations, it is very easy to make an error in the account of the reflections and absorptions which ultimately give one the net radiant loss. Thus, in this section we develop a generalized method for calculating net radiant losses, the electric network method, which will be used in radiation problems of an intermediate degree of difficulty. As defined in our Method Classification Scheme earlier, these are the problems involving less than about five surfaces and infinitudes of reflections.

Consider the enclosure shown in Fig. 4.34, which consists of n gray surfaces that can see themselves as well as each other. Of course in less general circumstances, some of these n surfaces may be black rather than gray. Some may not see themselves or see every other surface in the enclosure. Some may not be true physical surfaces, but instead may be openings that are behaving as laboratory black bodies at the temperature of the interior sides of the large enclosures into which they open. However, regardless of how general the surfaces making up the enclosure are, one must identify and deal with surfaces which form an enclosure, both for the electric network method being developed here and for the absorption factor method to be discussed in the next section. In Fig. 4.34, the surface of interest i, for which the net radiant loss is required, and some of the other $n-1$ surfaces such as surfaces 1, 2, and the general surface j are shown. Recall that the irradiation G_i of surface i was previously defined as the total rate at which radiant energy from all sources (including i itself) strikes or impinges on surface i per unit area of surface i. The irradiation G_i does not vary on surface i, because we have established that each surface is uniformly irradiated. The *radiosity* J_i of the ith surface is defined as the total rate at which radiant energy leaves surface i per unit area of i. Thus, the radiosity J_i includes both the emitted radiant energy and the reflected radiant energy from surface i. Since the total rate at which radiant energy from all sources strikes surface i is given by G_iA_i, and the total rate at which radiant energy (both emitted and reflected) leaves surface i is given by J_iA_i (by virtue of the definition of the radiosity J_i), the difference between the rate at which radiant energy leaves and strikes the surface is the net radiant loss q_i, since all radiant energy must be taken into account. Hence,

$$q_i = A_i(J_i - G_i). \tag{4.74}$$

However, since the net radiant loss q_i is the difference between the rate W_iA_i at which surface i emits radiant energy and the rate $\epsilon_iG_iA_i$ at which it absorbs radiant energy from all sources (in terms of the unknown irradiation), q_i can also be written

$$q_i = A_i(W_i - \epsilon_iG_i). \tag{4.75}$$

(Recall that the absorptivity α_i of a gray surface is by Kirchhoff's law equal to the emissivity ϵ_i. This is why the total absorption rate at the ith surface can be written as $\epsilon_iG_iA_i$.)

The emissive power W_i of a gray surface can be written in terms of the emissive power of a black surface at the same temperature by using the total hemispheric emissivity ϵ_i. Hence,

$$W_i = \epsilon_i\sigma T_i^4 = \epsilon_iW_{bi}. \tag{4.76}$$

Substituting W_i from Eq. (4.76) into Eq. (4.75) yields

$$q_i = \epsilon_iA_i(W_{bi} - G_i). \tag{4.77}$$

Solving for G_i from Eq. (4.77) gives

$$G_i = W_{bi} - \frac{q_i}{\epsilon_iA_i}. \tag{4.78}$$

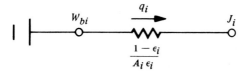

Figure 4.35 A basic circuit element of the electric network method.

Eliminating G_i from Eq. (4.74) by using Eq. (4.78) leads to

$$q_i = \frac{W_{bi} - J_i}{(1 - \epsilon_i)/(A_i\epsilon_i)} . \tag{4.79}$$

Equation (4.79) for the net radiant loss from surface i is in a form which has the appearance of Ohm's law for DC circuits and is readily interpreted in terms of a network element, a resistor of resistance $(1 - \epsilon_i)/(A_i\epsilon_i)$ through which a current q_i passes because of the potential difference $W_{bi} - J_i$ across the resistor. The actual electric network element is shown in Fig. 4.35, where W_{bi} is a known potential and is shown being held at this value by a battery. The element shown in Fig. 4.35 is insufficient to determine the net radiant loss, since the radiosity J_i is not known. However, there is an element like the one shown for every surface in the enclosure, and all that remains is to discover how these unknown radiosities are related. Since the total rate (emitted and reflected) at which radiant energy leaves surface i is A_iJ_i, F_{i1} of this directly strikes surface 1 (the angle factors apply to both the emitted and the reflected radiant energy since both of them are diffuse) and, by the same reasoning, $F_{1i}A_1J_1$ of the total radiant energy leaving surface 1 will directly strike surface i. The net rate of radiant exchange between surfaces i and 1 is the difference between these two quantities; that is,

$$q_{i1} = F_{i1}A_iJ_i - F_{1i}A_1J_1$$

or, using the reciprocity theorem,

$$q_{i1} = F_{i1}A_i(J_i - J_1). \tag{4.80}$$

This same type of expression can be written for the net exchange between surface i and every other surface, such as the jth surface, giving in general,

$$q_{ij} = F_{ij}A_i(J_i - J_j). \tag{4.81}$$

Summing these net radiant exchanges with the surfaces of the enclosure, the result must be q_i, since energy must be conserved. Hence,

$$q_i = \sum_{j=1}^{n} F_{ij}A_i(J_i - J_j). \tag{4.82}$$

Radiation Heat Transfer

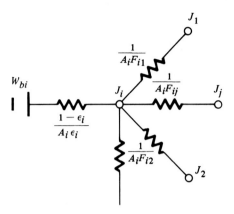

Figure 4.36 The form of the circuit elements connecting radiosities in the electric network method of radiation analysis.

After rearrangement of Eq. (4.81) as

$$q_{ij} = \frac{J_i - J_j}{1/F_{ij}A_i},$$

Eqs. (4.80–82) suggest connecting the radiosity node J_i with every other radiosity node through a resistance of value $1/F_{ij}A_i$. This is shown in Fig. 4.36 for three of the other nodes. Since i is a general node, connections like the one shown in Fig. 4.36 will occur at every node where the potential is the unknown radiosity. Thus it completes an electrical network which can be solved for the quantities of interest (the radiosities J) by standard electrical network methods such as Kirchhoff's current law [16].

Summarizing, an element like the one shown in Fig. 4.35 is sketched for every surface of the enclosure and then every radiosity node is connected to every other radiosity node by a resistance whose value is given by the general form $1/A_iF_{ij}$, or, because of reciprocity,

$$\frac{1}{A_jF_{ji}}. \tag{4.83}$$

The resulting electric network is solved for the radiosities of interest, and Eq. (4.79) is then used to compute the net radiant losses of interest.

Next, consider the problem of finding an expression for the net radiant loss from a gray cavity, surface 1, with emissivity ϵ_1 and temperature T_1. Surface 1 can see itself as well as an opening, surface 0, which behaves like a black surface at 0° absolute. The physical situation is shown in Fig. 4.37. Since gray surface 1 can see itself, this will give rise to an infinitude of reflections. Also, since there are two surfaces involved, the Method Classification Scheme suggests the use of the electric network method.

Since only two surfaces are involved, only two elements of the form shown in Fig. 4.35 are obtained, which are connected by a resistance of magnitude $1/A_1F_{10}$ as shown in Fig.

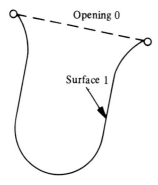

Figure 4.37 The gray cavity.

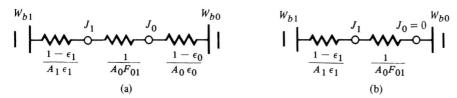

(a)   (b)

Figure 4.38 (a) The electric network. (b) Simplified network resulting from the fact that surface 0 is black.

4.38(a). However, because surface 0 is black, $\epsilon_0 = 1$ and the resistance to the far right in Fig. 4.38(a) is zero. Also, since surface 0 is at 0° absolute, $W_{b0} = 0$, and this equals J_0 since the resistance just mentioned is zero. With these changes, the circuit is redrawn as shown in Fig. 4.38(b). Since in this simple circuit, q_1 is the current through all parts of the circuit, the radiosity J_1 need not be found directly, since q_1 is equal to the overall voltage difference $W_{b1} - 0$ divided by the sum of the series resistances; that is,

$$q_1 = \frac{W_{b1}}{(1 - \epsilon_1)/(A_1\epsilon_1) + (1/A_0F_{01})}.$$

Noting that $A_0F_{01} = A_1F_{10}$ and $W_{b1} = \sigma T_1^4$, after some rearrangement the net radiant loss becomes

$$q_1 = \frac{A_1\epsilon_1\sigma T_1^4}{1 - \epsilon_1 + (\epsilon_1/F_{10})}. \tag{4.84}$$

The relationship among the angle factors in an enclosure yields $F_{11} + F_{10} = 1$; hence, $F_{10} = 1 - F_{11}$. Using this in Eq. (4.84) yields

$$q_1 = A_1\epsilon_1\sigma T_1^4\left[1 - \frac{\epsilon_1 F_{11}}{1 - F_{11}(1 - \epsilon_1)}\right]. \tag{4.85}$$

Radiation Heat Transfer   363

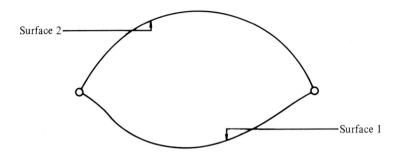

Surface 2

Surface 1

Figure 4.39 Two general gray surfaces forming an enclosure.

Recall that, as an illustration of the type of general problems the reflection method *could* be applied to, and to show the complexity of handling such problems, this problem was previously solved using the reflection method. The result was given by Eq. (4.64), which is, of course, the same as Eq. (4.85). However, comparing the effort required for each method, we can see that the network method is much simpler to apply, even in a problem as relatively easy and simple as this one is.

Two Gray Surfaces Forming an Enclosure

Consider two general gray surfaces that can see themselves as well as one another, and which together form an enclosure as shown in Fig. 4.39. Since this physical configuration, or degenerate versions of it, often may appear in practice, we would like to derive an expression for the net radiant loss from surface 1 when the temperatures, emissivities, areas, and angle factors are all known. Our Method Classification Scheme points to the electric network method.

Since there are only two surfaces, again only two elements of the form shown in Fig. 4.35 are required, and they are connected by a resistance of value $1/A_1F_{12}$ or $1/A_2F_{21}$ (which are equal by the reciprocity theorem). The relevant electric circuit is shown in Fig. 4.40. The current in the network is q_1 everywhere; hence,

$$q_1 = \frac{W_{b1} - W_{b2}}{\dfrac{1 - \epsilon_1}{A_1\epsilon_1} + \dfrac{1}{A_1F_{12}} + \dfrac{1 - \epsilon_2}{A_2\epsilon_2}} .$$

Using the Stefan-Boltzmann law for the black body emissive powers of surfaces 1 and 2 in this equation yields the result

$$q_1 = \frac{\sigma(T_1^4 - T_2^4)}{\dfrac{1 - \epsilon_1}{A_1\epsilon_1} + \dfrac{1}{A_1F_{12}} + \dfrac{1 - \epsilon_2}{A_2\epsilon_2}} . \tag{4.86}$$

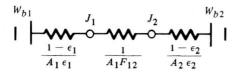

Figure 4.40 The electric network for an enclosure of two gray surfaces.

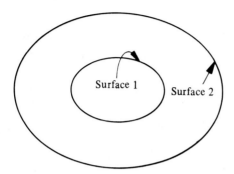

Figure 4.41 Two gray surfaces with $F_{11} = 0$.

A frequently encountered special case of Eq. (4.86) occurs when surface 1 cannot see itself so that $F_{12} = 1.0$ (see Fig. 4.39) and Eq. (4.86), when rearranged, takes the following form:

$$q_1 = \frac{A_1 \sigma (T_1^4 - T_2^4)}{\dfrac{1}{\epsilon_1} + \dfrac{A_1}{A_2}\left(\dfrac{1}{\epsilon_2} - 1\right)}, \tag{4.87}$$

which is known as *Christiansen's equation,* after the man who derived it in 1883 by the reflection method. Strictly speaking, it is only valid for two concentric spheres or cylinders or two parallel infinite planes, since only in these cases will the two surfaces be uniformly irradiated as required by one of the restrictions stated at the beginning of Sec. 4.3b. However, unless the eccentricity of the two bodies shown in Fig. 4.41 is very great—for example, body 1 practically touching the bottom portion of surface 2—Christiansen's equation is still used. This stems from the fact that in many circumstances the effect of nonuniform irradiation will be a second-order effect which does not cause much of a change in the value of the net radiant loss. Deciding whether or not the effect of nonuniform irradiation is important is another example of a situation where the engineer's judgement is important. As previously mentioned, the effects of nonuniform irradiation can be assessed by breaking the surfaces into a greater number of surfaces which are more nearly uniformly irradiated and repeating the calculation of the net radiant loss. Then, however, additional angle factors must be found, and since the electric network becomes more complicated as well, a decision must be made as to how much additional effort is justified for a particular problem.

Sometimes it happens that body or surface 1 has a relatively small surface area compared to surface 2, or that the total hemispherical emissivity, ϵ_2, of surface 2 is almost unity. When either of these situations prevails, the second term in the denominator of Eq. (4.87) is small compared to the first and can be neglected, giving us

$$q_1 = \epsilon_1 A_1 \sigma (T_1^4 - T_2^4). \tag{4.88}$$

This *simplified Christiansen's equation,* Eq. (4.88), is perhaps the most widely used simple equation in engineering radiation heat transfer calculations. It is valid when two gray surfaces form an enclosure with surface 1 not being able to see itself, $F_{11} = 0$, and when either $A_1 << A_2$ or when $\epsilon_2 \approx 1.0$.

Treatment of Surfaces in Radiant Balance

As was mentioned before, surfaces in radiant balance—that is, surfaces whose net radiant loss q is zero—commonly occur in engineering radiation heat transfer problems.

To develop the techniques used to handle these surfaces in the electric network method, consider the enclosure made up of three gray surfaces in Fig. 4.42. Surfaces 1 and 2 have known temperatures, T_1 and T_2, and emissivities, ϵ_1 and ϵ_2, while surface 3 has a known emissivity ϵ_3 and is also known to be in radiant balance because it cannot support a non-zero net radiant loss. After noting that the Method Classification Scheme indicates the use of the electric network method, we will develop the proper electric network for the problem and find the net radiant loss from surface 1, q_1, in terms of known quantities.

There are three elements of the general form shown in Fig. 4.35, and each of the radiosity nodes is connected to all the other radiosity nodes by a resistance of the form of Eq. (4.83), as shown in Fig. 4.43(a). However, by definition of a surface in radiant balance, q_3 is zero, and there is no current flow between J_3 and W_{b3}; hence, $W_{b3} = J_3$, an unknown potential. Thus the value of ϵ_3 is immaterial, since no current flows through that resistance, which constitutes an alternate proof of the statements previously made concerning the behavior of a surface in radiant balance using the reflection method of analysis. Since there is no current flow between W_{b3} and J_3, it is conventional to show the circuit without the resistance between J_3 and W_{b3} so that J_3 is a floating potential to be determined if the temperature of surface 3 is required. Thus, the circuit is redrawn in Fig. 4.43(b). To solve this circuit, note that the two resistances $1/A_1F_{13}$ and $1/A_2F_{23}$ are in series, so that their sum is the equivalent resistance R_1 in that leg of the circuit; that is,

$$R_1 = 1/A_1F_{13} + 1/A_2F_{23}. \tag{4.89}$$

Also let

$$R_2 = 1/A_1F_{12}. \tag{4.90}$$

Since the two legs of the circuit between J_1 and J_2 are in parallel, the equivalent resistance is

$$R_e = R_1R_2/(R_1 + R_2). \tag{4.91}$$

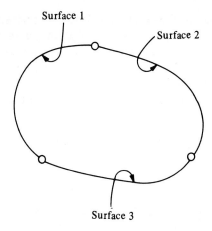

Figure 4.42 The enclosure of three surfaces.

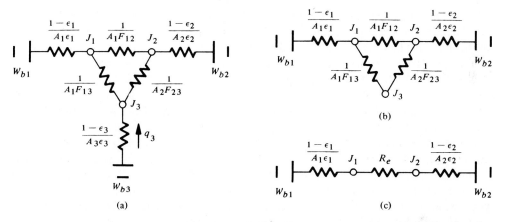

Figure 4.43 (a) The electric circuit for Fig. 4.42. (b-c) Successively simpler forms of the circuit due to surface 3 being in radiant balance and the application of standard network reduction techniques.

Using Eq. (4.91), the simple series circuit shown in Fig. 4.43(c) is obtained. Hence, solving for q_1,

$$q_1 = \frac{W_{b1} - W_{b2}}{(1 - \epsilon_1)/A_1\epsilon_1 + R_e + (1 - \epsilon_2)/A_2\epsilon_2}.$$

Since $W_{b1} = \sigma T_1^4$ and $W_{b2} = \sigma T_2^4$,

$$q_1 = \frac{\sigma(T_1^4 - T_2^4)}{(1 - \epsilon_1)/A_1\epsilon_1 + R_e + (1 - \epsilon_2)/A_2\epsilon_2}, \tag{4.92}$$

where R_e is found from Eqs. (4.89–91).

What if in addition to, or instead of, the net radiant loss from surface 1 in Fig. 4.42, one wants to find the unknown temperature T_3 of the surface in radiant balance? To do this, one can first compute q_1 from Eq. (4.92) and then use q_1, which is also $-q_2$ here, along with the *known* potentials W_{b1} and W_{b2} in Eq. (4.79) to find both radiosities, J_1 and J_2. With J_1 and J_2 now known, application of Kirchhoff's node law (the sum of all the currents into any node of an electric circuit, without capacitance at the node, must be zero) to radiosity node J_3 in Fig. 4.43(b) gives

$$\frac{J_1 - J_3}{\dfrac{1}{A_1 F_{13}}} + \frac{J_2 - J_3}{\dfrac{1}{A_2 F_{23}}} = 0.$$

Solving this equation for J_3, which is also W_{b3} when $q_3 = 0$, gives

$$J_3 = \sigma T_3^4.$$

This can now be solved for the temperature of the surface in radiant balance.

The main point of this section is not to present Eq. (4.92), although that certainly is useful for a configuration like Fig. 4.42. Rather, the point is to describe the procedure one uses when one or more surfaces in radiant balance occur in the enclosure for which a radiation solution is being found by the electric network method.

A General Four-Body Enclosure   Four general gray surfaces of known temperatures and emissivities form an enclosure. All areas are known and all angle factors are considered to have been already found. We would like to find the net radiant loss from surface 1, q_1. For this set of circumstances, the Method Classification Scheme calls for the use of the network method. Using the previously established rules which dictate that a circuit element of the form given in Fig. 4.35 is needed for each of the four surfaces, and that every radiosity node must be connected to all other radiosity nodes by a resistor of the form given by Eq. (4.83), we assemble the network shown in Fig. 4.44.

Examination of this network reveals that it cannot be collapsed into a simpler network by the application of standard electric network reduction techniques. So the net radiant loss from surface 1 is the current in the leg of the circuit connected to the battery W_{b1}. Hence, by either direct reference to the appropriate circuit element in Fig. 4.44 or by use of Eq. (4.79), one has

$$q_1 = \frac{W_{b1} - J_1}{(1 - \epsilon_1)/A_1 \epsilon_1} \tag{4.93}$$

However, the radiosity J_1 must be determined before q_1 can be found from Eq. (4.93). Thus, Kirchhoff's node law is applied to radiosity node 1 by summing the currents into node 1 and setting such currents equal to zero. Hence,

$$\frac{W_{b1} - J_1}{(1 - \epsilon_1)/A_1 \epsilon_1} + \frac{J_3 - J_1}{1/A_1 F_{13}} + \frac{J_2 - J_1}{1/A_1 F_{12}} + \frac{J_4 - J_1}{1/A_1 F_{14}} = 0. \tag{4.94}$$

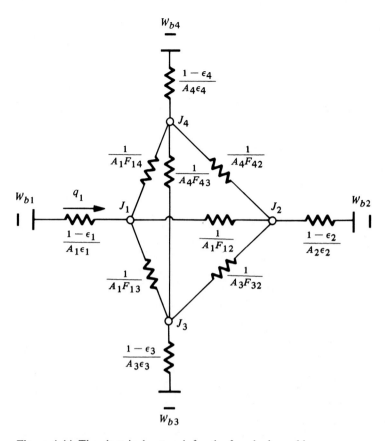

Figure 4.44 The electrical network for the four-body problem.

However, this has introduced the additional unknown radiosities J_2, J_3, and J_4. The three additional equations are found by summing the currents into the remaining three nodes and setting the sums equal to zero. Hence, for node 3,

$$\frac{W_{b3} - J_3}{(1 - \epsilon_3)/A_3\epsilon_3} + \frac{J_1 - J_3}{1/A_1F_{13}} + \frac{J_2 - J_3}{1/A_3F_{32}} + \frac{J_4 - J_3}{1/A_4F_{43}} = 0; \qquad \textbf{(4.95)}$$

for node 2,

$$\frac{W_{b2} - J_2}{(1 - \epsilon_2)/A_2\epsilon_2} + \frac{J_1 - J_2}{1/A_1F_{12}} + \frac{J_3 - J_2}{1/A_3F_{32}} + \frac{J_4 - J_2}{1/A_4F_{42}} = 0; \qquad \textbf{(4.96)}$$

for node 4,

$$\frac{W_{b4} - J_4}{(1 - \epsilon_4)/A_4\epsilon_4} + \frac{J_1 - J_4}{1/A_1F_{14}} + \frac{J_2 - J_4}{1/A_2F_{24}} + \frac{J_3 - J_4}{1/A_4F_{43}} = 0. \qquad \textbf{(4.97)}$$

Equations (4.94–97) can be solved simultaneously for the radiosities J_1, J_2, J_3, and J_4, and J_1 can then be used in Eq. (4.93) to obtain the net radiant loss q_1.

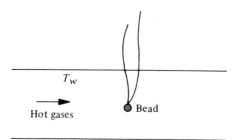

Figure 4.45 The duct and thermocouple.

Thermocouple Radiation Error

Figure 4.45 shows a small thermocouple bead situated at the centerline of a large duct in which hot gases at unknown temperature T_f are flowing. The inside wall of the duct is at known temperature T_w, and the temperature of the thermocouple bead has been calculated from the potentiometer readings as T_1. The emissivity of the bead is ϵ_1, while the surface coefficient of heat transfer between the bead surface and the flowing gas is h. If conduction through the thermocouple leads is negligible, and steady-state conditions prevail, we would like to develop an expression for the unknown temperature of the gas, T_f, which the thermocouple is attempting to measure.

First, one may ask why the measured temperature T_1 is *not* the gas temperature. The known bead temperature is not the gas temperature that the thermocouple is trying to measure because the bead can exchange radiant energy with the duct walls at temperature T_w, and this net radiant loss must then be made up by convection from the gas if steady-state conditions are to prevail. However, Newton's cooling law shows that if there is convection between the gas and the bead, there must also be a temperature difference between them. Hence, the source of the thermocouple error is exposed. Taking the entire bead as the control volume and using Eq. (1.8), noting that $R_{gen} = R_{stor} = 0$, yields $R_{in} = R_{out}$. Now, R_{out} is the expression for the net radiant loss q_1, and the convection is assumed to the control volume so that

$$R_{in} = hA_s(T_f - T_1) = q_1 = R_{out} \qquad (4.98)$$

An expression must still be found for the bead's net radiant loss q_1. This can be done quite easily by the application of the electric network method to the two-gray-surface enclosure formed by the bead surface and the duct surface. However, this was done earlier and resulted in Eqs. (4.86) through (4.88). The conditions in this problem satisfy the requirements for the use of the simplified Christiansen's equation, Eq. (4.88); hence,

$$q_1 = A_s\epsilon_1\sigma(T_1^4 - T_w^4).$$

Substituting this into Eq. (4.98) and rearranging, the unknown gas temperature is

$$T_f = T_1 + (\epsilon_1/h)\sigma(T_1^4 - T_w^4). \qquad (4.99)$$

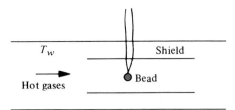

Figure 4.46 Shielded thermocouple.

Equation (4.99) shows that radiation error can be reduced by making the second term on the right negligibly small. A small bead emissivity ϵ_1 would accomplish this; however, even if ϵ_1 is made small, for example by polishing, in actual operation it would most likely become dirty, eroded, and/or corroded from the gas, thus causing the emissivity to rise to about 0.8 or 0.9. If the surface coefficient h is large, the error will be small, and h can be made larger than it would ordinarily be in the situation pictured in Fig. 4.45 by use of a collecting device to accelerate the gas locally in the vicinity of the bead. Such a device is called an aspirating, or suction, pyrometer, and is described in Jakob [12].

However, the usual way to reduce the radiation error of a thermocouple is to insert one or more radiation shields, such as a circular cylinder located as shown in Fig. 4.46, between the bead and the wall. The shield then comes to some temperature between T_f and T_w, which becomes the effective T_w to be used in Eq. (4.99), provided that the shield design is such that Eq. (4.88) is still the correct form to use to calculate q_1. The temperature which the shield comes to is determined by an energy balance on the shield taken as a control volume.

Gray Radiation Shield

Earlier, a black radiation shield interposed between two surfaces in a vacuum to reduce the net radiant loss was studied by the reflection method of analysis. It was found that the single black shield cuts the net radiant loss from either surface in half. At the time, it was mentioned that a gray shield with $\epsilon < 1$ would perform considerably better. The infinitude of reflections that a gray shield could necessitate calls for the application of the electric network method in the analysis of the effect of such a shield on the net radiant loss of the surface being shielded when both are in a vacuum.

Figure 4.47 shows a cross section of a long duct. Surfaces 1 and 2 are gray at known temperatures T_1 and T_2, with emissivities ϵ_1 and ϵ_2, respectively. Material a, interposed between surfaces 1 and 2, is a thin sheet of good conducting gray material with emissivity ϵ_a, which acts as a shield and neither adds nor subtracts any net radiant energy; that is, it acts as a resistance to radiation heat transfer between surfaces 1 and 2. The inside and outside surfaces of a are surfaces 3 and 4, respectively. Since a is a very good conductor and is relatively thin, it can be said that $T_3 = T_4$, which is an unknown temperature at present. We wish to solve for the net radiant loss from surface 1, q_1, by the electric network method.

Two separate networks are set up for the two separate enclosures, one for surfaces 1 and 3 and another for surfaces 4 and 2 as shown in Fig. 4.48(a). However, it is known

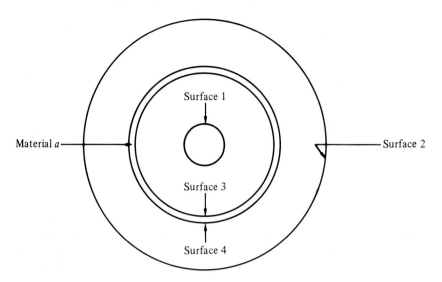

Figure 4.47 The annular duct with a thin concentric cylinder of material a between its surfaces.

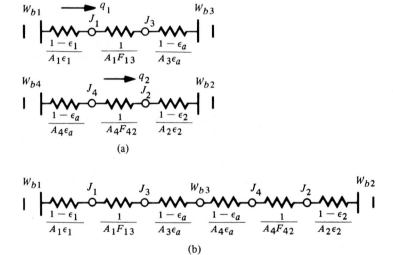

Figure 4.48 (a) Electric networks for the two enclosures. (b) Form of the network when surfaces 3 and 4 are at the same temperature.

Chapter Four

that since $T_4 = T_3$, W_{b4} and W_{b3} are equal in magnitude and, furthermore, since the shield a is in thermal balance, the current q_1 for the first enclosure must equal the current q_2 for the second enclosure (since the shield contributes no net amount of energy). This suggests that the batteries W_{b3} and W_{b4} be "removed" and the two ends connected, since the batteries contribute no net radiant energy and the node at which the two circuits are joined, from Fig. 4.48(a), behaves somewhat like a surface in radiant balance. The resulting circuit is shown in Fig. 4.48(b). Ohm's law is applied to this circuit to obtain the net radiant loss with the shield between the surfaces; that is, using $W_{b1} = \sigma T_1^4$ and $W_{b2} = \sigma T_2^4$,

$$q_1 = \frac{\sigma(T_1^4 - T_2^4)}{\dfrac{1 - \epsilon_1}{A_1\epsilon_1} + \dfrac{1}{A_1 F_{13}} + \dfrac{1 - \epsilon_a}{\epsilon_a A_3} + \dfrac{1 - \epsilon_a}{\epsilon_a A_4} + \dfrac{1}{A_4 F_{42}} + \dfrac{1 - \epsilon_2}{A_2\epsilon_2}}. \qquad (4.100)$$

The effect of the shield in decreasing the net radiant loss can be clearly seen from the middle four resistances in Eq. (4.100). Even when the shield is black, the first and last resistances of the middle four in Eq. (4.100) cause a reduction in q_1 over the case of no shield at all. Also, when $\epsilon_a < 1$, the middle two resistances can make very significant contributions to a reduction in q_1.

EXAMPLE 4.9

A soldering iron has a tip of area 0.0013 m² and emissivity $\epsilon = 0.88$, and the tip cannot see itself. The soldering is being done in an enclosure whose surfaces are all at 27°C. Steady-state conditions prevail, and power is supplied to the tip at the rate of 20 W. Calculate the surface temperature of the tip (a) if the soldering is being done in a vacuum; (b) if the soldering is taking place in a 27°C argon atmosphere with a surface coefficient of heat transfer of $h = 11.4$ W/m² °C between the tip and the argon.

Solution
In part a, since the soldering is being done in an evacuated enclosure, the only way the tip can lose or gain energy is by thermal radiation (neglecting conduction away from the tip along the body of the iron). Using the energy balance, Eq. (1.8), on the tip as a control volume yields, in the steady state,

$$R_{in} + R_{gen} = R_{out}.$$

Now, $R_{gen} = 20$ W; hence, $R_{out} - R_{in} = 20$ W also from this energy balance. But $R_{out} - R_{in}$, in the absence of convection and conduction, is simply the net radiant loss from the tip; call it q_1. Therefore, $q_1 = 20$ W. To relate this known net radiant loss to the unknown tip temperature T_1, one could use the Method Classification Scheme to select the proper method and then apply it to this configuration. However, the enclosure consisting of the tip and the other room surfaces constitutes a two-gray-surface enclosure problem whose solution has already been found in Eqs. (4.86) to (4.88). Actually, since the tip area of 0.0013 m² is presumably very small compared to the room surface area,

Eq. (4.88), the simplified Christiansen's equation, is applicable. Hence, inserting known quantities,

$$20 = 0.88(0.0013)5.669 \times 10^{-8} [T_1^4 - (273 + 27)^4].$$

Therefore, one gets

$$T_1 = 750°K = 477°C. \tag{a}$$

In part b, the procedure is modified by including the convection term in the energy balance, Eq. (1.8), giving,

$$20 = \epsilon_1 A_1 \sigma(T_1^4 - T_2^4) + hA_1(T_1 - T_2). \tag{4.101}$$

The temperature T_2 is also used in the convection term, since in this particular problem the temperature of the argon atmosphere is the same as that of the solid walls of the room. Since Eq. (4.101) cannot be solved in an explicit analytical fashion for T_1, it will be rearranged and solved iteratively for T_1. (Newton's method can also be employed to solve Eq. [4.101] if one so chooses.) Factoring as shown next,

$$T_1^4 - T_2^4 = (T_1^2 - T_2^2)(T_1^2 + T_2^2)$$
$$= (T_1 - T_2)(T_1 + T_2)(T_1^2 + T_2^2),$$

and inserting into Eq. (4.101) gives, after rearrangement,

$$T_1 = T_2 + \frac{20/hA_1}{1 + \frac{\epsilon_1 \sigma}{h}(T_1 + T_2)(T_1^2 + T_2^2)}. \tag{4.102}$$

Inserting the given known values into Eq. (4.102) gives

$$T_1 = 300 + \frac{1349.53}{1 + 0.4376 \times 10^{-8}(T_1 + 300)(T_1^2 + 90,000)}. \tag{4.103}$$

The solution procedure for Eq. (4.103) will be to put an initial estimate of T_1 into the right-hand side and calculate T_1 on the left. Then we will use this as a corrected estimate of T_1 on the right, recompute T_1 on the left, substitute it into the right-hand side, and so on, until two successive values of T_1 agree closely enough. As a first estimate of T_1, a value between 300°K and the 750°K temperature of part a would be chosen, since the effect of the convection is to reduce the tip temperature needed to dissipate 20 W across its surface. Hence, the answer of part a, $T_1 = 750°K$, is used as the first estimate on the right side of Eq. (4.103). This results in

$$T_1 = 300 + \frac{1349.53}{1 + 0.4376 \times 10^{-8}(750 + 300)(750^2 + 90,000)}$$
$$= 637.5°K.$$

Next, the iteration proceeds by using $637.5°K$ as T_1 on the right of Eq. (4.103) giving $T_1 = 744°K$. When this value is used on the right of Eq. (4.103), the result is $T_1 = 642.5°K$. Hence, a slow oscillatory type of convergence is apparently taking place. In order to speed up the process, the next T_1 value was taken to be the arithmetic average of the last two values, $642.5°K$ and $744°K$, and the iterative process continued until a converged value was reached,

$$T_1 = 690.5°K = 417.5°C. \tag{b}$$

Comparing the answers to parts a and b, one sees that the effect of convection to the argon is to reduce the tip temperature by $59.5°C$.

This problem is a typical example of the rather commonly occurring situation in which one has to account for the fact that both modes of heat transfer, radiation and convection, are acting simultaneously.

EXAMPLE 4.10

Figure 4.49 shows an enclosure consisting of a 0.3 m radius hemisphere, which is termed surface 1, capped by a disk divided into two equal surfaces called 2 and 3. Surface 1 is gray with $T_1 = 600°K$ and $\epsilon_1 = 0.35$, surface 2 is black at $T_2 = 350°K$, and surface 3 is gray, but is in radiant balance. Find the net radiant loss from surface 1 as well as the temperature, T_3, of the surface in radiant balance.

Solution
There are infinitudes of reflections here (most obviously because surface 1 is gray and sees itself) and three surfaces involved in the enclosure so that by the Method Classification Scheme, the electric network method is the most easily used method for the problem. The angle factors can be put down by inspection as $F_{21} = F_{31} = 1.0$ and $F_{23} = 0$. The gray surface in radiant balance acts like a floating potential at $J_3 = W_{b3}$ which, with the rest of the network, is shown in Fig. 4.50(a). Note that $J_2 = W_{b2}$ because surface 2 is black, but the battery W_{b2} must be shown because there *is* current in the zero resistance leg between J_2 and W_{b2}. Recall that a battery need not be shown *only* for a surface in radiant balance, since by definition its net radiant loss is zero. As a result of this, there is no current between J_3 and W_{b3}, and thus W_{b3} is not shown.

Figure 4.49 The capped hemisphere of Ex. 4.10.

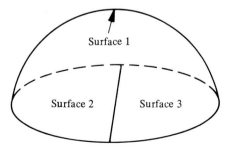

Surface 1

Surface 2   Surface 3

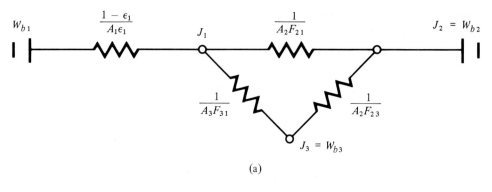

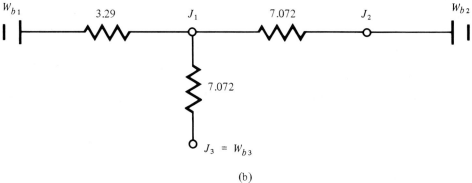

Figure 4.50 (a) The electric network of Ex. 4.10. (b) Simplified network resulting from $F_{23}=0$.

Now $A_2 = A_3 = \pi(0.3)^2/2 = 0.1414 \text{ m}^2$, $A_1 = 2\pi(0.3)^2 = 0.565 \text{ m}^2$, $(1 - \epsilon_1)/A_1\epsilon_1$ $= (1 - 0.35)/0.565(0.35) = 3.29$, $1/A_2F_{21} = 1/0.1414(1) = 7.072$, $1/A_3F_{31} = 7.072$ also, and $1/A_2F_{23} \to \infty$, since $F_{23} = 0$. This last result requires no connection, an open circuit, between nodes 2 and 3 in Fig. 4.50(a), and the simplified network that results is shown in Fig. 4.50(b).

Solving the electric network shown in Fig. 4.50(b) for q_1, which is the current between W_{b1} and J_1, gives

$$q_1 = \frac{5.669 \times 10^{-8}[600^4 - 350^4]}{3.29 + 7.072} = 627 \text{ W}.$$

To find the temperature T_3 of the surface in radiant balance, Kirchhoff's current law is applied to radiosity node 3, yielding $W_{b3} = J_1$. Ohm's law, written between W_{b1} and J_1, gives

$$627 = \frac{5.669 \times 10^{-8}(600)^4 - J_1}{3.29},$$

so that

$$J_1 = 5284 = W_{b3} = 5.669 \times 10^{-8}T_3^4.$$

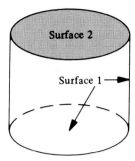

Figure 4.51 The right circular cylinder of Ex. 4.11.

Hence, $T_3 = 552.5°K$ is the temperature of the surface in radiant balance, while the net radiant loss from surface 1 was found to be 627 W.

This problem demonstrates how to find the temperature of the surface in radiant balance and how to treat black surfaces when using the electric network method. For a black surface, $J = W_b$, because of the fact that $(1 - \epsilon)/A\epsilon$ is zero for such a surface, or simply because of the definition of the radiosity as the total rate per unit area at which emitted *and* reflected radiant energy come off a surface. Since nothing can be reflected from a black surface, its total emissive power and its radiosity are the same.

EXAMPLE 4.11

Figure 4.51 shows the inside of a right circular cylinder made of a gray material with $\epsilon = 0.50$, which is held at 1000°R both on the curved inside surface and the bottom circular disk. The top of the cylinder is open to a surrounding that is at so low a temperature, relative to 1000°R, that the top, labelled surface 2, behaves like a black surface at 0°R. The 1000°R surface is called surface 1. If the cylinder diameter and height are both 2 ft, calculate the net radiant loss through the opening, surface 2. Next, in order to gauge the importance of the effects of nonuniform irradiation of the 1000°R surface, divide it into two surfaces which are more nearly uniformly irradiated: surface 3, the bottom circular disk, and surface 4, the lateral curved surface.

Solution

Surface 1, being gray and seeing itself, gives rise to an infinitude of reflections; hence the Method Classification Scheme points to the network method. However, in the first part of this problem, one is dealing with a two-body enclosure, and hence Eq. (4.86) may be used once the subscripts are interpreted properly. Therefore,

$$q_2 = \frac{W_{b2} - W_{b1}}{[(1 - \epsilon_1)/A_1\epsilon_1] + (1/A_2F_{21})}.$$

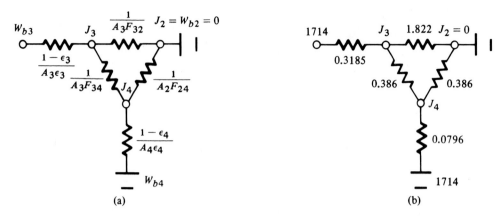

Figure 4.52 (a) The electric network of Ex. 4.11. (b) The network with numerical values shown.

Now, $W_{b2} = 0$, $W_{b1} = \sigma T_1^4 = 0.1714(\frac{1000}{100})^4 = 1714$, $F_{21} = 1$, $A_2 = \frac{1}{4}\pi D_2^2 = \frac{1}{4}\pi(2)^2 = 3.14$ ft^2, $A_1 = \frac{1}{4}\pi D_2^2 + \pi D_2 L = \frac{3.14}{4}(2)^2 + 3.14(2)(2) = 15.72$ ft^2, $(1 - \epsilon_1)/A_1\epsilon_1 = \frac{1 - 0.5}{(15.72)(0.5)} = 0.0636$, and $1/A_2F_{21} = \frac{1}{3.14(1)} = 0.3185$. Solving for q_2 yields

$$q_2 = -4490 \text{ Btu/hr.}$$

Now, the problem will be reworked by treating surface 1 as two surfaces, the bottom disk being surface 3 and the curved part of the cylinder surface 4. This will be done in order to work with surfaces which are more nearly uniformly irradiated than is the original surface 1, since uniformly irradiated surfaces are called for in our earlier listing of assumptions to be used in our radiation analyses.

Using the chart of Fig. 4.17 and the relationship for angle factors in an enclosure, the angle factors are $F_{32} = F_{23} = 0.175$, and $F_{34} = F_{24} = 0.825$. The electrical network for this three-surface situation is shown in Fig. 4.52(a). The various resistances and emissive powers are $W_{b3} = W_{b4} = \sigma T^4 = 0.1714(\frac{1000}{100})^4 = 1714$, $A_3 = \frac{1}{4}\pi(2)^2 = 3.14$ ft^2, $(1 - \epsilon_3)/\epsilon_3 A_3 = \frac{1 - 0.5}{0.5(3.14)} = 0.3185$, $(1 - \epsilon_4)/\epsilon_4 A_4 = \frac{1 - 0.5}{0.5(12.58)} = 0.0796$, $1/A_3F_{34} = \frac{1}{3.14(0.825)} = 0.386$, $1/A_2F_{24} = \frac{1}{3.14(0.825)} = 0.386$, and $1/A_3F_{32} = \frac{1}{3.14(0.175)} = 1.822$. The network is redrawn in Fig. 4.52(b) with the numerical values for resistances and for black body emissive powers shown. Summing the currents into node 4 gives

$$\frac{J_3 - J_4}{0.386} + \frac{1714 - J_4}{0.0796} + \frac{0 - J_4}{0.386} = 0. \qquad (4.104)$$

Summing the currents into node 3 yields

$$\frac{1714 - J_3}{0.3185} + \frac{J_4 - J_3}{0.386} + \frac{0 - J_3}{1.822} = 0. \qquad (4.105)$$

Equations (4.104–5) can be solved simultaneously to yield the radiosities as $J_3 = 1430$ and $J_4 = 1423$. Since J_3 and J_4 are so close, the surfaces involved are very close to being uniformly irradiated, and hence, q_2 should not be very different from the value found using surface 1. Now, q_2 is found by setting the sum of the currents entering node 2 equal to zero. Hence,

$$q_2 = -\left(\frac{1430 - 0}{1.822} + \frac{1423 - 0}{0.386}\right) = -4475 \text{ Btu/hr.}$$

For this case, the effect of nonuniform irradiation is so slight it is negligible.

The results of this problem, which show that nonuniform irradiation acts here as a second-order effect, seem to carry over into most practical engineering radiation heat transfer problems. As long as the different surfaces of an enclosure are not very highly reflective (and they usually are not) and are not arranged in a manner that promotes nonuniform irradiation of various surfaces, the effect of actual nonuniform irradiation will be so slight that it can be ignored in calculations of net radiant loss. Additional support for this conclusion is given in Gebhart [13].

4.3e Absorption Factor Method

The radiation calculation technique which is recommended in the Method Classification Scheme for the most difficult radiation heat transfer problems will be presented and discussed in this section. This absorption factor method, due to Gebhart [14], would be used in problems involving infinitudes of reflections and five or more surfaces forming an enclosure. In addition, as mentioned in the Method Classification Scheme, the absorption factor method would be used when the same *geometric* configuration must be solved a number of times, even if there are fewer surfaces than five. For example, consider the situation of four gray surfaces discussed earlier, whose electric network was given in Fig. 4.44, for which the net radiant loss from surface 1 can be found after Eqs. (4.94) through (4.97) are solved simultaneously for the radiosities.

If the net radiant loss from surface 1 must be found many times because the temperatures of the surfaces involved change, either during a transient or as a result of different steady-state operating conditions, the set of Eqs. (4.94–97) must be solved for each new set of temperatures, since the surface temperatures appear in the equation set in terms like W_{b1}, W_{b2}, etc. This difficult procedure can be averted by putting the equations in matrix form and calculating the inverse of the coefficient matrix. The same inverse coefficient matrix applies to each new set of temperatures as long as the emissivities (as well as the areas and the spatial relationship of each surface to the others) remain the same as the specified surface temperatures are changed [4].

However, in putting the equations of the electric network method into matrix form, one in effect loses the electric network itself, and thus also loses the advantage of the method, namely, the use of standard circuit reduction techniques to collapse the original circuit into a simpler one. The formulation of the absorption factor method, on the other hand, is such that it can easily and directly handle problems in which the same geometric configuration must be re-solved a number of different times.

The recommended limitation to less than five surfaces for the electric network method stems from two considerations. First of all, it becomes increasingly difficult to clearly draw the network when one has five or more wires emanating from *each* radiosity node, each wire having to cross many other wires while going to other radiosity nodes. Secondly, with this many batteries, nodes, and wires, the resulting circuit cannot be reduced or collapsed into significantly simpler circuits. One usually ends up writing Kirchhoff's node law at every radiosity node, thus generating a set of simultaneous algebraic equations for the unknown radiosities. The sum total of all this is that the network itself simply becomes an intermediate step which really does not help simplify the problem in any way. Actually, it is not even needed, since in the absorption factor method one writes down the appropriate algebraic equations directly without any need for intermediate steps.

Development of the Absorption Factor Method

Figure 4.53 shows an enclosure consisting of n gray surfaces with all angle factors, areas, and emissivities known. The net radiant loss from any of the n surfaces, say surface j, is to be calculated. By definition, q_j is the rate at which surface j emits radiant energy $W_j A_j$ minus the rate at which surface j absorbs radiant energy from all sources including itself, both on direct strikes of emitted radiant energy and by reflection and re-reflection of emitted radiant energy from all surfaces. Gebhart defines the absorption factor B_{ij} as the fraction of the radiant energy *emitted* by the ith surface which is eventually *absorbed* by the jth surface after a complex reflection pattern. Note the difference between the absorption factor B_{ij} and the angle factor F_{ij}. The angle factor F_{ij} is simply a geometrical entity which is the fraction of surface i's emitted radiant energy which directly strikes surface j. In general, all of F_{ij} is not absorbed at j, and some returns by reflections. The absorption factor B_{ij}, on the other hand, is defined as the fraction of the radiant emission from i which is eventually absorbed by surface j. Hence, in terms of the absorption factors, the net radiant loss for the jth surface is

$$q_j = W_j A_j - B_{1j} W_1 A_1 - B_{2j} W_2 A_2 - \cdots - B_{jj} W_j A_j - \cdots \\ - B_{ij} W_i A_i - \cdots - B_{nj} W_n A_n.$$

Since B_{1j} is the fraction of the emission from surface 1 which is eventually absorbed by surface j, the rate at which surface j absorbs radiant energy emitted by surface 1 is $B_{1j} W_1 A_1$. Similar reasoning gives the other terms. Using summation notation, the preceding equation becomes

$$q_j = W_j A_j - \sum_{i=1}^{n} B_{ij} W_i A_i. \tag{4.106}$$

Analytical expressions for the absorption factors $B_{1j}, B_{2j}, \ldots$, must be found. This can be begun by attempting to find the value of B_{1j}, the fraction of surface 1's emission which is eventually absorbed by surface j. The reflection method is used, concentrating only on the emission from surface 1 that is eventually absorbed by the jth surface. Thus,

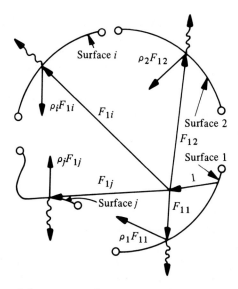

Figure 4.53 Some of the n gray surfaces in an enclosure with the radiant emission of surface 1 shown in detail as an aid in deriving the absorption factor B_{1j}.

in Fig. 4.53, all rays are a fraction of $W_1 A_1$. As shown, surface 1 emits the fraction 1 which perhaps strikes all the surfaces of the enclosure, not just the few shown, and the fraction of surface 1's emission which strikes the various surfaces directly is, by definition, the angle factors between surface 1 and the various surfaces. On the direct strike of surface 1's emission at surface j, the fraction $\epsilon_j F_{1j}$ is absorbed and the remainder $\rho_j F_{1j}$ is reflected and has a chance to strike surface j again. Hence, the fraction $\epsilon_j F_{1j}$ is *part* of the absorption factor B_{1j}, and B_{1j} can be written as

$$B_{1j} = \epsilon_j F_{1j} + \text{other terms yet to be found.} \qquad (4.107)$$

(Note that the reflectivity ρ_j is still equal to $1 - \epsilon_j$ by the eight restrictions of Sec. 4.3b; however, to conserve space, it is written as ρ_j until it must be actually determined.)

The emission from surface 1 which is absorbed by surfaces other than j on the direct strike is of no consequence, since it cannot return to j and therefore cannot contribute to the absorption factor B_{1j} of interest. However, the portion of surface 1's emission that is reflected off the surfaces is of interest, since this energy still has a chance of being absorbed by j, and hence can contribute to the absorption factor B_{1j}. Thus, the reflections of the direct strike from surface 1 are shown on all the surfaces with their strength. If it were known how much of the reflection from surface 1 would eventually be absorbed by surface j, what fraction of the reflection from surface 2 would eventually be absorbed by surface j, what fraction of the reflection from surface j would eventually be absorbed by surface j, etc., the respective fractions could be multiplied by the strengths of the reflections shown in Fig. 4.53 and the sum of these products added to $\epsilon_j F_{1j}$ in Eq. (4.107) to complete the expression for the absorption factor B_{1j}. But the fraction of the reflection from surface 1 which is eventually absorbed by j is the absorption factor B_{1j}, because the reflected radiant energy behaves exactly like an emission (by the restrictions of Sec. 4.3b), and the absorption factors, although defined in terms of emission, must apply to reflections as well, since they cannot distinguish between them. Hence, $B_{1j}F_{11}\rho_1$ is the fraction of the reflected radiant energy from surface 1 which is eventually absorbed by surface j, and $B_{2j}F_{12}\rho_2$ is the fraction of the reflected portion from surface 2 which originated at surface 1, which is eventually absorbed by surface j, etc. Adding these fractions to $\epsilon_j F_{1j}$ in Eq. (4.107), the expression for the absorption factor B_{1j} is

$$B_{1j} = \epsilon_j F_{1j} + B_{1j}F_{11}\rho_1 + B_{2j}F_{12}\rho_2 + \cdots + B_{ij}F_{1i}\rho_i$$
$$+ B_{nj}F_{1n}\rho_n. \tag{4.108}$$

However, Eq. (4.108) contains the other unknown absorption factors. Using the same reasoning that led to Eq. (4.108), the equations for the unknown absorption factors are

$$B_{2j} = \epsilon_j F_{2j} + B_{1j}F_{21}\rho_1 + B_{2j}F_{22}\rho_2 + \cdots + B_{ij}F_{2i}\rho_i$$
$$+ \cdots + B_{nj}F_{2n}\rho_n$$
$$B_{3j} = \epsilon_j F_{3j} + B_{1j}F_{31}\rho_1 + B_{2j}F_{32}\rho_2 + B_{3j}F_{33}\rho_3 + \cdots$$
$$+ B_{nj}F_{3n}\rho_n \tag{4.109}$$
$$\cdot$$
$$\cdot$$
$$\cdot$$
$$B_{nj} = \epsilon_j F_{nj} + B_{1j}F_{n1}\rho_1 + B_{2j}F_{n2}\rho_2 + \cdots + B_{nj}F_{nn}\rho_n.$$

Equations (4.108–9) constitute n simultaneous equations for the n unknown absorption factors. Now, since products such as $F_{21}\rho_1$, $F_{32}\rho_2$, etc. appear in almost every term, the single symbol for the product is defined for convenience as

$$K_{rs} = F_{rs}\rho_s \tag{4.110}$$

so that $F_{21}\rho_1 = K_{21}$, $F_{32}\rho_2 = K_{32}$, etc. Introducing Eq. (4.110) into Eqs. (4.108–9), combining like terms (i.e., B_{1j} appears twice in Eq. [4.108], etc.) and putting the terms which do not contain the absorption factors, such as $\epsilon_j F_{3j}$, on one side by themselves, the equation set for the n unknown absorption factors is

$$(K_{11} - 1)B_{1j} + K_{12}B_{2j} + K_{13}B_{3j} + \cdots + K_{1n}B_{nj} = -\epsilon_j F_{1j}$$
$$K_{21}B_{1j} + (K_{22} - 1)B_{2j} + K_{23}B_{3j} + \cdots + K_{2n}B_{nj} = -\epsilon_j F_{2j}$$
$$K_{31}B_{1j} + K_{32}B_{2j} + (K_{33} - 1)B_{3j} + \cdots + K_{3n}B_{nj} = -\epsilon_j F_{3j}$$
$$\cdot \tag{4.111}$$
$$\cdot$$
$$\cdot$$
$$K_{n1}B_{1j} + K_{n2}B_{2j} + K_{n3}B_{3j} + \cdots + (K_{nn} - 1)B_{nj} = -\epsilon_j F_{nj}.$$

Thus, Eq. (4.111) constitutes the n simultaneous equations for the n unknown absorption factors, which, when found, can be used in Eq. (4.106) to obtain the net radiant loss from the jth surface, the surface of interest.

Gebhart has also developed the relations

$$A_i B_{ij} \epsilon_i = A_j B_{ji} \epsilon_j, \tag{4.112}$$

$$\sum_{j=1}^{n} B_{ij} = 1. \tag{4.113}$$

Equations (4.112–13) are analogous to the reciprocity theorem and the relationship among angle factors in an enclosure, respectively.

In any particular problem, Eq. (4.111) can be solved for the n unknown absorption factors by any of the appropriate methods for sets of linear algebraic equations. Recall that such solution methods were given in Chapter 2 when finite difference methods were being presented. Thus, Eq. (4.111) can be written as a square coefficient matrix $[K'_{rs}]$ mapping a single column matrix $\{B_{sj}\}$ into another single column matrix $\{-\epsilon_j F_{rj}\}$; that is,

$$[K'_{rs}] \{B_{sj}\} = \{-\epsilon_j F_{rj}\} \tag{4.114}$$

for each value j. Equation (4.114) is rewritten with the elements of each matrix shown as

$$\begin{bmatrix} (K_{11} - 1) & K_{12} & K_{13} & \cdots & K_{1n} \\ K_{21} & (K_{22} - 1) & K_{23} & \cdots & K_{2n} \\ K_{31} & K_{32} & (K_{33} - 1) & \cdots & K_{3n} \\ \cdot & & & & \\ \cdot & & & & \\ \cdot & & & & \\ K_{n1} & K_{n2} & K_{n3} & \cdots & (K_{nn} - 1) \end{bmatrix} \begin{bmatrix} B_{1j} \\ B_{2j} \\ B_{3j} \\ \cdot \\ \cdot \\ \cdot \\ B_{nj} \end{bmatrix} = \begin{bmatrix} -\epsilon_j F_{1j} \\ -\epsilon_j F_{2j} \\ -\epsilon_j F_{3j} \\ \cdot \\ \cdot \\ \cdot \\ -\epsilon_j F_{nj} \end{bmatrix}. \tag{4.115}$$

Equation (4.114) can now be solved for the one-column, n-rowed matrix of absorption factors as

$$\{B_{sj}\} = [K'_{rs}]^{-1} \{-\epsilon_j F_{rj}\}, \tag{4.116}$$

where $[K'_{rs}]^{-1}$ is the inverse of the square matrix $[K'_{rs}]$. The matrix form of the equations can be used to solve for the absorption factors with the aid of a digital computer using subroutines appropriate for such equations when the number of surfaces involved is very large. If the number of surfaces involved is large and a computer is not available, Eq. (4.111) can still be solved by relaxation analysis as discussed in Chapter 2 when finite differences were used to arrive at steady-state temperature distributions.

When some of the surfaces are in radiant balance and the net radiant loss from one of the other surfaces is to be found, Gebhart recommends treating the surfaces in radiant balance as perfect reflectors. Thus, if surface a is in radiant balance, $\rho_a = 1$ and $\epsilon_a = 0$,

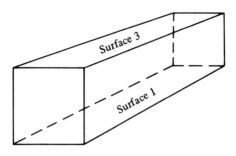

Figure 4.54 The food freezer of Ex. 4.12.

giving $W_a = 0$ in Eq. (4.106) so that the absorption factor B_{aj} need not be found. However, if the temperature of a surface a in radiant balance is to be found, it is recommended that it be treated as a black surface in radiant balance so that $\epsilon_a = 1$, $\rho_a = 0$, and $q_a = 0$. By Eq. (4.106), this gives, after the absorption factors are found, a single equation for the unknown temperature T_a.

Gebhart's method allows quick and easy calculation of the net radiant loss q_j for various sets of surface temperatures (as long as the absorption factors do not depend significantly upon the temperature level) by obtaining the absorption factors from Eq. (4.111) once and then letting the effect of changing the set of surface temperatures make itself felt on q_j only through the emissive powers W_1, W_2, . . . , W_i in Eq. (4.106).

EXAMPLE 4.12

The covered food freezer shown in Fig. 4.54 is 3 ft high, 3 ft wide and 10 ft long. The floor of the freezer, surface 1, is to be held at $-5°$F and has an emissivity of 0.93. The sides of the freezer are in radiant balance, and the four sides are considered as a single surface, called surface 2. The top of the freezer is at 70°F, has an emissivity of 0.7, and is called surface 3. Assuming that all surfaces are gray, find the rate at which the refrigerant must extract energy from the bottom surface of the freezer to maintain steady-state conditions, considering radiation only. In addition, calculate the energy transfer rate to the refrigerant at an off-design point which can occur in the event that the air conditioning system for the store, in which the freezer is located, fails on a hot day. If this happens, the freezer top will be at 95°F instead of 70°F, yet the freezer floor (actually the surface of the frozen food packages stored in the freezer) must still be maintained at $-5°$F.

Solution
Since gray surface 3 sees gray surface 1, there are an infinite number of reflections in this three-surface enclosure. Normally, this would mean, according to the Method Classification Scheme, that the electric network method should be used here. However, we want the net radiant loss from the floor, surface 1, at two different operating points (surface 3 at 70°F and when surface 3 is at 95°F) for the same geometrical configuration. Referring back to this point in the Method Classification Scheme and to the discussion at the beginning of this absorption factor section, we conclude that the absorption factor method should be used here in spite of the fact that the enclosure consists of less than

five surfaces. Using an energy balance, we find the rate at which the refrigerant must extract energy from surface 1 to be the net radiant loss q_1 from surface 1. In the absorption factor method, Eq. (4.106) becomes, with $n = 3$ and the surface of interest (surface j) called surface 1,

$$q_1 = W_1A_1 - B_{11}W_1A_1 - B_{21}W_2A_2 - B_{31}W_3A_3. \tag{4.117}$$

The surface in radiant balance, surface 2, is treated as a perfect reflector. Therefore, we assign to it the following values of emissivity and reflectivity: $\epsilon_2 = 0$ and $\rho_2 = 1$. Since $\epsilon_2 = 0$, it follows that $W_2 = \epsilon_2\sigma T_2^4 = 0$ as well. Because of this, the term containing $B_{21}W_2$ in Eq. (4.117) is zero, even though B_{21} is *not* zero. With this, Eq. (4.117) can be rewritten as follows:

$$q_1 = W_1A_1(1 - B_{11}) - B_{31}W_3A_3. \tag{4.118}$$

Now, $A_1 = 3(10) = 30$ ft^2, $A_3 = 3(10) = 30$ ft^2, and $A_2 = 3(10)(2) + 3(3)(2) = 78$ ft^2. Since $W_1 = \epsilon_1\sigma T_1^4 = 0.93(0.1714)(\frac{-5 + 460}{100})^4 = 68.2$, $W_1A_1 = 68.2(30) = 2045$; also since $W_3 = \epsilon_3\sigma T_3^4 = 0.7(0.1714)(\frac{70 + 460}{100})^4 = 94.5$, $W_3A_3 = 94.5(30) = 2840$. Hence, Eq. (4.118) becomes

$$q_1 = 2045(1 - B_{11}) - 2840B_{31}. \tag{4.119}$$

The absorption factors are now determined from Eq. (4.111) with $n = 3$ and $j = 1$; hence,

$$\begin{align}
(K_{11} - 1)B_{11} + K_{12}B_{21} + K_{13}B_{31} &= -\epsilon_1F_{11}, \\
K_{21}B_{11} + (K_{22} - 1)B_{21} + K_{23}B_{31} &= -\epsilon_1F_{21}, \\
K_{31}B_{11} + K_{32}B_{21} + (K_{33} - 1)B_{31} &= -\epsilon_1F_{31}.
\end{align} \tag{4.120}$$

(Note that although B_{21} is not required in Eq. [4.119], it is still an unknown and must be retained in Eq. [4.120].)

The angle factors are also required. Surfaces 1 and 3 are two equal-sized, parallel, directly-opposing planes, and the angle factor for this situation can be found from the chart of Fig. 4.16 as $F_{13} = 0.32$ and, by symmetry, $F_{31} = 0.32$. By the relationship among angle factors in an enclosure, $F_{12} = 0.68 = F_{32}$. By inspection, $F_{11} = F_{33} = 0$. By the reciprocity theorem, $A_1F_{12} = A_2F_{21}$; hence, $F_{21} = \frac{30}{78}(0.68) = 0.262$, $F_{23} = 0.262$, and $F_{22} = 0.476$.

Here, $\epsilon_1 = 0.93$, $\rho_1 = 0.07$, $\epsilon_3 = 0.7$, $\rho_3 = 0.3$, $\epsilon_2 = 0$, $\rho_2 = 1$, $\epsilon_1F_{11} = 0.93(0) = 0$, $\epsilon_1F_{21} = 0.93(0.262) = 0.244$, and $\epsilon_1F_{31} = 0.93(0.32) = 0.298$. As a result, $K_{11} = F_{11}\rho_1 = 0$, $K_{12} = F_{12}\rho_2 = 0.68$, $K_{13} = F_{13}\rho_3 = 0.32(0.3) = 0.096$, $K_{21} = F_{21}\rho_1 = 0.262(0.07) = 0.0184$, $K_{22} = F_{22}\rho_2 = 0.476(1) = 0.476$, $K_{23} = F_{23}\rho_3 = 0.262(0.3) = 0.0786$, $K_{31} = F_{31}\rho_1 = 0.32(0.07) = 0.0224$, $K_{32} = F_{32}\rho_2 = 0.68(1) = 0.68$, and $K_{33} = F_{33}\rho_3 = 0(0.3) = 0$. Substituting these values into Eq. (4.120) yields

$$\begin{align}
-B_{11} + 0.68B_{21} + 0.096B_{31} &= 0, \\
0.0184B_{11} - 0.524B_{21} + 0.0786B_{31} &= -0.244, \\
0.0224B_{11} + 0.68B_{21} - B_{31} &= -0.298.
\end{align} \tag{4.121}$$

Writing equation set (4.121) in matrix form gives,

$$\begin{bmatrix} -1.0 & 0.68 & 0.096 \\ 0.0184 & -0.524 & 0.0786 \\ 0.0224 & 0.68 & -1.0 \end{bmatrix} \begin{bmatrix} B_{11} \\ B_{21} \\ B_{31} \end{bmatrix} = \begin{bmatrix} 0.0 \\ -0.244 \\ -0.298 \end{bmatrix} \quad \textbf{(4.122)}$$

Solving Eq. (4.122) by use of a standard library computer program for simultaneous equations gives the following values of the absorption factors:

$$B_{11} = 0.472, \ B_{21} = 0.593, \ B_{31} = 0.712. \quad \textbf{(4.123)}$$

Using these results in Eq. (4.119) yields

$$q_1 = 2045 \ (1 - 0.472) - 2840(0.712) = -940 \ \text{Btu/hr},$$

the rate at which the refrigerant must extract energy from the bottom of the freezer. The minus sign is a consequence of the convention used, indicating that the net radiant loss is actually a gain, with energy entering the bottom surface at this net rate and then being carried away by the refrigerant. This value of $q_1 = -940$ Btu/hr is the refrigeration load when the freezer top, surface 3, is at $70°$F.

Next we would like to compute q_1 at the off-design point where the freezer top is at $95°$F instead, due to a failure in the store's air conditioning system. Since the only change is the temperature of surface 3, Eq. (4.118) applies again, as long as the new value of W_3 is used. Looking at equation set (4.120), which determines the values of the absorption factors needed in Eq. (4.118), we see that the geometric configuration is unchanged. Therefore, all previous values of the angle factors still apply, and the emissivities and reflectivities have not changed (as long as ϵ_3 has about the same value at $95°$F as it has at $70°$F). Hence, all the values of the K's are the same, and thus the absorption factor values are unchanged as well. Equation (4.123) gives the absorption factors. Now, the new value of W_3 is given by

$$W_3 = \epsilon_3 \sigma T_3^4 = 0.7(.1714)\left[\frac{95 + 460}{100}\right]^4 = 113.84 \ \text{Btu/hr ft}^2.$$

With this, and previously calculated values from the first part of the problem, Eq. (4.118) gives

$$q_1 = 2045(1 - 0.472) - 113.84(30)(0.712) = -1352 \ \text{Btu/hr}.$$

Thus, the refrigeration system should be sized to have about 50% more capacity than is needed at the design point, where the freezer top is at $70°$F, in order to accommodate the off-design condition that could occur if the store's air conditioner did not function on a $95°$F day.

Example 4.12 illustrates how quickly and easily the absorption factor method can be used to compute the net radiant losses of interest when only the surface temperatures have changed in a problem which was previously solved with a different set of surface temperatures.

It is also, perhaps, worthwhile to stress the way the surface in radiant balance, surface 2, was treated in this problem. It was stated earlier, during the development of the absorption factor method, that when one is trying to find the net radiant loss from a surface which is not in radiant balance, the fastest and simplest procedure is to treat all surfaces in radiant balance as being perfect reflectors, $\rho = 1$ and $\epsilon = 0$. This was done in this problem. If one instead had decided to treat the surface in radiant balance as being black, which of course is certainly permissible, the work to get q_1 would be effectively doubled. The use of $\epsilon_2 = 1.0$ instead of 0.0 would eventually lead to a set of absorption factors B_{11}, B_{21}, and B_{31} whose values would be different from those given in Eq. (4.123). But when these were substituted into Eq. (4.117) with the different value of W_2, they would yield exactly the same value of q_1. However, to get W_2 in the first place would require that Eq. (4.106) also be written out for $j = 2$ so that the use of $q_2 = 0$ would lead to the value of W_2. But this requires that the absorption factors for surface 2, B_{12}, B_{22}, and B_{32}, also be found. Hence, equation set (4.111) would have to be solved for *both* $j = 2$ and $j = 1$, thus doubling the work to get q_1.

4.4 NET RADIANT LOSS FROM NONGRAY SURFACES

The reflection, network, and absorption factor methods for the calculation of the net radiant loss have been utilized only for the case where the surfaces behave, approximately at least, as if they were gray. That is, we have worked with a surface when its monochromatic hemispheric emissivity ϵ_λ is independent of wavelength and is, therefore, equal to its total hemispheric emissivity ϵ. When this is true, Kirchhoff's law gives the relation that $\alpha = \epsilon$. However, situations also occur in which some of the surfaces involved are nongray, and their monochromatic hemispheric emissivity ϵ_λ is a known function of wavelength λ. Kirchhoff's law remains valid in the sense of the monochromatic values; that is, at each wavelength λ, $\alpha_\lambda = \epsilon_\lambda$. Theoretically, at least, any of the methods already developed could be used and the net radiant loss calculated from the jth surface *per unit wavelength,* around the wavelength λ; call if $q_{j\lambda}$. Then the net radiant loss from the jth surface over all wavelengths is

$$q_j = \int_0^\infty q_{j\lambda} \, d\lambda. \tag{4.124}$$

In using any of these methods to obtain $q_{j\lambda}$, quantities such as W_λ and ϵ_λ are dealt with rather than the total (over all wavelengths) corresponding quantities W and ϵ. Because of the variation of ϵ_λ with wavelength λ, the integration indicated in Eq. (4.124) must be done numerically.

To solve Eq. (4.124), *band approximation* is usually used. The entire wavelength interval is broken up into a finite number of bands, in each of which the monochromatic emissivities of all the surfaces are approximately constant, and then the net radiant loss is calculated for each band of finite width in wavelength by the reflection, network, or absorption factor method. The net radiant loss from the jth surface in each of the finite bands is then added to obtain the total (over all wavelengths) net radiant loss from surface j. This procedure, while perhaps seeming pretty straightforward, is greatly complicated by the presence of one or more surfaces in radiant balance, since these surfaces are in radiant balance over *all* wavelengths rather than on a monochromatic basis. See Sparrow and Cess [4] for a discussion of this point.

Semi-gray analysis, a procedure which is simpler than band approximation but also less accurate, can also be used. This procedure uses the total emissivity ϵ_j of surface j which is determined by the temperature of the surface T_j, but also uses an absorptivity for this same surface α_j that is determined by the temperature of the source of the incident radiant energy. Hence, if more than one source is involved, a different α_j is used for each source. Suppose some of the incident radiant energy on surface j is coming from surface 3 at temperature T_3; then, the absorptivity $\alpha_{j(3)}$ of surface j for radiant energy emitted by surface 3 at temperature T_3 is determined by setting $\alpha_{j(3)}$ equal to the emissivity of surface j at temperature T_3, so that $\alpha_{j(3)} = \epsilon_j (T_3)$. This is only an approximation, yet it gives far better results with nongray surfaces than an analysis which assumes the surfaces are all gray. It often gives results very close to those of a more precise analysis using many bands in the band approximation model. The reason for the relative success of the semi-gray analysis seems to lie in the fact that the monochromatic hemispheric absorptivity α_λ is a much stronger function of the wavelength of the incident radiation than of the surface temperature itself [4], [6], [7], [8], [15].

4.5 GAS RADIATION

We have assumed, according to the restrictions of Sec. 4.3b, that gases between the surfaces of interest were completely transparent to thermal radiation. That is, the gases were assumed neither to emit, absorb, nor reflect radiant energy. Many common gases, such as dry air, O_2, N_2, and H_2 are almost completely transparent except at very high temperatures. However, other common gases, such as water vapor, CO_2, SO_2, and hydrocarbons emit and absorb appreciable amounts of radiant energy even at moderate temperature levels. But these gases, unlike solids which emit and absorb radiant energy over the entire spectrum of wavelengths, emit and absorb radiant energy selectively in relatively narrow wavelength bands, as is shown qualitatively in Fig. 4.55. The shaded portions are the wavelength bands in which the gas emits and absorbs. This is plotted at a fixed temperature, and, for comparison purposes, the emissive power of a black surface at the same temperature is shown as the solid line in Fig. 4.55. Hence, the emission from a gas certainly does not even approach the behavior of a gray surface (i.e., ϵ_λ independent of λ) and, in fact, is highly nongray. Furthermore, the emissive power for the gas shown in Fig. 4.55 depends upon both the size and the shape of the gas body, since emission and absorption of radiant energy are not primarily surface phenomena, as was the case for the opaque solids previously discussed in this chapter. Hence, the apparent emissivity and absorptivity of a gas body, even if it is at constant temperature throughout, depends upon the size and the shape of the gas body. In some cases, the temperature variation within the gas volume is significant enough to complicate matters even further.

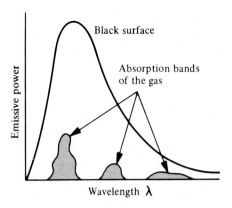

Figure 4.55 The monochromatic emissive power of gas, and, for comparison, of a black surface, versus wavelength at a single temperature.

Because of the complexities of gas radiation, as well as some difficulties not discussed concerning the thermal radiation properties of a gas, meaningful calculations of net radiant losses with participating gases are extremely involved and will not be presented here (see [4], [6], [7], [8], [9], and [11]). Instead, a brief description of the approximate *mean beam length method* will be given to allow an estimate to be made of gas radiation in some simple situations.

4.5a Mean Beam Length Method

This method was introduced by Hottel and is described in detail in the chapter he wrote in McAdams [19]. Basically, experimental measurements were made of the emissivity ϵ_g of a radiating gas at temperature T_g when the gas had the shape of a hemisphere of radius L. These measurements were made at the center of the base of the hemisphere. It was found that the gas emissivity depends upon the type of radiating gas, on the gas temperature T_g, on the total pressure P_T for gas mixtures, and on the product of length L and the partial pressure P_p of the emitting gas in the mixture; that is, $\epsilon_g = \epsilon_g(T_g, P_T, P_pL)$. These emissivity functions from [19] are presented here as Figs. 4.56 through 4.59 for water vapor and carbon dioxide, the most important emitting gases in industrial situations since they are present in products of combustion and in air. Thus, to determine the emissivity ϵ_{H_2O} for water vapor, for example, one would first find the value of $\epsilon^1_{H_2O}$ from Fig. 4.56 and then use Fig. 4.57 to find the factor C_{H_2O}. This, when multiplied by $\epsilon^1_{H_2O}$, corrects for total pressures other than one atmosphere and for a partial pressure other than very low ones. Hence, one has

$$\epsilon_{H_2O} = \epsilon^1_{H_2O} \, C_{H_2O}. \tag{4.125}$$

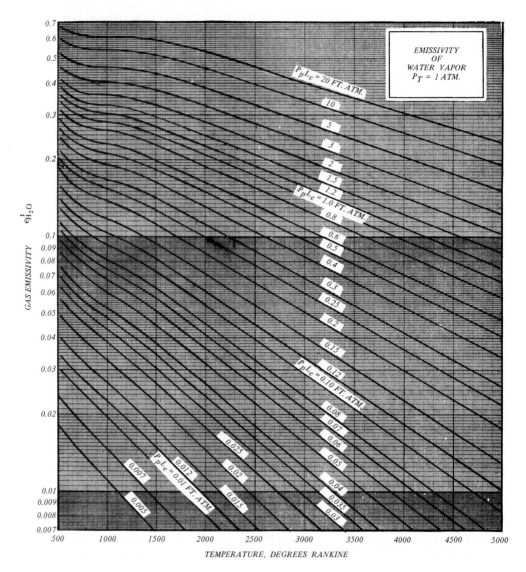

Figure 4.56 Emissivity of water vapor. (From *Heat Transmission,* 3d ed. by W. H. McAdams. Copyright © 1954 by William H. McAdams. Used with permission of the McGraw-Hill Book Company.)

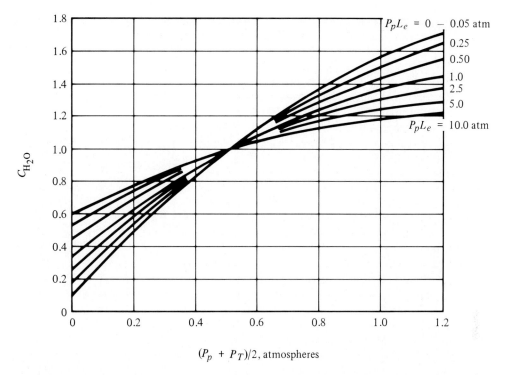

Figure 4.57 Correction factor C_{H_2O} for converting emissivity of H_2O to values of P_p and P_T other than 0 and 1 atm, respectively. (From McAdams, 1954. Used with permission.)

In a similar fashion, Figs. 4.58 and 4.59 give the emissivity of carbon dioxide as

$$\epsilon_c = \epsilon_c^1 C_c. \tag{4.126}$$

At this point, the gas emissivities strictly apply only to the situation in which a hemispherical gas body is radiating to the center of its base. However, Hottel [19] outlines the development of an equivalent mean beam length, L_e. For any gas shape, this is the radius that an equivalent hemisphere must have so that the gas radiates to the center of the hemisphere base, per unit area, at the same rate as the gas in the arbitrary shape radiates, per unit area, over some area A of interest. Hottel finds that at very small products $P_p L_e$, the L_e for radiation to the entire surface enclosing the gas is $4 V_0/A$, where V_0 is the volume of the enclosed gas and A is the total surface area of the container. However, for the average values of the product $P_p L_e$, suggestions in [9], [19], and [20], point toward a value of L_e which is about 90% of the value at small $P_p L_e$. We will use the following value for the mean beam length L_e for a gas radiating to the entire enclosure surface area A:

$$L_e = 3.6 \, V_0/A. \tag{4.127}$$

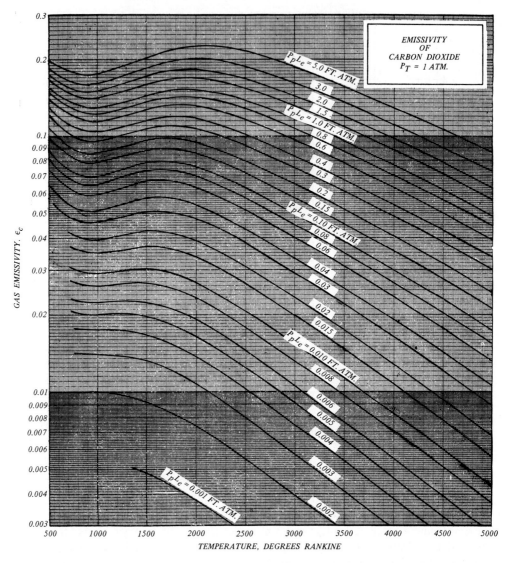

Figure 4.58 Emissivity of carbon dioxide. (From McAdams, 1954. Used with permission.)

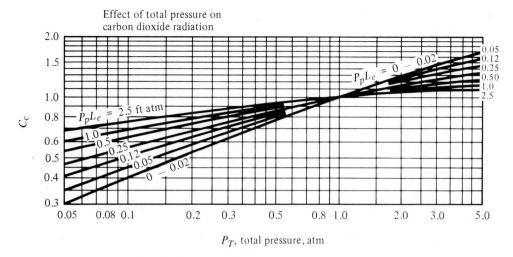

Effect of total pressure on
carbon dioxide radiation

Figure 4.59 Correction factor C_c for converting emissivity of CO_2 at 1 atm total pressure to emissivity at P_T atm. (From McAdams, 1954. Used with permission.)

Table 4.2
Mean Beam Lengths for Gas Radiation to the Entire Surface of Various Configurations

Configuration	Characterizing Dimension	Mean Beam Length L_e
Arbitrary shape	V_0/A	$3.6\,V_0/A$
Sphere	Diameter D	$0.6D$
Infinite cylinder	Diameter D	$0.9D$
Circular cylinder with height = diameter	Diameter D	$0.6D$
Infinite parallel planes	Plane spacing L	$1.8L$
Cube	Side length L	$0.6L$

Equation (4.127) was used to calculate the mean beam length for some configurations of interest, and the results are shown in Table 4.2.

Considered next will be the simplest case of radiation exchange between an emitting and absorbing gas and a solid surface. This is the radiant interaction of a black surface of area A and temperature T_w with an essentially isothermal gas body at temperature T_g which the black surface completely encloses.

The net radiant exchange between the two is the rate at which the radiant energy emitted by the gas is absorbed by the surface minus the rate at which the gas absorbs radiant energy emitted by the surface. Hence, we have that,

$$q_{wg} = \sigma A \left[\epsilon_g T_g^4 - \alpha_g T_w^4 \right]$$

where α_g is the absorptivity of the gas body at temperature T_g for radiation from the black surface at temperature T_w. It depends not only on the quantities that determine ϵ_g, but also on the black surface's temperature T_w. Hottel [19] reports that α_g can be found from the gas emissivity charts, Figs. 4.56–59, after making certain changes and using an additional correction factor. First, one evaluates the emissivity from Fig. 4.56 or 4.58 at the surface temperature of the solid, T_w, and uses $P_p L_e \left(\dfrac{T_w}{T_g} \right)$ in place of the parameter $P_p L_e$ of those figures. The correction factors C_{H_2O} and C_c are evaluated as was done for the gas emissivity, namely at P_T and $P_p L_e$. Finally, after incorporating a temperature ratio correction factor, the expressions for the absorptivities of water vapor and carbon dioxide become the following:

$$\alpha_{H_2O} = \epsilon^1_{H_2O} \left(T_w, P_p L_e \frac{T_w}{T_g} \right) C_{H_2O} \left[\frac{T_g}{T_w} \right]^{0.45} \tag{4.128}$$

$$\alpha_c = \epsilon^1_c \left(T_w, P_p L_e \frac{T_w}{T_g} \right) C_c \left[\frac{T_g}{T_w} \right]^{0.65} \tag{4.129}$$

In the event that one is dealing with a mixture of carbon dioxide and water vapor, Hottel provides additional charts in [19] to account for the mutual absorption which takes place in the mixture. However, Kreith [21] indicates that usually this correction is not appreciable and so, in mixtures of CO_2 and H_2O, the gas mixture emissivity and absorptivity will be taken to be the sum of the separate emissivities and absorptivities for the H_2O and CO_2 in the mixture.

For the more complex gas radiation problems involving gray surfaces, significant temperature variation within the gas body, and solid enclosing surfaces at different temperatures, one can consult Hottel [19], as well as [4], [6], and [9].

4.6 THE RADIATION SURFACE COEFFICIENT OF HEAT TRANSFER

In general, a solid surface exchanges some net amount of radiant energy with other solid surfaces (the net radiant loss) and also exchanges energy by free or forced convection with an adjacent, nonparticipating (transparent) gas. Hence, the total net rate at which the solid surface—call it surface 1—loses energy is the sum of the net radiant loss q_1 and the convection heat transfer rate to the adjacent gas at temperature T_f, called q_c. Hence, the total heat transfer rate q from the surface is

$$q = q_1 + q_c. \tag{4.130}$$

Using Newton's law of cooling for the convection heat transfer rate, Eq. (4.130) becomes

$$q = q_1 + hA_1(T_1 - T_f), \tag{4.131}$$

where h is the ordinary surface coefficient of heat transfer associated with the free or forced convection. The net radiant loss q_1 is, as has been seen in this chapter, a function of the radiation properties, areas, angle factors, and temperatures of all the surfaces involved. In some situations, it is convenient to have both terms on the right of Eq. (4.131) look similar in form. For this purpose, the radiation surface coefficient of heat transfer is defined as

$$q_1 = h_r A_1 (T_1 - T_f).$$ (4.132)

Note that the correct net radiant loss q_1 must still be found in every case by one of the methods presented in this chapter, and once it is found, then, and only then, can h_r be determined. For example, suppose surface 1 is gray and cannot see itself, but can see one other gray surface 2, which with surface 1 forms an enclosure. In this type of situation, an expression for the net radiant loss has previously been found as Christiansen's equation:

$$q_1 = \frac{A_1 \sigma (T_1^4 - T_2^4)}{1/\epsilon_1 + (A_1/A_2)(1/\epsilon_2 - 1)}.$$ [4.87]

Now, equating Eqs. (4.87) and (4.132) and solving for the radiation surface coefficient of heat transfer for this particular situation,

$$h_r = \frac{\sigma (T_1^4 - T_2^4)}{(T_1 - T_f)[1/\epsilon_1 + (A_1/A_2)(1/\epsilon_2 - 1)]}.$$

When Eqs. (4.131–32) are combined and simplified, the result is

$$q = (h_r + h) A_1 (T_1 - T_f).$$ (4.133)

The quantity $h_r + h$, the sum of the radiation surface coefficient of heat transfer and the ordinary convection surface coefficient of heat transfer, is often termed the *"total" surface coefficient of heat transfer* or the *total unit conductance*. Equations (4.132–33) have their greatest utility and value in situations where (1) the radiation loss and convection loss are of the same order of magnitude (as is often true for objects free convecting in air), and (2) the expression for the net radiant loss q_1 of the object is a relatively simple one (such as when q_1 is given by either Eq. [4.87] or [4.88]) which in turn gives a relatively simple expression for the radiation surface coefficient h_r by the use of Eq. (4.132).

4.7 PROBLEMS

4.1   A 100 W bulb is designed using a tungsten filament operating at 2750°K. At this temperature the total hemispheric emissivity of the tungsten is approximately $\epsilon = 0.30$. If the bulb is evacuated, calculate the minimum surface area of tungsten filament needed.

4.2   The total hemispherical emissivity of a material is to be measured by placing a sphere of the material, in which an electric resistance heater is buried, into an evacuated enclosure that has walls held at a very low temperature. The sphere has a surface area of 0.023 m² and a steady-state surface temperature 555°K which is maintained by dissipating electrical energy at the rate of 73.5 W. From this data, estimate the total hemispherical emissivity of the material.

4.3 Electrical heating panels in a space station are to be kept at a maximum temperature of 60°C to prevent serious burns. If the surface emissivity of the panels is 0.93, calculate the required heater size, in watts, for panels of area 0.2 m².

4.4 To protect workers from inhalation of the toxic vapor of a metal, a surface grinding operation on a thin sheet of this highly conductive metal is performed in an evacuated enclosure whose wall temperature is relatively low. The exposed surface area of the metal part is 1.4 ft² and the emissivity is 0.80. The grinder is dissipating mechanical energy at the rate of 1100 Btu/hr and, since the grinder is made of material with a much lower thermal conductivity than the metal, all of this energy goes into the metal piece. Find the steady-state surface temperature of the metal.

4.5 An oxidized nickel plate has electrical energy dissipated within it at the rate of 2000 Btu/hr while the plate is in a vacuum and can see only surfaces at very low temperature. If the surface area of the plate is 1 ft² and its emissivity varies linearly from 0.32 at 0°F to 0.66 at 1000°F, calculate the steady-state plate surface temperature.

4.6 If a black body is at 1110°K, determine the fraction of its emitted radiant energy (a) beyond a wavelength of 10 μm and (b) at wavelengths less than 10 μm.

4.7 Repeat Prob. 4.6 for a gray surface with emissivities of (a) 0.85 and (b) 0.25.

4.8 In Ex. 4.1, it was shown that a tungsten filament at 2778°K emits about 5.86% of its radiant energy in the visible range from 0.3 to 0.7 μm. If the filament temperature is raised to 3600°K, find the percentage by which the radiant energy emitted in the visible range increases.

4.9 Show that at relatively large values of λT, Planck's distribution law degenerates to

$$W_b = (C_1/\lambda^5)(\lambda T/C_2).$$

(This is called *Rayleigh's distribution law.*)

4.10 Find the wavelength at which the maximum emissive power per unit wavelength occurs for either black or gray surfaces at (a) 3000°R and (b) 1800°R.

4.11 Calculate the percentage of the total emission from a black or gray surface that lies in a wavelength band ± 1 μm about the value λ_m, at which the monochromatic emissive power is a maximum, when the surface is at (a) 3000°R and (b) 1800°R. (c) What do you conclude from the comparison of the answers in (a) and (b)?

4.12 A nongray material's monochromatic hemispheric emissivity at 555°K is 0.25 between 1 and 3 μm, 0.60 between 10 and 17 μm, and 0.15 between 90 and 100 μm, and is 0 for all other wavelengths. Calculate its total hemispheric emissivity at this temperature.

4.13 Find the total absorptivity and reflectivity for the material in Prob. 4.12 when it is at 555°K, for black body radiation from a 555°K source.

4.14 A long 5-cm-diameter cylinder of glass is used in the wall of a jet engine test cell to view the area near the nozzles in the combustor. This is equivalent to viewing a black body source at 2200°K. The transmissivity of the glass is zero except between the wavelengths 0.3 μm and 0.8μm, where its value is 0.35. Calculate the fraction of the incident radiant energy from the combustion which is transmitted through the glass.

4.15 The oven of an electric stove is operating at 200°C and its inside surfaces have an emissivity of 0.80. A defective oven door allows an opening of total area 0.002 m² along the perimeter of the door. Calculate the rate at which radiant emission from the oven crosses this opening into the room.

4.16 A thin-walled door of area 1 ft² is accidentally left open in a large heat-treating oven operating at 2000°F. The room into which the radiation escapes is 10 ft by 30 ft by 50 ft, and the air in the room is initially at 80°F and is at atmospheric pressure. If the door is left open for 15 min, find the maximum possible temperature change of the room air.

4.17 A 2.5 cm square hole is opened in the thin-walled door of a furnace whose inside walls are at 650°C. The total hemispheric emissivity of the walls is 0.73. Find the total rate at which radiant energy escapes from the furnace through the hole to the room.

4.18 A spherical satellite, whose outside surface is made of a very good conductor whose emissivity is ϵ, is in orbit and can receive radiant energy from the sun. At this point in orbit, a plane surface normal to the sun's rays receives radiant energy from the sun at the known rate H_s per unit area of plane surface. If the absorptivity of the satellite material for solar radiation is α_s and none of the satellite's emitted radiant energy ever returns, derive the following expression for the surface temperature of the satellite when the internal power supply is off and steady-state conditions prevail:

$$T = \left(\frac{\alpha_s}{4\epsilon} \frac{H_s}{\sigma} \right)^{1/4}.$$

4.19 Derive an expression for the net radiant loss of a small object, which is shaped such that its own radiant emission does not strike it directly, when the object is at surface temperature T and is in a large enclosure whose surfaces are at the temperature T_0. The emissivity of the object is ϵ and its absorptivity for black body radiation from a source at temperature T_0 is α_0.

4.20 A sphere of surface area 0.1 ft² and temperature 140°F is situated in a large evacuated enclosure whose wall temperature is 40°F. The total hemispheric emissivity ϵ of the sphere surface at 140°F is 0.85, while the absorptivity of the sphere at this temperature for black body radiation coming from a source at 40°F is $\alpha = 0.79$. If the sphere is in the steady state, compute the rate at which electrical energy is being dissipated by the resistor inside the sphere.

4.21 By following the procedure used in Sec. 4.2e, derive an expression relating the total emissive power W to the intensity I, when the emitter is nondiffuse and emits such that the local intensity is $I = I_0 \cos \phi$, where I_0 is the normal intensity at $\phi = 0$ and the angle ϕ is shown in Fig. 4.12. Show that the result is

$$I = \frac{3}{2} (W/\pi) \cos \phi,$$

where W is the total emissive power per unit area of surface.

4.22 Figure 4.60 shows two long circular cylinders, with the inner one of diameter D_1 and the outer one of diameter D_2. If the long axis of each cylinder is parallel to the other, find all the angle factors involved.

Figure 4.60

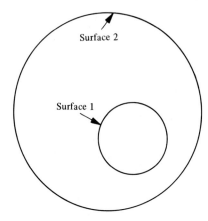

4.23 Explain why the surfaces for Eq. (4.28), even though relatively small and far apart, must be plane to give the angle factor directly.

4.24 Figure 4.61 shows a small plane area dA_1 which interacts with a sphere, surface 2. The normal to dA_1 passes through the center of the sphere. Find the angle factor from dA_1 to the sphere in terms of known quantities.

Figure 4.61

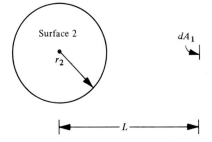

4.25 Find the value of all the angle factors involved in the enclosure shown in Fig. 4.62.

Figure 4.62

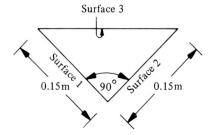

Chapter Four

4.26 Using exact analytical means, show that the angle factor between surface dA_1 and plane 2, which is parallel to it, as shown in Fig. 4.63, is given by

$$F_{dA_1 \rightarrow 2} = \frac{1}{2\pi} \left[\frac{X}{\sqrt{1 + X^2}} \tan^{-1} \left(\frac{Y}{\sqrt{1 + X^2}} \right) + \frac{Y}{\sqrt{1 + Y^2}} \right.$$
$$\left. \tan^{-1} \left(\frac{X}{\sqrt{1 + Y^2}} \right) \right],$$

where $X = a/L$ and $Y = b/L$.

Figure 4.63

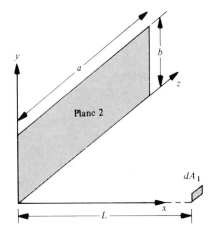

4.27 Figure 4.64 shows a very small area dA_1 interacting with the rectangular surface 2. Noting that dA_1 lies in the x, z plane and surface 2 lies in the x, y plane, derive an expression for the angle factor from dA_1 to 2 in terms of known quantities, such as a, h, and w.

Figure 4.64

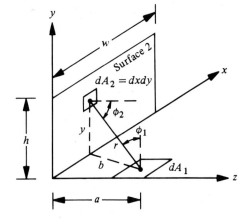

Radiation Heat Transfer

4.28 Find both angle factors F_{12} and F_{21} for the situation shown in Fig. 4.65; that is, two rectangles meeting at right angles and sharing a common edge.

Figure 4.65

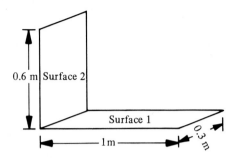

0.6 m | Surface 2

Surface 1

1 m

0.3 m

4.29 Figure 4.66 shows a hemisphere capping a circular disk of radius R where one-quarter of the disk has been cut away and is actually an opening which is called "surface" 3. Find the angle factor F_{23}.

Figure 4.66

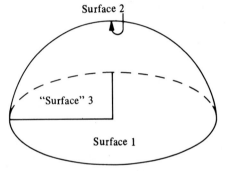

Surface 2

"Surface" 3

Surface 1

4.30 Find the angle factor between two disks, 1 and 2, when the normal to each of their centers coincides, if disks 1 and 2 have radii of 0.3 m and 0.6 m, respectively, when the disks are separated by a distance of 0.3 m.

4.31 Find the angle factors F_{12}, F_{13}, and F_{14} of Fig. 4.67.

Figure 4.67

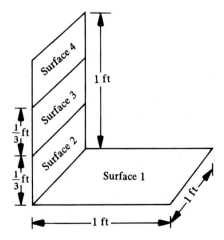

Surface 4

Surface 3

Surface 2

Surface 1

1 ft

$\frac{1}{3}$ ft

$\frac{1}{3}$ ft

1 ft

1 ft

400

4.32 Figure 4.68 shows a long cavity open at the top. Find the angle factor between the cavity and the opening.

Figure 4.68

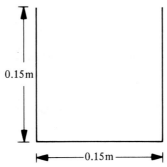

0.15 m

0.15 m

4.33 An annular area of outer radius 1 ft and inner radius 3/4 ft is capped by a 1-ft radius hemispherical bowl. Find the angle factor between (a) the bowl and the annular area and (b) the bowl and the circular disk of radius 3/4 ft.

4.34 Find the value of the angle factor between two directly opposed plane parallel plates, 0.3 m apart, if their shape is rectangular with dimensions 0.15 m by 0.3 m.

4.35 Figure 4.69 shows surface 1, a circular disk of diameter d, and a pipe of diameter d and length L, open at both ends. The interior surface of this pipe is called surface 2. Explain how to obtain F_{12}, the angle factor from the disk to the interior of the pipe, using equations already derived.

Figure 4.69

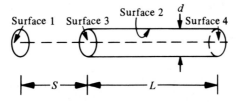

Surface 1   Surface 3   Surface 2 d Surface 4

S L

4.36 Find the angle factor F_{12} for the configuration shown in Fig. 4.70. Surfaces 1 and 2 are plane and parallel and corresponding edges are also parallel.

Figure 4.70

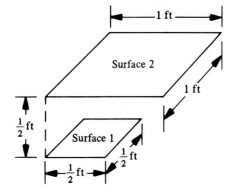

1 ft

Surface 2

1 ft

$\frac{1}{2}$ ft

Surface 1

$\frac{1}{2}$ ft

$\frac{1}{2}$ ft

4.37 Find the angle factor between the small disk called dA_1 in Fig. 4.71 and the ring-shaped disk 2 of inner radius r_1 and outer radius r_2 in terms of known quantities.

Figure 4.71

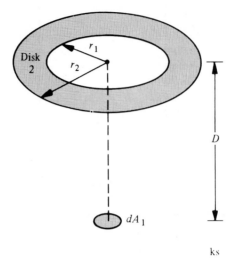

ks

4.38 Consider the room with the dimensions shown in Fig. 4.72. The floor and ceiling are surfaces 1 and 2, respectively, while the combination of the front and left-side wall is called surface 3 and the combination of the back and right-side wall is called surface 4. Find the angle factors F_{12}, F_{13}, and F_{14}.

Figure 4.72

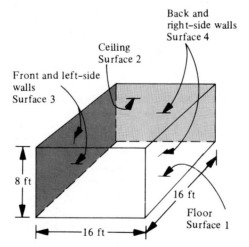

4.39 Does a black surface, for radiation calculation purposes, have to satisfy the requirement of being uniformly irradiated? Explain your answer.

4.40 Consider the geometrical surface 1 made of a single material at a constant temperature as shown in Fig. 4.73. All three surfaces emit diffusely. Determine if this is a proper surface for radiation calculations and explain.

Figure 4.73

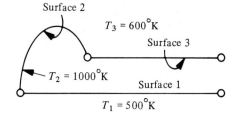

Surface 2

$T_3 = 600°K$

Surface 3

$T_2 = 1000°K$

Surface 1

$T_1 = 500°K$

4.41 Refer to the two cylinders shown in Fig. 4.60. Determine if the net radiant loss from the inner cylinder, surface 1, depends on its position within the outer cylinder if both have constant, but different, values of the temperature. Explain your answer.

4.42 Figure 4.74 shows an end view of a very long duct with all surfaces black and with known temperatures $T_1 = 1000°R$, $T_2 = 800°R$, and $T_3 = 700°R$. Calculate the net radiant loss from surface 1 per foot of duct.

Figure 4.74

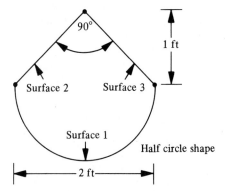

90°

1 ft

Surface 2   Surface 3

Surface 1   Half circle shape

2 ft

4.43 The inside surface of a hemispherical bowl of radius 0.15 m serves as the surface of an electric heater which dissipates electrical energy at the rate of 1500 W. The circular opening of the hemisphere into a room behaves like a black surface at 21°C. If the inside of the hemisphere is also black, calculate its steady-state surface temperature.

4.44 Consider two general black surfaces 1 and 2 that can see each other as well as other black surfaces. Set up an expression for the direct radiant exchange between these two surfaces, and then by noting that this expression must hold even when $T_1 = T_2$, conclude that $A_1F_{12} = A_2F_{21}$. (This is an alternate proof of the reciprocity theorem.) Discuss why the result of this problem, the reciprocity theorem, holds true even if the surfaces are not black.

4.45 An enclosure is formed by n black surfaces. Surface 1 is the surface of an electric resistance heater and its net radiant loss is known to be q_1. If the remaining surfaces all have known specified temperatures, derive an expression for the temperature of surface 1 in terms of known quantities. All angle factors and areas are considered known.

4.46 Two spacecraft have 10-ft by 10-ft black panels for emergency exchange of energy from one ship to another. One panel is to be at 2000°R while the cooler panel is to be at 500°R. All emitted energy hitting neither panel is assumed lost to space behaving as a black body at 0°R. If the net radiant loss of the cooler panel must be at least −1,360,000 Btu/hr, find the greatest possible spacing between the panels during energy exchange.

4.47 A hole is drilled into a plate with a 2.5-cm-diameter drill so that it is 2.5 cm deep. If all surfaces of this drilled cavity are at the same temperature and have an emissivity of 0.50, calculate (a) the angle factor between the cavity and the opening, (b) the apparent emissivity of the opening, and (c) the depth the hole would have to be drilled if the desired apparent emissivity is to be 0.98.

4.48 Referring to Eq. (4.47) and Fig. 4.26, disk 1 is kept in the steady state by a cooling liquid flowing over the underside of the disk, giving a surface coefficient of heat transfer of 56 W/m² °K between the disk and the fluid. Find the temperature of the cooling fluid.

4.49 In Fig. 4.75, all surfaces are black with surface 1 at 1500°R, surface 3 at 1000°R, and surface 2, the inside of the hemisphere, in radiant balance. If the radius of the hemisphere is 1 ft, find the temperature of the surface in radiant balance. Surface 3 is 1/4 of the circular disk.

Figure 4.75

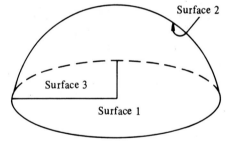

Surface 2

Surface 3

Surface 1

4.50 The room pictured in Fig. 4.72 has a radiant heating system installed in the ceiling, surface 2, causing the ceiling temperature to be 85°F. The floor, surface 1, is at 70°F. Three of the four walls are inside walls and behave as a single surface, called surface 3, which is in radiant balance. The remaining wall, surface 4, is an outside wall and is at 60°F. If all surfaces are black, find the net radiant loss from the ceiling.

4.51 Consider an enclosure of n black surfaces with one surface a in radiant balance. All angle factors and areas are known. Find an expression for the net radiant loss from surface i, when i is *not* surface a.

4.52 Figure 4.76 shows four surfaces forming a very long duct. All areas and angle factors are known. Surface 1 is black and has a temperature T_1, while surface 2 is gray with emissivity ϵ_2 and known temperature T_2. Surface 3 is a gray surface in radiant balance with an emissivity ϵ_3 and surface 4 is black at known temperature T_4. (a) By use of the reflection method, set up an expression in terms of known quantities for the temperature of surface 3. (b) Derive a relation for the net radiant loss from surface 1.

Figure 4.76

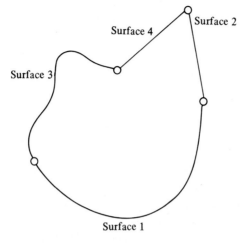

4.53 Figure 4.77 shows an enclosure consisting of an opening e through which no radiation enters, only leaves, a re-radiating surface a, a black surface of known temperature, surface 3, and two gray surfaces, 1 and 2, for which ϵ_1, T_1, ϵ_2, and T_2 are known. Assume all angle factors have been found. Compute the net radiant loss through the opening in terms of known quantities by use of the reflection method of analysis.

Figure 4.77

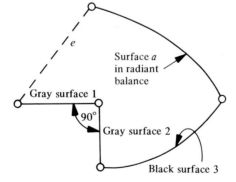

4.54 Figure 4.78 shows a 0.3-m-radius sphere whose inside surfaces are gray with emissivity 0.71 and are at constant temperature T_1. A small portion of the sphere is flattened, and it is proposed that a hole through this plane surface be drilled which will give an apparent emissivity of the hole of $\epsilon_0 = 0.995$. Find the required hole diameter for this proposed laboratory black body.

Figure 4.78

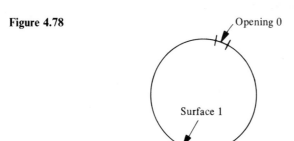

Opening 0

Surface 1

4.55 Surfaces 1 and 2 form an enclosure in which surface 1 is gray while surface 2 is black and cannot see itself. Find an expression for the net radiant loss from surface 1 using the reflection method of analysis.

4.56 Part of the surface of an otherwise flat plate is covered by roughly hemispherical cavities of average radius of 1/8 in. There are nine of these cavities to the square inch. If the actual emissivity of the surface material is 0.6, compute the apparent emissivity which should be used to treat the surface as being perfectly flat.

4.57 Consider the surfaces shown in Fig. 4.79. All radiation not directly striking any surface is assumed lost to space, which acts as a perfect absorber. Surface 1 is at 800°R, surface 2 is at 600°R, and surface 3 is in radiant balance. The angle factors and emissivities are $F_{13} = F_{23} = 0.60$, $F_{33} = 0.50$, $A_1 = A_2 = 1$ ft², $F_{32} = F_{31} = 0.10$, $\epsilon_1 = 0.7$, $A_3 = 6$ ft², $\epsilon_2 = 0.5$, and $\epsilon_3 = 0.22$. Compute the temperature of the surface in radiant balance.

Figure 4.79

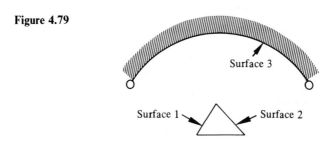

Surface 3

Surface 1

Surface 2

4.58 Figure 4.80 shows three parallel infinite planes. Plane 1 is black and at known temperature T_1 and plane 3 is also black at known temperature T_3. The plane separating them, plane 2, is not opaque and has known values of emissivity ϵ_2, transmissivity τ_2, and reflectivity ρ_2. Plane 2 is held at known temperature T_2. Using the reflection method of analysis, find the expression for the net radiant loss per unit area from plane 1 in terms of known quantities. All of material 2 is at known temperature T_2.

Figure 4.80

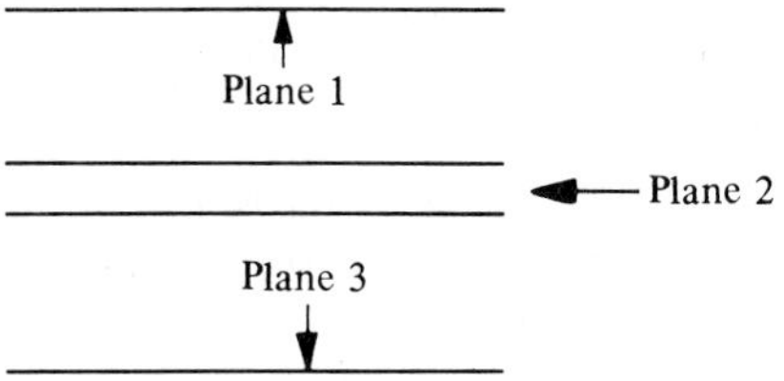

Plane 1

Plane 2

Plane 3

4.59 Consider now the simplest possible case, as far as reflections are concerned, of two gray surfaces interacting with one another. Neither of the surfaces can see themselves—that is, $F_{11} = F_{22} = 0$—and any radiant energy coming off either surface which does not strike the other surface is lost to the surroundings. Also, the surroundings contribute no radiant energy to either surface, because the effective temperature of the surroundings is much lower than either T_1 or T_2. Use the reflection method of analysis to derive an expression for the net radiant loss from surface 1.

4.60 Consider two gray circular disks, surfaces 1 and 2. Surface 1 has a radius of 1 m, $\epsilon_1 = 0.8$, and $T_1 = 830°$K. Surface 2 has a radius of 1 m, $\epsilon_2 = 0.90$, and $T_2 = 555°$K. These disks are interacting radiatively in outer space with the angle factor $F_{12} = 0.85$. Compute the net radiant loss from disk 1.

4.61 A circular hole 6 in. in diameter is in the side of a 3-in.-thick furnace wall as shown in Fig. 4.81. The interior sides of the furnace are at 3000°R while the curved surface of the hole in the furnace wall is in radiant balance and is gray with $\epsilon = 0.51$. The circular opening at the outside of the wall is surrounded by room surfaces at so low a temperature relative to 3000°R that the surface behaves essentially as if it were black at 0°R. Calculate the net radiant loss from surface 1.

Figure 4.81

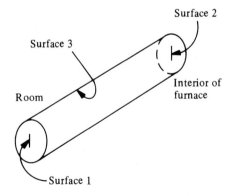

Surface 2

Surface 3

Interior of
furnace

Room

Surface 1

4.62 An electric heater whose surface temperature is not to exceed 280°C is used in a spacecraft compartment which has been evacuated and whose walls are at $-7°C$. The surface area of the heater is 0.1 m², its emissivity is 0.92, and the heater surface cannot see itself. Find the maximum rate at which electrical energy should be dissipated within this heater.

4.63 A thermos bottle is constructed with two concentric circular cylindrical walls with the space between them evacuated. The diameter of the inner wall is 4 in. and of the outer wall 4.5 in. Both walls have an emissivity of 0.10. If the inside wall is at 160°F because hot soup is in the bottle and the outside wall is essentially at the 70°F room temperature, calculate the energy transfer rate from the bottle if it is 12 in. high.

4.64 The thermos bottle of Prob. 4.63 initially contained hot soup at 160°F. Assuming that this soup has the properties of water, estimate the shortest possible time in which the soup will cool to 110°F.

4.65 Rework Prob. 4.50, calculating the net radiant loss from the ceiling of the room, surface 2, using the network method. (Note that the physical situation is shown in Fig. 4.72. All surfaces are black, with the ceiling, surface 2, at 85°F, the floor, surface 1, at 70°F, three inside walls, surface 3, in radiant balance, and the remaining wall, surface 4, at 60°F. From Prob. 4.50, $A_1 = A_2 = 256$ ft², $A_3 = 384$ ft², $A_4 = 128$ ft², $F_{12} = F_{21} = 0.42$, $F_{24} = 0.145$, $F_{23} = 0.435$, $F_{13} = 0.435$, and $F_{43} = 0.42$.)

4.66 Two gray circular disks are interacting radiatively in outer space. Disk 1 has a 3 ft radius with $\epsilon_1 = 0.80$ and $T_1 = 1500°R$, while disk 2 also has a 3 ft radius with $\epsilon_2 = 0.90$ and $T_2 = 1000°R$. The radiant energy not striking either disk is assumed to be lost to space acting as a black body at 0°R. Find the net radiant loss from disk 1 by the network method. $F_{12} = F_{21} = 0.85$.

4.67 An enclosure consists of four gray surfaces. Surfaces 1 and 2 have temperatures T_1 and T_2, and emissivities ϵ_1 and ϵ_2, respectively. Surfaces 3 and 4 have emissivities ϵ_3 and ϵ_4, respectively, and are in radiant balance. Assuming that all angle factors and areas are known, sketch the appropriate electric network.

4.68 Rework Prob. 4.55 using the network method of analysis.

4.69 Rework Prob. 4.52 using the network method of analysis.

4.70 Rework Prob. 4.53 using the network method.

4.71 If the panels referred to in Prob. 4.46 are both gray with emissivity $\epsilon = 0.70$, while all other conditions are the same as described in Prob. 4.46, find the greatest possible spacing between the panels during energy exchange.

4.72 Rework Prob. 4.61 using the network method.

4.73 The sphere shown in Fig. 4.82 is constructed of two hemispheres which are perfectly insulated along their contact area. The upper hemisphere, surface 1, has a temperature of 450°K and an emissivity of 0.80, while the lower hemisphere, surface 2, has a temperature of 330°K and an emissivity of 0.60. If the sphere radius is 0.3 m, calculate the net radiant loss from surface 1 using the network method.

Figure 4.82

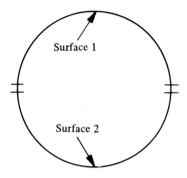

4.74 The inside of a hemispherical bowl, surface 1, is at 800°R with an emissivity of 0.60. The disk capping the bowl is divided into two surfaces, 2 and 3, of equal area split along a diameter. Surface 2 is held at 1000°R and has an emissivity of 0.80, while surface 3 is in radiant balance. If the radius of the bowl is 1 ft, calculate the net radiant loss from surface 2 using the network method.

4.75 A thermocouple is used to measure the temperature of dry air flowing in a large diameter pipe. The thermocouple bead has an emissivity of 0.80 while the emissivity of the pipe wall is 0.87. The surface coefficient of heat transfer between the surface of the bead and the gas has been estimated to be 10 Btu/hr ft² °F. If the pipe wall temperature is 500°F and the bead temperature is 1000°F, estimate the actual temperature of the dry air, neglecting conduction through the leads.

4.76 A household iron with a 50 W heating element has an exposed surface area of 0.046 m². The surrounding air is at 20°C, and the surface coefficient of heat transfer is approximately 8.5 W/m² °C. The emissivity of the iron's surface is 0.25. The walls, ceiling, and floor of the room in which the iron is located are all at 20°C. Estimate the average steady-state surface temperature of the iron.

4.77 Figure 4.83 shows two hemispherical caps of radius 0.3 m separated by a thin disk. The joint between the caps and the disk is made in such a way as to be essentially a perfect insulator. The spaces between the caps and disk have been evacuated. Cap 1 is black and is held at 500°K, while cap 2 is also black and is held at 320°K. The disk has an emissivity of 0.2 and is made of a good conducting material so that both the left and right faces of the disk have the same temperature. Compute the steady-state temperature of the disk between the two caps.

Figure 4.83

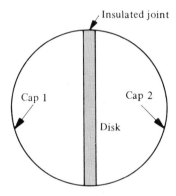

4.78 An electric transmission cable 1 in. in diameter and 30 ft long passes through a large work-room whose surfaces are at about 70°F. To prevent burns to the workmen if they accidentally brush against the cable, the outside surface temperature of the cable is limited to 130°F. If the emissivity of the outside surface of the cable is 0.80 and the convection coefficient of heat transfer between the cable surface and room air at 70°F is 1.2 Btu/hr ft² °F, find the maximum allowable current in the cable if the total resistance is 1 ohm. (Note that 1 watt = 3.413 Btu/hr, and the expression for the dissipation rate of electrical energy is I^2R.)

4.79 Frost forms on an apple when its temperature reaches 0°C. At night the apple is assumed to be completely surrounded by the night sky acting as a black body at −45°C. The surface coefficient of heat transfer between the surrounding air and the apple is 17 W/m² °C while the emissivity of the apple is approximately 0.93. Find the lowest temperature the air can reach during the night without frost formation on the apple. (Note that the apple is in steady state with no generation, and neglect energy conduction up the stem.)

4.80 A steam pipe at a temperature of 90°C with an emissivity of 0.80 passes through a small tank containing air. The inside tank surface is at 37°C and has an emissivity of 0.60. The area of the steam pipe is 6.5 m², while the tank area is 18.6 m². The surface coefficient of heat transfer between the steam pipe and the air is 11.3 W/m² °C, while the average surface coefficient between the tank surface and the air is 5.7 W/m² °C. Compute (a) the steady-state temperature of the air and (b) the total heat loss rate from the steam pipe.

4.81 A cylindrical shield of emissivity 0.40 is interposed between the thermocouple and duct wall of Prob. 4.75. After installation of the shield, the gas temperature is changed until the thermocouple bead again has a temperature of 1000°F. Calculate the true gas temperature in this case. The surface area of the thin shield is 10 times the surface area of the bead. All other conditions are the same as in Prob. 4.75.

4.82 Figure 4.84 shows a long solid cylinder, surface 1, at 800°R with a surface area of 1 ft². A resistor inside cylinder 1 dissipates electrical energy at the rate of 400 Btu/hr, and surface 1 has an emissivity of $\epsilon_1 = 0.5$. Surrounding surface 1 is surface 2, which has an area of 20 ft², an emissivity of 0.9, and a temperature of 500°R. A gas flows between the two surfaces with a surface coefficient of heat transfer of 3 Btu/hr ft² °F with surface 1. Determine (a) the gas temperature for steady-state conditions, and (b) the initial rate at which surface 1's temperature changes with time when the plug is pulled on the resistor heater if material 1 has a mass of 1 lbm and $c_p = 0.1$ Btu/lbm °R. (Assume that at any instant of time, all of material 1 is essentially at the same temperature.)

Figure 4.84

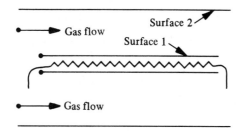

4.83 Rework Ex. 4.12 using the network method.

4.84 Using the network method, calculate the net radiant loss from a gray infinite plane, surface 1, which is at 280°C and has an emissivity of 0.6, which interacts with another gray infinite plane of emissivity 0.8 and temperature 60°C.

4.85 Two thin, infinite planes are spaced between the planes described in Prob. 4.84 to act as radiation shields and reduce the net radiant loss. If the emissivity of the shields is 0.30, calculate (a) the net radiant loss from surface 1 and (b) the temperature of the shield closest to surface 1.

4.86 Draw the electric network and determine the equations whose solution would yield the various radiosities for an enclosure consisting of five gray surfaces, all of which have their temperatures specified. All emissivities, areas, and angle factors are known.

4.87 Figure 4.85 shows a hemispheric gray surface, surface 3, of emissivity $\epsilon_3 = 0.70$ which is in radiant balance. The hemisphere is capped by a 0.3-m-radius circular disk divided into two surfaces by a diameter, as shown in Fig. 4.85. Surface 1 is gray with $T_1 = 550°$K and $\epsilon_1 = 0.35$, while surface 2 is black at 320°K. Using the absorption factor method, find the temperature of surface 3, the surface in radiant balance.

Figure 4.85

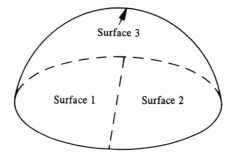

4.88   The physical situation of Prob. 4.87 must also operate in the steady state at another design point where $T_1 = 820°K$ and $T_2 = 550°K$. If the emissivities remain unchanged, calculate the temperature of surface 3, the surface in radiant balance.

4.89   The inside of a right circular cylinder is made of a gray material with $\epsilon = 0.50$ and is held at 1000°R both on the curved inside surface and the bottom circular disk, where this combined surface is called surface 1. The top of the cylinder is open to surroundings whose temperature is so low relative to 1000°R that the opening, called surface 2, behaves as if it were black at 0°R. The cylinder diameter and height are both 2 ft. Find the net radiant loss from the opening, surface 2, by the absorption factor method.

4.90   Calculate the net radiant loss from surface 1 in Prob. 4.87.

4.91   Using the absorption factor method, rework Prob. 4.90, treating surface 3 as a perfect reflector, and calculate the net radiant loss from surface 1.

4.92   Rework Prob. 4.84 using the absorption factor method.

4.93   Rework Prob. 4.57 using the absorption factor method.

4.94   Rework Prob. 4.49 using the absorption factor method.

4.95   Derive Eq. (4.64) using the absorption factor method.

4.96   Rework Prob. 4.60 using the absorption factor method.

4.97   Rework Prob. 4.61 using the absorption factor method.

4.98   Derive Christiansen's equation, Eq. (4.87), using the absorption factor method.

4.99   Rework Prob. 4.65 using the absorption factor method.

4.100 Consider the enclosure shown in Fig. 4.86, which consists of two disks, 1 and 2, and a cylindrical surface 3. Disk 1 has a temperature of 500°K and an emissivity of 0.80, while disk 2 has a temperature of 300°K and an emissivity of 0.40. The cylindrical surface joining the two disks is in radiant balance, and the angle factor F_{31} is 0.308. Using the absorption factor method of making radiant exchange calculations, compute the net radiant loss from surface 1.

Figure 4.86

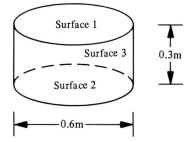

4.101 Show directly from Eq. (4.111) that the absorption factors become the angle factors when all surfaces of an enclosure are black.

REFERENCES

1. Jakob, M. *Heat Transfer.* Vol. 1. New York: Wiley, 1949.
2. Dunkle, R. V. "Thermal Radiation Tables and Applications." *Transactions of the A.S.M.E.,* Vol. 76 (1954), pp. 549–52.
3. Schornhorst, J. R., and R. Viskanta. "An Experimental Examination of the Validity of the Commonly Used Methods of Radiation Heat Transfer Analysis." *Journal of Heat Transfer,* Vol. 90, No. 4 (1968), pp. 429–36.
4. Sparrow, E. M., and R. D. Cess. *Radiation Heat Transfer.* Washington, D.C.: Hemisphere, 1978.
5. Planck, M. *The Theory of Heat Radiation.* New York: Dover Publications, 1959.
6. Siegel, R., and J. R. Howell. *Thermal Radiation Heat Transfer.* 2d ed. Washington, D.C.: Hemisphere, 1981.
7. Love, T. J. *Radiative Heat Transfer.* Columbus, Ohio: Charles E. Merrill, 1968.
8. Wiebelt, J. A. *Engineering Radiation Heat Transfer.* New York: Holt, Rinehart, and Winston, 1966.
9. Hottel, H. C., and A. F. Sarofim. *Radiative Transfer.* New York: McGraw-Hill, 1967.
10. Kreith, F. *Radiation Heat Transfer.* Scranton, Pa.: International Textbook Co., 1962.
11. Eckert, E. R. G., and R. M. Drake, Jr. *Analysis of Heat and Mass Transfer.* 2d ed. New York: McGraw-Hill, 1972.
12. Jakob, M. *Heat Transfer.* Vol. 2. New York: Wiley, 1957.
13. Gebhart, B. *Heat Transfer.* 2d ed. New York: McGraw-Hill, 1971.
14. Gebhart, B. "Unified Treatment for Thermal Radiation Processes—Gray, Diffuse Radiators and Absorbers," *A.S.M.E. Paper 57–A-34,* 1957.
15. Thomas, L. *Fundamentals of Heat Transfer.* Englewood Cliffs, N.J.: Prentice-Hall, 1980.
16. Oppenheim, A. K. "Radiation Analysis by the Network Method." *Transactions of the A.S.M.E.,* Vol. 78 (1956), pp. 725–35.
17. Hamilton, D. C., and W. R. Morgan. "Radiant Interchange Configuration Factors." *NACA TN 2836* (1952).
18. Gubareff, G. G., J. E. Janssen, and R. H. Torborg. *Thermal Radiation Properties Survey.* 2d ed. Minneapolis: Honeywell Research Center, Minneapolis-Honeywell Regulator Company, 1960.
19. McAdams, W. H. *Heat Transmission.* 3d ed. New York: McGraw-Hill, 1954.
20. Eckert, E. R. G., and R. M. Drake, Jr. *Heat and Mass Transfer.* 2d ed. New York: McGraw-Hill, 1959.
21. Kreith, F. *Principles of Heat Transfer.* 3d ed. New York: Intext Press, 1973.
22. Karlekar, B. V., and R. M. Desmond. *Heat Transfer.* 2d ed. St. Paul, Minn.: West, 1982.
23. Incropera, F. P., and D. P. Dewitt. *Fundamentals of Heat Transfer.* New York: Wiley, 1981.
24. Howell, J. R. *Radiation Configuration Factors.* New York: McGraw-Hill, 1982.

FORCED CONVECTION HEAT TRANSFER

5.1 INTRODUCTION

In Chapter 1, convection heat transfer was defined as energy transfer due to temperature differences between the surface of a solid and the bounding fluid which may be moving relative to the solid surface. If the fluid motion is induced by a pump, fan, blower, or reservoir of fluid under pressure or at a high elevation, the resulting energy transfer process is called *forced convection*. If the fluid motion is induced by density differences caused by the temperature field within the fluid itself, the energy transfer process is called *free* or *natural convection*. This chapter deals with forced convection heat transfer, and the following chapter is concerned with free convection heat transfer.

In a convection heat transfer situation, the energy transfer rate between the solid surface and the bounding fluid with which it interacts is, as in previous chapters, usually computed by use of Newton's cooling law,

$$q = h \, A \, \Delta T. \tag{1.4}$$

Thus far, whenever values of the surface coefficient of heat transfer h were required in the use of Eq. (1.4), these values were supplied. *The study of convection heat transfer is ultimately concerned with finding the values of the surface coefficient of heat transfer in terms of the physical quantities which are usually known or are given in an actual problem.* Values of the surface coefficient of heat transfer are arrived at by using the following general approaches: (1) Exact analytical solutions of the governing partial differential equations. Because of the mathematics required, this approach is generally treated only in graduate heat transfer courses. Only a brief explanation of the mathematical difficulties along with some final results of this approach will be presented here. (2) Approximate analytical solutions to the governing equations in integral form. This approach is the primary analytical method presented for prediction of the local surface coefficients of heat transfer; however, the bulk of the solutions of this type are also usually reserved for advanced courses because of different types of complexities from those which restricted the use of (1). Only a few of the simpler solutions obtainable by this approach are presented here. (3) Taking advantage of the similarity between the transport of heat and the transport of momentum. This is a very important technique whose principal application is to the difficult problem of finding surface coefficients of heat transfer in situations where the flow is turbulent. (4) Correlations in equation, graphical, or tabular form of the results of experiments designed to measure the surface coefficient of heat transfer.

We will first study the analytical and semi-analytical approaches, methods 1, 2, and 3, which yield expressions for the surface coefficient of heat transfer h in terms of the more fundamental known quantities upon which its value depends. This entails a discussion of what the starting point is for analytical prediction of the surface coefficient, and that leads to a review and study of the needed fluid mechanics. This is followed by sections which use exact solution techniques, approximate integral techniques, and semi-analytical "analogies" or similarity to derive the equations for the surface coefficients of heat transfer.

Important advances have been made in the analytical and semi-analytical methods, and the number of different situations that can be handled by these three predictive approaches has grown. Nevertheless, practicing engineers often compute surface coefficients of heat transfer from *experimental correlations,* since these are frequently more reliable, easier to use, and handle situations that are too complicated for any of the first three predictive techniques. Hence the last sections of the chapter will be devoted to the so-called "experimental correlations." These represent the results of extensive measurement, in the laboratory, of values of the surface coefficient as a function of the independent nondimensional groups which determine its value. They usually appear in easy-to-use equation form, but sometimes also appear in graphical or even tabular form. These experimental correlations, which are so often used for design and analysis in forced convection heat transfer situations, will be presented for a number of different geometries of practical importance, such as flow inside a circular tube, flow through tube banks, and external flow over flat plates, cylinders, and spheres.

5.2 ANALYSIS: THE FUNDAMENTAL PROBLEM

In attempting to use one of the first three methods to predict the local surface coefficient of heat transfer, we must determine where to begin the analysis. We must determine what is really needed, since the surface coefficient itself does not appear explicitly in any of the conservation equations. Consider Fig. 5.1, in which a solid surface is shown. At a point of interest on the surface, the surface temperature T_w is known, while the fluid far away from the surface is at the temperature T_s. The coordinate n perpendicular to the surface is measured outward from the surface as shown. A sketch of a typical temperature distribution within the moving fluid is also shown. At the point on the surface at T_w, the heat transfer rate per unit area by convection to the fluid is given by Newton's cooling law as $q'' = q/A = h\Delta T$. The temperature difference between the surface and the fluid some distance from the surface is $\Delta T = T_w - T_s$; hence,

$$q'' = h(T_w - T_s). \tag{5.1}$$

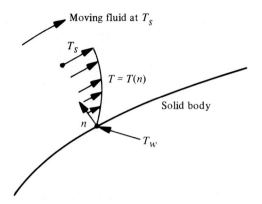

Figure 5.1 The temperature distribution in a fluid moving over a solid body.

Equation (5.1) is one expression for the heat transfer rate per unit area to the moving fluid. As noted in Chapter 1, another expression for this same quantity can be found by recognizing that the fluid velocity in the first few molecular layers against the solid surface has to be the same (for continuum fluid flow) as the velocity of the solid surface itself. Hence, in these first few molecular layers of fluid bounding the solid surface, there is no relative motion between the fluid and the solid, and the energy must first be transferred by pure conduction across these fluid layers before it reaches the moving fluid where it is then also carried away by the fluid motion. Therefore, Fourier's exact law of conduction, Eq. (1.3), also gives the "convection" heat transfer rate per unit area when it is applied to the fluid layers adjacent to the solid surface. Hence, another expression for q'' is given by Eq. (1.3) evaluated in the fluid at the solid surface, at $n = 0$, as

$$q'' = -k\frac{\partial T}{\partial n}\bigg|_{n=0}. \tag{5.2}$$

Equating Eqs. (5.1) and (5.2) and solving for h gives a general expression for the local surface coefficient of heat transfer as

$$h = \frac{-k\dfrac{\partial T}{\partial n}\bigg|_{n=0}}{T_w - T_s}. \tag{5.3}$$

It can be seen from Eq. (5.3) that to predict the local surface coefficient of heat transfer h, the temperature distribution as a function of the space coordinates and time within the *moving fluid* must be determined.

Thus, once one has predicted analytically the temperature profile within the moving fluid, Eq. (5.3) yields the quantity which the heat transfer analyst is interested in, the surface coefficient of heat transfer. As a result, the prediction of the temperature distribution becomes, as was also the case in conduction, the fundamental problem in convective heat transfer analysis.

As was pointed out in Chapter 1, the most distinguishing characteristic of convective heat transfer is the *motion of the fluid,* that is, the velocity field within the fluid. So even without seeing the specific form of the governing equations for the temperature distribution within the moving fluid, one's intuition about the nature of convective transport would cause one to conclude that the needed temperature field depends in part upon the motion itself. This means that the velocity field within the moving fluid—that is, the velocity vector as a function of space and time—must be predicted before one can find the temperature field, or it must be predicted simultaneously with the temperature field. Hence, at its root, every convective heat transfer problem is also a fluid mechanics problem. Therefore, the next section will be concerned with the fluid mechanics material and solution techniques required to predict the velocity field. With this result, the temperature field can be found and inserted into Eq. (5.3) to give the surface coefficient of heat transfer.

The analytical prediction of the velocity field, and of the temperature field which is coupled to it, often is relatively complicated, even for steady-state situations. Thus for our work in this book, only steady-state convection will be considered; the temperature and velocity distributions will depend only upon space and not upon time. Transient convection problems, when they arise, will be dealt with only by use of the quasi-steady approximation; that is, use of the steady-state relations at each instant of time, with all time-varying quantities inserted directly into the expression for the steady-state surface coefficient.

5.3 REVIEW OF FLUID MECHANICS

Since the velocity field is needed in order to find the temperature field and, therefore, the surface coefficient of heat transfer, this section will present the necessary fluid mechanics, much of which should be in the nature of a review of a previous course in fluids. The turbulent transport process which increases the shear stress in a fluid above that which would occur if the flow was laminar, also increases the energy transport rate in the same fluid. The similarities that can sometimes occur between the turbulent transport of momentum in the velocity field and of thermal energy for the temperature field make it appropriate to present the turbulent transport of heat in our discussion of fluid mechanics. In addition, sometimes the solution techniques for the governing equations of the velocity field can also be employed on the governing equation for the temperature field. When this is the case, these mathematical solution techniques (such as the powerful approximate integral method) will be developed carefully and in detail in this fluid mechanics section. They will be used again in later sections to solve for temperature fields.

5.3a Governing Equations for the Velocity Field

Two of the conservation laws of nature required to predict the velocity field as a function of space and time within a fluid are the *law of conservation of mass* and the *momentum theorem.* The momentum theorem, for a control volume, is the form taken by Newton's second law of motion. (More properly, Newton's second law is Euler's first law when considering a continuous distribution of mass rather than a particle, a mass point [4].) The

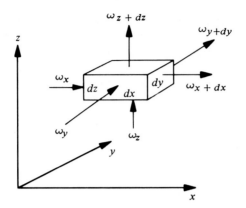

Figure 5.2 A differential control volume showing mass flow rates across its surfaces.

word form of these laws for a control volume can be found by starting with the known forms for a system, a region of fixed identity mass, and by doing an appropriate reinterpretation. This is done, for example, in [1], [2], and [5], and only the resulting word forms are presented here.

The *law of conservation of mass* states that the rate at which mass enters the control volume equals the rate at which mass leaves the control volume plus the time rate of change of stored mass within the control volume itself.

The *momentum theorem* states that the sum of all the forces on the control volume in any particular direction equals the momentum flux out (in that direction) minus the momentum flux in (in that direction) plus the time rate of change of control volume momentum in that direction. The momentum flux across a differential area dA on a control surface is the mass rate of flow $d\omega$ across that area times the velocity component in the direction in which the momentum flux is required.

Application of these two laws to a control volume yields, depending upon the choice of control volume, either partial differential or integral equations. The velocity vector, ordinarily expressed in terms of its components in a coordinate system, is a dependent variable, and space coordinates and time are independent variables.

Next we need quantitative equation forms of these two laws in order to predict the details of the velocity field's dependence upon the space coordinates. To arrive at the governing differential equation which represents the law of conservation of mass, the preceding word statement for mass conservation will be applied to the differential control volume, *dxdydz*, shown in Fig. 5.2. This control volume is immersed in a flowing compressible fluid with space points located in a rectangular cartesian coordinate system.

Mass crosses the face of area $dydz$ located at x at the rate ω_x and leaves across the face of area $dydz$, located at $x + dx$, at the rate ω_{x+dx}, where ω is the mass flow rate. Similarly, mass enters the control volume across the face at y at the rate ω_y, and leaves across the face at $y + dy$ at the rate ω_{y+dy}. Similarly for the faces at z and $z + dz$. Hence, the rate at which mass enters the control volume is the sum $\omega_x + \omega_y + \omega_z$ and the rate at which mass leaves the control volume is the sum $\omega_{x+dx} + \omega_{y+dy} + \omega_{z+dz}$. If the mass within the control volume at this instant of time is called m, then the time rate of change of stored mass within the control volume is $\partial m/\partial t$. Using these rates in the statement of the law of conservation of mass yields

$$\omega_x + \omega_y + \omega_z = \omega_{x+dx} + \omega_{y+dy} + \omega_{z+dz} + \frac{\partial m}{\partial t}. \qquad (5.4)$$

However, ω_{x+dx} can be expressed in terms of ω_x by expanding the function for the local mass flow rate in the x direction, at fixed y and z, in a Taylor series expansion about x across to $x + dx$. Neglecting higher order terms, this yields

$$\omega_{x+dx} = \omega_x + \frac{\partial \omega_x}{\partial x} \, dx. \qquad (5.5)$$

Similarly,

$$\omega_{y+dy} = \omega_y + \frac{\partial \omega_y}{\partial y} \, dy, \qquad (5.6)$$

$$\omega_{z+dz} = \omega_z + \frac{\partial \omega_z}{\partial z} \, dz. \qquad (5.7)$$

Substituting Eqs. (5.5–7) into Eq. (5.4), cancelling out terms that appear on both sides of the equality, and rearranging yields

$$\frac{\partial \omega_x}{\partial x} \, dx + \frac{\partial \omega_y}{\partial y} \, dy + \frac{\partial \omega_z}{\partial z} \, dz + \frac{\partial m}{\partial t} = 0. \qquad (5.8)$$

Since it is desirable to have the velocity components of the fluid at position x, y, and z appear as dependent variables in Eq. (5.8), the mass flow rates ω_x, ω_y, and ω_z must be related to the velocity components u, v, and w, where u is the local x component, v is the local y component, and w is the local z component of the velocity vector in the rectangular cartesian coordinate system. From fluid mechanics ([1], [2], or [5]), the mass flow rate across a differential area is the product of the fluid density ρ, area dA, and velocity component V_n normal to dA. Hence,

$$\omega_x = \rho u \, dzdy, \quad \omega_y = \rho v \, dxdz, \quad \omega_z = \rho w \, dxdy. \qquad (5.9)$$

The mass m within the control volume is

$$m = \rho \, dxdydz. \qquad (5.10)$$

We substitute Eqs. (5.9–10) into Eq. (5.8), noting that terms such as dx, dy, dz or their products can be brought outside the differential operators. After dividing through by a volume element, the partial differential equation describing mass conservation in rectangular cartesian coordinates is

$$\frac{\partial}{\partial x} (\rho u) + \frac{\partial}{\partial y} (\rho v) + \frac{\partial}{\partial z} (\rho w) + \frac{\partial \rho}{\partial t} = 0. \tag{5.11}$$

If the flow is steady, Eq. (5.11) reduces to

$$\frac{\partial}{\partial x} (\rho u) + \frac{\partial}{\partial y} (\rho v) + \frac{\partial}{\partial z} (\rho w) = 0, \tag{5.12}$$

since in the steady state, properties at any space point do not vary with time: hence, $\partial \rho / \partial t = 0$.

If the flow is incompressible—that is, the density is essentially constant—Eq. (5.11) reduces to

$$\frac{\partial u}{\partial x} + \frac{\partial v}{\partial y} + \frac{\partial w}{\partial z} = 0, \tag{5.13}$$

regardless of whether the flow is steady or unsteady, since if ρ is constant, $\partial \rho / \partial t = 0$ and ρ can be factored out of the remaining differential operators and cancelled. Equations (5.11–13) are all often referred to as the *continuity equation*. The various forms of the continuity equation, Eqs. (5.11–13), are, of course, for the rectangular cartesian coordinate system.

In order to derive a general equation for mass conservation which is independent of any particular coordinate system, one can use geometric vector notation in applying the word form of the law of conservation of mass to the arbitrarily shaped control volume of finite size, which is immersed in a flowing compressible fluid. The physical situation is shown in Fig. 5.3, and we will use the convention that the unit vector normal to a surface is positive when it points out of the volume which the surface is helping to envelop.

The unit normal vector to the area of magnitude dA on the control surface is n, and V is the local velocity vector with its direction such that mass enters the control volume across surface dA (see Fig. 5.3). Since only the normal component of velocity causes transport of mass across dA, the mass flow rate across dA is

$$d\omega = -\rho V \cdot n \, dA, \tag{5.14}$$

where the dot between vectors symbolizes the inner product in a geometric linear vector space, and the minus sign is required to ensure that the mass flow rate into the control volume is positive. The angle between V and n used in the dot product is between $90°$ and $180°$; hence, its cosine is negative and the minus sign is inserted to compensate for the one resulting from the cosine. The total net rate at which mass enters the control volume (the difference between the rate at which mass enters and the rate at which mass

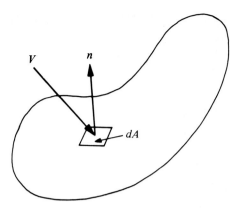

Figure 5.3 Differential area on an arbitrarily shaped finite control volume.

leaves the control volume) is determined by summing up all contributions of the type of Eq. (5.14) over the entire control surface in the definite integral

$$\omega = -\int_A \rho V \cdot n \, dA. \tag{5.15}$$

This must equal the time rate of change of mass within the control volume, that is, dm/dt.

The small amount of mass within a differential control volume dV_0, inside the finite-size control volume shown in Fig. 5.3, is given as the product of the local density and the volume element dV_0,

$$\rho \, dV_0. \tag{5.16}$$

The total mass at any instant of time within the control volume is found by adding all contributions of the type of Eq. (5.16) over the entire control volume in the definite integral

$$m = \int_{V_0} \rho \, dV_0.$$

Hence, the storage term is

$$\frac{dm}{dt} = \frac{d}{dt} \int_{V_0} \rho \, dV_0. \tag{5.17}$$

Finally, equating Eqs. (5.15) and (5.17), a general equation representing the conservation of mass for any control volume and for any coordinate system is

$$-\int_A \rho V \cdot n \, dA = \frac{d}{dt} \int_{V_0} \rho \, dV_0. \tag{5.18}$$

Next, the partial differential equation implied by Eq. (5.18) will be developed. Thus, the use of the Leibnitz rule for differentiating an integral with respect to a variable t which is not the integration variable, yields for the right-hand side of Eq. (5.18), after noting that the limit is independent of time,

$$\frac{d}{dt} \int_{V_0} \rho \, dV_0 = \int_{V_0} \frac{\partial \rho}{\partial t} \, dV_0.$$

Hence, after rearrangement, Eq. (5.18) becomes

$$\int_A \rho V \cdot n \, dA + \int_{V_0} \frac{\partial \rho}{\partial t} \, dV_0 = 0. \tag{5.19}$$

However, by Gauss' theorem (i.e., the divergence theorem) the surface integral on the left of Eq. (5.19) can be transformed into the volume integral

$$\int_A \rho V \cdot n \, dA = \int_{V_0} \text{div} \, (\rho V) \, dV_0. \tag{5.20}$$

Combining Eqs. (5.19–20) under the same integral yields

$$\int_{V_0} \left[\text{div} \, (\rho V) + \frac{\partial \rho}{\partial t} \right] dV_0 = 0. \tag{5.21}$$

But Eq. (5.21) must be true regardless of the size of the volume element V_0; hence, the integrand must equal zero, which gives the desired general partial differential equation for conservation of mass in any coordinate system as

$$\text{div} \, (\rho V) + \frac{\partial \rho}{\partial t} = 0. \tag{5.22}$$

The divergence operator is explicitly shown in all the common coordinate systems in any standard vector analysis textbook; hence, once the coordinate system is chosen, the expansion of Eq. (5.22) gives the partial differential equation for conservation of mass in that coordinate system. In particular, for the rectangular cartesian coordinate system,

$$\text{div} \, (\rho V) = \frac{\partial}{\partial x} \, (\rho u) + \frac{\partial}{\partial y} \, (\rho v) + \frac{\partial}{\partial z} \, (\rho w). \tag{5.23}$$

Substituting Eq. (5.23) into Eq. (5.22) yields

$$\frac{\partial}{\partial x} \, (\rho u) + \frac{\partial}{\partial y} \, (\rho v) + \frac{\partial}{\partial z} \, (\rho w) + \frac{\partial \rho}{\partial t} = 0,$$

which is the same as Eq. (5.11).

Note that any of the forms of the continuity equation, such as Eqs. (5.11–13) or Eq. (5.22), show that the velocity field cannot be obtained from the continuity equation alone, since even in its simplest form, Eq. (5.13), there are three unknowns, the velocity components u, v, and w, and only a single partial differential equation. Hence, additional equations are needed to solve for the velocity field. The momentum theorem yields three additional equations. In using the momentum theorem, say on the control volume shown in Fig. 5.2, each of the six faces has three forces acting upon it: two viscous shearing forces and a normal force that is usually divided into two parts, the part due to the thermodynamic pressure P and the part due to the normal viscous stress acting on the area. In addition, there are the three components of a body force; that is, a force whose effect is distributed throughout the material in the control volume, such as the weight or a magnetic attractive force. The body force per unit volume in the x, y, and z directions will be denoted by f_x, f_y, and f_z, respectively. The surface stresses on each face which cause the surface forces can be put in terms of various derivatives of the velocity components by using the *constitutive equation* for a linearly viscous fluid. This constitutive equation is also called *Stokes' law of friction* and, for simple cases, this equation for the shear-stress components is called *Newton's law of viscosity*. This procedure for deriving the momentum equations in the x, y, and z directions will not be carried out in detail for the fairly general three-dimensional situation being described here. In forthcoming sections, however, we will need to apply the momentum theorem in its word form, as given earlier, to control volumes in more restricted and less general situations. But at this point we simply present, *for a constant property fluid* (ρ and μ constant throughout the fluid), the final quantitative forms of the momentum equations for a laminar, three-dimensional flow in the rectangular cartesian coordinate system. These are called the *Navier-Stokes equations*, and the actual details of their derivation, which we have just outlined, can be found in [2], [6], [7], or [8]. The final equations are presented here:

$$\rho \left(\frac{\partial u}{\partial t} + u\frac{\partial u}{\partial x} + v\frac{\partial u}{\partial y} + w\frac{\partial u}{\partial z} \right)$$
$$= -\frac{\partial P}{\partial x} + \mu \left(\frac{\partial^2 u}{\partial x^2} + \frac{\partial^2 u}{\partial y^2} + \frac{\partial^2 u}{\partial z^2} \right) + f_x, \qquad (5.24)$$

$$\rho \left(\frac{\partial v}{\partial t} + u\frac{\partial v}{\partial x} + v\frac{\partial v}{\partial y} + w\frac{\partial v}{\partial z} \right)$$
$$= -\frac{\partial P}{\partial y} + \mu \left(\frac{\partial^2 v}{\partial x^2} + \frac{\partial^2 v}{\partial y^2} + \frac{\partial^2 v}{\partial z^2} \right) + f_y, \qquad (5.25)$$

$$\rho \left(\frac{\partial w}{\partial t} + u\frac{\partial w}{\partial x} + v\frac{\partial w}{\partial y} + w\frac{\partial w}{\partial z} \right)$$
$$= -\frac{\partial P}{\partial z} + \mu \left(\frac{\partial^2 w}{\partial x^2} + \frac{\partial^2 w}{\partial y^2} + \frac{\partial^2 w}{\partial z^2} \right) + f_z. \qquad (5.26)$$

(See [1], [2], [3], and [6].) Also, for constant property flow, the continuity equation takes the form of

$$\frac{\partial u}{\partial x} + \frac{\partial v}{\partial y} + \frac{\partial w}{\partial z} = 0. \tag{5.13}$$

Hence, *if the body forces f_x, f_y, and f_z are independent of temperature,* Eqs. (5.13) and (5.24–26) constitute a set of four partial differential equations in four dependent variables: the velocity field components u, v, and w and the pressure field P, as functions of x, y, z, and t. When the velocity field is completely determined by the solution to these four equations independent of the temperature distribution in the moving fluid, the velocity field is said to be *uncoupled* from the temperature field. (The converse, as will be seen in the following section, is definitely not true.) The fact that the velocity field can be determined independent of the temperature field is, perhaps, the major characteristic of constant property flow.

Thus, for a constant property flow with body forces independent of temperature, the governing partial differential equations, whose solution leads to the velocity field within the fluid, are Eqs. (5.13) and (5.24–26). The task in a number of the upcoming sections is to find solutions to less general versions of these equations. However, because of the nonlinearity of Eqs. (5.24) to (5.26), as manifested particularly by terms such as $u(\partial u/\partial x)$ on the left-hand side, very few exact analytical solutions are available. The two general types of flow for which most of the exact solutions have been found are (1) steady, two-dimensional, laminar, constant property, fully-developed internal flow in ducts and tubes, and (2) steady, two-dimensional, laminar, constant property, external boundary layer-type flow over bodies immersed in the fluid.

EXAMPLE 5.1

Two engineering assistants are to measure the three components of the velocity vector in a three-dimensional incompressible flow field, fit the data with equations, and report their results. When the results are reported, they realize that the z component of velocity w was not measured, and only the results for u and v have been obtained. In addition, the experimental rig has been torn down and cannot be rebuilt for a month. If, for the shape under consideration, $w = 0$ at $z = 0$, estimate the z component of velocity that can be used in rough calculations until the rig is rebuilt. Their results are

$$u = 100(1 + 0.5x^2y^2 - 0.17x^4y^4z^3), \quad v = 13(1 - x^2y^2 \cos z).$$

Solution
If this experimental data is valid and accurate, it must satisfy the continuity equation for incompressible three-dimensional flow; Eq. (5.13). From the given velocity components,

$$\frac{\partial u}{\partial x} = 100(xy^2 - 0.68x^3y^4z^3),$$

$$\frac{\partial v}{\partial y} = -26x^2y \cos z.$$

Substituting these into Eq. (5.13) and solving for $\partial w/\partial z$,

$$\frac{\partial w}{\partial z} = -100(xy^2 - 0.68x^3y^4z^3) + 26x^2y \cos z.$$

Integrating both sides of the equation partially with respect to z,

$$w = -100(xy^2z - 0.17x^3y^4z^4) + 26x^2y \sin z + F(x, y), \qquad \textbf{(5.27)}$$

where $F(x, y)$ is the function of integration resulting from integrating partially with respect to z, which can be determined from the boundary condition that $w = 0$ at $z = 0$. Substituting $z = 0$ into Eq. (5.27) and setting $w = 0$ yields $F(x, y) = 0$; hence, the rough solution for the z component of velocity is

$$w = -100(xy^2z - 0.17x^3y^4z^4) + 26x^2y \sin z.$$

Note that this result must be considered a rough estimate because the derivatives of two equations that were a fit to the experimental data were used. Although such equations can give back the measured velocity components at each x, y, and z, the slopes are suspect because they may not be close to the correct ones.

5.3b Fully Developed, Laminar, Two-Dimensional, Constant Property Flow in Tubes and Ducts

In this section, solutions for the velocity field and other hydrodynamic parameters which are derivable from the velocity field, such as wall shear stress τ_w and the friction factor f, will be presented for the simplest class of viscous flow problems. These include situations in which the flow is steady, constant property, laminar, two-dimensional, and fully developed. You should recall from fluid mechanics that a laminar flow is an orderly, layer-type flow with no bulk motion, motion on a macroscopic level, perpendicular to the main flow direction. There is, of course, molecular motion transverse to the main flow direction which causes the shearing stress within a laminar flow.

By *fully developed* is meant that the shape of the velocity profile remains unchanged in the general flow direction, which is taken here as the x direction. For this fully developed condition to occur, one must have a finite expanse of fluid, such as occurs within a duct, and a constant property fluid flowing. To illustrate how the fully developed condition arises, consider the duct shown in Fig. 5.4. In the case shown, the fluid enters the duct at $x = 0$ with an essentially uniform x component of velocity u which equals u_m for all values of y. However, as the real, viscous fluid contacts the walls, the no-slip condition at the wall, for a continuum flow, forces the x component of velocity to be zero at the walls, and the shear force exerted by the wall on the fluid slows down the fluid in the vicinity of the wall. This can be seen by examining the velocity profile at the second station, where some of the fluid—that which is in the layer of viscous flow against the wall—now has a nonuniform distribution of velocity in the y direction. The remaining fluid, in the so-called

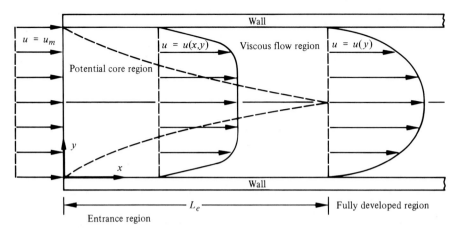

Figure 5.4 Development of the velocity profile in a duct.

potential core region, still has not been affected by the presence of the wall, and thus retains its uniform velocity profile in y, although conservation of mass considerations have caused a lengthening of the velocity in the x direction. Obviously the shape of the velocity profile is different at the second station than it is at the first one, the entrance. So u, the x component of velocity, depends upon both x and y in the entrance region, which is also termed the hydrodynamic, or the developing, entrance region. In this entrance region, $0 \leq x \leq L_e$, u is continually changing, lengthening or contracting in response to the shear stress which is being developed farther and farther away from the wall. Finally, when the viscous flow region grows to the centerline or centerplane of the duct, at $x = L_e$, the x component of velocity, u, no longer changes with x and depends only upon y, $u = u(y)$ for $x > L_e$. This relatively simple flow regime is the fully developed region.

From a purely mathematical viewpoint, fully developed means that $\partial u/\partial x = 0$ for all x and y. Also, when this is the case, the shear stress in a laminar flow is given by Newton's law of viscosity,

$$\tau = \mu \frac{\partial u}{\partial y} . \tag{5.28}$$

In Eq. (5.28), τ is the local shear stress in the x direction and μ, it will be recalled, is the dynamic, or absolute, viscosity, a transport property whose value for some common fluids is given as a function of temperature in Appendix A.

Of importance is the extent, L_e, of the entrance region. That represents how long a duct must be in order for the flow to be in the simpler regime, the fully developed region.

For laminar, constant property flow between two parallel plates spaced a distance b units apart in the y direction, Sparrow [9] has derived the following expression:

$$\frac{L_e}{b} = 0.026 \, N_{\mathrm{Re}_b}, \tag{5.29}$$

Chapter Five

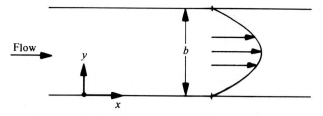

Figure 5.5 Laminar fluid flow between parallel plates.

where

$$N_{Re_b} = \rho U_m b / \mu \qquad (5.30)$$

is the Reynolds number based upon the mass average velocity U_m and the plate spacing b.

When the flow is inside a circular tube of diameter D and is laminar with constant properties, the entrance length is given by the following expression presented by Langhaar [10]:

$$\frac{L_e}{D} = 0.058 \, N_{Re_D} \qquad (5.31)$$

$$N_{Re_D} = \rho U_m D / \mu \qquad (5.32)$$

Flow Between Two Parallel Plates

Consider steady, laminar, constant property, fully developed flow between two parallel plates, spaced a distance b units apart as shown in Fig. 5.5. The mass average velocity of the fluid is U_m, the pressure is not significantly dependent upon y, and body forces are negligible. In order to eventually solve the convective heat transfer problem for this duct, we need to first solve for the velocity field, as has been previously explained.

Since the flow is two-dimensional (that is, there is no z velocity component), every term in Eq. (5.26) is zero. Also, Eq. (5.13) reduces to

$$\frac{\partial u}{\partial x} + \frac{\partial v}{\partial y} = 0.$$

However, because of the fully developed condition, $\partial u / \partial x = 0$; hence,

$$\frac{\partial v}{\partial y} = 0. \qquad (5.33)$$

Equation (5.33) can be integrated partially with respect to y to yield

$$v = F(x); \tag{5.34}$$

but, at $y = 0$, the bottom plate, and at $y = b$, the upper plate, $v = 0$ for all x since the plates are solid and impervious to mass. Thus, $F(x) = 0$ in Eq. (5.34), which yields $v = 0$ for all x and y. Using this information along with the fact that $\partial P/\partial y$ and f_y are zero, every term in Eq. (5.25) is zero. Since

$$\frac{\partial^2 u}{\partial x^2} = \frac{\partial^2 u}{\partial z^2} = \frac{\partial u}{\partial t} = u\frac{\partial u}{\partial x} = v\frac{\partial u}{\partial y} = w\frac{\partial u}{\partial z} = f_x = 0, \tag{5.35}$$

Eq. (5.24) becomes

$$0 = -\frac{\partial P}{\partial x} + \mu\frac{\partial^2 u}{\partial y^2}, \quad \text{or} \quad \frac{\partial^2 u}{\partial y^2} = \frac{1}{\mu}\frac{\partial P}{\partial x}. \tag{5.36}$$

Now, the fully developed condition $\partial u/\partial x = 0$ for all x and y can be integrated to yield

$$u = u(y).$$

As a result, it can be concluded that u depends upon y alone; hence, the second derivative of u with respect to y can at most depend upon y, certainly not upon x. However, in the problem statement it was stated that P did not depend upon y, only, at most, upon x. Hence, it must be concluded from Eq. (5.36) that both sides of the equation are simply a constant, and since u is a function only of one variable, Eq. (5.36) can be written as

$$\frac{d^2 u}{dy^2} = \frac{1}{\mu}\frac{\partial P}{\partial x}.$$

Since $\partial P/\partial x$ is a constant, integration of both sides twice with respect to y yields

$$u = \frac{1}{2\mu}\frac{\partial P}{\partial x}y^2 + c_1 y + c_2. \tag{5.37}$$

The boundary conditions are

$$\text{at } y = 0, u = 0, \tag{5.38}$$

$$\text{at } y = b, u = 0. \tag{5.39}$$

Applying the boundary condition of Eq. (5.38) to Eq. (5.37) yields $c_2 = 0$. Applying Eq. (5.39) to Eq. (5.37) leads to

$$0 = \frac{1}{2\mu}\frac{\partial P}{\partial x}b^2 + c_1 b, \quad \text{or} \quad c_1 = -\frac{1}{2\mu}\frac{\partial P}{\partial x}b.$$

With these values of c_1 and c_2, Eq. (5.37) becomes

$$u = \frac{1}{2\mu}\frac{\partial P}{\partial x}(y^2 - by).$$ (5.40)

This equation is not yet the final result, since $\partial P/\partial x$ is not ordinarily a known quantity. However, here the average velocity U_m is known; hence, it is logical to try to relate the average velocity and the pressure gradient in order to eliminate the latter. The actual total mass flow rate past any station x per unit in the z direction is

$$\omega = \rho A U_m = \rho b(1) U_m.$$ (5.41)

The small mass flow rate across the area $dy \times 1$ unit at position x is

$$d\omega = \rho u \, dy.$$

Using Eq. (5.40) in the preceding equation yields

$$d\omega = \frac{\rho}{2\mu}\frac{\partial P}{\partial x}(y^2 - by)dy.$$ (5.42)

The total mass flow rate can be found by summing up all contributions of the type of Eq. (5.42) over the entire cross section in the definite integral from $y = 0$ to $y = b$. Hence,

$$\int d\omega = \omega = \int_0^b \frac{\rho}{2\mu}\frac{\partial P}{\partial x}(y^2 - by)dy.$$

Integrating and inserting the limits leads to

$$\omega = \frac{\rho}{2\mu}\frac{\partial P}{\partial x}\left(-\frac{b^3}{6}\right).$$

Equating this and Eq. (5.41) and solving for $\partial P/\partial x$ yields

$$\frac{\partial P}{\partial x} = -\frac{12\mu\, U_m}{b^2}.$$ (5.43)

Equation (5.43) indicates that $\partial P/\partial x$ is negative; that is, the pressure is dropping in the flow direction. This agrees with what is to be expected, since the pressure drop is the sole driving potential for fluid flow in the given situation.

Substituting Eq. (5.43) into Eq. (5.40) yields the required velocity distribution in terms of known quantities as

$$u = 6 U_m\left(\frac{y}{b} - \frac{y^2}{b^2}\right).$$ (5.44)

Also,

$$v = 0. \qquad\qquad (5.45)$$

Equation (5.45) was established in the discussion that followed Eq. (5.34). Together, Eqs. (5.44) and (5.45) give the velocity field for this fully developed parallel plate duct flow. These relations are needed, and will be used, to predict analytically the local surface coefficient of heat transfer, h_x, between the duct wall and the flowing fluid.

Flow in a Circular Pipe

Next we will discuss steady, laminar, constant property, fully developed, two-dimensional flow in a horizontal circular pipe of radius R when body forces are negligible. We would like to find the velocity distribution function as well as the wall shear stress and the friction factor, parameters which are related to the pressure drop in the pipe and to the loss in available energy due to viscous effects.

Since the most appropriate coordinate system to be used here is the circular cylindrical, not the rectangular cartesian, we will not begin with the continuity equation and the Navier-Stokes equations in the form given by Eqs. (5.13) and (5.24–26). Instead, the governing equations for the velocity field will be derived by direct application of the word forms of the momentum theorem and the law of conservation of mass to the appropriate control volume. We will make the simplifications, such as steady flow and fully developed flow, which are valid for this flow problem, during the course of the derivation.

Take as the control volume the circular cylindrical shell of inner radius r, outer radius $r + dr$, and length dx as shown in Fig. 5.6. The momentum theorem applied to the control volume in the x direction requires that the sum of all the forces on the control volume in the x direction be found. There is a pressure force on the left-hand face, and one on the right-hand face where the Taylor series expansion of the pressure in the x direction has already been made and higher-order terms neglected. There are shear forces as well, shown on the faces at r and $r + dr$. Hence, the sum of all the forces in the x direction is

$$\Sigma F_x = P\,2\pi r dr - \left(P + \frac{\partial P}{\partial x}dx\right)2\pi r dr - \tau 2\pi r dx$$
$$- 2\pi dx \frac{\partial}{\partial r}(\tau r)dr + \tau 2\pi r dx.$$

Expanding the products and combining terms yields

$$\Sigma F_x = \frac{\partial P}{\partial x}2\pi r dx dr - 2\pi dx dr \frac{\partial(\tau r)}{\partial r}. \qquad (5.46)$$

To prove that the fully developed condition $\partial u/\partial x = 0$ implies that $V_r = 0$, just as in the case of parallel plates, the law of conservation of mass for the control volume shown in Fig. 5.6 is written. Mass enters the left-hand face at x at the rate $\rho u 2\pi r dr$, and leaves the right-hand face at $x + dx$ at the rate $\rho[u + (\partial u/\partial x)dx]2\pi r dr$. Mass enters at the

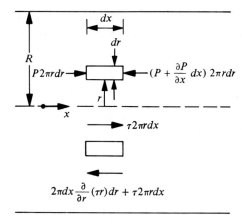

Figure 5.6 Forces on the differential control volume of a pipe.

face at r at the rate $\rho 2\pi r dx V_r$, and leaves at the face at $r + dr$ at the rate $\rho 2\pi r dx V_r + [\partial(\rho 2\pi r V_r dx)/\partial r]dr$. Since the rate at which mass enters the control volume must equal the rate at which mass leaves the control volume in the steady state,

$$\rho u 2\pi r dr + \rho 2\pi r dx V_r = \rho u 2\pi r dr + \rho \frac{\partial u}{\partial x} 2\pi r dx dr + \rho 2\pi r V_r dx$$

$$+ \rho 2\pi \frac{\partial}{\partial r}(r V_r)dxdr. \qquad (5.47)$$

Cancelling like terms and noting that the fully developed condition requires that $\partial u/\partial x = 0$, Eq. (5.47) becomes $(\partial/\partial r)(r V_r) = 0$. Integrating partially with respect to r yields

$$r V_r = F(x). \qquad (5.48)$$

However, at $r = R$, the pipe surface, there can be no relative motion between the fluid and the pipe since the pipe is not porous, and this condition must hold for all x. Hence, $F(x) = 0$ in Eq. (5.48), leading to the conclusion that the radial components of velocity are also zero, that is, $V_r = 0$. Since $V_r = 0$, there can be no mass flow rate across the surfaces at r and at $r + dr$ and hence no x-momentum flux can cross those surfaces. The x-momentum flux in across the surface at x is given by $\rho u 2\pi r dr\, u$, and the x-momentum flux out across the surface at $x + dx$ can be put in terms of the x-momentum flux at x by expanding the x-momentum flux in a Taylor series about x across to $x + dx$ and neglecting the higher-order terms. As a result, the x-momentum flux out at $x + dx$ becomes $\rho u 2\pi r dr\, u + \rho 4\pi r u(\partial u/\partial x)dxdr$. Since $\partial u/\partial x = 0$, the difference between the x-momentum flux out of the control volume and the x-momentum flux entering the control volume is zero and, for this case, the momentum theorem requires that $\sum F_x = 0$. (Recall that there is no time rate of change of control volume momentum because of the steady-state condition.) Therefore, Eq. (5.46) becomes, after setting it equal to zero and dividing by $2\pi rdrdx$,

$$\frac{1}{r}\frac{\partial}{\partial r}(\tau r) = -\frac{\partial P}{\partial x}.$$ (5.49)

Since the governing differential equation for the velocity distribution is required, the shear stress can be related to the velocity by Newton's law of viscosity

$$\tau = \mu\frac{\partial u}{\partial y}.$$ [5.28]

However, since the radial coordinate r is measured positively toward the wall rather than away from it, a positive value of fluid shear stress at the wall will be insured by writing Eq. (5.28) as $\tau = -\mu(\partial u/\partial r)$. Using this, Eq. (5.49) becomes

$$\frac{1}{r}\frac{\partial}{\partial r}\left(-\mu\frac{\partial u}{\partial r}r\right) = -\frac{\partial P}{\partial x},$$

or, after dividing through by $-\mu$, the governing differential equation for the velocity distribution in steady, laminar, constant property, fully developed flow in a circular tube is

$$\frac{1}{r}\frac{\partial}{\partial r}\left(r\frac{\partial u}{\partial r}\right) = \frac{1}{\mu}\frac{\partial P}{\partial x}.$$ (5.50)

The left-hand side has been left in the form shown, which is simple to integrate directly with respect to r.

To solve Eq. (5.50), note that since the flow is fully developed, $\partial u/\partial x = 0$; therefore, u is a function of r alone. Hence, total derivatives rather than partial derivatives can be used in Eq. 5.50. In addition, since P is not a function of r and the left-hand side of Eq. (5.50) cannot depend on x (since u depends upon r, but not x), it can be concluded that $\partial P/\partial x$ is constant. Hence, Eq. (5.50) becomes

$$\frac{1}{r}\frac{d}{dr}\left(r\frac{du}{dr}\right) = \frac{1}{\mu}\frac{\partial P}{\partial x}.$$

Multiplying by rdr yields

$$d\left(r\frac{du}{dr}\right) = \frac{1}{\mu}\frac{\partial P}{\partial x}rdr,$$

and integrating both sides leads to

$$r\frac{du}{dr} = \frac{1}{\mu}\frac{\partial P}{\partial x}\frac{r^2}{2} + c_1.$$ (5.51)

However, one of the boundary conditions on the velocity profile in a tube is the radial symmetry about $r = 0$, leading to a zero slope in the velocity profile at $r = 0$; hence, at $r = 0$, $\partial u/\partial r = 0$. Using this in Eq. (5.51) gives $c_1 = 0$, and Eq. (5.51) becomes

$$\frac{du}{dr} = \frac{1}{2\mu}\frac{\partial P}{\partial x}r.$$

Multiplying by dr and integrating both sides of the preceding equation yields

$$u = \frac{1}{\mu}\frac{\partial P}{\partial x}\frac{r^2}{4} + c_2. \tag{5.52}$$

The remaining boundary condition is the no-slip condition at the wall; that is, at $r = R$, $u = 0$. Using this, Eq. (5.52) becomes

$$0 = \frac{1}{\mu}\frac{\partial P}{\partial x}\frac{R^2}{4} + c_2, \quad \text{or} \quad c_2 = -\frac{1}{\mu}\frac{\partial P}{\partial x}\frac{R^2}{4}.$$

Substituting this into Eq. (5.52) and rearranging yields

$$u = \frac{1}{4\mu}\frac{\partial P}{\partial x}(r^2 - R^2). \tag{5.53}$$

The unknown pressure gradient $\partial P/\partial x$ can be written in terms of the average flow velocity U_m. The mass flow rate at any x through the tube cross section is

$$\omega = \rho\pi R^2 U_m. \tag{5.54}$$

The mass flow rate across a differential area $2\pi r dr$ is

$$d\omega = \rho 2\pi r dr\, u. \tag{5.55}$$

The total mass flow rate must equal the sum of all contributions of the type shown in Eq. (5.55) over the entire cross section as given by the definite integral

$$\omega = \int_0^R \rho 2\pi r u \, dr.$$

Substituting the velocity distribution Eq. (5.53) into the preceding equation leads to

$$\omega = \frac{2\pi\rho}{4\mu}\frac{\partial P}{\partial x}\int_0^R (r^3 - rR^2)dr.$$

Integrating and inserting the limits yields

$$\omega = \frac{-\pi\rho}{8\mu}\frac{\partial P}{\partial x}R^4. \qquad (5.56)$$

Equating Eqs. (5.56) and (5.54) and solving for $(1/4\mu)(\partial P/\partial x)$ leads to

$$\frac{1}{4\mu}\frac{\partial P}{\partial x} = \frac{-2U_m}{R^2}.$$

Substituting this into Eq. (5.53) and rearranging, the final form of the velocity distribution is

$$u = 2U_m\left(1 - \frac{r^2}{R^2}\right). \qquad (5.57)$$

The radial velocity component, V_r, was shown earlier to be equal to zero because of the fully developed and constant property conditions. Thus,

$$V_r = 0.$$

To compute the wall shear stress τ_w, Newton's law of viscosity, Eq. (5.28), can be used for the conditions of this problem after rewriting it in terms of r, as was shown in the derivation of Eq. (5.50). Hence,

$$\tau = -\mu\frac{\partial u}{\partial r}. \qquad (5.58)$$

Differentiating Eq. (5.57) with respect to r yields

$$\frac{\partial u}{\partial r} = -\frac{4U_m r}{R^2}. \qquad (5.59)$$

The wall shear stress is given by Eq. (5.58) evaluated at $r = R$; hence, $\tau_w = -\mu(\partial u/\partial r)_R$. From Eq. (5.59), $(\partial u/\partial r)_R = -4U_m/R$, and

$$\tau_w = \frac{4\mu U_m}{R}. \qquad (5.60)$$

The *Moody friction factor, f,* is a nondimensional wall shear stress defined as follows:

$$f = 4\tau_w/\tfrac{1}{2}\rho U_m^2. \qquad (5.61)$$

Combining Eqs. (5.60) and (5.61) gives,

$$f = \frac{4\tau_w}{\frac{1}{2}\rho U_m^2} = \frac{4\left(\frac{4\mu U_m}{R}\right)}{\frac{1}{2}\rho U_m^2} = \frac{32\mu}{\rho U_m R}.$$ (5.62)

The quotient μ/ρ is called the kinematic viscosity ν (or, the molecular diffusivity of momentum). The usual engineering units of ν are m^2/s or ft^2/hr, the same as for the thermal diffusivity α (which is also called the molecular diffusivity of heat).

Conventionally the diameter D, rather than the radius R, is used in an equation such as Eq. (5.62). Since $D = 2R$, Eq. (5.62) becomes

$$f = 64\frac{\mu}{\rho U_m D}.$$

The nondimensional group $\rho U_m D/\mu$, which is a measure of the ratio of "inertia" forces to viscous forces, is called the *Reynolds number*, denoted by N_{Re}. Hence,

$$N_{Re} = \frac{\rho U_m D}{\mu}.$$ (5.63)

Thus, in terms of the Reynolds number, the final result for the Moody friction factor is

$$f = \frac{64}{N_{Re}}.$$ (5.64)

The so-called *Fanning friction factor* or the *skin friction coefficient* C_f is also in wide use and is defined by

$$C_f = \frac{\tau_w}{\frac{1}{2}\rho U_m^2}.$$ (5.65)

Since C_f is one-fourth as large as the Moody friction factor,

$$C_f = \frac{16}{N_{Re}}.$$ (5.66)

The development which led to Eqs. (5.57), (5.58), (5.60), and (5.64) reviews the sequence of steps one normally follows in a fluid mechanics problem involving viscous flow when an exact analytical solution is possible. These steps are: to select, or derive, the governing differential equation forms of the momentum theorem and the law of conservation of mass; to solve these equations for the components of the velocity field as a function of the space coordinates; and, finally, to use the velocity field in the relation connecting the shear stress to the velocity gradients—in our case, Newton's law of viscosity—in order to predict the wall shear stress and the friction factor or skin friction coefficient.

5.3c Turbulent Flow and Heat Transfer

Thus far, only laminar flow has been discussed; that is, orderly layer-type flow. In a laminar flow the only motion transverse to the velocity vector is on the molecular scale. In turbulent flow, which is less ordered, it is found that even when extreme precautions are taken to keep the flow steady, there is still a basic unsteadiness in the flow caused by the motion of the turbulent eddies. An *eddy,* a macroscopic entity, can be viewed as a conglomerate of millions or billions of molecules moving more or less together as a single particle for some time period before being broken up. Because these eddies can move in all directions within a fluid, they constitute an additional mechanism by which transportable quantities, such as momentum and energy, can be transferred within a fluid. *Hence, in a turbulent flow situation, the heat transfer rates and the apparent shear stress are higher than those in an equivalent laminar flow situation.*

If a highly sensitive velocity probe is inserted into a turbulent flow in which everything possible is being done to keep the flow steady, the instantaneous x component of velocity u measured by the probe, when plotted as a function of time t, gives a plot that typically looks like that shown in Fig. 5.7. That is, the local x component of velocity fluctuates about some average value $\bar{u}$, generally with a relatively small amplitude and at a relatively high frequency. The quantity u' which must be added to the average velocity at any instant of time is called the *fluctuation component of velocity* and has a time-averaged value of zero. Hence, the local x component of velocity in a turbulent flow can be written as

$$u = \bar{u} + u'. \tag{5.67}$$

A similar result is found for the y and z components of velocity as well as for quantities such as the temperature and pressure. Hence,

$$v = \bar{v} + v', \tag{5.68}$$

$$w = \bar{w} + w', \tag{5.69}$$

$$T = \bar{T} + T', \tag{5.70}$$

$$P = \bar{P} + P'. \tag{5.71}$$

When the turbulent flow is steady, on the average, the barred quantities in Eqs. (5.67–71) are independent of time and are a function only of the spatial coordinates. In transient turbulent flow, if Eqs. (5.67–71) are used, the barred quantities depend upon time.

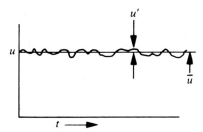

Figure 5.7 Measured fluctuating velocity in turbulent flow as a function of time.

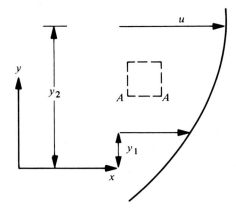

Figure 5.8 Diagram of an instantaneous velocity distribution in a flowing fluid as an aid in the explanation of turbulent transport.

Shear Stress in Turbulent Flow

To see how an additional apparent shear stress is generated by the fluctuation components of velocity which exist in steady (on the average), turbulent flow, consider the flow field shown in Fig. 5.8. For the present, the flow is laminar and all velocity vectors u are oriented in the x direction, although they do, as shown, have different magnitudes. A control volume is shown whose bottom surface of area dA is called surface A-A. In this laminar flow, the x-momentum flux across A-A is zero, because there is no mass flow rate through surface A-A as a result of the fact that the y component of velocity v is zero. Now suppose that conditions have changed so that the flow is *turbulent* and steady (on the average) with all the *average* velocity vectors pointing in the x direction. This means that $\bar{v}$ and $\bar{w}$ are zero, but the fluctuation components v', w', and u' are not zero except on the average. Therefore, since there is a y component of the fluctuation velocity which is not, in general, zero, the instantaneous mass flow rate across plane A-A is

$$\omega' = \rho dA\, v'. \tag{5.72}$$

This fluid which crosses surface A-A has, at any instant of time, the x component of velocity

$$u = \overline{u} + u', \tag{5.67}$$

so that the instantaneous x-momentum flux H_x being transported across A-A is given by the product of Eqs. (5.72) and (5.67) as

$$H_x = \rho dA \, v'\overline{u} + \rho dA \, u'v'. \tag{5.73}$$

In steady (on the average), turbulent flow, average quantities rather than instantaneous quantities are dealt with. (This is part of the reason for using Eqs. [5.67–71].) Therefore, Eq. (5.73) is time-averaged to obtain the average x-momentum flux across plane A-A. The time average of some time-dependent function g is called $\overline{g}$ and is defined by

$$\overline{g} = \frac{1}{t_2 - t_1} \int_{t_1}^{t_2} g \, dt. \tag{5.74}$$

(Note that in this chapter, t, rather than τ, is used to denote time, since here, τ is used to denote shear stress.) Because of the linearity of the integral operator in Eq. (5.74), it is readily seen that the time average of a sum is the sum of the time averages; hence, Eq. (5.73) becomes, after time-averaging both sides,

$$\overline{H}_x = \rho dA \, \overline{v'\overline{u}} + \rho dA \, \overline{u'v'} \,. \tag{5.75}$$

By Eq. (5.74),

$$\overline{v'\overline{u}} = \frac{1}{t_2 - t_1} \int_{t_1}^{t_2} \overline{u}v' dt = \frac{\overline{u}}{t_2 - t_1} \int_{t_1}^{t_2} v' dt,$$

where $\overline{u}$ is independent of time in steady (on the average), turbulent flow and

$$\frac{1}{t_2 - t_1} \int_{t_1}^{t_2} v' dt = \overline{v'} = 0$$

by the definition of the fluctuation component as a function of time whose time average is zero. Hence, only the second term on the right of Eq. (5.75) remains; that is,

$$\rho dA \, \overline{u'v'} = \frac{\rho dA}{t_2 - t_1} \int_{t_1}^{t_2} u'v' dt. \tag{5.76}$$

Although both elements of the product $u'v'$ in Eq. (5.76) have zero time averages separately, their product, in general, does not have a zero time average. This is borne out by experiment and by the following argument:

An eddy of fluid arriving at plane A-A, which originated from a value of y below A-A (therefore having a positive y component of the fluctuation velocity v') has, on the average, a lower x component of velocity than the fluid at A-A (observe the length of the average x velocity vectors in Fig. 5.8). Hence, if this eddy "collides" with the fluid at A-A, it tends to slow it down; that is, it tends to induce a negative x component of velocity u'. Therefore, on the average, it may be said that a positive y component of the fluctuation velocity v' is usually associated with a negative x component of the fluctuation velocity u', and vice versa. So the product $u'v'$ is ordinarily negative and hence would not have a zero average value.

As yet another way of looking at this point, suppose u' happened to be 10 sin t and v' happened to be 3 sin t at some location in the flow. Now, even though the time average of u' and of v' are both zero, the time average of their product, 30 sin^2 t, is obviously *not* zero. As a result of the nonzero time average of the product $u'v'$, Eq. (5.75) for the x-momentum flux transported across plane A-A by the y component of the fluctuation velocity is (on the average)

$$\overline{H}_x = \rho dA \, \overline{u'v'}, \tag{5.77}$$

where $\overline{u'v'}$ is called the *cross-correlation of the fluctuation components of velocity* u' and v'.

Thus, when applying the momentum theorem to the control volume shown in Fig. 5.8 and time-averaging all terms, a momentum flux contribution as given by Eq. (5.77) appears which is not present in laminar flow. However, rather than treat Eq. (5.77) as what it is, an additional x-momentum flux, it is common practice to insert a minus sign in front of it and put it on the side of the momentum equation containing the forces. Therefore, the term $-\rho dA \, \overline{u'v'}$ is treated as an additional apparent force on the control volume in the x direction. Or, dividing by the area dA, the term

$$-\rho \, \overline{u'v'} \tag{5.78}$$

is treated as an additional shear stress τ_t called the *turbulent shear stress* or the *Reynolds stress* or the *apparent shear stress*. Hence,

$$\tau_t = -\rho \, \overline{u'v'} . \tag{5.79}$$

The total shear stress on the face is taken to be the sum of the laminar shear stress due to molecular motion and the turbulent shear stress due to the motion of eddies of macroscopic size. Thus,

$$\tau = \tau_{\text{laminar}} + \tau_t. \tag{5.80}$$

The laminar shear stress is presumed to have the same constitutive equation as in a purely laminar flow, except that derivatives of the time-averaged components of velocity are involved; that is, when Newton's law of viscosity is valid,

$$\tau_{\text{laminar}} = \mu \frac{\partial \overline{u}}{\partial y} .$$

Using this along with Eq. (5.79) in Eq. (5.80) yields

$$\tau = \mu \frac{\partial \overline{u}}{\partial y} - \rho \overline{u'v'}.$$

Factoring out ρ and noting that $\nu = \mu/\rho$,

$$\tau = \rho \left(\nu \frac{\partial \overline{u}}{\partial y} - \overline{u'v'} \right). \tag{5.81}$$

Lack of knowledge of the detailed structure of turbulent flow does not permit the prediction of the quantity $-\overline{u'v'}$ from first principles. A constitutive equation for τ_t, partly because of a rough analogy between the motion of the turbulent eddies and the motion of molecules in a gas, is sometimes written as

$$\tau_t = -\rho \overline{u'v'} = \rho \epsilon_m \frac{\partial \overline{u}}{\partial y}, \tag{5.82}$$

where ϵ_m is called the *eddy diffusivity of momentum* or *turbulent diffusivity of momentum* and is a turbulent transport coefficient which is analogous to the molecular transport coefficient ν. (Compare the first term in Eq. [5.81] with Eq. [5.82].) However, ϵ_m, unlike the molecular transport coefficient ν, is not a property of the fluid, but rather depends upon such conditions as the local mean velocity gradient, possibly some distance from a nearby solid surface, and also upon the previous history of the flow. These dependencies are, in general, very complex and ϵ_m can vary from point to point in the flow. One approach to the entire problem of turbulent flow consists of trying, by a combination of theory and experiment, to formulate expressions for ϵ_m for different types of turbulent flows that have a wide range of validity.

Substituting Eq. (5.82) into Eq. (5.81) and rearranging gives for the total shear stress

$$\tau = \rho(\nu + \epsilon_m) \frac{\partial \overline{u}}{\partial y}. \tag{5.83}$$

The expression for the shear stress written in this fashion explicitly shows the increased apparent shear due to the motion of the turbulent eddies. Experiments have shown that in many situations ϵ_m is 50 to 100 times greater than the molecular transport quantity ν. This factor of 50 to 100 or more would occur at positions in the flow that are not close to the wall; in addition, the actual numerical value of this factor depends strongly upon the Reynolds number of the flow. Close to a wall itself, the turbulence is damped or constrained by the presence of the wall (the no-slip condition in the x direction for the viscous continuum flow of a fluid past a solid surface and the zero y component of velocity required right at the wall because of the impenetrability of the wall material). Therefore, ϵ_m has much smaller values near the wall, with its value going to zero itself right at the wall.

We can see, from this discussion of the size of ϵ_m, that even though the largest values of u' may be only a few percent of $\bar{u}$ in many cases, its effect on the shearing stress is significant and must be accounted for.

Pipe Flow   Experiment shows that the most significant indicator of whether or not fully developed pipe flow is laminar or turbulent is the Reynolds number, based on pipe diameter, previously given as Eq. (5.63). When the Reynolds number is less than about 2000, the pipe flow will be laminar. Hence, the expression developed in an earlier section for the Moody friction factor f in laminar, fully developed pipe flow, Eq. (5.64), is valid only if $N_{Re_D} < 2000$.

When $2000 < N_{Re_D} < 4000$, the pipe flow is in a critical or transition region between laminar flow and turbulent flow. This is a region where the phenomena occurring are very poorly understood. Hence, no analytical or semi-analytical results for f are available. Even experimental evidence is very scanty and widely scattered, so that the uncertainty as to the values of f in this region dictates that one design away from this region.

Finally, for pipes ordinarily used in industrial situations, the pipe flow is generally turbulent when the Reynolds number is greater than about 3000 or 4000; we will use the value 4000 here. Hence, the criteria for fully developed pipe flow are

$$N_{Re_D} < 2000 \quad \text{(laminar flow)},$$
$$N_{Re_D} > 4000 \quad \text{(turbulent flow)}.$$

For fully developed turbulent flow in both smooth and rough pipes, the friction factors have been measured and plotted in Fig. 5.9, the Moody diagram from [11]. The parameter on the chart is relative roughness, the mean height of the roughness elements divided by the pipe diameter. This chart will subsequently prove useful for finding values of the surface coefficient of heat transfer by the analogy between the transport of heat and momentum.

The friction factors for turbulent flow given in Fig. 5.9 are also approximately valid for fully developed flow in ducts of noncircular cross section if the hydraulic diameter D_H is used in the Reynolds number and in the relative roughness. (Note that D_H is equal to the ratio of four times the cross-sectional area A perpendicular to the flow, to the wetted perimeter, the portion of the perimeter over which the major shear stress acts.)

Since the friction factors given in Fig. 5.9 are for fully developed flows, expressions for the hydrodynamic entrance length L_e are given next. These are the analogues, for turbulent flow, of the relations for laminar flow, Eqs. (5.29) and (5.31).

For turbulent flow in a pipe, an entrance length expression first developed by Latzko [12] can be given as follows:

$$\frac{L_e}{D} = 0.69 \, N_{Re_D}^{1/4} . \tag{5.84}$$

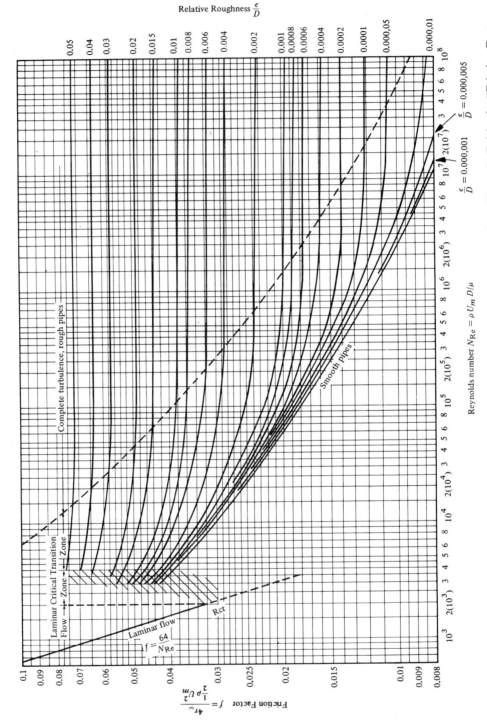

Figure 5.9 Moody diagram. Friction factor and relative roughness for fully developed flow in pipes. (From L. F. Moody, "Friction Factors for Pipe Flow," *Transactions of the A.S.M.E.*, Vol. 66 [1944], pp. 671–84.)

For turbulent flow in a parallel plate channel with walls spaced a distance b units apart, the hydrodynamic entrance length can be given by

$$\frac{L_e}{b} = 1.85 \, N_{Re_b}^{1/4}. \tag{5.84a}$$

Entrance lengths in turbulent flow are subject to more ambiguity than are the corresponding ones for laminar flow. This is due to the dependence of L_e in turbulent flow on the type of duct entrance (sharp lip, bellmouth, etc.), and also on the sense in which *fully developed* is meant. Is it defined in terms of the velocity profile, or the local wall shear, or the average friction factor over length, attaining the ultimate, fully developed value? Some authors recommend that the entrance length in turbulent flow be estimated as about 10 diameters, or as between 10 and 50 diameters. Here, Eq. (5.84) will be used to indicate whether a pipe is long enough to justify the use of the fully developed expressions for friction factor and for velocity profile throughout the vast majority of the pipe's length.

EXAMPLE 5.2

Consider the flow of 20°C water with density $\rho = 998$ kg/m³ and absolute viscosity $\mu = 1.01 \times 10^{-3}$ kg/ms, through a 5-cm-diameter pipe with a mass average velocity of 0.21 m/s. Calculate the actual value of the wall shear stress and compare it to the value that would occur if the flow could be laminar at the same Reynolds number. The pipe is long enough for fully developed flow conditions to prevail.

Solution

The Moody diagram, Fig. 5.9, plots the friction factor versus the diameter Reynolds number at various values of the relative roughness. Since no mention was made of roughness and the pipe material is not given, we will assume that the pipe is hydraulically smooth. Thus, the Reynolds number is now computed as

$$N_{Re_D} = \frac{\rho U_m D}{\mu} = \frac{998(0.21)(0.05)}{1.01 \times 10^{-3}} = 10,375.$$

This value indicates that the flow is turbulent. From Fig. 5.9, for a smooth pipe at this Reynolds number, one obtains $f = 0.031$. From Eq. (5.61), after rearrangement,

$$\tau_w = \frac{f}{8} \rho U_m^2 = \frac{0.031}{8}(998)(0.21)^2 = 0.171 \text{ N/m}^2.$$

If the flow could be maintained as laminar, even at a diameter Reynolds number of 10,375, then Eq. (5.64) would give the friction factor as

$$f_{lam} = 64/10,375 = 0.00617.$$

Thus, the shear stress at the wall would be smaller by the ratio of the friction factors, 0.00617/0.031, and would be 0.034 N/m² instead.

Hence, the effect of turbulence is quite pronounced in this problem; it causes the actual wall shear stress to be about five times the value that would be obtained if the flow could somehow be kept laminar.

Heat Transfer Rates in Turbulent Flow

It has been pointed out that in turbulent flow the heat transfer rates are higher than those which occur when the flow is laminar. To demonstrate this, consider the physical situation shown in Fig. 5.8 again with all the velocity vectors pointing in the x direction and, for the present, laminar flow. If there is a temperature distribution in the y direction, the heat transfer rate across surface A-A of area dA must be by pure conduction, since there is no bulk motion of the fluid in the y direction. Hence, the heat transfer rate q_L across plane A-A when the flow is laminar is, by the exact law of conduction,

$$q_L = -kdA \frac{\partial \overline{T}}{\partial y}.$$

Using the fact that $k/\rho c_p = \alpha$, the thermal diffusivity or molecular diffusivity of heat, the heat transfer rate per unit area is

$$q_L'' = -\rho c_p \alpha \frac{\partial \overline{T}}{\partial y}. \tag{5.85}$$

Suppose that the flow is now turbulent and steady (on the average) with all the average velocity vectors pointing in the x direction. Even though the flow is turbulent, energy still crosses plane A-A in Fig. 5.10 by molecular conduction at the rate given by Eq. (5.85). In addition, the fluctuation component of velocity v' in the y direction causes eddies of macroscopic size to move across plane A-A, carrying with them the energy they had before they crossed the plane. Hence, at some instant of time, an eddy crosses the surface A-A from below at the mass flow rate ω' and carries with it the energy it had when it was formed at some value of y below surface A-A. The energy carried across plane A-A per unit mass is approximately equal to $c_p T_1$, where T_1 is the temperature of fluid at the y location where the eddy emanated. At the same time, on the average, an eddy must cross plane A-A from above in order that mass be conserved (since $\overline{v} = 0$ in this case). The mass flow rate across A-A from above is also ω' (so that mass is conserved) and the energy per unit mass carried by the eddy crossing from above is $c_p T_2$, where T_2 is the temperature at the y location above A-A from which the eddy came. Hence, the *net* energy transfer across the plane in the $+y$ direction is the difference between the rate at which energy crosses the surface A-A from below and the rate at which it crosses surface A-A from above; namely, $\omega' c_p(T_1 - T_2)$. Hence, the heat transfer rate in the y direction is accentuated in turbulent flow because of the physical transport of energy across various planes by the motion of macroscopic particles of fluid, the eddies. The primary purpose of this

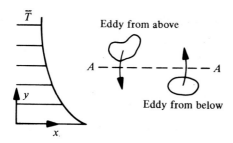

Figure 5.10 Eddy motion in a temperature field causing turbulent transport of energy.

qualitative description is to explain the mechanism by which the heat transfer rate increases in turbulent flow over that in laminar flow. To obtain the equation for the heat transfer rate across A-A due to the motion of the eddies, consider the instantaneous energy transfer rate across plane A-A due to the eddy motion, $\omega' c_p T$. However, $\omega' = \rho dA v'$ and by Eq. (5.70), the instantaneous temperature is $T = \overline{T} + T'$. Hence,

$$\omega' c_p T = \rho dA \, v' c_p \overline{T} + \rho dA \, c_p v' T'. \tag{5.86}$$

The energy transfer rate (on the average) across A-A due to the eddy motion is found by time-averaging the terms in Eq. (5.86) by applying Eq. (5.74) to each term. It is seen that the first term on the right has a time average of zero, since the only quantity depending upon time is the y component of the fluctuation velocity, and by definition its time average is zero; that is, $\overline{v'} = 0$. Time-averaging the second term of Eq. (5.86) leads to

$$\rho dA \, c_p \, \overline{v'T'}. \tag{5.87}$$

Although the time averages of v' and T' are individually zero, the time average of their product, in general, is not zero. Despite the fact that the energy transfer rate in Eq. (5.87) is due to bulk motion of the fluid, it is treated as if it were an apparent conduction term and is added to the actual molecular or laminar conduction term, Eq. (5.85). Dividing first by the area in Eq. (5.87), the heat transfer rate q_t'' in the y direction per unit area as a result of the eddy motion is

$$q_t'' = \rho c_p \, \overline{v'T'}.$$

Combining this with Eq. (5.85), the total "conduction" heat transfer rate per unit area in the y direction is

$$q'' = -\rho c_p \alpha \frac{\partial \overline{T}}{\partial y} + \rho c_p \, \overline{v'T'}. \tag{5.88}$$

Forced Convection Heat Transfer

Using the same reasoning that led to Eq. (5.82) for the apparent shear stress due to the eddy motion, the eddy diffusivity of heat ϵ_H, a turbulent transport quantity analogous to the molecular diffusivity of heat α, is defined as

$$\rho c_p \overline{v'T'} = -\rho c_p \epsilon_H \frac{\partial \overline{T}}{\partial y}.$$

Using this with Eq. (5.88), the expression for the apparent conduction heat transfer rate in a turbulent flow situation in the y direction becomes

$$q'' = -\rho c_p (\alpha + \epsilon_H) \frac{\partial \overline{T}}{\partial y}. \qquad (5.89)$$

In this section, we discussed the mechanism—the mixing motion of the turbulent eddies caused by the fluctuation components of velocity—by which the shearing stress and the heat transfer rate are increased in a turbulent flow. A consideration of the transfer rates caused by the fluctuation velocities and temperature led eventually to Eqs. (5.83) and (5.89), which show explicitly the increased transport rates, due to turbulence, by the presence of the eddy diffusivities ϵ_m and ϵ_H. One way to use these two equations is to employ various models of turbulence to replace the diffusivities ϵ_m and ϵ_H with their equivalents in terms of more fundamental quantities, such as distance from the nearest wall, the local velocity, the local velocity gradient, and so on. When this is done, these equations are inserted into the momentum theorem and into the law of conservation of energy, both in time-averaged form. The result is a set of nonlinear partial differential equations which are then solved by finite difference methods. This is far beyond the scope of this text, and the interested reader can consult [7], [13], and [14] for discussions of the various facets of this approach.

In this text, we will use Eqs. (5.83) and (5.89) in connection with the similarity between heat and momentum transfer. This leads to expressions for the surface coefficient h, under certain conditions, in terms of the friction factor f, which is already available, from experiment, in the Moody diagram, Fig. 5.9.

5.3d Velocity Distributions for Fully Developed Turbulent Flow in a Pipe

The velocity profile for fully developed *laminar* flow in a tube has been derived analytically and is given by Eq. (5.57). A qualitative sketch of the profile shape is given in Fig. 5.11. Also predicted analytically was the friction factor for laminar, fully developed pipe flow, which is found in Eq. (5.64). However, our lack of understanding of the details of the physics of turbulent flow does not allow us to predict, in a truly analytical fashion based on first principles alone, the corresponding quantities in a turbulent flow. In turbulent flow, we must resort to experiment to either measure rather directly the velocity profiles and wall shear stress, or to measure other quantities which are not fundamental properties, but which ultimately lead to velocity distributions and friction factors. This experimental approach in turbulent flow has already led to the friction factors of Fig. 5.9,

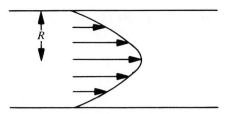

Figure 5.11 A typical fully developed laminar velocity profile in a pipe.

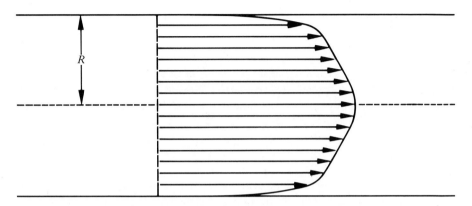

Figure 5.12 A typical fully developed turbulent velocity profile in a pipe.

which were discussed in the previous section. In this section, experimental evidence concerning velocity distributions in fully developed turbulent pipe flow will be presented and discussed. The entrance length expressions that allow one to decide whether the flow is fully developed have been given previously as Eqs. (5.84) and (5.84a).

Measurements of the fully developed turbulent velocity profile in a circular pipe indicate a shape that typically looks as shown in Fig. 5.12. Notice that the velocity profile for turbulent flow is fuller than that for laminar flow; that is, its average velocity is closer to the centerline velocity than in laminar flow. For laminar pipe flow, the average velocity is equal to 50% of the centerline velocity, while in turbulent pipe flow the average velocity is anywhere from about 80% to 90% of the centerline velocity, depending upon the Reynolds number. Curve fits of the experimental data show that the fully developed turbulent velocity profile in pipe flow can, in the range of Reynolds numbers between about 10,000 and 100,000, be adequately represented by the so-called *one-seventh power law profile,*

$$u = U_0(y/R)^{1/7}, \qquad (5.90)$$

where R is the pipe radius, y is the coordinate measured positively in the direction away from the wall, and U_0 is the centerline velocity. Henceforth, unless otherwise indicated, the bars will be dropped from u and T even in turbulent flow, and these quantities will be interpreted as time-average values. Note that this has already been done in Eq. (5.90).

In a region of fluid extremely close to the wall, the low fluid velocities combined with the viscosity μ serve to dampen any turbulent fluctuations and keep the fluid laminar. This region is called the *viscous sublayer,* and Eq. (5.90) is not valid in this region. In particular, Eq. (5.90) does not predict the correct slope which, although very steep, is not infinite as predicted by Eq. (5.90).

Actually, a better and more universally applicable characterization of the velocity distribution in turbulent flow can be arrived at by considering the physical differences between various regions of any cross section of the flow, which are revealed by the relative contribution of laminar and turbulent effects on the local shear stress.

From Eq. (5.80),

$$\tau = \tau_{\text{laminar}} + \tau_t \qquad\qquad [5.80]$$

where τ is the local shear stress and the terms on the right side represent the laminar and turbulent contributions to it, respectively. By use of Newton's law of viscosity and the eddy diffusivity of momentum, ϵ_m, this was rewritten as Eq. (5.83).

$$\tau = \rho(\nu + \epsilon_m)\frac{\partial u}{\partial y} \qquad\qquad [5.83]$$

As was stated in the previous section, near the wall one expects laminar effects to dominate, since ϵ_m goes to zero right at the wall and is smaller than ν at least for some short distance away from the wall. This portion of the flow cross section where laminar (molecular) effects are most dominant is called the *inner layer* or the *viscous layer.* Experimental evidence indicates that far away from the wall $\epsilon_m >> \nu$, and at far enough distances ν can be neglected in comparison to the size of ϵ_m. This region, where turbulent effects completely dominate and determine the value of the local shear in Eq. (5.83), is called the *outer layer* or the *wake region.* As one might expect, between the inner layer where molecular effects are primary, and the outer layer where turbulent effects are primary, there is a region or layer where both laminar and turbulent effects are important and must be considered. This is called the *overlap layer.*

Shear Stress Distribution Over a Pipe Cross Section

Before discussing the form of the velocity profile in the inner, overlap, and outer layers, the shear stress distribution in the radial r direction or the y direction will be investigated for steady, constant property, fully developed turbulent flow in a pipe. This shear stress distribution will be of aid in discussing certain aspects of velocity profiles and will also be used in the theories that relate turbulent heat and momentum transfer.

Consider the control volume of finite size shown in Fig. 5.13. For fully developed flow, it was previously shown that there is no net change in momentum in the x direction; hence, the x-momentum flux out in the x direction equals the x-momentum flux in, and the time rate of change of stored x momentum within the control volume is zero as a result of the

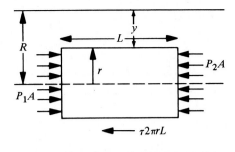

Figure 5.13 Forces on a control volume.

steady-state condition. (Recall that the x-momentum flux across any surface due to the motion of the eddies, if the flow is turbulent, is treated as part of the total shear force on that surface and is accounted for in this way.) For this case, the momentum theorem reduces to the sum, which must be set equal to zero, of all the forces on the control volume in the x direction. These forces are shown in Fig. 5.13. Hence,

$$\sum F_x = P_1\pi r^2 - P_2\pi r^2 - \tau 2\pi rL = 0.$$

Note that in fully developed pipe flow, $\partial u/\partial x = 0$ means that u depends only upon y; hence, dP/dx is a constant and the shear stress does not vary with x. This allows the use of a control volume of finite length L rather than length dx. Solving for τ,

$$\tau = \frac{(P_1 - P_2)}{2L}\, r. \tag{5.91}$$

Setting $r = R$ and then $\tau = \tau_w$, the wall shear stress, τ_w becomes

$$\tau_w = \frac{(P_1 - P_2)}{2L}\, R. \tag{5.92}$$

Dividing Eq. (5.91) by Eq. (5.92) and replacing r by $R - y$, the required variation of shear stress over the cross section is

$$\tau = \tau_w\left(1 - \frac{y}{R}\right). \tag{5.93}$$

We can see that the shear stress in fully developed pipe flow varies linearly over the cross section of the pipe, regardless of whether the flow is laminar or turbulent.

Inner and Overlap Layer Velocity Profiles

Deep within the inner or viscous layer, very close to the wall, $\epsilon_m \ll \nu$, and the part of the inner layer where this is true is called the *viscous sublayer*. Hence, combining the shear distribution of Eq. (5.93) with Eq. (5.83) when ϵ_m can be neglected gives

$$\tau_w \left(1 - \frac{y}{R}\right) = \rho\nu \frac{\partial u}{\partial y} . \tag{5.94}$$

However, deep within the inner layer, practically at $y = 0$ in Fig. 5.13, $y \ll R$, causing Eq. (5.94) to take the following form in the viscous sublayer:

$$\frac{\partial u}{\partial y} = \frac{\tau_w}{\rho\nu} . \tag{5.95}$$

Since τ_w, ρ, and ν do not depend upon y, partial integration with respect to y yields

$$u = \frac{\tau_w}{\rho\nu} y + c(x). \tag{5.96}$$

But at $y = 0$ (the pipe wall), the no-slip condition requires that $u = 0$; hence $c(x) = 0$, and Eq. (5.96) becomes

$$u = \frac{\tau_w}{\rho\nu} y. \tag{5.97}$$

One way that this result can be put into nondimensional form is by defining the *friction velocity* or *shear velocity* $u*$ as

$$u* = \sqrt{\frac{\tau_w}{\rho}} , \tag{5.98}$$

where τ_w is the wall shear stress. The velocity u and the distance from the wall y can be put into nondimensional form using the definitions

$$u^+ = u/u*, \tag{5.99}$$

$$y^+ = yu*/\nu. \tag{5.100}$$

Here u^+ and y^+ are the nondimensional velocity and position coordinate, respectively, and ν is the kinematic viscosity. The shear velocity is defined by Eq. (5.98) as $u* = \sqrt{\tau_w/\rho}$, so $u*^2 = \tau_w/\rho$, and Eq. (5.97) becomes

$$u = \frac{u*^2}{\nu} y, \quad \text{or} \quad \frac{u}{u*} = \frac{u*}{\nu} y. \tag{5.101}$$

However, by Eqs. (5.99–100), $u/u^* = u^+$ and $u^*y/\nu = y^+$. Hence, Eq. (5.101) becomes

$$u^+ = y^+. \qquad\qquad (5.102)$$

The velocity profile in the viscous sublayer is given by Eq. (5.102). We would also like to know how the velocity is distributed within the rest of the inner layer, within the overlap layer, and within the outer layer. Equation (5.102) is as far as pure analysis from first principles can carry us, since in the rest of the pipe cross section, $\epsilon_m \neq 0$.

Experiments indicate that, for fully developed turbulent pipe flow, the overlap layer extends out practically to the pipe centerline; that is, there is no significant separate outer, or wake, region in this type of flow. Furthermore, the experimental data for the inner layer and for the overlap layer, when made nondimensional using Eqs. (5.98), (5.99), and (5.100) are found to plot as a single curve in the u^+ versus y^+ plane, regardless of the diameter Reynolds number. Prandtl had concluded this from a theoretical basis as one consequence of his famous mixing length theory, an account of which is given in [6] and [7].

Thus, it is found that there is a unique relationship of the form

$$u^+ = F(y^+). \qquad\qquad (5.103)$$

Equations of the form of Eq. (5.103) are called the *law of the wall* or the *universal velocity distribution* (although it is certainly not an ordinary velocity distribution, due to the fact that a value of y does not yield a value of u directly because of the presence of the shear stress at the wall τ_w, as shown by Eqs. [5.98–100]). Many different forms have been proposed for Eq. (5.103) which fit the experimental data very well, and seven of these forms are given in [15]. However, only the form presented by von Kàrmàn is given here. In von Kàrmàn's expression for the law of the wall, three different regions of flow at any cross section are defined, and there is a corresponding $F(y^+)$ for each of these regions. The three different regions used by von Kàrmàn were the viscous sublayer, the buffer region—which is the remainder of the earlier-defined inner layer—and the overlap layer. Thus, von Kàrmàn's *law of the wall* is, for flow through a smooth pipe,

$$u^+ = y^+, \text{ where } 0 < y^+ < 5 \text{ (viscous sublayer)};$$
$$u^+ = 5.0 \ln y^+ - 3.05, \text{ where } 5 < y^+ < 30 \text{ (buffer region)}; \qquad (5.104)$$
$$u^+ = 2.5 \ln y^+ + 5.5, \text{ where } y^+ > 30 \text{ (overlap layer)}.$$

As mentioned earlier, there is no significant outer or wake layer in fully developed turbulent pipe flow. Suffice it to say at this point that in the outer layer, u^+ is *not* a function of y^+ alone, and it exhibits a separate dependence upon a pressure gradient parameter.

Thus, as soon as the wall shear stress τ_w is known (recall that it is contained in both u^+ and y^+ by the definitions, Eqs. [5.98–100]), Eq. (5.104) becomes the velocity profile for the turbulent pipe flow. One way to arrive at the needed wall shear stress is by use of the smooth pipe friction factors from the Moody diagram, Fig. 5.9. The other way to arrive at the needed wall shear stress is from Eq. (5.104) itself, as will be shown next.

One of the important implications of the law of the wall in fully developed turbulent duct flow is that it is also an expression for determining the wall shear stress τ_w. This results from the fact that a plot of experimental data of u^+ vs. y^+ must also yield the shear stress, since both u^+ and y^+ contain u^*, which in turn contains τ_w. Hence, the wall shear stress should be obtainable from the law of the wall. In a range of Reynolds numbers from about 10,000 to 100,000, the form of the law of the wall which correlates the experimental data fairly well is

$$u^+ = 8.7(y^+)^{1/7}. \tag{5.105}$$

Whereas Eqs. (5.104) hold for all Reynolds numbers for which the pipe flow is turbulent, Eq. (5.105) is a good approximation to it in the limited Reynolds number range indicated. Because the functional form of (5.105) is simpler than (5.104), the algebra involved in extracting the wall shear stress is simpler. However, the overall procedure to be followed to find τ_w is the same regardless of whether one is using Eq. (5.105) or (5.104).

First, Eqs. (5.99) and (5.100) are inserted into Eq. (5.105) to get the friction velocity, u^*, into the picture so as to eventually find τ_w. Thus,

$$\frac{u}{u^*} = 8.7 \left(\frac{yu^*}{\nu}\right)^{1/7}, \quad \text{or} \quad \frac{u\nu^{1/7}}{8.7y^{1/7}} = u^{*8/7}. \tag{5.106}$$

However, $u^* = \sqrt{\tau_w/\rho}$; hence,

$$(u^*)^{8/7} = \left(\frac{\tau_w}{\rho}\right)^{4/7}. \tag{5.107}$$

Substituting Eq. (5.107) into Eq. (5.106) and raising both sides to the 7/4 power yields

$$\frac{\tau_w}{\rho} = \frac{\nu^{1/4}u^{7/4}}{44y^{1/4}}. \tag{5.108}$$

At the tube centerline, $y = D/2$ and $u = U_0$. Hence, Eq. (5.108) becomes

$$\frac{\tau_w}{\rho} = \frac{\nu^{1/4}U_0^{7/4}}{44(D/2)^{1/4}}. \tag{5.109}$$

The centerline velocity U_0 is not known, but the mass average velocity U_m is ordinarily a known quantity, and U_0 can be related to U_m as will now be shown.

Dividing Eq. (5.105) by the form it has when y is set equal to $D/2$ and u is set equal to U_0 gives, after using Eqs. (5.99) and (5.100),

$$u = U_0 (y/R)^{1/7}. \tag{5.110}$$

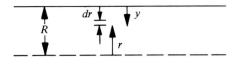

Figure 5.14 The differential area used in the calculation of average velocity.

In terms of U_m, the mass flow rate is given by

$$\omega = \rho \, \pi R^2 U_m.$$ (5.111)

The mass flow through the differential annular area of inner radius r and outer radius $r + dr$ shown in Fig. 5.14 is the product of a density ρ, the area $2\pi r dr$, and the velocity u normal to the area. Hence,

$$d\omega = \rho \, 2\pi r dr \, u,$$

or, in terms of y, since $r = R - y$,

$$d\omega = \rho \, 2\pi (R - y) dy \, u.$$ (5.112)

Summing up all contributions of the type shown in Eq. (5.112) in the definite integral from $y = 0$ to $y = R$ gives the mass flow rate ω, which is also given by Eq. (5.111). Substituting Eq. (5.110) into Eq. (5.112) for u, integrating, and setting the result equal to Eq. (5.111) yields

$$\rho \, \pi R^2 U_m = 2\pi \rho \int_0^R U_0 \left(\frac{y}{R}\right)^{1/7} (R - y) dy,$$

or

$$R^2 U_m = 2U_0 \int_0^R \left(y^{1/7} R^{6/7} - \frac{y^{8/7}}{R^{1/7}}\right) dy$$
$$= 2U_0 \left(\frac{7}{8} y^{8/7} R^{6/7} - \frac{7}{15} \frac{y^{15/7}}{R^{1/7}}\right)\Big|_0^R.$$

Substitution of the limits and rearrangement leads to

$$U_m = 0.81 \, U_0.$$ (5.113)

Eliminating U_0 between Eqs. (5.113) and (5.109) and rearranging the result so that the Reynolds number appears leads to the following expression for the wall shear stress:

$$\tau_w = 0.039 \frac{\rho \, U_m^2}{N_{Re}^{0.25}}.$$ (5.114)

The friction factor is $f = 4\tau_w / \frac{1}{2}\rho U_m^2$. Therefore, using Eq. (5.114) for τ_w yields $f = 0.312/N_{\text{Re}}^{0.25}$. This is within 1.5% of a good curve fit to the experimental friction factor data for turbulent pipe flow at Reynolds numbers less than about 100,000.

Using essentially the same procedure as that which led to Eq. (5.114), the friction factor f that can be determined from von Kàrmàn's form of the law of the wall, Eq. (5.104), is one that is valid over the entire range of turbulent Reynolds numbers in fully developed flow through a smooth pipe. One way the additional algebraic complexity of the more general Eq. (5.104) shows up is in the expression that corresponds to Eq. (5.114), in that the expression must be solved iteratively for τ_w. However, once this is done, the resulting expression for τ_w, when substituted into Eq. (5.104), gives the velocity distribution $u = u(y)$ in the turbulent pipe flow.

The simplest way, however, of having Eq. (5.104) give the velocity profile $u = u(y)$ is to simply find f from the Moody diagram, Fig. 5.9, convert this into a value of τ_w, using the definition of f, and use this τ_w in Eq. (5.104).

EXAMPLE 5.3

In Ex. 5.2, it was found that, when water was flowing in a fully developed fashion through a 5-cm-diameter pipe at a Reynolds number of 10,375, the value of the wall shear was $\tau_w = 0.171$ N/m^2. Also given was the density and viscosity, $\rho = 998$ kg/m^3 and $\mu = 1.01 \times 10^{-3}$ kg/ms, respectively. Find the thickness of the viscous sublayer and compute the percentage of the radius occupied by this sublayer thickness.

Solution
From Eq. (5.104), we can see that the viscous sublayer ends at $y^+ = 5.0$. Using the definition of y^+ given by Eq. (5.100) and solving for y_s at the sublayer edge where y^+ is 5 gives

$$y_s = 5\,\nu / \sqrt{\frac{\tau_w}{\rho}}$$

since u^* is given by Eq. (5.98). Hence,

$$y_s = 5(1.01 \times 10^{-3}/998)/\sqrt{0.171/998} = 0.00039 \text{ m}$$

is the thickness of the viscous sublayer. The percentage of the radius occupied by this sublayer is

$$100(y_s/R) = 100(0.00039/0.025) = 1.56\%.$$

These specific numerical results tend to validate an assumption made earlier in the development of Eq. (5.102); namely, that the thickness of the viscous sublayer is so small that τ is about equal to τ_w by virtue of Eq. (5.93).

5.3e External Flow Over Bodies Immersed in the Flow

In previous sections, the emphasis, as far as solutions and experiments to give the velocity field are concerned, has been on internal flow situations such as flow within a pipe. Only the simplest class of internal flow situations was discussed, namely those internal flows which are fully developed hydrodynamically. At the time we noted that these fully developed internal flows are also the simplest type of viscous flows, in all of fluid mechanics, to handle analytically.

In this section we will consider the class of external viscous flows, such as flow over an airplane wing or over a sphere. Actually, even the viscous flow in the developing entrance region of a duct (that is, in the region for which $x < L_e$ in Fig. 5.4) could be considered to be in the external flow category, because the governing differential equations and boundary conditions are the same as for an external flow because of the presence of the potential core.

Since the fully developed condition has no meaning, or counterpart, in an external flow, external viscous flow problems are far more difficult to handle analytically than are fully developed internal flow problems. Most external viscous flow problems are found to be of the thin boundary layer type which, though allowing major simplifications in the governing partial differential equations, still leave *nonlinear* partial differential equations to be solved for the velocity field. The differential equations for the velocity field in a thin boundary layer–type flow will be derived next by two different approaches.

Derivation of the Boundary Layer Equations

Consider a body such as that shown in Fig. 5.15 immersed in a flowing fluid whose velocity far away from the body is U_s in the indicated direction. Furthermore, suppose the flow is planar two-dimensional, so that the figure is in the x, y plane, and is steady, and constant property. Then, if the flow around the body is laminar, the velocity field is given by the solution to the continuity equation and the Navier-Stokes equations, Eqs. (5.13) and (5.24–26), after dropping the terms containing z and w. This is a formidable set of equations to solve. Fortunately, conditions are often such that the viscous effects (i.e., the shear stresses) are concentrated within a fairly thin layer of fluid along the body called the *boundary layer,* sometimes also called the *velocity boundary layer* or the *hydrodynamic*

Figure 5.15 Hydrodynamic boundary layer development on a body immersed in a flowing fluid.

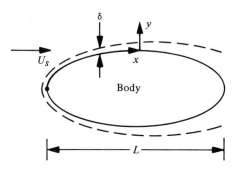

boundary layer. The local boundary layer thickness is denoted by δ and a sketch is shown in Fig. 5.15, with a dashed line indicating the outer extremity of the boundary layer. The region of fluid flow outside the boundary layer which is essentially free of viscous effects is called the *free stream,* the *external flow,* the *invisid flow,* or the *potential flow field.* The equations governing the velocity distribution in the free stream are Eqs. (5.13) and (5.24–26), with the viscosity μ set equal to zero. The resulting equations, with the exception of the continuity equation, are called *Euler's equations of motion.* These equations are the ones solved in classical hydrodynamics, so that any solutions required for the free stream velocity distribution will be assumed known. Now, the equations to be used for finding the velocity field within the boundary layer itself must be found. This concept of a thin boundary layer is due to Prandtl, and *the criterion that the layer is thin, the fundamental boundary layer assumption, is that δ/L << 1, where L is a characteristic length of the body in the general flow direction and δ is the boundary layer thickness at any point on the body.* Assuming that there are no body forces, Eqs. (5.13) and (5.24–26) for steady-state conditions become, respectively, after dropping out all z and w terms,

$$\frac{\partial u}{\partial x} + \frac{\partial v}{\partial y} = 0, \tag{5.115}$$

which is the continuity equation for these conditions,

$$\rho \left(u\frac{\partial u}{\partial x} + v\frac{\partial u}{\partial y} \right) = -\frac{\partial P}{\partial x} + \mu \left(\frac{\partial^2 u}{\partial x^2} + \frac{\partial^2 u}{\partial y^2} \right), \tag{5.116}$$

which is the conservation of momentum in the x direction, and

$$\rho \left(u\frac{\partial v}{\partial x} + v\frac{\partial v}{\partial y} \right) = -\frac{\partial P}{\partial y} + \mu \left(\frac{\partial^2 v}{\partial x^2} + \frac{\partial^2 v}{\partial y^2} \right), \tag{5.117}$$

which is momentum conservation in the y direction. To determine the form these equations take when the boundary layer is thin—when δ/L << 1—use the natural coordinates shown in Fig. 5.15, where x is measured along the body and y is measured normally outward from the body surface. The x, y coordinate system shown in Fig. 5.15 is *not* a rectangular cartesian coordinate system, which is the system for which Eqs. (5.115–17) were developed, but is a general curvilinear coordinate system for which the correct analogues of Eqs. (5.115–17) are much more complicated, containing what are called *curvature terms.* The origin of these additional terms can be most simply explained by saying that they arise from the fact that the unit vectors in the x, y system of Fig. 5.15 do not have a fixed direction, but have directions which change from point to point in the flow field. Fortunately, most bodies of interest have curvatures which are sufficiently gentle to neglect the fact that the unit vectors change direction and, hence, Eqs. (5.115–17) constitute the starting point of two-dimensional, steady, constant property, planar, boundary layer flows. (For additional information on this curvature effect see [4] and [6].)

The implication of $\delta << L$ is that the x derivatives (along the general direction of flow) are much smaller than the y derivatives (transverse to the general direction of flow) of the same quantities, and thus some of the terms in Eqs. (5.115–17) can be neglected in comparison to the remaining ones.

To illustrate this, consider the order of magnitude of the x derivative $\partial u/\partial x$. An estimate of the order of magnitude of $\partial u/\partial x$ is found by noting that u can change from U_s (the approach velocity) in the vicinity of $x = 0$ to a very low value, say zero, at $x = L$, so that

$$\frac{\partial u}{\partial x} \approx \frac{U_s - 0}{L} = \frac{U_s}{L}. \tag{5.118}$$

Now that the order of magnitude of $\partial u/\partial x$ has been established, consider the order of magnitude of the y derivative $\partial u/\partial y$, of the same quantity u. The velocity u can change from zero at the wall to U_s at $y = \delta$, so that roughly,

$$\frac{\partial u}{\partial y} \approx \frac{U_s - 0}{\delta} = \frac{U_s}{\delta}. \tag{5.119}$$

Comparing Eqs. (5.118) and (5.119), we see that when $\delta << L$, the y derivative is much greater than the x derivative; that is, $\partial u/\partial y >> \partial u/\partial x$. Using the same reasoning, it can be shown that

$$\frac{\partial^2 u}{\partial x^2} = \frac{\partial}{\partial x}\left(\frac{\partial u}{\partial x}\right) \approx \frac{U_s/L}{L} = \frac{U_s}{L^2},$$

$$\frac{\partial^2 u}{\partial y^2} = \frac{\partial}{\partial y}\left(\frac{\partial u}{\partial y}\right) \approx \frac{U_s/\delta}{\delta} = \frac{U_s}{\delta^2}.$$

Hence, $\partial^2 u/\partial y^2 >> \partial^2 u/\partial x^2$, and since these two terms appear added together on the right-hand side of Eq. (5.116), $\partial^2 u/\partial x^2$ can be neglected. Hence, Eq. (5.116) reduces to

$$\rho\left(u\frac{\partial u}{\partial x} + v\frac{\partial u}{\partial y}\right) = -\frac{\partial P}{\partial x} + \mu\frac{\partial^2 u}{\partial y^2}. \tag{5.120}$$

Note that, as yet, nothing can be done to the left-hand side of this equation, even though $\partial u/\partial y >> \partial u/\partial x$, because each is multiplied by a different quantity for which the order of magnitude has not yet been determined.

Thus far, the order of magnitude study of Eq. (5.116) for a thin boundary layer–type flow has shown that one of the terms in Eq. (5.116) is negligible for such a flow, and therefore the simpler result, Eq. (5.120), applies. Next, we continue the order of magnitude study to see whether any of the terms on the left side of Eq. (5.120) can be dropped because of being very small.

By Eq. (5.115),

$$\frac{\partial u}{\partial x} + \frac{\partial v}{\partial y} = 0. \tag{5.115}$$

The order of magnitude of $\partial u / \partial x$ has already been determined to be

$$\frac{\partial u}{\partial x} \approx \frac{U_s}{L} .$$

In addition,

$$\frac{\partial v}{\partial y} \approx \frac{V_s - 0}{\delta} ,$$

since v changes from 0 at $y = 0$, the body surface, to some value V_s at $y = \delta$. Using Eq. (5.115), it is seen that $\partial v / \partial y \approx U_s / L$; hence, $V_s / \delta \approx U_s / L$, or $V_s \approx (\delta / L) U_s$. The order of magnitude of v is then $v \approx (\delta / L) U_s$.

Since the order of magnitude of u is about U_s, the order of magnitude of the left-hand side of Eq. (5.120) is

$$\rho \left(u \frac{\partial u}{\partial x} + v \frac{\partial u}{\partial y} \right) \approx \rho \left[U_s \left(\frac{U_s}{L} \right) + \frac{\delta U_s}{L} \left(\frac{U_s}{\delta} \right) \right] .$$

(Recall that $\partial u / \partial y \approx U_s / \delta$ was found previously.) Hence, the order of magnitude of the left-hand side is

$$\rho \left(\frac{U_s^2}{L} + \frac{U_s^2}{L} \right)$$

and, since both terms $u(\partial u / \partial x)$ and $v(\partial u / \partial y)$ are of the same order of magnitude, both must be retained on the left-hand side of Eq. (5.120).

The question now arises as to whether the entire left-hand side of Eq. (5.120) is much different in order of magnitude from the terms on the right-hand side. Since at present there appears to be no way to establish the order of magnitude of the viscosity, it can only be said that the viscous term should be retained. If it is not, this is equivalent to saying the shear stresses are not important for the calculation of the flow field in the boundary layer. Yet it is reasonable to anticipate that they are important; hence, the term is retained. Equation (5.120) with μ set equal to zero is assumed to govern the flow in the free stream outside the boundary layer; that is, outside of the region where the important viscous effects are concentrated. Thus, for $y > \delta$,

$$\rho \left(u \frac{\partial u}{\partial x} + v \frac{\partial u}{\partial y} \right) = -\frac{\partial P}{\partial x} .$$

But when $y > \delta$, $\partial u / \partial y$ is very small, for if it were not, an appreciable shear stress could arise by Newton's law of viscosity. Hence, for $y > \delta$,

$$\rho u \frac{\partial u}{\partial x} = -\frac{\partial P}{\partial x} .$$

The order of magnitude of the left-hand side could conceivably be about the same as for the like term within the boundary layer, so that $\partial P/\partial x$ could be of the same order of magnitude as the terms on the left-hand side of Eq. (5.120), and they will be retained in that equation.

On the strength of these arguments, Eq. (5.120) cannot be further simplified. Therefore, it becomes the form of the x-momentum equation valid for thin boundary layer flows. Next, an order of magnitude study will be done for the y-momentum equation, Eq. (5.117), to see what simpler form it takes for a thin boundary layer.

It has been found that $u \approx U_s$ and $v \approx \delta U_s/L$. Hence,

$$\frac{\partial v}{\partial x} \approx \frac{\delta U_s/L}{L} = \frac{\delta U_s}{L^2},$$

$$\frac{\partial^2 v}{\partial x^2} = \frac{\partial}{\partial x}\left(\frac{\partial v}{\partial x}\right) \approx \frac{\delta U_s/L^2}{L} = \frac{\delta U_s}{L^3},$$

$$\frac{\partial^2 v}{\partial y^2} = \frac{\partial}{\partial y}\left(\frac{\partial v}{\partial y}\right) \approx \frac{U_s/L}{\delta} = \frac{U_s}{\delta L}.$$

Thus, on an order of magnitude basis, Eq. (5.117) becomes

$$\rho\left[U_s\left(\frac{\delta U_s}{L^2}\right) + \frac{\delta U_s}{L}\left(\frac{U_s}{L}\right)\right] = -\frac{\partial P}{\partial y} + \mu\left(\frac{\delta U_s}{L^3} + \frac{U_s}{\delta L}\right).$$

On the right-hand side, $\delta U_s/L^3 << U_s/\delta L$ when $\delta << L$, so that the second term on the right can be dropped. Following the same reasoning used in the preceding discussion, we conclude that the remaining terms on the right-hand side are of the same order of magnitude as those on the left-hand side, $\rho(\delta U_s^2/L^2)$. Hence every term in Eq. (5.117), the y-momentum equation, is of the order $(\delta/L)[\rho(U_s^2/L)]$. But earlier it was shown that every term in the x-momentum equation, Eq. (5.120), was of the order $\rho(U_s^2/L)$. Since $\delta << L$, *every term in the y-momentum equation is much smaller than every term in the x-momentum equation;* hence, the former will be neglected in comparison with the latter. Since every term in the y-momentum equation can be neglected, $\partial P/\partial y \approx 0$; that is, the pressure does not change significantly through the thin boundary layer and is equal to the pressure in the free stream just outside the boundary layer. But, as mentioned earlier, the free stream conditions are considered known from classical hydrodynamics solutions for the ideal (frictionless, no viscosity) flow about the bodies of interest. Hence, P as a function of x is considered to be specified. Thus the order of magnitude study shows that for a thin boundary layer, in which $\delta << L$, the y-momentum equation is unnecessary. In view of this, the so-called *boundary layer equations* become

$$\frac{\partial u}{\partial x} + \frac{\partial v}{\partial y} = 0, \qquad\qquad\qquad \textbf{(5.121)}$$

which is the continuity equation, and

$$\rho \left(u\frac{\partial u}{\partial x} + v\frac{\partial u}{\partial y} \right) = -\frac{\partial P}{\partial x} + \mu\frac{\partial^2 u}{\partial y^2} , \qquad (5.122)$$

which is x momentum conservation.

As soon as one has the pressure gradient which appears in Eq. (5.122), $\partial P/\partial x$, from the potential or invisid solution for flow over the body of interest, Eqs. (5.121) and (5.122) constitute a set of two mutually coupled nonlinear partial differential equations whose solution is the velocity field $u = u(x, y)$ and $v = v(x, y)$ in steady, laminar, constant property, two-dimensional planar, thin boundary layer–type flow.

More rigorous and detailed derivations of the boundary layer equations are available in [6], [16], and [17].

Alternate Derivation of the Boundary Layer Equations

We have just shown how the boundary layer equations, Eqs. (5.121) and (5.122), may be derived from the more complicated and more general Navier-Stokes relations, Eqs. (5.24) to (5.26), and the continuity relation, Eq. (5.13), by studying the order of magnitude of all terms in these equations while considering the condition for a thin boundary layer, $\delta/L \ll 1$. Now we would like to show the development of Eqs. (5.121) and (5.122) by use of first principles. That is, we will apply the word form of the conservation laws, mass and momentum, to a differential control volume within the boundary layer. This procedure has already been carried out for the law of conservation of mass and culminated in Eq. (5.13) which, when $\partial w/\partial z$ is set equal to zero in a two-dimensional flow, becomes Eq. (5.121). Hence, we are left with the task of deriving the x-momentum equation for a thin boundary layer–type flow.

In going through the derivation of Eq. (5.122) by first principles, it then should be clear how the more general Navier-Stokes results, Eqs. (5.24) to (5.26), were derived, since the procedure for doing so was only outlined earlier in the chapter. In addition, the mechanics of this derivation are essentially the same as one must employ when the momentum theorem is applied to other control volumes, such as ones which are differential in extent in one direction and of finite extent in all other directions.

Shown in Fig. 5.16(a) is a differential control volume within a steady, laminar, constant property, two-dimensional planar, thin boundary layer, to which the momentum theorem will be applied.

In using the momentum theorem, the sum of all the forces on the control volume in the x direction is required. Hence, Fig. 5.16(b) shows the pressure forces and shear forces after the appropriate Taylor series expansions have been completed. Because of the thin boundary layer, it was shown that $\partial u/\partial y \gg \partial u/\partial x$. Thus, no normal force due to viscosity is shown in the x direction because the stress which causes this force is proportional to $\partial u/\partial x$, and this is negligible compared to $\partial u/\partial y$, which the shear stress τ depends upon.

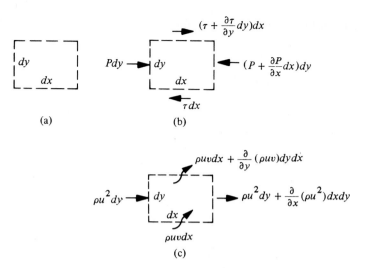

Figure 5.16 (a) The control volume within the boundary layer. (b) Forces on the control surface. (c) Momentum flux across the control surface.

Therefore,

$$\Sigma F_x = Pdy + \left(\tau + \frac{\partial \tau}{\partial y}\, dy\right) dx - \tau dx - \left(P + \frac{\partial P}{\partial x}\, dx\right) dy,$$

or, simplifying,

$$\Sigma F_x = \left(\frac{\partial \tau}{\partial y} - \frac{\partial P}{\partial x}\right) dxdy. \tag{5.123}$$

The momentum theorem also requires finding the x-momentum flux out of the control volume minus the x-momentum flux into the control volume. In general, there is x-momentum flux crossing every portion of the control surface which mass crosses. Hence, in Fig. 5.16(c), the x-momentum flux crossing the left-hand face into the control volume is the product of the mass flow rate in times the local x component of velocity, or,

$$\rho u dy\, u = \rho u^2 dy.$$

There is also x-momentum flux leaving at the right-hand face at $x + dx$. This can be put in terms of that crossing the face at x by a Taylor series expansion of $\rho u^2 dy$ about x across to dx as

$$\rho u^2 dy + \frac{\partial}{\partial x}(\rho u^2)\, dydx.$$

Also, x-momentum flux enters the bottom face at y and is

$$\rho v dx\, u = \rho u v dx,$$

and x-momentum flux leaves at the upper face with the value, by a Taylor series expansion about y,

$$\rho u v dx + \frac{\partial}{\partial y}(\rho u v)\, dy dx.$$

Hence, the expression for the x-momentum flux out minus the x-momentum flux in is

$$\rho u v dx + \rho \frac{\partial}{\partial y}(uv)\, dx dy + \rho u^2 dy + \rho \frac{\partial}{\partial x}(u^2)\, dy dx$$
$$- \rho u v dx - \rho u^2 dy,$$

which, after cancelling like terms, becomes

$$\rho \left[\frac{\partial}{\partial y}(uv) + \frac{\partial}{\partial x}(u^2) \right] dx dy. \qquad (5.124)$$

Since the momentum theorem requires that in the steady state the net force on the control volume in the x direction, Eq. (5.123), equal the net momentum flux out in the x direction, Eqs. (5.123) and (5.124) may be equated to yield

$$\rho \left[\frac{\partial}{\partial x}(u^2) + \frac{\partial}{\partial y}(uv) \right] = -\frac{\partial P}{\partial x} + \frac{\partial \tau}{\partial y}. \qquad (5.125)$$

However, by Newton's law of viscosity, τ is equal to $\mu(\partial u/\partial y)$. [More correctly, $\tau = \mu(\partial u/\partial y + \partial v/\partial x)$; but, the thin boundary layer yields $\partial v/\partial x \ll \partial u/\partial y$; hence the form used.] Thus,

$$\frac{\partial \tau}{\partial y} = \mu \frac{\partial^2 u}{\partial y^2}.$$

Substituting this into Eq. (5.125) and expanding the left-hand side yields

$$\rho \left(2u\frac{\partial u}{\partial x} + u\frac{\partial v}{\partial y} + v\frac{\partial u}{\partial y} \right) = -\frac{\partial P}{\partial x} + \mu \frac{\partial^2 u}{\partial y^2}.$$

Rearranging the quantity in brackets gives

$$\rho \left[u\frac{\partial u}{\partial x} + v\frac{\partial u}{\partial y} + u\left(\frac{\partial u}{\partial x} + \frac{\partial v}{\partial y} \right) \right] = -\frac{\partial P}{\partial x} + \mu \frac{\partial^2 u}{\partial y^2}.$$

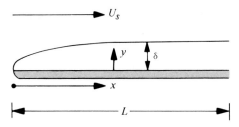

Figure 5.17 Hydrodynamic boundary layer development on a flat plate.

But by the continuity equation, Eq. (5.121), $\partial u / \partial x + \partial v / \partial y = 0$, so that the term in parentheses vanishes and the final result is

$$\rho \left(u \frac{\partial u}{\partial x} + v \frac{\partial u}{\partial y} \right) = -\frac{\partial P}{\partial x} + \mu \frac{\partial^2 u}{\partial y^2} . \qquad \text{[5.122]}$$

This is the same equation derived earlier by reducing the Navier-Stokes equation using an order of magnitude analysis which employed the fundamental boundary layer theory assumption, that is, $\delta \ll L$.

Now that Eqs. (5.121) and (5.122) have been developed to determine the velocity field in thin boundary layer–type flows, the next step is to seek solutions to these equations. Some solution techniques for them will be discussed in the next two sections.

5.3f An Exact Solution to the Boundary Layer Equations

Consider the steady, laminar, constant property, planar two-dimensional flow over a flat plate at zero angle of attack to the flow, as shown in Fig. 5.17. If the flow is of the boundary layer type, it is desirable to determine the x and y components of the velocity as functions of the space coordinates, and also the skin friction coefficient C_f, defined by

$$C_f = \frac{\tau_w}{\frac{1}{2} \rho U_s^2} . \qquad (5.126)$$

The governing equations are Eqs. (5.121–22), the boundary layer equations with $\partial P / \partial x = 0$. The pressure distribution impressed on the boundary layer is essentially[1] that produced by invisid flow over the body; but, since the flat plate has essentially no thickness,

---

1. The word *essentially* is inserted because the slowing down of the fluid due to viscosity in the boundary layer actually presents a body which appears to be of slightly different shape and thickness to the external potential flow, and hence modifies the potential flow pressure and velocity distribution. This effect, usually small, can be taken into account by adding the local displacement thickness to the body and recalculating the potential flow distribution and, from this, the flow in the boundary layer, using an iterative procedure until the apparent body shape stops changing. Because of the effect of displacement thickness, even the flat plate sustains a small induced pressure gradient. More detailed information can be found in [6] and [18].

it does not disturb an invisid flow, and hence produces no pressure variation or variation in the free stream velocity U_s. Thus, Eqs. (5.121–22) become

$$\frac{\partial u}{\partial x} + \frac{\partial v}{\partial y} = 0, \quad \rho\left(u\frac{\partial u}{\partial x} + v\frac{\partial u}{\partial y}\right) = \mu\frac{\partial^2 u}{\partial y^2}, \tag{5.127}$$

with boundary conditions,

$$\text{at } x = 0 \text{ and } 0 < y < \infty, \quad u = U_s,$$
$$\text{at } y = 0 \text{ and } 0 < x < L, \quad u = 0 \text{ and } v = 0,$$
$$\text{as } y \rightarrow \infty, \quad\quad\quad\quad u \rightarrow U_s.$$

The exact analytical method for solving certain sets of nonlinear partial differential equations subject to certain special forms of the boundary conditions is called a *similarity transformation*. In effect, this method is a way to reduce, using a new independent variable called the *similarity variable,* the set of nonlinear partial differential equations to a set of nonlinear ordinary differential equations which are amenable to solution by numerical methods. The finding of this similarity variable, if indeed one even exists, is still partly an art. (See [19] for the free-parameter technique, which is one of the best general methods for determining whether a similarity transformation exists and, if so, the form of the similarity variable.) Only the results of applying this method to Eqs. (5.127) will be discussed. It can be shown that using

$$u = U_s f'(\eta) = U_s \frac{df}{d\eta}, \quad \eta = y\sqrt{\frac{U_s}{vx}},$$

Eqs. (5.127) and the boundary conditions reduce to the following problem involving an ordinary differential equation (see [19]):

$$ff'' + 2f''' = 0,$$
$$\text{at } \eta = 0, \quad f = 0 \text{ and } f' = 0, \tag{5.128}$$
$$\text{as } \eta \rightarrow \infty, \quad f' \rightarrow 1.$$

(Note that primes indicate differentiation of the quantity with respect to the similarity variable η.) These equations were first solved by Blasius (1908), and subsequently by Howarth [20]. Only the result for the boundary layer thickness and the skin friction coefficient are given here. Although the solution for u extends quite far in the y direction, it reaches 99% of U_s very quickly and this distance from the wall, at any x, at which $u = 0.99U_s$ is defined as the *boundary layer thickness* δ. The result is

$$\delta = \frac{5.0x}{\sqrt{N_{Re_x}}}, \tag{5.129}$$

where N_{Re_x}, the Reynolds number based on the distance x away from the leading edge, is

$$N_{Re_x} = \frac{\rho\, U_s x}{\mu}. \tag{5.130}$$

The result for the local skin friction coefficient, defined by Eq. (5.126), is

$$C_f = \frac{0.664}{\sqrt{N_{Re_x}}}. \tag{5.131}$$

Often in a fluid mechanics analysis, the average skin friction coefficient $\overline{C}_f$ is needed so that one can, in a direct fashion, compute the total drag force on the plate. The value $\overline{C}_f$ is defined as follows, where $\overline{\tau}_w$ is the average wall shear stress over the plate surface:

$$\overline{C}_f = \frac{\overline{\tau}_w}{\frac{1}{2}\rho U_s^2}. \tag{5.132}$$

The total viscous force on the plate per unit of length in the z direction is the average shear stress $\overline{\tau}_w$ multiplied by the plate area, $L \times 1$ unit; hence,

$$F = \overline{\tau}_w L. \tag{5.133}$$

Another expression for the total force is obtained by adding all the small shear forces dF which act on areas dx in the definite integral $F = \int dF$. But, dF is the local shear stress τ_w times the area dx over which it acts; thus,

$$F = \int_0^L \tau_w\, dx.$$

Equating this with Eq. (5.133) yields

$$\overline{\tau}_w L = \int_0^L \tau_w\, dx. \tag{5.134}$$

The local skin friction coefficient, Eq. (5.131), is

$$C_f = \frac{\tau_w}{\frac{1}{2}\rho U_s^2} = \frac{0.664}{\sqrt{N_{Re_x}}} = \frac{0.664 x^{-1/2}}{\sqrt{\rho U_s/\mu}}.$$

Solving for τ_w leads to

$$\tau_w = \tfrac{1}{2}\,\rho U_s^2 \left(\frac{0.664}{\sqrt{\rho U_s/\mu}}\right) x^{-1/2}.$$

Substituting this into Eq. (5.134) and rearranging gives

$$\overline{C}_f = \frac{\overline{\tau}_w}{\tfrac{1}{2}\rho U_s^2} = \frac{0.664}{L\sqrt{\rho U_s/\mu}} \int_0^L x^{-1/2}\, dx.$$

Integrating and rearranging, the final result for the average skin friction coefficient over the plate is

$$\overline{C}_f = \frac{\overline{\tau}_w}{\tfrac{1}{2}\rho U_s^2} = \frac{1.328}{\sqrt{N_{\mathrm{Re}_L}}}. \tag{5.135}$$

A procedure which parallels that which was just used to go from the local skin friction coefficient, Eq. (5.131), to the average skin friction coefficient, Eq. (5.135), will also need to be used in forthcoming sections, both in this and other chapters, to find the average surface coefficient of heat transfer for a body when the local surface coefficient is known as a function of position on the body.

Criterion for a Thin Boundary Layer

In reducing the Navier-Stokes equations to the boundary layer equations, the condition $\delta << L$ for a thin boundary layer was invoked and used to determine the order of magnitude of the individual terms in the equation. It was noted that this condition was satisfied in a large number of practical situations, but it was not determined why $\delta << L$ is correct. From Eq. (5.129), *it is seen that for $\delta << L$, the Reynolds number N_{Re_L} must not be too small; that is, $N_{\mathrm{Re}_L} >> 1$ for $\delta << L$.* Actually, to ensure that the boundary layer approximations are valid locally at every x, it is required that $\delta << x$, and it is seen, by rearranging Eq. (5.129) in the form

$$\frac{\delta}{x} = \frac{5.0}{\sqrt{N_{\mathrm{Re}_x}}},$$

that near $x = 0$, the leading edge of the plate, δ can be much greater than x. Hence, in this leading edge region, the boundary layer approximations are not valid. The x direction derivatives are not small compared to the y direction and cannot be neglected as was done to arrive at Eq. (5.122). This can also be seen directly from the physics of the situation. Referring to Fig. 5.17, fluid at a value of y almost zero and just upstream of the leading edge is moving at about the speed U_s when, after entering the boundary layer, it must decrease its x component of velocity almost to zero in practically zero distance; hence,

$\partial u/\partial x$ is very large in the leading edge region and could not be neglected in comparison to, say $\partial u/\partial y$. Fortunately, in most cases the leading edge region is an extremely small fraction of the total plate length L, and the fact that the boundary layer solution is not strictly correct in that region is of negligible consequence elsewhere along the vast majority of the plate surface.

This section was an introduction to exact solution techniques for the boundary layer equations. Further study of this solution technique is beyond the scope of topics normally covered in a first course in heat transfer. The bulk of the effort to solve the boundary layer equations analytically will, in this book, be devoted to the approximate integral methods which are presented in the next section. However, exact analytical solutions of the boundary layer equations by the method of the present section, the similarity transformation, have been found for some situations of interest other than the flat plate. These solutions are available in [6], [17], [18], and [19].

5.3g Integral Methods

The approximate analytical methods for solving boundary layer–type problems in fluid flow and heat transfer are usually called *integral methods*.[2] These methods are important because in a great many cases they yield accurate results with a reasonable amount of effort and can be a powerful tool for the engineer. Also, the rudiments of the method can be learned rather easily compared with, say, the method of similarity transformations.

Integral methods will be carefully developed in some detail in this section, not only because of their application to the prediction of velocity fields in thin boundary layer–type flows, but also because they will be applied again later in the chapter, to predict the temperature field within the moving fluid which leads to the local surface coefficient of heat transfer by using Eq. (5.3).

Derivation of the Integral Equations

The first step in the integral technique is to derive the governing integral forms of the conservation equations. This is done by applying the conservation laws to a control volume. For planar two-dimensional thin boundary layer problems, the most useful control volume is one that is differential in extent in the x direction (i.e., the principal flow direction) and finite in extent in the y direction, encompassing all of the boundary layer in the y direction. Such a control volume is shown on a general body in Fig. 5.18, where the boundary layer thickness has been greatly exaggerated, relative to other dimensions, in order to more clearly show the details. The left-hand face of the control volume at position x is higher than the local boundary layer thickness δ at x and, therefore, extends slightly into the free stream, while the right-hand face at $x + dx$ has a height equal to the boundary layer thickness at $x + dx$. One unit of z direction (perpendicular to the plane of the

---

2. These integral methods are one type of approximate procedure which can be classed under the "method of weighted residuals" [21]. As used in heat transfer (the von Kàrmàn-Pohlhausen procedure), the integral methods are a hybrid mix of other approximate techniques that fall under the method of weighted residuals, such as collocation and the method of moments.

Free stream

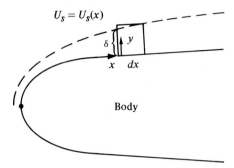

Figure 5.18 Control volume for the derivation of the integral equations of the hydrodynamic boundary layer on a general body.

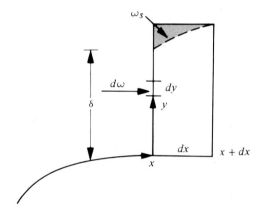

Figure 5.19 Details of the control volume used to derive the integral form of a continuity equation.

figure) is considered and, since the body has an arbitrary shape, the pressure will be a function of x as is the local velocity outside the boundary layer $U_s = U_s(x)$. The flow is constant property and steady.

We will first consider the derivation of the integral equation form of the law of conservation of mass. The control volume is shown in more detail in Fig. 5.19. In the steady state, the rate at which mass enters the control volume must equal the rate at which mass leaves the control volume. Mass enters the control volume at the left-hand face from *within* the boundary layer at the total rate of ω units of mass per unit time. Now ω can be put in terms of the local x component of velocity at the face x by first finding the rate $d\omega$ at which mass enters the control volume through the differential area dy *within* the boundary layer.

$$d\omega = \rho \, dy \, u \tag{5.136}$$

Since ω is the sum of all contributions of the type shown in Eq. (5.136), integrating from $y = 0$ to $y = \delta$ yields

$$\omega = \int_0^\delta \rho u \, dy. \tag{5.137}$$

Mass leaves the control volume through the boundary layer at the right-hand face at $x + dx$ at the rate ω_{x+dx}. But ω_{x+dx} can be put in terms of ω at x by a Taylor series expansion of the function ω (the expression for the local mass flow rate at any x through the boundary layer) about x across to $x + dx$. Neglecting higher-order terms, the result is

$$\omega_{x+dx} = \omega + \frac{d\omega}{dx} \, dx. \tag{5.138}$$

Using Eq. (5.137) in Eq. (5.138) gives

$$\omega_{x+dx} = \int_0^\delta \rho u \, dy + \frac{d}{dx}\left(\int_0^\delta \rho u \, dy\right) dx. \tag{5.139}$$

Integrals appear in Eq. (5.139) because the integral in Eq. (5.137) represents the function of x that is the local mass flow rate within the boundary layer.

No mass crosses the bottom face, because in this case the body surface is impervious to mass flow. In general, mass would be entrained by the boundary layer at the upper face, between x and $x + dx$, and the net rate at which mass enters the boundary layer from the free stream between x and $x + dx$ is called ω_s. Hence, equating the rate at which mass enters the control volume of Fig. 5.19 to the rate at which it leaves yields

$$\omega_s + \int_0^\delta \rho u \, dy = \int_0^\delta \rho u \, dy + \frac{d}{dx}\left(\int_0^\delta \rho u \, dy\right) dx.$$

Cancelling like terms, the integral form of the law of conservation of mass for the control volume in Fig. 5.19 for steady, constant property, planar two-dimensional flow is

$$\omega_s = \frac{d}{dx}\left(\int_0^\delta \rho u \, dy\right) dx. \tag{5.140}$$

Equation (5.140) is valid regardless of whether the flow is laminar or turbulent, and since ρ was never taken out of the integral or the differential operators, it is valid even for the flow of a compressible fluid.

Next, the integral form of the x-momentum equation will be derived by applying the word form of the momentum theorem to the control volume in Fig. 5.19. First the forces on the control volume in the x direction are summed. These forces are shown in Fig. 5.20 as the shear force on the lower face, a pressure force on the left face (note that P does

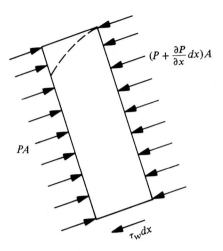

Figure 5.20 Forces on the control volume used to derive the integral form of a momentum equation.

not vary with y within the boundary layer as a result of the layer being so thin that $\partial P / \partial y \approx 0$ as discussed earlier), and another pressure force on the right face. There is no shear force shown on the upper face because this face is just outside the boundary layer in the free stream where the viscous effects are negligible by definition of the boundary layer; also, no normal stresses due to viscosity are shown either on the face at x or at $x + dx$ as a result of their being very small due to $\partial u / \partial x \ll \partial u / \partial y$ and, therefore, neglecting $\partial u / \partial x$ by comparison. There is a component of the weight of the material inside the control volume in the x direction, but this force ordinarily can be neglected in comparison to the other forces. Hence,

$$\sum F_x = PA - \left(P + \frac{\partial P}{\partial x} dx \right) A - \tau_w \, dx$$
$$= -A \frac{\partial P}{\partial x} \, dx - \tau_w \, dx. \qquad (5.141)$$

There is x-momentum flux into the control volume from *within* the boundary layer at the left face, at x, as a result of mass entering with an x component of velocity. The momentum flux entering through the differential area dy is

$$\rho u \, dy \, u = \rho u^2 dy, \qquad (5.142)$$

as is shown in Fig. 5.21. The total momentum flux H_x entering the left face from within the boundary layer, found by summing small contributions of the type shown in Eq. (5.142) in the definite integral from $y = 0$ to $y = \delta$, is

$$H_x = \int_0^\delta \rho u^2 dy. \qquad (5.143)$$

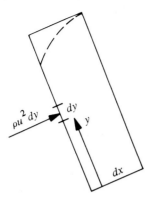

Figure 5.21 Momentum flux across the differential area.

The x-momentum flux leaving the control volume across the right face at $x + dx$, called H_{x+dx}, can be put in terms of H_x by a Taylor series expansion of the function for the x-momentum flux within the boundary layer at any value of x. Thus,

$$H_{x+dx} = H_x + \frac{dH_x}{dx}\,dx. \tag{5.144}$$

Using Eq. (5.143) in Eq. (5.144) leads to

$$H_{x+dx} = \int_0^\delta \rho u^2 dy + \frac{d}{dx}\left(\int_0^\delta \rho u^2 dy\right) dx. \tag{5.145}$$

Now, x-momentum flux also enters the control volume at the upper surface, between x and $x + dx$, because the mass which enters at the rate ω_s carries with it the x component of velocity it had before it crossed the control surface. That x component of velocity is approximately equal to U_s, so that the x-momentum flux entering at the upper surface is

$$\omega_s U_s. \tag{5.146}$$

The difference between the x-momentum flux leaving and entering the control volume is Eq. (5.145) minus the sum of Eqs. (5.146) and (5.143); that is,

$$\frac{d}{dx}\left(\int_0^\delta \rho u^2 dy\right) dx - \omega_s U_s.$$

Setting this equal to the sum of all the forces on the control volume, Eq. (5.141), yields

$$\frac{d}{dx}\left(\int_0^\delta \rho\, u^2 dy\right) dx - \omega_s U_s = -A\frac{\partial P}{\partial x}\,dx - \tau_w\,dx. \tag{5.147}$$

From Eq. (5.140),

$$\omega_s = \frac{d}{dx} \left(\int_0^\delta \rho u \, dy \right) dx.$$

Substituting this into Eq. (5.147) and dividing by dx and ρ yields

$$\frac{d}{dx} \int_0^\delta u^2 dy - U_s \frac{d}{dx} \int_0^\delta u \, dy = -\frac{1}{\rho} \frac{\partial P}{\partial x} A - \frac{\tau_w}{\rho}. \qquad (5.148)$$

Now, $\partial P/\partial x$ can be related to $U_s(x)$ by the x-momentum equation outside the boundary layer, where the effects of viscosity are so small that μ can be set equal to zero. Setting $\mu = 0$ in Eq. (5.122), and neglecting $\partial u/\partial y$ for $y > \delta$ yields

$$\rho U_s \frac{\partial U_s}{\partial x} = -\frac{\partial P}{\partial x}.$$

Substituting this into Eq. (5.148) and rearranging gives

$$U_s \frac{d}{dx} \int_0^\delta u \, dy - \frac{d}{dx} \int_0^\delta u^2 dy + U_s \frac{dU_s}{dx} A = \frac{\tau_w}{\rho}, \qquad (5.149)$$

and, since $A = \delta + d\delta$, the term containing $d\delta$ will be much smaller than any other in Eq. (5.149); hence, $A \approx \delta$, and this can be rewritten as

$$A = \int_0^\delta dy.$$

Thus, Eq. (5.149) becomes

$$U_s \frac{d}{dx} \int_0^\delta u \, dy - \frac{d}{dx} \int_0^\delta u^2 dy + U_s \frac{dU_s}{dx} \int_0^\delta dy = \frac{\tau_w}{\rho}. \qquad (5.150)$$

This form is acceptable, but it can be made more compact by additional operations. First, from the identity

$$\frac{d}{dx} \left(U_s \int_0^\delta u \, dy \right) = U_s \frac{d}{dx} \int_0^\delta u \, dy + \frac{dU_s}{dx} \int_0^\delta u \, dy,$$

it can be seen that

$$U_s \frac{d}{dx} \int_0^\delta u \, dy = \frac{d}{dx} \left(U_s \int_0^\delta u \, dy \right) - \frac{dU_s}{dx} \int_0^\delta u \, dy, \qquad (5.151)$$

and, since U_s depends only upon x and not upon y, it can be brought inside the integral operator of the first term on the right of Eq. (5.151) and also in the third term on the left of Eq. (5.150). Then, by replacing the first term on the left of Eq. (5.150) by Eq. (5.151) and combining, the final form of the integral momentum equation for a steady, planar two-dimensional, constant property, boundary layer–type flow is

$$\frac{d}{dx} \int_0^\delta u(U_s - u)\, dy + \frac{dU_s}{dx} \int_0^\delta (U_s - u)\, dy = \frac{\tau_w}{\rho}. \tag{5.152}$$

At this point the integral x-momentum equation, Eq. (5.152), is valid for either laminar or turbulent flow and is just as "exact" as the differential form of the x-momentum equation, Eq. (5.122), since no additional approximations were required, other than those used to arrive at Eq. (5.122). (Note that Eq. [5.152] can be derived directly from Eq. [5.122] by multiplying every term of Eq. [5.122] by dy and integrating from 0 to δ. After the use of the continuity equation, integration by parts, Leibnitz' rule, and considerable rearrangement, Eq. [5.152] is obtained, with no approximations.) It is as a result of the solution techniques employed that "integral methods" generally have the word "approximate" associated with them.

The wall shear stress in Eq. (5.152) is, by Newton's law of viscosity,

$$\tau_w = \mu \left(\frac{\partial u}{\partial y} \right)_{y=0},$$

and, since the kinematic viscosity is defined by $\nu = \mu/\rho$, Eq. (5.152) becomes

$$\frac{d}{dx} \int_0^\delta u(U_s - u)\, dy + \frac{dU_s}{dx} \int_0^\delta (U_s - u)\, dy = \nu \left(\frac{\partial u}{\partial y} \right)_{y=0}. \tag{5.153}$$

Equation (5.153) is the form of the integral x-momentum equation which is most suitable for laminar thin boundary layer–type flow problems. In this form it is seen that Eq. (5.153) represents a single equation (the continuity equation has already been used to obtain Eq. [5.153]) in the two dependent variables $\delta = \delta(x)$ and $u = u(x, y)$. Note that $U_s(x)$ is a known quantity from the potential flow solution for the body of interest. The next task is to solve Eq. (5.153) for the velocity field, u and v, as a function of x and y.

Approximate Solution Method for the Integral Equations

One way to solve Eq. (5.153) is to insert a reasonable, though approximate, velocity profile $u = u(x, y)$, so that the equation becomes an ordinary differential equation for δ as a function of x. The approximate profile for u must satisfy the known boundary conditions that the actual profile satisfies. It may also be forced to satisfy some additional conditions, other than boundary conditions, that the true velocity profile is known to satisfy. Finally, it must satisfy the integral form of the momentum equation, Eq. (5.153), on the average in the y direction between 0 and δ by insertion of the approximate velocity profile into

Eq. (5.153). For integral equations of the general type of Eq. (5.153), a Kantorovich profile [22] is used as the approximating profile for $u = u(x, y)$. The general form of the Kantorovich profile for Eq. (5.153) is

$$u = a_0(x)f_0(y) + a_1(x)f_1(y) + a_2(x)f_2(y) + \cdots, \tag{5.154}$$

where $a_0(x)$, $a_1(x)$, $a_2(x)$, etc., are functions of x that must be determined, and $f_0(y)$, $f_1(y)$, $f_2(y)$, etc., are *specified* functions of y. In some cases, the engineer has information or insight which allows an immediate choice for the function or functions of y. When information about the possible form of the functions $f_0(y)$, $f_1(y)$, $f_2(y)$, etc., is not available, they are chosen to be elements of a linearly independent, complete set of functions. Linearly independent means that the only way

$$c_0 f_0(y) + c_1 f_1(y) + c_2 f_2(y) + \cdots + c_n f_n(y) = 0$$

can be satisfied, throughout some y domain, is to choose *all* the c's equal to zero. By complete is meant that by taking a sufficiently great number of elements of the set, $f_0(y)$, $f_1(y)$, $f_2(y)$, etc., any function of y, say $G(y)$, can be approximated as close as is necessary by a sequence of the $f(y)$.[3] (See [23] and [24].) Although there are many sets of functions that satisfy these conditions, only the set that is simplest to work with is used herein; that is

$$f_0(y) = 1, f_1(y) = y, f_2(y) = y^2, \ldots, \text{or}$$
$$1, y, y^2, y^3, y^4, y^5, \ldots, y^n. \ldots \tag{5.155}$$

Hence, the approximating sequence for the velocity profile becomes

$$u = a_0 + a_1 y + a_2 y^2 + a_3 y^3 + a_4 y^4 + \cdots, \tag{5.156}$$

where the a's are functions of x that must be determined. The number of terms to take in Eq. (5.156) depends upon whether higher accuracy is desired with the use of a greater number of terms even though computational effort is greater. Ordinarily, in heat transfer and fluid flow problems of the boundary layer type, three or four terms suffice, although in some situations, such as near a laminar separation point, more may be required.

---

3. The reasons for these two conditions on the function $f(y)$ are perhaps more readily understood by viewing the elements of the set of all continuous functions of y as abstract vectors in a linear vector space, a function space. Then, by analogy with the ordinary geometric vector space, the functions (vectors) $f_0(y), f_1(y), f_2(y)$, etc., are the base vectors which can be combined in a linear fashion to yield any other vector, such as $G(y)$, in the space. It then becomes obvious why the set of functions (base vectors) must be linearly independent and complete, for no attempt would be made to represent a geometric vector, which has both x and y components, by two vectors which are both pointing in the x direction (need for linear independence). Nor can one represent a geometric vector that has nonzero components in all three directions, x, y, and z, by only two unit vectors, one in the y and one in the x direction, because the z component could not be taken into account.

Once a decision is made as to the number of terms to retain in the approximating sequence, such as the first four terms of Eq. (5.156) as suggested in the discussion just completed, then the various coefficients a_0, a_1, a_2, and a_3 must be found as functions of x in general. Some of these functions, these various a's, are found or are related to the remaining unknown a's by forcing the approximate profile, Eq. (5.156), to satisfy the boundary conditions which the true solution function for u, whatever it is, also satisfies. In addition, one or more other unknown $a(x)$ functions may be found or related to one final unknown coefficient by forcing the approximating sequence, Eq. (5.156), to satisfy other conditions, besides boundary conditions, which the true solution for u also satisfies. When this is done and only *one* of the $a(x)$ functions remains as an unknown, the last step is to insert the approximating sequence into the integral equation, such as Eq. (5.153), to force the sequence to satisfy the integral form of the governing conservation law *on the average* throughout the boundary layer. This is also a condition that the true solution function satisfies, and this condition determines the last unknown $a(x)$. With this done, one then has an *approximate* analytical solution for the velocity profile $u(x, y)$.

In short, what is being done here in this approximate solution method is to start with an approximating sequence for u which contains a number of adjustable parameter functions, the various coefficients $a(x)$ in Eq. (5.156). These are then determined by forcing the approximate profile to satisfy a number of conditions that the true, or exact, solution function also satisfies, even though we do not know what the exact solution is. In this way, one is attempting to "twist" or "force" the approximate profile toward the true profile.

Flat Plate Solution

The approximate integral solution methods just described will now be demonstrated in detail by a comprehensive application to a situation involving a thin hydrodynamic boundary layer.

Consider the same problem for which an exact analytical solution by the method of the similarity transformation was presented earlier; that is, the laminar, steady, constant property, planar two-dimensional, boundary layer–type flow over a flat plate at zero angle of attack to the flow. The physical situation is shown again in Fig. 5.22 with a qualitative sketch of the boundary layer also shown. Integral methods will be used to predict the velocity field $u = u(x, y)$ and $v = v(x, y)$, the local skin friction coefficient $C_f = C_f(x)$, and the local hydrodynamic boundary layer thickness $\delta = \delta(x)$.

To solve for the velocity field and the shear stress from the integral momentum equation, Eq. (5.153), a velocity profile is needed; the one we have chosen is of the form of Eq. (5.156). The first four terms are used to approximate u; hence,

$$u = a_0 + a_1 y + a_2 y^2 + a_3 y^3. \qquad (5.157)$$

Recall that a_0, a_1, a_2, and a_3 are, in general, functions of x that must be determined. First, this approximate velocity profile is forced to satisfy the known boundary conditions that the true velocity profile satisfies in the y direction. Obviously, at $y = 0$, $u = 0$; hence, from Eq. (5.157), $0 = a_0 + 0 + 0 + 0$, or

$$a_0 = 0, \qquad (5.158)$$

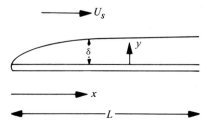

Figure 5.22 Boundary layer development on a flat plate.

and Eq. (5.157) becomes

$$u = a_1 y + a_2 y^2 + a_3 y^3. \tag{5.159}$$

At the edge of the boundary layer, $y = \delta$, and u, by definition, must equal U_s; hence,

$$\text{at } y = \delta, \quad u = U_s. \tag{5.160}$$

Also, since $\partial u/\partial y \approx 0$ outside the boundary layer, $\partial u/\partial y$ must be 0 at $y = \delta$ so that the velocity profile for $y \leq \delta$, Eq. (5.159), blends smoothly into the profile for $y \geq \delta$, $u = U_s$. Thus,

$$\text{at } y = \delta, \quad \partial u/\partial y = 0. \tag{5.161}$$

Thus, once boundary conditions, Eqs. (5.160–61), are imposed on Eq. (5.159), which already satisfies the condition $u = 0$ at $y = 0$, the approximate velocity profile will satisfy the same boundary conditions as does the true velocity profile. However, if Eqs. (5.160–61) are used in Eq. (5.159), we find that there are two unknown functions of x in the velocity profile, δ and one of the a's, but only one equation, Eq. (5.153), the integral form of the x-momentum equation, remains. Hence, another condition or equation is required. Additional conditions can be found from the partial differential form of the x-momentum equation, Eq. (5.122), with the $\partial P/\partial x$ term set equal to zero because of the flat plate geometry. Thus,

$$\rho \left(u \frac{\partial u}{\partial x} + v \frac{\partial u}{\partial y} \right) = \mu \frac{\partial^2 u}{\partial y^2}. \tag{5.162}$$

Additional conditions on the approximate velocity profile can be found by collocation, that is, by forcing the approximate profile, Eq. (5.159), to satisfy the exact partial differential equation, Eq. (5.162), at various space points in the boundary layer. (Recall that the exact solution for u satisfies Eq. [5.162] at *all* space points in the boundary layer.) Since only one more condition is required because of the number of terms used in the original approximation, Eq. (5.157), only one space point can be picked at which the approximating sequence for u will exactly satisfy the partial differential equation. Perhaps the simplest point to use is at $y = 0$ on the boundary. It is also an important point because the accurate prediction of the wall shear stress depends upon how good the approximate velocity profile is near the wall. At $y = 0$, for a solid impervious wall, $u = 0$ and $v = 0$, so that Eq. (5.162) reduces to the form $\partial^2 u / \partial y^2 = 0$. Hence, the additional condition that the approximate profile is forced to satisfy is

$$\text{at } y = 0, \quad \frac{\partial^2 u}{\partial y^2} = 0. \tag{5.163}$$

This has been called, by some authors, a *compatibility condition*. Now, Eqs. (5.160–61) and (5.163) are imposed on the approximate profile, Eq. (5.159). Thus, since $u = a_1 y + a_2 y^2 + a_3 y^3$,

$$\frac{\partial u}{\partial y} = a_1 + 2a_2 y + 3a_3 y^2, \tag{5.164}$$

$$\frac{\partial^2 u}{\partial y^2} = 2a_2 + 6a_3 y. \tag{5.165}$$

Using Eq. (5.163) in Eq. (5.165) yields $a_2 = 0$; hence, Eq. (5.164) becomes

$$\frac{\partial u}{\partial y} = a_1 + 3a_3 y^2. \tag{5.166}$$

Using Eq. (5.161) in Eq. (5.166) gives $0 = a_1 + 3a_3 \delta^2$ or $a_3 = -a_1/3\delta^2$. Using this, along with $a_2 = 0$, in Eq. (5.159) yields

$$u = a_1 \left(y - \frac{y^3}{3\delta^2} \right). \tag{5.167}$$

Using Eq. (5.160) in Eq. (5.167) leads to

$$U_s = a_1 \left(\delta - \frac{\delta^3}{3\delta^2} \right) = \frac{2}{3} \delta a_1,$$

from which $a_1 = 3U_s/2\delta$. Substituting this into Eq. (5.167) and rearranging yields

---

$$u = U_s \left(\frac{3}{2} \frac{y}{\delta} - \frac{1}{2} \frac{y^3}{\delta^3} \right). \tag{5.168}$$

Equation (5.168) is not yet the local x component of velocity within the boundary layer as a function of x and y, because δ still must be found as a function of x by inserting the approximate profile, Eq. (5.168), into the integral form of the momentum equation, Eq. (5.153). This will force the approximate velocity profile to satisfy the x-momentum equation (on the average) throughout the boundary layer in the y direction.

Since $\partial P/\partial x = 0$ for a flat plate, dU_s/dx is also zero (cf. discussion just prior to Eq. [5.149]). Hence, Eq. (5.153) becomes

$$\frac{d}{dx} \int_0^\delta u(U_s - u)dy = \nu \left(\frac{\partial u}{\partial y}\right)_{y=0}. \tag{5.169}$$

Now, from the approximate velocity profile, Eq. (5.168),

$$\frac{\partial u}{\partial y} = U_s \left(\frac{3}{2\delta} - \frac{3y^2}{2\delta^3}\right),$$

or at $y = 0$,

$$\left(\frac{\partial u}{\partial y}\right)_{y=0} = \frac{3U_s}{2\delta}.$$

Substituting this into the right-hand side of Eq. (5.169) and inserting u from Eq. (5.168) into the left-hand side yields

$$\frac{d}{dx} \int_0^\delta U_s \left(\frac{3}{2}\frac{y}{\delta} - \frac{1}{2}\frac{y^3}{\delta^3}\right)\left[U_s - U_s \left(\frac{3}{2}\frac{y}{\delta} - \frac{1}{2}\frac{y^3}{\delta^3}\right)\right] dy = \nu \frac{3U_s}{2\delta}.$$

Expanding the integrand of the left-hand side, a polynomial in y is obtained to integrate over y from 0 to δ (recall that δ does not depend upon y, only upon x). Performing the indicated operations yields

$$\frac{39}{280} U_s^2 \frac{d\delta}{dx} = \frac{3}{2} \frac{U_s \nu}{\delta}. \tag{5.170}$$

Separating the variables δ and x in Eq. (5.170) leads to

$$\delta d\delta = \frac{140}{13} \frac{\nu}{U_s} dx.$$

Integrating both sides from $x = 0$, where $\delta = 0$, and the general position x, where the boundary layer thickness is δ, yields

$$\frac{\delta^2}{2} = \frac{140}{13} \frac{\nu}{U_s} x.$$

Taking the square root,

$$\delta = 4.64 \sqrt{\nu x/U_s}.$$

Multiplying numerator and denominator by x and introducing the local length Reynolds number, N_{Re_x}, the expression for the local boundary layer thickness becomes

$$\delta = \frac{4.64x}{\sqrt{N_{Re_x}}}. \qquad \textbf{(5.171)}$$

Substituting Eq. (5.171) into Eq. (5.168) yields the approximate x-velocity component as a function of x and y, while Eq. (5.171) alone gives the local boundary layer thickness as a function of x.

Now, expressions for the skin friction coefficient and the y component of the velocity will be developed. The definition of the skin friction coefficient is

$$C_f = \frac{\tau_w}{\frac{1}{2}\rho U_s^2}, \qquad \textbf{[5.126]}$$

and τ_w is

$$\tau_w = \mu \left(\frac{\partial u}{\partial y}\right)_{y=0}.$$

From Eq. (5.168), u is

$$u = U_s \left(\frac{3y}{2\delta} - \frac{y^3}{2\delta^3}\right);$$

hence,

$$\left(\frac{\partial u}{\partial y}\right)_{y=0} = \frac{3U_s}{2\delta}.$$

Therefore,

$$\tau_w = \frac{3\mu U_s}{2\delta}. \qquad \textbf{(5.172)}$$

However, from Eq. (5.171), δ for this case is

$$\delta = 4.64 \sqrt{\frac{\nu x}{U_s}} = 4.64 \sqrt{\frac{\mu x}{\rho U_s}}. \qquad \textbf{(5.173)}$$

Combining Eqs. (5.172) and (5.173) yields

$$\tau_w = 0.323 \frac{\mu^{1/2}\rho^{1/2}U_s^{3/2}}{x^{1/2}}. \qquad \textbf{(5.174)}$$

Multiplying numerator and denominator of the right-hand side of Eq. (5.174) by $\rho^{1/2}\, U_s^{1/2}$ introduces the local length Reynolds number, and the expression for the wall shear stress becomes

$$\tau_w = 0.323 \, \frac{\rho U_s^2}{\sqrt{N_{Re_x}}}.$$

Using this result in Eq. (5.126), the expression for the local skin friction coefficient for the flat plate by use of the approximate integral method is

$$C_f = \frac{0.646}{\sqrt{N_{Re_x}}}. \tag{5.175}$$

Now we have found a solution for the x component of velocity $u(x, y)$, namely, the combination of Eqs. (5.168) and (5.171). The simplest way of arriving at the y component of velocity, v, is to use the continuity equation, Eq. (5.121), which is the connection required by the law of conservation of mass between these two velocity components. Thus,

$$\frac{\partial u}{\partial x} + \frac{\partial v}{\partial y} = 0. \tag{5.121}$$

Transposing terms and integrating partially with respect to y from 0 to y yields

$$v = -\int_0^y \frac{\partial u}{\partial x} \, dy + F(x). \tag{5.176}$$

However, for an impervious surface, $v = 0$ at $y = 0$; hence, $F(x) = 0$ for all x on the surface, and Eq. (5.176) becomes

$$v = -\int_0^y \frac{\partial u}{\partial x} \, dy. \tag{5.177}$$

From Eq. (5.168),

$$u = U_s \left(\frac{3y}{2\delta} - \frac{y^3}{2\delta^3} \right). \tag{5.168}$$

Differentiating this with respect to x, noting that only δ depends upon x, gives

$$\frac{\partial u}{\partial x} = U_s \left(-\frac{3}{2} \frac{y}{\delta^2} + \frac{3}{2} \frac{y^3}{\delta^4} \right) \frac{d\delta}{dx} = \frac{3}{2} U_s \frac{d\delta}{dx} \left(-\frac{y}{\delta^2} + \frac{y^3}{\delta^4} \right).$$

Substituting this into Eq. (5.177), integrating with respect to y (recalling that δ and $d\delta/dx$ are independent of y), and inserting the limits yields

$$v = \frac{3}{2} U_s \frac{d\delta}{dx} \left(\frac{y^2}{2\delta^2} - \frac{y^4}{4\delta^4} \right). \tag{5.178}$$

From Eq. (5.173),

$$\delta = 4.64 \sqrt{\nu x/U_s}\,; \tag{5.179}$$

hence, $d\delta/dx = \frac{1}{2}(4.64) \sqrt{\nu/U_s}\, x^{-1/2}$. Substituting this into Eq. (5.178),

$$v = 3.48 \sqrt{\frac{\nu U_s}{x}} \left(\frac{y^2}{2\delta^2} - \frac{y^4}{4\delta^4} \right). \tag{5.180}$$

Since δ is given by Eq. (5.179) as a known function of x, Eq. (5.180) constitutes an approximate expression for v as a function of x and y within the boundary layer on a flat plate.

Equation (5.168) combined with (5.171); Eq. (5.180) combined with (5.171); Eq. (5.175); and Eq. (5.171) give the approximate analytical integral method solution for the x component of velocity, the y component of velocity, local skin friction coefficient, and local boundary layer thickness, respectively, for steady, laminar, constant property, two-dimensional, thin boundary layer–type flow over a flat plate at zero angle of attack.

Compare this approximate solution to the exact solution for the same problem for the boundary layer thickness and the skin friction coefficient, Eqs. (5.129) and (5.131). One finds that the integral method differs by about 7% in the prediction of boundary layer thickness and by about 3% for the skin friction coefficient. This is quite good agreement indeed when one looks at the relative ease with which the approximate integral solution was found, compared with the exact solution by the similarity transformation. Of course, the integral method would ordinarily be applied to problems for which we are unable to develop an exact solution. Its power and importance derives from the fact that the engineer can apply it, in a relatively easy and straightforward manner, to such problems and usually get results of acceptable engineering accuracy.

Summary Review and Explanation of Integral Methods

Let us review the integral method to see why it works well. Begin with an integral equation, such as Eq. (5.153), which is exact, and construct an approximate profile for the dependent variable in the argument of the integrands, in this case, the x component of velocity u. This approximate profile is chosen to be physically reasonable. That is, the function $\sin(30\pi y/\delta)$ would not be used alone, because if u were chosen as $u = a_0 \sin(30\pi y/\delta)$, a plot of this at some fixed value of x would give an x component of velocity which oscillates with y between $y = 0$ and $y = \delta$; for the flat plate, the actual velocity profile would not exhibit such behavior. The approximate profile is then forced to satisfy all the boundary conditions in y that the true profile satisfies. It may also be forced to satisfy some other conditions satisfied by the true profile, such as the compatibility condition in the flat plate problem. Finally the approximate profile for u is forced to satisfy the integral momentum equation (on the average) throughout the boundary layer. In view of what the approximate profile is forced to do, it is not unreasonable to expect the final result for this profile to be relatively close to the true profile. (We can also see why these methods might work in terms of the difference, between the true velocity profile and the approximation to it, being a vector in function space whose components in the various abstract "directions" are to be made as small as possible, in some sense, thus making the resultant error vector as small as possible.) For a more formal mathematical discussion of the convergence, in the mean, of the method of weighted residuals in certain types of problems, see [24], [25], and [26].

EXAMPLE 5.4

If the velocity profile for the flat plate boundary layer is plotted as a function of y at a given value of x, it would appear as shown in Fig. 5.23. If Eq. (5.168), from which the sketch could be made, was unavailable, and yet a sketch of the profile was required, the sketch would probably look much like Fig. 5.23, partly because of the knowledge of the shapes of velocity profiles in laminar, fully developed tube flow discussed earlier. The profile appears much like a portion of a sine wave. This suggests the use of the approximate profile

$$u = U_s \sin(\pi y / 2\delta). \tag{5.181}$$

A check will show that this approximate profile satisfies all the boundary conditions and the compatibility condition. Use Eq. (5.181) to derive an expression for the local boundary layer thickness as a function of x for the same conditions which led to Eq. (5.171).

Solution
The same integral momentum equation must be solved; that is,

$$\frac{d}{dx} \int_0^\delta u(U_s - u)\, dy = \nu \left(\frac{\partial u}{\partial y} \right)_{y=0}. \tag{5.169}$$

From Eq. (5.181), $(\partial u / \partial y)_{y=0} = \pi U_s / 2\delta$. Using this on the right-hand side of Eq. (5.169) and using Eq. (5.181) on the left yields, after some rearrangement,

$$U_s^2 \frac{d}{dx} \int_0^\delta \left(\sin\frac{\pi y}{2\delta} - \sin^2\frac{\pi y}{2\delta} \right) dy = \frac{\nu \pi U_s}{2\delta}.$$

Performing the indicated integration with respect to y, inserting the limits, and simplifying yields

$$\left(\frac{2}{\pi} - \frac{1}{2} \right) \frac{d\delta}{dx} = \frac{\nu \pi}{2 U_s \delta}.$$

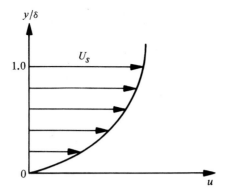

Figure 5.23 The flat plate laminar velocity profile.

The variables can be separated and the equation integrated to yield finally

$$\frac{\delta}{x} = \frac{4.78}{\sqrt{N_{\mathrm{Re}_x}}}.$$

This is quite close to both the result of Eq. (5.171) using different profile functions and the exact result, Eq. (5.129).

Flow Criterion

Thus far, the solutions obtained in this section for the boundary layer thickness, the velocity field, and the skin friction coefficient have been for *laminar* external flow over a body immersed in the flow. The flow in the boundary layer of bodies immersed in an external stream usually begins as laminar, but then may undergo a transition, somewhere on the body, to turbulent flow within the boundary layer. The prediction of the location of this transition point is, at present, one of the least understood and yet one of the most important areas of boundary layer theory. The location of this transition point depends upon surface roughness, the size and algebraic sign of the pressure gradient, whether the surface is being heated or cooled and how strongly, injection or suction of mass at the surface, the intensity of turbulence in the free stream, the curvature of the body, any large-scale disturbances in the free stream such as pulsations, and the ratio of inertial to viscous forces. A detailed discussion of some of these factors can be found in [6]. Reference 7 also discusses these factors and presents a number of experimental and semi-experimental correlations to estimate the position of the transition point of boundary layer–type flows.

Experimental data for flow over a smooth flat plate with no pressure gradient generally indicates a transition to turbulent boundary layer–type flow at values of the local length Reynolds number N_{Re_x} defined in Eq. (5.130), between about 300,000 and 500,000. However, transition can also occur at values of the Reynolds number outside this range, depending upon the intensity of turbulence in the free stream outside the boundary layer.

The value of 300,000 will be used here because this corresponds to a value, 360, of the more general momentum thickness Reynolds number which indicates turbulent flow in fully developed pipe flow. This point is discussed in [17].

For present purposes, unless other information is provided, the boundary layer will be assumed turbulent once a local length Reynolds number of 300,000 is reached. Hence, a *laminar boundary layer* exists when

$$N_{Re_x} < 300,000 \qquad \textbf{(5.182a)}$$

and a *turbulent external boundary* layer is implied when

$$N_{Re_x} \geq 300,000 \qquad \textbf{(5.182b)}$$

where

$$N_{Re_x} = \rho U_s x / \mu. \qquad \textbf{[5.130]}$$

For completeness, the flow criterion discussed earlier for *fully developed pipe flow* will be repeated here. *Laminar fully developed pipe flow* exists when

$$N_{Re_D} < 2000, \qquad \textbf{(5.182c)}$$

and *turbulent fully developed pipe flow* occurs when

$$N_{Re_D} > 4000 \qquad \textbf{(5.182d)}$$

where

$$N_{Re_D} = \rho U_m D / \mu. \qquad \textbf{[5.63]}$$

EXAMPLE 5.5

Water is flowing at 1.2 ft/sec over a 1-ft-long flat plate. Compute the boundary layer thickness δ at the end of the plate, and compare it with the plate length to see if the fundamental assumption of boundary layer theory is satisfied under these conditions. For water, $\rho = 62.2$ lbm/ft^3 and $\mu = 2.0$ lbm/hr ft.

Solution

If the flow in the boundary layer is laminar, the thickness at any x can be calculated from Eq. (5.129). The flow will be laminar to the end of the plate if the local length Reynolds number at $x = L$ is less than 300,000; hence, a computation must be made to determine N_{Re_L} at $x = L$. Since $U_s = 1.2$ ft/sec $\times$ 3600 sec/hr $= 4320$ ft/hr,

$$N_{Re_L} = \frac{\rho U_s L}{\mu} = \frac{62.2(4320)(1)}{2.0} = 134,300;$$

Re_x = N_{Re_L}
our book
this book

hence, the flow is completely laminar. From Eq. (5.129),

$$\delta = \frac{5x}{\sqrt{N_{Re_x}}} = \frac{5(1)}{\sqrt{134,000}} = 0.01368 \text{ ft} = 0.164 \text{ in.}$$

The fundamental assumption from which the boundary layer simplifications arise is $\delta << L$. In this case, $\delta/L = 0.01368/1$, and since δ is just over 1% of L, application of boundary layer theory is valid.

5.3h Turbulent External Boundary Layers

When the boundary layer over a body becomes turbulent, the integral x-momentum equation in the form of Eq. (5.153) still applies. However, Eq. (5.153) contains $(\partial u/\partial y)_{y=0}$, and in external turbulent boundary layers it is found, as for fully developed turbulent pipe flow, that the velocity gradient is very steep close to the wall and relatively small everywhere else. It is extremely difficult to represent this type of profile accurately near the wall using an approximate velocity profile; hence, in turbulent flow problems, the integral x-momentum equation in the form of Eq. (5.152) is used in which the wall shear stress appears explicitly as τ_w. Experiments yield a relationship between the local wall shear stress τ_w and the local boundary layer thickness δ which, when substituted into Eq. (5.152) along with a velocity profile, permits the calculation of the boundary layer thickness as a function of x.

Velocity Profile and Shear Stress

Experiments show that turbulent external boundary layers have at least part of their velocity's y dependence, at a fixed value of x, given by the law of the wall, $u^+ = F(y^+)$, Eq. (5.103). In fact, the explicit functional form of the law of the wall is essentially the same as it was for fully developed pipe flow, namely, Eq. (5.104). These same forms persist, again for at least part of the boundary layer thickness, even with pressure gradients, demonstrating a type of universality of the law of the wall. In the outer extremities of the boundary layer, when the Reynolds number becomes high enough, it is found that u^+ does not depend upon y^+ alone anymore. The flow is found to have a wake-like character (like the flow downstream of a cylinder at high enough Reynolds numbers in cross flow) and the velocity distribution in this region is given by Coles' *law of the wake* [27]. For a flat plate the wake portion of the flow is not a significant percentage of the local boundary layer thickness until N_{Re_x} exceeds about 10^6. In heat transfer calculations in which the law of the wall is used, it is generally assumed that the law of the wall is valid across the entire layer, and for the most part very good results have been obtained because the temperature gradient at the wall, upon which the heat transfer coefficient depends, is most sensitive to the velocity distribution near the wall. This region near the wall is where the law of the wall is valid, and thus fairly good results are obtained even without employing the law of the wake. For a more detailed discussion of the law of the wake, see [15] and [7].

For turbulent flow over flat plates or bodies producing very mild pressure gradients, experiments show that a fairly good approximation to the velocity profile is given by the one-seventh power law in the following form for length Reynolds numbers less than about 10^6 or 10^7:

$$u = U_s \left(\frac{y}{\delta} \right)^{1/7}, \qquad\qquad (5.183)$$

where δ is the local boundary layer thickness and U_s is the free-stream velocity just outside the boundary layer.

When the conditions are satisfied which allow Eq. (5.183) to be an adequate representation of the velocity profile within a turbulent boundary layer, experiments give the following expression for the local skin friction coefficient [6]:

$$C_f = \frac{\tau_w}{\frac{1}{2} \rho U_s^2} = 0.045 \left(\frac{\nu}{U_s \delta} \right)^{1/4}. \qquad\qquad (5.184)$$

Actually this result, Eq. (5.184), can be derived from the experimental correlation for the friction factor f in fully developed turbulent pipe flow. Replacing the diameter D in Eq. (5.114) by twice the boundary layer thickness, relating U_m to U_s through Eq. (5.183), and then rearranging Eq. (5.114) leads to Eq. (5.184). This is not completely unexpected, in view of the fact that essentially the same form of the law of the wall applies to *both* fully developed turbulent pipe flow and turbulent thin boundary layer–type flow.

Integral Method Solution for Turbulent Flow Over a Flat Plate

As we have discussed, experiment indicates that the velocity profile and skin friction factor for turbulent flow over a smooth flat plate with no pressure gradient are given, with reasonable accuracy for Reynolds numbers less than about 10^7, by Eqs. (5.183) and (5.184). However, these expressions are not yet complete, because the unknown local boundary layer thickness δ appears in them. To determine δ, Eq. (5.183) is used as the approximating expression for the velocity profile and Eq. (5.184) as the corresponding one for the shear stress, and these are inserted into the appropriate integral form of the x-momentum equation, Eq. (5.152).

Since the body is a flat plate, $dU_s/dx = 0$, and Eq. (5.152) becomes

$$\frac{d}{dx} \int_0^\delta u(U_s - u)\,dy = \frac{\tau_w}{\rho}.$$

Using Eq. (5.183) for u and Eq. (5.184) for τ_w yields

$$U_s^2 \frac{d}{dx} \int_0^\delta \left[\left(\frac{y}{\delta}\right)^{1/7} - \left(\frac{y}{\delta}\right)^{2/7}\right]dy = \frac{0.045}{2}\left(\frac{\nu}{U_s}\right)^{1/4} \delta^{-1/4} U_s^2.$$

Performing the integration on the left yields

$$\frac{7}{72}\frac{d\delta}{dx} = \frac{0.0225}{1}\left(\frac{\nu}{U_s}\right)^{1/4} \delta^{-1/4}.$$

Separating the variables δ and x leads to

$$\delta^{1/4}d\delta = \frac{72}{7}(0.0225)\frac{\nu^{1/4}}{U_s^{1/4}}\,dx.$$

Integrating from 0 to δ on the left and from 0 to x on the right (this assumes that the boundary layer thickness is zero at $x = 0$; that is, the boundary layer is turbulent over the entire plate) gives

$$\frac{4}{5}\delta^{5/4} = \frac{72}{7}(0.0225)\frac{\nu^{1/4}}{U_s^{1/4}}x.$$

Raising both sides to the $\frac{4}{5}$ power and solving for δ gives

$$\delta = 0.37\left(\frac{\nu}{U_s}\right)^{1/5} x^{4/5}. \tag{5.185}$$

It is seen from Eq. (5.185) that the turbulent boundary layer on a flat plate grows at a greater rate with x than does a laminar boundary layer, since δ is proportional to $x^{4/5}$ rather than $x^{1/2}$ as in a laminar boundary layer.

By multiplying the right-hand side of Eq. (5.185) by $x^{1/5}/x^{1/5}$, the local length Reynolds number is introduced and Eq. (5.185) becomes

$$\delta = \frac{0.37x}{(N_{Re_x})^{1/5}}. \tag{5.186}$$

It will be recalled that in the solution it was assumed that the boundary layer was turbulent over the entire plate. This can occur if the leading edge of the plate is rough enough to "trip" the boundary layer and cause it to be turbulent over the entire plate.

Also, for high enough Reynolds numbers, the initial laminar portion of the boundary layer will cover a small enough fraction of the total plate length that it can be neglected and the entire layer assumed turbulent. Situations can arise, of course, in which the boundary layer development will have to be calculated using some combination of Eq. (5.129) for the laminar portion and Eq. (5.186) for the turbulent portion.

The local skin friction coefficient is now found by inserting the expression for δ just derived, Eq. (5.186), into Eq. (5.184), giving

$$C_f = 0.045 \left[\frac{\nu N_{\mathrm{Re}_x}^{1/5}}{U_s(0.37x)} \right]^{1/4} = 0.0576 \left(\frac{N_{\mathrm{Re}_x}^{1/5}}{N_{\mathrm{Re}_x}} \right)^{1/4} = \frac{0.0576}{(N_{\mathrm{Re}_x})^{1/5}} .$$

Hence,

$$C_f = \frac{0.0576}{N_{\mathrm{Re}_x}^{1/5}} . \tag{5.187}$$

EXAMPLE 5.6

A portion of a ship's hull can be idealized as a flat plate, 30 m long, moving with the ship's speed of 3 m/s through water with properties $\rho = 1000$ kg/m^3 and $\mu = 1.6 \times 10^{-3}$ kg/ms. Calculate the maximum thickness of the boundary layer.

Solution
The maximum boundary layer thickness occurs at the end of the hull where $x = L = 30$ m. To see whether the boundary layer flow is laminar or turbulent, the Reynolds number criteria, Eqs. (5.182a) and (5.182b), are checked. Thus,

$$N_{\mathrm{Re}_L} = \rho U_s L / \mu = 1000(3)(30)/1.6 \times 10^{-3} = 5.63 \times 10^7.$$

Since this is greater than 300,000, the flow is turbulent for the part of the hull downstream of the position, call it x_c, at which the *local* length Reynolds number is 300,000. We will treat it as turbulent flow over the entire hull and make a check on the suitability of this later. Hence, by Eq. (5.186) evaluated at $x = L = 30$ m,

$$\delta = \frac{0.37L}{N_{\mathrm{Re}_L}^{1/5}} = \frac{0.37(30)}{(5.63 \times 10^7)^{1/5}} = 0.313 \text{ m}.$$

(Note that, strictly speaking, the length Reynolds number is outside the range of validity of Eq. (5.186), but since the value is just a little above 10^7, it will be assumed accurate enough.)

This computed boundary layer thickness of 0.313 m does not seem very "thin," but the boundary layer approximations rely not on the boundary layer being thin in some absolute sense, but rather thin relative to the length of the body in the flow direction.

That is, $\delta << L$. In this problem $\delta/L = 0.313/30 \approx 0.01$, so the maximum boundary layer thickness is only about 1% of the total length.

Now, in computing $\delta = 0.313$ m, it was assumed that the flow was turbulent over the *entire* hull, not just the section downstream of x_c, where the local Reynolds number is greater than 300,000. To see if this is justified, let us compute the percentage of the total hull length L which might be covered by a laminar boundary layer which begins at the leading edge and would be of length x_c. Thus,

$$\frac{\rho U_s x_c}{\mu} = 300{,}000 = \frac{1000(3)x_c}{1.6 \times 10^{-3}} \text{, or } x_c = 0.16 \text{ m.}$$

Thus, only $(0.16/30)100 = 0.53\%$ of the hull might be covered by a laminar boundary layer; thus, the assumption of turbulent flow over the entire hull is validated.

This completes our formal study and review of fluid mechanics.

It will be recalled that the primary aim of the study of convection heat transfer is finding the value of the surface coefficient of heat transfer. The surface coefficient h depends upon the temperature gradient in the fluid at the solid surface, which in turn depends upon the motion of the fluid past the solid surface. Hence, to find surface coefficients of heat transfer in any analytical or semi-analytical approach, the velocity field in the moving fluid must be determined. In constant property flows, the principal subject of this section, the velocity field is independent of the temperature field and can be solved for initially. In subsequent sections, this information about the velocity field will be used in some form of the thermal energy equation to solve for the temperature distribution in the moving fluid, which will, by Eq. (5.3), yield the desired local surface coefficient of heat transfer. Also, the work in subsequent sections will, at times, entail the use of integral techniques to predict the temperature field in the moving fluid. We have studied the use of integral methods in this section so that they can be used with relatively little explanation henceforth.

5.4 ANALYTICAL AND SEMI-ANALYTICAL SOLUTIONS TO FORCED CONVECTION PROBLEMS

With a number of solutions to the velocity field already in hand from Sec. 5.3, we are now in a position to confront the major problem of convection heat transfer, namely, the prediction of the local surface coefficient of heat transfer h_x as a function of position on a body or within a duct. To predict h_x analytically, we must find the temperature distribution. Hence the appropriate governing differential equations and integral equation, whose solution yields the temperature distribution, will be derived by use of the energy balance, Eq. (1.8). Exact solutions will then be found for the simplest convection problems, which are the ones in which the velocity and temperature field are both fully developed within a laminar duct flow. Turbulent, fully developed duct flow will be then treated using the similarity between the transport of momentum and of thermal energy. Finally, solutions for the surface coefficient will be found analytically for laminar, external boundary layer flows, for the most part by use of the powerful approximate integral method.

5.4a Fully Developed Laminar Flow Within Tubes and Ducts

In this section, the problem of predicting the local surface coefficient of heat transfer in steady, laminar, constant property, fully developed (thermally and hydrodynamically) flow within tubes and ducts is considered. In the previous section, it was seen that *fully developed hydrodynamically* means that the velocity profile remains unchanged in the direction of the duct or tube axis; that is, $\partial u/\partial x = 0$. *Fully developed thermally* means that a particular nondimensional temperature excess ratio remains unchanged in the direction of the duct axis. Precisely what ratio is meant will be explained later; but at this point it must be emphasized that, ordinarily, fully developed thermally does not mean simply that $\partial T/\partial x = 0$. As in the case of a fully developed hydrodynamic flow, the entrance region of a duct in which the profile of temperature and velocity do change substantially with x must be excluded. This is the so-called "developing entrance region." Relations for the length of the entrance region and expressions for the surface coefficient of heat transfer in the entrance region will be presented later in the chapter.

Bulk Mean Temperature

Consider now fully developed, constant property flow through a duct of arbitrary cross-sectional shape, a shape that does not vary with position down the duct and which is of total area A_c. The perimeter of this cross section, at the position x shown in Fig. 5.24, is at the constant temperature T_w. A view of the duct cross section along the flow direction is shown in Fig. 5.25. Once the temperature distribution is found in the moving fluid, Eq. (5.3) permits a determination of the local surface coefficient of heat transfer h_x. That is,

$$h_x = \frac{-k \left.\dfrac{\partial T}{\partial n}\right|_{n=0}}{T_w - T_s}.$$ [5.3]

The denominator is the ΔT in Newton's cooling law, the "appropriate" driving potential difference between the surface and the fluid. However, in thermally fully developed duct flow, there is not, at any cross section, some large mass of fluid at some easily identified temperature, such as there is, for example, away from the surface of a submarine: namely, the temperature of the water it is passing through. Hence, a decision must be made as to what fluid temperature T_s to use in Eq. (5.3) since, in general, the fluid temperature will have some distribution over the entire cross section shown in Fig. 5.25. Naturally, any temperature in the cross section can be defined as T_s, since Eq. (5.3) defines h_x, and one of the more logical choices would seem to be the temperature at the "center" of the cross section. But for an asymmetrical cross section as shown in Fig. 5.25, there is no point that could be clearly identified as the center. The temperature of the fluid, at any cross section of a duct, that is most logical is the temperature which characterizes the energy per unit mass of all the fluid passing through the cross section; in particular, its enthalpy per unit mass. This temperature is referred to as the *bulk mean temperature of the fluid,* the *bulk temperature,* the *mixing cup temperature,* or, simply, the *average fluid temperature,* and is given the symbol T_B. (The term *mixing cup temperature* comes from the fact that if the fluid passing through a cross section is suddenly placed into an adiabatic cup without

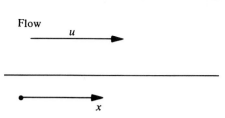

Figure 5.24 Flow through a duct of arbitrary cross section.

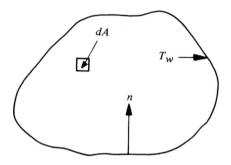

Figure 5.25 Cross section of an asymmetrical duct showing the differential area and normal to the wall.

disturbing its velocity, and is allowed to mix thoroughly, the temperature eventually arrived at is T_B.)

To derive a general expression for the bulk mean temperature T_B at any cross section, first the total rate at which energy is carried across the surface is computed. This is done by summing up all contributions passing through every differential area dA on A_c and setting that sum equal to the rate at which the energy would be carried across if all the fluid were at the same temperature, the mixing cup temperature.

The rate at which energy crosses the differential area dA (except for the kinetic energy) is given by the mass flow rate $d\omega$ through dA times the enthalpy per unit mass, which is approximated by $c_p T$, where T is the local fluid temperature at the position on the cross section where dA is located. Thus this energy transfer rate is

$$d\omega \, c_p T. \tag{5.188}$$

But,

$$d\omega = \rho u \, dA, \tag{5.189}$$

where u is the local x component of velocity, the velocity component normal to dA. Hence, Eq. (5.188) becomes

$$\rho c_p u T \, dA. \tag{5.190}$$

The total rate at which energy passes through the entire cross section is given by summing contributions of the type shown in Eq. (5.190) over the entire cross section in the definite integral

$$\int_{A_c} \rho c_p u T \, dA. \tag{5.191}$$

The total mass flow rate through the entire cross section of area A_c is given by the definite integral of Eq. (5.189) over the cross section as

$$\omega = \int_{A_c} \rho u \, dA. \tag{5.192}$$

If all the fluid passing the cross section is at the bulk mean temperature T_B, its energy per unit mass is $c_p T_B$, and the total rate at which energy passes the cross section is

$$\omega c_p T_B. \tag{5.193}$$

Equating Eqs. (5.193) and (5.191), substituting Eq. (5.192), and cancelling the term ρc_p, since the fluid is constant property, the general expression for the bulk mean temperature in a duct of arbitrary cross section is

$$T_B = \frac{\displaystyle\int_{A_c} u T \, dA}{\displaystyle\int_{A_c} u \, dA}. \tag{5.194}$$

For fully developed duct flow, the temperature T_s in Newton's cooling law and in the defining equation for h_x, Eq. (5.3), is the bulk mean temperature T_B, defined mathematically by Eq. (5.194).

Sometimes, even in thermally undeveloped flow where there is a core of fluid in the central region of the duct at a known temperature T_s, the bulk mean temperature T_B might be used as the fluid temperature in Newton's cooling law instead of T_s. In addition, for any type of internal flow, be it fully developed thermally or not, the bulk mean temperature is often a quantity of interest, because it characterizes the energy per unit mass possessed by the fluid at any cross section in the duct. Often, the function of the duct is to heat or cool the fluid—that is, to increase or decrease its energy per unit mass. Hence the expression for the bulk mean temperature, Eq. (5.194), has value and implications beyond the topic of flows that are fully developed thermally.

Governing Differential Equation for the Temperature Field

In order to predict the local surface coefficient of heat transfer from Eq. (5.3), the steady-state temperature distribution as a function of the space coordinates within the moving fluid must be determined. For conduction heat transfer within solids, the temperature distribution was found by solving the governing partial differential equation representing

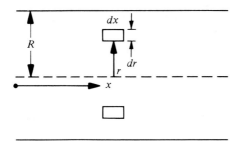

Figure 5.26 Annular control volume for a tube.

conservation of energy. In forced convection heat transfer as well, a partial differential equation for the temperature distribution can be found by making an energy balance on a stationary control volume within the flowing fluid.

Circular Tube   The governing partial differential equation for the temperature distribution in steady, laminar, constant property, fully developed flow in a circular tube will be derived by applying the energy balance, Eq. (1.8), to the control volume shown in Fig. 5.26.

For steady-state conditions and no generation, Eq. (1.8) becomes $R_{in} = R_{out}$. Since the flow is fully developed, $\partial u/\partial x = 0$, and, from the previous section, this means that the radial velocity component is zero; hence, no energy is carried into or out of the control volume by bulk motion of the fluid in the r direction because there is no bulk fluid motion in the radial direction. Energy is conducted into the control volume across the surface at r at the rate q_r and conducted out at $r + dr$ at the rate $q_r + (\partial q_r/\partial r)dr$ (using a Taylor series expansion for the conduction heat transfer rate in the r direction). The various energy transfer terms are shown in Fig. 5.27. Energy is also conducted into the control volume in the x direction at the rate q_x and conducted out at the right face $x + dx$ at the rate $q_{x+dx} = q_x + (\partial q_x/\partial x)dx$. Energy comes into the control volume at the left-hand face not only by conduction, which has already been accounted for in the term q_x, but also because every unit mass that crosses the face, by virtue of its motion in the x direction, carries its own internal energy, kinetic energy, and the flow work done on it by the upstream fluid to force it into the control volume. The internal energy and flow work are combined in enthalpy i, and the total rate at which energy enters the control volume due to mass motion in the x direction is the mass flow rate times the total energy per unit mass, or,

$$\rho u 2\pi r\, dr(i + \tfrac{1}{2}u^2). \tag{5.195}$$

The rate at which energy leaves by the same mechanism at the face $x + dx$ can be put in terms of Eq. (5.195) by a Taylor series expansion to yield

$$\rho u 2\pi r\, dr(i + \tfrac{1}{2}u^2) + \frac{\partial}{\partial x}\left[\rho u 2\pi r\, dr(i + \tfrac{1}{2}u^2)\right] dx.$$

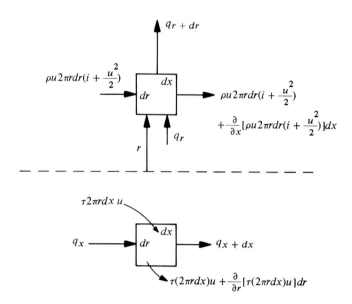

Figure 5.27 Energy transfer rates for the control volume of Fig. 5.26.

Energy also enters the control volume because of shear work done by the faster-moving material just outside the control surface at r on the control volume at the rate $\tau(2\pi r\,dx)u$, the shear force times the velocity. By the same mechanism, energy leaves the control volume at the surface $r + dr$ at the rate $\tau(2\pi r\,dx)u + (\partial/\partial r)\,[\tau(2\pi r\,dx)u]\,dr$. Collecting the R_{in} and R_{out} terms, the energy balance becomes

$$q_r + q_x + \rho u 2\pi r\,dr(i + \tfrac{1}{2}u^2) + \tau 2\pi r\,dx\,u$$
$$= q_r + \frac{\partial q_r}{\partial r}\,dr + q_x + \frac{\partial q_x}{\partial x}\,dx + \rho u 2\pi r\,dr(i + \tfrac{1}{2}u^2)$$
$$+ \frac{\partial}{\partial x}\,[\rho u 2\pi r\,dr(i + \tfrac{1}{2}u^2)]\,dx + \tau 2\pi r\,dx\,u$$
$$+ \frac{\partial}{\partial r}\,(\tau 2\pi r\,dx\,u)\,dr.$$

Cancelling terms leads to

$$0 = \frac{\partial q_r}{\partial r}\,dr + \frac{\partial q_x}{\partial x}\,dx + \frac{\partial}{\partial x}\,[\rho u 2\pi r\,dr(i + \tfrac{1}{2}u^2)]\,dx$$
$$+ \frac{\partial}{\partial r}\,(2\pi r\tau\,dx\,u)\,dr. \tag{5.196}$$

By Fourier's exact law of conduction,

$$q_r = -k2\pi r\,dx(\partial T/\partial r), \quad q_x = -k2\pi r\,dr(\partial T/\partial x).$$

Using these in Eq. (5.196), noting that the term involving $\frac{1}{2}u^2$ disappears because of the fully developed condition $\partial u/\partial x = 0$, cancelling the term $2\pi r\, dr\, dx$, and rearranging gives

$$\frac{k}{r}\frac{\partial}{\partial r}\left(r\frac{\partial T}{\partial r}\right) + k\frac{\partial^2 T}{\partial x^2} = u\rho\frac{\partial i}{\partial x} + \frac{1}{r}\frac{\partial}{\partial r}(r\tau u). \tag{5.197}$$

The enthalpy per unit mass for a constant property fluid is $i = c_p T + P/\rho$. Using this in Eq. (5.197) and expanding the second term on the right yields

$$\frac{k}{r}\frac{\partial}{\partial r}\left(r\frac{\partial T}{\partial r}\right) + k\frac{\partial^2 T}{\partial x^2} = u\rho c_p\frac{\partial T}{\partial x} + u\frac{\partial P}{\partial x} + \frac{u}{r}\frac{\partial}{\partial r}(\tau r) + \frac{\partial u}{\partial r}\tau. \tag{5.198}$$

This equation can be simplified through the aid of the x-momentum equation for this situation:

$$\frac{1}{r}\frac{\partial}{\partial r}(\tau r) = -\frac{\partial P}{\partial x}. \tag{5.49}$$

Multiplying both sides by u and rearranging yields

$$\frac{u}{r}\frac{\partial}{\partial r}(\tau r) + u\frac{\partial P}{\partial x} = 0.$$

This is called the *mechanical energy balance*, and as a result, the second and third terms on the right of Eq. (5.198) sum to zero. Hence, Eq. (5.198) becomes

$$\frac{k}{r}\frac{\partial}{\partial r}\left(r\frac{\partial T}{\partial r}\right) + k\frac{\partial^2 T}{\partial x^2} = \rho c_p u\frac{\partial T}{\partial x} + \tau\frac{\partial u}{\partial r}. \tag{5.199}$$

Note that Eq. (5.199) is referred to as the *thermal energy equation.*

The last term on the right of Eq. (5.199) is the *viscous dissipation term*, the rate at which mechanical energy is degraded to random thermal energy as a result of the irreversible action of the viscous shear stresses. This viscous dissipation term is roughly analogous to the mechanical energy dissipated when sandpapering a piece of lumber, which appears as a heating effect. More will be said later about convection with appreciable viscous dissipation; for the present, only situations in which it can be neglected will be considered, and these occur in what is usually termed *low-speed flow*, flow in which the Eckert number, a dimensionless quantity, is much less than unity [16]. That is,

$$\frac{U_m^2}{c_p(T_w - T_B)} < 1. \tag{5.200}$$

When Eq. (5.200) holds, Eq. (5.199) becomes, after noting that $\alpha = k/\rho c_p$,

$$\frac{1}{r}\frac{\partial}{\partial r}\left(r\frac{\partial T}{\partial r}\right) + \frac{\partial^2 T}{\partial x^2} = \frac{u}{\alpha}\frac{\partial T}{\partial x}.$$ (5.201)

That the temperature field is coupled to the velocity field is seen from Eq. (5.201), since unless u is known as a function of r and x, the equation could not be solved for $T = T(r, x)$.

Depending upon whether the flow is low speed or high speed, the differential equation which must be solved to give the temperature profile within the moving fluid in a circular tube is either Eq. (5.201) or (5.199), respectively.

Implications of the Fully Developed Thermally Condition

Flow within a duct that is fully developed thermally will now be explored. Earlier, we noted that this ordinarily did not mean that $\partial T/\partial x = 0$, for if the fluid flowing is being heated by the duct, its bulk mean temperature, if no phase change occurs, must rise in the flow direction. Therefore, axial temperature gradients will exist. However, in a constant property flow it is expected that after the entrance region effects have decayed, the temperature profile, in some sense, will begin to appear similar at the various x stations. Indeed, when the ducts are long enough, this similarity takes the form of the local (with x) nondimensional temperature excess ratio

$$\frac{T - T_B}{T_w - T_B} \neq f(x)$$ (5.202)

being independent of x and a function only of the space coordinates transverse to x that lie in the flow cross section itself. Recall that, in general, T_B, T_w, and T can depend upon x; it is only when they are combined in the form of Eq. (5.202) that there is an independence of x. When this occurs, the duct flow is said to be fully developed thermally. *When this is true, the entire problem of predicting the temperature distribution is simplified, because Eq. (5.202) implies a particular form for $\partial T/\partial x$ which depends upon the boundary condition at the duct surface.*

To demonstrate this, consider the specific case of fully developed (thermally and hydrodynamically) flow in a circular tube of radius R when the heat flux at the wall is a known constant q''_w. This type of boundary condition can easily arise if the tube is wrapped with electric resistance heating tape; it is also well approximated by events in parts of nuclear reactors. The special form of $\partial T/\partial x$ that the fully developed condition implies, Eq. (5.202), must be found. If steady-state conditions prevail, there will be radial symmetry, and the nondimensional temperature excess ratio of Eq. (5.202) will depend only upon the general radial coordinate r; hence,

$$\frac{T - T_B}{T_w - T_B} = F(r).$$ (5.203)

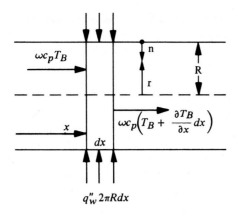

$$\omega c_p T_B$$

$$\omega c_p \left(T_B + \frac{\partial T_B}{\partial x} dx \right)$$

$$q_w'' 2\pi R dx$$

Figure 5.28 Control volume with energy transfer rates for fully developed flow with constant tube wall flux.

In order to develop an expression for $\partial T/\partial x$, the logical starting place is the differentiation, with respect to x, of both sides of Eq. (5.203). Thus,

$$\frac{(\partial T/\partial x) - (\partial T_B/\partial x)}{T_w - T_B} - \frac{(T - T_B)[(\partial T_w/\partial x) - (\partial T_B/\partial x)]}{(T_w - T_B)^2} = 0.$$

Rearranging and solving for $\partial T/\partial x$ yields

$$\frac{\partial T}{\partial x} = \frac{\partial T_B}{\partial x} + \frac{(T - T_B)[(\partial T_w/\partial x) - (\partial T_B/\partial x)]}{(T_w - T_B)}. \tag{5.204}$$

To obtain information about $\partial T_B/\partial x$, the energy balance, Eq. (1.8), is applied to the control volume dx long and extending over the entire cross section of the tube, shown in Fig. 5.28. With $R_{gen} = R_{stor} = 0$, $R_{in} = R_{out}$. The various energy transfer terms are shown in Fig. 5.28, so that the energy balance becomes

$$\omega c_p T_B + q_w'' 2\pi R \, dx = \omega c_p T_B + \omega c_p \frac{\partial T_B}{\partial x} dx,$$

where ω is the mass flow rate. Simplifying and solving for $\partial T_B/\partial x$ yields

$$\frac{\partial T_B}{\partial x} = \frac{q_w'' 2\pi R}{\omega c_p}.$$

However, $\omega = \rho \pi R^2 U_m$; hence,

$$\frac{\partial T_B}{\partial x} = \frac{2q_w''}{\rho c_p U_m R}. \tag{5.205}$$

Hence, since q_w'' is a constant by the constant flux boundary condition of this problem, it is seen that $\partial T_B/\partial x$ is also a constant.

Since the energy goes across the first few molecular layers by pure conduction,

$$q_w'' = -k(1) \left. \frac{\partial T}{\partial n} \right|_{n=0},\tag{5.206}$$

or, since $n = R - r$,

$$\left. \frac{\partial T}{\partial n} \right|_{n=0} = - \left. \frac{\partial T}{\partial r} \right|_{r=R},$$

and Eq. (5.206) becomes, after rearrangement,

$$\left. \frac{\partial T}{\partial r} \right|_{r=R} = + \frac{q_w''}{k}.\tag{5.207}$$

Hence, $(\partial T/\partial r)_{r=R}$ is also a constant independent of x.

Consider the expression

$$\frac{\partial}{\partial r} \left(\frac{T - T_B}{T_w - T_B} \right) = \frac{\partial}{\partial r} [F(r)].\tag{5.208}$$

Now, T_w certainly does not depend upon r, since it is the temperature at a specific value of r, that is, at $r = R$, and T_B does not depend upon r since it characterizes the energy per pound mass passing the cross section; hence, performing the differentiations in Eq. (5.208) yields

$$\frac{\partial T/\partial r}{T_w - T_B} = F'(r).$$

This expression can be evaluated at $r = R$ from Eq. (5.207); hence,

$$\frac{(\partial T/\partial r)_{r=R}}{T_w - T_B} = F'(R).\tag{5.209}$$

Substituting Eq. (5.207) into Eq. (5.209) and solving for $T_w - T_B$ yields

$$T_w - T_B = \frac{q_w''/k}{F'(R)}.\tag{5.210}$$

The right-hand side of Eq. (5.210) is a constant for the boundary condition being considered so that the difference $T_w - T_B$ is independent of x. Differentiating both sides of

Eq. (5.210) with respect to x (since this will relate some of the derivatives which appear in Eq. [5.204]) gives

$$\frac{\partial T_w}{\partial x} - \frac{\partial T_B}{\partial x} = 0.$$

Using this and Eq. (5.204), the entire second term on the right of Eq. (5.204) vanishes, and using Eq. (5.205) for the remaining term on the right gives, for $\partial T/\partial x$ in fully developed tube flow with the constant flux boundary condition,

$$\frac{\partial T}{\partial x} = \frac{2q_w''}{\rho c_p U_m R}. \qquad (5.211)$$

Hence, as expected, the fully developed condition simplifies this problem, since the axial temperature gradient is a known constant.

This consequence of the flow being fully developed thermally, Eq. (5.211), *with the heat flux q_w'' at the wall being a specified constant,* considerably simplifies the solution of the governing differential equation for the temperature profile, as will be evident in the next section.

For the case in which the tube wall temperature T_w is the known condition, instead of the surface heat flux being known and being constant, $\partial T/\partial x$ has a different, more complicated form than in Eq. (5.211). However, this form still allows the governing differential equation to be solved much more easily than is possible if the flow is not fully developed thermally and hydrodynamically. More details concerning cases of constant wall temperature and fully developed flow are available in [17].

Solution for Constant Flux Tube Wall

Now that the simplification for a fully developed flow has been found, as Eq. (5.211), for a circular tube of radius R when the surface heat flux q_w'' is a known constant, we will seek the temperature distribution within the moving fluid and the local surface coefficient of heat transfer h_x for the same situation. The flow is steady, laminar, low speed, and constant property, and at the inlet, $x = 0$, the fluid bulk mean temperature T_{B_1} would be known. (Note that this "inlet," $x = 0$, is far enough downstream of the actual tube entrance so that the flow and temperature fields are already fully developed.)

The governing partial differential equation for the temperature distribution has been previously derived for flow with radial symmetry as Eq. (5.201) for the low-speed situation. In addition, the fully developed nature of the flow causes Eq. (5.211) to be the expression for $\partial T/\partial x$, and this leads to $\partial^2 T/\partial x^2 = 0$ for the constant flux wall condition. Hence, Eq. (5.201) becomes

$$\frac{1}{r}\frac{\partial}{\partial r}\left(r\frac{\partial T}{\partial r}\right) = \frac{u}{\alpha}\frac{\partial T}{\partial x}. \qquad (5.212)$$

The boundary conditions are:

1. The temperature derivative in the radial direction is zero at the centerline because of the radial symmetry; hence, at $r = 0$ for all x, $\partial T / \partial r = 0$.
2. At the wall, the heat flux into the fluid is specified as a known constant q_w''; hence, $q_w'' = -k(\partial T / \partial n)_{n=0}$. However, $n = R - r$, where n is the coordinate measured in a positive sense perpendicular to and away from the tube wall. Thus,

$$\text{at } r = R \text{ for all } x, \quad (\partial T / \partial r)_R = q_w''/k.$$

3. At $x = 0$, $T_B = T_{B_1}$.

It has been shown that a fully developed temperature profile for this case implies that

$$\frac{\partial T}{\partial x} = \frac{2q_w''}{\rho c_p U_m R}, \tag{5.211}$$

or that $\partial T / \partial x$ is a known constant, say G_0. Hence,

$$\frac{\partial T}{\partial x} = G_0 = \frac{2q_w''}{\rho c_p U_m R}. \tag{5.213}$$

Before Eq. (5.212) can be solved for the temperature distribution, the velocity profile $u = u(r, x)$ must be determined. But, for the conditions of this problem, the velocity profile was obtained previously as

$$u = 2U_m \left(1 - \frac{r^2}{R^2} \right). \tag{5.57}$$

Substituting Eqs. (5.213) and (5.57) into Eq. (5.212) yields

$$\frac{1}{r} \frac{\partial}{\partial r} \left(r \frac{\partial T}{\partial r} \right) = \frac{2U_m G_0}{\alpha} \left(1 - \frac{r^2}{R^2} \right).$$

This equation can be integrated partially with respect to r to obtain

$$r \frac{\partial T}{\partial r} = \frac{2U_m G_0}{\alpha} \left(\frac{r^2}{2} - \frac{r^4}{4R^2} \right) + F(x).$$

Applying boundary condition 1 yields $F(x) = 0$; hence,

$$\frac{\partial T}{\partial r} = \frac{2U_m G_0}{\alpha} \left(\frac{r}{2} - \frac{r^3}{4R^2} \right). \tag{5.214}$$

The second boundary condition is also on the radial temperature derivative, and since there are no more functions or constants of integration at this point, Eq. (5.214) must satisfy boundary condition 2 as it stands. Thus, at $r = R$,

$$\left.\frac{\partial T}{\partial r}\right|_R = \frac{U_m G_0 R}{2\alpha},$$

or, using the definition of G_0, Eq. (5.213), Eq. (5.214) becomes at $r = R$ (noting that $\alpha = k/\rho c_p$)

$$\left.\frac{\partial T}{\partial r}\right|_R = \frac{R U_m (2 q_w''/\rho c_p U_m R)}{2k/\rho c_p} = \frac{q_w''}{k}.$$

This is the same result as boundary condition 2, and the reason it is automatically satisfied here is clarified by looking at the steps that led to the special expression of Eq. (5.211) for $\partial T/\partial x$ when the temperature profile is fully developed with a constant heat flux boundary condition. The constant heat flux boundary condition has already been used to obtain Eq. (5.211), which is being used in the present problem.

Integrating Eq. (5.214) with respect to r once more gives

$$T = \frac{2 U_m G_0}{\alpha}\left(\frac{r^2}{4} - \frac{r^4}{16 R^2}\right) + f(x). \tag{5.215}$$

The form of $f(x)$ can be found by integrating Eq. (5.213) partially with respect to x to yield

$$T = G_0 x + Q(r). \tag{5.216}$$

Since Eqs. (5.215–16) represent the same temperature distribution, $Q(r)$ must be within a constant of the function of r shown in Eq. (5.215), while $f(x)$ must be within a constant of the function of x shown in Eq. (5.216); hence, Eq. (5.215) can be written as

$$T = \frac{2 U_m G_0}{\alpha}\left(\frac{r^2}{4} - \frac{r^4}{16 R^2}\right) + G_0 x + C_1. \tag{5.217}$$

The constant C_1 is determined by boundary condition 3. The general expression for the bulk mean temperature is

$$T_B = \frac{\displaystyle\int_{A_c} u T \, dA}{\displaystyle\int_{A_c} u \, dA}. \tag{5.194}$$

In this case, $dA = 2\pi r \, dr$, so Eq. (5.194) becomes at $x = 0$, after substituting Eq. (5.217) and the velocity distribution, Eq. (5.57),

$$T_{B_1} = \frac{\int_0^R 2U_m \left(1 - \frac{r^2}{R^2}\right)\left[\frac{2U_m G_0}{\alpha}\left(\frac{r^2}{4} - \frac{r^4}{16R^2}\right) + C_1\right] 2\pi r \, dr}{\int_0^R 2U_m \left(1 - \frac{r^2}{R^2}\right) 2\pi r \, dr}.$$

Performing the indicated integrations and solving for C_1 yields

$$C_1 = T_{B_1} - \frac{14}{96}\frac{G_0 R^2 U_m}{\alpha}.$$

Substituting this into Eq. (5.217), the temperature distribution as a function of r and x for the given conditions is

$$T(r, x) = \frac{2U_m G_0}{\alpha}\left(\frac{r^2}{4} - \frac{r^4}{16R^2} - \frac{7}{96}R^2\right) + G_0 x + T_{B_1}. \qquad (5.218)$$

The local surface coefficient of heat transfer h_x is now computed from Eq. (5.3) as

$$h_x = \frac{-k(\partial T/\partial n)_{n=0}}{T_w - T_B} = \frac{+k(\partial T/\partial r)_{r=R}}{T_w - T_B}.$$

Boundary condition 2 requires that $(\partial T/\partial r)_R = q_w''/k$. From the temperature distribution, Eq. (5.218), the wall temperature T_w at any x can be found, by setting $r = R$, as

$$T_w(x) = \frac{22}{96}\left(\frac{U_m G_0 R^2}{\alpha}\right) + G_0 x + T_{B_1}. \qquad (5.219)$$

In Eq. (5.3), the local bulk mean temperature $T_B(x)$ is also needed. It has been shown that for these conditions,

$$\frac{\partial T_B}{\partial x} = \frac{2q_w''}{\rho c_p U_m R} = G_0.$$

Integrating and noting that $T_B = T_{B_1}$ at $x = 0$ yields

$$T_B(x) = G_0 x + T_{B_1}, \qquad (5.220)$$

where, from Eq. (5.213),

$$G_0 = \frac{2q_w''}{\rho c_p U_m R}. \qquad [5.213]$$

Using boundary condition 2 and Eqs. (5.219–20) in Eq. (5.3) yields

$$h_x = \frac{k(q_w''/k)}{(22/96)(U_m G_0 R^2/\alpha)}.$$

Using Eq. (5.213) and $\alpha = k/\rho c_p$ yields $h_x = (96/44)(k/R)$, or, introducing the diameter $D = 2R$,

$$h_x = 4.36(k/D). \tag{5.221}$$

Thus, Eq. (5.221) is the predicted local surface coefficient of heat transfer for steady, laminar, low speed, constant property, fully developed (both thermally and hydrodynamically) flow within a circular tube with negligible axial conduction in the fluid and a known constant surface heat flux q_w'' to the fluid. It is standard practice to put relations of the type shown in Eq. (5.221) in nondimensional form by rearranging in the manner

$$N_{Nu} = h_x D/k = 4.36. \tag{5.222}$$

The nondimensional group on the left is called the *local Nusselt number*.

Equations (5.221–22) show one of the major thermal characteristics of fully developed duct flow; namely, a local surface coefficient of heat transfer which is *not* a function of position. Note also that the local surface coefficient does not even depend upon the Reynolds number, as long as the Reynolds number is less than 2000, giving laminar flow through the tube. This will not be the case for fully developed turbulent flow, where it will be found that the Nusselt number does depend upon the Reynolds number. Hence, not only the procedure leading up to Eq. (5.222), but also the final result, Eq. (5.222), reflect the rather simple character of the convective heat transfer problem when fully developed conditions exist for both the velocity and the temperature field. With this, it should now be apparent why problems of this nature were earlier referred to as the simplest class of problems to handle analytically.

EXAMPLE 5.7

Air which will be used in an experiment is to be heated first from 21°C to 95°C in a 2.5-cm-diameter tube which is 28 m long and is wrapped with electrical resistance heating tape. The mass flow rate of the air is 0.0008 kg/s, and its properties at some suitable "average" temperature are given as $k = 0.03$ W/m °C, $\rho = 0.998$ kg/m³, $\mu = 2.075 \times 10^{-5}$ kg/ms, and $c_p = 1000$ J/kg °C. Calculate (a) the constant heat flux in W/m² which must be dissipated by the tape, and (b) the wall temperature at the tube exit.

Solution

If the flow is laminar and fully developed, then results such as Eq. (5.213) and (5.221) will be applicable here. Hence, first determine if the flow is laminar by computing and checking the diameter Reynolds number, Eq. (5.63).

$$N_{Re_D} = \rho U_m D / \mu \qquad\qquad\qquad [5.63]$$

The value U_m is found from the known mass flow rate:

$$\omega = \rho U_m \frac{\pi}{4} D^2$$

so,

$$U_m = \frac{4(0.0008)}{\pi(0.998)(0.025)^2} = 1.633 \text{ m/s.}$$

Thus, from Eq. (5.63),

$$N_{Re_D} = \frac{0.998(1.633)(0.025)}{2.075 \times 10^{-5}} = 1964.$$

By Eq. (5.182c), the flow criterion, this indicates that laminar flow conditions prevail, though just barely. At this point, some idea of whether or not the flow is fully developed hydrodynamically can be found by comparing the actual L/D of the tube to the one for the entrance length, which is given by Eq. (5.31).

So

$$L_e/D = 0.058(1964) = 114,$$

while

$$L/D = 28/0.025 = 1120.$$

Hence, we can judge that L is larger than L_e by a large enough factor, practically 10, to cause the velocity field to be fully developed throughout most of the tube. (A criterion for a *thermal* entrance length will be discussed shortly in connection with the constant wall temperature solution, and it is assumed to be satisfied.)

Thus, the wall flux will be found from Eq. (5.213) as soon as G_0 is determined. However, G_0 can be found from the axial distribution of the bulk mean temperature, from Eq. (5.220), since both the inlet and the exit bulk mean temperatures are known in the present problem. So, $95 = G_0(28) + 21$, or $G_0 = 2.64$. With this and Eq. (5.213),

$$q_w'' = \frac{0.998(1000)(1.633)(0.025/2)(2.64)}{2} = 26.9 \text{ W/m}^2. \qquad\qquad \textbf{(a)}$$

To find the tube wall temperature at the exit, one can either use Eq. (5.219) or work through the local surface coefficient of heat transfer from Eq. (5.221). The latter procedure will be used here, since this is what the surface coefficient is used for in the first place, namely, finding the wall temperature when there is a known flux at the tube wall.

So, from Eq. (5.221),

$$h_x = 4.36(0.03/0.025) = 5.23 \text{ W/m}^2 \text{ °C}.$$

Since, by Newton's cooling law, one has that

$$q_w'' = h_x (T_w - T_B),$$

and at the tube exit, $T_B = 95°C$, it follows from this that

$$T_w(L) = 95 + \frac{26.9}{5.23} = 100°C. \tag{b}$$

Note that in a problem such as the one just worked, where the surface heat flux is known, the basic question faced by the heat transfer analyst is to find the wall temperature distribution. In particular, the maximum wall temperature is often the quantity of most interest.

Results for Ducts of Various Cross Sections

Another common and important type of boundary condition for circular ducts and for ducts of other cross-sectional shapes as well, occurs when the duct surface is being held at a uniform known temperature rather than being subjected to a known surface flux. These constant wall temperature cases are more involved than the constant flux case. Most duct cross-sectional shapes, except the duct composed of two parallel planes, are more difficult to handle mathematically than is the circular duct, regardless of what the thermal boundary condition is at the wall. The general procedure to be used on these more complicated cases is outlined in [28] and [17]. Some of these results are presented here—along with the circular tube case already considered and with some other cases solved by Clark and Kays [29]—in Table 5.1.

From any of the cases listed in Table 5.1, the effect of *thermal history* on the local surface coefficient can be seen by noting the difference between the surface coefficients of heat transfer when everything is the same except the thermal boundary condition at the wall. For a circular tube, in Table 5.1, the surface coefficient of heat transfer for the constant wall flux condition is about 19% higher than that for the constant wall temperature condition. This results from the fact that two different types of thermal boundary conditions give rise to different temperature distributions within the moving fluid and, hence, different surface coefficients of heat transfer. Thus, if the wall temperature varies sinusoidally down the tube, a value different from those given in Table 5.1 would be expected for h_x.

Table 5.1
Local Nusselt Numbers

Description of Duct Cross-sectional Shape	Cross Section Hydraulic Diameter D_H	Local Nusselt Number $= \dfrac{h_x D_H}{k}$	
		Constant Flux q_w''	Constant Wall Temperature T_w
Case 1 Circular tube of diameter D	D	4.36	3.66
Case 2 Rectangle of width a and height $b : a = b$	a	3.63	2.98
Case 3 Rectangle of width a and height $b : a = 2b$	$\dfrac{2}{3} a$	4.11	3.39
Case 4 Rectangle of width a and height $b : a = 4b$	$\dfrac{2}{5} a$	5.35	4.44
Case 5 Parallel plates spaced a distance b apart	$2b$	8.22	7.54
Case 6 Equilateral triangle of side length b	$\dfrac{2}{3} b$	3.00	

Based on the hydraulic diameter D_H, for steady laminar, low speed, constant property, fully developed (both thermally and hydrodynamically) flow within ducts when axial conduction is negligible. Constant wall flux solutions are for a constant axial flux q_w'', but with perimeter of any cross section having constant temperature.

As indicated, the local Nusselt numbers, and therefore the local surface coefficients of heat transfer, are those for hydrodynamically and thermally fully developed laminar flow for the boundary conditions of either constant known flux or constant known wall temperature. The characteristic dimension D_H to be used in the Nusselt numbers given in Table 5.1 is the previously defined *hydraulic diameter,* namely,

$$D_H = 4A/P, \tag{5.223}$$

where A is the cross-sectional area through which the fluid is flowing and P is the wetted perimeter, the portion of the duct perimeter which has shear stress on it.

A criterion whereby one can check to see whether the flow is fully developed hydrodynamically or not has already been given as Eq. (5.31) for laminar flow. However, an additional criterion is needed to see what the thermal entrance length L_t is equal to. The value L_t is the thermal counterpart of the hydrodynamic entrance length L_e shown in Fig. 5.4. That is, L_t is the distance down the duct that is required so that the nondimensional temperature distribution does not depend upon position, x, down the duct. In other words,

L_t is the first position down the duct at which Eq. (5.202), which represents the fully developed thermally condition, is satisfied. By inspecting the exact solution for the local Nusselt number, in a constant temperature tube, in the general case in which the flow is *not* fully developed thermally, one can conclude that the Nusselt number becomes constant, and the temperature fully developed, at about [17]:

$$L_t/D = 0.05 \, N_{Re_D} \, N_{Pr}. \tag{5.224}$$

$$N_{Pr} = \mu c_p/k \tag{5.225}$$

Because of its importance and frequent appearance in convection heat transfer, the Prandtl number, N_{Pr}, as defined in Eq. (5.225), is listed in the property tables in Appendix A for some common heat transfer fluids.

Though Eq. (5.224) is valid, strictly speaking, only for a constant wall temperature tube, it can also be used as an estimate for the constant flux tube wall thermal condition.

Logarithmic Mean Temperature Difference

Since the local surface coefficient of heat transfer h_x is constant with x for fully developed flow in constant cross-sectional area ducts, as can be seen from Table 5.1, we can expect that the average value of the surface coefficient from $x = 0$ to $x = L$ is also equal to the same constant. Hence, the total heat transfer rate over the length L can be computed from Newton's law of cooling as

$$q = h_x P L \Delta T, \tag{5.226}$$

where P is the perimeter of the duct and ΔT is the appropriate driving temperature difference. If the duct wall is at the known temperature T_w, then ΔT must be interpreted as T_w minus some "average" temperature of the fluid. The "average" fluid temperature which comes to mind most quickly, perhaps, is the arithmetic average of the inlet and outlet bulk mean temperatures,

$$\overline{T}_s = \tfrac{1}{2}(T_{B_1} + T_{B_2}).$$

However, no guessing about the correct ΔT is required in such a situation, since the law of conservation of energy determines the correct ΔT to use.

Figure 5.29 shows a general duct of constant perimeter P and length L. Its walls are held at the constant known temperature T_w and the flow is fully developed, giving a constant value of the local surface coefficient of heat transfer h_x. Since its value is constant for all x, this is also the average value of the surface coefficient from $x = 0$ to $x = L$. Other known quantities are the mass flow rate ω, specific heat c_p, and the inlet and outlet bulk mean temperatures T_{B_1} and T_{B_2} of the fluid. We would like to derive an expression for the correct driving potential difference, ΔT, to use in Eq. (5.226) for the total heat transfer rate to or from the fluid over the length L of the duct.

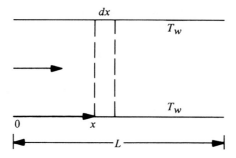

Figure 5.29 Differential element and duct.

Taking the entire duct from $x = 0$ to $x = L$ as the control volume, Eq. (1.8) reduces to $R_{in} = R_{out}$ for steady-state conditions and no generation. Energy enters by convection from the duct walls at the rate given by Newton's cooling law as

$$q = h_x PL\Delta T, \tag{5.227}$$

where ΔT is to be found in terms of known temperatures in the problem and where h_x is also the average surface coefficient, since h_x does not vary with x. Energy also enters the control volume because of mass entering the face at $x = 0$ by virtue of its motion in the x direction at the rate

$$\omega c_p T_{B_1}. \tag{5.228}$$

(The kinetic energy term is not shown because of the assumption of low-speed flow.) Energy leaves the control volume, the duct, at $x = L$ by virtue of its motion at the rate

$$\omega c_p T_{B_2}. \tag{5.229}$$

(Axial conduction is considered negligible; see the restrictions on the use of Table 5.1.) Substituting Eqs. (5.227–29) into the energy balance yields

$$h_x PL\Delta T + \omega c_p T_{B_1} = \omega c_p T_{B_2}.$$

Solving for ΔT gives

$$\Delta T = \frac{\omega c_p}{h_x PL}(T_{B_2} - T_{B_1}). \tag{5.230}$$

Now in an effort to eliminate all quantities in Eq. (5.230) except temperatures and temperature differences, apply Eq. (1.8) to the differential control volume of length dx extending over the entire cross section of the duct, shown in Fig. 5.29. Again, $R_{in} = R_{out}$.

Energy is transferred from the wall to the fluid at the rate

$$h_x P \, dx \, (T_w - T_B),$$
$$(5.231)$$

where T_B is the local bulk mean temperature. There is no ambiguity regarding the ΔT in Eq. (5.231) because it was chosen to be $T_w - T_B$ on a local basis by the reasoning presented earlier. Energy enters the control volume at x by virtue of the fluid motion, carrying it across the control surface at the rate

$$\omega c_p T_B.$$
$$(5.232)$$

Energy leaves the right-hand face at $x + dx$ at the rate given by the Taylor series expansion of Eq. (5.232) about x across to $x + dx$; that is, neglecting the higher-order terms,

$$\omega c_p T_B + \omega c_p \frac{dT_B}{dx} \, dx.$$
$$(5.233)$$

(Note that ω is not a function of x.) Substituting Eqs. (5.231–33) into the energy balance yields

$$h_x P \, dx \, (T_w - T_B) + \omega c_p T_B = \omega c_p T_B + \omega c_p \frac{dT_B}{dx} \, dx.$$

Cancelling terms and rearranging,

$$h_x P \, (T_w - T_B) = \omega c_p \frac{dT_B}{dx}.$$
$$(5.234)$$

Now the variables T_B and x can be separated, leading to the definite integral

$$\frac{\omega c_p}{h_x P} \int_{T_{B_1}}^{T_{B_2}} \frac{dT_B}{T_w - T_B} = \int_0^L dx.$$

Integration yields

$$\frac{\omega c_p}{h_x P} \ln \left(\frac{T_w - T_{B_1}}{T_w - T_{B_2}} \right) = L.$$

Solving for $\omega c_p / h_x P L$ yields

$$\frac{\omega c_p}{h_x P L} = \frac{1}{\ln \left(\dfrac{T_w - T_{B_1}}{T_w - T_{B_2}} \right)}.$$

Using this to eliminate the same quantity in Eq. (5.230), the appropriate driving potential difference is

$$\Delta T_L = \frac{T_{B_2} - T_{B_1}}{\ln\left(\dfrac{T_w - T_{B_1}}{T_w - T_{B_2}}\right)}.\tag{5.235}$$

This is called the *logarithmic mean temperature difference,* or simply the L.M.T.D., and thus the subscript L has been placed on the ΔT term. This is the appropriate driving potential difference to use in Newton's law of cooling for fully developed flow in a duct of constant cross-sectional area when its wall is being held at the constant temperature T_w. This result approaches the ΔT based upon the arithmetic mean temperature difference, namely, $T_w - \frac{1}{2}(T_{B_1} + T_{B_2})$, when $T_{B_1} - T_{B_2}$ is small in magnitude when compared to $T_{B_1} - T_w$ or $T_{B_2} - T_w$.

Equation (5.235), though developed in the present section on laminar, fully developed flow, can be applied in a number of different situations. For example, it can be applied directly to fully developed *turbulent* flow through a constant temperature duct, since in this case also, h_x is independent of distance down the duct. In addition, if one uses the average surface coefficient $\overline{h}_L$ over the length L instead of the local h_x in Eq. (5.226), then Eq. (5.235) can be used even in situations where the velocity field, temperature field, or both are not fully developed, as long as the duct wall temperature is a constant. In Chapter 8, a derivation will be presented which allows a modified version of Eq. (5.235) to be used when the wall temperature and h_x both vary with position down the duct.

Bulk Mean Temperature Distribution for Duct of Constant Wall Temperature

Previously, it has been shown that, for fully developed duct flow, the bulk mean temperature varies linearly with x *when the heat flux at the duct wall is constant.* Since very often the function of a single duct, or one of many ducts in a heat exchanger, is to increase or decrease the bulk mean temperature of the fluid within the duct, it is of interest to know how the bulk mean temperature varies with x in a constant wall temperature duct. This can be arrived at quite easily by reference to the development of Eq. (5.235).

We can see that Eq. (5.234), which was derived by an energy balance on the fluid flowing in the duct, is the differential equation for the local bulk mean temperature distribution with x.

$$h_x P\,(T_w - T_B) = \omega c_p \frac{dT_B}{dx} \qquad [5.234]$$

If the situation is limited to fully developed flow, so that h_x is a constant, and to a constant duct wall temperature so that T_w is independent of x, Eq. (5.234) is rearranged and integrated from $x = 0$, where $T_B = T_{B_1}$, a known value, to the general position x

down the duct where the bulk mean temperature has its general value, $T_B(x)$. This leads to

$$\int_{T_{B_1}}^{T_B} \frac{dT_B}{T_B - T_w} = -\frac{h_x P}{\omega c_p} \int_0^x dx, \quad \text{or} \quad \ln\left(\frac{T_B - T_w}{T_{B_1} - T_w}\right) = -\frac{h_x P}{\omega c_p} x.$$

Taking the antilogarithm and rearranging yields

$$T_B = T_w + (T_{B_1} - T_w) e^{-h_x Px/\omega c_p}. \tag{5.236}$$

From Eq. (5.236), we can see that the bulk mean fluid temperature varies exponentially with x, and not linearly. In view of this, it is not expected that the arithmetic mean temperature difference is the correct one to use in Newton's cooling law. Indeed, it is not, as was demonstrated in the derivation of Eq. (5.235) for the constant wall temperature condition.

Limiting Values of the Surface Coefficient

The expressions for the surface coefficients of heat transfer given in Table 5.1 are valuable not only for the fully developed laminar flows to which they apply, but also for placing a bound on problems where the flow is not fully developed or even turbulent. In the last section of this chapter, the local and average surface coefficients of heat transfer will be found to be generally higher than those given in Table 5.1 when the flow is not fully developed and/or turbulent. Thus, use of the relations in Table 5.1 gives a conservative estimate (lower bound) of the surface coefficient in these other cases. This lower bound can be calculated quite quickly and easily from Table 5.1, whereas the actual values of the surface coefficients in undeveloped and/or turbulent flow come from relations, given in another section of this chapter, that are more complicated than those in Table 5.1. In some design situations, it may be useful to establish some of the bounds on a problem, such as the greatest length a particular pipe must be, very quickly at the beginning of the problem.

However, in some cases, such as with fully developed turbulent flow, the actual values of the surface coefficient h are so much greater than the limiting values given in Table 5.1 that an estimate, using the values in Table 5.1, may be so ultra-conservative that its effective use in design and in analysis is somewhat limited.

EXAMPLE 5.8

Twenty lbm/hr of an oil is to cool from 200°F to 100°F by flowing through a 1/2-in.-diameter tube which is 15 ft long. Since oils have a highly temperature-dependent viscosity, the flow is never fully developed in the sense of Table 5.1. In the temperature range from 100°F to 200°F, the viscosity can easily change by a factor of 5 or more. Make a first estimate of the tube wall temperature required to accomplish the cooling by assuming that the properties of the oil are constant at the values $\rho = 54$ lbm/ft^3, $c_p = 0.47$ Btu/lbm °F, $k = 0.075$ Btu/hr ft °F, and $\mu = 19$ lbm/hr ft.

Solution

Since the oil is to be treated as having constant properties, Case 1 of Table 5.1 for the constant wall temperature condition can be used if the flow is laminar. To check the Reynolds number, first the average oil velocity is obtained from

$$\omega = \rho \, \tfrac{1}{4} \, \pi D^2 U_m, \quad \text{or} \quad 20 = 54 \, \tfrac{1}{4} \, \pi(\tfrac{1}{24})^2 \, U_m, \quad \text{or} \quad U_m = 272 \text{ ft/hr.}$$

Hence,

$$N_{\text{Re}} = \frac{\rho \, U_m D}{\mu} = \frac{54(272)(\tfrac{1}{24})}{19} = 32.2,$$

and the flow is laminar.

An energy balance on the entire pipe yields

$$h_x PL \Delta T = \omega c_p (T_{B_2} - T_{B_1}), \tag{5.237}$$

where h_x can be found from Case 1 of Table 5.1 for the constant wall temperature condition. Thus, since $(h_x D/k) = 3.66$, or $[h_x(\tfrac{1}{24})]/0.075 = 3.66$,

$$h_x = 6.59 \text{ Btu/hr ft}^2 \, {}^\circ\text{F.}$$

Also, $P = \pi D = 3.14/24 = 0.131$.

The appropriate driving potential difference to use here is the logarithmic mean temperature difference

$$\Delta T_L = \frac{T_{B_2} - T_{B_1}}{\ln\left(\dfrac{T_w - T_{B_1}}{T_w - T_{B_2}}\right)}. \tag{5.235}$$

Substituting this into Eq. (5.237) and cancelling common terms yields

$$\frac{h_x PL}{\ln\left(\dfrac{T_w - T_{B_1}}{T_w - T_{B_2}}\right)} = \omega c_p.$$

Substituting known values yields

$$\frac{6.59(0.131)15}{\ln\left(\dfrac{T_w - 200}{T_w - 100}\right)} = 20(0.47), \quad \text{or} \quad \ln\left(\frac{T_w - 200}{T_w - 100}\right) = 1.377.$$

Taking the antilogarithm leads to

$$\frac{T_w - 200}{T_w - 100} = 3.96, \quad \text{or} \quad T_w = 66.2^\circ\text{F.}$$

In Ex. 5.8, even though we felt that the properties of the oil were so temperature dependent that the flow and temperature fields could not be fully developed, Table 5.1 was employed to make an initial estimate of the h_x and, therefore, of the required tube wall temperature. Thus, the entrance length expressions, Eqs. (5.31) and (5.224), were not even checked. The Reynolds number of 32.2, when inserted into Eq. (5.31), leads to

$$L_e/D = 1.87.$$

Since $N_{Pr} = \mu c_p/k = 19(0.47)/0.075 = 119$, Eq. (5.224) gives

$$L_t/D = 192.$$

The actual L/D is $15(12)/0.5 = 360$. Hence, even if the oil was a more or less constant property fluid, it would not be thermally fully developed. Hence, our estimate of the surface coefficient is in error, not only because of the variation of properties with temperature, but also because the tube is too short for fully developed flow of even a constant property fluid. Thus, the wall temperature of about 66°F is only a rough first estimate to be used for preliminary design or feasibility studies.

5.4b Similarity Between the Transport of Heat and Momentum

In the previous section, attention focused on finding exact solutions for the surface coefficient of heat transfer in laminar, fully developed duct flow. Now we look at the problem of predicting the local surface coefficient of heat transfer in fully developed *turbulent* flow of a fluid in a duct. However, as mentioned in our review of fluid mechanics, we are not able to develop exact solutions to turbulent flow and heat transfer problems from first principles alone. As a result of this, our "analytical" predictions and "theory" of turbulent heat transfer are actually semi-analytical and semi-theoretical approaches. They emphasize, in this text, the similarities which exist under certain conditions between energy transfer and momentum transfer, and they make use of the experimental information which, in effect, is contained within the law of the wall, Eq. (5.104).

In this section two of the simpler theories, due to Reynolds and von Kàrmàn, are presented for the similarity between the transport of heat due to temperature gradients, and the transport of momentum due to velocity gradients.

The basis for the similarity theory (not to be confused with the method of the similarity transformation discussed briefly in Sec. 5.3) lies partly in the exact equivalence of the governing partial differential equations and the boundary conditions for both the velocity field and the temperature field in certain laminar flow situations [15], and partly in the physical observation that, in gas flow, the same molecules that cause the shear stress because of transport of momentum on a molecular level also cause conduction heat transfer because of the transport of energy on the molecular level. Under a certain condition, the molecular transport coefficients to which these motions give rise, the kinematic viscosity v and the thermal diffusivity α, are equal. This in turn, in certain laminar flows, causes the appropriate nondimensional velocity profile and temperature profile to be the same function of the space coordinates, that is, to be "similar." This physical observation can be extended to turbulent flow, where there are additional apparent shear stresses and increased conduction heat transfer rates due to the motion of the turbulent eddies as described in Sec. 5.3. The same eddy that transports momentum across any plane also transports energy; thus, it might be expected that the turbulent transport coefficients ϵ_m and ϵ_H caused by this eddy motion are practically equal to each other. The ratio of ϵ_m to ϵ_H is called the *turbulent Prandtl number* N_{Pr_t}. Thus,

$$N_{Pr_t} = \epsilon_m/\epsilon_H.$$

As a result, if the turbulent transport coefficients are equal, the turbulent Prandtl number is unity. In fully developed pipe flow and boundary layers, measurements have indicated average turbulent Prandtl numbers of approximately 0.7, while for jets and wake-like flows, values are closer to 0.5 [15], [30]. The use of $N_{Pr_t} = 1$ has, however, generally given acceptable results and will be used herein.

The ratio of the molecular diffusivity of momentum v (the kinematic viscosity) to the molecular diffusivity of heat α (the thermal diffusivity) is called the *molecular Prandtl number*, or simply the *Prandtl number* N_{Pr}. Hence,

$$N_{Pr} = \frac{v}{\alpha} = \frac{\mu c_p}{k}. \tag{5.238}$$

The second part of the equality in Eq. (5.238) can be proven by recourse to the definitions of v and α as μ/ρ and $k/\rho c_p$, respectively. The Prandtl number is a fluid property weakly dependent upon pressure and more strongly dependent, at least for liquids, on temperature. Values of the Prandtl number for selected fluids are given in Appendix A.

The Reynolds Analogy

Consider now the problem of finding the local surface coefficient of heat transfer in steady, fully developed, turbulent, constant property flow in a circular tube of radius R when the tube wall is at the constant temperature T_w. Use y as the general position coordinate transverse to x, as shown in Fig. 5.30, where $y = R - r$. The local shear stress at any value y and the local heat transfer rate per unit area in the y direction are given by the equations developed in Sec. 5.3; that is,

$$\tau = \rho(v + \epsilon_m) \frac{\partial u}{\partial y}, \tag{5.83}$$

$$q'' = -\rho c_p(\alpha + \epsilon_H) \frac{\partial T}{\partial y}. \tag{5.89}$$

Also in Sec. 5.3, it was shown that the local shear stress varies linearly away from the wall according to

$$\tau = \tau_w \left(1 - \frac{y}{R}\right). \tag{5.93}$$

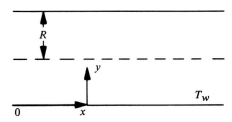

Figure 5.30 Details of the duct used in the derivation of the Reynolds analogy.

Substituting Eq. (5.93) into Eq. (5.83), rearranging, and using Eq. (5.89) leads to

$$\frac{q''}{\rho} = -c_p(\alpha + \epsilon_H) \frac{\partial T}{\partial y}, \tag{5.239}$$

$$\frac{\tau_w}{\rho}\left(1 - \frac{y}{R}\right) = (\nu + \epsilon_m) \frac{\partial u}{\partial y}. \tag{5.240}$$

Assume that the fluid flowing has a Prandtl number of 1; that is, from Eq. (5.238),

$$\nu = \alpha. \tag{5.241}$$

(Most diatomic gases, such as air, have Prandtl numbers of about 0.70, which for present purposes is close enough to unity.) Reynolds also assumed that $\epsilon_m = \epsilon_H$ by the arguments just used for the similarity between the transport of heat and momentum by the motion of the same turbulent eddies. Hence,

$$\epsilon_m = \epsilon_H, \tag{5.242}$$

and, from Eqs. (5.241–42), it is concluded that

$$\alpha + \epsilon_H = \nu + \epsilon_m. \tag{5.243}$$

Hence, dividing Eq. (5.239) by Eq. (5.240) and using Eq. (5.243),

$$\frac{q''}{\tau_w\left(1 - \frac{y}{R}\right)} = -c_p \frac{\partial T}{\partial u}. \tag{5.244}$$

In order to develop an equation for the local surface coefficient of heat transfer, q'' at $y = 0$, say q''_w, is required in Eq. (5.244), rather than the general heat flux q'' at any y position. Because of the similarity between the transport of momentum and heat on the molecular level, Eq. (5.241), and on the macroscopic level (level of the turbulent eddies),

Eq. (5.242), it is not unreasonable to expect that the local heat flux will have a distribution in y which is essentially the same as the local shear stress, Eq. (5.93). Hence, assume that

$$q'' = q''_w \left(1 - \frac{y}{R}\right).$$

Substituting this into Eq. (5.244) and cancelling the common quantity yields

$$\frac{q''_w}{c_p \tau_w} = -\frac{\partial T}{\partial u}.$$

Separating the variables u and T gives

$$\frac{q''_w}{c_p \tau_w} \, du = -dT. \tag{5.245}$$

This equation is integrated at constant x in the y direction from $y = 0$, where $u = 0$ and $T = T_w$, to the y coordinate, where $u = U_m$, the average flow velocity, and where T is assumed to be equal to the bulk mean temperature T_B. This assumption—that the average velocity and bulk mean temperature occur at the same value of y—is not ordinarily a poor one, since from experiment it is known that the velocity profile has steep gradients near the wall with only relatively small changes over the central part of the pipe cross section. Because of the expected similarity, the temperature profile will also be flat near the center of the pipe. Even if the local temperature at the point where $u = U_m$ is not T_B, it would not be expected to be very different than T_B; hence, little error is incurred by this assumption (for turbulent flows).

Since Eq. (5.245) is to be integrated at constant x, the parameter $q''_w / c_p \tau_w$ behaves as a constant during the integration. Therefore,

$$\frac{q''_w}{c_p \tau_w} \int_0^{U_m} du = -\int_{T_w}^{T_B} dT, \quad \text{or} \quad \frac{q''_w U_m}{c_p \tau_w} = T_w - T_B. \tag{5.246}$$

But, by Newton's cooling law the local surface coefficient of heat transfer can be written

$$h_x = \frac{q''_w}{T_w - T_B}.$$

Substituting this into Eq. (5.246) and rearranging yields

$$h_x = \frac{c_p \tau_w}{U_m}. \tag{5.247}$$

Ordinarily, the local skin friction coefficient

$$C_f = \frac{\tau_w}{\frac{1}{2} \rho U_m^2} \tag{5.65}$$

is introduced; thus, $\tau_w = \rho U_m^2 \frac{1}{2} C_f$. Substituting this into Eq. (5.247) and rearranging yields

$$\frac{h_x}{\rho c_p U_m} = \frac{C_f}{2} .$$

The nondimensional group on the left containing the local surface coefficient of heat transfer is called the *Stanton number* N_{St_x}, and the final result is

$$N_{St_x} = \frac{h_x}{\rho c_p U_m} = \frac{C_f}{2} . \tag{5.248}$$

This simplest of the theories for the similarity between the transport of heat and momentum is called the *Reynolds analogy*. It can also be put in terms of the Moody friction factor f, which for pipe flow is $4C_f$; hence, Eq. (5.248) can be written

$$N_{St_x} = \frac{h_x}{\rho c_p U_m} = \frac{C_f}{2} = \frac{f}{8} . \tag{5.249}$$

This is the expression for the local surface coefficient of heat transfer in steady, turbulent, low-speed, fully developed, constant property flow in a circular pipe when the wall temperature is a known constant, the fluid Prandtl number is 1 (or near 1), and when axial conduction is negligible. This result must be classified as semi-analytical, because the predicted h_x depends upon the friction factor, which cannot be determined from first principles alone. However, a result such as Eq. (5.249) is extremely useful because the friction factors have been measured and presented in the Moody diagram, Fig. 5.9, by researchers primarily interested in pressure loss calculations, and their efforts can now be used to obtain surface coefficients of heat transfer as well.

An examination of Eq. (5.249) shows that, similar to fully developed laminar flows of the type represented by the results in Table 5.1, fully developed turbulent pipe flows also yield a local value of the surface coefficient which does not vary with position. Hence, by earlier reasoning, the logarithmic mean temperature difference, Eq. (5.235), should be used when calculating the heat transfer rate over a finite length of pipe.

A major limitation on the usefulness of the Reynolds analogy, Eq. (5.249), is the restriction, employed in the derivation, to a molecular Prandtl number of about unity. However, on the basis of an examination of experimental data for turbulent pipe flow, Colburn [31] inserted the Prandtl number dependence of the Stanton number. This gives the relationship which follows and is valid for Prandtl numbers between about 0.6 and 100:

$$N_{St_x} N_{Pr}^{2/3} = \frac{f}{8} , \tag{5.250}$$

where

$$N_{\text{St}_x} = h_x/\rho c_p U_m. \tag{5.251}$$

Equation (5.250) applies to steady, constant property, thermally and hydrodynamically fully developed turbulent flow through a pipe held at constant wall temperature, and is referred to as either the *Modified Reynolds Analogy* or as *Colburn's Analogy*.

By a simple rearrangement, the Stanton number N_{St_x} can be related to the nondimensional group containing the surface coefficient of heat transfer which we worked with previously. This is the local Nusselt number N_{Nu}, defined in Eq. (5.222).

$$N_{\text{St}_x} = N_{\text{Nu}}/N_{\text{Re}_D}N_{\text{Pr}}. \tag{5.252}$$

By fitting a curve through experimental friction factor data for fully developed turbulent flow in smooth-walled pipes and using this expression, along with Eq. (5.252), in Eq. (5.250), Colburn [31] arrived at the following result:

$$N_{\text{Nu}} = 0.023 \, N_{\text{Re}_D}^{0.8} N_{\text{Pr}}^{1/3}. \tag{5.253}$$

The nondimensional groups in Eq. (5.253) have been previously defined in Eqs. (5.222), (5.63), and (5.225), respectively. Comparison of this semi-analytical result, Eq. (5.253) (which was derived in part on the basis of the arguments that led to the Reynolds analogy, Eq. [5.249]), with experimental heat transfer measurements in turbulent, fully developed pipe flow indicates reasonable agreement between the two.

Noncircular Cross Sections   In fully developed turbulent flow, Eq. (5.250) applies to ducts of noncircular cross section with reasonable accuracy, as long as the hydraulic diameter D_H defined in Eq. (5.223) is used as the length coordinate in the Reynolds number when obtaining the friction factors from the Moody diagram, Fig. 5.9. (Fully developed turbulent flow is, in some respects, more universal than laminar flow. For example, the same form of the law of the wall as applies over the entire cross section in fully developed flow also applies over part of any external, turbulent, boundary layer-type flow—see Sec. 5.3.) However, the use of the hydraulic diameter D_H is not proper in those cross sections whose shape is such that an appreciable percentage of the cross section is experiencing laminar flow, regions which can be described as "corners," while the central portion of the cross section is experiencing turbulent flow. A cross section which possibly could behave in this manner is shown in Fig. 5.31, where the probable regions of laminar flow are shaded.

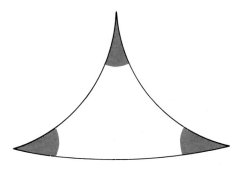

Figure 5.31 A cross section with regions of laminar flow (shown as shaded) in the corners.

von Kàrmàn's Analogy

T. von Kàrmàn [32] devised an improved similarity theory that removes the major restriction, that the Prandtl number be unity, from the Reynolds analogy. The basic procedure of von Kàrmàn for fully developed, turbulent, steady (on the average), constant property tube flow is now briefly outlined. More detailed discussions can be found in [17], [7], and [32]. The same two basic equations used at the beginning of the derivation of the Reynolds analogy are used with Eq. (5.93):

$$\tau = \rho(\nu + \epsilon_m) \frac{\partial u}{\partial y},$$ [5.83]

$$q'' = -\rho c_p(\alpha + \epsilon_H) \frac{\partial T}{\partial y},$$ [5.89]

$$\tau = \tau_w \left(1 - \frac{y}{R}\right).$$ [5.93]

As in the Reynolds analogy, it is assumed that the turbulent transport coefficients are equal, that is, $\epsilon_m = \epsilon_H$. Using this and introducing Eq. (5.93) into Eq. (5.83) yields

$$\tau_w \left(1 - \frac{y}{R}\right) = \rho(\nu + \epsilon_m) \frac{\partial u}{\partial y},$$ (5.254)

$$q'' = -\rho c_p(\alpha + \epsilon_m) \frac{\partial T}{\partial y}.$$ (5.255)

The similarity assumption

$$q''/\tau = q_w''/\tau_w$$ (5.256)

is also used. The temperature and flow fields are broken up into the three regions of the laminar sublayer, the buffer region, and the fully turbulent core, to correspond to the three different regions identified by von Kàrmàn for his form of the law of the wall, Eq. (5.104).

Thus,

$$u^+ = y^+ \text{ when } 0 < y^+ < 5 \text{ in the viscous sublayer,}$$

$$u^+ = 5.0 \ln y^+ - 3.05 \text{ for } 5 < y^+ < 30 \text{ in the buffer region,} \quad \textbf{[5.104]}$$

$$u^+ = 2.5 \ln y^+ + 5.5 \text{ when } y^+ > 30 \text{ in the overlap layer,}$$

where, by the definitions in Sec. 5.3,

$$u^+ = u/u^*, \qquad\qquad\qquad \textbf{[5.99]}$$

$$y^+ = yu^*/\nu, \qquad\qquad\qquad \textbf{[5.100]}$$

$$u^* = \sqrt{\tau_w/\rho} \ . \qquad\qquad\qquad \textbf{[5.98]}$$

The viscous sublayer is so thin and the velocities within it are so small that it is assumed that the local heat flux q'' in the y direction is about constant at the value for the wall q_w''. Also, in the viscous sublayer, the turbulent exchange coefficient ϵ_m is essentially zero, so that within the viscous sublayer, Eq. (5.255) becomes

$$q_w'' = -\rho c_p \alpha \frac{\partial T}{\partial y} \ . \qquad\qquad\qquad \textbf{(5.257)}$$

In reality, the eddy diffusivity of momentum ϵ_m, though small in the viscous sublayer, is not zero except right at the wall itself, as was discussed in Sec. 5.3d. This fact has important implications for *high* Prandtl number fluids, since much of the resistance to heat transfer for such fluids is located in the viscous sublayer, where ϵ_m, in the present analysis, is being neglected. The use of $\epsilon_m = 0$ in the viscous sublayer limits the accuracy of the final result somewhat for N_{Pr} greater than about 5. Additional information concerning this point, as well as a discussion of the use of a nonzero value of ϵ_m in the viscous sublayer, can be found in [17]. Separating the variables and integrating from $y = 0$, where $T = T_w$, to $y = y_L$, the edge of the viscous sublayer where $T = T_L$, the unknown fluid temperature at the edge of the viscous sublayer, leads to

$$\frac{q_w'' y_L}{\rho c_p \alpha} = T_w - T_L. \qquad\qquad\qquad \textbf{(5.258)}$$

According to the law of the wall, Eq. (5.104), the edge of the viscous sublayer occurs at

$$y^+ = 5. \qquad\qquad\qquad \textbf{(5.259)}$$

Using Eqs. (5.100) and (5.98) in Eq. (5.259) yields

$$y_L = \frac{5\nu}{\sqrt{\tau_w/\rho}} \ . \qquad\qquad\qquad \textbf{(5.260)}$$

Substituting this into Eq. (5.258) and noting that the ratio of ν to α is the Prandtl number gives

$$\frac{5q_w'' N_{\text{Pr}}}{\rho c_p \sqrt{\tau_w/\rho}} = T_w - T_L. \tag{5.261}$$

This by itself does not give q_w'' in terms of τ_w, and therefore not in terms of C_f or f, because the unknown temperature at the edge of the sublayer must be determined. Thus, another expression such as Eq. (5.261) for the buffer region and another for the overlap layer must be developed. In this way, the unknown temperatures at the edge of the sublayer T_L and at the edge of the buffer region T_0 can be eliminated to yield q_w'' as a function of τ_w and, therefore, of either C_f or f.

In the buffer region too, von Kàrmàn assumes that the region is thin enough and the velocities low enough so that the local heat flux q'' in the y direction does not differ too much from its value at the wall; thus, $q'' \approx q_w''$ in the buffer region. The buffer region is the portion of the flow cross section in which molecular and turbulent transport coefficients are of the same order of magnitude; hence, Eq. (5.255) becomes, for the buffer region,

$$q_w'' = -\rho c_p (\alpha + \epsilon_m) \frac{\partial T}{\partial y}. \tag{5.262}$$

Although q_w'' is essentially constant over the buffer region, Eq. (5.262) cannot yet be integrated directly, as was Eq. (5.257) in the sublayer, because the eddy diffusivity of momentum ϵ_m can be a function of y. In order to determine how ϵ_m varies with y, consider Eq. (5.254). The buffer region is still so close to the wall at $y = 0$ that the term y/R can be neglected in comparison to unity, and Eq. (5.254) becomes

$$\tau_w/\rho = (\nu + \epsilon_m)\,(\partial u/\partial y).$$

The left-hand side is, by Eq. (5.98), equal to u^{*2}, and hence can be rearranged as

$$1 = \left(1 + \frac{\epsilon_m}{\nu}\right)\left(\frac{\nu}{u^{*2}}\right)\frac{\partial u}{\partial y}. \tag{5.263}$$

Since the partial differentiation indicated by Eq. (5.263) is performed while holding x constant, the quantities u^* and u^*/ν can be brought inside the differential operator to yield

$$1 = \left(1 + \frac{\epsilon_m}{\nu}\right)\left[\frac{\partial(u/u^*)}{\partial(yu^*/\nu)}\right]. \tag{5.264}$$

However, by Eqs. (5.99–100), Eq. (5.264) becomes

$$1 = \left(1 + \frac{\epsilon_m}{\nu}\right)\frac{du^+}{dy^+}. \tag{5.265}$$

Only the ordinary derivative is required in Eq. (5.265), since $u^+ = F(y^+)$ where the law of the wall is valid. From Eq. (5.265),

$$\epsilon_m = \left(\frac{1}{du^+/dy^+} - 1 \right) \nu. \tag{5.266}$$

But, from Eq. (5.104), in the buffer region, the law of the wall is $u^+ = 5.0 \ln y^+ - 3.05$; hence,

$$\frac{du^+}{dy^+} = \frac{5.0}{y^+}.$$

With this, Eq. (5.266) gives the eddy diffusivity of momentum within the buffer region as a known function of the coordinate y^+,

$$\epsilon_m = \nu \left(\frac{y^+}{5} - 1 \right). \tag{5.267}$$

Using Eq. (5.267), Eq. (5.262) can be integrated if the y variable is changed to y^+ by multiplying numerator and denominator by u^*/ν and rearranging to yield

$$q_w'' = -\rho c_p (\alpha + \epsilon_m) \frac{u^*}{\nu} \frac{\partial T}{\partial (yu^*/\nu)} = -\rho c_p \left(\frac{1}{N_{\text{Pr}}} + \frac{\epsilon_m}{\nu} \right) u^* \frac{\partial T}{\partial y^+}.$$

Separating the variables and integrating at constant x from $y^+ = 5$ where $T = T_L$ to $y^+ = 30$ where $T = T_0$, and using Eq. (5.267) for ϵ_m yields

$$\frac{q_w''}{\rho c_p u^*} \int_5^{30} \frac{dy^+}{\left(\dfrac{1}{N_{\text{Pr}}} + \dfrac{y^+}{5} - 1 \right)} = - \int_{T_L}^{T_0} dT.$$

Integrating, inserting the limits and rearranging yields, after using $u^* = \sqrt{\tau_w/\rho}$,

$$\frac{5q_w''}{\rho c_p \sqrt{\tau_w/\rho}} \ln (5 N_{\text{Pr}} + 1) = T_L - T_0. \tag{5.268}$$

Consideration of the viscous sublayer and of the buffer region—that is, of the entire inner or viscous layer of Sec. 5.3d—has yielded Eqs. (5.261) and (5.268) for the three unknowns q_w'', T_L, and T_0. A third equation will be found by treating the overlap region of the cross section next. Here it cannot be assumed that $q'' = q_w''$, since the significant fluid motion in the x direction is physically transporting away the energy, thus causing smaller and smaller values, relative to q_w'', of q'' as the distance y from the wall increases. Here the similarity assumption, that $q''/\tau = q_w''/\tau_w$, Eq. (5.256), is used instead. Also, in this overlap region, molecular conduction is considered negligible in comparison to transport of energy by the motion of the turbulent eddies, as manifested by the value of ϵ_m;

hence, α is neglected in comparison to ϵ_m. Note that this is not the case at low Prandtl numbers, where α may have very large values. Thus, this neglect of α limits the analysis to Prandtl number values greater than about 0.5.

Proceeding next in a manner that is very much the same as was just discussed to arrive at Eq. (5.268) for the buffer region, one can use Eqs. (5.254–56), with $\alpha = 0$, in conjunction with the form of the law of the wall in the overlap layer, from Eq. (5.104), to derive an equation relating q_w'', T_0, and T_B. Since the steps required parallel so closely those used to develop Eq. (5.268), the details will be omitted. Readers, if they are interested, can attempt the development themselves or else consult [32] or [33]. Using this overlap layer relation along with Eqs. (5.261) and (5.268), one can eliminate the unknown temperatures T_L and T_0, and the final result in terms of the Stanton number is

$$N_{St} = \frac{f/8}{1 + 5\sqrt{f/8} \; \{N_{Pr} - 1 + \ln[1 + \frac{5}{6}(N_{Pr} - 1)]\}}. \qquad (5.269)$$

Note that *von Kàrmàn's analogy*, Eq. (5.269), reduces to the Reynolds analogy when $N_{Pr} = 1$. Comparison of Eq. (5.269) with experimental results for fully developed turbulent tube flow yields good agreement for the Prandtl number range from 0.5 to about 5 or 10 for a smooth-walled, constant surface temperature tube. As was also the case for the less general Reynolds analogy, Eq. (5.249), or Colburn's modified, improved version of it, Eq. (5.250), the logarithmic mean temperature difference, Eq. (5.235), is the temperature difference to be used in Newton's cooling law when the h_x given by Eq. (5.269) is being used to calculate the heat transfer rate across the total inside area of the tube. The concept of the hydraulic diameter, Eq. (5.223), allows the application of Eq. (5.269) to ducts of other than circular cross section.

Strictly speaking, Eqs. (5.249), (5.250), and (5.269) are valid only if the thermal boundary condition at the duct wall is a constant, uniform temperature. In general one might expect that, as was the case of fully developed laminar flow in a tube, the effect of thermal history (the influence of the wall temperature distribution on the surface coefficient of heat transfer) is to cause a change in the surface coefficient of heat transfer if the duct wall is not at a single temperature T_w. However, in turbulent flow, as long as the Prandtl number of the fluid is not too low—not far below 0.5—thermal history effects are not nearly as important as in laminar flow. Thus the surface coefficient of heat transfer for the constant flux boundary condition is only a few percent higher than Eq. (5.269) gives for the constant-temperature tube. Hence, as is also pointed out by Petukhov [34], expressions such as Eq. (5.269) can also be used for the constant flux boundary condition, *as long as the flow is turbulent.*

The derivation which led to von Kàrmàn's analogy, Eq. (5.269), is important not only because of the final result for the surface coefficient of heat transfer, but also because the ideas, procedures, and concepts employed for these are quite alike, in many respects, to those used in analyses of turbulent heat transfer which are considered to be more sophisticated and refined than von Kàrmàn's analysis. Hence the fundamental steps of von Kàrmàn's analysis should provide enough groundwork for readers to understand not only the approach which uses the so-called "temperature law of the wall" (which really is not all that different from von Kàrmàn's basic approach), but also the approach which uses more sophisticated eddy diffusivity of momentum distributions such that equations, such as Eq. (5.89), must be integrated numerically rather than analytically across the boundary layer [17], [34].

Finally, it may be remarked that von Kàrmàn's analogy, Eq. (5.269), is of the same basic form as the much later and more sophisticated development by Petukhov [34], which will be presented in a later section. The basic difference between one of Petukhov's results, which predicts the experimental data very well indeed, and Eq. (5.269) is the function of the Prandtl number which appears in the second term of the denominator.

EXAMPLE 5.9

In a particular design situation, 2000 lbm/hr of water must be raised from a temperature of 60°F to 120°F while passing through a pipe whose inside wall temperature is 150°F. Because of space considerations, a single pipe 15 ft long must be used. It is expected that the flow will be turbulent, and the water properties are constant at the average values $c_p = 1.0$ Btu/lbm °F; $\rho = 61.7$ lbm/ft^3, $N_{Pr} = 3.64$, $\mu = 1.35$ lbm/hr ft, and $k = 0.371$ Btu/hr ft °F. Find the required pipe diameter.

Solution
An energy balance on the entire pipe taken as the control volume yields

$$h_x PL\Delta T = \omega c_p (T_{B_2} - T_{B_1}). \tag{5.270}$$

From Eq. (5.235), the logarithmic mean temperature difference is

$$\Delta T_L = \frac{T_{B_2} - T_{B_1}}{\ln\left(\dfrac{T_w - T_{B_1}}{T_w - T_{B_2}}\right)} = \frac{120 - 60}{\ln\left(\dfrac{150 - 60}{150 - 120}\right)} = 54.6°\text{F}.$$

Assuming that the flow is turbulent (this cannot be checked until the diameter is found), von Kàrmàn's analogy will be used to find the surface coefficient of heat transfer. From Eq. (5.269),

$$\frac{h_x}{\rho c_p U_m} = \frac{f/8}{1 + 5\sqrt{f/8}\,\{3.64 - 1 + \ln[1 + \frac{5}{6}(3.64 - 1)]\}}$$
$$= \frac{f/8}{1 + 6.72\sqrt{f}}\ ;$$

therefore,

$$h_x = \frac{\rho c_p U_m f/8}{1 + 6.72\sqrt{f}} = \frac{61.7(1)U_m f/8}{1 + 6.72\sqrt{f}} = \frac{7.71\,U_m f}{1 + 6.72\sqrt{f}}. \tag{5.271}$$

At this point, f is unknown because it depends upon the unknown diameter D, and U_m is unknown for the same reason, since $\omega = \rho \frac{1}{4} \pi D^2 U_m$, or $2000 = 61.7 \frac{1}{4} \pi D^2 U_m$; hence,

$$41.3/D^2 = U_m. \tag{5.272}$$

Using Eq. (5.272), U_m can be eliminated from Eq. (5.271) in favor of the diameter D; hence,

$$h_x = \frac{318\ f/D^2}{1 + 6.72\sqrt{f}}. \tag{5.273}$$

The Reynolds number is

$$N_{\mathrm{Re}} = \frac{\rho\ U_m D}{\mu} = \frac{61.7(41.3/D^2)D}{1.35} = \frac{1890}{D}. \tag{5.274}$$

Inserting the known quantities into Eq. (5.270) yields

$$h_x\ \pi D(15)(54.6) = 2000(1)(120 - 60),$$

and rearrangement yields

$$D = 46.6/h_x. \tag{5.275}$$

Combining Eqs. (5.273) and (5.275) and rearranging yields

$$D = \frac{6.83f}{1 + 6.72\sqrt{f}}. \tag{5.276}$$

Thus, there are three unknowns—D, f, and N_{Re}—and three equations: Eqs. (5.274) and (5.276) and the Moody diagram of Fig. 5.9, which relates f to N_{Re}. A solution is obtained by estimating a value of D, using it to obtain N_{Re} and then f. The check is made when f is substituted into Eq. (5.276) to see if it yields the estimated value of D; if not, the procedure is repeated.

As a first estimate, assume $D = \frac{1}{6}$ ft $= 0.167$ ft. From Eq. (5.274),

$$N_{\mathrm{Re}} = \frac{1890}{0.167} = 1.13 \times 10^4.$$

Note that since $N_{\mathrm{Re}} > 4000$, the criterion for turbulent flow, Eq. (5.182d), is satisfied at the value of the diameter being used at this point. From the chart of Fig. 5.9 at $N_{\mathrm{Re}} = 1.13 \times 10^4$, $f = 0.03$. Substituting this into Eq. (5.276) yields

$$D = \frac{6.83(0.03)}{1 + 6.72\sqrt{0.03}} = 0.096 \text{ ft.}$$

Since this is not in agreement with the initial guess, a new estimate must be made in an attempt to balance Eq. (5.276). Take 0.096 feet as the second estimate. From Eq. (5.274),

$$N_{Re} = \frac{1890}{0.096} = 1.97 \times 10^4.$$

Using Fig. 5.9, $f = 0.026$. Substituting this into Eq. (5.276) yields

$$D = \frac{6.83(0.026)}{1 + 6.72\sqrt{0.026}} = 0.0856 \text{ ft.}$$

Using 0.0856 ft as the new estimate, the iteration cycle is continued until one gets $D = 0.084$ ft, which balances Eq. (5.276). The corresponding values of N_{Re} and f are 2.25×10^4 and 0.0255, respectively. We can see that the Reynolds number does indicate that the flow is turbulent, thus making Eq. (5.269) appropriate, from that standpoint.

Thus, the required pipe diameter is

$$D = 0.084 \text{ ft} = 1.01 \text{ in.} \approx 1.0 \text{ in.}$$

Some indication of whether or not the flow is fully developed may be gleaned by a check of the hydrodynamic entrance length estimate as given by Eq. (5.84). Thus,

$$L_e/D = 0.69[2.25 \times 10^4]^{1/4} = 8.5.$$

Since the actual $L/D = 15/(1/12) = 180$, it would seem that hydrodynamic development for most of the tube's length is assured. A criterion to be used to see if a turbulent pipe flow is thermally fully developed will be given in an upcoming section and is simply that $L/D > 60$ for the average surface coefficient to be close enough to the fully developed value of h_x. This condition is also satisfied by the actual ratio of L to D in Ex. 5.9.

5.4c Analytical Solutions for Heat Transfer in External Boundary Layer–Type Flows

So far, analytical and semi-analytical solutions for the local surface coefficient of heat transfer have been presented for fully developed internal flows in both the laminar and turbulent flow regimes. As was also the case in Sec. 5.3 where velocity fields had to be predicted, the fully developed, internal flow heat transfer problems represent the simplest class of problems to handle analytically. In this section, the more difficult heat transfer problems involving external, thin, boundary layer–type flow over immersed bodies will be considered. First the governing differential and integral equations for the temperature field will be derived. Then one exact solution for the important case of the flat plate will be very briefly outlined. This will be followed by the use of the powerful integral methods to find approximate analytical solutions for the temperature field and, eventually, for the local surface coefficient of heat transfer. We will discuss some of the ways in which available constant property solutions and experimental results can be modified to take approximate account of the fact that the thermophysical properties do vary with fluid

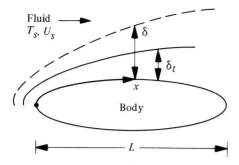

Figure 5.32 Thermal and velocity boundary layer development on a general body.

temperature. Finally, Ambrok's method will be presented. This allows one to fairly easily find approximate analytical solutions for arbitrary thermal history—that is, for any wall temperature distribution along the body, in both laminar and turbulent flows.

Overview of the Problem

Figure 5.32 shows a body, with a specified wall temperature distribution $T_w(x)$, immersed in a fluid which is flowing over the body and which has known temperature T_s in the undisturbed part of the flow far away from the body surface. Also shown as a dashed line is the relatively thin velocity boundary layer, of thickness δ at x. (The thickness is exaggerated for clarity.) When conditions lead to a relatively thin velocity boundary layer ($N_{Re_x} \gg 1$, from Sec. 5.3), there is generally a relatively narrow region of fluid adjacent to the body surface in which the major temperature gradients are concentrated. Outside of this thin layer, the fluid is essentially at the undisturbed temperature T_s everywhere. This layer, in which the major temperature gradients are concentrated adjacent to the body surface, is called the *thermal boundary layer,* and its local thickness is δ_t. The outer edge of the thermal boundary layer is sketched in Fig. 5.32 as a solid line. In general, δ_t does not equal δ; rather, their ratio depends upon the Prandtl number and other factors.

The concept of a thin thermal boundary layer means that

$$\delta_t \ll L, \tag{5.277}$$

and, in analogy with the concept of a thin velocity boundary layer discussed in Sec. 5.3, it can be concluded that when Eq. (5.277) is valid, temperature derivatives in the general flow direction, the x direction, are small compared to the y derivative at the same point, the direction transverse to the main flow direction. This can be shown by use of Eq. (5.277) in the same overall type of order of magnitude studies which led to the boundary layer Eqs. (5.121) and (5.122) in Sec. 5.3e. Doing this, we can derive the following two inequalities when the thermal boundary layer is thin—that is, when Eq. (5.277) is satisfied:

$$\frac{\partial T}{\partial x} \ll \frac{\partial T}{\partial y}, \quad \frac{\partial^2 T}{\partial x^2} \ll \frac{\partial^2 T}{\partial y^2}. \tag{5.278}$$

The second inequality permits neglecting x direction conduction in boundary layer–type flow over a body immersed in the flow.

The basic problem is still the prediction of the local surface coefficient of heat transfer and, by Eq. (5.3), the temperature distribution within the flowing fluid must be found as a function of the space coordinates. Thus, a governing partial differential equation or an integral equation is required before the temperature distribution can be determined.

Governing Differential Equations

Consider steady, laminar, constant property, planar two-dimensional, boundary layer–type flow over a body whose shape is gentle enough in the x direction that the natural coordinate system (x measured along the body surface and y perpendicular to the body surface) can be treated as a rectangular cartesian coordinate system. (Refer to Sec. 5.3 for a discussion of this point.) By applying the law of conservation of energy, Eq. (1.8), to a differential control volume dx by dy within the thermal boundary layer, the partial differential equation which governs the temperature distribution in the fluid will be derived when the caloric equation for the enthalpy, i, is as follows:

$$i = c_p T + P/\rho. \tag{5.279}$$

The differential control volume is shown in Fig. 5.33. For steady flow with no generation, Eq. (1.8) reduces to $R_{in} = R_{out}$. Because of the assumption of a thin thermal boundary layer, the x direction conduction can be neglected; however, in the y direction, there is conduction into the control volume at y at the rate q_y and conduction out of the control volume at $y + dy$ at the rate $q_{y+dy} = q_y + (\partial q_y/\partial y)\,dy$, which we find after using a Taylor series expansion for the conduction heat transfer rate in the y direction. Energy is also carried into the control volume across the lower face at y by virtue of the mass entering the control volume at this face. The rate at which energy enters is the product of the mass flow rate and the energy per unit mass, consisting of the sum of the enthalpy and kinetic energy per unit mass. This energy transfer rate is

$$\rho v\,dx\,[i + (u^2 + v^2)/2]. \tag{5.280}$$

Energy leaves the control volume at the upper face $y + dy$ by the same mechanism, and by a Taylor series expansion, the rate can be put in terms of Eq. (5.280) as

$$\rho v\,dx\left(i + \frac{u^2 + v^2}{2}\right) + \frac{\partial}{\partial y}\left[\rho v\,dx\left(i + \frac{u^2 + v^2}{2}\right)\right]dy.$$

Energy crosses the left-hand face at x by virtue of the mass crossing the face and transporting in its energy per unit mass; hence, the rate is

$$\rho u\,dy\,[i + (u^2 + v^2)/2]. \tag{5.281}$$

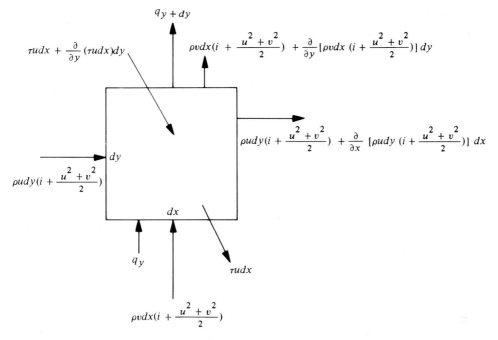

Figure 5.33 Energy transfer rates across the differential control volume.

In a similar fashion, energy leaves the right-hand face at the rate, given by the Taylor series expansion of Eq. (5.281) at $x + dx$,

$$\rho u \, dy \left(i + \frac{u^2 + v^2}{2}\right) + \frac{\partial}{\partial x}\left[\rho u \, dy \left(i + \frac{u^2 + v^2}{2}\right)\right] dx.$$

Energy also leaves the control volume at the lower face x because of the shear work done by the faster-moving control volume fluid on the outside fluid at the rate $\tau u \, dx$, and enters at the upper face by the same mechanism at the rate $\tau u \, dx + (\partial/\partial y)(\tau u \, dx)dy$. In Sec. 5.3, it was shown that for a thin boundary layer $v << u$; hence, $v^2 << u^2$, and v^2 can be neglected compared with u^2 in the kinetic energy terms. Also, because $v << u$, the shear work done on the faces at x and $x + dx$ is neglected and is not shown in Fig. 5.33.

Setting the rate at which energy of all forms enters the control volume R_{in} equal to the rate at which energy of all forms leaves the control volume R_{out} yields

$$\rho u \, dy(i + \tfrac{1}{2} u^2) + \rho v \, dx(i + \tfrac{1}{2} u^2) + \tau u \, dx + \frac{\partial}{\partial y}(\tau u \, dx) \, dy + q_y$$

$$= \rho u \, dy(i + \tfrac{1}{2} u^2) + \frac{\partial}{\partial x}[\rho u \, dy(i + \tfrac{1}{2} u^2)] \, dx + q_y + \frac{\partial q_y}{\partial y} \, dy$$

$$+ \rho v \, dx(i + \tfrac{1}{2} u^2) + \frac{\partial}{\partial y}[\rho v \, dx(i + \tfrac{1}{2} u^2)] \, dy + \tau u \, dx.$$

Substituting for q_y the result from Fourier's exact law of conduction, $q_y = -kdx(\partial T/\partial y)$, and for τ, $\tau = \mu(\partial u/\partial y)$, cancelling like terms, replacing i (the enthalpy per unit mass) by its equivalent from Eq. (5.279), and dividing by $dx\,dy$ yields

$$\mu \frac{\partial}{\partial y}\left(u\,\frac{\partial u}{\partial y}\right) = \rho\frac{\partial}{\partial x}\left[u\left(c_p T + \frac{P}{\rho} + \frac{u^2}{2}\right)\right]$$
$$+ \rho\frac{\partial}{\partial y}\left[v\left(c_p T + \frac{P}{\rho} + \frac{u^2}{2}\right)\right] - k\frac{\partial^2 T}{\partial y^2}.$$

Using the chain rule,

$$\mu \frac{\partial}{\partial y}\left(u\,\frac{\partial u}{\partial y}\right) = \rho\left(c_p T + \frac{P}{\rho} + \frac{u^2}{2}\right)\left(\frac{\partial u}{\partial x} + \frac{\partial v}{\partial y}\right)$$
$$+ \rho u\frac{\partial}{\partial x}\left(c_p T + \frac{P}{\rho} + \frac{u^2}{2}\right)$$
$$+ \rho v\frac{\partial}{\partial y}\left(c_p T + \frac{P}{\rho} + \frac{u^2}{2}\right) - k\frac{\partial^2 T}{\partial y^2}. \qquad \textbf{(5.282)}$$

Note that the quantity $\partial u/\partial x + \partial v/\partial y$ must be zero by the continuity equation for this situation, Eq. (5.115); thus, the entire first term on the right of Eq. (5.282) is zero. Performing the remaining differentiations and noting that because of the thin boundary layer, $\partial P/\partial y = 0$ (see the discussion preceding Eq. [5.121]), after some rearrangement the final result, called the *total energy equation,* is

$$\rho c_p\left(u\frac{\partial T}{\partial x} + v\frac{\partial T}{\partial y}\right) + u\frac{\partial P}{\partial x} + \rho u^2\frac{\partial u}{\partial x} + \rho uv\frac{\partial u}{\partial y}$$
$$= k\frac{\partial^2 T}{\partial y^2} + \mu\left(\frac{\partial u}{\partial y}\right)^2 + \mu u\frac{\partial^2 u}{\partial y^2}. \qquad \textbf{(5.283)}$$

You may notice, by studying Eq. (5.283), that the last three terms on the left-hand side and the last term on the right-hand side form the x-momentum equation, Eq. (5.122), of the boundary layer if the velocity u is factored out. This allows a further simplification of Eq. (5.283). This product of the local x-component of velocity u and the x-momentum equation (5.122), is called the *mechanical energy balance,* and upon multiplying u by Eq. (5.122), one arrives at

$$\rho u^2\frac{\partial u}{\partial x} + \rho uv\frac{\partial u}{\partial y} = -u\frac{\partial P}{\partial x} + \mu u\frac{\partial^2 u}{\partial y^2}. \qquad \textbf{(5.284)}$$

Subtracting this mechanical energy balance from the total energy balance, Eq. (5.283), yields

$$\rho c_p\left(u\frac{\partial T}{\partial x} + v\frac{\partial T}{\partial y}\right) = k\frac{\partial^2 T}{\partial y^2} + \mu\left(\frac{\partial u}{\partial y}\right)^2. \qquad \textbf{(5.285)}$$

Equation (5.285) is called the *thermal energy equation* and is the form ordinarily used (rather than Eq. [5.283]) when solving the partial differential equation for the temperature distribution within the moving fluid in the boundary layer. The last term on the right-hand side of Eq. (5.285) is called the *dissipation function* or the *dissipation term* and represents the rate at which mechanical energy is degraded by the work of the viscous shear stresses into random thermal energy. If the flow is relatively low speed—that is, if the Eckert number is much less than 1—

$$N_{Ec} = \frac{U_s^2}{c_p(T_w - T_s)} < 1. \tag{5.286}$$

Then the dissipation term can be neglected.

Physically, the Eckert number is proportional to the ratio of the maximum change in kinetic energy that can be experienced by a fluid particle to the maximum change in static enthalpy; thus, when the number is small, the temperature rises that could result by total degradation of the kinetic energy remain small compared to the impressed temperature difference $T_w - T_s$ across the thermal boundary layer. The Eckert number was earlier introduced, in connection with flow inside a tube, in Eq. (5.200), along with some discussion of viscous dissipation which one may wish to refer back to.

When the inequality, Eq. (5.286), is satisfied so that the viscous dissipation of energy term, $\mu(\partial u/\partial y)^2$, is negligible, Eq. (5.285) reduces to the following low-speed thermal energy equation for laminar, constant property, two-dimensional planar, thin boundary layer–type flow:

$$\rho c_p \left(u\frac{\partial T}{\partial x} + v\frac{\partial T}{\partial y} \right) = k\frac{\partial^2 T}{\partial y^2}. \tag{5.287}$$

Thus, depending upon whether one is dealing with low-speed flow or high-speed flow, flow with appreciable viscous dissipation, the solution to Eqs. (5.287) or (5.285), respectively, will yield the temperature distribution within the moving fluid. This leads, via Eq. (5.3), to the prediction of the local surface coefficient of heat transfer.

Exact Solution for Laminar Flow Over an Isothermal Plate

Here we consider steady, laminar, low-speed, constant property, two-dimensional planar, thin boundary layer–type flow over a constant surface temperature flat plate which is at zero angle of attack to the flow. This is a configuration of great practical importance because of the different physical systems which can be adequately idealized as being flow over such a plate.

We seek an exact solution for the fluid temperature as a function of the space coordinates. From this, we would like to predict both the local surface coefficient h_x and the average surface coefficient of heat transfer $\overline{h}_L$ over the entire plate.

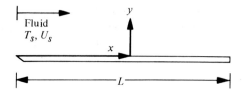

Figure 5.34 Diagram of a flat plate.

The physical situation is depicted in Fig. 5.34, and for low-speed flow, the governing differential equation is given by Eq. (5.287).

$$\rho c_p \left(u \frac{\partial T}{\partial x} + v \frac{\partial T}{\partial y} \right) = k \frac{\partial^2 T}{\partial y^2} . \tag{5.288}$$

The boundary conditions are

$$
\begin{array}{lll}
\text{at } x = 0 \text{ and } y > 0, & T = T_s, \\
\text{at } y = 0 \text{ and } 0 < x < L, & T = T_w, \\
\text{as } y \to \infty, & T \to T_s.
\end{array}
$$

The boundary conditions can be put into simpler form by introducing the temperature excess ratio

$$\phi = \frac{T - T_w}{T_s - T_w} . \tag{5.289}$$

Using this, Eq. (5.288) becomes

$$\rho c_p \left(u \frac{\partial \phi}{\partial x} + v \frac{\partial \phi}{\partial y} \right) = k \frac{\partial^2 \phi}{\partial y^2} , \tag{5.290}$$

and the boundary conditions become

$$
\begin{array}{lll}
\text{at } x = 0 \text{ and } y > 0, & \phi = 1, \\
\text{at } y = 0 \text{ and } 0 < x < L, & \phi = 0, & \tag{5.291} \\
\text{as } y \to \infty, & \phi \to 1.
\end{array}
$$

An exact analytical solution for the velocity field of this problem by the method of the similarity transformation was discussed in Sec. 5.3, where the x component of velocity was given as $u = U_s f'(\eta)$, with $\eta = y\sqrt{U_s/vx}$. The same similarity variable η that transformed the x-momentum equation, (5.122), into an ordinary differential equation also transforms Eq. (5.290) into the ordinary differential equation

$$\frac{d^2\phi}{d\eta^2} + \frac{N_{Pr}}{2} f(\eta) \frac{d\phi}{d\eta} = 0. \tag{5.292}$$

Recall that $f(\eta)$ is a known function of η that resulted from the solution of the x-momentum equation for the velocity field of this same problem. Pohlhausen [35] solved Eq. (5.292) subject to the boundary conditions of Eq. (5.291), and arrived at an equation of the form

$$\phi = \phi(\eta). \tag{5.293}$$

For the same reasons discussed in Sec. 5.3, only the results of this exact analytical solution by the method of the similarity transformation are presented herein. When $\phi = \phi(\eta)$ is known, then, by Eq. (5.289) and the expression for η, $T = T(x, y)$ is known. Hence, using Eq. (5.3), $h_x = h_x(x)$ can be predicted. When arranged in the nondimensional Nusselt number form, the result for the flat plate is

$$\frac{h_x x}{k} = 0.332\, F(N_{Pr})\sqrt{N_{Re_x}}\,, \tag{5.294}$$

where, for the Prandtl number range from 0.6 to 15, it is found that the function of the Prandtl number is well represented by

$$F(N_{Pr}) = N_{Pr}^{1/3}. \tag{5.295}$$

Hence, Eq. (5.294) becomes

$$\frac{h_x x}{k} = 0.332\, N_{Pr}^{1/3}\sqrt{\frac{\rho U_s x}{\mu}}\,, \tag{5.296}$$

or

$$N_{Nu_x} = 0.332\, N_{Pr}^{1/3}\, N_{Re_x}^{1/2}. \tag{5.297}$$

The total heat transfer rate from the plate to the fluid, per unit length of plate normal to the x, y plane, is

$$q = \bar{h}_L L(T_w - T_s), \tag{5.298}$$

where $\bar{h}_L$ is the average surface coefficient of heat transfer over the entire plate from $x = 0$ to $x = L$. This total heat transfer rate can also be found by summing up all the local heat transfer rates across differential areas dx. Thus, the local heat transfer rate from the plate to the fluid from x to $x + dx$ is

$$dq = h_x\, dx(T_w - T_s). \tag{5.299}$$

The sum of all contributions of the type shown in Eq. (5.299) from $x = 0$ to $x = L$, in a definite integral, is the total heat transfer rate q from the plate to the fluid; that is,

$$q = \int_0^L h_x(T_w - T_s)dx = (T_w - T_s) \int_0^L h_x \, dx. \tag{5.300}$$

(Note that $T_w - T_s$ is not a function of x here.) Now, from Eq. (5.296),

$$h_x = 0.332 \, k \, N_{Pr}^{1/3} \sqrt{\rho U_s/\mu} \; x^{-1/2}.$$

Substituting this into Eq. (5.300) and integrating yields

$$q = (T_w - T_s)(0.332 \, k \, N_{Pr}^{1/3} \sqrt{\rho U_s/\mu}) \, 2x^{1/2}\big|_0^L \; .$$

Substituting the limits and rearranging yields

$$q = (T_w - T_s)(0.664 \, k \, N_{Pr}^{1/3} \sqrt{\rho U_s L/\mu}).$$

Equating this and Eq. (5.298) and rearranging to introduce the average Nusselt number over the plate length yields

$$\overline{h}_L L/k = 0.664 \, N_{Pr}^{1/3} \sqrt{\rho U_s L/\mu} \; , \tag{5.301}$$

or

$$\overline{N}_{Nu_L} = 0.664 \, N_{Pr}^{1/3} \, N_{Re_L}^{1/2} \; . \tag{5.302}$$

By comparing Eqs. (5.301) and (5.296), it is seen that the average surface coefficient of heat transfer from $x = 0$ to $x = L$ is twice the local value at $x = L$. Using similar reasoning, we note that the average surface coefficient *from $x = 0$ to any* value of $x \le L$, say $x = b_0$, is simply twice the local value, given by Eq. (5.296), at $x = b_0$. Note that this two-to-one ratio between average and local coefficients is valid only for the conditions of the physical situation being discussed here and *not* for other situations, such as turbulent flow over a plate or laminar flow over a cylinder.

As noted before, because of the use of Eq. (5.295) for $F(N_{Pr})$, the expressions for the surface coefficients, Eqs. (5.297) and (5.302), are valid only for Prandtl numbers in the range $0.6 < N_{Pr} < 15$. This range of Prandtl numbers encompasses the usual heat transfer fluids such as air and other gases, and water. However, most heavy oils have Prandtl numbers far in excess of 15. At the other end of the spectrum, the liquid metals have very low Prandtl numbers, typically less than 0.03.

Churchill [36] has found an analytical expression which reproduces the actual function $F(N_{Pr})$ which appears in Eq. (5.294), over the full range of Prandtl numbers from zero to infinity. This leads to the following two equations valid for the laminar, constant property flow over a constant temperature flat plate for all values of the Prandtl number:

$$N_{Nu_x} = \frac{0.3387 \, N_{Pr}^{1/3} \, N_{Re_x}^{1/2}}{[1 + (0.0468/N_{Pr})^{2/3}]^{1/4}} \tag{5.303}$$

$$\overline{N}_{Nu_L} = \frac{0.6774 \, N_{Pr}^{1/3} \, N_{Re_L}^{1/2}}{[1 + (0.0468/N_{Pr})^{2/3}]^{1/4}} \tag{5.304}$$

Integral Equation for the Thermal Boundary Layer

Exact analytical solutions to forced convection heat transfer problems can be found only in a very small number of different cases. This, it will be recalled, was also the situation earlier when dealing with solutions for the velocity field. Thus, the powerful approximate integral method is often used to get analytical solutions to problems for which exact solutions cannot be developed. So, in the present section, the integral equation whose solution gives the temperature distribution as a function of the space coordinates in the flow will be derived.

For steady, constant property, low-speed, two-dimensional planar, thin boundary layer–type flow over a body of arbitrary shape, as shown in Fig. 5.35, the integral form of the low-speed thermal energy equation is found by applying the law of conservation of energy, Eq. (1.8), to an appropriate control volume.

The control volume is shown in Fig. 5.36. Conservation of energy for steady-state conditions and no generation is $R_{in} = R_{out}$. Energy enters the control volume at $y = 0$ by conduction across the first few molecular layers of fluid against the plate at the rate

$$q_w'' \, dx, \tag{5.305}$$

where q_w'' is the local heat flux at the wall, at $y = 0$. Energy enters the control volume from within the thermal boundary layer at x, the left-hand face, by virtue of the motion of the fluid crossing the face, carrying with it the energy it had before it crossed. The low-speed assumption permits us to neglect the kinetic energy; hence, the energy carried in across the differential area dy at x is

$$\rho u \, dy \, c_p T.$$

Figure 5.35 The body and control volume used to derive the integral form of the energy equation.

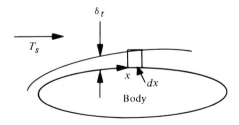

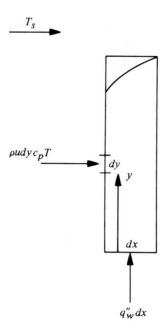

T_s

$\rho u \, dy \, c_p T$

dy

y

dx

$q_w'' \, dx$

Figure 5.36 Details of the control volume of Fig. 5.35 showing some of the energy transfer rates.

The total rate at which energy enters the control volume across the left face from within the thermal boundary layer is

$$\int_0^{\delta_t} \rho u c_p T \, dy. \tag{5.306}$$

The total rate at which energy leaves the control volume at $x + dx$, through the thermal boundary layer, can be put in terms of Eq. (5.306) by a Taylor series expansion as

$$\int_0^{\delta_t} \rho u c_p T \, dy + \frac{d}{dx}\left(\int_0^{\delta_t} \rho u c_p T \, dy\right) dx. \tag{5.307}$$

Conduction in the x direction is neglected because the flow is of the boundary layer type. There is no conduction in the y direction at the upper face near $y = \delta_t$, because this upper surface lies outside the thermal boundary layer and there, by definition, the temperature gradients are negligible. Energy does enter the control volume, however, from outside the thermal boundary layer between x and $x + dx$ by virtue of the net mass flow rate entering and carrying its enthalpy with it. The rate at which energy enters between x and $x + dx$ from outside the thermal boundary layer is

$$\omega_0 c_p T_s, \tag{5.308}$$

where ω_0 is the net mass flow rate into the thermal boundary layer between x and $x + dx$. Since this mass is essentially at the temperature T_s of the material outside the thermal boundary layer, its enthalpy per unit mass is equal to $c_p T_s$. Now, ω_0 can be found by application of the law of conservation of mass to the control volume. This has already been done in Sec. 5.3 for the hydrodynamic boundary layer of thickness δ. The only difference between that case and this is that the layer thickness is δ_t. Hence, replacing δ by δ_t in Eq. (5.140) gives

$$\omega_0 = \frac{d}{dx}\left(\int_0^{\delta_t} \rho u\, dy\right) dx. \tag{5.309}$$

Using this, and noting that the product $c_p T_s$ is constant and can be brought inside both the differential and integral operators of Eq. (5.309), Eq. (5.308) becomes

$$\frac{d}{dx}\left(\int_0^{\delta_t} \rho u c_p T_s\, dy\right) dx. \tag{5.310}$$

Substituting Eqs. (5.310), (5.307), (5.306), and (5.305) into the energy balance yields

$$q_w''\, dx + \int_0^{\delta_t} \rho u c_p T\, dy + \frac{d}{dx}\left(\int_0^{\delta_t} \rho u c_p T_s\, dy\right) dx$$
$$= \int_0^{\delta_t} \rho u c_p T\, dy + \frac{d}{dx}\left(\int_0^{\delta_t} \rho u c_p T\, dy\right) dx.$$

Cancelling like terms, combining, and rearranging yields

$$\frac{d}{dx}\int_0^{\delta_t} u(T_s - T)\, dy = -\frac{q_w''}{\rho c_p}. \tag{5.311}$$

This is the integral form of the low-speed thermal energy equation and, in its present form, *is applicable to both laminar and turbulent boundary layers* so long as u and T are interpreted as time-averaged values.

The heat flux q_w'' at the wall is given by Fourier's exact law of conduction within the fluid at $y = 0$ as

$$q_w'' = -k\left(\frac{\partial T}{\partial y}\right)_{y=0}.$$

Substituting this into Eq. (5.311) yields

$$\frac{d}{dx}\int_0^{\delta_t} u(T_s - T)\, dy = \alpha\left(\frac{\partial T}{\partial y}\right)_{y=0}, \tag{5.312}$$

where $\alpha = k/\rho c_p$ is the thermal diffusivity.

Equations (5.311) and (5.312) are reasonably general forms of the low-speed, integral energy equation which can be applied to laminar or to turbulent flow over a planar body of arbitrary shape. It will also be noted, by recalling the comments after our development of the integral form of the x-momentum equation, Eq. (5.152), that these, Eqs. (5.311–12) are *exact* within the framework of their development. The approximate nature of the "integral method" is a consequence of the solution technique used, and solutions of the integral energy equation are the next subject of discussion.

Heat Transfer Solutions by Integral Methods

The procedure needed to solve the integral form of the energy equation, Eq. (5.311) or (5.312), by approximate integral methods is essentially the same as was already used in Sec. 5.3g. This procedure is discussed in great detail in the subsections entitled "Approximate Solution Methods for the Integral Equations" and "Flat Plate Solution." In the work of this section, these procedures are extended and used to solve the convective heat transfer problem—namely, the prediction of local surface coefficients of heat transfer once Eqs. (5.311) or (5.312) have been solved for the temperature profile within the moving fluid.

The Constant Temperature Flat Plate   First, let us deal with steady, laminar, constant property, low-speed, two-dimensional planar, thin boundary layer flow over a flat plate with an *unheated starting length*. Specifically, the thermal condition on the surface of the flat plate pictured in Fig. 5.34 is as follows. For values of x less than x_0, $0 \leq x \leq x_0$, the surface temperature is the *same* as in the undisturbed stream, T_s; hence there is no heat transfer across this unheated portion of the plate. On the other hand, for values of x greater than x_0, $x_0 < x \leq L$, the plate is held at a *constant* temperature T_w which is different than T_s. We would like to derive an expression for the surface coefficient of heat transfer.

To approximately solve integral Eq. (5.312), profiles of both velocity and temperature are required. Hence, Eq. (5.312) clearly shows, as did Eq. (5.287), that the temperature field is coupled to the velocity field, since $u = u(x, y)$ appears explicitly in the equation for the temperature field $T = T(x, y)$. In constant property problems, the velocity field is found by solution of the integral form of the x-momentum equation, Eq. (5.152) or (5.153). However, this has already been done for the case of the flat plate, and the velocity field is given by Eqs. (5.168) and (5.180) coupled with (5.179). Next, an approximating sequence for the temperature profile must be specified, and this is ordinarily taken to be of the same general type as shown in Eq. (5.156), with T replacing u. To keep the algebra manageable, Eq. (5.156) is truncated after the fourth term; hence, the approximate profile is

$$T = a_0 + a_1 y + a_2 y^2 + a_3 y^3, \tag{5.313}$$

where a_0, a_1, a_2, and a_3 can be functions of x. This assumed temperature profile is then forced to satisfy the boundary conditions and, perhaps, an additional auxiliary condition.

In the present problem containing the unheated starting length x_0, $T(x, y)$ is simply T_s for $x < x_0$, so we consider the determination of a_0, a_1, etc., in Eq. (5.313) only for $x > x_0$ where the wall temperature is the known constant T_w. Hence,

$$\text{at } y = 0, T = T_w. \tag{5.314}$$

By definition of the thermal boundary layer,

$$\text{at } y = \delta_t, T = T_s. \tag{5.315}$$

Also by definition of the thermal boundary layer as the region where the major temperature gradients are concentrated,

$$\text{at } y = \delta_t, \partial T/\partial y = 0, \tag{5.316}$$

and thus, at its outer edge, the gradients must be zero.

Another condition on the temperature profile is found by forcing the profile to satisfy the exact partial differential equation, Eq. (5.287), at some space point. Since the local surface coefficient of heat transfer is of primary interest and, by Eq. (5.3), the temperature distribution near the wall determines h_x, $y = 0$ is chosen as the point at which Eq. (5.313) should exactly satisfy Eq. (5.287). (Essentially the same type of reasoning has already been used in the discussion following Eq. [5.162].) Thus, in Eq. (5.287) at $y = 0$, $u = v = 0$, so that Eq. (5.287) becomes

$$\text{at } y = 0, \partial^2 T/\partial y^2 = 0. \tag{5.317}$$

Equations (5.314–17) are now applied to Eq. (5.313). It is seen that Eq. (5.314) yields

$$a_0 = T_w. \tag{5.318}$$

From Eq. (5.313),

$$\partial T/\partial y = a_1 + 2a_2 y + 3a_3 y^2, \tag{5.319}$$

$$\partial^2 T/\partial y^2 = 2a_2 + 6a_3 y. \tag{5.320}$$

Equation (5.320) along with Eq. (5.317) indicates that

$$a_2 = 0. \tag{5.321}$$

Equation (5.319) with Eq. (5.316) yields

$$0 = a_1 + 3a_3 \delta_t^2, \quad \text{or} \quad a_3 = -a_1/3\delta_t^2. \tag{5.322}$$

Finally, using Eqs. (5.321–22) and (5.318) in Eq. (5.313) and then using Eq. (5.315) leads to

$$T_s = T_w + a_1\delta_t - \frac{a_1\delta_t}{3};$$

hence,

$$a_1 = \frac{3}{2}\frac{T_s - T_w}{\delta_t}. \qquad (5.323)$$

Using this in Eq. (5.322) yields

$$a_3 = \frac{1}{2}\frac{T_s - T_w}{\delta_t^3}. \qquad (5.324)$$

Substituting Eqs. (5.323–24), Eq. (5.318), and Eq. (5.321) into Eq. (5.313), the temperature distribution is

$$T = T_w + (T_s - T_w)\left(\frac{3}{2}\frac{y}{\delta_t} - \frac{1}{2}\frac{y^3}{\delta_t^3}\right). \qquad (5.325)$$

One parameter, the local thermal boundary layer thickness $\delta_t = \delta_t(x)$, remains to be determined by substituting the approximate temperature profile into the integral form of the low-speed thermal energy equation, Eq. (5.312). This equation also requires the x component of the velocity profile which has been found previously, for $0 \le y \le \delta$, as

$$u = U_s\left(\frac{3}{2}\frac{y}{\delta} - \frac{1}{2}\frac{y^3}{\delta^3}\right), \qquad [5.168]$$

where

$$\delta = 4.64x/\sqrt{N_{\mathrm{Re}_x}}. \qquad [5.171]$$

Now, $(\partial T/\partial y)_{y=0}$ is required on the right-hand side of Eq. (5.312). From Eq. (5.325),

$$\frac{\partial T}{\partial y} = (T_s - T_w)\left(\frac{3}{2\delta_t} - \frac{3y^2}{2\delta_t^3}\right).$$

Hence, at $y = 0$,

$$\left(\frac{\partial T}{\partial y}\right)_{y=0} = \frac{3(T_s - T_w)}{2\delta_t}. \qquad (5.326)$$

Substituting Eqs. (5.326), (5.168), and (5.325) into Eq. (5.312) yields

$$\frac{d}{dx}\int_0^{\delta_t} U_s\left(\frac{3}{2}\frac{y}{\delta} - \frac{1}{2}\frac{y^3}{\delta^3}\right)\left\{T_s - \left[T_w + (T_s - T_w)\left(\frac{3}{2}\frac{y}{\delta_t} - \frac{1}{2}\frac{y^3}{\delta_t^3}\right)\right]\right\} dy$$

$$= \frac{3\alpha(T_s - T_w)}{2\delta_t}.$$

(Note that at this point, a restriction has been introduced by the use of Eq. [5.168] as the velocity distribution; that is, $\delta_t \leq \delta$, since, if $\delta_t > \delta$, u would be given by U_s rather than by Eq. [5.168].) Factoring out and cancelling $T_s - T_w$ and rearranging gives

$$\frac{d}{dx} \int_0^{\delta_t} \left(\frac{3}{2} \frac{y}{\delta} - \frac{1}{2} \frac{y^3}{\delta^3} \right) \left(1 - \frac{3}{2} \frac{y}{\delta_t} + \frac{y^3}{2\delta_t^3} \right) dy = \frac{3\alpha}{2\delta_t U_s} .$$

Performing the multiplication indicated in the integrand, integrating with respect to y (noting that δ and δ_t are functions only of x), inserting the limits, and defining $\eta = \delta_t/\delta$ yields

$$\frac{3}{20} \frac{d}{dx} [\delta(\eta^2 - \tfrac{1}{14} \eta^4)] = \frac{3\alpha}{2\delta\eta U_s} . \tag{5.327}$$

Assuming that η is small compared to 1, $\tfrac{1}{14} \eta^4 < \eta^2$; that is, the $\tfrac{1}{14} \eta^4$ term can be neglected compared with η^2. (Note that η does not have to be much smaller than 1, since even at $\eta = 1$, $\tfrac{1}{14} \eta^4$ is only $\tfrac{1}{14}$ of η^2.) Hence, Eq. (5.327) becomes

$$\frac{d}{dx} (\delta\eta^2) = \frac{10\alpha}{\delta\eta U_s} .$$

Expanding the left-hand side gives

$$2\delta\eta \frac{d\eta}{dx} + \frac{d\delta}{dx} \eta^2 = \frac{10\alpha}{\delta\eta U_s} .$$

Multiplying through by η and noting that $\eta^2 \, d\eta/dx = \tfrac{1}{3}(d\eta^3/dx)$ yields

$$\frac{2\delta}{3} \frac{d}{dx} (\eta^3) + \frac{d\delta}{dx} \eta^3 = \frac{10\alpha}{\delta U_s} .$$

Defining $\eta^3 = z$ and rearranging yields

$$\frac{dz}{dx} + \frac{3}{2\delta} \frac{d\delta}{dx} z = \frac{15\alpha}{\delta^2 U_s} . \tag{5.328}$$

From Eq. (5.171),

$$\delta^2 = \frac{21.5x}{\rho U_s/\mu} = \frac{280}{13} \frac{x\nu}{U_s} , \quad \frac{d\delta}{dx} = \frac{4.64}{\sqrt{\rho U_s/\mu}} \left(\frac{x^{-1/2}}{2} \right) ;$$

hence, $(3/2\delta)(d\delta/dx) = (3/4x)$, and Eq. (5.328) becomes

$$\frac{dz}{dx} + \frac{3z}{4x} = \frac{0.696}{x N_{\text{Pr}}} .$$

Solving this first-order, linear, ordinary differential equation by the method of the integrating factor yields

$$z = \frac{C_1}{x^{3/4}} + \frac{0.929}{N_{Pr}},$$ (5.329)

where $z = (\delta_t/\delta)^3$. In this problem, δ has been developing from $x = 0$, while δ_t has been 0 from $x = 0$ to $x = x_0$; hence, at $x = x_0$, $z = 0$. From Eq. (5.329), C_1 can be determined as

$$0 = \frac{C_1}{x_0^{3/4}} + \frac{0.929}{N_{Pr}}, \quad \text{or} \quad C_1 = \frac{-0.929 \, x_0^{3/4}}{N_{Pr}}.$$

Using this, Eq. (5.329) becomes

$$z = \frac{0.929}{N_{Pr}} \left[1 - \left(\frac{x_0}{x} \right)^{3/4} \right].$$

Solving for δ_t/δ by taking the cube root of both sides of this equation, since $(\delta_t/\delta)^3 = z$, yields

$$\frac{\delta_t}{\delta} = \frac{[1 - (x_0/x)^{3/4}]^{1/3}}{1.026 \, N_{Pr}^{1/3}}.$$ (5.330)

Equation (5.330) together with Eqs. (5.171) and (5.325) give the temperature distribution as a function of x and y. The local surface coefficient of heat transfer is found from Eq. (5.3) with the coordinate n identified as y. Therefore,

$$h_x = \frac{-k(\partial T/\partial y)_{y=0}}{T_w - T_s}.$$ (5.331)

Substituting the value of $(\partial T/\partial y)_{y=0}$ from Eq. (5.326) into Eq. (5.331) yields

$$h_x = \frac{3k}{2\delta_t},$$ (5.332)

where δ_t is given by Eq. (5.330) and δ by Eq. (5.171). Using these values in Eq. (5.332) and rearranging to give the local Nusselt number yields

$$\frac{h_x x}{k} = \frac{0.332 \, N_{Pr}^{1/3} \, N_{Re_x}^{1/2}}{[1 - (x_0/x)^{3/4}]^{1/3}}.$$ (5.333)

Equation (5.333) is the integral method solution for the local Nusselt number for steady, laminar, constant property, thin boundary layer–type flow over a flat plate with an unheated starting length—that is, when the plate surface temperature is the free-stream temperature T_s for $x < x_0$ and is the known constant T_w for $x_0 < x \leq L$. The accuracy of this approximate solution will be discussed shortly. At this point, we note that the problem under discussion cannot be solved in an exact fashion by the method of the similarity transformation, because of the unheated starting length x_0. Yet we were able to solve it in a relatively easy fashion by the approximate integral method.

Entire Plate at T_w   If the entire plate, from $x = 0$ to $x = L$, is being held at the known constant temperature T_w, the solution for the local Nusselt number and for the local thermal boundary layer thickness δ_t can be found by simply letting the unheated starting length x_0 approach zero in Eqs. (5.333) and (5.330), respectively. Thus, one arrives at

$$\frac{h_x x}{k} = 0.332\, N_{\mathrm{Pr}}^{1/3}\, N_{\mathrm{Re}_x}^{1/2}\,. \tag{5.334}$$

A comparison with the exact solution to this same problem using the method of the similarity transformation, Eq. (5.297), reveals that the results are identical in the Prandtl number range of 0.6 to 15. This "exact" agreement between the results of the approximate integral method and the exact analytical solution is the exception rather than the rule. Ordinarily, if the integral method produces a result within 10% to 15% of the exact result for the surface coefficient of heat transfer, it is considered good enough for many engineering calculations.

The original assumption of $\delta_t \leq \delta$ should now be checked for its implications and restrictions on the solution. From Eq. (5.330),

$$\frac{\delta_t}{\delta} = \frac{1}{1.026\, N_{\mathrm{Pr}}^{1/3}}\,. \tag{5.335}$$

Hence, for $\delta_t \leq \delta$, the Prandtl number should be greater than about 1. For Prandtl numbers as low as 0.7 or 0.6, the result is still correct, since the neglected term $\frac{1}{14}\eta^4$ still does not make an appreciable contribution to the final result. The restriction of $N_{\mathrm{Pr}} > 0.6$ is not really severe, because the two most common heat transfer fluids, air and water, satisfy it. However, liquid metals do not, and they are discussed later in the text.

Since the expression for the local surface coefficient of heat transfer h_x is the same as the exact analytical result, the average surface coefficient of heat transfer for the plate will be the same as given by Eq. (5.302); that is

$$\overline{N}_{\mathrm{Nu}_L} = 0.664\, N_{\mathrm{Pr}}^{1/3}\, N_{\mathrm{Re}_L}^{1/2}\,. \tag{5.336}$$

The Constant-Flux Flat Plate   Here we will consider steady, laminar, low-speed, constant property, thin boundary layer–type flow over a flat plate which is providing a constant, known energy transfer rate to the fluid over the entire surface of the plate. Hence, there is a known surface heat flux, q_w'', from the plate to the fluid, rather than a known constant surface temperature as for the case completed before. By use of approximate integral methods, we mean to derive an expression for the local Nusselt number and also for the unknown wall temperature variation with distance x down the plate.

Since the surface heat flux is known, the form of the integral energy equation given by Eq. (5.311) is most easily used. After replacing ρc_p by its equivalent, k/α, Eq. (5.311) becomes

$$\frac{d}{dx} \int_0^{\delta_t} u(T_s - T)\, dy = -\alpha \frac{q_w''}{k}. \tag{5.337}$$

The velocity distribution is given by Eqs. (5.168) and (5.171). (Recall that for constant properties, in forced convection, the velocity field is independent of the temperature field, and hence the same velocity field can be used with different thermal loadings, i.e., boundary conditions, as long as the same body—in this case the flat plate—is involved.) The temperature profile is taken to be of the general form of Eq. (5.156), where the functions of x, a_0, a_1, a_2, and a_3 are to be determined. Hence,

$$T = a_0 + a_1 y + a_2 y^2 + a_3 y^3. \tag{5.338}$$

The boundary conditions which the temperature profile must satisfy are as follows: At $y = 0$, the surface temperature T_w is not specified, and hence is unknown; however, the heat flux is known, so that Fourier's exact law of conduction, applied right at the wall where the fluid is stationary, gives the fluid temperature derivative at the wall as

$$(\partial T/\partial y)_{y=0} = -q_w''/k. \tag{5.339}$$

At the outer edge of the thermal boundary layer, the temperature must be T_s and the y derivative equals zero; hence,

$$\text{at } y = \delta_t, \partial T/\partial y = 0, \tag{5.340}$$

$$\text{at } y = \delta_t, T = T_s. \tag{5.341}$$

One more condition is required and, for the reasons discussed preceding Eq. (5.317), the temperature profile, Eq. (5.338), is forced to exactly satisfy the governing partial differential equation, Eq. (5.287), at $y = 0$. Since $u = v = 0$ at $y = 0$,

$$\text{at } y = 0, \partial^2 T/\partial y^2 = 0. \tag{5.342}$$

Next, these four conditions, Eqs. (5.339–42), are forced on the approximating sequence, Eq. (5.338), in the same basic manner as was used to arrive at Eq. (5.325) earlier. This gives

$$T = T_s + \frac{2}{3} \frac{q_w'' \delta_t}{k} - \frac{q_w''}{k} y + \frac{q_w''}{3k} \frac{y^3}{\delta_t^2}$$

$$= T_s + \frac{q_w''}{k} \left(\frac{2\delta_t}{3} - y + \frac{y^3}{3\delta_t^2} \right).$$

(5.343)

The last parameter in the temperature profile, the thermal boundary layer thickness as a function of x, is found by substituting Eqs. (5.343) and (5.168) into the integral energy equation, Eq. (5.337). (Recall that using Eq. [5.168] implies that $\delta_t \leq \delta$, just as in the previous solution.) Hence,

$$\frac{d}{dx} \int_0^{\delta_t} U_s \left(\frac{3}{2} \frac{y}{\delta} - \frac{1}{2} \frac{y^3}{\delta^3} \right) \frac{q_w''}{k} \left(-\frac{2}{3} \delta_t + y - \frac{y^3}{3\delta_t^2} \right) dy$$

$$= -\frac{q_w'' \alpha}{k}.$$

(5.344)

Performing the indicated multiplications, integrating, inserting the limits, defining $\eta = \delta_t/\delta$, and rearranging leads to

$$\frac{d}{dx} \left[\frac{\delta^2}{10} (\eta^3 - \frac{1}{14} \eta^5) \right] = \frac{\alpha}{U_s}.$$

(5.345)

This can be integrated directly, and the fact that $\delta = 0$ at $x = 0$ forces the definite integral on the left-hand side to have the limits 0 and $[(\delta^2/10)(\eta^3 - \frac{1}{14} \eta^5)]$; hence,

$$\frac{1}{10} \delta^2 (\eta^3 - \frac{1}{14} \eta^5) = (\alpha/U_s) x.$$

(5.346)

Since $\eta \leq 1$, it is assumed that $\frac{1}{14} \eta^5 << \eta^3$, and $\frac{1}{14} \eta^5$ will be neglected compared with η^3. Using this approximation, Eq. (5.346) becomes

$$\frac{1}{10} \delta^2 \eta^3 = (\alpha/U_s)x.$$

(5.347)

To eliminate δ^2, Eq. (5.171) is squared; that is,

$$\delta^2 = \frac{(4.64)^2 x^2}{N_{Re_x}} = \frac{21.6 x^2}{U_s x/\nu} = \frac{21.6 \nu x}{U_s}.$$

Substituting this into Eq. (5.347), cancelling x/U_s, noting that $N_{Pr} = \nu/\alpha$, and rearranging yields

$$\eta^3 = \frac{1}{2.16 N_{Pr}}.$$

Since $\eta = \delta_t/\delta$, extracting the cube root gives

$$\frac{\delta_t}{\delta} = \frac{1}{1.293\ N_{\mathrm{Pr}}^{1/3}}. \tag{5.348}$$

To determine the implication of the assumption $\delta_t \le \delta$, Eq. (5.348) is examined. If $\delta_t/\delta \le 1$, then

$$\frac{1}{1.293\ N_{\mathrm{Pr}}^{1/3}} \le 1, \quad \text{or} \quad N_{\mathrm{Pr}}^{1/3} \ge \frac{1}{1.293}.$$

Solving this inequality yields the condition that $N_{\mathrm{Pr}} \ge 0.462$ in order that $\delta_t \le \delta$. This is not a severe restriction, because the two most common heat transfer fluids, air and water, satisfy this condition over a wide range of temperatures and pressures.

By Eq. (5.3), with n replaced by y, the expression for the local surface coefficient of heat transfer is

$$h_x = \frac{-k(\partial T/\partial y)_{y=0}}{T_w - T_s}.$$

Using q_w'' in the numerator yields

$$h_x = q_w''/(T_w - T_s). \tag{5.349}$$

Now, T_w is found from the temperature distribution, Eq. (5.343), by setting $y = 0$, or

$$T_w = T_s + \tfrac{2}{3}(q_w''/k)\,\delta_t. \tag{5.350}$$

Inserting δ_t from Eq. (5.348) into Eq. (5.350) gives

$$T_w = T_s + \frac{0.515\delta}{N_{\mathrm{Pr}}^{1/3}}\frac{q_w''}{k}. \tag{5.351}$$

By Eq. (5.171), $\delta = 4.64x/N_{\mathrm{Re}_x}^{1/2}$. Substituting this into Eq. (5.351) and rearranging, the final result for the wall temperature distribution as a function of x is

$$T_w = T_s + \frac{2.39\ q_w''x}{kN_{\mathrm{Pr}}^{1/3}\ N_{\mathrm{Re}_x}^{1/2}}. \tag{5.352}$$

The value $T_w - T_s$ from Eq. (5.352) is now inserted into the denominator of Eq. (5.349), and the result is rearranged to give the following expression for the local Nusselt number for the case in which the flux q_w'' is a known constant on the surface of the plate:

$$h_x x/k = 0.419\ N_{\mathrm{Pr}}^{1/3}\ N_{\mathrm{Re}_x}^{1/2}. \tag{5.353}$$

In a more exact solution to this problem (see Kays [33]) the constant is 0.453 instead of 0.419, and hence the approximate integral method predicts a result which, in this problem, is 7.5% too low. This is still considered close agreement for most types of engineering calculations as far as surface coefficients of heat transfer are concerned, and the error here is more typical of what must be expected than is the zero error in Eq. (5.334).

An equation which represents essentially the exact solution over the full range of Prandtl numbers from zero to infinity, for the known constant surface heat flux, was arrived at by Churchill and Ozoe [37], [36]. Their result is as follows:

$$N_{Nu_x} = \frac{0.4637 \, N_{Pr}^{1/3} \, N_{Re_x}^{1/2}}{[1 + (0.02052/N_{Pr})^{2/3}]^{1/4}}. \tag{5.354}$$

Combining Eq. (5.354) with Eq. (5.349) gives the wall temperature distribution on the constant surface flux flat plate as the following:

$$T_w(x) = T_s + \frac{2.157 q_w'' x \, [1 + (0.02052/N_{Pr})^{2/3}]^{1/4}}{k \, N_{Pr}^{1/3} \, N_{Re_x}^{1/2}}. \tag{5.355}$$

For the utmost in accuracy over the entire Prandtl number range, Eqs. (5.354) and (5.355) are recommended rather than the simpler-to-use but less accurate results of the integral method, Eqs. (5.352) and (5.353), which are restricted to Prandtl numbers greater than about 0.46.

Hence, for steady, laminar, constant property, low-speed flow over a flat plate with a known constant heat flux q_w'' at its surface, the local Nusselt number and wall temperature distribution are given by Eqs. (5.354) and (5.355).

5.4d Thermal History Effect

Upon comparing the constant flux solution, Eq. (5.354), to the constant wall temperature solution, Eq. (5.334), for the flat plate, the effect of different thermal histories shows up as a surface coefficient for the constant flux case which is almost 40% higher than the one for the constant temperature condition, at least for Prandtl numbers equal to or greater than unity. Thus the thermal history effect—the effect which surface temperature variations have on the surface coefficient of heat transfer—is quite significant in this particular laminar flow situation. Therefore, it must certainly be accounted for in calculations. Thus one would *not* use the constant wall temperature solution for h_x from Eq. (5.334) to calculate, from Newton's cooling law $q_w'' = h_x(T_w - T_s)$, the local wall temperature for a plate where the flux q_w'' is being maintained constant along it. Instead, Eq. (5.355), which incorporates the h_x from Eq. (5.354), would be used. You will recall that the thermal history effect was noted earlier in connection with the results for the local Nusselt number in laminar, fully developed tube flow, as given in Table 5.1, where almost a 20% difference was noted between the h_x for the constant flux and for the constant wall temperature cases. In addition, the discussion in connection with Eq. (5.269) indicated that this equation could be used with high accuracy for the constant surface flux case as well, when one is dealing with *turbulent* fully developed flow within a tube. Thus, thermal history effects are not nearly as important in turbulent flow (as long as the Prandtl number is not extremely low; see [17]) as they are in laminar flow.

In an upcoming section, a rather general approximate technique, due to Ambrok, will be presented which handles arbitrary surface temperature variations in the downstream direction, *x*, for both laminar and turbulent flow regimes and therefore yields the thermal history effect in fairly general situations.

5.4e Corrections for the Effect of Temperature-Dependent Fluid Properties

All of the analytical and semi-analytical results presented so far for the surface coefficient of heat transfer, in both external boundary layer–type flows and internal flows within a duct, and for turbulent flow as well as laminar flows, have been for *constant property* flows. Actual fluids, in general, have temperature-dependent fluid properties. The density, viscosity, and thermal conductivity of most gases are fairly temperature dependent, and the viscosity of most liquids is a strong function of temperature, while their thermal conductivity is a much weaker function of the temperature.

In addition to the fact that most analytical predictions of the surface coefficient are for constant property conditions, it turns out that many experiments performed to measure the surface coefficient of heat transfer and to establish the working relations referred to as experimental correlations, involved small temperature differences between the fluid and the solid—small enough that, in effect, they also represent constant property results.

As a result of this, what we need is a way to at least approximately correct the available constant property solutions so that they reflect the influence of the actual property variation with temperature. That is, we would like to incorporate some sort of after-the-fact correction factor in these various constant property results. With this we can generalize them so that they take into approximate account the properties' temperature dependence. An attempt to find such a correction factor is complicated by the fact that the temperature dependence of ρ, μ, and k are each important for most gases (though often the dependence upon T of ρ and μ tends to cancel), whereas the viscosity μ is most important for most liquids. Also, for gases, μ increases with increasing T, while for liquids, μ usually decreases with increasing T. This sometimes leads to the necessity of using different types of corrections for gases and liquids. Perhaps two different types of corrections might be needed for each, depending upon whether the fluid is being heated or cooled by the solid surface with which it exchanges energy by convection.

These schemes to correct the effect of the temperature-dependent fluid properties originate either from the results of direct experiments, which employ large temperature differences and use fluids with properties which are fairly sensitive to temperature, or, more recently, they come from solutions to problems that include the effect of property variation. These problems usually involve finite difference methods. Using these variable property experiments or solutions, one can then investigate how best to modify the available constant property results so that they reproduce fairly closely the variable property results. Once a correction scheme is deduced in this way, it is then assumed that the same scheme applies to other constant property results for which variable property results are not available.

The simplest such correction factor scheme is the one which employs a "film temperature," T_f, which is used to evaluate all temperature-dependent fluid properties. This frequently used scheme is applied to both liquids and gases, but works better for gases (see Eckert [38]). For this particular reference temperature scheme which uses the film temperature, the *film temperature* itself is given by the following expressions for external and internal flows:

$$T_f = (T_w + T_s)/2 \quad \textit{External boundary layer flow} \tag{5.356}$$

$$T_f = \frac{T_w + (T_{B_1} + T_{B_2})/2}{2} \quad \textit{Internal flows} \tag{5.357}$$

The values T_{B_1} and T_{B_2} are the inlet and exit bulk mean temperatures, respectively, T_w is the surface temperature, and T_s is the undisturbed temperature of the fluid outside the thermal boundary layer.

For the flow of a gas, the appropriate film temperature, Eq. (5.356) or (5.357), may be applied to any of the results derived so far, such as Eq. (5.303) for example, in order to approximately account for the effect of the temperature-dependent properties. This film temperature can also be used for such liquids as water when the temperature differences are moderate. If, however, liquids are used whose viscosity μ is extremely temperature dependent, as is the case for many oils, or if the liquids are subjected to very large temperature differences, then there are better ways to account for temperature-dependent fluid properties.

Another way of correcting constant property results is to evaluate all properties in the constant property results at the undisturbed stream temperature, T_s, for external boundary layer–type flows, or at the arithmetic average bulk mean temperature T_{B_a}, where

$$T_{B_a} = (T_{B_1} + T_{B_2})/2 \tag{5.358}$$

for internal flows in ducts. Next, a multiplying factor is put on the right side of the Nusselt or Stanton number relationship. This multiplying factor may be the ratio of the viscosity at T_s or T_{B_a}, as the case may be, to the viscosity evaluated at the wall temperature, with the ratio raised to some power which is normally between about 0.11 and 0.14. This is a common correction factor scheme for liquids (see Sieder and Tate [39] and Petukhov [34]). Another often-used multiplying factor is the ratio of the appropriate temperature T_s or T_{B_a} to the wall temperature T_w when the ratio is raised to a power whose value depends upon whether the fluid is a liquid or a gas and whether the fluid is being heated or cooled. The value of the power may also be dependent upon the Prandtl number (see Kays and Crawford [17]).

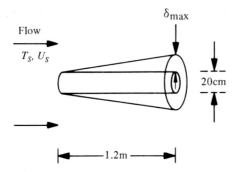

Figure 5.37 The boundary layer on the cylinder of Ex. 5.10.

In the analytical, semi-analytical, and experimental results for the Nusselt or Stanton number to be presented later in this chapter, we will see that all three schemes just mentioned for taking into account property variation will be used, depending upon which one does the best job of correlating the variable property results. Each separate result will have a specification with it as to the way the effect of variable thermophysical properties should be handled.

EXAMPLE 5.10

Air at one atmosphere pressure and 27°C flows at a speed of 3 m/s along a circular cylindrical electric resistance heater 20 cm in diameter and 1.2 m long. As shown in Fig. 5.37, the air flow is in the direction of the long axis, and it is assumed that it flows onto the cylinder smoothly. If the maximum surface temperature of the cylinder is not to exceed 127°C and the cylinder heater is designed to deliver a constant flux across its surface, find the maximum allowable surface heat flux and the total electrical energy dissipation rate if the net radiant loss can be safely neglected.

Solution
This particular geometry has not been treated. However, referring to Fig. 5.37, it should be reasonable to treat the cylinder as a flat plate of length L equal to 1.2 m and width of $\pi D = \pi(0.2) = 0.628$ m, if the boundary layer thickness at $x = L$ is just a small fraction of the cylinder radius of 0.1 m. If the boundary layer thicknesses are small compared to the radius, the effect of the curvature can be expected to be small. (Remember, the ratio of the boundary layer thickness to the "radius" of an actual flat plate is zero.)

First, the length Reynolds number is computed to see if the flow is laminar or turbulent and to compute the boundary layer thicknesses. The temperature-dependent fluid properties will be accounted for by use of the film temperature in the form applicable to external boundary layer–type flow, Eq. (5.356). Thus, at the end of the cylinder, $x = L$,

$$T_f = (T_w + T_s)/2 = (127 + 27)/2 = 77°C = 350°K.$$

The value T_w was evaluated at $x = L$, since by Eq. (5.355), the maximum surface temperature occurs at $x = L$ because T_w increases as $x^{1/2}$ if the flow is laminar.

From the property tables for air at $T_f = 350°K$, $\rho_f = 0.998$ kg/m³, $\mu_f = 2.075 \times 10^{-5}$ kg/ms, $k_f = 0.03$ W/m °C, and $N_{Pr_f} = 0.697$.

From Eq. (5.130), the length Reynolds number can be computed as

$$N_{Re_L} = \frac{\rho_f U_s L}{\mu_f} = \frac{0.998(3)(1.2)}{2.075 \times 10^{-5}} = 173,000.$$

By the criterion, Eq. (5.182a), the flow along the cylinder is laminar. Note that if the pressure were not one standard atmosphere or if the density were not listed in the property tables, the density of the air would have been calculated from the ideal gas equation of state.

Since the flow is laminar and the thermal boundary condition is one of constant flux, the approximate integral solution for the thermal boundary layer thickness is given by Eq. (5.348), which gives $\delta_t/\delta = 1/1.293\, N_{Pr}^{1/3}$. Hence,

$$\delta_t/\delta = 1/1.293(0.697)^{1/3} = 0.872.$$

Thus, it is seen that $\delta > \delta_t$, so that it is the hydrodynamic boundary layer thickness δ which should be compared to the cylinder radius to see if the cylinder can be satisfactorily idealized as a flat plate. Now, from Eq. (5.171), $\delta = 4.64x/N_{Re_x}^{1/2}$, and the maximum value of δ occurs at $x = L = 1.2$ m. Therefore,

$$\delta_{max} = 4.64(1.2)/(173,000)^{1/2} = 0.0134 \text{ m}.$$

The cylinder radius is $R = 0.1$ m,

$$\delta_{max}/R = 0.0134/.1 = 0.134.$$

Normally, ratios of this type (such as the ratio δ/L which is used to see if the boundary layer is thin) if less than 0.1, are found to be small enough so that replacing the cylinder by a flat plate is a reasonable approximation. Actually, more detailed information about this problem available in White's book [7] indicates that the boundary layer development *for our conditions* is essentially the same as for over a flat plate.

Next, the maximum allowable flux is found by inserting the maximum wall temperature of 127°C which occurs at the downstream end of the cylinder, $L = 1.2$ m, into Eq. (5.355). So,

$$127 = 27 + \frac{2.157 q_w''(1.2) \, [1 + (0.02052/0.697)^{2/3}]^{1/4}}{0.03(0.697)^{1/3} (173,000)^{1/2}},$$

or

$$q_w'' = 417.8 \text{ W/m}^2.$$

The total electrical energy dissipation rate from the heater is the surface flux times the area, so one has

$$q_w'' \pi D L = 417.8\pi(0.2)(1.2) = 315 \text{ W}.$$

EXAMPLE 5.11

The first inch of a particular gas turbine blade is idealized as a flat plate over which air at 1200°F flows at 100 ft/sec. If the blade surface is held at 800°F, compute the heat transfer rate for a 2-in.-span blade. The pressure is 60 psia.

Solution

The film temperature is $T_f = \frac{1}{2}(800 + 1200) = 1000°F$. From the property tables in Appendix A at this temperature, $\mu_f = 0.088$ lbm/hr ft, $k = 0.0337$ Btu/hr ft °F, and $N_{Pr_f} = 0.690$. The density at the film temperature can be found from the ideal gas equation of state. Thus,

$$\rho_f = \frac{P}{RT_f} = \frac{60(144)}{53.3(1460)} = 0.111 \text{ lbm/ft}^3.$$

Since $U_s = 100(3600)$ ft/hr $= 3.6 \times 10^5$ ft/hr,

$$N_{Re_L} = \frac{0.111(3.6 \times 10^5)(1/12)}{0.088} = 0.378 \times 10^5,$$

and the flow is laminar.

The appropriate expression to use for the Nusselt number and, therefore, for the surface coefficient of heat transfer, depends upon the type of thermal condition at the plate surface because of the previously discussed thermal history effect. Here the surface is at a constant known temperature, and since the total heat transfer rate is required, Eq. (5.302) will be used. (Equation [5.304], basically, is the same as Eq. [5.302] at Prandtl numbers fairly close to unity. Since Eq. [5.302] is simpler from the algebraic standpoint, it will be used in our calculation.)

Therefore,

$$\overline{N}_{Nu_L} = 0.664 \, N_{Pr}^{1/3} \, N_{Re_L}^{1/2} = 0.664(0.69)^{1/3} \, (3.78 \times 10^4)^{1/2} = 114,$$

or

$$\frac{\overline{h}_L(1/12)}{0.0337} = 114,$$

from which

$$\overline{h}_L = 46.2 \text{ Btu/hr ft}^2 \text{ °F}.$$

The heat transfer rate is

$$q = \overline{h}_L A(T_w - T_s).$$

Here, since $A = LW = (\frac{1}{12})(\frac{2}{12}) = \frac{1}{72}$ ft^2,

$$q = 46.2(\tfrac{1}{72})(800 - 1200) = -257 \text{ Btu/hr}.$$

The negative sign indicates that the heat transfer is from the fluid to the blade. This energy either has to be conducted down the blade to a cooler region or carried away by air circulating inside the blade.

5.4f Arbitrary Surface Temperature and Free Stream Velocity Variation

In general, the prediction of the local surface coefficient of heat transfer for a planar two-dimensional body of arbitrary shape which gives rise to an arbitrary free stream velocity as a function of x, $U_s(x)$, and with arbitrary but specified surface temperature variation $T_w(x)$, is considerably more difficult than the cases for a constant temperature or constant surface heat flux flat plate (U_s = constant) worked so far. However, as we have pointed out already, the thermal history effect is significant, particularly in laminar flows. In addition, different free stream velocity variations, $U_s(x)$, can also cause large changes in the surface coefficient from the form and value appropriate to flow over a flat plate at zero angle of attack where $U_s(x)$ is just a constant.

The integral methods we have discussed can be applied directly to such problems to obtain an approximate analytical solution. This is more complicated than the flat plate cases treated earlier, but certainly can be done, at least for laminar flows. The situation in the case of turbulent flow, like turbulent flow itself, is not as straightforward as far as solutions by integral methods are concerned.

Only the simplest method of attack on the general heat transfer problem, involving arbitrary $U_s(x)$ and $T_w(x)$, will be developed here. This is the semi-analytical approach of Ambrok [40]. This approach, while not as accurate as the more sophisticated methods, possesses the advantages that it is easy to use, is fairly accurate, and has the capability of handling laminar and turbulent flows.

Development of Ambrok's Method

For steady, constant property, low-speed, planar two-dimensional, boundary layer–type flow, either laminar or turbulent, the integral form of the thermal energy equation after multiplying through by -1 is

$$\frac{d}{dx} \int_0^{\delta_t} u(T - T_s)\, dy = \frac{q_w''}{\rho c_p}. \qquad [5.311]$$

(Recall that the wall heat flux q_w'' is not generally a constant.) Multiplying the integrand by $U_s(T_w - T_s)/U_s(T_w - T_s)$, where both U_s and T_w can be arbitrary functions of x, and then factoring out $U_s(T_w - T_s)$ in the numerator of the integral yields

$$\frac{d}{dx}\left[U_s(T_w - T_s) \int_0^{\delta_t} \frac{u}{U_s}\left(\frac{T - T_s}{T_w - T_s}\right) dy \right] = \frac{q_w''}{\rho c_p}. \qquad (5.359)$$

Inspection of the integral indicates that it has the dimensions of a length and is called the *enthalpy thickness* δ_t^*. Therefore,

$$\delta_t^* = \int_0^{\delta_t} \frac{u}{U_s}\left(\frac{T - T_s}{T_w - T_s}\right) dy. \tag{5.360}$$

Using this definition, Eq. (5.359) becomes

$$\frac{d}{dx}[U_s(T_w - T_s)\,\delta_t^*] = \frac{q_w''}{\rho c_p}.$$

Dividing both sides by $U_s(T_w - T_s)$ yields

$$\frac{1}{U_s(T_w - T_s)}\frac{d}{dx}[U_s(T_w - T_s)\,\delta_t^*] = \frac{q_w''}{\rho c_p U_s(T_w - T_s)}. \tag{5.361}$$

By Newton's law of cooling,

$$q_w''/(T_w - T_s) = h_x, \tag{5.362}$$

and using this in Eq. (5.361) gives on the right the nondimensional group identified earlier as the local Stanton number; that is,

$$\frac{h_x}{\rho c_p U_s} = N_{\mathrm{St}_x}. \tag{5.363}$$

Using Eqs. (5.362–63) in Eq. (5.361) and performing the indicated differentiations on the left side of Eq. (5.361) gives

$$\frac{d\delta_t^*}{dx} + \left[\frac{1}{U_s}\frac{dU_s}{dx} + \left(\frac{1}{T_w - T_s}\right)\frac{d(T_w - T_s)}{dx}\right]\delta_t^* = N_{\mathrm{St}_x}. \tag{5.364}$$

Equation (5.364) is another general form of the integral energy equation, Eq. (5.311).

For a moment, the discussion will be restricted to the flat plate in laminar flow at constant wall temperature T_w, where the exact analytical solution for the local Nusselt number has been found to be

$$h_x x/k = 0.332\, N_{\mathrm{Pr}}^{1/3} N_{\mathrm{Re}_x}^{1/2}. \tag{5.297}$$

Ambrok notes that this equation is of the general form

$$N_{\mathrm{Nu}_x} = A\, N_{\mathrm{Re}_x}^{n}, \tag{5.365}$$

where, for laminar flow over the plate,

$$A = 0.332 \, N_{\mathrm{Pr}}^{1/3}, \quad n = \frac{1}{2}. \tag{5.366}$$

Experiments indicate that the same form as Eq. (5.365) holds for turbulent flow over a flat plate at constant surface temperature, but with values of A and n differing from those given in Eq. (5.366) for laminar flow. The relationship between the Nusselt and Stanton numbers was earlier shown to be (replacing D everywhere by x)

$$N_{\mathrm{St}_x} = \frac{N_{\mathrm{Nu}_x}}{N_{\mathrm{Re}_x} N_{\mathrm{Pr}}}. \tag{5.252}$$

Eliminating the Nusselt number between Eqs. (5.252) and (5.365) yields

$$N_{\mathrm{St}_x} = \frac{A}{N_{\mathrm{Pr}}} N_{\mathrm{Re}_x}^{n-1}. \tag{5.367}$$

For a flat plate at constant wall temperature, dU_s/dx and $(d/dx)(T_w - T_s)$ are both zero and, for the flat plate, Eq. (5.364) becomes, after using Eq. (5.367) and writing out the local length Reynolds number,

$$\frac{d\delta_t^*}{dx} = \frac{A}{N_{\mathrm{Pr}}} \left(\frac{\rho U_s}{\mu} \right)^{n-1} x^{n-1}.$$

Integrating this from $x = 0$, where $\delta_t^* = 0$, by Eq. (5.360), to x, where the enthalpy thickness is δ_t^*, gives

$$\delta_t^* = \frac{A}{n \, N_{\mathrm{Pr}}} \left(\frac{\rho U_s}{\mu} \right)^{n-1} x^n.$$

Solving for x in terms of the enthalpy thickness,

$$x = \left(\frac{n N_{\mathrm{Pr}}}{A} \right)^{1/n} \left(\frac{\mu}{\rho U_s} \right)^{\frac{n-1}{n}} \delta_t^{*1/n}.$$

Substituting this into Eq. (5.367) to eliminate x in favor of the enthalpy thickness δ_t^* and rearranging leads to

$$N_{\mathrm{St}_x} = \left(\frac{n^{n-1}A}{N_{\mathrm{Pr}}} \right)^{1/n} \left(N_{\mathrm{Re}_{\delta_t^*}} \right)^{\frac{n-1}{n}}, \tag{5.368}$$

where

$$N_{\text{Re}_{\delta_t^*}} = \frac{\rho U_s \delta_t^*}{\mu}.$$ (5.369)

The term $N_{\text{Re}_{\delta_t^*}}$ is called the *enthalpy thickness Reynolds number.* Ambrok reasoned that the enthalpy thickness is a better indicator of local conditions, such as h_x, than is the length x from some point to the point of interest, and that, in effect, Eq. (5.368) should hold not only for the flat plate, but for an arbitrarily shaped body with an arbitrary surface temperature distribution. *That is, the local Stanton number is assumed to be a universal function of the local enthalpy thickness Reynolds number, with the exception that A and n have one set of values in a laminar flow and a different set of values in a turbulent flow.* Before Eq. (5.368) can be of use for a general body, the enthalpy thickness as a function of x must be found for an arbitrarily shaped body with arbitrary surface temperature distribution. This can be determined by solving Eq. (5.364). Substituting Eq. (5.368) into the right-hand side of Eq. (5.364) yields

$$\frac{d\delta_t^*}{dx} + \left[\frac{1}{U_s}\frac{dU_s}{dx} + \left(\frac{1}{T_w - T_s}\right)\frac{d(T_w - T_s)}{dx}\right]\delta_t^*$$

$$= \left(\frac{n^{n-1}A}{N_{\text{Pr}}}\right)^{1/n}\left(\frac{\rho U_s}{\mu}\right)^{\frac{n-1}{n}}\delta_t^{*\frac{n-1}{n}}.$$

Multiplying by $\delta_t^{*\frac{1-n}{n}}$ gives

$$\delta_t^{*\frac{1-n}{n}}\frac{d\delta_t^*}{dx} + \left[\frac{1}{U_s}\frac{dU_s}{dx} + \left(\frac{1}{T_w - T_s}\right)\frac{d(T_w - T_s)}{dx}\right]\delta_t^{*1/n}$$

$$= \left(\frac{n^{n-1}A}{N_{\text{Pr}}}\right)^{1/n}\left(\frac{\rho U_s}{\mu}\right)^{\frac{n-1}{n}}.$$ (5.370)

Noting that

$$\frac{d}{dx}\left(\delta_t^{*1/n}\right) = (1/n)\delta_t^{*\frac{1}{n}-1}\left(\frac{d\delta_t^*}{dx}\right) = (1/n)\delta_t^{*\frac{1-n}{n}}\frac{d\delta_t^*}{dx},$$

Eq. (5.370) becomes a linear, first-order, ordinary differential equation in the new dependent variable $\delta_t^{*(1/n)}$. This equation can be solved by the method of the integrating factor to yield a general expression for δ_t^* which can then be substituted into Eq. (5.368). Hence,

$$N_{\text{St}_x} = \left(\frac{n^{n-1}A}{N_{\text{Pr}}}\right)^{1/n} \frac{\left[\left(\frac{A}{nN_{\text{Pr}}}\right)^{1/n}\int_0^x \frac{U_s(T_w - T_s)^{1/n}}{\nu}dx\right]^{n-1}}{(T_w - T_s)^{\frac{n-1}{n}}}.$$ (5.371)

In solving Eq. (5.370), the point $x = 0$ coincided with the point where either $\delta_t^* = 0$ (sharp leading edge of a body), $U_s = 0$ (stagnation point of a body) or $(T_w - T_s) = 0$ (beginning of a heated section of a body).

Making use of Eq. (5.252), a general form for the local Nusselt number can be developed as

$$
N_{\mathrm{Nu}_x} = \frac{N_{\mathrm{Pr}}\left[\dfrac{n^{n-1}A}{N_{\mathrm{Pr}}}\right]^{1/n}\left[\left(\dfrac{A}{nN_{\mathrm{Pr}}}\right)^{1/n}\displaystyle\int_0^x \frac{U_s(T_w - T_s)^{1/n}}{\nu}\,dx\right]^{n-1}}{(T_w - T_s)^{\frac{n-1}{n}}} N_{\mathrm{Re}_x}. \tag{5.372}
$$

Substituting the values for A and n from Eq. (5.366) for *laminar flow* into Eq. (5.372) gives

$$
N_{\mathrm{Nu}_x} = \frac{0.332 N_{\mathrm{Pr}}^{1/3} N_{\mathrm{Re}_x}(T_w - T_s)}{\left[\displaystyle\int_0^x \frac{U_s(T_w - T_s)^2\,dx}{\nu}\right]^{1/2}}. \tag{5.373}
$$

For *turbulent flow*, flat plate experiments yield the values $A = 0.0288 N_{\mathrm{Pr}}^{1/3}$ and $n = 0.8$. Substituting these values into Eq. (5.372), the general expression for the local Nusselt number on a body of arbitrary shape and arbitrary surface temperature distribution is

$$
N_{\mathrm{Nu}_x} = \frac{0.0288 N_{\mathrm{Pr}}^{1/3} N_{\mathrm{Re}_x}(T_w - T_s)^{0.25}}{\left[\displaystyle\int_0^x \frac{U_s(T_w - T_s)^{1.25}}{\nu}\,dx\right]^{0.2}}. \tag{5.374}
$$

Ambrok's fundamental assumption that Eq. (5.368) is valid regardless of the body shape or surface temperature distribution holds more closely in turbulent flow than in laminar flow, and hence Eq. (5.374) is more reliable than is its laminar flow counterpart, Eq. (5.373).

Ambrok's result, Eq. (5.372), can also be applied, as an approximation, in the transitional flow regime between the point on the body where the flow is just still completely laminar and the point at which the flow is just barely completely turbulent. One needs an expression of the form of Eq. (5.365) to determine the proper expressions for A and n in this flow regime. Gauntner and Sucec [41], using experimental results given in Ambrok [40], have put forth a tentative expression for the local surface coefficient in the transition region along with rough estimates of the beginning and end of transition.

For present purposes, unless more accurate results in the form of exact or approximate analytical solutions or experimental correlations are available, Eqs. (5.373) and (5.374) for laminar and turbulent flow, respectively, will be used to predict the local surface coefficient of heat transfer in steady, constant property (variable properties are taken into approximate account by use of the film temperature T_f), low-speed, planar two-dimensional flow of a fluid at undisturbed temperature T_s over a body of arbitrary shape [which gives rise to $U_s = U_s(x)$] and arbitrary specified surface temperature distribution $T_w(x)$.

Application of Ambrok's Result

This section will be concerned with the application of Ambrok's results to both laminar and turbulent heat transfer problems. Predictions will be compared to more nearly exact solutions to assess the accuracy one might typically expect when using Ambrok's method, and also to help decide upon the overall value of the method.

Laminar Flow Over a Flat Plate with Linear Wall Temperature Distribution Steady, constant property, laminar, boundary layer, low-speed flow over a flat plate will be considered for the case when the plate surface temperature varies linearly with x according to the following prescription:

$$T_w = T_s + bx. \tag{5.375}$$

In Eq. (5.375), b is a known constant which might be positive or negative, x is the length coordinate along the plate measured from the plate's leading edge, and T_s is the thermal free stream temperature outside the thermal boundary layer. An expression for the local Nusselt number for this particular surface temperature variation will be derived.

Equation (5.373) is used with U_s a constant and $T_w - T_s = bx$. Substitution yields

$$N_{Nu_x} = \frac{0.332 N_{Pr}^{1/3} N_{Re_x} (bx)}{\left(\dfrac{U_s}{\nu} \displaystyle\int_0^x b^2 x^2 \, dx\right)^{1/2}}.$$

Performing the integration in the denominator yields

$$N_{Nu_x} = \frac{0.332 N_{Pr}^{1/3} N_{Re_x} (bx)}{[(U_s/\nu) b^2 (x^3/3)]^{1/2}}.$$

Rearranging yields the result

$$N_{Nu_x} = 0.574 N_{Pr}^{1/3} N_{Re_x}^{1/2}. \tag{5.376}$$

This result, Eq. (5.376), using Ambrok's method is 7.3% higher than the more nearly exact solution for the same case as reported in [17]. Hence the accuracy of the method, at least on this problem, is quite acceptable.

Compare Eq. (5.376) or the more nearly exact solution to the case of the *same* flat plate at a *constant* (instead of linear) wall temperature, Eq. (5.297). Note that the local surface coefficient is 61% higher for the linear variation of wall temperature than for the constant wall temperature case. This illustrates once again the important effect of different thermal histories on the surface coefficient in laminar flow heat transfer problems. This was seen earlier for the flat plate and for ducts by comparing the constant temperature results with the constant surface flux results.

Turbulent Flow Over a Flat Plate with Linear Wall Temperature Distribution   For turbulent flow, we would like to predict the local Nusselt number variation with x in steady, constant property, two-dimensional, low-speed boundary layer flow over the plate when the plate surface temperature varies linearly as given in Eq. (5.375).

Thus, substituting Eq. (5.375) into the result of Ambrok for turbulent flow, Eq. (5.374), one has

$$N_{Nu_x} = \frac{0.0288 \, N_{Pr}^{1/3} \, N_{Re_x} \, (bx)^{0.25}}{\left[\dfrac{U_s}{\nu} \displaystyle\int_0^x b^{1.25} x^{1.25} \, dx \right]^{0.2}} \, .$$

Performing the indicated integration and rearranging gives

$$N_{Nu_x} = 0.0339 \, N_{Pr}^{1/3} \, N_{Re_x}^{0.8} \, . \tag{5.377}$$

When we compare the results of Ambrok's method, Eq. (5.377), with results (considered practically exact) which can be derived with the aid of [17], we find that Ambrok's prediction is only about 3.75% in error. When the error of Ambrok's method for a nearly identical problem in which the flow is laminar is compared to the error here where the flow is turbulent, we can see that Ambrok's method is more accurate for turbulent than laminar flow, all other things being equal. This is in accord with what we noted earlier about Ambrok's fundamental assumption: that Eq. (5.368) is valid in very general circumstances, being more nearly satisfied in turbulent flows than in laminar flows.

Experiments for a *constant temperature* flat plate when the flow is turbulent indicate that the Nusselt number has the form shown in Eq. (5.377), with the coefficient being 0.0288 instead of the value of 0.0339 for the linear wall temperature variation. Thus we can see that the linear wall temperature distribution of Eq. (5.375) causes, in turbulent flow, the local Nusselt number to be about 11.9% higher than the value for the constant temperature flat plate. Recalling that for the same case, except for the flow being laminar, the corresponding factor was 61%, we see that although the thermal history effect is important here, it is nowhere near as important as it is in laminar flow. We presented other evidence of this when we compared h_x for constant surface temperature and constant surface flux for both laminar and turbulent flows.

Thus, the thermal history effect is more important in laminar flow than in turbulent flow, as long as the Prandtl number is not too much lower than about 0.5 [17].

The two solutions just presented by Ambrok's method could give one a distorted view of the accuracy of the method, since the errors involved were 7.3% and 3.75%. It can be demonstrated that, in some cases, the error is substantially greater. For example, one "worst case" situation is laminar flow at the stagnation point of a cylinder, where the error involved in using Ambrok's method is about 17.5%. But usually for laminar flow it will not be as large an error as this, and as we have seen already, the error in the method is considerably less for turbulent flows. From our examples using Ambrok's procedure, it should be evident that the method involves relatively little effort. One simply inserts a particular $T_w(x)$ and $U_s(x)$ into Eq. (5.373) or (5.374) and performs the indicated integrations. Even when they cannot be performed analytically, they can be very easily done numerically on the digital computer, or even graphically at one's desk. Hence, if we define an "efficiency," of sorts, of prediction methods for the surface coefficient, as being the ratio of the desired effect (a relatively accurate value of h_x) to the price paid to achieve it (amount of effort expended to make the prediction of h_x), Ambrok's method would easily be the most efficient.

The performance of Ambrok's method, particularly for turbulent flows, has earned it high grades [17], [7], [41].

EXAMPLE 5.12

A constant temperature body immersed in a laminar flow has a shape such that the free stream velocity variation is $U_s = U_0 + cx^{1/3}$. When the shape is such that $c > 0$, the flow accelerates along the body, whereas, if the shape yields $c < 0$, the flow decelerates along the body. Predict the local Nusselt number as a function of x.

Solution
For constant wall temperature, $T_w - T_s$ is a constant and can be cancelled from Eq. (5.373). Substituting the free stream velocity U_s as a function of x gives

$$N_{\mathrm{Nu}_x} = \frac{0.332 N_{\mathrm{Pr}}^{1/3} N_{\mathrm{Re}_x}}{\left[\int_0^x \dfrac{(U_0 + cx^{1/3})}{\nu}\, dx\right]^{1/2}} = \frac{0.332 N_{\mathrm{Pr}}^{1/3} N_{\mathrm{Re}_x}}{\left(\dfrac{U_0 x + \frac{3}{4} c x^{4/3}}{\nu}\right)^{1/2}}. \qquad (5.378)$$

The denominator of Eq. (5.378) can be rearranged by multiplying inside the bracket by U_s/U_s and then factoring x to yield

$$\left[\left(\frac{U_s x}{\nu}\right)\left(\frac{U_0 + \frac{3}{4}cx^{1/3}}{U_0 + cx^{1/3}}\right)\right]^{1/2} = N_{\mathrm{Re}_x}^{1/2}\left(\frac{U_0 + \frac{3}{4}cx^{1/3}}{U_0 + cx^{1/3}}\right)^{1/2}.$$

Using this in Eq. (5.378) leads to

$$N_{\mathrm{Nu}_x} = 0.332 N_{\mathrm{Pr}}^{1/3} N_{\mathrm{Re}_x}^{1/2}\left(\frac{U_0 + cx^{1/3}}{U_0 + \frac{3}{4}cx^{1/3}}\right)^{1/2}. \qquad (5.379)$$

From this result it can be seen that if $c > 0$, accelerated flow, the Nusselt number is greater than for constant velocity flow, $c = 0$; and, if $c < 0$, the Nusselt number is less than that for a constant velocity flow.

When the free stream velocity takes the simpler form $U_s = cx^{1/3}$, Eq. (5.379) reduces to

$$N_{\mathrm{Nu}_x} = 0.382 N_{\mathrm{Pr}}^{1/3} N_{\mathrm{Re}_x}^{1/2} \,,$$

which is 13.2% lower than the exact solution given in [17].

5.5 EXPERIMENTAL CORRELATIONS FOR FORCED CONVECTION SITUATIONS

Earlier in this chapter, it was mentioned that many forced convection heat transfer situations are so complex that exact analytical analysis has thus far been precluded, and the surface coefficients of heat transfer are found from nondimensional correlations of experimental data for different geometries and boundary conditions. One very important example of this complexity, since it is encountered routinely in actual practice, is *flow separation*. Since separation is caused by certain types of free stream velocity variations, $U_s(x)$, it might be thought that Ambrok's procedure could be used here. However, neither Eq. (5.373) or Eq. (5.374) is valid in or near a region of separated flow. Flow separation can occur when the external velocity U_s outside the boundary layer decreases in the flow direction. Separation can thus occur on the rear portion of spheres and on cylinders normal to the flow. It occurs when the low-kinetic-energy fluid in the boundary layer cannot negotiate the adverse pressure gradient caused by the shape of the bodies themselves or produced by one body's interaction with another. This low-energy boundary layer fluid can be likened to a ball rolling on a horizontal plane which originally has a high enough speed to roll to the top of a hill (at this point, the ball is similar to a fluid particle *outside* the boundary layer in the free stream), but is slowed down on its way to the hill because of the frictional resistance of the air and the surface (similar to the fluid particle entering the boundary layer and having its velocity decreased by the viscous effects, i.e., the shear stresses). Hence, when it reaches the hill, its speed is so low that it can only go part way up until its speed decreases to zero. It then begins to roll back down the hill. (The low-velocity fluid particle in the boundary layer reaches a portion of the body where the pressure is rising in the flow direction, and the particle must convert its kinetic energy to "pressure energy"—flow work—to get up the "pressure hill," but it does not have sufficient kinetic energy and begins a reverse flow at the separation point.) The flow and heat transfer picture in the separated region is very complex, and average surface coefficients of heat transfer between a body surface and the external stream, when a separated flow region is present, will be found by the experimental correlations which will be presented later in this section.

Even when an exact analytical solution is available, an experimental correlation for the situation may at times be much easier to use to calculate an actual value of the surface coefficient of heat transfer. This might be the case when the exact solution is given by a slowly convergent infinite series in which a great many terms of the series must be retained and evaluated. It sometimes may also be that the eigenvalues needed in the individual terms of the series are no small task to evaluate from a complicated transcendental equation. Thus, under these conditions, an available experimental correlation might be preferred for use over an available exact solution.

Correlations of the experimental data for surface coefficients of heat transfer will be given for the following: for external, turbulent flow over bodies, including some shapes which cause flow separation; for internal turbulent flow, both fully developed and developing; for laminar flows in the developing entrance region of tubes; and for laminar or turbulent flow over bluff bodies, such as cylinders and spheres, which have separated regions of flow.

With an experimental correlation, say for the Nusselt number as a function of the Reynolds and Prandtl number, or the Stanton number as a function of the Reynolds and Prandtl number, some range of the variables will be presented wherein the correlation is valid. We will also see how the problem of temperature-dependent fluid properties should be handled. Care should be taken to adhere to the instructions relevant to a particular correlation.

5.5a Turbulent Boundary Layer–Type Flow Over a Constant Temperature Flat Plate

In Sec. 5.4b, we noted that after an examination of experimental data, Colburn [31] proposed that the Reynolds analogy, Eq. (5.249), be modified for the effect of Prandtl numbers other than unity by rewriting it as

$$N_{St}N_{Pr}^{2/3} = \frac{1}{2}\,C_f. \tag{5.380}$$

The product of the Stanton and Prandtl number to the two-thirds power is often called *Colburn's j factor* or simply the *j factor,* and the modified Reynolds analogy, Eq. (5.380), is called *Colburn's analogy.*

But of what use is this relation, derived for fully developed tube flow, in the problem of external turbulent boundary layer flow over a flat plate? Well, for steady, low-speed, constant property, laminar or turbulent, planar two-dimensional, boundary layer–type flow over a flat plate at zero angle of attack and constant wall temperature T_w, the essential ideas embodied in the development of the Reynolds analogy are satisfied [15]; thus, Eq. (5.380) is assumed to be valid. In the local length Reynolds number range from 3×10^5 to about 10^7—that is, fully turbulent flow over a flat plate—the local skin friction coefficient is

$$C_f = 0.0576/N_{Re_x}^{1/5}. \tag{5.187}$$

Substituting Eq. (5.187) into Eq. (5.380), the expression for the local Stanton number in turbulent flow over a flat plate is

$$N_{St_x} = 0.0288\,N_{Pr}^{-2/3}\,N_{Re_x}^{-1/5}. \tag{5.381}$$

Using Eq. (5.252), Eq. (5.381) can be rearranged to yield

$$N_{Nu_x} = 0.0288\,N_{Pr}^{1/3}\,N_{Re_x}^{0.8} = h_x x/k. \tag{5.382}$$

Equations (5.381–82) agree with experimental data for turbulent flow over a flat plate for Prandtl numbers from approximately 0.5 to 50, provided that the flow is turbulent over the entire plate. Temperature-dependent fluid properties should be evaluated at the film temperature previously defined as $T_f = (T_w + T_s)/2$.

Now, Eq. (5.382) can be used to develop an expression for the average surface coefficient of heat transfer over the entire plate of length L. We do this by following exactly the same procedure as was used earlier in connection with the corresponding average surface coefficient for laminar flow over a constant temperature plate, Eq. (5.301). This leads to the following result:

$$\frac{\overline{h_L} L}{k} = \overline{N_{Nu_L}} = 0.036 \, N_{Pr}^{1/3} \, N_{Re_L}^{0.8} . \qquad (5.383)$$

Equations (5.382) and (5.383) give the local and average surface coefficients of heat transfer, respectively, over a *constant temperature* flat plate when the flow is *turbulent* over the *entire* plate and when the Prandtl number is in the range $0.5 < N_{Pr} < 50$.

Sometimes, because of a boundary layer trip wire, natural roughness, or a slight lip or step near the leading edge of the plate, turbulent flow is realized practically from the leading edge of the plate, and flow remains turbulent over the entire plate. However, it is more common for the flow to be laminar from $x = 0$ to $x = x_c$, at which point turbulent flow begins and continues to the end of the plate at $x = L$. If $x_c << L$, little error occurs if Eq. (5.383) is used for the entire plate. If L is only slightly greater than x_c, Eq. (5.302) should be used. But if the length of the plate x_c over which the flow is laminar is roughly the same order of magnitude as the length of plate $L - x_c$ over which the flow is turbulent, this must be accounted for in a relation for the average Nusselt number.

This can be done by using Eq. (5.296) for the local surface coefficient in the laminar region and then using Eq. (5.382) for the local value of the surface coefficient, h_x, in the turbulent region. This assumes that the h_x for $x > x_c$, where x_c is the distance down the plate at which the transition from laminar to turbulent flow occurs, is essentially the same as that which would have occurred if the boundary layer had been turbulent from $x = 0$. By use of $N_{Re_x} = 300,000$ as the transition Reynolds number (see the discussion of Eqs. [5.182a–b]), we will now derive an expression for the average surface coefficient over the plate when the turbulent boundary layer is preceded by a laminar one.

A general expression for the average surface coefficient over an *isothermal body* can be arrived at by equating Eqs. (5.298) and (5.300) and rearranging to yield

$$\overline{h_L} = \frac{1}{L} \int_0^L h_x dx. \qquad (5.384)$$

Because of the different functional form for h_x in the laminar and turbulent portions of the plate, Eq. (5.384) must be split into the two integrals

$$\overline{h_L} = \frac{1}{L} \int_0^{x_c} h_x \, dx + \frac{1}{L} \int_{x_c}^L h_x \, dx.$$

Using Eqs. (5.296) and (5.382) to obtain h_x for the first and second integrals, respectively, leads to

$$\overline{h}_L = \frac{0.332}{L} k \, N_{\text{Pr}}^{1/3} \left(\frac{\rho U_s}{\mu}\right)^{1/2} \int_0^{x_c} x^{-1/2} \, dx$$
$$+ \frac{0.0288}{L} k \, N_{\text{Pr}}^{1/3} \left(\frac{\rho U_s}{\mu}\right)^{0.8} \int_{x_c}^L x^{-0.2} \, dx.$$

Performing the integrations and rearranging yields

$$\overline{h}_L L / k = 0.664 \, N_{\text{Pr}}^{1/3} N_{\text{Re}_c}^{1/2} + 0.036 \, N_{\text{Pr}}^{1/3} N_{\text{Re}_L}^{0.8} - 0.036 \, N_{\text{Pr}}^{1/3} N_{\text{Re}_c}^{0.8},$$

or

$$\overline{N_{\text{Nu}_L}} = 0.036 N_{\text{Pr}}^{1/3} (N_{\text{Re}_L}^{0.8} - N_{\text{Re}_c}^{0.8} + 18.43 \, N_{\text{Re}_c}^{0.5}),$$

where $N_{\text{Re}_c} = \rho U_s x_c / \mu$, and x_c is the position where laminar flow ends. Using $N_{\text{Re}_c} = 3 \times 10^5$, the final result for the average Nusselt number is

$$\overline{N_{\text{Nu}_L}} = 0.036 \, N_{\text{Pr}}^{1/3} (N_{\text{Re}_L}^{0.8} - 14{,}100). \tag{5.385}$$

Note that Eq. (5.385) is to be used only for $N_{\text{Re}_L} > 300{,}000$, *when the flow is laminar up to this value of local length Reynolds number,* for a constant temperature plate (strictly speaking), and for $0.5 < N_{\text{Pr}} < 50$.

The effect of temperature-dependent fluid properties is approximately accounted for in Eqs. (5.382), (5.383), and (5.385) by employing the film temperature $T_f = (T_w + T_s)/2$.

Whitaker [42] gives the following correlation of experimental data which extends into the higher Prandtl number range and thus correlates the data for highly viscous oils better than Eq. (5.385). The result, modified slightly for a transition local length Reynolds number of 300,000, is as shown.

$$\overline{N_{\text{Nu}_L}} = 0.036 \, N_{\text{Pr}}^{0.43} [N_{\text{Re}_L}^{0.8} - 14{,}100] \, (\mu_s/\mu_w)^{0.25} \tag{5.386}$$

In Eq. (5.386), μ_s is the dynamic viscosity evaluated at the undisturbed stream temperature T_s, μ_w is the same quantity evaluated at the wall temperature T_w, and all fluid properties which appear in $\overline{N_{\text{Nu}_L}}$, N_{Re_L}, and N_{Pr} are to be evaluated at T_s. Note that Eq. (5.386) is easily modified for the case in which the flow is turbulent over the entire plate by replacing the 14,100 by zero. Equation (5.386) is recommended in preference to Eq. (5.385) for the very high Prandtl number fluids such as hydrocarbon oils.

Equation (5.382) is best used in the indicated Prandtl number range, $0.5 < N_{\text{Pr}} < 50$. Churchill [36] provides the following correlating equation for turbulent flow over an isothermal flat plate for the entire range of Prandtl numbers:

$$N_{\text{Nu}_x} = \frac{0.032 \, N_{\text{Pr}}^{8/15} \, N_{\text{Re}_x}^{0.8}}{[1 + (0.0468/N_{\text{Pr}})^{2/3}]^{0.4}} \,. \tag{5.387}$$

In this expression for the local Nusselt number, temperature-dependent fluid properties are to be evaluated at the film temperature. An examination of Eqs. (5.382), (5.383), and (5.385) makes apparent the modifications needed to arrive at average values of the surface coefficient for Eq. (5.387).

Constant Flux Flat Plate in Turbulent Flow

Thus far, all results in this section are, strictly speaking, valid only for the constant wall temperature flat plate. However, as has been pointed out before, the effect of thermal history in turbulent flow is not as important as it is in laminar flow. In particular, the constant flux thermal condition at the wall *in turbulent flow* causes the local surface coefficient to be only about 4% higher than the corresponding one for the constant temperature plate [17]. Hence, the expression for the local surface coefficient for the turbulent flow over a *constant flux* flat plate is given by

$$N_{\text{Nu}_x} = 0.03 \, N_{\text{Pr}}^{1/3} \, N_{\text{Re}_x}^{0.8} \,. \tag{5.388}$$

The quantity of most interest for a constant flux flat plate is the local wall temperature variation with x caused by the known, constant flux, q_w''.

Since, by Newton's cooling law,

$$q_w'' = h_x(T_w - T_s), \tag{5.389}$$

h_x can be found from Eq. (5.388), inserted into Eq. (5.389), and the result solved for $T_w(x)$ to yield

$$T_w(x) = T_s + \frac{33.33 \, x \, q_w''}{k \, N_{\text{Pr}}^{1/3} \, N_{\text{Re}_x}^{0.8}} \,. \tag{5.390}$$

EXAMPLE 5.13

A thin sheet of metal 3 m by 3 m is to be heat-treated in an oven and the average surface temperature, in time, of the metal during the heat-treating operations is expected to be about 500°K. Oven air, at standard atmospheric pressure, is to be blown over the plate along one of the 3 m dimensions at a speed of 12 m/s and at a temperature of 800°K. Estimate the value of the average surface coefficient of heat transfer between the oven air and the plate.

Solution

We first establish whether the flow is laminar or turbulent by calculating the value of the local length Reynolds number, Eq. (5.130), for this external flow situation. Since most of the results presented thus far, for both laminar and turbulent flow over flat plates, use the film temperature T_f to calculate the temperature-dependent fluid properties, this is found first.

$$T_f = (T_w + T_s)/2 = (500 + 800)/2 = 650°\text{K}.$$

With this as the film temperature, the property tables for air yield $\rho_f = 0.543 \text{ kg/m}^3$ (since the pressure is one standard atmosphere), $\mu_f = 3.177 \times 10^{-5} \text{ kg/ms}$, $k_f = 0.0495$ W/m °K, and $N_{Pr_f} = 0.682$. Thus, from Eq. (5.130),

$$N_{Re_L} = \rho_f U_s x / \mu_f = 0.543(12)(3)/3.177 \times 10^{-5} = 615,000.$$

Thus the flow is turbulent, since N_{Re_L} exceeds 300,000, at least over the last part of the plate. Since no mention was made of any projection, lip, or roughness that could trip the boundary layer and cause it to be turbulent right from the leading edge, we assume that the boundary layer is laminar up to a local Reynolds number of 300,000 and then turbulent the rest of the way. Hence, Eq. (5.385) should be used, since the Prandtl number in this problem is also in the proper range. Substituting known values into Eq. (5.385) yields

$$\frac{\overline{h}_L(3)}{0.0495} = 0.036(0.682)^{1/3} [(615,000)^{0.8} - 14,100].$$

Thus, $\overline{h}_L = 15 \text{ W/m}^2$ °K.

This value of $\overline{h}_L$ can now be used, for example, in solving for the time it takes to raise these thin sheets to a particular temperature of interest with the aid of the Heisler's charts in Chapter 3 on transient conduction problems.

5.5b Turbulent Flow in Tubes

Thermally and hydrodynamically fully developed turbulent flow in tubes was introduced in an earlier section, where the similarity between the transport of heat and momentum led to Reynolds' analogy. This was then generalized by Colburn to Colburn's, or the extended Reynolds, analogy, Eq. (5.250). With the insertion of an equation which correlates the experimental friction factor data into Eq. (5.250), the working form of Colburn's analogy becomes Eq. (5.253), which is repeated here for easy reference.

$$N_{Nu} = hD/k_f = 0.023 N_{Pr_f}^{1/3} N_{Re}^{0.8}, \tag{5.391}$$

where $N_{Re} = \rho_f U_m D / \mu_f$.

When we use Eq. (5.391), all temperature-dependent fluid properties are to be evaluated at the *film temperature, T_f*, given by Eq. (5.357) for tube flow. The diameter Reynolds number should be greater than about 10^4, and the Prandtl number should lie in the range $0.7 < N_{Pr} < 100$. Also, the value of tube length to diameter, *L/D*, should be large enough to insure the fully developed nature of both the velocity field and the temperature field. Generally, it is accepted that this is the case, as far as the average surface coefficient is concerned [39], when this ratio exceeds 50 to 60; so,

$$L/D \geq 60. \tag{5.392}$$

So, when inequality (5.392) is satisfied, the expressions for the *average* surface coefficient in *fully developed turbulent* tube flow can be used safely.

Equation (5.391) is essentially of the same form as an earlier, direct correlation of experimental data due to Dittus and Boelter [43], and this constitutes additional evidence of the adequacy of the semi-analytical result, Eq. (5.391), in representing experimental data for *smooth* tubes.

For fully developed turbulent tube flow of fluids whose properties are strongly temperature dependent, such as oils, the following equation is recommended by Sieder and Tate [39] where, again, $L/D > 60$:

$$hD/k_a = 0.023 N_{Pr}^{1/3} N_{Re}^{0.8} (\mu_a/\mu_w)^{0.14}. \tag{5.393}$$

The subscript *a* refers to the arithmetic average of the fluid inlet bulk mean temperature and outlet bulk mean temperature; that is,

$$T_{B_a} = \tfrac{1}{2}(T_{B_1} + T_{B_2}). \tag{5.394}$$

All temperature-dependent fluid properties are evaluated at T_{B_a} except for the denominator of the viscosity ratio, which is subscripted *w* to indicate that this should be evaluated at the wall temperature.

Von Kàrmàn's analogy, Eq. (5.269), is also valid for fully developed turbulent flow, but Eqs. (5.391–93) are usually preferred because they are easier to use.

For the utmost in accuracy of the *local* surface coefficient of heat transfer, with this increase in accuracy over equations such as Eqs. (5.391) and (5.393) purchased at the price of increased complexity of the calculation, the various equations presented by Petukhov [34] can be used for developed turbulent tube flow. Petukhov's base equation represents a close curve fit to his semi-analytical results which arise from numerical solution, by computer, of Eq. (5.89) once a distribution of eddy diffusivity of heat ϵ_H has been inserted and q'' has been related to q_w'' by an energy balance. Though his solutions are for constant wall flux, they can also be used for constant wall temperature in turbulent flow, since the two different values of the local surface coefficient differ by only a few percent as long as $N_{Pr} > 0.6$ [17]. Petukhov's expressions correlate the experimental data better than the expressions presented so far.

For liquids

$$N_{Nu} = \frac{(f/8)N_{Pr}N_{Re_D}}{1.07 + 12.7\sqrt{f/8}\,(N_{Pr}^{2/3} - 1)}\left(\frac{\mu_B}{\mu_w}\right)^n \tag{5.395}$$

$n = 0.11$ for heating, $T_w > T_B$
$n = 0.25$ for cooling, $T_w < T_B$

For gases

$$N_{Nu} = \frac{(f/8)N_{Pr}N_{Re_D}F_G}{1.07 + 12.7\sqrt{f/8}\,(N_{Pr}^{2/3} - 1)} \tag{5.396}$$

$$F_G = \left(\frac{T_w}{T_B}\right)^n$$

$$n = -0.3\log_{10}(T_w/T_B) + 0.36 \tag{5.397}$$
$$\text{for } \textit{heating, } T_w > T_B$$

$$F_G = 1.27 - 0.27(T_w/T_B) \tag{5.398}$$
$$\text{for } \textit{cooling, } T_w < T_B$$

In using these results, the following ranges of Reynolds and Prandtl numbers should be observed:

$$10^4 < N_{Re_D} < 5 \times 10^6, \quad 0.5 < N_{Pr} < 2000.$$

The friction factor in smooth pipes can either be found from the Moody diagram, Fig. 5.9, or from the following relationship given by Petukhov:

$$f = 1/(1.82\log_{10} N_{Re_D} - 1.64)^2. \tag{5.399}$$

All temperature-dependent fluid properties are to be evaluated at the local bulk mean temperature T_B, except for μ_w, when it appears, which is to be evaluated at the wall temperature.

Equations (5.395) and (5.396) are for the *local* surface coefficient of heat transfer in developed turbulent tube flow when the surface heat flux q_w'' is a constant or, with a few percent error, when the wall temperature is a constant. When compared to experimental

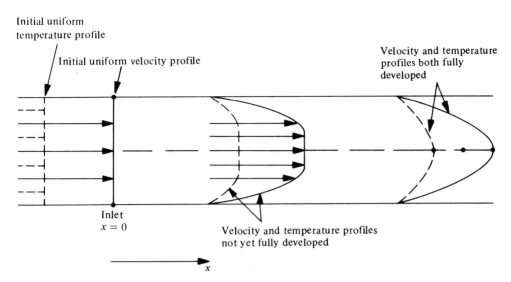

Initial uniform
temperature profile

Initial uniform velocity profile

Velocity and temperature
profiles both fully
developed

Inlet
$x = 0$

Velocity and temperature profiles
not yet fully developed

x

Figure 5.38 Developing velocity and temperature profiles in the entrance region of a tube.

data for the local surface coefficient, the data scatter by less than about $\pm 10\%$ about these two equations. On the other hand, for a simpler correlation such as Eq. (5.393), the experimental data lie $\pm 15\%$ to $\pm 20\%$ about the equation.

In order to have an expression for the average surface coefficient of heat transfer for the entire tube from Petukhov's results, it is suggested that all properties except μ_w in Eqs. (5.395) and (5.396) be evaluated at the arithmetic average bulk mean temperature given by Eq. (5.394).

Turbulent Heat Transfer in the Tube Entrance Region

In the entrance region of tubes, $L/D < 60$, the profiles of velocity and temperature are changing from their (perhaps) uniform initial forms at the tube entrance to their fully developed forms. This is shown in Fig. 5.38, where at $x = 0$, both profiles are uniform over the cross section. At the next station, both profile shapes have changed to something between the initial uniform distribution and the ultimate fully developed distributions shown at the last station in Fig. 5.38. The average surface coefficients of heat transfer are higher in this entrance region than in the fully developed region, and they depend upon the L/D ratio. For tubes having a sharp-edged entrance (as opposed to a smooth nozzle or bellmouth) with a flow Reynolds number (based on diameter) greater than 10^4, and for fluid Prandtl numbers between 0.7 and 120, McAdams [44] recommends the use of the equation

$$\frac{hD}{k_a} = 0.023 N_{\text{Pr}}^{1/3} \left(\frac{\rho U_m D}{\mu_a}\right)^{0.8} \left(\frac{\mu_a}{\mu_w}\right)^{0.14} \left[1 + \left(\frac{D}{L}\right)^{0.7}\right]. \qquad (5.400)[4]$$

---

4. From *Heat Transmission* by William H. McAdams. Copyright 1954 by William H. McAdams. Used with permission of McGraw-Hill Book Company.

As can be seen from Eq. (5.400), all properties except μ_w should be evaluated at the average bulk mean temperature T_{B_a}, which is calculated from Eq. (5.394).

Additional information, both from direct experiments and from the results of semi-analytical approaches, concerning the surface coefficient of heat transfer in the entry regions in turbulent flow is available in [45] and [17]. The information appears both in graphical form and as eigenvalues and coefficients to be used in an infinite series solution.

After Eqs. (5.269), (5.391), (5.393), (5.400), or appropriately modified versions of Eqs. (5.395) and (5.396) are used to determine the average surface coefficient of heat transfer to be used in Newton's cooling law, the appropriate driving potential difference is the logarithmic mean temperature difference; that is,

$$\Delta T_L = \frac{T_{B_2} - T_{B_1}}{\ln\left(\dfrac{T_w - T_{B_1}}{T_w - T_{B_2}}\right)}.$$

[5.235]

The basis for doing this, it will be recalled, was discussed during the derivation of Eq. (5.235).

EXAMPLE 5.14

Water at 40°F flows at 1000 lbm/hr through a sharp-edged 1/2-in.-diameter tube which is 10 in. long. If the tube wall temperature is 240°F, compute the exit temperature of the water.

Solution
An energy balance applied to the entire tube gives

$$h\pi DL\Delta T_L = \omega c_p(T_{B_2} - T_{B_1}).$$

(5.401)

Since $D = 1/24$ ft and $L = 10/12$ ft, $L/D = 20$. Since $L/D < 60$, the flow is not fully developed, and if turbulent, Eq. (5.400) should be used to calculate h.

The average bulk mean temperature is required to evaluate most of the temperature-dependent fluid properties in Eq. (5.400), but this contains the unknown exit temperature of the water T_{B_2}; that is,

$$T_{B_a} = \tfrac{1}{2}(40 + T_{B_2}).$$

(5.402)

A value of T_{B_2} is estimated and used to find h, and then T_{B_2} is solved for from Eq. (5.401); if the new value of T_{B_2} does not agree with the estimated value, the new value is used and the procedure repeated until there is satisfactory agreement between successive estimates.

Initially assume $T_{B_2} = 100°F$; therefore, $T_{B_a} = 70°F$ by Eq. (5.402). From the property tables in Appendix A, $k_a = 0.347$ Btu/hr ft °F, $N_{Pr_a} = 6.81$, $\mu_a = 2.37$ lbm/hr ft, and $c_p \approx 1.0$ Btu/lbm °F. At $T_w = 240°F$, $\mu_w = 0.588$ lbm/hr ft. From the continuity equation,

$$\rho U_m = \frac{4}{\pi} \frac{\omega}{D^2} = \frac{4}{\pi} \frac{1000}{(1/24)^2} = 7.34 \times 10^5;$$

hence,

$$N_{Re} = \frac{\rho U_m D}{\mu_a} = \frac{7.34 \times 10^5 (1/24)}{2.37} = 12,900.$$

Substituting these values into Eq. (5.400) yields

$$\frac{h(1/24)}{0.347} = 0.023(6.81)^{1/3}(12,900)^{0.8} \left(\frac{2.37}{0.588}\right)^{0.14} [1 + (1/20)^{0.7}],$$

or

$$h = 959 \text{ Btu/hr ft}^2 \text{ °F}.$$

Now, from Eq. (5.235),

$$\Delta T_L = \frac{T_{B_2} - 40}{\ln\left(\dfrac{240 - 40}{240 - T_{B_2}}\right)} = \frac{T_{B_2} - 40}{\ln\left(\dfrac{200}{240 - T_{B_2}}\right)}.$$

Substitution of ΔT_L and h along with other known quantities into Eq. (5.401) and cancelling the common factor $T_{B_2} - 40$ leads to

$$\frac{959(1/24)\pi(10/12)}{\ln\left(\dfrac{200}{240 - T_{B_2}}\right)} = 1000(1), \quad \text{or} \quad 0.1045 = \ln\left(\frac{200}{240 - T_{B_2}}\right).$$

Therefore,

$$1.11 = \frac{200}{240 - T_{B_2}}, \quad \text{or} \quad T_{B_2} \approx 60°F.$$

Since the calculations were based on an estimated value of T_{B_2} of 100°F, the problem must be reworked using the value of $T_{B_2} = 60°F$. Thus, $T_{B_a} = \frac{1}{2}(40 + 60) = 50°F$, $\mu_a = 3.16$ lbm/hr ft, $k_a = 0.334$ Btu/hr ft °F, $N_{Pr_a} = 9.49$, and $\mu_w = 0.588$. As a result,

$$N_{Re} = \frac{7.34 \times 10^5 (1/24)}{3.16} = 9700.$$

This value of N_{Re} permits the use of the correlation, Eq. (5.400)(because 9700 is so close to 10,000). Hence,

$$\frac{h(1/24)}{0.334} = 0.023(9.49)^{1/3}(9700)^{0.8}(3.16/0.588)^{0.14}\,[1\,+\,(1/20)^{0.7}],$$

or $\quad h = 860$ Btu/hr ft^2 °F.

Substituting this into Eq. (5.401) along with the other known quantities yields

$$0.094 = \ln\left(\frac{200}{240 - T_{B_2}}\right), \quad \text{or} \quad 1.0985 = \frac{200}{240 - T_{B_2}},$$

or $\quad T_{B_2} = 58.3°$F.

This is sufficiently close to the value 60°F used to find the properties, so that it can be considered to be the exit temperature of the water.

In addition, further iterations are probably not warranted at this point, simply because of the uncertainty in the value of h being used as a consequence of what may be a ±15% or more scatter of the experimental data about the correlating relation, Eq. (5.400).

5.5c Fully Developed Flow in Rough Tubes

The relations developed thus far are for use in tubes that can be considered smooth. Experiment has shown that while surface roughness in tubes increases the Stanton number, it does not usually increase it in the same proportion as the increased friction factor f [47]. This increase in the Stanton number or, equivalently, the Nusselt number, and the resulting increase in the surface coefficient of heat transfer as a result of roughness, is obtained only for *turbulent* flow. Furthermore, the increase in the surface coefficient depends upon the type of roughness element—such as three-dimensional sand grains, two-dimensional square-edged ribs, wire ribs, screw threads, etc.; the relative heights of the elements; and the surface spacing of the elements. The experimental data also suggest that the surface coefficient of heat transfer ultimately reaches a ceiling. That is, as the friction factor f increases due to increasing roughness, a value of the ratio f/f_s (where f_s is the friction factor if the pipe were smooth) is reached beyond which the surface coefficient of heat transfer no longer increases.

As a result of the inherent complexity of the phenomenon of turbulent heat transfer in a rough pipe, there is no single relation, for the surface coefficient's dependence upon roughness parameters, which is adequate for all roughness conditions.

A crude first estimate of the surface coefficient when roughness is present is to simply use results such as von Kàrmàn's analogy, Eq. (5.269), Colburn's modification of Reynolds' analogy, Eq. (5.250), or Petukhov's results, Eqs. (5.395) and (5.396), and insert the rough pipe friction factor from Fig. 5.9 directly into these equations. This procedure, however, usually gives a surface coefficient which is higher than the actual one, sometimes two or three times too high. Hence, this may indeed result in a crude estimate.

In order to obtain a much better estimate, we can use the results of Norris [47], which appear in a simple, easy-to-use form. Norris presents the results of various experiments on rough pipe heat transfer as the ratio of the rough pipe Nusselt number to the smooth pipe Nusselt number N_{Nu_s}, when all other conditions are the same, as a function of the ratio f/f_s, where f is the actual rough pipe friction factor and f_s is the one for a smooth pipe under the same conditions. Norris finds that when f/f_s reaches a value of about 4 to 6, depending upon the type of roughness elements present, the Nusselt number no longer increases with increasing roughness as measured by increasing f. Below this limiting, or ceiling, value, the rough pipe heat transfer can be correlated by the following relation:

$$\frac{N_{Nu}}{N_{Nu_s}} = \left(\frac{f}{f_s}\right)^n. \tag{5.403}$$

The value of n in Eq. (5.403) depends to some degree on the type of roughness element and on the Prandtl number of the fluid. At a Prandtl number of 0.7, typical of air and other diatomic gases, n ranges from about 0.5 to 0.63 for different types of roughness elements. Here, as an estimate of the roughness effect, we suggest the use of an average value of 0.57, regardless of the type of roughness. Thus,

$$f/f_s \leq 4, \quad N_{Pr} = 0.7, \quad n = 0.57 \tag{5.404}$$

in Eq. (5.403).

Again neglecting the effect of different types of roughness on the Nusselt number, the following expression can be used for higher Prandtl numbers:

$$f/f_s \leq 4, \quad 1 < N_{Pr} < 6, \quad n = 0.68 \, N_{Pr}^{0.215}. \tag{5.405}$$

If $f/f_s > 4$, use $f/f_s = 4$ in Eq. (5.403) along with the appropriate value of n from Eq. (5.404) or (5.405).

More accurate predictions of the influence of roughness on heat transfer require that one take the type of roughness element into account by use of the appropriate plots of experimental data or the more complicated correlation equations which apply to specific types of roughness. For these, the reader can consult [46], [47], [48], and [45].

5.5d Laminar Flow in the Entrance Region of Tubes

For relatively high Prandtl number fluids, the velocity profile reaches a somewhat fully developed form much more quickly (in a much smaller length of tube) than does the temperature profile. This can be seen by noting that at the entrance of the tube, as shown in Fig. 5.38, a boundary layer–type flow persists for some distance (for a sharp leading edge). There is growth, from zero thickness, of both the velocity boundary layer and thermal boundary layer of revolution. Near the very beginning of the tube where the boundary layer thicknesses are small relative to the tube radius, if the flow is laminar, they develop almost like those on a constant temperature flat plate. It might be expected that in this region near the entrance the ratio of the boundary layer thicknesses is given quantitatively for some distance and qualitatively for a greater distance by the flat plate relation

$$\frac{\delta_t}{\delta} = \frac{1}{1.026 N_{Pr}^{1/3}} . \qquad \text{[5.335]}$$

Thus, for $N_{Pr} > 1$, $\delta > \delta_t$ at any x from the entrance, and when $\delta = R$, the tube radius, the velocity profile is fairly well developed. However, if the Prandtl number is large enough, the temperature profile will still be essentially undeveloped. For this case, where the velocity profile in laminar tube flow develops in a relatively short distance, Hausen [49] developed the following relation, which fits the experimental data and the exact solution by Graetz [12]:

$$N_{Nu} = \frac{hD}{k} = 3.66 + \frac{0.0668[N_{Re}N_{Pr}(D/L)]}{1 + 0.04[N_{Re}N_{Pr}(D/L)]^{2/3}} . \qquad \textbf{(5.406)}$$

This is the average Nusselt number, with the appropriate driving temperature difference being the logarithmic mean temperature difference and the Reynolds number based on diameter.

For oils, it is expected that Hausen's relation will better represent the data if a Sieder-Tate type of correction is made [16], [39]. In this case,

$$\frac{hD}{k_a} = \left\{ 3.66 + \frac{0.0668[N_{Re}N_{Pr}(D/L)]}{1 + 0.04[N_{Re}N_{Pr}(D/L)]^{2/3}} \right\} \left(\frac{\mu_a}{\mu_w} \right)^{0.14} . \qquad \textbf{(5.407)}$$

All properties in Eq. (5.407), except μ_w, are evaluated at the average bulk mean temperature $T_{B_a} = \frac{1}{2}(T_{B_1} + T_{B_2})$, while μ_w is the value of the viscosity at the tube wall temperature.

An experimental correlation which is easier to use than Hausen's is given by Sieder and Tate [39], and more recently this has been endorsed by Whitaker [42] in his review of experimental data and recommendation of correlating equations for a number of different geometries. This relation represents the experimental data for laminar flow of water and oils in particular, and even for gases, as long as the following inequality is satisfied [51], [52]:

$$N_{Re}N_{Pr} \frac{D}{L} > 10. \qquad \textbf{(5.408)}$$

When Eq. (5.408) is satisfied, the Sieder-Tate relationship is

$$\frac{hD}{k_a} = 1.86 \, (N_{Re}N_{Pr})^{1/3} \left(\frac{D}{L} \right)^{1/3} \left(\frac{\mu_a}{\mu_w} \right)^{0.14} . \qquad \textbf{(5.409)}$$

All properties, except μ_w, are evaluated at the average bulk mean temperature; this average value of h over the length L for laminar flow is based on the driving temperature difference for Newton's cooling law,

$$\Delta T = T_w - \tfrac{1}{2}(T_{B_1} + T_{B_2}).\qquad(5.410)$$

Hence, unlike all previous relations for tube flow which are based on the logarithmic mean temperature difference, the Sieder-Tate relation, Eq. (5.409), is based on the arithmetic average temperature difference between the surface and the fluid.

If the condition of Eq. (5.408) is satisfied for the laminar flow of liquids in tubes, Eq. (5.409) is usually used because it is simpler to use than Hausen's relation, Eq. (5.407). If Eq. (5.408) is not satisfied, then Eq. (5.407) is used.

For gases like air, whose Prandtl number is near 0.70 over an extremely wide range of temperatures, Eq. (5.335) shows that the velocity and temperature profiles in the entrance of a tube will tend to develop at the same rate, so that Eqs. (5.406) and (5.409) are not expected to be valid. In this case, the velocity field and the temperature field are developing simultaneously, so that a combined hydrodynamic and thermal entrance length problem must be handled. Solutions to this problem have been arrived at by numerical finite difference solution of the governing partial differential equation; tabular and graphical results for the Nusselt number are available in [17] and [45]. An alternative is to compute an estimate of the average surface coefficient for fluids—such as air, whose Prandtl number is low enough to cause simultaneous development of the velocity and temperature profiles—by using either Eq. (5.406) or (5.409). When applying Eq. (5.406) to air and other gases, it is recommended that all temperature-dependent fluid properties be evaluated at the film temperature, Eq. (5.357).

EXAMPLE 5.15

In Ex. 5.8, an oil, whose properties were given as constant at average values, flowed through a 1/2-in.-diameter tube which was 15 ft long. The solution for the wall temperature, based on the fully developed (thermally and hydrodynamically) expression for the surface coefficient, was 66°F. At 66°F, this same oil has a viscosity of 190 lbm/hr ft. Assuming that the property values in Ex. 5.8 are those at the average bulk mean temperature, make a better estimate of the average surface coefficient of heat transfer and recalculate the wall temperature.

Solution

From Ex. 5.8, $N_{Re} = 32.2$, so that the oil flow is laminar and either Eq. (5.409) or Eq. (5.407) should be used. The criterion of Eq. (5.408) is checked to see if Eq. (5.409) may be used. From Ex. 5.8, $N_{Pr} = \mu c_p/k = \frac{19(0.47)}{0.075} = 119$, and $D/L = \frac{1}{24}/15 = \frac{1}{360}$. Hence,

$$N_{Re}N_{Pr}\frac{D}{L} = \frac{32.2(119)}{360} = 10.6.$$

Thus, the criterion of Eq. (5.408) is just satisfied, so that Eq. (5.409) can be employed. Substitution into Eq. (5.409) yields

$$\frac{hD}{k_a} = \frac{h(\frac{1}{24})}{0.075} = 1.86[32.2(119)]^{1/3} \left(\frac{1}{360}\right)^{1/3} \left(\frac{19}{190}\right)^{0.14},$$

or $h = 5.34$ Btu/hr ft^2 °F.

Note that this is lower than the fully developed result of 6.59 Btu/hr ft^2 °F determined in Ex. 5.8. The fact that D/L is not zero causes the surface coefficient to rise, but the effect of the temperature-dependent fluid viscosity, which is taken into account more properly, causes a net lowering of the surface coefficient. Substituting this new value of h into Eq. (5.237), noting that when using Eq. (5.409), $\Delta T = \frac{1}{2}(T_{B_1} + T_{B_2}) - T_w = 150 - T_w$, leads to

$$5.34 \frac{1}{24} \pi(15)(150 - T_w) = 20(0.47)(200 - 100),$$

or $T_w = 60°$ F.

Notice that, in Ex. 5.15, the high dynamic viscosity at the wall, $\mu_w = 190$ lbm/hr ft, relative to the much lower viscosity of the bulk of the fluid away from the wall, $\mu_{B_a} = 19$ lbm/hr ft, causes a significantly lower h, because of the factor $(19/190)^{0.14}$, than would occur if all the fluid had the same viscosity. The high viscosity of the fluid near the wall, compared to the bulk of the fluid away from the wall, distorts the velocity profile by causing lower velocities in the region near the wall. As a consequence, it also causes lower energy transfer rates, since the energy cannot be carried away as quickly by this slower-moving fluid. If this oil were being heated instead of cooled, the opposite behavior would occur, and a surface coefficient of heat transfer higher than the constant property value would result.

EXAMPLE 5.16

Engine oil, whose mass flow rate is 0.032 kg/s, is to be cooled from 120°C to 40°C while passing through 1.25-cm-diameter tubing whose inside surface will be maintained at 20°C. One of the steps in the preliminary design process for a heat exchanger to do this duty is to calculate the length of tubing necessary, L.

Solution

The basic equation which must be solved to find L is the energy balance which follows from Eq. (1.8) applied to the tube. Thus,

$$h\pi DL \, \Delta T = \omega c_p \, (T_{B_2} - T_{B_1}). \tag{5.411}$$

To find the value of h, we first check to see if the flow is laminar or turbulent, and we then select the appropriate analytical, semi-analytical, or experimental correlation result to give the surface coefficient. In the present problem, we have oil, a highly viscous fluid, flowing in a relatively small diameter tube. So probably the diameter Reynolds number will be less than 2000, indicating laminar flow. Since oils have highly temperature-dependent viscosities, the flow will probably never be fully developed thermally, and either Hausen's result, Eq. (5.407), or the Sieder-Tate relation, Eq. (5.409), will in all likelihood be used. In either case, all properties, except μ_w, are to be evaluated at the arithmetic average bulk mean temperature T_{B_a}. So, from Eq. (5.394),

$$T_{B_a} = (120 + 40)/2 = 80°C.$$

With this, the temperature-dependent properties from Appendix A are $k_a = 0.138$ W/m °C, $c_{p_a} = 2131$ J/kg °C, $N_{Pr_a} = 490$.

$$\mu_a = N_{Pr_a} k_a/c_{p_a} = 490(0.138)/2131 = 0.0317 \text{ kg/ms}$$

Similarly, $\mu_w = 0.8021$ kg/ms.
Next, the Reynolds number is found to be

$$N_{Re_D} = \rho_a U_m D/\mu_a = 4\omega/\pi D\mu_a = 4(0.032)/\pi(0.0125)(0.0317)$$
$$= 102.8.$$

Hence the flow is laminar, and the Sieder-Tate relation, Eq. (5.409), will be used, provided that the criterion, Eq. (5.408), is satisfied. This check, however, must wait until we calculate L on the assumption that Eq. (5.409) can be used.

Thus, inserting known values into Eq. (5.409),

$$\frac{h(0.0125)}{0.138} = 1.86 \ [102.8(490)]^{1/3}(0.0125/L)^{1/3}(0.0317/0.8021)^{0.14},$$

or

$$h = 112/L^{1/3}. \tag{5.412}$$

In the energy balance, Eq. (5.411), the appropriate ΔT is the arithmetic average temperature difference given by Eq. (5.410); thus,

$$\Delta T = 20 - (1/2)(120 + 40) = -60°C.$$

Substituting known quantities into Eq. (5.411) gives

$$\frac{112}{L^{1/3}} \pi(0.0125)L(-60) = 0.032(2131)(40 - 120), \quad \text{or} \quad L = 94 \text{ m}.$$

Now the criterion, Eq. (5.408), must be checked to see if the use of Eq. (5.409) for h was justified.

$$N_{Re_D} N_{Pr_a} \frac{D}{L} = 102.8(490)(0.0125/94) = 6.7$$

This value, 6.7, does not satisfy Eq. (5.408). Therefore, Hausen's relationship, Eq. (5.407), must be used instead to compute the needed length of tubing. Substituting known quantities and rearranging Eq. (5.407) gives

$$h = 7.023 \left\{ 3.66 + \frac{42.1/L}{1 + (2.94/L^{2/3})} \right\}. \tag{5.413}$$

When using Hausen's relationship, the ΔT in Newton's cooling law and also on the left side of Eq. (5.411) is the logarithmic mean temperature difference, Eq. (5.235); hence,

$$\Delta T_L = (40 - 120)/\ln \left[\frac{20 - 120}{20 - 40} \right] = -49.7°C.$$

Substituting this, Eq. (5.413), and other known quantities into Eq. (5.411) yields, after rearranging,

$$L = \frac{398}{\left\{ 3.66 + \frac{42.1/L}{1 + (2.94/L^{2/3})} \right\}}.$$

This equation can be solved by trial and error by iterating as follows: First, the contribution of the function of L on the right is neglected, and L is computed on the left. Then, this is used as L on the right side to compute a new value of L on the left, and the process is repeated until successive values of L are deemed sufficiently close to one another.

Hence, one has, after neglecting the function of L on the right,

$$L = \frac{398}{3.66} = 108.7 \text{ m.}$$

Using this on the right-hand side of the equation gives

$$L = \frac{398}{3.66 + \frac{42.1/108.7}{1 + (2.94/108.7^{2/3})}} = 99.4 \text{ m.}$$

Using this value of 99.4 m in the right side of the equation leads to

$$L = 98.7 \text{ m.}$$

This is close enough to the previous value so that the length of tubing required is 98.7 m.

The length required is such that a single straight tube this long would not be a practical solution as a heat exchanger. Instead, the ultimate design of the exchanger will incorporate a larger number of shorter tubes or a single tube which is coiled so that its greatest straight linear dimension is far less than 99 m.

5.5e Forced Convection Over Cylinders, Spheres, and Tube Banks in Crossflow

These geometries are the primary ones referred to earlier when mention was made of boundary layer separation and regions of separated flow. Some analytical work can be done for the portions of the objects which have unseparated flow, but for an average surface coefficient of heat transfer over the entire surface of these objects, experimental correlations provide the most applicable information.

Crossflow Over Circular Cylinders

It may be recalled from fluid mechanics that the flow pattern over a right circular cylinder in crossflow is often, depending upon the Reynolds number, exceedingly complicated, and may include regions of laminar boundary layer flow, turbulent boundary layer flow, and separated flow. The Reynolds number used here is formed with the cylinder diameter D as the characteristic dimension and the undisturbed oncoming velocity far upstream of the cylinder, U_0, as the velocity scale. The shape of the cylinder causes the free stream velocity, outside boundary layers or separated flow regions, to vary with distance x along the cylinder's circumference, $U_s = U_s(x)$.

At extremely low Reynolds numbers, the flow around the cylinder is laminar, non-boundary layer–type flow, which stays attached until the rear stagnation point. This is the creeping flow regime. As the Reynolds number increases, a number of different flow regimes evolve, sometimes with very large and abrupt changes in the fluid mechanics picture. Subcritical flow occurs just beyond the creeping flow regime. It is characterized by laminar boundary layer flow from about the front stagnation point to about the 80° angle from the front stagnation point, along the circumference, where separation takes place and the separated zone covers the remaining downstream portion of the cylinder. As the Reynolds number increases further, the critical flow regime occurs where the flow beyond the laminar separation point re-attaches to the cylinder as a turbulent boundary layer which itself then separates at about the 140° angle. Beyond this, at higher Reynolds numbers, the supercritical regime is reached where the laminar boundary layer undergoes natural transition (without separation) to a turbulent boundary layer flow which delays the first, and only, separation until about an angle of 130°. Finally, at the highest Reynolds numbers, the transcritical regime is encountered, in which the natural transition point of the laminar boundary layer has shifted very close to the front stagnation point, and the resultant turbulent boundary layer is separating at about the 115° angle.

These fluid mechanics phenomena are reflected in distributions of the local skin friction coefficient $C_f(x)$ and local Nusselt number $h_x D/k$, which depend strongly on position x and on the Reynolds number. More complete descriptions of the flow and experimental results for skin friction, local Nusselt number, drag coefficient, and average Nusselt number are available in Achenbach [53] and in Zukauskas [54].

In spite of the complex dependence of the *local* Nusselt number on the Reynolds number because of the large, rather abrupt, and drastic changes in the flow field, the *average* Nusselt number over the cylinder circumference is relatively uninfluenced by these changes. It displays a much simpler dependence upon Reynolds number than do the local values, as pointed out by Achenbach [53] and by Churchill and Bernstein [55].

The preceding remarks, though made explicitly for crossflow over a circular cylinder, also apply in a general, qualitative sense to other geometries which lead to flow separation, such as the sphere and crossflow over banks of tubes.

Stagnation Point Nusselt Number for the Cylinder The heat transfer problem at the forward stagnation point of a circular cylinder in crossflow can be handled in an exact analytical fashion as described, for example, in Kays and Crawford [17]. These exact solutions utilized the variable free stream velocity, predicted by potential flow theory, for the stagnation point of a cylinder, namely,

$$U_s(x) = 2U_0 x / R, \tag{5.414}$$

where U_0 is the undisturbed upstream velocity and R is the radius of the cylinder. Churchill and Bernstein [55] fit the following correlating equation to the theoretical solutions *for all values of the Prandtl number*. This result can be used for *either constant wall temperature or constant flux*, since one implies the other at the stagnation point region as a consequence of h_x being independent of x in this region.

$$N_{Nu} = \frac{1.276 \, N_{Re}^{1/2} \, N_{Pr}^{1/3}}{[1 + (0.412/N_{Pr})^{2/3}]^{1/4}} \tag{5.415}$$

Where

$$N_{Nu} = hD/k_f, \quad N_{Re} = \rho_f U_0 D / \mu_f,$$

all temperature-dependent fluid properties are to be calculated at the film temperature $T_f = (T_w + T_s)/2$, and U_0 is the undisturbed velocity of the fluid upstream of the cylinder and appears in the free stream velocity variation with x as shown in Eq. (5.414).

Evidence in [55] and [56] indicates that Eq. (5.415) can be used not only at the front stagnation point of the cylinder, but also for about the first 15° of angle, measured along the circumference from this stagnation point, with very little error.

Average Nusselt Number for Crossflow Over a Circular Cylinder For air and other gases whose Prandtl number is about 0.7, Morgan [57] calculated corrected values of correlating constants for extensive experimental data taken earlier by Hilpert [58] for crossflow over a *constant temperature* circular cylinder. The average Nusselt number for the entire cylinder, when all properties are evaluated at the film temperature, is given for air as

$$\frac{hD}{k_f} = c \left(\frac{\rho_f U_0 D}{\mu_f} \right)^n . \tag{5.416}$$

Table 5.2
The Values c and n for Use in Eq. (5.416)

$N_{\text{Re}} = \dfrac{\rho_f U_0 D}{\mu_f}$			c	n
10^{-4}	$\rightarrow$	4×10^{-3}	0.437	0.0895
4×10^{-3}	$\rightarrow$	9×10^{-2}	0.565	0.136
9×10^{-2}	$\rightarrow$	1	0.800	0.280
1	$\rightarrow$	35	0.795	0.384
35	$\rightarrow$	5000	0.583	0.471
5000	$\rightarrow$	50,000	0.148	0.633
50,000	$\rightarrow$	200,000	0.0208	0.814

Ref. 71 suggests the following result:

200,000	$\rightarrow$	2×10^6	0.023	0.80

The values of c and n depend upon the Reynolds number range and are listed in Table 5.2.

Churchill and Bernstein [55] correlate many investigators' experimental data for the average surface coefficient of heat transfer over a wide range of Prandtl and Reynolds numbers. Their resultant experimental correlation is as follows:

$$\frac{hD}{k_f} = 0.3 + \frac{0.62\, N_{\text{Re}}^{1/2}\, N_{\text{Pr}}^{1/3}}{[1 + (0.4/N_{\text{Pr}})^{2/3}]^{1/4}}\left[1 + \left(\frac{N_{\text{Re}}}{282{,}000}\right)^{5/8}\right]^{4/5}. \tag{5.417}$$

Equation (5.417) is valid as long as the Peclet number, the product of the Reynolds and the Prandtl numbers, is greater than 0.2; that is,

$$N_{\text{Re}}\, N_{\text{Pr}} > 0.2. \tag{5.418}$$

The equation $N_{\text{Re}} = \rho_f U_0 D/\mu_f$ and all properties in Eq. (5.417) are to be evaluated at the film temperature $T_f = (T_w + T_s)/2$. It is recommended that Eq. (5.417) be used for gases and liquids when the temperature differences involved are moderate ones. Some of the experimental data correlated by Eq. (5.417) was taken for essentially constant surface heat flux, rather than constant wall temperature. Thus, as a first approximation, Eq. (5.417) gives an h that allows one to estimate the *average* cylinder surface temperature when the surface flux being constant is the thermal boundary condition at the cylinder's surface (also see Zukauskas [54]).

For very large temperature differences between the cylinder wall temperature and the undisturbed stream temperature, or for fluids with extremely strong dependence of one

Table 5.3
The Values b and p for Use in Eq. (5.419)

N_{Re}			b	p
1	$\rightarrow$	40	0.75	0.4
40	$\rightarrow$	1000	0.51	0.5
1000	$\rightarrow$	200,000	0.26	0.6
200,000	$\rightarrow$	10^6	0.076	0.7

or more of the properties upon temperature (such as oils), the following experimental correlation due to Zukauskas [54] is recommended:

$$N_{Nu} = b \, N_{Re}^{p} \, N_{Pr}^{0.37} \left(\frac{N_{Pr}}{N_{Pr_w}} \right)^{0.25}, \tag{5.419}$$

where the values of b and p are given in Table 5.3.

If $N_{Pr} > 10$, then Zukauskas recommends that the power on the Prandtl number be changed to 0.36.

All temperature-dependent fluid properties in Eq. (5.419) should be evaluated at the undisturbed free stream temperature T_s except for N_{Pr_w}, which is to be evaluated at the cylinder wall temperature T_w.

Figure 5.39 shows, as a continuous solid line, the correlating curve to which Zukauskas fit the experimental correlation equation given as Eq. (5.419). The various symbols (dots, squares, etc.) represent the actual experimentally measured values of many different investigators and many different fluids, such as air, water, and transformer oil. The symbol on the ordinate is defined as

$$K_f^1 = N_{Nu} \, N_{Pr}^{-0.36} \, (N_{Pr}/N_{Pr_w})^{-0.25},$$

while the abscissa is the Reynolds number defined by

$$N_{Re_f} = \rho_s U_0 D / \mu_s$$

Note that Zukauskas uses the subscript symbol f not to denote film conditions, but rather to denote the undisturbed main stream conditions.

Figure 5.39 shows scattering of experimental data about the correlating equation or curve, which is typical of many different forced convection situations besides the crossflow over the circular cylinder being considered. From this, the reader can get an appreciation for the uncertainty to be expected when calculating values of the surface coefficient of heat transfer from an experimental correlation. For a more complete discussion of this, see [54] and [55].

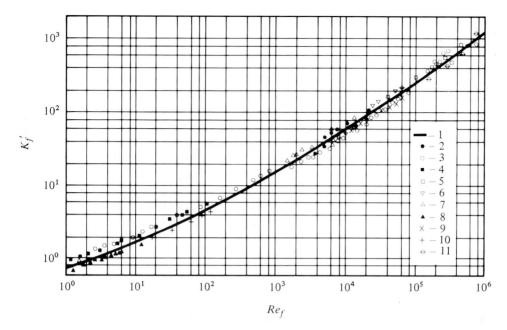

Figure 5.39 Comparison of experimental data for the average heat transfer coefficient in crossflow over a circular cylinder, and the proposed solid correlating curve, Eq. (5.419). (From A. Zukauskas, "Heat Transfer from Tubes in Crossflow." In J. P. Hartnett and T. F. Irvine, Jr., [Eds.], *Advances in Heat Transfer*, Vol. 8. Copyright © 1972 by Academic Press. Used with permission.)

EXAMPLE 5.17

A thin wire, 0.01 in. in diameter, is to be used as a speed measuring device for the 80°F air in crossflow over the wire. Measurements indicate that electrical energy is dissipated at the rate of 1.2 W in the 2-in.-long wire, with its surface temperature being 120°F when the wire is immersed in the 80°F airstream. Estimate the airstream speed.

Solution

The electrical energy dissipated is 1.2 W $\times$ 3.413 Btu/hr W = 4.09 Btu/hr. An energy balance on the entire wire taken as the control volume gives, in the steady state, $R_{in} + R_{gen} = R_{out}$, or $0 + 4.09 = h\pi DL(T_w - T_s)$, or $4.09 = h\pi(\frac{0.01}{12})(\frac{2}{12})(120 - 80)$, or

$$h = 234 \text{ Btu/hr ft}^2 \text{ °F.}$$

This known value of the surface coefficient can now be related by Eq. (5.416) to the unknown airspeed U_0. Since $T_f = \frac{1}{2}(T_w + T_s) = \frac{1}{2}(120 + 80) = 100°F$, then

$\mu_f = 0.0459$ lbm/hr ft, $k_f = 0.0157$ Btu/hr ft °F, and $\rho_f = P/RT_f = \frac{14.7(144)}{53.3(460 + 100)} = 0.0711$ lbm/ft^3. Hence,

$$\frac{hD}{k_f} = \frac{234(\frac{0.01}{12})}{0.0157} = 12.5.$$

Since the unknown is the air velocity U_0, the Reynolds number can be viewed as the unknown. It appears explicitly in Eq. (5.416) and implicitly as well, since c and n depend upon the Reynolds number. So an iterative procedure is necessary whereby we will first estimate the value of the Reynolds number, use it to find the corresponding values of the parameters c and n from Table 5.2, and then insert these into Eq. (5.416) along with the known Nusselt number of 12.5 and solve for N_{Re}. This computed value of N_{Re} will be used as the new estimate and therefore will determine the values of c and n. If they are different from the previous values, the procedure is continued. If not, the computed value of N_{Re} can be solved for U_0.

Taking N_{Re} between 5000 and 50,000 as a first estimate for this small-diameter wire, one has, from Table 5.2, that $c = 0.148$ and $n = 0.633$. Hence, using 12.5 as the value of N_{Nu} in Eq. (5.416), one has

$$12.5 = 0.148 \, N_{Re}^{0.633}, \quad \text{or} \quad N_{Re} = 1106.$$

This Reynolds number, however, does not lie in the range for which c and n have the values just used. Hence, new values of c and n are chosen for the range in which $N_{Re} = 1106$ does lie, namely, 35 → 5000 in Table 5.2. This gives $c = 0.583$ and $n = 0.471$. Again using Eq. (5.416) with 12.5 as the Nusselt number,

$$12.5 = 0.583 \, N_{Re}^{0.471}, \quad \text{or} \quad N_{Re} = 670.5.$$

This Reynolds number is in the proper range, 35 → 5000, for the values of c and n used; hence, this is the solution for the Reynolds number. Now, since

$$N_{Re_D} = \frac{\rho_f U_0 D}{\mu_f}, \quad 670.5 = \frac{0.0711 \, U_0 \, (0.01/12)}{0.0459},$$

the air speed is

$$U_0 = 519,425 \text{ ft/hr} = 144.3 \text{ ft/sec}.$$

Implicit in the solution to this problem, up to this point, is the assumption that the net radiant loss from the wire to the surrounding solid surfaces at about the air temperature, is negligible compared to the forced convection. This is easily verified by applying the simplified version of Christiansen's equation from Chapter 4 to this situation.

EXAMPLE 5.18

An electric resistance heater is 2.5 cm in diameter and 1 m long, and is immersed in water which is moving with a speed of 3 m/s and which has an undisturbed temperature of 20°C. If the maximum average temperature over the surface of the heater is not to exceed 60°C, calculate the highest allowable electrical energy dissipation rate for the heater.

Solution
An energy balance, Eq. (1.8), applied to the entire heater as a control volume in the steady state yields $R_{in} + R_{gen} = R_{out}$, or

$$R_{gen} = h\pi DL(T_w - T_s), \qquad (5.420)$$

where R_{gen} is the required electrical energy dissipation rate. The surface coefficient h of heat transfer for a liquid in crossflow over a cylinder is given by Eq. (5.417). Now, $T_f = (20 + 60)/2 = 40°C$. The properties of water at this film temperature are found from Appendix A to be $\nu_f = 0.658 \times 10^{-6}$ m²/s, $k_f = 0.628$ W/m °C, and $N_{Pr_f} = 4.34$. Therefore,

$$N_{Re} = \frac{U_0 D}{\nu_f} = \frac{3(0.025)}{0.658 \times 10^{-6}} = 1.14 \times 10^5.$$

Also, $N_{Re} N_{Pr} = 1.14 \times 10^5 (4.34) = 4.95 \times 10^5$, so condition (5.418) is satisfied. Thus, substituting into Eq. (5.417), one has

$$\frac{h(0.025)}{0.628} = 0.3 + \frac{0.62[1.14 \times 10^5]^{0.5} (4.34)^{1/3}}{[1 + (0.4/4.34)^{2/3}]^{1/4}}$$

$$\left[1 + \left(\frac{114,000}{282,000}\right)^{5/8}\right]^{4/5},$$

or $h = 11,741$ W/m² °C.

Using this and other known quantities in Eq. (5.420) gives

$$R_{gen} = 11,741\pi(0.025)(1)(60 - 20) = 36,885 \text{ W}.$$

Spheres

Whitaker [42] has developed a correlating equation for the experimental data of other investigators concerned with the flow of fluids, such as air, water, and oils, over a constant temperature sphere. In his search for a suitable form for the correlating equation, he was guided by the following considerations: First, the expression for the Nusselt number should reduce to the pure conduction solution for a sphere in an infinite medium as the fluid velocity, and therefore also the Reynolds number, goes to zero. In addition, there should be some dependence upon the square root of the Reynolds number, $N_{Re}^{1/2}$, since both analytical and experimental work indicate this functional dependence for many different shapes on the portion covered by a laminar boundary layer. Evidence also exists which suggests that the heat transfer in a separated region of flow depends upon $N_{Re}^{2/3}$ and, since a portion of the flow over the sphere is separated, this factor was included explicitly as an additive quantity. Finally, of course, the correlating equation should pass as close as possible through the experimental data points for a wide range of Prandtl and Reynolds numbers.

This finally led to the result which follows for the average Nusselt number over a constant surface temperature sphere. The domain of validity of the correlation is represented by the indicated Prandtl and Reynolds number ranges within which the experimental data points lie.

$$\frac{hD}{k} = 2 + [0.4\, N_{Re}^{1/2} + 0.06\, N_{Re}^{2/3}]\, N_{Pr}^{0.4} \left(\frac{\mu}{\mu_w}\right)^{1/4} \tag{5.421}$$

$$0.71 < N_{Pr} < 380, \quad 3.5 < N_{Re} < 7.6 \times 10^4,$$

$$N_{Re} = \rho U_0 D / \mu \tag{5.422}$$

The value U_0 is the undisturbed speed of the fluid, far upstream of the sphere, moving toward the sphere.

All temperature-dependent fluid properties are evaluated at the undisturbed stream temperature T_s except for μ_w, which is to be evaluated at the sphere wall temperature T_w.

Crossflow Through Tube Banks

One common type of crossflow heat exchanger has one fluid passing inside circular tubes arranged in either an in-line or a staggered pattern in a bank, as shown in Fig. 5.40, while another fluid flows over the outside of the tubes in crossflow. This type of exchanger is frequently used in the boiler and condenser sections of power plants, where the phase change process may cause the tube wall to be approximately isothermal. In the more general case, the lack of a controlling resistance for heat transfer at the inside of a tube, or conditions which cause an appreciable bulk mean temperature change of the fluid inside the tubes, will lead to a variable outside tube wall temperature. However, the starting point of the analysis of even these more general tube bank problems is still the consideration of crossflow of a fluid over a bank of constant temperature tubes, and this is considered next.

Banks of tubes of diameter D are usually arranged either in-line, as shown in Fig. 5.40(a), or staggered, as shown in Fig. 5.40(b). Now S_L and S_T, called the longitudinal and transverse tube spacings, respectively, are shown in Figs. 5.40(a–b). For air flowing over tube banks 10 or more rows deep (in the direction of flow), Grimison [63] correlated the data of Huge [62] and Pierson [61] for heat transfer to tube banks as

$$hD/k_f = B(\rho U_m D / \mu_f)^m, \tag{5.423}$$

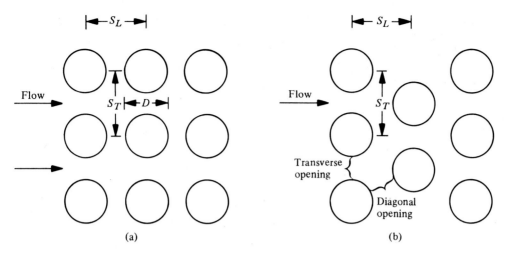

Figure 5.40 (a) In-line and (b) staggered tube banks.

where B and m are given for various longitudinal and transverse spacing ratios (S_L/D and S_T/D) in Table 5.4. The data were obtained in the Reynolds number range from 2000 to 40,000, the properties are to be evaluated at the film temperature, and the logarithmic mean temperature difference should be used in Newton's cooling law. *The product ρU_m is the maximum value which occurs in the minimum area section, either the transverse opening between tubes or the diagonal opening,* both of which are depicted in Fig. 5.40(b).

Equation (5.423) can be used for the flow of liquids over tubes by adding a Prandtl number dependence, as has occurred in many problems in this chapter. Equation (5.423) as it stands is for air with a Prandtl number of about 0.70. Multiplying the right-hand side of Eq. (5.423) by $N_{\text{Pr}}^{1/3}/(0.7)^{1/3}$ yields

$$hD/k_f = 1.13B(\rho U_m D/\mu_f)^m N_{\text{Pr}}^{1/3} . \tag{5.424}$$

Again, B and m are given in Table 5.4.

For banks with more than 10 rows, Eqs. (5.423–24) are valid, but for banks less than 10 rows deep, the average surface coefficient of heat transfer decreases somewhat. A correction factor for staggered banks of tubes with less than 10 rows is given by Kays, London, and Lo [64]. It is presented here as Table 5.5, where h is the average surface coefficient for a bank with the number of rows indicated, while h_{10} refers to the average surface coefficient for a 10-row or more bank and is calculated from either Eq. (5.423) or Eq. (5.424) in conjunction with Table 5.4.

If the diameter Reynolds number, based on the maximum velocity in the minimum area section, is outside the range for which the values of B and m of Table 5.4 were established, or if there are large temperature differences between the tubes and the bulk of the fluid, or if the fluid is an oil with a strongly temperature-dependent viscosity, then the results of Zukauskas [54] can be used. These are given next. These expressions are for the average surface coefficient of the entire 10 or more row tube bank.

Table 5.4
Constants B and m for Use in Either Eq. (5.423) or (5.424)
for Tube Banks 10 or More Rows Deep

Bank Design	Longitudinal Spacing Ratio S_L/D	Transverse Spacing Ratio S_T/D							
		1.25		1.5		2.0		3.0	
		B	m	B	m	B	m	B	m
Staggered	0.600	—	—	—	—	—	—	0.213	0.636
	0.900	—	—	—	—	0.446	0.571	0.401	0.581
	1.000	—	—	0.497	0.558	—	—	—	—
	1.125	—	—	—	—	0.478	0.565	0.518	0.560
	1.250	0.518	0.556	0.505	0.554	0.519	0.556	0.522	0.562
	1.500	0.451	0.568	0.460	0.562	0.452	0.568	0.488	0.568
	2.000	0.404	0.572	0.416	0.568	0.482	0.556	0.449	0.570
	3.000	0.310	0.592	0.356	0.580	0.440	0.562	0.421	0.574
In-line	1.250	0.348	0.592	0.275	0.608	0.100	0.704	0.0633	0.752
	1.500	0.367	0.586	0.250	0.620	0.101	0.702	0.0678	0.744
	2.000	0.418	0.570	0.299	0.602	0.229	0.632	0.198	0.648
	3.000	0.290	0.601	0.357	0.584	0.374	0.581	0.286	0.608

From Grimison, "Correlation and Utilization of New Data on Flow Resistance and Heat Transfer for Crossflow of Gases over Tube Banks," *Transactions of the A.S.M.E.,* Vol. 59 (1937), pp. 583–94.

Table 5.5
Ratio of the Average Heat Transfer Coefficient to That for 10 Rows for Staggered
Tube Banks

Number of Rows	4	5	6	7	8	9	10
h/h_{10}	0.88	0.92	0.94	0.97	0.98	0.99	1.0

From W. M. Kays, A. L. London, and R. K. Lo, "Heat Transfer and Friction Characteristics for Gas Flow Normal to Tube Banks—Use of a Transient Test Technique," *Transactions of the A.S.M.E.,* Vol. 76 (April 1954), pp. 387–96.

For $10 < N_{Re} < 100$,

$$\frac{hD}{k} = c\, N_{Re}^{0.4}\, N_{Pr}^{0.36}\, (N_{Pr}/N_{Pr_w})^{0.25}, \tag{5.425}$$

where $c = 0.8$ for in-line arrangement, and $c = 0.9$ for staggered
arrangement.

For $1000 < N_{Re} < 2 \times 10^5$,

$$\frac{hD}{k} = 0.27 \, N_{Re}^{0.63} \, N_{Pr}^{0.36} \, (N_{Pr}/N_{Pr_w})^{0.25} \qquad (5.426)$$

for in-line arrangement and $S_T/S_L > 0.7$.

$$\frac{hD}{k} = 0.35(S_T/S_L)^{0.2} \, N_{Re}^{0.60} \, N_{Pr}^{0.36} \, (N_{Pr}/N_{Pr_w})^{0.25} \qquad (5.427)$$

for staggered arrangement and $S_T/S_L < 2$.

$$\frac{hD}{k} = 0.40 \, N_{Re}^{0.60} \, N_{Pr}^{0.36} \, (N_{Pr}/N_{Pr_w})^{0.25} \qquad (5.428)$$

for staggered arrangement and $S_T/S_L > 2$.

For $N_{Re} > 2 \times 10^5$,

$$\frac{hD}{k} = a \, N_{Re}^{0.84} \, N_{Pr}^{0.36} \, (N_{Pr}/N_{Pr_w})^{0.25} \qquad (5.429)$$

where $a = 0.021$ for in-line banks, $a = 0.022$ for staggered banks.

In Eqs. (5.425) through (5.429), all fluid properties except N_{Pr_w} should be evaluated at the arithmetic average bulk mean temperature $T_{B_a} = (T_{B_1} + T_{B_2})/2$, where T_{B_1} is the bulk mean temperature of the fluid entering the bank and T_{B_2} is the exit bulk mean temperature. Again, N_{Pr_w} should be evaluated at the tube wall temperature. The Reynolds number is formed using U_m as the *maximum* velocity in the bank. This occurs, as mentioned above, in the minimum area section of the bank, either the transverse or diagonal openings.

EXAMPLE 5.19

Water at 50°F, moving at 0.80 ft/sec, flows over 1-in.-diameter tubes in a staggered tube bank in which the tubes are at a temperature of 210°F due to steam condensing inside them. The transverse and longitudinal spacings are both 2 in. in this bank, which is 6 rows deep in the flow direction, 10 ft high in the transverse direction, contains 60 tubes in the transverse direction, and is 10 ft wide in the remaining direction. If the latent heat of vaporization of the steam in the tubes is 950 Btu/lbm, calculate the condensation rate of the steam in lbm/hr.

Solution
An energy balance on the steam in all the tubes yields

$$hA_s(T_w - T_s) = \omega_0 h_{fg}, \qquad (5.430)$$

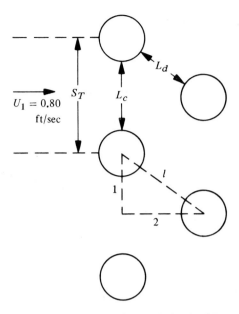

Figure 5.41 Portion of the tube bank of Ex. 5.19.

where ω_0 is the unknown condensation rate of the steam, h_{fg} is its latent heat of vaporization, h is the average surface coefficient of heat transfer between the outside of the tubes and the water flowing over them, A_s is the total outside surface area of the tubes, and, in using $T_w - T_s$ as the driving temperature difference, it has been assumed that the temperature change of the water will be negligible. This assumption will be checked later and, if necessary, modified.

There are 60 tubes per row and 6 rows, giving a total of 360 tubes. Since each has a diameter of 1 in. and is 10 ft long, $A_s = 360\pi(\frac{1}{12})10 = 943$ ft², and Eq. (5.430) becomes $h(943)(210 - 50) = \omega_0 950$, or

$$\omega_0 = 159h. \tag{5.431}$$

The surface coefficient of heat transfer for a 10 or more row bank (in the flow direction) is, for liquids flowing over the bank, given by Eq. (5.424). The film temperature is $T_f = \frac{1}{2}(T_w + T_s) = \frac{1}{2}(210 + 50) = 130°F$, and from the property tables for water at $130°F$, $\mu_f = 1.24$ lbm/hr ft, $k_f = 0.375$ Btu/hr ft °F, and $N_{Pr} = 3.29$. To compute the product of ρU_m in the minimum area section, which will be either L_c or L_d, a few tubes in the first two rows are shown in Fig. 5.41. Hence, $L_c = S_T - D = 2 - 1 = 1$ in. Also, L_d can be found by first finding the length l of the triangle shown in the figure. Thus, $l^2 = (1)^2 + (2)^2$, or $l = 2.23$ in. Hence, $L_d = l - D = 2.23 - 1 = 1.23$ in. Therefore, the minimum area section occurs not in the diagonal openings, but in the transverse opening for these spacing ratios. The continuity equation applied to the water flow in the channel shown gives $\rho_1 U_1 S_T = \rho U_m L_c$. Using this and $U_1 = 0.80(3600) = 2880$ ft/hr, ρU_m can be determined to be $\rho U_m = 62.4(2880)(\frac{2}{1}) = 62.4(5760)$. Hence,

$$N_{\text{Re}} = \frac{\rho U_m D}{\mu} = \frac{62.4(5760)(\frac{1}{12})}{1.24} = 24,200.$$

From the geometry, $S_L/D = S_T/D = \frac{2}{1} = 2$. As a result, from Table 5.4, for staggered tubes with both spacing ratios equal to 2, $B = 0.482$ and $m = 0.556$. Substituting known values into Eq. (5.424),

$$\frac{h(1/12)}{0.375} = 1.13(0.482)(24,200)^{0.556}(3.29)^{1/3}, \quad \text{or}$$

$$h = 997 \text{ Btu/hr ft}^2 \text{ °F}.$$

This would be the value of h for a 10-row deep bank. For a 6-row bank, Table 5.5 indicates that $h/h_{10} = 0.94$, so the surface coefficient for the 6-row bank is $h = 0.94(997) = 937$ Btu/hr ft^2 °F. Substituting this into Eq. (5.431) yields $\omega_0 = 159(937) = 149,000$ lbm/hr for the steam condensation rate.

Now the assumption of negligible temperature rise of the cooling water should be checked. Since the energy lost by the steam must have been gained by the cooling water, the law of conservation of energy requires that

$$\omega_0 h_{fg} = \omega c_p(T_{B_2} - T_{B_1}). \tag{5.432}$$

Now, $\omega = \rho_1 U_1 A_1 = 62.4(2880)(10 \times 10) = 1.8 \times 10^7$ lbm/hr, and, for water, $c_p = 1.0$ Btu/lbm °F; hence, Eq. (5.432) becomes

$$(149,000)(950) = (1.8 \times 10^7)(1)(T_{B_2} - 50), \quad \text{or} \quad T_{B_2} = 57.9°\text{F}.$$

The new film temperature would differ by less than 4°F from the value used here, and the logarithmic mean temperature difference is within 4°F of the ΔT used, so we need not recalculate. The condensation rate of the steam is $\omega_0 = 149,000$ lbm/hr.

5.6 SUMMARY OF FORCED CONVECTION EXPRESSIONS

In this chapter, we have dealt with exact analytical, approximate analytical, and semi-analytical methods, as well as the all-important experimental correlations to arrive at numerical values of the surface coefficient of heat transfer in various forced convection situations of importance.

In order to start readers on their way to the proper choice of the most appropriate relation for the surface coefficient, and to reduce the time needed to make the proper choice, Table 5.6 is supplied. This provides an overview and brief description of the various relations for h. Note that the description of the conditions of validity of the various equations listed in Table 5.6 is, in general, not complete. Therefore, once a tentative choice of an equation is made based on Table 5.6, the next step is to read the more detailed description of the conditions and restrictions on the equation in this chapter. This is also the place to see how the temperature-dependent fluid properties are to be evaluated and, for internal flows, what driving temperature difference should be used in Newton's cooling law.

Table 5.6
Summary of Expressions for the Forced Convection Heat Transfer Coefficient

Geometry of Body	Comments and Conditions	Equation Number
Flow inside circular tubes	Laminar, constant T_w, fully developed	Table 5.1
	Laminar, constant q_w'', fully developed	(5.222)
	Laminar, constant T_w, undeveloped	
	Gases; liquids with moderate temperature differences	(5.406)
	Oils; liquids with large temperature differences	(5.407)
	$N_{Re} N_{Pr} D/L > 10$. Simpler to use than Eqs. (5.406) or (5.407)	(5.409)
	Turbulent, constant T_w or q_w'', fully developed, $L/D \geq 60$	
	Moderate temperature differences	
	$0.5 \leq N_{Pr} \leq 10$	(5.269)
	$0.5 \leq N_{Pr} \leq 100$	(5.250) or (5.391)
	Oils; liquids with large temperature differences	
	$N_{Pr} \geq 0.5$	(5.393)
	$0.5 \leq N_{Pr} \leq 2000$. Use when utmost accuracy is needed. More complicated to use than previous results.	
	Gases	(5.396)
	Liquids	(5.395)
	Turbulent, constant T_w, undeveloped	
	$0.5 \leq N_{Pr} \leq 120$. Average heat transfer coefficient over the length L.	(5.400)
Flow inside noncircular ducts	Laminar, constant T_w, fully developed	Table 5.1
	Laminar, constant q_w'', fully developed	Table 5.1
	Turbulent, constant T_w or q_w'', fully developed	
	Use relations above for turbulent flow through a circular duct with the hydraulic diameter, D_H, defined in Eq. (5.223), in place of D.	

Table 5.6
Continued

Geometry of Body	Comments and Conditions	Equation Number
Flow in rough tubes	Turbulent, constant T_w or q_w'', fully developed	
	$0.5 \leq N_{Pr} < 6$	(5.403)
Flow parallel to a flat plate	Laminar, constant T_w	
	$\quad 0.5 \leq N_{Pr} \leq 15$	
	$\qquad$ Local	(5.297)
	$\qquad$ Average	(5.302)
	$\quad 0 < N_{Pr} < \infty$	
	$\qquad$ Local	(5.303)
	$\qquad$ Average	(5.304)
	Laminar, constant T_w, unheated starting length	
	$\quad 0.5 \leq N_{Pr} \leq 15$	
	$\qquad$ Local	(5.333)
	Laminar, constant q_w''	
	$\quad 0.5 \leq N_{Pr} \leq 15$	
	$\qquad$ Local	(5.353)
	$\quad 0 < N_{Pr} < \infty$	
	$\qquad$ Local	(5.354)
	Turbulent, constant T_w	
	$\quad 0.5 < N_{Pr} \leq 50$	
	$\qquad$ Local	(5.382)
	$\qquad$ Average	(5.383)
	$\quad 0 < N_{Pr} < \infty$	
	$\qquad$ Local	(5.387)
	Laminar until $N_{Re_x} = 3 \times 10^5$, then turbulent for the rest of the plate	
	$\quad 0.5 < N_{Pr} < 50$, gases; liquids with moderate temperature differences, average.	(5.385)
	$\quad N_{Pr} > 0.5$, oils; liquids with large temperature differences, average	(5.386)
	Turbulent, constant q_w''	
	$\quad 0.5 < N_{Pr} \leq 50$	
	$\qquad$ Local	(5.388)

Table 5.6
Continued

Geometry of Body	Comments and Conditions	Equation Number
Flow over body of arbitrary shape with no separation	Laminar, arbitrary variation of T_w with x	(5.373)
	Turbulent, arbitrary variation of T_w with x	(5.374)
Stagnation point of circular cylinder	Laminar, constant T_w or q_w'' $0 < N_{Pr} < \infty$	(5.415)
Crossflow over circular cylinder	Constant T_w, air $N_{Pr} \approx 0.7$, average $10^{-4} < N_{Re} < 2 \times 10^5$	(5.416)
	Constant T_w, $N_{Re}\,N_{Pr} > 0.2$, average, gases, liquids with moderate temperature differences	(5.417)
	Constant T_w, $1 < N_{Re} < 10^6$, average, oils, liquids with large temperature differences	(5.419)
Crossflow over sphere	Constant T_w, $0.5 < N_{Pr} < 380$, $3.5 < N_{Re} < 7.6 \times 10^4$, average	(5.421)
Crossflow through tube banks	Constant T_w, average for banks of 10 or more rows	
	$2000 < N_{Re} < 40,000$	(5.424)
	$10 < N_{Re} < 100$	(5.425)
	$1000 < N_{Re} < 2 \times 10^5$	
	In-line	
	$S_T/S_L > 0.7$	(5.426)
	Staggered	
	$S_T/S_L < 2$	(5.427)
	$S_T/S_L > 2$	(5.428)
	$N_{Re} > 2 \times 10^5$	(5.429)

5.7 PROBLEMS

5.1   Determine if the velocity distribution

$$u = 37x^3 \sin y, \quad v = +111(\cos y)(x^2)$$

is possible for a two-dimensional, steady, constant property flow.

5.2   In a certain two-dimensional, constant property, steady flow situation, the y component of velocity has been found to be $v = 75 \sin x \sin y$. Find the form of the x component of velocity as a function of x and y.

5.3   Consider steady, laminar, constant property flow between two parallel plates spaced a distance b apart. If the lower plate is stationary and the upper plate moves in the x direction with constant speed V_1, show by solution of the Navier-Stokes equations that the velocity distribution is $u = V_1(y/b)$. (This is called *Couette flow*.)

5.4   Rework Prob. 5.3 for the more general case in which there is a pressure gradient in the x direction, and show that the solution is

$$u = V_1 \left(\frac{3y^2}{b^2} - \frac{2y}{b} \right) - 6U_m \left(\frac{y^2}{b^2} - \frac{y}{b} \right).$$

5.5   Consider laminar, steady, constant property, vertically downward flow between two parallel plates spaced a distance b apart. If the pressure gradient is zero, show, by applying the momentum theorem to a control volume of fluid dx by dy, that the governing differential equation for the velocity distribution is

$$\mu \frac{\partial^2 u}{\partial y^2} = -\rho \frac{g}{g_0},$$

where ρ is the mass density, g the local acceleration of gravity, and g_0 the gravitational constant.

5.6   Show that the velocity distribution for the conditions given in Prob. 5.5 is

$$u = \frac{\rho g}{2\mu g_0} (by - y^2).$$

5.7   By applying the momentum theorem to the circular cylindrical control volume which is dx long and of general radius r, shown in Fig. 5.42, for steady, laminar, fully developed, and constant property flow in a circular tube of radius R, derive the governing equation for the velocity distribution as

$$\frac{\partial u}{\partial r} = \frac{1}{2\mu} \frac{\partial P}{\partial x} r,$$

and show that it is equivalent to Eq. (5.50).

Figure 5.42

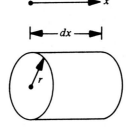

5.8   Use the result of Prob. 5.7 to verify that the velocity distribution in steady, laminar, fully developed, constant property, horizontal flow through an annulus of inside radius R_i and outside radius R is

$$u = -\frac{1}{4\mu} \frac{dP}{dx} \left[R^2 - r^2 + \frac{R^2 - R_i^2}{\ln(R_i/R)} \ln \left(\frac{R}{r} \right) \right].$$

5.9   Water at 20°C flows with an average velocity of 3 m/s through a 15-cm-diameter cast iron pipe. Compute the value of the Moody friction factor.

5.10 For turbulent flow of a fluid between two parallel planes spaced a distance $2s$ apart, the velocity profile

$$u = U_0 \left(\frac{y}{s} \right)^{1/7}$$

is found to be adequate in a certain Reynolds number range. Find the relation between the average velocity U_m and the centerline velocity U_0.

5.11 For the Reynolds number range in which Eq. (5.90) is valid, find the distribution of the eddy diffusivity of momentum ϵ_m over the cross section as a function of y. The pipe flow is turbulent and fully developed, and the average velocity U_m is known.

5.12 If the order of magnitude of all the terms in Eq. (5.122) is the same, find an expression for the order of magnitude of the viscosity μ, and from this expression deduce what must be true for a boundary layer to be thin, that is, for $\delta/L \ll 1$.

5.13 It is known that the potential flow around a certain body is given by $U_s = U_0 \sin x$. If the flow is constant property, compute the pressure gradient $\partial P/\partial x$ impressed upon the boundary layer as a function of x.

5.14 Equation (5.129) gives the exact analytical solution for the local boundary layer thickness δ as a function of x for steady, laminar, constant property, planar two-dimensional flow over a flat plate. Suppose that $\delta \ll L$ means that δ should not exceed 10% of L. Compute the lowest plate Reynolds number that can be tolerated.

5.15 Water at 40°C is flowing over the top of a 0.6-m-long flat plate with a speed of 0.3 m/s. Calculate the maximum boundary layer thickness.

5.16 Consider a flat plate over which an incompressible fluid of density ρ is flowing in a steady fashion. Because of the plate's porosity, mass is being injected into the hydrodynamic boundary layer at a constant rate $\dot{m}_s$ per unit length. This injected fluid is the same type as the main stream. Take the appropriate control volume and derive the integral form of the law of conservation of mass for this situation. The physical situation is shown in Fig. 5.43.

Figure 5.43

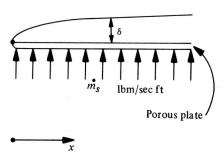

5.17 An incompressible fluid is flowing in a steady fashion *into a circular pipe* of radius r_0. In the so-called entrance region, the major transverse velocity gradients take place within a boundary layer of local thickness δ. Derive the integral form of the law of conservation of mass for the boundary layer in the entrance region. A convenient coordinate system is shown in Fig. 5.44, where r is the general radial coordinate.

Figure 5.44

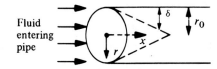

Fluid
entering
pipe

5.18 Figure 5.45 shows a very thin-walled, sharp-edged pipe with an incompressible fluid flowing at a velocity U_∞ parallel to the long axis of the pipe. As can be seen, the hydrodynamic boundary layer builds up on the outside of the pipe from zero thickness at the leading edge to the general thickness δ a distance x from the leading edge. The pipe has radius R, and the general coordinate from the pipe surface to anywhere within the boundary layer in a radial direction is called y, as shown. There is no pressure gradient over the outside of the pipe. The pipe wall is solid and the flow is steady and laminar. The local x component of velocity is denoted by u. Derive the integral form of the conservation of mass equation for the *outside surface* of the pipe.
(*Hint:* Do not treat the outside of the pipe as a flat plate in your derivations.)

Figure 5.45

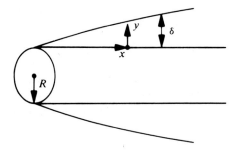

5.19 Consider steady, laminar, constant property, planar two-dimensional, boundary layer–type flow over a flat plate. Another constraint on an approximate velocity profile, for the integral method, can be found by collocating the x-momentum equation, Eq. (5.122), at $y = \delta$. Show that this leads to the condition that at $y = \delta$, $\partial^2 u/\partial y^2 = 0$.

5.20 By using only the first two terms of Eq. (5.156), prove that the approximate velocity profile for steady, laminar, constant property, planar two-dimensional flow over a flat plate is

$$u = U_s(y/\delta).$$

5.21 Using the first three terms of Eq. (5.156), construct the approximate velocity profile for the conditions described in Prob. 5.20 and show that the result is

$$u = U_s\left(\frac{2y}{\delta} - \frac{y^2}{\delta^2}\right).$$

5.22 Find an expression for the local boundary layer thickness δ for laminar, steady, constant property, planar two-dimensional flow over a flat plate by using the relatively crude linear velocity profile given in Prob. 5.20 and substituting it into the appropriate integral form of the x-momentum equation, Eq. (5.169).

5.23 Use the velocity profile given in Prob. 5.21 in the appropriate integral form of the x-momentum equation for a flat plate to obtain $\delta/x = 5.48/\sqrt{N_{Re_x}}$.

5.24 Using the results of Prob. 5.23, derive an expression for the local skin friction coefficient C_f defined by $C_f = \tau_w / \frac{1}{2} \rho U_s^2$.

5.25 Based upon the integral analysis of the steady, laminar, incompressible flow over a flat plate, it was found that an approximate expression for the x component of velocity within the boundary layer is

$$u = U_s[\tfrac{3}{2}(y/\delta) - \tfrac{1}{2}(y^3/\delta^3)],$$

where $\delta = 4.64\sqrt{xv/U_s}$. (a) Find the function which expresses the total mass rate of flow *within* the boundary layer at any x. (b) Show how this can be used to calculate the influx of mass from the free stream between any two values of x.

5.26 Air at 40°C and 101,000 N/m² flows over a 1-m-long flat plate with a free stream velocity of 16 m/s. Estimate the boundary layer thickness at the end of the plate.

5.27 By making use of Eq. (5.187), derive the following expression for the average skin friction coefficient in turbulent boundary layer–type flow over a flat plate of length L:

$$\overline{C_f} = 0.072/N_{Re_L}^{1/5} .$$

5.28 By the use of the continuity equation, derive an expression for the local y component of velocity in a steady, turbulent, constant property, two-dimensional flow over a flat plate.

5.29 A blade of a gas turbine engine is idealized as a 1-in.-long flat plate. The gas has a velocity of 1000 ft/sec relative to the blade, with $\mu = 0.09$ lbm/hr ft and ρ calculated to be 0.20 lbm/ft³. Calculate the maximum thickness of the boundary layer if (a) the boundary layer is completely turbulent over the entire blade (which is very likely the case) and (b) the boundary layer can be maintained as laminar over the entire plate.

5.30 Consider simple Couette flow between two parallel plates with the lower plate stationary and the upper plate moving at a constant speed of V_1 ft/sec. For steady, laminar, constant property, fully developed flow, the velocity profile was previously found to be $u = V_1(y/b)$ (see Prob. 5.3). The physical situation is shown in Fig. 5.46. A constant flux q_w'' is entering the fluid at the lower plate, and the temperature profile is developed in such a manner that the same constant flux q_w'' leaves at the upper plate. Neglecting axial conduction, derive the expression for the local Nusselt number at the lower plate, based on the hydraulic diameter, as $h_x D_H/k = 3.0$. (Note that the bulk mean temperature of the fluid was used in the temperature difference in the defining equation for h_x.)

Figure 5.46

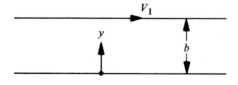

Chapter Five

5.31 Consider steady, laminar, constant property, fully developed, low-speed flow between two parallel plates, as shown in Fig. 5.47. Neglect axial conduction as being small compared to conduction in the y direction. Derive the governing partial differential equation for the temperature distribution by using Eq. (1.8), the energy balance, on the control volume $dx \times dy$ shown in Fig. 5.47.

Figure 5.47

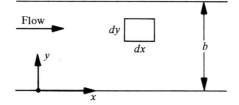

5.32 Consider steady, laminar, low-speed, constant property, fully developed (both thermally and hydrodynamically) flow between the two parallel plates of Fig. 5.47 when axial conduction can be neglected and when the surfaces at $y = 0$ and $y = b$ are heated such that the flux into the fluid at both surfaces is equal to the known constant q_w''. By the same reasoning that preceded Eq. (5.213), it can be shown that the thermally fully developed condition implies that

$$\frac{\partial T}{\partial x} = \frac{\partial T_B}{\partial x} = \frac{2q_w''}{\rho U_m c_p b}.$$

The bulk mean temperature of the fluid at $x = 0$ is T_{B_1}. Find the temperature distribution as a function of x and y, the local surface coefficient of heat transfer, and the local Nusselt number.

5.33 Consider steady, laminar, low-speed, constant property, fully developed (both thermally and hydrodynamically) flow between two parallel plates a distance s apart. The upper plate is perfectly insulated, and the lower plate has a known constant flux q_w'' crossing it going into the fluid. Axial conduction can be neglected. It is argued that the results of Prob. 5.32 apply here as long as the identification that $b = 2s$ is made. The physical situation is pictured in Fig. 5.48. The argument proceeds by showing that the governing partial differential equation and the boundary conditions on it are the same here as they were for Prob. 5.32. That is,

$$\frac{\partial^2 T}{\partial y^2} = \frac{u}{\alpha}\frac{\partial T}{\partial x}$$

subject to the boundary conditions that because of the perfect insulation,

$$\text{at } y = s = \tfrac{1}{2}b, \quad \partial T/\partial y = 0,$$

Figure 5.48

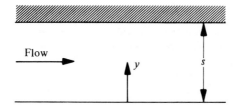

(note that in Prob. 5.32, $\partial T/\partial y = 0$ because of the symmetry about the center line)

$$\text{at } y = 0, \quad -k\,(\partial T/\partial y) = q_w'',$$
$$\text{at } x = 0, \quad T_B = T_{B_1}.$$

Determine if the argument is correct.

5.34 Water flows through a 2-in.-diameter tube at the rate of 30 lbm/hr. Electrical resistance tape wrapped around the tube dissipates energy at the constant rate of 100 W/ft² of tube surface. If the water temperature rise is to be 100°F, calculate the length of tube needed and also the local surface coefficient of heat transfer. The properties of water are assumed constant at the values $c_p = 1.0$ Btu/lbm°F, $k = 0.36$ Btu/hr ft°F, $\mu = 1.60$ lbm/hr ft, and $\rho = 62$ lbm/ft³.

5.35 Air is flowing in a fully developed fashion through a 0.6-cm-diameter tube at an average speed of 3 m/s. The average values of the properties are $\rho = 1.18$ kg/m³, $\mu = 1.98 \times 10^{-5}$ kg/ms, and $k = 0.026$ W/m °C. Find the value of the local surface coefficient of heat transfer.

5.36 If the air in Prob. 5.35 enters the tube at 10°C and the tube is 0.6 m long and at 70°C, compute the exit temperature of the air.

5.37 An equilateral triangular duct is wrapped with electrical resistance tape which imparts a constant flux of 10,700 W/m² to the water flowing in a laminar fashion through the duct. The side length of the triangular cross section is 1.25 cm, and the water leaves the duct at 100°C. Find the maximum wall temperature of the duct if $k = 0.63$ W/m °C for the water.

5.38 Suppose the triangular duct of Prob. 5.37 is used to heat air, initially at 27°C, instead of water, but with the same power setting. Assuming that the air flow is laminar and $k = 0.026$ W/m °C for air, (a) estimate the duct wall temperature and (b) decide whether or not the duct should be used for heating air.

5.39 Engine oil flowing at 500 lbm/hr at 260°F enters a 1/2-in.-diameter tube 100 ft long whose wall temperature is 175°F. Estimate the exit temperature of the oil, whose properties are to be treated as being constant at the values $\rho = 52.4$ lbm/ft³, $\mu = 41.3$ lbm/hr ft, $k = 0.08$ Btu/hr ft °F, and $c_p = 0.53$ Btu/lbm °F.

5.40 Oil ($k = 0.14$ W/m °C) is to be heated to 60°C in a 1.25-cm-diameter tube wrapped with electrical resistance tape. If the tube wall temperature is not to exceed 500°C anywhere and if the oil flow is laminar, compute the maximum allowable heat flux q_w'' that the tapes should be allowed to dissipate.

5.41 Water flowing at 6 kg/s must be cooled from 100°C to 40°C. An old heat exchanger consisting of a 5-cm-diameter pipe 6 m long, which can be held as low as 20°C, is available. Can this exchanger definitely do the required cooling?

5.42 Calculate the mass flow rate of water that definitely could be cooled from 100°C to 40°C in the heat exchanger of Prob. 5.41.

5.43 A rectangular duct 4 in. high and 8 in. wide carries heated air from a furnace whose temperature is 150°F. The 20-ft-long duct passes through a cellar where the surroundings cause the duct walls to be at 70°F. The properties of the air are assumed constant at the values $c_p = 0.24$ Btu/lbm °F, $k = 0.016$ Btu/hr ft °F, $\mu = 0.046$ lbm/hr ft, and $\rho = 0.07$ lbm/ft³. If the air flow is 40 lbm/hr, calculate the bulk mean temperature of the fluid at the exit of the duct.

5.44 In a simple heat exchanger with inside walls of 1 ft length, 1/4-in.-diameter tubes are held at 212°F by steam condensing on the outside of the tubes. Water is to enter the tubes at 50°F and leave at 150°F. The water properties are assumed constant at the values $c_p = 1.0$ Btu/lbm °F, $\rho = 62$ lbm/ft^3, $k = 0.36$ Btu/hr ft °F, and $\mu = 1.65$ lbm/hr ft. (a) Calculate the mass flow rate of water per tube that can be accommodated. (b) Find the number of tubes needed if 100 lbm/hr must be processed.

5.45 Water, flowing at 3.2 kg/s, is to be heated in a single pipe exchanger from 20°C to 60°C. The pipe is 5 cm in diameter and the wall temperature is 100°C. Calculate the required length of pipe.

5.46 Air at 70°C and 104,000 N/m² flows at 0.25 kg/s through a square duct, 15 cm on a side and 10 m long. If the duct wall temperature is 30°C, estimate the exit temperature of the air.

5.47 Water flowing at 3000 lbm/hr is to be raised from 70°F to 130°F by passing through a 1-in.-diameter tube whose inside surface is held at 150°F. The properties of the water are constant at the average values $\rho = 61.6$ lbm/ft^3, $c_p = 1.0$ Btu/lbm °F, $\mu = 1.30$ lbm/hr ft, $N_{Pr} = 3.45$, and $k = 0.373$ Btu/hr ft °F. Determine the length of tube required.

5.48 When Eq. (5.325) is substituted into Eq. (5.312) to solve for h_x, determine if the result is valid for all body shapes, so long as the wall temperature is constant at T_w, since in arriving at Eq. (5.325) the actual shape of the body was not specified (other than that it is two-dimensional and planar). Explain your answer.

5.49 Consider a vertical flat plate 1 ft high by 1 ft deep into the paper. The plate is held at 130°F, and 70°F air is being blown over the plate at a speed of 50 ft/sec. Compute the *convective* heat transfer rate from one side of the plate to the air. (Note that the pressure is atmospheric.)

5.50 Water at 60°F flows past a body being held at 200°F. Compute the maximum free stream velocity the water may have and still be considered a "low-speed" flow.

5.51 Consider steady, laminar, constant property, low-speed, planar two-dimensional, boundary layer–type flow of a fluid at temperature T_s over a flat plate whose wall temperature varies linearly with position along the plate according to $T_w = T_s + mx$, where m is a constant. By the use of *integral* methods, derive an expression for the local Nusselt number for this situation.
(*Hint:* Begin with the approximate temperature profile $T = a_0 + a_1 y + a_2 y^2 + a_3 y^3$, and force it to satisfy the boundary conditions and the "additional condition" $\partial^2 T / \partial y^2 = 0$ at $y = 0$; then substitute it into the integral form of the energy equation, Eq. [5.312].)

5.52 Find the Prandtl number restriction for the results of Prob. 5.51 if the assumption that $\delta_t / \delta < 1$ was made and resulted in $\delta_t / \delta = 1/1.47 N_{Pr}^{1/3}$. Is the resultant restriction severe?

5.53 Rework the development which led to Eq. (5.334) using the same temperature profile, Eq. (5.325), but using the simple linear velocity profile $u = U_s(y/\delta)$, where $\delta = 3.46\sqrt{\nu x/U_s}$.

5.54 Air at 40°C and 1.013×10^5 N/m² is being blown over the top of a flat plate which is 0.3 m long and 0.6 m wide. The air velocity is 4.5 m/s and the plate temperature is 100°C. Calculate (a) the total heat transfer rate from plate to air, and (b) the fraction of the total heat transfer which is given up by the rear half of the plate.

Forced Convection Heat Transfer

5.55 Consider steady, laminar, low-speed, constant property flow over a flat plate of length L. Derive a relationship for the average surface coefficient of heat transfer over the last two-thirds of the plate.

5.56 Thin plates 2.5 ft long are placed in a heat-treating oven and 1000°F air is blown over them at 20 ft/sec. If the average surface temperature of the plates is to be 600°F, find the average surface coefficient of heat transfer between the plate surface and the air.

5.57 The plate described in Prob. 5.56 is taken out of the oven when its temperature is 800°F and is exposed to a 100°F air stream with a speed of 25 ft/sec until the plate temperature reaches 200°F. In calculating the plate transient, Heisler's charts would be used with some "average" value of h over the time involved. To see how much deviation the surface coefficient goes through, compute the surface coefficient of heat transfer between the plate and the air when the plate is (a) at 800°F and (b) again when it is at 200°F.

5.58 Plates which are 0.6 m long and are at 60°C are to be cooled by setting them on a floor and blowing 20°C air over them with a speed of 3 m/s. Because of the position of the fans, this air first runs over 1 m of floor before reaching the leading edge of the plate. (a) Compute the surface coefficient of heat transfer at the end of the plate. (b) Compare the answer to part a to the coefficient of heat transfer that would have been erroneously calculated if the starting length were not properly taken into account.

5.59 Water at 70°F is flowing at 0.5 ft/sec over a plate 3 ft long in the direction of flow and 1 ft wide. Design requirements call for the plate to transfer energy at the rate of 12,000 Btu/hr to the water. Find the required plate surface temperature.

5.60 Derive the appropriate forms of Eqs. (5.352) and (5.353) for the case in which the plate is at the undisturbed stream temperature T_s from $x = 0$ to $x = x_0$, and then has a constant known flux q_w'' for $x > x_0$.

5.61 A flat plate has a fluid flowing over it in a laminar, boundary layer–type fashion. The plate is L units long, but the first half of it is at the fluid temperature and only its last half is held at a different temperature. Find the ratio of the correct local surface coefficient at the end of the plate where $x = L$ to that computed by the use of Eq. (5.334).

5.62 Air at 1 atm pressure and 60°F flows at a speed of 10 ft/sec over a circular cylindrical electric resistance heater 6 in. in diameter and 3 ft long. The air flow is in the direction of the long axis, and fairing has been provided so that it flows along the cylinder smoothly. If the maximum cylinder temperature is not to exceed 220°F, estimate the maximum heat flux and the total electrical energy dissipation rate.

5.63 A flat plate electric resistance heater is 6 in. long in the direction of the water flow over it. The water is at 60°F and is moving at 3 ft/sec over the heater. If the maximum surface temperature of the heater is to be 200°F, calculate the maximum electrical energy dissipation rate per unit area q_w'' if it is constant over the heater surface.

5.64 A heat sink for some electronic components is in the shape of a flat plate 7.5 cm long and 15 cm wide. Because of the dissipation of electrical energy within the components, energy must cross the heat sink surface at the rate of 2000 W. Water at 20°C is available for cooling. If the maximum heat sink temperature is not to exceed 80°C, calculate the speed with which the water must be forced over the plate heat sink.

Chapter Five

5.65 Show that the integral form of the thermal energy equation that applies *throughout* the entrance region of a circular tube for low-speed, constant property, laminar, steady flow is

$$\frac{d}{dx} \int_{R-\delta_t}^{R} \rho u c_p (T_s - T) \, r \, dr = -kR \left(\frac{\partial T}{\partial r} \right)_R.$$

5.66 In a rough calculation, a gas turbine blade 1.25 in. long in the flow direction is idealized as a flat plate with 1000°F gases passing over both surfaces of the blade at 300 ft/sec. The gas, which has the properties of air, has a density of 0.21 lbm/ft³. If the blade is 4 in. wide in span and the surface is to be held at 600°F, calculate the heat transfer rate to the blade.

5.67 Using Eq. (5.382), find the average surface coefficient of heat transfer between a flat plate at constant wall temperature and the constant property fluid flowing over the plate in a steady, turbulent, boundary layer–type, planar two-dimensional fashion.

5.68 Air at -23°C is blowing at 25 m/s parallel to the 2 m dimension of a 2 m by 1.25 m picture window. If the outside surface of the window is at 0°C, compute the convective heat transfer rate from the window.

5.69 A heat exchanger surface is a 0.3 m by 0.3 m plate and is designed to transfer energy at the rate of 30,000 W to the 20°C water flowing over it at 3.5 m/s. Estimate the average surface temperature of the exchanger.

5.70 A heat transfer surface is at 300°F and consists of a flat plate 1 ft long in the direction of the 100°F air flowing over it and 2 ft wide. If energy must be transferred across this surface at a rate of 4000 Btu/hr, calculate the required air speed.

5.71 A heat exchanger surface is to have condensing steam on one side and cooling water flowing at 20 ft/sec over the other side. The surface is in the shape of a flat plate 3 ft wide and is at 230°F. The required condensation rate of the steam will cause the heat transfer from the plate to the 70°F water to be 1,500,000 Btu/hr. Find the length the plate should be in the direction of the water flow.

5.72 The pressure surface of a turbine blade is to be idealized as a 5-cm-long flat plate over which the hot gases flow at 275 m/s. The fluid properties evaluated at the film temperature are $\mu_f = 2.85 \times 10^{-5}$ kg/ms, $k_f = 0.044$ W/m °C, $N_{Pr_f} = 0.69$, and $\rho_f = 14$ kg/m³. If the flow is expected to be turbulent over the entire surface, compute the average surface coefficient of heat transfer between the blade surface and the hot gases.

5.73 A heater plate at 180°F has dimensions 1 ft by 1 ft, and water at 60°F is flowing over it in a turbulent fashion over the entire plate due to boundary layer trip wires installed at the leading edge. The water speed is $U_s = 7.2$ ft/sec. Compute the average surface coefficient of heat transfer and the heat transfer rate from the plate to the fluid.

5.74 An air preheater is in the shape of a flat plate 15 cm long in the direction of flow and 0.6 m wide. The electrical energy dissipation rate is constant over the surface at the rate of 3200 W/m². The air, which is at 40°C and moves with a speed of 15 m/s, passes over a 0.3-m-long surface before reaching the preheater. Estimate the maximum temperature of the surface of the heater. (*Hint:* See Prob. 5.60.)

5.75 Show that Ambrok's relation, Eq. (5.373), for laminar, constant property flow over a flat plate which is held at the temperature T_s of the stream from 0 to x_0 and then at T_w for $x > x_0$, reduces to

$$N_{Nu_x} = 0.332 \frac{N_{Pr}^{1/3} N_{Re_x}^{1/2}}{[1 - (x_0/x)]^{1/2}}.$$

Compare this with the more nearly exact result for this situation, Eq. (5.333).

5.76 Find an expression for the total heat transfer rate from a flat plate, whose wall temperature varies linearly as $T_w = T_s + bx$, to a constant property stream at temperature T_s flowing in a laminar, steady fashion over the plate. The plate is of length L and width W.
(*Hint:* The local surface coefficient for this situation is given by Eq. [5.376].)

5.77 Consider steady, constant property crossflow over a circular cylinder of radius R. On the forward portion of the cylinder, the boundary layer is laminar, and the external velocity is known from the potential flow solution to be $U_s = 2 U_0 \sin(x/R)$ where U_0 is the undisturbed upstream velocity. Using Ambrok's method, derive an expression for the local Nusselt number when the cylinder surface is at the constant temperature T_w.

5.78 Show that in steady, turbulent, constant property, low-speed flow over a flat plate which has a surface temperature equal to the stream temperature T_s from $x = 0$ to $x = x_0$ and then equal to T_w for $x > x_0$, the local Nusselt number predicted by Ambrok's method is

$$N_{Nu_x} = \frac{0.0288 N_{Pr}^{1/3} N_{Re_x}^{0.8}}{[1 - (x_0/x)]^{0.2}}.$$

The factor in the denominator should be compared with the more-or-less correct result

$$\frac{1}{[1 - (x_0/x)^{0.9}]^{1/9}}.$$

5.79 Water at $60°$F flows over a 1-ft-long plate also at $60°$F, and then flows over a second plate of length 1 ft which is at $140°$F and which immediately follows the first plate. If the water speed is 20 ft/sec and the boundary layer is turbulent over both plates, estimate the value of the local surface coefficient of heat transfer at the far end of the second plate.
(*Hint:* The correction for the unheated starting length can be found in Prob. 5.78.)

5.80 For steady, turbulent, constant property, low-speed flow over a flat plate with the surface temperature variation $T_w = T_s + bx^{0.2}$, where T_s is the stream temperature and b is a known positive constant, (a) find the local Nusselt number as a function of x, and (b) determine why this is an important surface temperature variation.
(*Hint:* Using the results of part a, see how the local surface flux varies.)

5.81 Water at $20°$C flows at 3 m/s over a flat plate electric resistance heater which dissipates electrical energy at a rate per unit area which is independent of location. If the heater temperature is not to exceed $100°$C, calculate the maximum allowable energy dissipation rate from the 0.3 m by 0.3 m heater.

5.82 Air at $100°$F passes over a 1-ft-wide electric resistance flat plate heater which must dissipate electrical energy into the air at the constant, over the surface, rate of 2500 W/ft², with the heater temperature not to exceed $800°$F. If the air flows at 100 ft/sec, find how long the heater must be in the flow direction to satisfy these requirements.

5.83 Consider steady, turbulent, low-speed, constant property flow over a body, at constant wall temperature, whose shape gives rise to the variation of velocity outside the boundary layer $U_s = cx^{1/3}$, where $c > 0$. Show that the local Nusselt number is $N_{Nu_x} = 0.03045 N_{Pr}^{1/3} N_{Re_x}^{0.8}$.

5.84 By comparing the result of Prob. 5.83 with the case in which U_s is a constant, and with the results for the equivalent laminar flow situations (see Ex. 5.12), draw a conclusion concerning the relative importance of free stream velocity variation on the constant velocity surface coefficient in laminar and turbulent flow.

5.85 Because of the internal cooling scheme inside the leading edge of a turbine vane, the temperature of the outside surface in the vicinity of the stagnation point is expected to vary linearly according to $T_w = T_0 + bx$. The free stream, at temperature T_s, has the following velocity variation in the neighborhood of the stagnation point: $U_s = cx$, and the flow is laminar. Find an expression for the local Nusselt number.

5.86 A two-dimensional nozzle is designed such that the flow is planar and the velocity outside the boundary layer on the nozzle walls varies with x in the manner $U_s = U_0 + cx$. If the nozzle walls are at constant temperature and the flow is turbulent throughout, derive an expression for the variation of the local Nusselt number.

5.87 Water at $140°F$ flows through the nozzle described in Prob. 5.86 with a speed of $U_s = 5 + 10x$ ft/sec. If the nozzle walls are held at $60°F$, calculate the local surface coefficient of heat transfer 2 ft from the nozzle entrance.

5.88 Rework Prob. 5.86 for the case in which the nozzle flow is laminar, and obtain the result

$$ N_{Nu_x} = 0.332 N_{Pr}^{1/3} N_{Re_x}^{1/2} \left(\frac{U_0 + cx}{U_0 + \frac{1}{2} cx} \right)^{1/2} . $$

5.89 It is desired to cool 1.5 lbm/s of water from $85°F$ to $75°F$ in a 2-in.-diameter tube which is 10 ft long. Find the required tube wall temperature.

5.90 Water flowing at 3 kg/s at $20°C$ enters a 0.6-m-long, 7.5-cm-diameter tube whose inside surface is at $100°C$. Calculate the exit temperature of the water.

5.91 How long must the tube of Prob. 5.90 be to heat the water to $60°C$?

5.92 Air flowing at 900 lbm/hr at $250°F$ leaves the first stage of a compressor and enters an intercooler consisting of a 6-in.-diameter pipe held at $60°F$. If the air is to be cooled to $150°F$, calculate the length of the intercooler.

5.93 Calculate the required length of the intercooler of Prob. 5.92 if a 2-in.-diameter pipe is used instead of the 6-in. pipe.

5.94 Air flowing at 0.125 kg/s at $20°C$ enters a 0.3-m-diameter pipe that is 2 m long. If the pipe wall temperature is held at $100°C$, calculate the exit temperature of the air.

5.95 Air is flowing through a 3-in.-diameter galvanized iron pipe in a fully developed fashion with a diameter Reynolds number of 100,000. The film temperature is $100°F$. Estimate the value of the surface coefficient of heat transfer.

5.96 In a 3-in.-diameter pipe which is 20 ft long, 265 lbm/hr of air must be heated from $70°F$ to $130°F$. Find the temperature at which the pipe wall must be held to accomplish this.

5.97 Water at 20°C enters a 1.25-cm-diameter tube at the rate of 0.0025 kg/s. If the water is to leave at 50°C when the tube wall is held at 100°C, compute the required length of tubing.

5.98 An engine oil cooler is constructed of 1/2-in.-diameter tubes 20 ft long and held at 104°F. The oil enters the tubes at 270°F and leaves at 154°F. Find the mass flow rate of oil per tube.

5.99 Oil, flowing at 0.0063 kg/s at 85°C, enters a 5-cm-diameter, 1.2-m-long tube and is to leave at 75°C. Compute the wall temperature required to accomplish this.

5.100 A simple internal cooling scheme for a gas turbine vane can be idealized as air, at 800°F, entering a 1/8-in.-diameter, 3-in.-long tube at the rate of 0.4 lbm/hr. If the tube wall temperature is 1200°F, compute the exit temperature of the air, when the properties are given as $k = 0.034$ Btu/hr ft °F, $\mu = 0.09$ lbm/hr ft, $c_p = 0.26$ Btu/lbm °F, and $N_{Pr} = 0.69$.

5.101 Air flowing at 0.0015 kg/s at 110°C enters an intercooler consisting of a 5-cm-diameter pipe, 0.6 m long. If the air is to leave at 80°C, determine the temperature at which the pipe wall should be held.

5.102 Air is to be heated from 50°F to 150°F in a heat exchanger made up of 1/4-in.-diameter tubes. If each tube is at 500°F and must process 1.3 lbm/hr of air, calculate the required length of the tube.

5.103 Oil flowing at 250 lbm/hr is to be cooled from 200°F to 152°F in a 1/2-in.-diameter tube whose inside surface temperature is 68°F. Calculate the required length of tube.

5.104 If the conditions of Prob. 5.103 remain the same except that the oil enters the tube at 248°F and leaves at 104°F, calculate the required length of 1/2-in.-diameter tubing.

5.105 The leading edge of a gas turbine blade is a two-dimensional stagnation region with the external velocity equal to $U_s = 24,000x$ ft/sec. The property values are assumed constant at $\mu = 0.09$ lbm/hr ft, $N_{Pr} = 0.69$, $k = 0.034$ Btu/hr ft °F, and $\rho = 0.30$ lbm/ft³. Compute the local surface coefficient of heat transfer between the hot gases and the blade in the vicinity of the stagnation point.

5.106 On a particular gas turbine vane, the external gas velocity in the vicinity of the leading edge of the vane (which is actually a stagnation point) is given as $U_s = 40,000x$ m/s. Estimate the local surface coefficient of heat transfer at the leading edge if the gas properties are $\rho = 4$ kg/m³, $\mu = 4.15 \times 10^{-5}$ kg/ms, $k = 0.06$ W/m °C, and $N_{Pr} = 0.70$.

5.107 A 5 cm outside diameter steam line 6 m long runs between two buildings. Compute the convective heat transfer rate from the line when a 25 m/s wind at 30°C blows in crossflow over the 180°C steam line.

5.108 An electrical resistance heater is in the shape of a cylinder 1 m long and 2.5 cm in diameter. If its surface temperature is not to exceed 300°C when air at 30°C blows over it in crossflow at a speed of 12 m/s, compute the maximum allowable electrical energy dissipation rate.

5.109 Bars 2.5 cm in diameter are taken out of a heat-treating oven and cooled by having 60°C oil move over the bars at 0.6 m/s in crossflow. If the average bar surface temperature is 160°C, compute the average surface coefficient of heat transfer between the bar and the oil.

5.110 Consider a pipe 7.5 m long whose outside diameter is 2.5 cm and whose outside surface temperature is 40°C. Water flows inside the pipe at a mass rate of flow of 0.25 kg/s and it leaves the pipe 10°C hotter than it was when it came in. Neglecting radiant losses, compute the velocity U_s at which 150°C air at 10^5 N/m² would have to be blown across the *outside* of the pipe to accomplish this heating of the water.

5.111 An electric resistance heater, in the shape of a 1-in.-diameter and 3-ft-long right circular cylinder, dissipates electrical energy at the rate of 2000 watts to air at 80°F which is being blown over the cylinder in crossflow at a speed of 100 ft/sec. Calculate the surface temperature of the heater.

5.112 During part of the cooling period after heat treatment, a long 4-in.-diameter bar, whose average surface temperature during the cooling period is 176°F, has 248°F oil flowing over it with a velocity of 0.5 ft/sec. The oil has the properties listed in Appendix A. Find the average surface coefficient of heat transfer between the bar and the oil.

5.113 Water at 48°C is flowing with a speed of 3 m/s in crossflow over a right circular cylinder of diameter 5 cm and surface temperature of 20°C. If the cylinder is 3 m long, compute the heat transfer rate to the cylinder.

5.114 A tube which encloses thermocouple wires is 0.20 in. in diameter. It is situated in a large duct where 100°F air flows over it in crossflow at 10 ft/sec. Estimate the value of the average surface coefficient of heat transfer between the tube and the air.

5.115 Water flows inside 1-m-long, 0.6-cm-diameter tubes in a heat exchanger. The tubes, which are at 80°C, are cooled by having 20°C air blow over them in crossflow at 30 m/s. If the water temperature must decrease by 30°C when passing through the tubes, calculate the mass flow rate of water that can be processed per tube.

5.116 Repeat Prob. 5.115 where water is used as the coolant and flows over the tubes at 1.5 m/s.

5.117 A thermocouple bead 0.15 cm in diameter is used to measure the temperature in an air stream at atmospheric pressure at about 150°C. To calculate the radiation error of the bead, the average surface coefficient of heat transfer between the bead and the air must be found. Compute its value when the air speed is 3 m/s.

5.118 After being taken from a heat-treating oven, a ball bearing 1/2 in. in diameter is being cooled by having 80°F air blown over it at 20 ft/sec. If the average surface temperature of the ball bearing is 320°F, compute the average surface coefficient of heat transfer between the bearing surface and the air.

5.119 Six-in.-diameter spheres, having an average surface temperature of 320°F, are cooling in oil at 212°F which is passing over the spheres at 0.25 ft/sec. Estimate the average surface coefficient of heat transfer.

5.120 A sphere suspended in an air stream is used as a speed-measuring device. A 1/2-in. sphere, when suspended in a 90°F air stream, maintains a surface temperature of 110°F while dissipating electrical energy at the rate of 0.60 W. From this information, calculate the air speed.

5.121 After a machining process, 1.25-cm-diameter ball bearings at 90°C are cooled by passing 10°C water at 0.3 m/s over them. Find the value of the average surface coefficient of heat transfer between the ball bearings and the water.

5.122 Fuel, squirted from a nozzle in the combustor of a jet engine, is in the form of droplets, roughly spherical in shape. The rate at which the droplets evaporate or reach the ignition point depends, for one thing, on the average surface coefficient of heat transfer between the droplet and the surrounding gas. Estimate the value of this surface coefficient for a 0.0038-cm-diameter droplet traveling at 6 m/s with the surrounding gas having the properties $N_{Pr} = 0.70$, $k = 0.09$ W/m °C, $\mu = 5.7 \times 10^{-5}$ kg/ms, and $\rho = 4.8$ kg/m³.

5.123 In the tube bank of Ex. 5.19, provision has been made for forcing 50°F air over the tubes at the same Reynolds number as for the water in the event of failure of the water source or pumps. Estimate the percentage decrease in the condensation rate of the steam if the bank must be operated with air instead of water as the coolant.

5.124 Atmospheric air at 20°C flows across an in-line tube bank whose 2.5-cm-diameter tubes are spaced 7.5 cm apart in both the longitudinal and transverse directions. The tubes are held at 50°C and the air velocity prior to entering the tube bank is 18 m/s. Assuming that the temperature rise of the air as it passes through the bank is negligible, calculate the average surface coefficient of heat transfer between the bank and the air for a 10-row-deep (in the flow direction) bank.

5.125 The tube bank in Prob. 5.124 is 1.5 m high (i.e., the dimension in the transverse direction) and has 20 tubes per row in the transverse direction. The surface coefficient of heat transfer in Prob. 5.124 was based on a negligible rise in the air temperature. For the tube bank just described, compute the temperature of the air at the bank exit.

5.126 Water is to be cooled by 50°F while passing inside the 1/4-in. outside diameter tubes of an in-line tube bank. Air enters the bank at 80°F with a speed of 100 ft/sec and passes over the outside surface of the 3-ft-long tubes, which have a surface temperature of 180°F. There are 10 rows of tubes in the flow direction, spaced 1/2 in. apart in that direction, and each row contains 48 tubes spaced, transversely, also 1/2 in. apart in a total height of 2 ft. Find (a) the mass flow rate of water which is processed by the tube bank, and (b) the temperature of the air leaving the bank.

5.127 Rework Prob. 5.126 if all conditions are the same except that the bank is constructed in a staggered arrangement.

5.128 A tube bank of 5-cm-diameter tubes at 120°C serves as a power plant condenser. There are 20 rows of tubes in the flow direction set in a staggered fashion with 15 cm longitudinal spacings and 10 cm transverse spacings. Each row consists of 15 tubes, 1.5 m long, which are spaced in the total transverse height of 1.5 m. Water at 10°C and 0.15 m/s is supplied to flow over the tubes and provide the cooling. Calculate the rate at which saturated steam at 120°C is condensed by this tube bank.

REFERENCES

1. Parker, J. D., J. H. Boggs, and E. F. Blick. *Introduction to Fluid Mechanics and Heat Transfer.* Reading, Mass.: Addison-Wesley, 1969.
2. Shames, I. H. *Mechanics of Fluids.* New York: McGraw-Hill, 1962.
3. Gebhart, B. *Heat Transfer.* 2d ed. New York: McGraw-Hill, 1971.
4. Leigh, D. C. *Non-Linear Continuum Mechanics.* New York: McGraw-Hill, 1968.
5. Streeter, V. L., and E. B. Wylie. *Fluid Mechanics.* 7th ed. New York: McGraw-Hill, 1979.
6. Schlichting, H. *Boundary Layer Theory.* 6th ed. New York: McGraw-Hill, 1968.
7. White, F. M. *Viscous Fluid Flow.* New York: McGraw-Hill, 1974.
8. White, F. M. *Fluid Mechanics.* New York: McGraw-Hill, 1979.
9. Sparrow, E. M. "Analysis of Laminar Forced Convection Heat Transfer in Entrance Region of Flat Rectangular Ducts." *National Advisory Committee for Aeronautics (NACA) Technical Note 3331* (January 1955).
10. Langhaar, H. L. "Steady Flow in the Transition Length of a Straight Tube." *Journal of Applied Mechanics,* Vol. 64 (1942).
11. Moody, L. F. "Friction Factors for Pipe Flow." *Transactions of the A.S.M.E.,* Vol. 66 (1944).
12. Jakob, M. *Heat Transfer.* Vol. 1. New York: Wiley, 1959.
13. Patankar, S. V., and D. B. Spalding. *Heat and Mass Transfer in Boundary Layers.* 2d ed. London: Intertext Books, 1970.
14. Launder, B. E., and D. B. Spalding. *Mathematical Models of Turbulence.* London: Academic Press, 1972.
15. Kestin, J., and P. D. Richardson. "Heat Transfer Across Turbulent Incompressible Boundary Layers." *International Journal of Heat and Mass Transfer,* Vol. 6 (1963), pp. 147–89.
16. Eckert, E. R. G., and R. M. Drake, Jr. *Analysis of Heat and Mass Transfer.* New York: McGraw-Hill, 1972.
17. Kays, W. M., and M. E. Crawford. *Convective Heat and Mass Transfer.* 2d ed. New York: McGraw-Hill, 1980.
18. Rosenhead, L., ed. *Laminar Boundary Layers.* New York: Oxford University Press, 1963.
19. Hansen, A. G. *Similarity Analyses of Boundary Value Problems in Engineering.* Englewood Cliffs, N.J.: Prentice-Hall, 1964.
20. Howarth, L. "On the Solution to the Laminar Boundary Layer Equations." *Proceedings of the Royal Society,* Series A, Vol. 164 (London, 1938), pp. 547–79.
21. Finlayson, B. A., and L. E. Scriven. "The Method of Weighted Residuals—A Review." *Applied Mechanics Reviews,* Vol. 19, No. 9 (September 1966), pp. 735–48.
22. Arpaci, V. S. *Conduction Heat Transfer.* Reading, Mass.: Addison-Wesley, 1966.
23. Hildebrand, F. B. *Methods of Applied Mathematics.* 2d ed. Englewood Cliffs, N.J.: Prentice-Hall, 1965.
24. Finlayson, B. A. *The Method of Weighted Residuals and Variational Principles.* New York: Academic Press, 1972.
25. Kantorovich, L. V., and V. I. Krylov. *Approximate Methods of Higher Analysis.* 3d ed. New York: Wiley/Interscience, 1964.
26. Pol'skii, N. I. "Projective Methods in Applied Mathematics." *Soviet Mathematics,* Vol. 3 (1962), pp. 488–91. English translation.
27. Coles, D. "The Law of the Wake in the Turbulent Boundary Layer." *Journal of Fluid Mechanics,* Vol. 1 (1956), pp. 191–226.
28. Rohsenow, W. M., and H. Y. Choi. *Heat, Mass and Momentum Transfer.* Englewood Cliffs, N.J.: Prentice-Hall, 1961.
29. Clark, S. H., and W. M. Kays. "Laminar Flow Forced Convection in Rectangular Tubes." *Transactions of the A.S.M.E.,* Vol. 75 (1953), pp. 859–66.
30. Schubauer, G. B., and C. M. Tchen. *Turbulent Flow.* Princeton, N.J.: Princeton University Press, 1961.
31. Colburn, A. P. "A Method of Correlating Forced Convection Heat Transfer Data and a Comparison with Fluid Friction." *Transactions of the American Institute of Chemical Engineers,* Vol. 29 (1933), pp. 174–210.

32. von Kàrmàn, T. "The Analogy Between Fluid Friction and Heat Transfer." *Transactions of the A.S.M.E.,* Vol. 61 (1939), pp. 705–11.

33. Kays, W. M. *Convective Heat and Mass Transfer.* New York: McGraw-Hill, 1966.

34. Petukhov, B. S. "Heat Transfer and Friction in Turbulent Pipe Flow with Variable Physical Properties." In *Advances in Heat Transfer,* Vol. 6, edited by J. P. Hartnett and T. F. Irvine, Jr. New York: Academic Press, 1970.

35. Pohlhausen, E. "Der Wärmeaustausch zwischen festen Körpern und Flüssigkeiten mit kleiner Reibung und kleiner Wärmeleitung." *Z. Angew. Math. Mech.,* Vol. 1 (1921), p. 115.

36. Churchill, S. W. "A Comprehensive Correlating Equation for Forced Convection from Flat Plates," *AIChE Journal,* Vol. 22, No. 2 (March 1976), pp. 264–68.

37. Churchill, S. W., and H. Ozoe. "Correlations for Laminar Forced Convection with Uniform Heating in Flow Over a Plate and in Developing and Fully Developed Flow in a Tube." *Journal of Heat Transfer,* Vol. 95, No. 1 (1973), pp. 78–84.

38. Eckert, E. R. G. "Engineering Relations for Friction and Heat Transfer to Surfaces in High Velocity Flow." *Journal of the Aero. Sciences* (August 1955), pp. 585–87.

39. Sieder, E. N., and G. E. Tate. "Heat Transfer and Pressure Drop of Liquids in Tubes." *Industrial and Engineering Chemistry,* Vol. 28, No. 12 (1936), pp. 1429–35.

40. Ambrok, G. S. "Approximate Solutions of the Equations for the Thermal Boundary Layer with Variations in Boundary Layer Structure." *Soviet Physics, Technical Physics,* Vol. 2, No. 9 (1957), pp. 1979–86.

41. Gauntner, D. J., and J. Sucec. "Method for Calculating Convective Heat Transfer Coefficients Over Turbine Vane Surfaces." *NASA Technical Paper 1134* (1978).

42. Whitaker, S. "Forced Convection Heat Transfer Correlations for Flow in Pipes, Past Flat Plates, Single Cylinders, Single Spheres, and for Flow in Packed Beds and Tube Bundles." *AIChE Journal,* Vol. 18, No. 2 (March 1972), pp. 361–71.

43. Dittus, F. W., and L. M. K. Boelter. "Heat Transfer in Automobile Radiators of the Tubular Type." *University of California Publications in Engineering,* Vol. 2, No. 13 (Berkeley, 1930), pp. 443–61.

44. McAdams, W. H. *Heat Transmission.* 3d ed. New York: McGraw-Hill, 1954.

45. Rohsenow, W. M., and J. P. Hartnett, eds. *Handbook of Heat Transfer.* New York: McGraw-Hill, 1973.

46. Dipprey, D. F., and R. H. Sabersky. "Heat and Momentum Transfer in Smooth and Rough Tubes at Various Prandtl Numbers." *International Journal of Heat and Mass Transfer,* Vol. 6 (1963), pp. 329–53.

47. Norris, R. H. "Some Simple Approximate Heat Transfer Correlations for Turbulent Flow in Ducts with Rough Surfaces." In *Augmentation of Convective Heat and Mass Transfer,* edited by A. E. Bergles and R. L. Webb. New York: American Society of Mechanical Engineers, 1970.

48. Han, J. C., L. R. Glicksman, and W. M. Rohsenow. "An Investigation of Heat Transfer and Friction for Rib Roughened Surfaces." *International Journal of Heat and Mass Transfer,* Vol. 21 (1978), pp. 1143–56.

49. Hausen, H. "Darstellung des Wärmeüberganges in Rohren durch verallgemeinerte Potenzbeziehungen." *Zeitschr. ver. deut. Ing. Beihefte Verfahrenstechnik,* No. 4 (1943), pp. 91–98.

50. Walz, A. *Boundary Layers of Flow and Temperature.* Cambridge, Mass.: The M.I.T. Press, 1969.

51. Knudsen, J. G., and D. L. Katz. *Fluid Dynamics and Heat Transfer.* New York: McGraw-Hill, 1958.

52. Holman, J. P. *Heat Transfer.* 5th ed. New York: McGraw-Hill, 1981.

53. Achenbach, E. "Total and Local Heat Transfer from a Smooth Circular Cylinder in Crossflow at High Reynolds Number." *International Journal of Heat and Mass Transfer,* Vol. 18 (1975), pp. 1387–96.

54. Zukauskas, A. "Heat Transfer from Tubes in Crossflow." In *Advances in Heat Transfer,* Vol. 8, edited by J. P. Hartnett and T. F. Irvine, Jr. New York: Academic Press, 1972.

55. Churchill, S. W., and M. Bernstein. "A Correlating Equation for Forced Convection from Gases and Liquids to a Circular Cylinder in Crossflow." *Journal of Heat Transfer,* Vol. 99, No. 2 (1977), pp. 300–306.

56. Kreith, F. *Principles of Heat Transfer.* 3d ed. New York: Intext Press, 1973.

57. Morgan, V. T. "The Overall Convective Heat Transfer from Smooth Circular Cylinders." In *Advances in Heat Transfer,* Vol. 11, edited by T. F. Irvine, Jr. and J. P. Hartnett. New York: Academic Press, 1975.

58. Hilpert, R. "Wärmeabgabe von geheizen Drähten und Rohren." *Forsh. Gebiete Ingenieurw.,* Vol. 4 (1933), pp. 215–24.

59. Fand, R. M. "Heat Transfer by Forced Convection from a Cylinder to Water in Crossflow." *International Journal of Heat and Mass Transfer,* Vol. 8 (1965), pp. 995–1010.

60. Vliet, G. C., and G. Leppert. "Forced Convection Heat Transfer from an Isothermal Sphere to Water." *Journal of Heat Transfer,* Vol. 83 (May 1961), pp. 163–75.

61. Pierson, O. L. "Experimental Investigation of the Influence of Tube Arrangement on Convection Heat Transfer and Flow Resistance in Crossflow of Gases Over Tube Banks." *Transactions of the A.S.M.E.,* Vol. 59 (1937), pp. 563–72.

62. Huge, E. C. "Experimental Investigation of Effects of Equipment Size on Convection Heat Transfer and Flow Resistance in Crossflow of Gases Over Tube Banks." *Transactions of the A.S.M.E.,* Vol. 59 (1937), pp. 573–82.

63. Grimison, E. D. "Correlation and Utilization of New Data on Flow Resistance and Heat Transfer for Crossflow of Gases Over Tube Banks." *Transactions of the A.S.M.E.,* Vol. 59 (1937), pp. 583–94.

64. Kays, W. M., A. L. London, and R. K. Lo. "Heat Transfer and Friction Characteristics for Gas Flow Normal to Tube Banks—Use of a Transient Test Technique." *Transactions of the A.S.M.E.,* Vol. 76 (April 1954), pp. 387–96.

65. Theoclitus, G. "Heat Transfer and Flow Friction Characteristics of Nine Pin-Fin Surfaces." *Journal of Heat Transfer,* Vol. 88, No. 4 (1966), pp. 383–90.

66. Succc, J., and W. W. Bowley. "Analytical Models for the Prediction of Total Pressure Loss in Staggered Banks of Tubes with Small Side Plate Spacings." *Pratt and Whitney Aircraft Report TGM-111* (August 1967).

67. Brand, R. S., and L. N. Persen. "Implications of the Law of the Wall for Turbulent Boundary Layers." *Acta Polytechnica Scandinavica,* Ph 30, UDC 532.526.4. Trondheim, Norway (1964).

68. Karlekar, B. V., and R. M. Desmond. *Heat Transfer.* 2d ed. St. Paul, Minn.: West, 1982.

69. Shah, R. K., and A. L. London. "Laminar Flow Forced Convection in Ducts." *Advances in Heat Transfer,* Supplement 1. New York: Academic Press, 1978.

70. Succc, J. "Extension of a Modified Integral Method to Boundary Conditions of Prescribed Surface Heat Flux." *International Journal of Heat and Mass Transfer,* Vol. 22 (1979), pp. 771–74.

71. Zukauskas, A. Personal communication, 1984.

FREE CONVECTION

6.1 INTRODUCTION

Chapter 5 discussed forced convection heat transfer, in which the fluid motion is caused by a pump, fan, or blower, or perhaps by the fluid coming from a reservoir at a higher elevation. One of the primary characteristics of forced convection was that the velocity field did not depend on the temperature field (for constant property situations), and hence, the velocity field could be solved independent of the temperature field. That is, the velocity field was not coupled to the temperature field, although the temperature field in the moving fluid was coupled to the velocity field. In *free* or *natural convection,* the fluid motion is a consequence of the temperature field inducing density gradients in the fluid which, in turn, are acted upon by a force field, such as the gravitational field. Since in free convection, the fluid motion is generated by the density differences produced by the temperature field, it seems reasonable to assume that the velocity field is coupled to the temperature field. Since the temperature field is also coupled to the velocity field, the fields are said to be *mutually coupled.* This mutual coupling of the temperature and velocity fields does not allow the velocity field to be determined first and then inserted into the energy equation to obtain the temperature field, as was the case for constant property forced convection problems. Because of this mutual coupling, solving a free convection problem by exact analytical or even approximate analytical techniques is considerably more difficult than an equivalent forced convection problem. For this reason, the details of the available exact analytical solutions will not be discussed. However, an approximate analytical solution using integral methods will be presented which will reveal the nature of the additional complications in free convection, relative to constant property forced convection. It will also introduce us to the parameters upon which the surface coefficient of heat transfer depends in free convection.

In this chapter, we will begin by deriving the governing partial differential equations, and then the governing integral equations, for the velocity field and the temperature field in natural convection situations. This will be followed by an approximate analytical solution, by use of integral methods, for the surface coefficient of heat transfer on a vertical

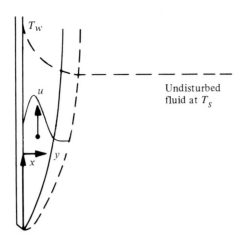

Figure 6.1 Representative velocity and temperature distributions in free convection boundary layer flow on a vertical plate.

plate. Presented next will be experimental correlations for the free convection surface coefficient for a number of important geometries, such as vertical and inclined plates, cylinders, spheres, and rectangular enclosures. Experimental correlations are, if anything, used more often in free convection calculations than in forced convection calculations. This occurs because there are fewer exact and approximate solutions to free convection problems. Another reason is that the predicted surface coefficients in laminar free convection tend to be smaller than the experimentally measured values because of rate-increasing disturbances, present in actual equipment, which the analyses do not account for. The chapter concludes with a discussion of mixed, or combined, convection, in which the influence of both free and forced convection must be dealt with.

6.2 GOVERNING EQUATIONS FOR VELOCITY AND TEMPERATURE FIELDS

Experiment and analysis have both shown that most free convection problems, like their forced convection counterparts, are of the thin boundary layer type [1]. That is, the thermal and viscous effects are concentrated in relatively thin layers adjacent to the solid surface and derivatives in the flow direction are much smaller than those in the direction perpendicular to the solid surface. Figure 6.1 shows representative velocity and temperature distributions for thermal and velocity boundary layers on a flat plate whose surface temperature T_w is greater than the undisturbed fluid temperature T_s far away from the plate. The velocity boundary layer and x component of the velocity distribution are shown as solid lines, while the thermal boundary layer and temperature distribution within it, at a single value of x, are shown as dashed lines. One obvious difference from forced convection over a flat plate is that the velocity begins at zero at the plate surface, reaches a maximum somewhere inside the velocity boundary layer, and then decreases to zero at the outer edge of the boundary layer, in the undisturbed fluid.

Most of the analytical solutions to free convection problems are for constant property situations. That is, the density variation is accounted for only in the term that actually causes the free convection currents and is considered small enough to neglect in other places, such as the continuity equation. Assuming then that density variation can be neglected except in the term that causes the motion in free convection, that all other properties are constant, and that the flow is steady, laminar, low speed, and of the thin boundary layer type, the governing equations for the velocity field and the temperature field in free convection will now be developed. First, by reference to Fig. 6.1, the governing *partial differential* equations will be derived, and, after that, the *integral* forms of the governing equations will be found.

6.2a Partial Differential Equations

Consider the free convection hydrodynamic and thermal boundary layer–type flows adjacent to the vertical plate shown in Fig. 6.1. The governing equations are, as was also the case for forced convection in the last chapter, the continuity equation, x-momentum equation (the thin boundary layer assumption makes the y-momentum equation unnecessary), and the thermal energy equation. The continuity equation for this situation is given by Eq. (5.13) with the last term equal to zero because the present situation is two dimensional. Hence,

$$\frac{\partial u}{\partial x} + \frac{\partial v}{\partial y} = 0. \tag{6.1}$$

The x-momentum equation is given by Eq. (5.24) with several terms dropped because of the thin boundary layer assumption and the two-dimensionality of the problem, just as in the case of forced convection over a flat plate. However, the body force f_x per unit volume must be retained, because in the present problem, the free convection currents are caused by gravitational attraction. (They could, in a different situation, for example, be caused by the "centrifugal" force field induced by rotation about an axis [2].) Hence, for the situation shown in Fig. 6.1, Eq. (5.24) becomes

$$\rho \left(u \frac{\partial u}{\partial x} + v \frac{\partial u}{\partial y} \right) = -\frac{\partial P}{\partial x} + \mu \frac{\partial^2 u}{\partial y^2} + f_x. \tag{6.2}$$

For low speed, constant property, two-dimensional, steady, laminar flow, the low-speed energy equation has exactly the same form as for the equivalent forced convection situation, Eq. (5.287). Hence,

$$\rho c_p \left(u \frac{\partial T}{\partial x} + v \frac{\partial T}{\partial y} \right) = k \frac{\partial^2 T}{\partial y^2}. \tag{6.3}$$

In order to determine what the body force per unit volume in the x direction equals, consider the forces, in the x direction, on a control volume of differential extent dx by dy. (Recall that the right-hand side of Eq. [6.2] represents the sum of all the forces on

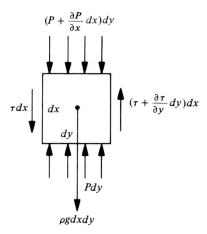

Figure 6.2 A differential control volume with forces during free convection shown.

the control volume in the x direction.) Figure 6.2 shows these forces, the shear forces and pressure forces, just as in the case of forced convection. In addition, the attraction of the earth for the mass contained in the control volume, the weight of the control volume material, is shown. The weight is the weight density, the product of the mass density and the local acceleration of gravity g, times the volume of material $dxdy$. Summing the forces in the x direction,

$$\Sigma F_x = P\,dy - \left(P + \frac{\partial P}{\partial x}\,dx\right)dy + \left(\tau + \frac{\partial \tau}{\partial y}\,dy\right)dx - \tau\,dx$$
$$- \rho g\,dxdy = \left(-\frac{\partial P}{\partial x} + \frac{\partial \tau}{\partial y} - \rho g\right)dxdy. \qquad (6.4)$$

Since for laminar, two-dimensional, boundary layer–type flow, $\tau = \mu(\partial u/\partial y)$, substitution into Eq. (6.4) and division by the volume element leads to

$$\Sigma F_x = -\frac{\partial P}{\partial x} + \mu\frac{\partial^2 u}{\partial y^2} - \rho g. \qquad (6.5)$$

Comparison of Eq. (6.5) with the right-hand side of Eq. (6.2) yields

$$f_x = -\rho g. \qquad (6.6)$$

Equation (6.4), the force balance for the x-momentum equation, is also valid at the same value of x with the control volume moved to the right into the undisturbed fluid at temperature T_s and density ρ_s. But $\partial^2 u/\partial y^2 = 0$ at this location because the point is outside the velocity boundary layer, and in addition, all momentum flux terms are zero at large values of y (the undisturbed fluid) because all velocities are zero in this region for a free convection process; hence, by the momentum theorem, $\Sigma F_x = 0$ in the undisturbed fluid. Therefore,

$$-(\partial P/\partial x)_s - \rho_s g = 0$$

or

$$-(\partial P/\partial x)_s = \rho_s g. \tag{6.7}$$

However, the assumption of a thin boundary layer implies that $\partial P/\partial y \approx 0$ (see Sec. 5.3); hence,

$$-(\partial P/\partial x)_s = -\partial P/\partial x = \rho_s g. \tag{6.8}$$

Substituting Eqs. (6.8) and (6.6) into Eq. (6.2) and rearranging yields

$$\rho \left(u \frac{\partial u}{\partial x} + v \frac{\partial u}{\partial y} \right) = \mu \frac{\partial^2 u}{\partial y^2} + g(\rho_s - \rho). \tag{6.9}$$

The last term on the right-hand side of Eq. (6.9) is the net upward force per unit volume, the difference between the buoyant force and the weight, which initiates and sustains the motion of the free convection currents in the boundary layer of the plate. The x-momentum equation, Eq. (6.9), will be valid as it stands for the case of $T_w < T_s$ also.

The net upward force per unit volume in Eq. (6.9) is usually put into a form which is linear in the difference between the local temperature T and the undisturbed temperature T_s by introducing the thermal coefficient of volume expansion β, where β is defined as

$$\beta = \frac{1}{\underline{v}} \left(\frac{\partial \underline{v}}{\partial T} \right)_P, \tag{6.10}$$

where $\underline{v}$ is the specific volume of the substance, T is the temperature, and P is the pressure.

To arrive at a form for $g(\rho_s - \rho)$, which is linear in the temperature difference and is, therefore, more easily handled when solutions to the governing equations are being sought, one must first get the density ρ into the expression for the thermal coefficient of volume expansion, β in Eq. (6.10). Since the specific volume $\underline{v} = 1/\rho$, we have

$$\left(\frac{\partial \underline{v}}{\partial T} \right)_P = \frac{\partial}{\partial T} \left(\frac{1}{\rho} \right)_P = -\frac{1}{\rho^2} \left(\frac{\partial \rho}{\partial T} \right)_P.$$

Thus, Eq. (6.10) becomes

$$\beta = -\frac{1}{\rho} \left(\frac{\partial \rho}{\partial T} \right)_P. \tag{6.11}$$

Separating the variables T and ρ in Eq. (6.11) permits integration (while holding P constant) in the y direction at constant x (since $\partial P/\partial y = 0$). Integrating from a point inside the boundary layer, where the variables have the values ρ and T, to a point at the same x in the undisturbed stream, where ρ and T take on the values ρ_s and T_s, yields

$$\int_T^{T_s} \beta dT = -\int_\rho^{\rho_s} \frac{d\rho}{\rho}.$$

(6.12)

If β does not vary much with temperature, or if a suitable average value of β is used, the left-hand side of Eq. (6.12) becomes

$$\beta(T_s - T).$$

(6.13)

Since the property variations are small, the ρ in the denominator cannot vary much between ρ and ρ_s and, therefore, can be factored, or considered as the proper "average" value and factored, giving for the right-hand side of Eq. (6.12),

$$-(\rho_s - \rho)/\rho.$$

(6.14)

Equating Eqs. (6.13) and (6.14) and solving for $g(\rho_s - \rho)$ yields

---

$$g(\rho_s - \rho) = g\beta\rho(T - T_s).$$

(6.15)

Using Eq. (6.15) in Eq. (6.9), the equations which describe the velocity and temperature field in this free convection situation are

---

$$\frac{\partial u}{\partial x} + \frac{\partial v}{\partial y} = 0,$$

[6.1]

$$\rho\left(u\frac{\partial u}{\partial x} + v\frac{\partial u}{\partial y}\right) = \mu\frac{\partial^2 u}{\partial y^2} + g\beta\rho(T - T_s),$$

[6.16]

$$\rho c_p\left(u\frac{\partial T}{\partial x} + v\frac{\partial T}{\partial y}\right) = k\frac{\partial^2 T}{\partial y^2}.$$

[6.3]

The *mutual coupling* between the velocity and temperature fields is clearly shown by these equations. The thermal energy equation, Eq. (6.3), cannot be solved for the temperature field $T = T(x, y)$ without knowledge of the velocity field or without simultaneously solving for it, since $u = u(x, y)$ and $v = v(x, y)$ appear in Eq. (6.3). This, however, was also the case for constant property forced convection. We can also see that *the x-momentum equation cannot be solved for the velocity field without finding the temperature field simultaneously, since it contains the unknown temperature field explicitly in the term $g\beta\rho(T - T_s)$.*

Equations (6.1), (6.16), and (6.3) have been solved exactly by the method of the similarity transformation for laminar flow and for various boundary conditions that admit a similarity transformation in [1], [3], and [4]. A presentation and short discussion of these exact solutions, as well as a more detailed development of the free convection boundary layer equations, Eqs. (6.1), (6.16), and (6.3), with special attention paid to the approximations involved in arriving at Eq. (6.15), is available in Gebhart [5].

As was mentioned earlier, rate-increasing disturbances existing with actual equipment cause the measured free convection surface coefficient of heat transfer to be higher (by about 15%) than the analytically predicted values for the laminar flow of gases, because these disturbances are not taken into consideration in the analytical solution. It is found that, under most conditions, the free convective velocities for gases in the laminar boundary layer are fairly small. Thus, any *small disturbance,* such as stray fluid currents outside the hydrodynamic boundary layer, can significantly influence the similarly small velocities within the boundary layer and therefore change the convective heat transfer rate. As another example of a small disturbance, we have equipment vibration, which could also have the same type of augmenting effect on the small velocities in the boundary layer. As a result of these rate-increasing disturbances, experimental correlations for the surface coefficient in laminar free convection of gases, such as air, whose Prandtl number is close to 1, should be utilized when they are available.

6.2b Integral Equations

To solve free convection heat transfer problems by the powerful approximate integral methods requires that we first develop, as also was the case for forced convection, the integral forms of the momentum equation and of the energy equation so that we can ultimately solve for the velocity field and the temperature field. We will begin by deriving the integral form of the x-momentum equation for the free convection hydrodynamic boundary layer shown in Fig. 6.1.

The appropriate control volume is shown in Fig. 6.3. When all the forces in the x direction are summed up, the shear force $\tau_w \, dx$ is present, as shown in Fig. 6.3. Also, there is a net upward force per unit volume $g\beta\rho(T - T_s)$ as in Eq. (6.15). Hence, the net upward buoyant force on the entire control volume is

$$\int_0^\delta g\beta\rho(T - T_s) \, dydx.$$

Thus,

$$\Sigma F_x = \int_0^\delta g\beta\rho(T - T_s) \, dydx - \tau_w \, dx.$$

The momentum fluxes into and out of the control volume in the x direction are calculated as in Chapter 5; that is,

$$\int_0^\delta \rho u^2 \, dy$$

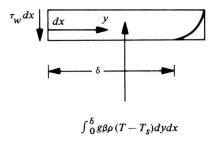

$$\int_0^\delta g\beta\rho\,(T-T_s)dydx$$

Figure 6.3 The control volume used in the derivation of the integral form of the momentum theorem with forces shown.

for the x-momentum flux in, and

$$\int_0^\delta \rho u^2\,dy + \frac{d}{dx}\left(\int_0^\delta \rho u^2\,dy\right)dx$$

for the x-momentum flux out. The x-momentum flux that enters from the free stream between x and $x + dx$ is zero since $U_s = 0$ in a free convection situation. (Note that ω_s is not zero, however.) Setting the sum of all the forces on the control volume in the x direction equal to the momentum flux out minus the momentum flux in, in the x direction, as required by the momentum theorem, yields

$$\frac{d}{dx}\int_0^\delta \rho u^2\,dy = -\tau_w + \int_0^\delta g\beta\rho(T-T_s)\,dy. \tag{6.17}$$

This is the appropriate integral form of the x-momentum equation for the free convection situation shown in Fig. 6.1.

The integral form of the x-momentum equation, Eq. (6.17), also clearly shows the dependence of the velocity distribution on the temperature distribution in a free convection situation, by virtue of the temperature difference in the last term on the right. Obviously then, Eq. (6.17) cannot be solved by itself to yield the velocity field, $u = u(x, y)$, since it also contains another dependent variable, the temperature $T = T(x, y)$, which is also unknown. Hence, even to find the velocity field in a constant property free convection situation, the governing equation for the temperature distribution is needed and must be used.

From the energy balance on the thermal boundary layer of local thickness δ_t that was carried out in Chapter 5 for forced convection, one sees that, as long as the free convection problem is one which is steady, low speed, constant property, and of the thin boundary layer type, the same integral energy equation can be used in free convection as was previously derived for forced convection. This is Eq. (5.312), which will be renumbered here and is shown next.

$$\frac{d}{dx} \int_0^{\delta_t} u(T_s - T)dy = \alpha \left(\frac{\partial T}{\partial y}\right)_{y=0} \tag{6.18}$$

Thus, the integral x-momentum equation, Eq. (6.17), and the integral form of the thermal energy equation, Eq. (6.18), constitute two *mutually* coupled integral equations which would have to be solved *simultaneously* for the velocity field and the temperature field in free convection, since both the velocity u and the temperature T appear as unknown dependent variables in both equations.

Notice that the set of partial differential equations, Eqs. (6.1), (6.16), and (6.3), and the set of integral equations just developed, Eqs. (6.17) and (6.18), though discussed in the framework of the vertical flat plate of Fig. 6.1, are also valid for other geometries. If body curvature is gentle enough, as discussed in Sec. 5.3, they will be valid for arbitrarily shaped, planar two-dimensional bodies when the fields of velocity and temperature are of a thin boundary layer character, once modified so that the component of the weight in the local x direction is used.

EXAMPLE 6.1

Figure 6.4 shows a flat plate which is inclined at an angle θ from the vertical. What modifications would be needed in the x-momentum equations, both the partial differential form, Eq. (6.16), and the integral form, Eq. (6.17), to allow them to be valid for this inclined plate, as long as the flow remains the thin boundary layer type along the length of the plate, which has its cold surface facing up?

Solution
The inclination of the plate causes the component of the weight of the control volume material, and the buoyant force on the control volume along the plate, the x direction, to

Figure 6.4 The inclined flat plate of Ex. 6.1.

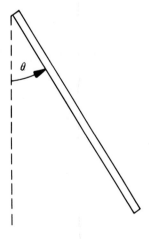

Chapter Six

be smaller than if the plate is vertical. This can be seen by reference to Fig. 6.2 and the development of the net buoyant force term in Eq. (6.16). Hence, if the plate were vertical, the net driving force per unit volume for free convection would be $g\beta\rho(T - T_s)$ as shown in Fig. 6.5. The inclination changes this net driving force to the value $g\beta\rho(T - T_s)$ $\cos\theta$, as is seen in Fig. 6.5. Hence, the effect of inclination, as long as the flow remains the thin boundary layer type, can be accounted for by multiplying the term $g\beta\rho(T - T_s)$ by $\cos\theta$ in both Eqs. (6.16) and (6.17). Later we will discuss the limiting values of θ for which this is a proper correction for inclination.

6.3 VERTICAL PLATE SOLUTION BY INTEGRAL METHODS

In this section, an approximate analytical solution of a free convection problem by the integral method, which was used extensively in Chapter 5, will be presented. The objective of the solution is to predict the local surface coefficient of heat transfer h_x as a function of quantities that would ordinarily be known in a free convection situation. By contrasting the work involved in this free convection solution with the work that was done to arrive at the forced convection counterpart in Chapter 5—namely, Eq. (5.334)—it will be clear that the mutual coupling of the velocity and temperature fields makes a free convection solution, all other things being equal, considerably more complicated than a solution for a forced convection situation. The approximate solution for the local surface coefficient will also serve to introduce us to the nondimensional groups that appear in free convection problems, not only for the case that will be dealt with here, but for other geometries and boundary conditions as well.

Consider steady, laminar, low speed, constant property, thin boundary layer, free convective flow over the vertical, constant wall temperature plate shown in Fig. 6.1. In order to analytically predict the local surface coefficient of heat transfer h_x, we must, as in the case of any convection problem, first predict the temperature field within the moving fluid. This, in the case of free convection, necessitates solving the x-momentum equation and the energy equation simultaneously. Hence, after using Newton's law of viscosity for the

Figure 6.5 The control volume for the inclined plate of Ex. 6.1 with some of the forces shown.

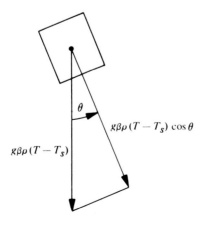

shear stress τ_w, and rearranging Eq. (6.17), the integral equations to be solved are as follows:

$$\int_0^\delta g\beta\rho(T - T_s)dy = \mu\left(\frac{\partial u}{\partial y}\right)_{y=0} + \frac{d}{dx}\int_0^\delta \rho u^2 dy \qquad (6.19)$$

$$\frac{d}{dx}\int_0^{\delta_t} u(T_s - T)dy = \alpha\left(\frac{\partial T}{\partial y}\right)_{y=0} \qquad [6.18]$$

Now, in free convection, since it is temperature differences which create or cause the velocity field in the first place, it seems reasonable to expect that the hydrodynamic boundary layer thickness δ, to which the velocity field is confined, is about equal to the thermal boundary layer thickness δ_t, where the temperature variation from T_w to the undisturbed temperature T_s takes place. Hence we assume on the basis of this argument that

$$\delta_t \approx \delta. \qquad (6.20)$$

This assumption, Eq. (6.20), was also made by Squire, who first applied integral methods to this problem in an unpublished work that is reported in [6].

To solve, in an approximate fashion, Eqs. (6.19) and (6.18), the procedure developed in detail in Chapter 5 is followed. Thus we now must put down approximating sequences for the velocity and temperature distributions which are physically reasonable and which satisfy the boundary conditions that the true solutions for u and for T satisfy. Hence, as in Chapter 5, we take polynomials in y as the approximating sequences as follows:

$$u = a_0 + a_1 y + a_2 y^2 + a_3 y^3 \qquad (6.21)$$
$$T = b_0 + b_1 y + b_2 y^2 + b_3 y^3 \qquad (6.22)$$

For the velocity profile, the boundary conditions to be satisfied are as follows:

$$y = 0, \qquad \text{all } x,\ u = 0$$
$$y = \delta, \qquad \text{all } x,\ \partial u/\partial y = 0 \qquad (6.23)$$
$$y = \delta, \qquad \text{all } x,\ u = 0$$

The last condition in equation set (6.23) is due to the free convective nature of the problem where the fluid far away from the plate, or outside the hydrodynamic boundary layer, is not moving.

Applying these conditions, Eq. (6.23), to Eq. (6.21) leads to the following:

$$u = a_1\left[y - \frac{2y^2}{\delta} + \frac{y^3}{\delta^2}\right]. \qquad (6.24)$$

In Eq. (6.24), *both* a_1 and δ are unknown functions of x to be determined eventually from the two governing equations, Eqs. (6.19) and (6.18).

The boundary conditions and an additional condition to be satisfied by the temperature profile, Eq. (6.22), are the same as in the forced convection counterpart of this problem in Chapter 5 and are shown in the steps leading to Eq. (5.325), which is also what Eq. (6.22) becomes in this free convection problem. Thus, because of Eq. (6.20), and the imposition of the conditions which led to Eq. (5.325), Eq. (6.22) becomes

$$T = T_w + (T_s - T_w)\left(\frac{3}{2}\frac{y}{\delta} - \frac{1}{2}\frac{y^3}{\delta^3}\right). \tag{6.25}$$

From Eq. (6.25), one has that

$$\left(\frac{\partial T}{\partial y}\right)_{y=0} = \frac{3}{2}\frac{(T_s - T_w)}{\delta}. \tag{6.26}$$

Inserting the approximating sequence for the velocity profile, Eq. (6.24), into the integral on the right side of Eq. (6.19) and performing the integration leads to

$$\int_0^\delta \rho u^2 dy = \rho\,\frac{a_1\delta^3}{105}. \tag{6.27}$$

In a similar fashion, using Eqs. (6.24) and (6.25) in the integral of Eq. (6.18) gives, after using Eq. (6.20),

$$\int_0^\delta u(T_s - T)dy = a_1(T_s - T_w)$$
$$\times \int_0^\delta \left[y - \frac{2y^2}{\delta} + \frac{y^3}{\delta^3}\right]\left[1 - \frac{3}{2}\frac{y}{\delta} + \frac{1}{2}\frac{y^3}{\delta^3}\right]dy \tag{6.28}$$
$$= \frac{4}{105}a_1(T_s - T_w).$$

Using Eq. (6.25) in the integral on the left of Eq. (6.19) gives

$$\int_0^\delta g\beta\rho(T - T_s)dy = g\beta\rho(T_w - T_s)\frac{3}{8}\delta. \tag{6.29}$$

Differentiating the approximate velocity profile, Eq. (6.24), and evaluating the result at $y = 0$ gives

$$\left(\frac{\partial u}{\partial y}\right)_{y=0} = a_1. \tag{6.30}$$

Next, Eqs. (6.26) through (6.30) are inserted, appropriately, into the integral x-momentum equation (6.19) and the integral energy equation, Eq. (6.18), to give the following equations:

$$g\beta\rho\Delta T \frac{3}{8}\delta = \mu a_1 + \frac{\rho}{105}\frac{d}{dx}(a_1^2\delta^3) \tag{6.31}$$

$$\frac{4}{105}\frac{d}{dx}(a_1\delta^2) = \frac{3}{2}\frac{\alpha}{\delta} \tag{6.32}$$

$$\Delta T = T_w - T_s \tag{6.33}$$

Equations (6.31) and (6.32) constitute a set of mutually coupled nonlinear ordinary differential equations for the two unknown functions of x, the boundary layer thickness δ and the parameter a_1. In general, one would have to solve these equations by numerical finite difference techniques. However, recall from Chapter 5 that the boundary layer thicknesses in the forced convection problems solved by integral methods turned out to be a power function in x. So, before resorting to a finite difference solution for Eqs. (6.31) and (6.32), let us try a power function solution for both δ and a_1 to see if that form satisfies these equations. If it does not, then finite difference methods will be used. Hence, the following become our candidate solutions. The exponents m and n and the coefficients A and B should be determined such that, if possible, Eqs. (6.31) and (6.32) are satisfied for all x on the vertical plate.

$$\delta = Ax^m \tag{6.34}$$

$$a_1 = Bx^n \tag{6.35}$$

Using Eqs. (6.34) and (6.35) in Eq. (6.31) yields

$$\frac{3}{8}g\beta\rho\Delta T \, Ax^m = \mu \, Bx^n + \frac{B^2A^3(2n+3m)}{105}x^{2n+3m-1}. \tag{6.36}$$

For Eq. (6.36) to be valid for all x on the vertical plate, the exponent on x must be the same in every term. Hence,

$$m = n = 2n + 3m - 1. \tag{6.37}$$

Solution of the two equalities of Eq. (6.37) gives the exponents or powers m and n as

$$m = n = 1/4. \tag{6.38}$$

With Eq. (6.38), the x cancels from Eq. (6.36), and after introduction of $\nu = \mu/\rho$, Eq. (6.36) becomes

$$\frac{3}{8}g\beta\Delta T \, A = \nu \, B + \frac{B^2A^3}{84}. \tag{6.39}$$

At this point, it is not yet guaranteed that the forms given by Eqs. (6.34) and (6.35) are correct, with $m = n = 1/4$, since these forms have not yet been checked to see if they satisfy Eq. (6.32) for all values of x. So using $\delta = Ax^{1/4}$ and $a_1 = Bx^{1/4}$ in Eq. (6.32), one finds that after performing the indicated operations, the x cancels from the equation and rearrangement of Eq. (6.32) gives

$$B = \frac{105}{2} \frac{\alpha}{A^3}. \tag{6.40}$$

Next, A is solved for by eliminating B by combining Eqs. (6.39) and (6.40). Thus,

$$A = \frac{3.44}{\left[\dfrac{g\beta\Delta T}{\nu^2} N_{Pr} \right]^{1/4}} \left[\frac{N_{Pr} + 5/8}{N_{Pr}} \right]^{1/4}. \tag{6.41}$$

If desired, B can now be found by using Eq. (6.41) in Eq. (6.40).

From Chapter 5, the expression for the local surface coefficient of heat transfer, h_x, in terms of the temperature profile within the moving fluid was developed in Eq. (5.3) as

$$h_x = \frac{-k \left(\dfrac{\partial T}{\partial y} \right)_{y=0}}{T_w - T_s}. \tag{6.42}$$

Using Eq. (6.26) in Eq. (6.42) gives

$$h_x = \frac{3}{2} \frac{k}{\delta}. \tag{6.43}$$

The value δ is given by Eq. (6.34) with $m = 1/4$ and A given by Eq. (6.41). Using this in Eq. (6.43) and multiplying both sides of Eq. (6.43) by x/k to form the local Nusselt number gives

$$N_{Nu_x} = \frac{h_x x}{k} = 0.436 \left[\frac{N_{Pr}}{N_{Pr} + 5/8} \right]^{1/4} \left[\frac{g\beta x^3 \Delta T}{\nu^2} N_{Pr} \right]^{1/4}. \tag{6.44}$$

Using the same procedure as was demonstrated in detail in Chapter 5, we can now use Eq. (6.44) to derive an expression for the average surface coefficient of heat transfer, and thus the average Nusselt number over the entire constant temperature vertical plate of height L. This is left as an exercise for the student to do—to show that

$$h = \frac{4}{3} h_{x=L}. \tag{6.45}$$

Equations (6.45) and (6.44) give the expression for the average Nusselt number as

$$N_{\text{Nu}} = \frac{hL}{k} = 0.581 \left[\frac{N_{\text{Pr}}}{N_{\text{Pr}} + 0.625}\right]^{1/4} \left[\frac{g\beta L^3 \Delta T}{\nu^2} N_{\text{Pr}}\right]^{1/4}. \qquad (6.46)$$

The nondimensional group multiplying the Prandtl number in the second bracket of Eq. (6.46) is termed the *Grashof number,* N_{Gr}.

$$N_{\text{Gr}} = g\beta L^3 \Delta T / \nu^2 \qquad (6.47)$$

In Eq. (6.47), g is the local acceleration of gravity, β is the thermal coefficient of volume expansion, L is a characteristic length of the surface, ΔT is the driving temperature difference between the surface and the undisturbed fluid (always taken as positive), and ν is the kinematic viscosity. The Grashof number can be interpreted as the ratio of the net buoyancy forces to the viscous forces and is sometimes referred to as the *free convection Reynolds number.* In an interferometric study of vertical plates freely convecting in air, Eckert and Soehngen [7] found that *transition from laminar to turbulent flow occurs at a value of the Grashof number about 4×10^8.*

The results of the experiments of [7] can be summarized as follows:

Free Convection Flow Criteria for Vertical Plates

Flow is laminar, but not of the thin boundary layer type.	$N_{\text{Gr}} < 10^4$
Flow is laminar and of the thin boundary layer type.	$10^4 < N_{\text{Gr}} < 10^8 \rightarrow 10^9$
Flow is turbulent and of the thin boundary layer type.	$10^8 \rightarrow 10^9 < N_{\text{Gr}}$

These criteria will, in general, be different for inclined or horizontal plates.

In free convection, the *product* of the Grashof and Prandtl numbers appears so often, as it does in the second bracket of Eq. (6.46), that it is called the *Rayleigh number,* with the symbol N_{Ra}.

$$N_{\text{Ra}} = N_{\text{Gr}} N_{\text{Pr}} \qquad (6.48)$$

Using Eqs. (6.48) and (6.47), Eq. (6.46) becomes

$$10^4 < N_{\text{Gr}} < 4 \times 10^8, \quad N_{\text{Nu}} = 0.581 \left[\frac{N_{\text{Pr}}}{N_{\text{Pr}} + 0.625}\right]^{1/4} N_{\text{Ra}}^{1/4}. \qquad (6.49)$$

Thus, in Eq. (6.49), we have an approximate, analytical solution, by the integral method, for the average surface coefficient of heat transfer in steady, laminar, constant property, thin boundary layer–type, free convection over a constant temperature vertical plate.

Comparing Eq. (6.49) with the *exact* analytical solution to the same problem as reported in [5], indicates that Eq. (6.49) is about 3.8% low at $N_{Pr} = 1$ and about 7.5% low at $N_{Pr} = 10$. Once again it is found, as was also the case for integral method solutions to flow and heat transfer problems in Chapter 5, that the integral method has yielded a result of acceptable accuracy for a modest amount of effort invested (compared to an exact solution, by the method of the similarity transformation, *when* it can be found).

Integral methods can also be applied to problems of *turbulent* free convection heat transfer, and solutions of this kind have been developed by a number of authors. Results and details of the procedure are available in Eckert and Jackson [8] and in Kays and Crawford [9].

Consider the solution procedure which led to Eq. (6.49) and compare the work involved in this free convection problem to that needed to arrive at the solution to the equivalent *forced* convection problem, Eq. (5.336). It is clear that the *mutual* coupling of the velocity and temperature fields causes free convection problems to be more complicated than forced convection problems. In addition to this insight provided by the analysis, the final form of the solution to this vertical plate problem, Eq. (6.49), also introduces us to the important nondimensional groups in free convection, the Grashof, Prandtl, and Rayleigh numbers. These same groups will appear both in analytical solutions and in experimental correlations for geometries other than the vertical flat plate.

6.4 WORKING RELATIONSHIPS FOR FREE CONVECTION HEAT TRANSFER

In Secs. 6.1 and 6.2a, we discussed the need to use experimental correlations even more extensively in free convection than was the case, in Chapter 5, for forced convection. This is due to the increased difficulty of finding analytical solutions in free convection because of the mutual coupling of the velocity and temperature fields, and because predictions are somewhat low when dealing with laminar free convection of a gas or other fluid whose Prandtl number is close to 1. Hence, in the present section, experimental correlations for the free convection surface coefficient of heat transfer will be presented for a variety of important geometries and boundary conditions. In addition, some analytical and semi-analytical results, which have been verified to some degree by reasonable comparison with experiments, will also be given.

In using expressions for free convection surface coefficients, perhaps the most striking difference between the relations for free and forced convection surface coefficients of heat transfer is that the free convection surface coefficient depends explicitly upon the driving temperature difference ΔT, usually to the 1/4 power for laminar flow and the 1/3 power for turbulent flow. The reason for the explicit appearance of ΔT in the relations for h is most evident by reference to Eq. (6.15), where it is clear that the net buoyant force per unit volume, which *causes* the motion in free convection, depends upon the temperature differences within the fluid.

6.4a Temperature-Dependent Fluid Properties

In free convection heat transfer, the modification of both analytical and experimental constant property relations for the surface coefficient of heat transfer, to take into approximate account the fact that the fluid properties actually vary with temperature, will ordinarily be done by use of the film temperature. This reference temperature, the film temperature T_f, was discussed in Chapter 5, and its definition will be repeated here for convenience.

$$T_f = (T_w + T_s)/2 \tag{6.50}$$

In Eq. (6.50), T_w is the surface temperature of the body and T_s is the undisturbed fluid temperature outside of the thermal boundary layer. The film temperature is recommended for use in the previously derived analytical relations, Eqs. (6.44) and (6.49), as well as in the correlations to be presented in this section. If temperature-dependent fluid properties are to be evaluated at some reference temperature other than the film temperature, this will be pointed out explicitly in the instructions governing the proper use of the correlating equation for h.

6.4b Calculation of the Rayleigh and Grashof Numbers

As was already mentioned, the various correlations and analytical expressions for the free convection surface coefficient contain the Rayleigh or the Grashof number, and hence one or both of these must be calculated. The Rayleigh number is given by Eq. (6.48) as the product of the Grashof and Prandtl numbers. Hence, substituting Eq. (6.47) for the Grashof number, $\mu c_p/k$ from Chapter 5 for the Prandtl number, and replacing the kinematic viscosity ν by its equivalent, μ/ρ, in Eq. (6.48) yields

$$N_{Ra} = \frac{g\beta\rho^2 L^3 \Delta T}{\mu^2}\left(\frac{\mu c_p}{k}\right) = \left(\frac{g\beta\rho^2 c_p}{\mu k}\right) L^3 \Delta T.$$

Notice that the last expression in parentheses, except for the local acceleration of gravity g, is dependent only upon the properties of the fluid. This term is commonly given the symbol a; that is,

$$a = \frac{g\beta\rho^2 c_p}{\mu k}. \tag{6.51}$$

This factor a, defined by Eq. (6.51), was apparently introduced and first used by King [10]. Thus, in terms of the shorthand symbol a, the expression for the Rayleigh number becomes as follows:

$$N_{Ra} = aL^3 \Delta T. \tag{6.52}$$

Using Eqs. (6.48) and (6.52), an expression for the Grashof number is found in terms of the factor a as

$$N_{Gr} = \frac{aL^3 \Delta T}{N_{Pr}} . \tag{6.53}$$

For a local acceleration of gravity equal to 32.2 ft/sec^2 (4.17 $\times$ 10^8 ft/hr^2) or 9.81 m/s^2 and standard atmospheric pressure, the term a, defined by Eq. (6.51), depends only upon temperature and the type of fluid. Its value is tabulated in Appendix A for both air and water as a function of temperature for the conditions just described. For liquid water at ordinary temperatures and pressures, the properties involved in the value of a are sensibly independent of pressure, and hence the values of a from Appendix A can be used for water at pressures other than 14.7 psia or 1.013 $\times$ 10^5 N/m^2.

6.4c Vertical Plates and Vertical Cylinders

For vertical plates and vertical cylinders whose diameter is not too small compared to their height, McAdams [11] correlated some of the experimental data for air. His recommended curve through the data points (not displayed) is shown in Fig. 6.6. Since the Prandtl number of air is fairly close to a value of 0.7 for a wide range of temperatures, the correlating curve of Fig. 6.6 should provide a reasonable estimate of the free convection surface coefficient for other gases and other fluids if their Prandtl number is fairly close to about 1.0. Equation fits can be made to the curve shown in Fig. 6.6 for various Rayleigh number ranges, and McAdams[1] recommends the use of

$$\frac{hL}{k_f} = N_{Nu} = 0.59 \, N_{Ra_f}^{1/4} \tag{6.54}$$

for $10^4 < N_{Ra_f} < 10^9$, and

$$\frac{hL}{k_f} = N_{Nu} = 0.13 \, N_{Ra_f}^{1/3} \tag{6.55}$$

for $10^9 < N_{Ra_f} < 10^{12}$.

Equation (6.54) corresponds to a laminar, boundary layer–type, free convection flow, while Eq. (6.55) corresponds to turbulent, boundary layer–type, free convection flows. The properties, as is indicated, are to be evaluated at the film temperature T_f, where

---

1. From *Heat Transmission* by McAdams. Copyright © 1954 by William H. McAdams. Used with permission of McGraw-Hill Book Company.

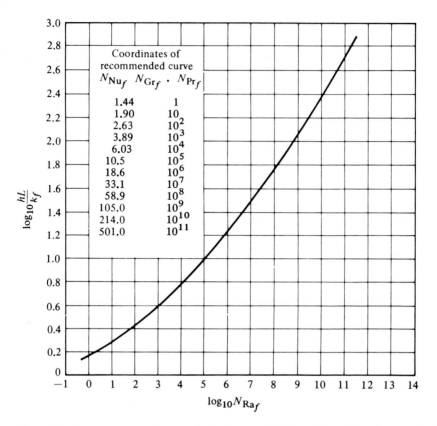

Figure 6.6 Free convection from vertical plates to air. (From *Heat Transmission* by W. H. McAdams. Copyright © 1954 by William H. McAdams. Used with permission of McGraw-Hill Book Company.)

$T_f = \frac{1}{2}(T_w + T_s)$ for a constant temperature surface at T_w immersed in a fluid of undisturbed temperature T_s. Since the Rayleigh number contains the driving temperature difference to the first power, Eq. (6.54) shows that the free convection film coefficient h is directly proportional to $(\Delta T)^{1/4}$, while from Eq. (6.55), for turbulent flow, it is proportional to $(\Delta T)^{1/3}$.

The discussion preceding Eq. (6.54) indicates that Eqs. (6.54) and (6.55) and the correlating curve, Fig. 6.6, are recommended for use on constant temperature vertical plates when the fluid is air or has a Prandtl number close to unity.

Churchill and Chu [12] provide the following correlating equations for the laminar and the turbulent free convection regimes on vertical constant temperature plates. Their equations fit the available exact solutions in the laminar regime and, in addition, correlate some of the available experimental data fairly well in the laminar and the turbulent regime. *They suggest that their equations are valid for all Prandtl numbers from 0 to ∞. It is recommended that the film temperature, Eq. (6.50), be used to account for the temperature dependence of the fluid properties.* These relations are for the *average* Nusselt number over the plate; the plate height L is to be used as the characteristic dimension in both the Nusselt and the Rayleigh numbers.

For $0 < N_{Ra_f} < 10^9$, $0 < N_{Pr_f} < \infty$

$$\frac{hL}{k_f} = N_{Nu} = 0.68 + \frac{0.670 \, N_{Ra_f}^{1/4}}{[1 + (0.492/N_{Pr_f})^{9/16}]^{4/9}} \tag{6.56}$$

for $N_{Ra_f} > 10^9$, $0.6 < N_{Pr_f} < \infty$

$$\frac{hL}{k_f} = N_{Nu} = \frac{0.15 \, N_{Ra_f}^{1/3}}{[1 + (0.492/N_{Pr_f})^{9/16}]^{16/27}} \tag{6.57}$$

for $N_{Ra_f} > 10^9$, $0 < N_{Pr_f} < 0.6$,

$$\frac{hL}{k_f} = N_{Nu} = \left\{ 0.825 + \frac{0.387 \, N_{Ra_f}^{1/6}}{[1 + (0.492/N_{Pr_f})^{9/16}]^{8/27}} \right\}^2 \tag{6.57a}$$

Equation (6.56) is recommended for fluids whose Prandtl number is not too close to unity or to that of air. Equation (6.57) is recommended for high Prandtl number fluids, while Eq. (6.57a) would be used on low Prandtl number fluids, particularly the liquid metals.

As implied earlier, Eqs. (6.54–55) and Fig. 6.6 also apply with good accuracy to vertical cylinders if the diameter D is not too small relative to the height. Sparrow and Gregg [13] derive a criterion which indicates the conditions under which the average Nusselt number, for a vertical cylinder, will not deviate by more than about 5% from the value derived by treating the vertical cylinder, of height L and diameter D, as a vertical flat plate of height L. By assuming that the Prandtl number effect on the Nusselt number can be approximated by $N_{Pr}^{1/4}$, their criterion can be put in the form

---

$$D/L \geq 34/N_{Ra_f}^{1/4} . \tag{6.58}$$

(Note that the Rayleigh number is based on the *height* L of the cylinder.)

If the diameter to height ratio of the vertical cylinder satisfies Eq. (6.58), then at least for laminar flow of fluids with Prandtl numbers near unity, the vertical plate relations can be used to calculate the free convection surface coefficient with an error of about 5%. In the absence of other information, it is suggested that Eq. (6.58) be used as a rough indicator for fluids whose Prandtl number deviates from unity. For vertical cylinders, such as thin wires, which do not satisfy Eq. (6.58), see Kyte et al. [14].

Constant Flux Vertical Plate

Sparrow and Gregg [3] present an analytical solution by the method of the similarity transformation for the Nusselt number in the case of a vertical flat plate of height L, with a constant flux boundary condition, freely convecting to a fluid in a laminar fashion. They also found that if the temperature difference between the fluid and plate at $x = \frac{1}{2} L$

was used as the driving potential difference in Newton's cooling law, the average Nusselt number for an isothermal flat plate could be used with little error. This fact, combined with the following expression from their paper for the surface temperature variation over the plate, permits the approximate solution of the constant flux vertical flat plate in laminar free convection. The surface temperature variation is

$$T_w - T_s = (T_w - T_s)_L (x/L)^{1/5}, \qquad\qquad (6.59)$$

where L is the total height of the plate, $(T_w - T_s)_L$ indicates the difference between the plate temperature and fluid temperature at $x = L$, and $T_w - T_s$ is the temperature difference at any x.

Churchill and Chu [12] also indicate that the expressions for the *average* surface coefficient over a vertical plate with *constant flux* at its surface are essentially the same as for the constant temperature plate—namely, Eqs. (6.56) and (6.57)—as long as the driving potential difference in Newton's cooling law is taken to be

$$\Delta T = T_{w_{x=L/2}} - T_s \qquad\qquad (6.60)$$

as was also recommended by Sparrow and Gregg [3]. The value $T_{w_{x=L/2}}$ means the plate surface temperature, for the constant flux plate, at a point midway between the top and bottom. In using Eqs. (6.56) and (6.57) for the constant flux plate, Churchill and Chu recommend changing the 0.492 to 0.437.

EXAMPLE 6.2

A large tank of water at 20°C is to be heated slightly by a flat plate, 0.3 m high by 0.3 m wide, which is imbedded in the side of the tank. If it is desired to transfer energy to the water at the rate of 3000 W, what should be the surface temperature T_w of the plate?

Solution
The plate must deliver energy at the rate of 3000 W to the water by the process of free convection. Hence, Newton's law of cooling applies and the unknown wall temperature appears within it as shown next.

$$q = hA\Delta T, \text{ or } 3000 = h(0.3 \times 0.3)(T_w - 20).$$

Hence, we have that

$$h(T_w - 20) = 33{,}333. \qquad\qquad (6.61)$$

We can see that the unknown wall temperature appears explicitly in Eq. (6.61) as well as implicitly, since the surface coefficient of heat transfer h depends upon T_w. The value T_w appears directly in the Rayleigh number, Eq. (6.52), and in the Grashof number, Eq.

(6.53), and one or the other appears in all of the relations for the free convective surface coefficient. In addition, the temperature-dependent fluid properties are to be evaluated at T_f, which, by Eq. (6.50), also contains T_w.

Since a vertical plate is being dealt with and a fluid, water, is being used whose Prandtl number is probably not very close to unity, either Eq. (6.56) or (6.57) is the appropriate relation for h. Assume that the Rayleigh number is greater than 10^9, allowing the use of Eq. (6.57). This assumption will be checked once T_w is found using that equation. Solving for h from Eq. (6.57), after making use of Eq. (6.52) for N_{Ra}, noting that the L cancels, and that $\Delta T = T_w - 20$, gives

$$h = \frac{0.15 k_f a_f^{1/3} (T_w - 20)^{1/3}}{[1 + (0.492/N_{Pr_f})^{9/16}]^{16/27}} .$$

Combining this equation with Eq. (6.61) to eliminate h and then solving for $T_w - 20$ by raising both sides of the result to the 0.75 power yields the following:

$$T_w - 20 = \frac{10,235 \, [1 + (0.492/N_{Pr_f})^{9/16}]^{0.444}}{k_f^{0.75} \, a_f^{0.25}} . \tag{6.62}$$

Next, Eq. (6.62) will be solved iteratively for T_w, since T_w appears implicitly on the right side in the properties evaluated at the film temperature T_f. From Eq. (6.50),

$$T_f = (T_w + T_s)/2 = (T_w + 20)/2. \tag{6.63}$$

As a first estimate of T_w, for the purposes of calculating the film temperature, a reasonable guess would seem to be about 60°C. (If T_w were greater than 100°C, this value would lead to boiling on the plate and the necessity to use the relations given in the chapter on boiling and condensation in order to calculate the surface coefficient.)

This first estimate of T_w will be used in Eq. (6.63) to find T_f, which will then lead to k_f, a_f, and N_{Pr_f} from the property tables. We will insert these into Eq. (6.62) and then solve for T_w. If this value differs from our original estimate, the procedure will be repeated until two successive values of T_w agree closely enough.

So, with $T_w = 60°C$, Eq. (6.63) yields $T_f = 40°C$. From the property tables,

$$a_f = 36.9 \times 10^9, \quad k_f = 0.633 \text{ W/m °C}, \quad \text{and} \quad N_{Pr_f} = 4.33.$$

Inserting these values into the right side of Eq. (6.62), T_w is determined to be

$$T_w = 56.9°C,$$

whereas our first estimate of it was 60°C. So now T_f is recomputed using this revised value of T_w, and one arrives at

$$T_f = (56.9 + 20)/2 = 38.45°C.$$

At this value of T_f, interpolation in the property tables for water in Appendix A gives

$$a_f = 34.85 \times 10^9, \quad k_f = 0.631 \text{ W/m °C}, \quad N_{Pr_f} = 4.5.$$

Using these property values in the right side of Eq. (6.62) and solving for the unknown T_w yields

$$T_w = 57.4°C.$$

This value is within a half degree Celsius of the previous iterate, and hence it is judged to be close enough so that additional iteration is not warranted.

Now the Rayleigh number must be checked, since the procedure used is predicated on the validity of Eq. (6.57). Thus, from Eq. (6.52),

$$N_{Ra_f} = a_f L^3 \Delta T = 38.45 \times 10^9 (0.3)^3 (57.4 - 20) = 2.58 \times 10^{10}.$$

This satisfies the requirement of Eq. (6.57), that $N_{Ra} > 10^9$. Hence, the plate surface must be held at $T_w = 57.4°C$ to accomplish the required heating function.

EXAMPLE 6.3

After a compressor is shut off, a compressed air storage tank holds air at 100 psia and 300°F while the tank side wall, which is a cylinder 3 feet in diameter by 5 feet high, is at 100°F. Compute the heat transfer rate to the wall at this instant of time, neglecting any enclosure effect on the surface coefficient of heat transfer.

Solution
By Newton's cooling law,

$$q = hA\Delta T = h\pi(3)(5)[300 - 100] = 9425h. \tag{6.64}$$

If the enclosure effect can be neglected, then, since we have air as the fluid, the vertical plate relations, Eq. (6.54) or (6.55), will apply provided that the inequality Eq. (6.58) is satisfied. This will be checked next by computation of the Rayleigh number based on the cylinder height of 5 feet.

Now, $T_f = \frac{1}{2}(300 + 100) = 200°F$, and from the property tables for air at 200°F, $k_f = 0.0181$ and $a_f = 0.59 \times 10^6$. But a_f is for a pressure of 14.7 psia, not 100 psia. To determine how this value can be modified for the effect of a pressure other than 14.7 psia, examine the definition of a,

$$a = \frac{g\beta\rho^2 c_p}{\mu k}, \tag{6.51}$$

where g, β, c_p, μ, and k are independent of pressure for air in the normal range of temperatures and pressures. The density ρ, however, depends upon the pressure P through the ideal gas equation of state, $\rho = P/RT$. The value of a tabulated in Appendix A is for standard atmospheric pressure 14.7 psia, and therefore the value of ρ used in a is ρ_0 and the standard atmospheric pressure is P_0. By the equation of state, the air density at any pressure P and film temperature T_f is related to the density ρ_0 at pressure P_0 and temperature T_0 by the equation

$$\rho = \rho_0(P/P_0)(T_0/T_f).$$

Using this for ρ in Eq. (6.51) gives, at any temperature and pressure,

$$a = \left(\frac{P}{P_0}\right)^2 \left(\frac{g\beta\rho_0^2 c_p}{\mu k}\right)\left(\frac{T_0}{T_f}\right)^2.$$

However, a was tabulated as a function of temperature; thus, $(T_0/T_f)^2$ is unity. The second factor is the value of a tabulated at 14.7 psia, while the first ratio takes into account the effect of a different pressure. *Hence, if P is in psia, the expression a_p for a at any pressure P becomes*

$$a_p = a_f\left(\frac{P}{14.7}\right)^2.$$

Thus, in this problem,

$$a_p = 0.59 \times 10^6 (100/14.7)^2 = 27.3 \times 10^6.$$

The Rayleigh number is found from Eq. (6.52) to be

$$N_{\mathrm{Ra}_f} = 27.3 \times 10^6 (5)^3 (300 - 100) = 6.83 \times 10^{11}.$$

This value of Rayleigh number indicates that the natural convection flow on the vertical plate is turbulent, whereas the criterion, Eq. (6.58), was derived for the case of laminar flow. However, in the absence of a corresponding inequality for turbulent flow, we will still use Eq. (6.58) to get some idea whether we are justified in treating this vertical cylinder as an equivalent vertical plate. Hence,

$$34/N_{\mathrm{Ra}_f}^{1/4} = 34/(6.83 \times 10^{11})^{1/4} = 0.0374.$$

The actual D/L is given by

$$D/L = 3/5 = 0.6.$$

Since $0.6 >> 0.0374$, the inequality, Eq. (6.58), is easily satisfied and it is assumed that this carries over to our turbulent case as well.

The value of the Rayleigh number dictates the use of Eq. (6.55), so that

$$\frac{h(5)}{0.0181} = 0.13[6.83 \times 10^{11}]^{1/3}, \quad \text{or} \quad h = 4.14 \text{ Btu/hr ft}^2 \text{ °F.}$$

Finally, using this value of h in Eq. (6.64) yields

$$q = 9425(4.14) = 39{,}020 \text{ Btu/hr.}$$

In this problem, note that since the air is not at standard atmospheric pressure, the value of a from the property tables, in order to be used to compute the Rayleigh number, must be modified for the strong dependence of a upon the pressure. In addition, it was found that the natural convective flow was turbulent, yet the criterion, Eq. (6.58), for deciding whether or not the vertical cylinder can be treated as a vertical plate, applies to laminar flows. But, since the actual D/L of 0.6 turned out to be so much greater than the criterion, Eq. (6.58), would require for a laminar flow, it seems reasonable to assume that the conclusion holds for turbulent flow as well. If the actual D/L had come out to be closer to the value given by the criterion, or even less than that value, then we would have had to refer to the work on free convection on thin vertical cylinders by Kyte [14].

6.4d Horizontal and Inclined Planes

In dealing with horizontal planes, or inclined planes (other than vertical) such as that depicted in Fig. 6.4, there are two basically different cases in regard to orientation of a hot surface or a cold surface. These two cases are (1) hot surface facing up or, equivalently, cold surface facing down; and (2) hot surface facing down or, equivalently, cold surface facing up. In case 1, the orientation is such that the convection currents are relatively unimpeded by the surface, since they are moving away from the surface. In case 2 the convection currents tend to be toward the surface, and hence the presence of the surface impedes them and gives rise to *lower* surface coefficients of heat transfer than are found in case 1. This is perhaps most easily seen in the case of a horizontal surface which, if the hot surface is facing up, has a freely upward flow of natural convection currents away from the surface carrying off the energy picked up at the surface. On the other hand, if the hot surface of the horizontal plate is facing down, the free convection currents which are trying to rise cannot do so because the plate blocks their motion. Hence, only the spilling away along the edges of the plate prevents a completely stagnant, thick layer of fluid from accumulating along the bottom of the plate.

As a consequence of these differences, there will be separate experimental correlations, when available, for the two different cases discussed.

Horizontal Square Plates

Experimental information on free convection from square constant temperature horizontal plates with hot side facing up, or cold side facing down, is available in a number of sources, including Fishenden and Saunders [15], Fujii and Imura [16], Goldstein, Sparrow, and Jones [17], Lloyd and Moran [18], Al-Arabi and El-Riedy [19], and Yousef, Tarasuk, and McKeen [20]. From an examination of these works, the following two expressions are recommended for the average surface coefficient of heat transfer.

Constant Temperature Surface with Hot Side Up or Cold Side Down

$$\frac{hL}{k} = N_{Nu} = 0.622 \, N_{Ra}^{1/4} \quad 10^4 < N_{Ra} < 10^8 \tag{6.65}$$

$$\frac{hL}{k} = N_{Nu} = 0.16 \, N_{Ra}^{1/3} \quad 10^8 < N_{Ra} < 10^{11} \tag{6.66}$$

Constant Temperature Surface with Hot Side Down or Cold Side Up

Fishenden and Saunders [15] recommend

$$\frac{hL}{k} = N_{Nu} = 0.25 \, N_{Ra}^{1/4} \quad 10^4 < N_{Ra} < 10^9 \tag{6.67}$$

Fujii and Imura [16] recommend

$$\frac{hL}{k} = N_{Nu} = 0.58 \, N_{Ra}^{1/5} \quad 10^6 < N_{Ra} < 10^{11} \tag{6.68}$$

In these four experimental correlations, the side length L of the square plate is to be used in both the Rayleigh and the Nusselt numbers, and all temperature-dependent fluid properties are to be evaluated at the film temperature.

Inspection of the Rayleigh number range of validity of Eqs. (6.67) and (6.68) indicates an overlap between Rayleigh numbers of 10^6 and 10^9, with Eq. (6.68) being about 10% higher than Eq. (6.67) at $N_{Ra} = 10^6$ and about 20% lower at $N_{Ra} = 10^9$. The experimental data which are correlated by Eq. (6.67) are for air, $N_{Pr} \approx 0.7$, while the data which led to Eq. (6.68) are for water in the Prandtl number range from about 1 to 10. This may serve as a basis for selection of one correlation or the other in the range of overlap of Rayleigh numbers.

For the most part, the experimental data upon which the correlations, Eqs. (6.65) through (6.67), are based has been taken using air as the fluid. However, some of the data correlated has been for water in a Prandtl number range of 1 to 10, and the techniques used in [17] and [18] correspond to Prandtl numbers of 2.5 and 2200, respectively. Hence, it would seem as if the aforementioned equations could be used on fluids whose Prandtl numbers are greater than about 0.6.

Other Horizontal Shapes

Lloyd and Moran [18] performed experiments with horizontal surfaces in the shape of squares, rectangles, circles, and right triangles, while Goldstein, Sparrow, and Jones worked with horizontal squares, rectangles, and circles. Both experimental investigations found that the results for the different shapes were correlated well by relations which used the following parameter, L^*, as the characteristic length in both the Rayleigh and Nusselt numbers:

$$L^* = A/P. \tag{6.69}$$

The value A is the surface area of one face of the horizontal shape, while P is the perimeter of that shape. Lloyd and Moran recommend the following experimental correlations:

Constant Temperature Horizontal Shape with Hot Surface Up or Cold Surface Down

$$\frac{hL^*}{k} = N^*_{Nu} = 0.54\, N^{1/4}_{Ra^*} \tag{6.70}$$

for $2.6 \times 10^4 < N_{Ra^*} < 8 \times 10^6$.

$$\frac{hL^*}{k} = N^*_{Nu} = 0.15\, N^{1/3}_{Ra^*} \tag{6.71}$$

for $8 \times 10^6 < N_{Ra^*} < 1.5 \times 10^9$.

Regarding the case of the hot side of the horizontal shape facing down, or its cold side facing up, Eqs. (6.67) and (6.68) are tentatively recommended, using L^* from Eq. (6.69) instead of L. It is recommended that temperature-dependent fluid properties be evaluated at the film temperature in these relations.

Inclined Planes

Discussed here will be the recommended correlations for the surface coefficient of heat transfer for a plate inclined at an angle θ to the *vertical* ($\theta = 0$ for a vertical plate) as was shown earlier in Fig. 6.4.

An additional complicating factor for the inclined planes, which is not shared by the vertical surface, occurs in the phenomenon of transition from laminar flow to turbulent flow. For inclined planes with hot side facing down or cold side facing up, the local Grashof number and local Rayleigh number for transition from laminar to turbulent flow increase. This is due to the stabilizing effect of the orientation of the net buoyant force, which now has a component tending to force the boundary layer fluid against the surface and a smaller component tending to drive the flow along the surface. On the other hand, if the inclined surface has its hot side facing up or its cold side down, the net buoyant force now causes transition to turbulent flow at a lower value of critical Grashof and Rayleigh numbers. In this case, the nonzero (for inclination angles other than $\theta = 0$) component of the buoyant force normal to the surface is tending to cause the fluid to lift away from the surface in a type of flow separation. When this does occur on the plate surface, experiment indicates that the flow will be turbulent beyond that point.

Constant Temperature Inclined Surfaces with Hot Side Down or Cold Side Up   Evidence from Vliet [21], Fujii and Imura [16], and Fussey and Warneford [22] indicates that, for angles of inclination up to 86° from the vertical, and for laminar flow, the *vertical plate relations,* Eqs. (6.54) and (6.56), *may be used as long as the Rayleigh number is multiplied by cos θ.* Thus, for example, modification of the vertical plate relation, Eq. (6.54), leads to the following:

$$\frac{hL}{k_f} = 0.59 \left[N_{\text{Ra}_f} \cos \theta \right]^{1/4} \tag{6.72}$$

for $10^4 < N_{\text{Ra}_f} \cos \theta < 10^9$.

A similar modification would be made to Eq. (6.56) for the case of fluids with Prandtl numbers not close enough to that of air.

This placement of the cos θ next to the Rayleigh number is equivalent to using the component of the net buoyant force along the plate as the driving force for free convection, as was discussed in some detail in Ex. 6.1.

For turbulent flow, Churchill and Chu [12], after analyzing the results of Vliet [21], find that the vertical plate relations, Eqs. (6.55) and (6.57), may be used *unmodified* for $N_{\text{Ra}_f} \cos \theta > 10^9$.

Constant Temperature Inclined Surfaces with Cold Side Down or Hot Side Up   Fujii and Imura [16] indicate that, for $60° < \theta \leq 90°$, the flow will usually be turbulent and the horizontal plate relation, Eq. (6.66), can be used as it stands. For other inclination angles, they combine laminar relations and turbulent relations to give a number of different experimental correlations, and these are available in [16].

Constant Flux Surfaces

For an inclined surface with a constant flux q_w'' maintained at the surface, the usual problem is to predict the surface temperature distribution along the plate and, particularly, the maximum surface temperature. Hence, presented here will be the *local* Nusselt number expressions, since the local surface temperature, $T_w(x)$, is connected to the local Nusselt number through Newton's cooling law as follows:

$$q_w'' = h_x(T_w(x) - T_s). \tag{6.73}$$

Since the flux q_w'' is known, $T_w(x)$ can be solved for from Eq. (6.73), as soon as h_x is known as a function of x.

A modified Rayleigh number is used in the experimental correlations for the constant flux surfaces. It is defined as

$$N_{\text{Ra}_{mx}} = \frac{g\beta x^4 q_w''}{k\alpha\nu}. \tag{6.74}$$

Constant Flux Inclined Surface with Hot Side Down   Fussey and Warneford [22] recommend the following results:

$$\frac{h_x x}{k} = N_{\text{Nu}_x} = 0.592 \left[N_{\text{Ra}_{mx}} \cos \theta \right]^{1/5} \tag{6.75}$$

for $0 \le \theta \le 86.5°$ and $N_{\text{Ra}_{mx}} < N_{\text{Ra}_c}$ where

$$N_{\text{Ra}_c} = 6.31 \times 10^{12} \, e^{0.0705\theta}, \tag{6.76}$$

$$\frac{h_x x}{k} = N_{\text{Nu}_x} = 0.889 \left[N_{\text{Ra}_{mx}} \cos \theta \right]^{0.205} \tag{6.77}$$

for $0 < \theta < 31.7°$ and $N_{\text{Ra}_{mx}} > N_{\text{Ra}_c}$.

Constant Flux Inclined Surface with Hot Side Up   The recommended relations here come from Vliet [21] and from Vliet and Liu [23].

$$\frac{h_x x}{k} = 0.6 \left[N_{\text{Ra}_{mx}} \cos \theta \right]^{1/5} \tag{6.78}$$

for $0 < \theta < 60°$ and $N_{\text{Ra}_{mx}} < N_{\text{Ra}_{mt}}$, where

$$N_{\text{Ra}_{mt}} = 0.3 \times 10^7 \, e^{0.18(90 - \theta)}, \tag{6.79}$$

$$\frac{h_x x}{k} = 0.568 \, N_{\text{Ra}_{mx}}^{0.22} \tag{6.80}$$

for $0 < \theta < 60°$ and $N_{\text{Ra}_{mx}} > N_{\text{Ra}_{mt}}$.

The influence of temperature-dependent fluid properties can be approximated by use of the film temperature for property evaluation in these constant flux relations.

Note also that many of these constant flux relations can also be used for a vertical plate, $\theta = 0$. Hence, they serve to complement the constant temperature vertical plate results given previously.

6.4e Horizontal Cylinders

For free convection between a constant temperature horizontal cylinder and a fluid, Morgan [24] recommends the following relation which, according to Fand, Morris, and Lum [25], agrees well with experimental data for air and water in the Prandtl number range $0.69 < N_{\text{Pr}} < 7$:

$$N_{\text{Nu}} = \frac{hD}{k_f} = B \, N_{\text{Ra}_{D_f}}^m \tag{6.81}$$

where B and m are given as functions of the Rayleigh number range in Table 6.1.

Table 6.1
The Values B and m as Functions of the Rayleigh Number Range

Rayleigh Number (N_{Ra_D})Range	B	m
10^{-10}–10^{-2}	0.675	0.058
10^{-2}–10^{2}	1.02	0.148
10^{2}–10^{4}	0.85	0.188
10^{4}–10^{7}	0.48	0.250
10^{7}–10^{12}	0.125	0.333

Using the values of B and m from Table 6.1 in Eq. (6.81) for the Rayleigh numbers most often encountered leads to the following two relations:

$$10^4 < N_{Ra_{D_f}} < 10^7, \quad \frac{hD}{k_f} = 0.48\, N_{Ra_{D_f}}^{0.25} \tag{6.82}$$

$$10^7 < N_{Ra_{D_f}} < 10^{12}, \quad \frac{hD}{k_f} = 0.125\, N_{Ra_{D_f}}^{0.333} \tag{6.83}$$

In Eqs. (6.81) to (6.83), all temperature-dependent fluid properties are to be evaluated at the film temperature, both the Nusselt and Rayleigh numbers use the cylinder diameter D as the characteristic dimension, and the Prandtl number range in which these correlations are recommended for use is $0.69 < N_{Pr} < 7$.

For the higher Prandtl numbers associated with oils, the following relation given by Fand, Morris, and Lum [25] is recommended. Actually they present two other expressions which correlate the data better. These use reference temperatures other than the film temperature to account for temperature-dependent fluid properties. However, it is felt that the one presented here, which uses the film temperature for property evaluation, is adequate for the high Prandtl number fluids.

$$\frac{hD}{k_f} = 0.474\, N_{Ra_{D_f}}^{0.25}\, N_{Pr_f}^{0.047} \tag{6.84}$$

for $0.7 \le N_{Pr} \le 3000$, $3 \times 10^2 \le N_{Ra_{D_f}} \le 2 \times 10^7$.

Churchill and Chu [26] recommend the next two correlations, Eqs. (6.85–86), for laminar and turbulent free convection, respectively, over the complete range of Prandtl numbers, $0 \le N_{Pr} \le \infty$.

$$\frac{hD}{k_f} = 0.36 + 0.518 \left\{ \frac{N_{Ra_{D_f}}}{[1 + (0.559/N_{Pr_f})^{9/16}]^{16/9}} \right\}^{1/4} \tag{6.85}$$

for $10^{-6} < N_{Ra_{D_f}} < 10^9$.

Equation (6.85) is particularly recommended for the low Prandtl number fluids, such as the liquid metals, rather than Eq. (6.81) or (6.84).

For turbulent natural convection over the horizontal cylinder, Churchill and Chu present the following:

$$\frac{hD}{k_f} = \left\{ 0.60 + 0.387 \left(\frac{N_{\mathrm{Ra}_{D_f}}}{[1 + (0.559/N_{\mathrm{Pr}_f})^{9/16}]^{16/9}} \right)^{1/6} \right\}^2 \qquad (6.86)$$

for $N_{\mathrm{Ra}_{D_f}} > 10^9$.

It is recommended that Eq. (6.86) be used for the very low Prandtl number fluids and also for the very high Prandtl number fluids when the Rayleigh number range of validity of Eq. (6.84) is exceeded.

Based on the limited evidence available, Churchill and Chu recommend that these equations also be used for the constant flux boundary with the surface temperature, T_w, being the temperature halfway up the cylinder at the 90° angle from the bottom.

EXAMPLE 6.4

A steam pipe, 6 in. in diameter and 20 ft long, at 130°F runs through a room where the air and the other surfaces are at 70°F. If the emissivity of the pipe surface is 0.85, calculate the heat transfer loss from the pipe.

Solution
The total heat transfer rate from the pipe surface is the sum of the free convection loss and the net radiant loss. The net radiant loss for this situation is given by Eq. (4.88). Now, $A = \pi DL = \pi(\frac{1}{2})(20) = 31.4$ ft², $T_w = 130 + 460 = 590°R$, and $T_s = 70 + 460 = 530°R$. Thus, substituting known values into Eq. (4.88) yields

$$q_r = 0.85(31.4)(0.1714)[(5.9)^4 - (5.3)^4] = 1913 \text{ Btu/hr.}$$

By Newton's cooling law, the free convection loss is

$$q_c = h(31.4)(130 - 70) = 1883h. \qquad (6.87)$$

Since the fluid being dealt with here is air, the recommended expression for the surface coefficient between it and the horizontal cylinder is Eq. (6.81). Next, the value of the Rayleigh number must be determined.

The film temperature and air properties at the film temperature are $T_f = \frac{1}{2}(130 + 70) = 100°F$, $a_f = 1.26 \times 10^6$, and $k_f = 0.0157$. The Rayleigh number is calculated using Eq. (6.52) as $N_{\mathrm{Ra}_f} = 1.26 \times 10^6(\frac{1}{2})^3(130 - 70) = 9.45 \times 10^6$. This Rayleigh number permits the use of Eq. (6.82). Substituting known values,

$$\frac{h(\frac{1}{2})}{0.0157} = 0.48(9.45 \times 10^6)^{0.25}$$

or

$$h = 0.836 \text{ Btu/hr ft}^2 \text{ °F.}$$

Substituting this into Eq. (6.87) yields

$$q_c = 1883(0.836) = 1574 \text{ Btu/hr.}$$

Hence the total heat transfer rate from the pipe is given by

$$q_c + q_r = 1574 + 1913 = 3487 \text{ Btu/hr.}$$

Note how close the value of h is to 1.0 Btu/hr ft² °F. This calculated value and numerous others give evidence which leads to an observation made earlier in the text, that *when objects of ordinary size and shape are free convecting to air at one atmosphere with ordinary temperature differences between the object surface and the air, the surface coefficient of heat transfer usually has a value in the vicinity of 1.0 Btu/hr ft² °F, or about 6 W/m² °C.*

In addition, this problem indicates a net radiant loss and a free convective loss to air which are of about the same size. As pointed out in Chapter 4, this is generally true in most problems involving free convection in air. Hence, because of its relative size, the net radiant loss must always be computed if the total heat transfer rate, rather than just the convective heat transfer rate, is desired from the surface of an object which is free convecting in air.

EXAMPLE 6.5

An electric resistive element in a water heater is a circular cylinder of diameter 0.03 m and length 0.65 m mounted in a horizontal position. If the heater element size is 3000 W and is to operate in 50°C water, calculate the surface temperature T_w of the resistive element.

Solution
By the energy balance, Eq. (1.8), the electrical energy dissipation rate from the element must be balanced off by convective heat transfer to the water when steady-state conditions prevail. Hence, using this fact and Newton's cooling law yields the following equation:

$$3000 = h\pi DL(T_w - T_s). \tag{6.88}$$

In this situation, the electrical energy dissipation may lead to an essentially constant flux condition at the element surface. But, according to the discussion just following Eq. (6.86), the expressions for the surface coefficient for an isothermal cylinder may be used as a good approximation as long as T_w is interpreted to be the temperature at the midpoint of the cylinder. Therefore, for this situation, Eq. (6.81) will be used for h.

Thus,

$$h = \frac{k_f}{D} B N_{Ra_{D_f}}^m$$

where

$$N_{Ra_{D_f}} = a_f D^3 (T_w - T_s).$$

Inserting these two equations into Eq. (6.88) gives

$$3000 = \pi L k_f B [a_f D^3]^m (T_w - T_s)^{m+1}. \tag{6.89}$$

The values B and m in Eq. (6.89) depend upon the unknown wall temperature T_w because of their dependence upon the Rayleigh number as given by Table 6.1. In addition, the film temperature T_f depends upon the unknown wall temperature and is needed to evaluate the fluid properties k_f and a_f. Thus an iterative solution of Eq. (6.89) is dictated. One possible iterative procedure is as follows: Estimate an initial value of T_w and use it to evaluate k_f, a_f, and $N_{Ra_{D_f}}$. This allows a selection of B and m from Table 6.1, and these values are now used in Eq. (6.89) to solve for T_w. This value of T_w is taken as the new estimate and the procedure continued until two successive iterates yield values of T_w close enough to one another.

As a first estimate, $70°C = T_w$ was chosen. Hence,

$$T_f = \frac{T_w + T_s}{2} = \frac{70 + 50}{2} = 60°C.$$

From the property tables for water at $T_f = 60°C$, $k_f = 0.654$ W/m °C, $a_f = 67.6 \times 10^9$, and $N_{Pr_f} = 3.04$. This value of the Prandtl number, since it is between 0.69 and 7.0, indicates that Eq. (6.81) is the relation to be used as recommended in the presentation and discussion of the various correlations. Also, since $D = 0.03$ m,

$$N_{Ra_{D_f}} = 67.6 \times 10^9 (0.03)^3 (70 - 50) = 3.65 \times 10^7.$$

With this value of Rayleigh number, the values of m and B are found from Table 6.1 to be

$$m = 0.333 \text{ and } B = 0.125.$$

Inserting these values, as well as $L = 0.65$ m, into Eq. (6.89) yields

$$3000 = \pi(0.65)(0.654)(0.125)[67.6 \times 10^9 (0.03)^3]^{0.333} \times (T_w - 50)^{1.333}.$$

Solving for T_w gives

$$T_w = 92.4°C.$$

Chapter Six

Since this value of wall temperature is significantly different from the estimate of 70°C used to arrive at m and B as well as T_f, the entire procedure is now repeated using $T_w = 92.4$°C.

With this, $T_f = 71$°C, or rounding to 70°C for T_f, one gets $k_f = 0.664$, $a_f = 85.4 \times 10^9$, and $N_{Pr_f} = 2.57$. With these, $N_{Ra_{D_f}} = 9.78 \times 10^7$. This value of Rayleigh number indicates the use of the same values of B and m as before, namely, $m = 0.333$ and $B = 0.125$. These new estimates of k_f, a_f, B, and m, when inserted into Eq. (6.89) yield the wall temperature, T_w, as the following:

$$T_w = 89.6°C.$$

Since this value of wall temperature yields virtually the same film temperature $T_f = 70$°C as was used to calculate it and also does not change the Rayleigh number enough to cause any change in the values of m and B, the iterative solution is complete with the resistive element's wall temperature being about 90°C.

If the calculation procedure had yielded a wall temperature significantly above the saturation temperature of the water at the pressure level in the water heater, we would have had to resort to the correlations presented in one of the later chapters because of the boiling that would be occurring locally around the heating element. That should not be the case here, since even at standard atmospheric pressure, T_w would have to exceed 100°C, and in all likelihood, the pressure level, and therefore also the saturation temperature, is higher than one standard atmosphere in a water delivery system.

6.4f Spheres

Yuge [27] presents the following correlation for free convection between spheres and air with the sphere diameter used as the characteristic dimension, the properties evaluated at the film temperature, and the Nusselt number being an average over the sphere surface:

$$N_{Nu} = 2 + 0.392 N_{Gr_f}^{0.25} \tag{6.90}$$

for $1 < N_{Gr_f} < 10^5$.

This relation can be extended to other Prandtl numbers by assuming that the Prandtl number dependence will enter into the second term on the right of Eq. (6.90) as $N_{Pr_f}^{0.25}$. Multiplying this term by $N_{Pr_f}^{0.25}/(0.72)^{0.25}$, where 0.72 has been used as an average Prandtl number for air, leads to

$$N_{Nu} = 2 + 0.426 N_{Ra_f}^{0.25} \tag{6.91}$$

for $1 < N_{Ra_f} < 10^5$.

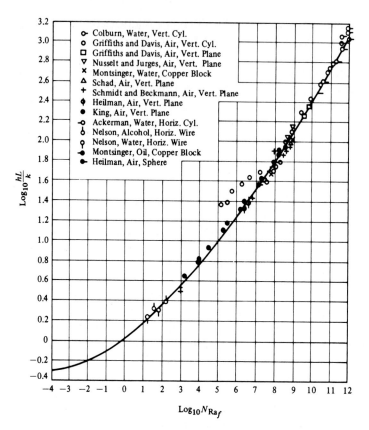

Figure 6.7 Correlation curve for free convection between various shapes and fluids. (From "The Basic Laws and Data of Heat Transmission III—Free Convection" by W. J. King. *Mechanical Engineering*, Vol. 54 [1932], p. 350.)

6.4g Additional Shapes

King [10] based the curve of Fig. 6.7 on experimental data for free convection from vertical plates, vertical cylinders, horizontal cylinders, spheres, and blocks to various fluids such as air, water, alcohol, and oil. The Nusselt number is the average over the surface, the temperature-dependent fluid properties are evaluated at the film temperature, and L is the characteristic linear dimension. Thus, for a vertical plate, L is the height, while for a long horizontal cylinder, L is the cylinder diameter. For situations in which it is suspected that both a vertical and a horizontal dimension influence the value of the surface coefficient, such as in the case, perhaps, of a short solid vertical cylinder, King recommends the following formula for estimating a characteristic length L:

$$\frac{1}{L} = \frac{1}{L_v} + \frac{1}{L_H}, \tag{6.92}$$

where L_v is a significant vertical dimension, such as the height of a short solid vertical cylinder, and L_H is a significant horizontal dimension, such as the diameter of a short solid vertical cylinder.

However, experimental work by Sparrow and Ansari [41], who used a vertical circular cylinder whose height equalled its diameter, does *not* support the choice of length L recommended by King and given by Eq. (6.92). King's length L from Eq. (6.92), when used in conjunction with his correlating curve in Fig. 6.7, leads to an overprediction of the surface coefficient by 40 to 50% for the finite vertical cylinder described here. Thus, the conclusion from [41] is that *Eq. (6.92) does not do an acceptable job in attempting to collapse free convection data from multidimensional bodies onto the correlating curve of King given in Fig. 6.7.* Hence, with this in mind, Eq. (6.92) will be used, in this text, only as a very rough estimate for multidimensional bodies when specific correlations for the shapes are not available.

If Fig. 6.7 is used for spheres, King recommends that the characteristic dimension to be used in both the Nusselt and Rayleigh numbers should be the *sphere radius,* since this allows the sphere data to be closer to the correlating curve of Fig. 6.7 than does the use of the diameter.

Figure 6.7 is expected to yield a fair estimate of the free convection coefficient for objects whose shape is unlike those of the horizontal cylinder and plate. In addition, Fig. 6.7 may also be used for the more common shapes when the Rayleigh number involved lies outside the range of the specific correlation for that shape.

EXAMPLE 6.6

A 6-in.-diameter sphere at $700°F$ is cooling in $100°F$ air after heat treatment. Estimate the free convection coefficient between the sphere and the air.

Solution

The film temperature is $T_f = \frac{1}{2}(700 + 100) = 400°F$. Hence, the property table for air in Appendix A gives $a_f = 0.18 \times 10^6$ and $k_f = 0.0225$. From Eq. (6.52), $N_{Ra_f} = 0.18 \times 10^6(\frac{1}{2})^3(700 - 100) = 13.5 \times 10^6$. Eq. (6.53) yields the Grashof number, after we note from Appendix A that $N_{Pr_f} = 0.681$: $N_{Gr_f} = (13.5 \times 10^6/0.681) = 1.98 \times 10^7$. This Grashof number is outside the range of the correlation for spheres freely convecting in air, Eq. (6.90). However, the curve in Fig. 6.7 represents various shapes, including spheres, *if L is taken to be the sphere radius.* Since the radius is one-half the diameter and appears cubed in the expression for the Rayleigh number, Eq. (6.52), the Rayleigh number based on the radius is $(\frac{1}{2})^3$ times that based on the diameter, or $N_{Ra} = 13.5 \times 10^6(\frac{1}{2})^3 = 1.69 \times 10^6$ and $\log_{10}(1.69 \times 10^6) = 6.228$. Using this value of the abscissa in Fig. 6.7 yields

$$\log_{10} N_{Nu} = 1.30, \quad \text{or} \quad N_{Nu} = \frac{hR}{k_f} = 19.9 = \frac{h(\frac{1}{4})}{0.0225},$$

$$\text{or} \quad h = 1.79 \text{ Btu/hr ft}^2 \text{ °F}.$$

EXAMPLE 6.7

A long 10-cm-high by 5-cm-wide wooden beam runs horizontally across a factory work space. If a fire in an adjacent room raises the temperature of the air around the beam to 1100°K, estimate the surface coefficient of heat transfer between the air and the 300°K beam.

Solution

There is no single correlation for this particular shape, since both the 10-cm height, L_v = 0.1 m, and the 5-cm width, L_H = 0.05 m, would seem to be important in determining the value of the convection coefficient h. Hence, the curve of Fig. 6.7, which is a result of tests on many different shapes, will be used with the characteristic dimension computed from Eq. (6.92) as a rough estimate in the absence of other data. Hence,

$$\frac{1}{L} = \frac{1}{L_v} + \frac{1}{L_H} = \frac{1}{0.1} + \frac{1}{0.05}$$

or, L = 0.0333 m.

Now, T_f = (300 + 1100)/2 = 700°K, and therefore the needed properties of air at this temperature are found from Appendix A as

$$k_f = 0.052 \text{ W/m °K}, \quad a_f = 2.188 \times 10^6,$$

$$N_{Ra_f} = a_f L^3 \Delta T = 2.188 \times 10^6 (0.0333)^3 (1100 - 300) = 64{,}635.$$

Thus, $\log_{10}(64{,}635) = 4.81$.

Using this as the abscissa value in Fig. 6.7 yields

$$\log_{10} N_{Nu} = 0.95, \quad \text{or} \quad \frac{hL}{k_f} = 8.91 = \frac{h(0.0333)}{0.052}.$$

Thus, h = 13.92 W/m² °K.

This value of the surface coefficient might be used in a solution, utilizing Heisler's charts from Chapter 3, to estimate how long it will take before this wooden beam ignites or how long it will take before its strength deteriorates appreciably due to high temperatures in the beam.

6.4h Free Convection in Fluids Confined in Enclosures by Horizontal and Vertical Plates

Consider a fluid confined between two large horizontal plates spaced a distance b apart. If the lower plate has a lower surface temperature than the upper plate, the process is one of pure conduction across the fluid layer, neglecting end effects, and, *if the driving temperature difference in Newton's cooling law is taken to be the difference in the plate temperatures,* it can be shown that the Nusselt number for this situation is unity; that is,

$$hb/k = 1. \tag{6.93}$$

For air confined between horizontal plates with the lower plate hotter than the upper plate, Jakob [28] presents the following correlations for the average Nusselt number, where the characteristic dimension is the plate spacing b, the properties are evaluated at the average of the two plate temperatures, and *the driving temperature difference in Newton's cooling law is the difference between the surface temperatures of the plates:*

$$N_{\text{Nu}} = 0.195 \, N_{\text{Gr}}^{1/4} \tag{6.94}$$

for $10^4 < N_{\text{Gr}} < 3.7 \times 10^5$, and

$$N_{\text{Nu}} = 0.068 \, N_{\text{Gr}}^{1/3} \tag{6.95}$$

for $3.7 \times 10^5 < N_{\text{Gr}} < 10^7$.

For liquids confined between two parallel horizontal plates with the hotter plate below, Globe and Dropkin [29] determined the following correlation for the data from their experiments in which they used water, silicone oils, and mercury:

$$N_{\text{Nu}} = 0.069 \, N_{\text{Ra}}^{1/3} \, N_{\text{Pr}}^{0.074} \tag{6.96}$$

for $1.5 \times 10^5 < N_{\text{Ra}} < 10^9$. In using Eq. (6.96), the properties are evaluated at the average of the two plate temperatures, the temperature difference to be used in Newton's cooling law is the difference between the plate temperatures, and the characteristic dimension is the plate spacing b.

For vertical air spaces between two plates spaced a distance b apart and of height L in the vertical direction, Jakob [28] has correlated the results of other investigators. For Grashof numbers based on the plate spacing b somewhat less than 2000, the process is pure conduction and the Nusselt number based on the temperature difference between the plates is

$$hb/k = 1.0 \tag{6.97}$$

for $N_{\text{Gr}} < 2000$,

$$N_{\text{Nu}} = \frac{0.18 \, N_{\text{Gr}}^{1/4}}{(L/b)^{1/9}} \tag{6.98}$$

for $2 \times 10^4 < N_{Gr} < 2 \times 10^5$, and

$$N_{Nu} = \frac{0.065 \, N_{Gr}^{1/3}}{(L/b)^{1/9}} \qquad (6.99)$$

for $2 \times 10^5 < N_{Gr} < 10^7$. In using Eqs. (6.97–99), the Nusselt number and Grashof number are formed using the plate spacing b as the characteristic dimension, the fluid properties are evaluated at the average of the two surface temperatures, the appropriate driving temperature difference, in Newton's cooling law, is the surface temperature difference, and it is required that

$$L/b > 3. \qquad (6.100)$$

For values of $L/b < 3$, it is expected that the surface coefficients for a vertical surface freely convecting in a large expanse of fluid apply, at least approximately, to each of the two surfaces.

Emery and Chu [30] have investigated enclosed vertical layers of fluids with Prandtl numbers between 3 and 30,000, and recommend

$$N_{Nu} = 1$$

for $N_{Ra} < 10^3$, and

$$N_{Nu} = \frac{0.280 \, N_{Ra}^{1/4}}{(L/b)^{1/4}} \qquad (6.101)$$

for $10^3 < N_{Ra} < 10^7$. The layer thickness b serves as the characteristic dimension in the Nusselt and Rayleigh numbers in Eq. (6.101).

Equation (6.101) is also supported by the finite difference calculations of MacGregor and Emery [31], who fit an equation virtually identical to Eq. (6.101) to their numerical calculations.

For the case of an enclosed vertical fluid layer with *constant flux* at one wall and constant temperature at the other wall, experiments by MacGregor and Emery [31] are correlated by the following equations:

$$N_{Nu} = \frac{hb}{k} = \frac{0.42 \, N_{Ra}^{0.25} \, N_{Pr}^{0.012}}{(L/b)^{0.30}} \qquad (6.102)$$

for $10^4 < N_{Ra} < 10^7$, $1 < N_{Pr} < 20{,}000$, $10 < L/b < 40$ and *constant flux;*

$$N_{Nu} = \frac{hb}{k} = 0.046 \, N_{Ra}^{1/3} \tag{6.103}$$

for $10^6 < N_{Ra} < 10^9$, $1 < N_{Pr} < 20$, $1 < L/b < 40$ and *constant flux.*

In these equations also, the plate spacing b is to be used as the characteristic dimension in both the Nusselt and the Rayleigh numbers, and the temperature-dependent fluid properties can be evaluated at the arithmetic average of the two plate temperatures. The Nusselt number is the average value over the entire height L of the vertical enclosure; the temperature difference to be used, in both Newton's cooling law and in the Rayleigh number, is the temperature of the constant flux wall (not known at the beginning of the analysis) minus the temperature of the cooler isothermal wall.

Finally, some evidence relating the effect of angle of inclination to the surface coefficient for rectangular enclosures is available in the work of Randall, Mitchell, and El-Wakil [32] and that of ElSherbiny, Raithby, and Hollands [40], which is particularly suited to solar collectors.

EXAMPLE 6.8

Compute the heat transfer rate per unit area across a 2-in.-thick horizontal air space between two floors when the lower surface is at 85°F and the upper surface is at 65°F. Both surfaces have an emissivity of 0.90.

Solution
Here the total heat transfer rate per unit area is the net radiant loss across the air space from the hotter lower floor plus the free convection loss. The net radiant loss from the lower floor, called 1, is given by Christiansen's equation, Eq. (4.87), with $A_1 = A_2$, where 2 is used as the subscript for the upper floor:

$$q_1 = \frac{A_1 \sigma (T_1^4 - T_2^4)}{(1/\epsilon_1) + (A_1/A_2)(1/\epsilon_2 - 1)}. \tag{4.87}$$

Here, $T_1 = 85 + 460 = 545°R$, $T_2 = 65 + 460 = 525°R$, $A_1 = A_2 = 1$, and $\epsilon_1 = \epsilon_2 = 0.9$. Hence,

$$q_1 = \frac{0.1714[(5.45)^4 - (5.25)^4]}{\frac{1}{0.9} + 1(\frac{1}{0.9} - 1)} = 17.62 \text{ Btu/hr ft}^2.$$

The free convection loss per square foot is

$$q_c = h(T_1 - T_2). \tag{6.104}$$

Depending upon the Grashof number, either Eq. (6.94) or (6.95) may be used for a horizontal air space when the lower plate is hotter. The film temperature is $T_f = \frac{1}{2}(85 + 65) = 75°F$. Hence, the air properties at this temperature are $a_f = 1.60 \times 10^6$, $N_{Pr_f} = 0.709$, and $k_f = 0.015$. The Rayleigh number is found from Eq. (6.52), using $L = \frac{2}{12}$ ft and $T_1 - T_2 = 20°F$, as $N_{Ra_f} = 1.60 \times 10^6 (\frac{2}{12})^3 (20) = 1.48 \times 10^5$. Using Eq. (6.53), the Grashof number is $N_{Gr_f} = 1.48 \times 10^5 / 0.709 = 2.09 \times 10^5$. Thus, Eq. (6.94) can be used. Therefore,

$$h \frac{2}{12} / 0.015 = N_{Nu} = 0.195(2.09 \times 10^5)^{1/4},$$

or $h = 0.374$ Btu/hr ft² °F.

Using this in Eq. (6.104) yields

$$q_c = 0.374(85 - 65) = 7.48 \text{ Btu/hr ft}^2.$$

Hence,

$$q = q_c + q_1 = 7.48 + 17.62 = 25.1 \text{ Btu/hr ft}^2.$$

EXAMPLE 6.9

A double-walled window is to be designed with a vertical air space such that the energy transfer (excluding radiation) across this space of thickness b, when the temperature difference is 16°C, is by pure conduction. Determine the largest value of plate spacing that will permit this when the inner window is at 17°C and the outer one is at 1°C.

Solution
The discussion of vertical air spaces leading to Eq. (6.97) indicates that the process across them will be pure conduction if the Grashof number, based upon b, is less than 2000. From Eq. (6.53),

$$N_{Gr} = \frac{a_f b^3 \Delta T}{N_{Pr_f}}. \qquad [6.53]$$

(Recall that b is to be used as the characteristic dimension, not the vertical height L.) The film temperature T_f is

$$T_f = (17 + 1)/2 = 9°C = 282°K.$$

Interpolation in the air tables at this temperature gives the following properties: $k_f = 0.0248$ W/m °K, $N_{Pr_f} = 0.713$, $a_f = 157.9 \times 10^6$.
    Thus the Grashof number becomes

$$N_{Gr} = \frac{157.9 \times 10^6 \, b^3 (17 - 1)}{0.713} = 3.54 \times 10^9 \, b^3.$$

Setting this equal to the pure conduction limit of 2000 and solving for b yields the following result:

$$b = 0.0083 \text{ m} = 0.83 \text{ cm}.$$

Thus value of b certainly represents a practical, reasonable spacing of the two panes (glazings) of such a double-walled window and actually lies in the range of advertised spacings for such windows.

6.4i Convection in Enclosures Between Concentric Cylinders and Spheres

Experimental results for the free convection heat transfer of various fluids, such as air, water, and oils, when they are confined in the space between concentric cylinders and spheres have been correlated by Raithby and Hollands [33] and are presented here.

The values D_i and D_o are the diameters of the inside and outside, respectively, of the long cylinders or the spheres. The values T_i and T_o are the corresponding surface temperatures, L is the length of the long cylinders, and b is the enclosure gap or thickness of the enclosed fluid layer.

Concentric Cylindrical Annuli

$$q/L = 2\pi k_{\text{eff}}(T_i - T_o)/\ln\left(\frac{D_o}{D_i}\right) \qquad (6.105)$$

$$k_{\text{eff}}/k = 0.386 \left[\frac{N_{\text{Pr}}}{0.861 + N_{\text{Pr}}}\right]^{1/4} N_{\text{Ra}_{cc}}^{1/4} \qquad (6.106)$$
when $100 < N_{\text{Ra}_{cc}} < 10^7$.

$$N_{\text{Ra}_{cc}} = \frac{\left[\ln\left(\frac{D_o}{D_i}\right)\right]^4 N_{\text{Ra}_b}}{b^3 \left\{\frac{1}{D_i^{3/5}} + \frac{1}{D_o^{3/5}}\right\}^5} \qquad (6.107)$$

Concentric Spherical Annuli

$$q = \pi k_{\text{eff}}(D_i D_o/b)(T_i - T_o) \qquad (6.108)$$

$$k_{\text{eff}}/k = 0.74 \left[\frac{N_{\text{Pr}}}{0.861 + N_{\text{Pr}}}\right]^{1/4} N_{\text{Ra}_{cs}}^{1/4} \qquad (6.109)$$
when $10 < N_{\text{Ra}_{cs}} < 10^6$.

$$N_{\text{Ra}_{cs}} = \frac{b \, N_{\text{Ra}_b}}{D_o^4 D_i^4 \left\{\frac{1}{D_i^{7/5}} + \frac{1}{D_o^{7/5}}\right\}^5} \qquad (6.110)$$

In both Eqs. (6.107) and (6.110), N_{Ra_b} is the Rayleigh number based on the thickness b of the annular fluid layer.

$$N_{\mathrm{Ra}_b} = ab^3 \left| T_i - T_o \right| \tag{6.111}$$

In addition, temperature-dependent fluid properties are to be evaluated at the arithmetic average of T_i and T_o, and Eqs. (6.106) and (6.109) are invalid if k_{eff} is found, from them, to be less than unity. If k_{eff} is calculated to be smaller than one, the process is one of pure conduction in the fluid; therefore, $k_{\mathrm{eff}} = k$ should be used.

6.5 MIXED FREE AND FORCED CONVECTION

In almost all forced convection situations, buoyancy forces are present which cause some additional fluid motion beyond that induced by the forces which are sustaining the forced convection flow. When the effects of the buoyancy forces are of the same order of magnitude as the forces causing the forced convection, the heat transfer regime is called *combined,* or *mixed free and forced convection.* Also, in free convection problems, a forced flow velocity which may seem very small in magnitude may cause a sufficiently large deviation in the convection coefficient so that the flow situation actually is mixed convection.

The primary purpose here is to present criteria which enable a determination of whether or not free convection effects are important in forced convection problems and whether forced convection effects are important in what is considered, at first glance, to be a free convection problem.

The value of the Nusselt number in the mixed convection regime depends upon the Grashof, Reynolds, and Prandtl numbers, and experimental results for the Nusselt number in this regime are still scarce. Because of the scarcity and uncertainty of the Nusselt number in the mixed convection regime, it is desirable, whenever possible, to design away from this heat transfer regime; that is, to design in such a way that either forced or free convection, rather than mixed convection, is assured. Thus, the emphasis will be placed upon determining when either a forced or free convection situation is close to being in the mixed convection regime.

Let us consider how to determine the basic form of the criterion, or criteria, which indicate whether one is dealing with free, forced, or mixed convection. Consider steady, laminar, upward flow over a thin vertical plate which is at a temperature higher than the external stream. This implies a forced convection situation. However, the fact that the plate temperature is higher than the stream temperature and that the plate is vertical indicates that there will be free convection currents as well. By comparing the relations for pure forced convection and pure free convection in this situation, we will deduce a qualitative criterion which will show when forced convection effects dominate, when free convection effects dominate, and when both effects have to be considered.

For steady, laminar, constant property flow over a flat plate, the average Nusselt number for forced convection over the plate was found in Chapter 5 to be

$$N_{\mathrm{Nu}_{\mathrm{forced}}} = 0.664 \, N_{\mathrm{Re}}^{0.5} \, N_{\mathrm{Pr}}^{1/3} . \tag{5.336}$$

For pure free convection in laminar flow on a vertical flat plate, the average Nusselt number is given by Eq. (6.54). After using Eq. (6.48), Eq. (6.54) becomes

$$N_{Nu_{free}} = 0.59 \, N_{Gr}^{0.25} \, N_{Pr}^{0.25} . \tag{6.112}$$

Dividing Eq. (6.112) by Eq. (5.336) gives

$$\frac{N_{Nu_{free}}}{N_{Nu_{forced}}} = \frac{0.889}{N_{Pr}^{0.083}} \frac{N_{Gr}^{0.25}}{N_{Re}^{0.5}} . \tag{6.113}$$

In order to qualitatively determine the relative importance of forced and free convection effects, let the factor $0.889/N_{Pr}^{0.083}$ equal unity, since it is very close to unity except for rather extreme values of the Prandtl number; hence, Eq. (6.113) becomes

$$\frac{N_{Nu_{free}}}{N_{Nu_{forced}}} \approx \frac{N_{Gr}^{0.25}}{N_{Re}^{0.5}} . \tag{6.114}$$

When forced convection effects dominate, the ratio shown in Eq. (6.114) must be very small compared with unity. This implies that

$$N_{Gr}^{0.25}/N_{Re}^{0.5} \ll 1.$$

Or raising both sides to the fourth power, *we see that forced convection effects will dominate when*

$$N_{Gr}/N_{Re}^2 \ll 1. \tag{6.115}$$

Using similar reasoning, we see that *free convection effects dominate when the ratio in Eq. (6.114) is much greater than unity; that is, when*

$$N_{Gr}/N_{Re}^2 \gg 1. \tag{6.116}$$

Finally, when the ratio in Eq. (6.114) is of the order of unity, we can expect that both free and forced convection effects are important; that is, *mixed convection occurs when*

$$N_{Gr}/N_{Re}^2 \approx 1. \tag{6.117}$$

In the absence of more specific information, Eqs. (6.115–17) can be used as rough criteria even for shapes other than the vertical flat plate, as long as the Grashof and Reynolds numbers are formed with the appropriate characteristic length.

6.5a Vertical Plates

Sparrow and Gregg [34] find that the effect of free convection on the average surface coefficient for laminar forced flow over a vertical flat plate is less than 5% when

$$N_{Gr} \leq 0.225 \, N_{Re}^2 \qquad \text{(6.118)}$$

for $0.01 \leq N_{Pr} \leq 10$.

From the results of Acrivos [35], it can be deduced that the effects of forced convection on free convection are less than about 10% for the local surface coefficient in laminar flow on a vertical flat plate when

$$N_{Gr_x}/N_{Re_x}^2 \geq 4 \qquad \text{(6.119)}$$

for $0.73 \leq N_{Pr} \leq 10$, and

$$N_{Gr_x}/N_{Re_x}^2 \geq 60 \qquad \text{(6.120)}$$

for $N_{Pr} \approx 100$.

6.5b Horizontal Plates

Mori [36] has shown that the effect of natural convection on the local heat transfer coefficient in laminar flow over a *horizontal* flat plate (either side) is less than about 10% for fluids whose Prandtl number is near 0.72 when

$$N_{Gr_x} \leq 0.0831 \, N_{Re_x}^{2.5} . \qquad \text{(6.121)}$$

6.5c Tubes

The free, forced, and mixed regimes for flow in both vertical and horizontal tubes are reported by Metais and Eckert [37]. Two of their figures are reproduced here. Figure 6.8 is for flow through vertical tubes, while Fig. 6.9 is for flow in horizontal tubes. Their symbols, UWT and UHF, stand for uniform wall temperature and uniform heat flux, respectively, and Re, Gr, and Pr stand for the Reynolds, Grashof, and Prandtl numbers, respectively, while Gz stands for the Graetz number, which is defined by

$$Gz = N_{Re}N_{Pr}(D/L). \qquad \text{(6.122)}$$

Both the Reynolds and Grashof numbers use the tube diameter D as their characteristic dimension, the temperature difference in the Grashof number is the tube wall temperature minus the fluid bulk mean temperature, and properties are evaluated at the film temperature.

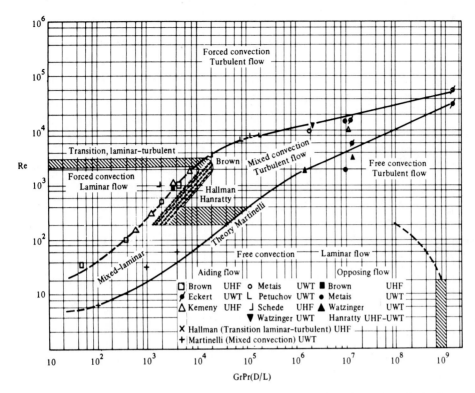

Figure 6.8 Regimes of free, forced, and mixed convection for flow through vertical tubes $[10^{-2} < Pr(d/L) < 1]$. (From "Forced, Mixed and Free Convection Regimes" by B. Metais and E. R. G. Eckert. *Journal of Heat Transfer, Transactions of the A.S.M.E.* [May 1964], pp. 295–96.)

6.5d Spheres and Horizontal Cylinders

Yuge [27] indicates that forced convection dominates in crossflow or parallel flow (basic free convection velocities at right angles to forced flow or in the direction of the forced flow, respectively) over a *sphere* if the following is satisfied:

$$N_{Re}^2 > 100 \, N_{Gr}. \tag{6.123}$$

Fand and Keswani [38] give the following criteria for *horizontal cylinders:*
Forced convection effects dominate, with no more than a 5% influence of free convection, when

$$\frac{N_{Gr}}{N_{Re}^2} < 0.5. \tag{6.124}$$

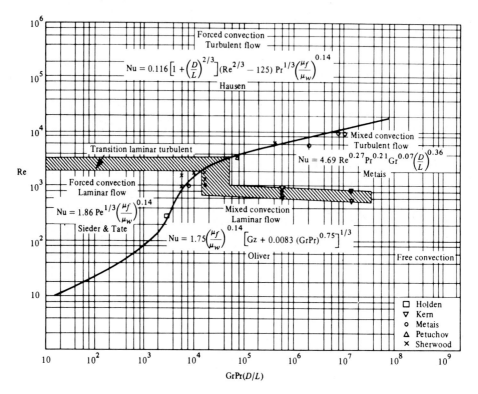

Figure 6.9 Regimes of free, forced, and mixed convection for flow through horizontal tubes $[10^{-2} < \Pr(d/L) < 1]$. (From Metais and Eckert, 1964.)

Natural convection effects dominate when the following condition holds:

$$\frac{N_{Gr}}{N_{Re}^2} > 40. \tag{6.125}$$

EXAMPLE 6.10

Suppose water is forced past a 1-ft-high vertical flat plate in such a manner that the plate Reynolds number is only 5×10^4. If the water temperature is 60°F, calculate the highest plate temperature that can be used without free convection effects causing a change greater than 5% in the average forced convection surface coefficient of heat transfer.

Solution

The effect of free convection on the average forced convection surface coefficient will be less than 5% if Eq. (6.118) is satisfied, since at the temperatures involved, the Prandtl number at the film temperature does not exceed 10.

Thus,

$$N_{Gr} \leq 0.225 \, N_{Re}^2, \quad \text{or} \quad N_{Gr} \leq 5.64 \times 10^8.$$

Using Eq. (6.53) along with $L = 1$ ft and $T_s = 60°F$ gives

$$\frac{a_f(1)^3(T_w - 60)}{N_{Pr_f}} \leq 5.64 \times 10^8. \tag{6.126}$$

Since a_f and the Prandtl number depend upon the unknown surface temperature, the solution will proceed by trial and error. Therefore, assume that $T_w = 140°F$; then $T_f = \frac{1}{2}(140 + 60) = 100°F$, $a_f = 547 \times 10^6$, and $N_{Pr_f} = 4.55$. Substituting these values into Eq. (6.126) yields

$$1.2 \times 10^8(T_w - 60) \leq 5.64 \times 10^8.$$

Clearly, $T_w = 140°F$ is too large and leads to a surface temperature only 5°F different from 60°F. Choosing a lower value as the new estimate, say $T_w = 80°F$, $T_f = \frac{1}{2}(80 + 60) = 70°F$, $a_f = 238 \times 10^6$, and $N_{Pr_f} = 6.81$. Substituting these values into Eq. (6.126) yields

$$0.35 \times 10^8(T_w - 60) \leq 5.64 \times 10^8, \quad \text{or} \quad T_w \leq 76°F.$$

The new film temperature, with $T_w = 76°F$, becomes $T_f = \frac{1}{2}(76 + 60) = 68°F$. Using linear interpolation, $a_f = 221 \times 10^6$ and $N_{Pr_f} = 7.05$. Substituting these values into Eq. (6.126) and solving for T_w yields $T_w \leq 78°F$. This is sufficiently close to the previous value. Hence, at this relatively low Reynolds number, a 1-ft vertical plate in 60°F water begins to show the influence of free convection on the forced convection coefficient at a temperature difference of 18°F, or at $T_w = 78°F$.

EXAMPLE 6.11

Water at an average temperature of 30°C is free convecting within a 7.5-cm-diameter, 38-cm-high vertical tube whose wall temperature is 90°C. Find the value of the forced convection Reynolds number which would just begin to cause mixed convection.

Solution
For the various heat transfer flow regimes within vertical tubes, Fig. 6.8 is available. Here,

$$T_f = \frac{1}{2}(90 + 30) = 60°C.$$

Hence, from the property tables for liquid water,

$$a_f = 6.76 \times 10^9, \quad N_{Pr_f} = 3.04.$$

First, the product $N_{\text{Pr}_f} \dfrac{D}{L}$ is checked to see if Fig. 6.8 can be used.

$$N_{\text{Pr}_f} \frac{D}{L} = 3.04(7.5/38) = 0.6$$

This is within the range of this parameter given on Fig. 6.8. Next, from Eq. (6.53), using the tube diameter, the Grashof number is

$$N_{\text{Gr}} = 67.6 \times 10^9 (0.075)^3 (90 - 30)/3.04 = 5.63 \times 10^8.$$

Hence,

$$N_{\text{Gr}} \, N_{\text{Pr}}(D/L) = 5.63 \times 10^8 (0.6) = 3.38 \times 10^8.$$

Use this as the value of the abscissa in Fig. 6.8, and move vertically until the lower curve, between the free convection and mixed convection regimes, is reached. At this point the ordinate is

$$N_{\text{Re}} \approx 17,000.$$

Thus, in this free convection situation, a substantial diameter Reynolds number due to forced flow is required before the free convection process is significantly influenced.

6.6 SUMMARY OF FREE CONVECTION EXPRESSIONS

Numerous expressions for the surface coefficient of heat transfer, in different free convection situations, have been presented in this chapter. The vast majority of these expressions represent correlations of experimental data, while the few remaining ones result from approximate analytical approaches.

Similar to the summary section of Chapter 5 on forced convection, this section will organize the various free convection relations in Table 6.2 in order to minimize the time and effort needed by the reader to make a proper selection of the most appropriate relation to be used in the problem at hand. The descriptions of the conditions for validity of the various equations for h, which are given in Table 6.2, are incomplete. Readers are strongly advised to study the more complete statements concerning the domain of applicability of equations tentatively selected from Table 6.2 by referring to the discussion of the equation where it is first presented in the chapter.

Table 6.2
Summary of Expressions for the Free Convection Heat Transfer Coefficient

Geometry of Body	Comments and Conditions	Equation Number
Vertical plates and cylinders	Laminar, constant T_w $D/L > 34/N_{Ra}^{1/4}$ for vertical cylinders. May also be used for constant wall flux if T_w is evaluated halfway up the plate.	
	$10^4 < N_{Gr} < 4 \times 10^8$	(6.49)
	$10^4 < N_{Ra} < 10^9$, Prandtl number close to unity. Air, other gases	(6.54)
	$0 < N_{Ra} < 10^9$, $0 < N_{Pr} < \infty$	(6.56)
	Turbulent, constant T_w Same comment as above for a constant flux surface.	
	$10^9 < N_{Ra} < 10^{12}$, Prandtl number close to unity. Air, other gases	(6.55)
	$N_{Ra} > 10^9$, $0.6 < N_{Pr} < \infty$	(6.57)
	$N_{Ra} > 10^9$, $0 < N_{Pr} < 0.6$	(6.57a)
Horizontal and inclined planes	Horizontal square plates, constant T_w Hot side up or cold side down	
	$10^4 < N_{Ra} < 10^8$, $N_{Pr} > 0.6$	(6.65)
	$10^8 < N_{Ra} < 10^{11}$, $N_{Pr} > 0.6$	(6.66)
	Hot side down or cold side up	
	$10^4 < N_{Ra} < 10^9$	(6.67)
	$10^6 < N_{Ra} < 10^{11}$	(6.68)
	Other horizontal shapes, constant T_w, $L^* = A/P$ used in N_{Ra^*} and N_{Nu}^* Hot side up or cold side down	
	$2.6 \times 10^4 < N_{Ra^*} < 8 \times 10^6$	(6.70)
	$8 \times 10^6 < N_{Ra^*} < 1.5 \times 10^9$	(6.71)
	Hot side down or cold side up	
	$10^4 < N_{Ra^*} < 10^9$	(6.67)
	$10^6 < N_{Ra^*} < 10^{11}$	(6.68)
	Inclined planes, constant T_w Hot side down or cold side up	
	$10^4 < N_{Ra} \cos \theta < 10^9$, Prandtl number close to unity. Air or other gases	(6.72)
	$0 < N_{Ra} \cos \theta < 10^9$, $0 < N_{Pr} < \infty$ Use $N_{Ra} \cos \theta$ in place of N_{Ra}	(6.56)
	$10^9 < N_{Ra} \cos < 10^{12}$, Prandtl number close to unity. Air, other gases	(6.55)
	$N_{Ra} \cos \theta > 10^9$, $0 < N_{Pr} < \infty$	(6.57–57a)
	Cold side facing down or hot side up	
	$60 \leq \theta \leq 90°$	(6.66)

Table 6.2
(Continued)

Geometry of Body	Comments and Conditions	Equation Number
	Inclined planes, constant flux *local* surface coefficient of heat transfer	
	Hot side down	
	$N_{\mathrm{Ra}_{mx}} < N_{\mathrm{Ra}_c}$, $0 \leq \theta \leq 86.5°$	(6.75)
	$N_{\mathrm{Ra}_{mx}} > N_{\mathrm{Ra}_c}$, $0 \leq \theta \leq 31.7°$	(6.77)
	Hot side up	
	$N_{\mathrm{Ra}_{mx}} < N_{\mathrm{Ra}_{mt}}$, $0 \leq \theta \leq 60°$	(6.78)
	$N_{\mathrm{Ra}_{mx}} > N_{\mathrm{Ra}_{mt}}$, $0 \leq \theta \leq 60°$	(6.80)
Horizontal cylinders	Constant T_w	
	Use for air and water and other moderate Prandtl number fluids	
	$10^{-10} < N_{\mathrm{Ra}} < 10^{12}$, $0.69 < N_{\mathrm{Pr}} < 7$	(6.81)
	$10^4 < N_{\mathrm{Ra}} < 10^7$, $0.69 < N_{\mathrm{Pr}} < 7$	(6.82)
	$10^7 < N_{\mathrm{Ra}} < 10^{12}$, $0.69 < N_{\mathrm{Pr}} < 7$	(6.83)
	Use for high Prandtl numbers	
	$3 \times 10^2 < N_{\mathrm{Ra}} < 2 \times 10^7$, $0.7 < N_{\mathrm{Pr}} < 3000$	(6.84)
	Use for low Prandtl numbers, with Eq. (6.86) also used for high Prandtl numbers	
	$10^{-6} < N_{\mathrm{Ra}} < 10^9$, $0 < N_{\mathrm{Pr}} < \infty$	(6.85)
	$N_{\mathrm{Ra}} > 10^9$, $0 < N_{\mathrm{Pr}} < \infty$	(6.86)
Spheres	Constant T_w, $1 < N_{\mathrm{Ra}} < 10^5$	(6.91)
Additional shapes	Constant T_w, vertical plates and cylinders, horizontal cylinders, spheres and blocks	Fig. 6.7
Enclosures	Horizontal plates	
	Lower plate cooler	(6.93)
	Lower plate hotter	
	$10^4 < N_{\mathrm{Gr}} < 3.7 \times 10^5$, air	(6.94)
	$3.7 \times 10^5 < N_{\mathrm{Gr}} < 10^7$, air	(6.95)
	$1.5 \times 10^5 < N_{\mathrm{Ra}} < 10^9$, liquids	(6.96)
	Vertical plates, constant temperature	
	$N_{\mathrm{Gr}} < 2000$, air	(6.97)
	$2 \times 10^4 < N_{\mathrm{Gr}} < 2 \times 10^5$, air	(6.98)
	$2 \times 10^5 < N_{\mathrm{Gr}} < 10^7$, air	(6.99)
	$10^3 < N_{\mathrm{Ra}} < 10^7$, $3 < N_{\mathrm{Pr}} < 30,000$	(6.101)

Table 6.2
(Continued)

Geometry of Body	Comments and Conditions	Equation Number
	Vertical plates, constant flux at one wall, constant T_w at the other. Average surface coefficient.	
	$10^4 < N_{Ra} < 10^7$, $1 < N_{Pr} < 20{,}000$, $10 < L/b < 40$	(6.102)
	$10^6 < N_{Ra} < 10^9$, $1 < N_{Pr} < 20$, $1 < L/b < 40$	(6.103)
	Concentric cylinders	
	$100 < N_{Ra_{cc}} < 10^7$	(6.106)
	Concentric spherical annuli	
	$10 < N_{Ra_{cs}} < 10^6$	(6.109)

6.7 PROBLEMS

6.1 Suppose that a pressurized space ship is moving at constant velocity in outer space. Would you expect the free convection coefficients on vertical surfaces to be higher or lower than those for the same surface on earth when the cabin pressure is 14.7 psia? Briefly explain your reasoning.

6.2 Show that the thermal coefficient of volume expansion for a gas that obeys the Clausius equation of state $P(v - b) = RT$, where v is the specific volume and b is a known constant, is given by

$$\beta = \frac{1}{(bP/R) + T}.$$

6.3 Explain why Eq. (6.17) is not valid for a free convection situation on the top of a heated horizontal flat plate.

6.4 During one stage of heat treatment, a thin plate, 1 ft high by 1 ft wide, with a surface temperature of 200°F, is cooling in 60°F water which is completely undisturbed far away from the plate. Estimate the average surface coefficient of heat transfer between the water and the plate surface.

6.5 Some of the plates being cooled, as described in Prob. 6.4, are only 1 in. high. Compute the value of the average surface coefficient of heat transfer between these plates and the water.

6.6 A heater panel in the shape of a vertical flat plate 0.6 m high by 0.6 m wide is mounted on the wall of a large room. If the room air and surfaces are at 20°C and the maximum heater temperature is 60°C, calculate the rate at which electrical energy is dissipated by the heater if its surface has an emissivity of 0.90.

6.7 The heating element for a water heater is a vertical flat plate 15 cm high. The element when operating at 95°C in 65°C water is to dissipate electrical energy at the rate of 3500 W. Compute the needed width of the element.

6.8 The heating element of a vertical water heater is 15 cm high and 15 cm wide and dissipates electrical energy at the rate of 3000 W. Determine at what pressure the 40°C water must be held to ensure that no boiling occurs in the vicinity of the heating element.

6.9 The vertical door of a large meat freezer is 2 m high and 1.3 m wide. Due to the low temperature maintained inside the freezer, the door is at a temperature of 15°C while the room air temperature is 30°C. Compute the heat transfer rate by free convection to the freezer door.

6.10 Suppose the meat freezer door of Prob. 6.9 was designed so that, when closed, it made an angle of 35° with the vertical. Calculate the free convection heat transfer rate.

6.11 A large tank of water at 80°F is to be heated slightly by a flat plate, 6 in. high and 1 ft wide, which is imbedded in the side of the tank. If it is desired to transfer energy to the water from this plate at the rate of 5000 Btu/hr, find the temperature at which the plate surface should be held.

6.12 A vertical heating panel 2 ft high by 2 ft wide is situated on the wall of a pressurized chamber. The chamber air and surfaces other than the heater are at 60°F and the air pressure is 60 psia. If the surface temperature of the heater is not to exceed 140°F, and its emissivity is 0.90, calculate the size, in watts, of the heater element.

6.13 An electric heating panel 0.3 m high by 0.6 m wide is to have an average temperature no greater than 325°K while freely convecting in 300°K air. The room surfaces are also at 300°K and the heater has an emissivity of 0.70. Estimate the total electrical energy dissipation rate of the heater.

6.14 In another design situation, where there is relatively little danger of being burned, the heater panel described in Prob. 6.13 will have its surface held at 750°K. (a) Compute the total electrical energy dissipation rate within the heater. (b) Is radiation relatively more or less important at this operating level compared with the operating level of Prob. 6.13?

6.15 A small vertical heater plate operates in air under conditions that permit only small excursions around a film temperature of 300°K. The pressure is 1 atm and the plate height is small enough so that laminar free convection is expected. Derive a simple equation for computing the free convection coefficient under these conditions for any given ΔT and L.

6.16 A steam pipe 0.3 m in diameter passes vertically through a room which is 2.3 m high. The outside pipe temperature is 350°K while the air temperature is 300°K and the air is at atmospheric pressure. The outside of the pipe has been coated so as to give it an emissivity of about 0.55. The room surfaces are at 295°K. Calculate the heat loss.

6.17 A large circular chimney of outer diameter 1.3 m is passing vertically through a factory work space. The outside of the chimney is made of building brick and the temperature of the outside chimney surface is 325°K. The room air is at 300°K while the walls, ceiling, and floor are at 295°K. The chimney is 3 m high and $\epsilon = 0.45$ for the brick. Calculate the total heat transfer loss from the chimney.

6.18 An electric hot water tank has a diameter of 1 m and a height of 1.6 m. If the heat transfer rate from the vertical curved surface is not to exceed 175 W when the surrounding air and surfaces are at 290°K, calculate the highest outside surface temperature of the tank that can be allowed. The surface emissivity is 0.90.

6.19 A large vertical plate heater operates in air in a temperature range which fluctuates slightly about a film temperature of 450°K. The free convection flow is expected to be turbulent. Develop a simplified equation for the surface coefficient h to be used in this specific set of circumstances.

6.20 A resistance heater in the wall of a 70°F water tank can be idealized as a flat plate 6 in. high and 3 ft wide. If its average temperature is not to exceed 90°F, compute the rate at which electrical energy can be dissipated within the heater.

6.21 Suppose the electrical energy dissipation in Prob. 6.20 was done in such a way that essentially a constant flux condition existed at the plate surface. If the energy dissipation rate remains the same, estimate the maximum heater temperature.

6.22 A vertical plate heater, 0.3 m wide, operates at 60°C in 30°C water. If the electrical energy dissipation rate is to be 1500 W, calculate the required height of the heater.

6.23 Oil at 124°F is heated by a 4-ft-high vertical plate which has a surface temperature of 300°F. The properties of the oil at a film temperature of 212°F are $a_f = 105 \times 10^6$, $k_f = 0.0741$, and $N_{Pr_f} = 117$. Compute the value of the average surface coefficient of heat transfer.

6.24 A block of ice, 0.6 m on a side, is at −8°C in a room where the air is at 27°C. Estimate the value of the surface coefficient of heat transfer between the top surface of the ice and the room air.

6.25 If the block of ice of Prob. 6.24 is suspended such that its top and bottom faces are horizontal, compute the value of the surface coefficient of heat transfer between its bottom face and the air.

6.26 A square hot plate 0.3 m by 0.3 m is in a horizontal position with its hot surface at 425°K facing upward. The undisturbed air temperature is 300°K and the room surfaces are also at 300°K. The emissivity of the hot plate is 0.90. Calculate the total heat transfer rate from the hot plate with steady-state conditions prevailing.

6.27 Design requirements are such that a heater plate at 500°F must transfer energy by free convection to 100°F air and yet have a width of 1 ft. Calculate the required vertical height of the plate if the energy transfer rate to the air from one side of the plate must be 200 Btu/hr.

6.28 A jet engine casing shaped like a thin cylindrical shell, 3 ft in diameter, is 1 ft high at 1000°F when taken out of a heat-treating oven. It is hung to cool with its 1-ft dimension in the vertical direction in 100°F air. Compute the average surface coefficient of heat transfer between the casing and the air.

6.29 If the casing in Prob. 6.28, instead of hanging vertically, is hanging canted at 30° to the vertical, estimate the value of the surface coefficient.

6.30 The horizontal top of a small laboratory furnace is a horizontal plate 2 ft on a side. The top is at 100°F and has an emissivity of 0.85. If the surrounding air and the room surfaces are at 70°F, calculate the heat transfer rate from the furnace through the top.

6.31 The top of a chest-type home food freezer can be idealized as a flat plate, 4 ft by 4 ft, at 70°F with a total hemispheric emissivity of 0.91. If the surrounding air and surfaces are at 80°F, compute the rate at which the refrigerant extracts energy due to heat transfer to the freezer top.

6.32 A home water heater is designed with a 9 in. by 9 in., 3500 W horizontal heater element located at the bottom of the tank. If the element switches on when the tank water is at 140°F, compute the steady-state surface temperature of the element.

6.33 If the water heater of Prob. 6.32 is located in a residential district where the water pressure is 70 psia, compute the maximum rate at which the heating element may dissipate energy without causing a phase change when the bulk water temperature is 140°F.

6.34 Assume that the top of a stove can be idealized as a horizontal square plate 1 m by 1 m. While the oven is on, the stove top maintains a temperature of 40°C while the room air is at 20°C and 1 atm. Calculate the heat transfer rate by convection from the stove top.

6.35 Rework Prob. 6.23 for the condition that the plate is horizontal and the surface coefficient between the upper surface and the oil must be estimated.

6.36 Small heating panels, 15 cm on a side, are to be installed in the ceiling of a room where the air and other surfaces are at 300°K. If the panel surface temperature is 350°K and its emissivity is 0.91, estimate the rate at which electrical energy is dissipated by each panel.

6.37 Derive the following expression for the average Nusselt number over both surfaces of a horizontal flat plate which is warmer than the surrounding fluid when $10^4 < N_{Ra_f} < 10^8$:

$$\overline{N}_{Nu} = 0.436 \, N_{Ra_f}^{1/4}.$$

(Note that both the top and bottom are at approximately the same temperature.)

6.38 Estimate the value of the surface coefficient of heat transfer between a large lake covered with ice at a surface temperature of 30°F and the surrounding 0°F air on a completely windless day.

6.39 The liquid surface in a coffee cup is 3 in. in diameter and is at 160°F. Estimate the value of the free convection surface coefficient between this surface and the 70°F air above it.

6.40 An object of total volume 1 ft³ and total surface area 10 ft² is initially at 1000°F throughout. The thermal conductivity of the object is $k = 150$ Btu/hr ft °F. At time $\tau = 0$, the object is placed into a liquid whose temperature is 500°F and the average surface coefficient between the object and the fluid in this free convection situation is $h = 12.5(T - 500)^{1/3}$ in Btu/hr ft² °F, where T is the surface temperature of the object at any time τ. The density of the object is 100 lbm/ft³ and $c_p = 0.5$ Btu/lbm °R. (a) Derive an expression for the temperature of the object as a function of time. (b) Find the time it takes for the object to cool to 700°F.

6.41 A 3-in.-diameter, 300°F horizontal steam pipe passes through the 124°F oil whose properties are given in Prob. 6.23. Compute the value of the heat transfer rate from the pipe to the oil if the pipe is 10 ft long.

6.42 One of the top surfaces on a stove is an oxidized iron plate, with dimensions 0.3 m by 0.3 m, at a temperature of 150°C. This plate is inclined at an angle of 60° to the vertical, and the surrounding room air is at 22°C. Calculate the heat transfer rate from this plate.

6.43 Part of a transition section in a hot air duct consists of a plate 1.5 ft wide by 2.0 ft long inclined at an angle of 45° to the horizontal. If the underside of this plate is at 120°F and the surrounding air is at 60°F, calculate the free convective heat transfer rate from the plate.

6.44 A house roof, inclined at 27° to the horizontal, has an outside surface temperature of 30°C while the outside air is at 35°C. Calculate the value of the free convection surface coefficient of heat transfer between the roof and air.

6.45 One of the electric resistance heating units in a vat of water is a square plate 1 ft by 1 ft inclined at 50° to the vertical. If the heater has a constant flux of 6000 Btu/hr ft² and the water temperature is 140°F, find the maximum surface temperature of the heater.

6.46 A 15-cm-diameter pipe in a hot air heating system runs horizontally across a cellar where the air and other surfaces are at 20°C. If the pipe is 5 m long, at 60°C, and with an emissivity of 0.92, calculate the total heat transfer loss from the pipe.

6.47 Determine the water pressure necessary to prevent boiling, if a horizontal, 2.5-cm-diameter, 0.6-m-long heating element dissipates electrical energy at the rate of 4500 W in 350°K water.

6.48 Four-in.-diameter aluminum shafts at 700°F are supported horizontally in 100°F air to cool. Calculate the value of the free convection surface coefficient between the shafts and the surrounding air.

6.49 An electric resistance heater in the shape of a horizontal cylinder 1 in. in diameter and 3 ft long is used in an experiment in which the net radiant loss of the heater is made zero by placing it in an oven whose walls are held at the surface temperature of the heater. If the heater dissipates electrical energy at the rate of 100 W, estimate the temperature at which the oven walls should be held if the oven air is at 100°F.

6.50 An electric resistance heater in a water heater is a 0.6-m-long, 2.5-cm-diameter horizontal cylinder. If the heater surface temperature is not to exceed 90°C when the water temperature is 60°C, calculate the size, in watts, of the heater element.

6.51 A small, simple condenser consists of a 6-in. horizontal pipe which is at 200°F because steam condenses inside. This 10-ft length of pipe passes through a tank containing 60°F water. If the latent heat of condensation of the steam is 978 Btu/lbm, estimate the rate at which steam is condensing.

6.52 Liquid mercury, for which $N_{Pr} = 0.025$ and $k = 5.02$ Btu/hr ft °F, is heated using a 1-in.-diameter horizontal tube heater. The average film temperature is 68°F, and the value of a_f, defined in Eq. (6.51), has been computed as 53×10^6. Estimate the value of the average surface coefficient of heat transfer between the cylinder and the mercury if the temperature difference is 100°F.

6.53 A 1.25-cm-diameter, 6-m-long, horizontal copper water pipe whose surface temperature is 10°C runs through a cellar where the surrounding air and surfaces are at 20°C. The emissivity of the water pipe is 0.75. Calculate the total heat transfer rate to the pipe.

6.54 0.6-cm-diameter steel pellets came from a furnace at 650°K and are cooling in 300°K air. Calculate the value of the free convection coefficient between the pellets and the air.

6.55 A thermocouple bead is 1/8 in. in diameter and has an emissivity of 0.90. It is used to measure the temperature of air in a large chamber in which the walls are at 150°F. If the thermocouple reading is 200°F, estimate the value of the air temperature.

6.56 A sphere, 5 cm in diameter, encloses instrumentation and is positioned in part of a river that is stagnant because of a flood control dam. If the sphere surface is to be maintained at 30°C and the river water is at 10°C, compute the electrical energy dissipation rate required within the sphere.

6.57 An electric motor can be idealized as a horizontal cylinder 15 cm in diameter and 25 cm long. The emissivity of its outside surface is about 0.85. When the motor runs under load in the steady state, its average outside surface temperature is 350°K. The room air and other surfaces in the room are at about 300°K. (a) Compute the total heat transfer rate from the motor surface. (b) If the motor is doing useful shaft work at the rate of 373 W, calculate its efficiency of conversion of electrical energy.

6.58 A thermocouple bead is 1/8 in. in diameter. Compute the surface coefficient of heat transfer between the bead and the 130°F water it is immersed in at a time when the bead temperature is 70°F.

6.59 An instrument balloon is 1.25 m in diameter and is initially at 325°K. The surrounding air is at 275°K. Estimate the value of the free convection coefficient between the balloon and the surrounding air.

6.60 A 0.75-mm-diameter, long horizontal wire at 60°C is surrounded by 20°C water. Compute the electrical energy dissipation rate per meter of wire.

6.61 Compute the heat transfer rate per meter from the wire of Prob. 6.60 if the fluid surrounding it is replaced by 20°C air and the wire emissivity is 0.80.

6.62 A vertical 2.5-cm-diameter pipe, 0.6 m high, carries steam to a radiator. The pipe surface is at 400°K, while the surrounding air is at 300°K. Compute the free convection heat transfer rate from the pipe.

6.63 A metal component shaped approximately like a 2.5-cm cube is at 500°K after a final soldering operation and is placed on a workbench in 300°K air to cool. Estimate the average surface coefficient of heat transfer between the component and the air.

6.64 A spherical container, 1 ft in diameter, holds electronic instruments near the bottom of 60°F lake water. The sphere surface is held at 80°F by the electrical energy dissipated within the sphere for the purpose of providing the right atmosphere for the instruments. Determine the rate at which electrical energy is being dissipated.

6.65 Water in a large tank at 60°F is heated by a 1-in.-diameter pipe at 200°F passing vertically through the water with a total height of 3 ft. Compute the heat transfer rate to the water.

6.66 A short, solid, vertical cylinder, 6 in. high and 6 in. in diameter, is at 500°F and is cooling in 100°F air. Estimate the value of the average surface coefficient of heat transfer between the *entire* outside surface of the solid cylinder and the surrounding air.

6.67 Estimate the value of the average surface coefficient of heat transfer between short horizontal bars 5 cm long and 2 cm in diameter if they are at 650°K and are cooling in 300°K air.

6.68 Estimate the average surface coefficient of heat transfer between a cubical instrument container which is 1 ft on a side and at 120°F and the surrounding 80°F water.

6.69 A long, 2-in.-high by 4-in.-wide beam runs horizontally across a work space. If a fire in an adjacent room raises the temperature of the air around the beam to 1500°F, estimate the average surface coefficient of heat transfer between the air and the 100°F beam.

6.70 Derive Eq. (6.93).

6.71 Large horizontal plates at 300°K are heated by a plate heater positioned 7.5 cm below the plate and parallel to it. The heater surface is at 550°K and both surfaces have an emissivity of 0.90. Calculate the total heat transfer rate per unit area to the plate being heated. Air is in the space between the heater and the plate.

6.72 Rework Prob. 6.71 when the heater is placed 7.5 cm above the plate to be heated.

6.73 A 1.25-cm horizontal space between two large plates is filled with water. If the lower plate is at 90°C while the upper plate is at 30°C, compute the heat transfer rate per unit area across the water layer.

6.74 Rework Prob. 6.73 when air fills the space between the two plates. The emissivity of both plates is 0.85.

6.75 Rework Prob. 6.73 when the layer is vertical and 0.3 m high.

6.76 Rework Prob. 6.73 when the layer is vertical, 0.3 m high, and filled with air. The emissivity of both surfaces is 0.85.

6.77 When installed, a storm window is 1 in. away from the inside window and 2 ft high. If the inside window is at 65°F and the outside window is at 35°F, compute the convection heat transfer rate per unit area across the window and the storm window.

6.78 The side wall of a building has a vertical air space 4 in. thick and 8 ft high. If the inner wall is at 85°F and the outer wall is at 115°F, compute the total heat transfer rate across the air space if the emissivity of both surfaces is 0.85.

6.79 When the oven is on, the back of a 1-m-high stove is at 310°K, while the kitchen wall 10 cm away is at 293°K. If both the wall and the stove have an emissivity of 0.80, calculate the heat transfer loss per unit area from the back of the stove.

6.80 Consider a vertical fluid layer whose height L is less than three times its thickness b. Derive an expression for the free convection coefficient, based on the difference between the temperatures of the two vertical planes bounding the layer, in terms of the free convection coefficient for either vertical plane in an infinite expanse of fluid.

6.81 A vertical fin, 2 ft high, at 300°F is 2 in. away from the vertical side of a housing which is at 100°F. The space between them is filled with an oil whose properties at the film temperature are $a = 105 \times 10^6$, $k_f = 0.0741$, and $N_{Pr} = 117$. Compute the heat transfer rate from the fin to the housing if the fin is 4 ft in width.

6.82 Derive the integral form of the x-momentum equation for mixed convection on a vertical flat plate.

6.83 In free convection heat transfer, statements such as the following can be found: (1) The free convection coefficient is independent of plate height if the height is greater than 0.6 m. (2) For turbulent free convection, even the use of the "wrong" characteristic dimension will give the correct result for the surface coefficient. Explain why these statements can be made.

6.84 Air at atmospheric pressure and 450°K is forced upward at 1 m/s past a vertical flat plate at 300°K. Calculate whether or not free convection effects are important in this forced convection problem. The plate height is 0.3 m.

6.85 Rework Prob. 6.84 for the case in which the plate is horizontal.

6.86 Air at standard atmospheric pressure and 300°F is freely convecting over a 1-ft-high vertical flat plate at 100°F. Calculate the magnitude of the extraneous external velocity that can be tolerated without appreciably affecting the free convection surface coefficient of heat transfer.

6.87 Water at 10°C is to be forced upward past a 15-cm-high heated vertical plate at a speed of 0.6 m/s. If free convection is not to appreciably influence the forced convection process, calculate the maximum plate temperature.

6.88 Water at an average temperature of 100°F is free convecting within a 3-in.-diameter, 15-in.-high vertical tube whose wall temperature is 200°F. Find the value of the forced convection Reynolds number which would just begin to cause mixed convection.

6.89 Air at an average bulk mean temperature of 100°F is flowing at a Reynolds number of 100 in a 1-in.-diameter tube which is 5 ft long and horizontal. If the wall temperature is 500°F, determine whether or not the flow is pure forced convection.

6.90 Water is being forced, in crossflow, over a 1-in.-diameter horizontal tube which is at 200°F. If the water is at 60°F, estimate whether or not free convection will appreciably influence the value of the forced convection coefficient if the forced velocity is 10 ft/sec.

6.91 Water at an average temperature of 70°C is flowing in a vertical 0.6-cm-diameter tube whose wall temperature is 30°C. Estimate the Reynolds number at which free convection effects begin to be important if the tube is 1 m high.

6.92 Rework Prob. 6.91 if the fluid flowing is replaced by air.

6.93 Air at 60°C in a hot air heating system is to flow through a 20-cm-diameter, 4.6-m-long horizontal duct. If the duct wall temperature is 30°C, find (a) the Reynolds number and (b) the flow velocity at which free convection effects will become important in what is viewed as a forced convection situation.

6.94 A 0.3-m-diameter sphere containing instrumentation is at 40°C in a 10°C river where the river velocity is about 0.3 m/s. Decide whether this situation falls into the free, forced, or mixed convection regime.

REFERENCES

1. Ostrach, S. "An Analysis of Laminar Free Convection Flow and Heat Transfer About a Flat Plate Parallel to the Direction of the Generating Body Force." *N.A.C.A. Report 1111* (1953).
2. Kreith, F. *Principles of Heat Transfer.* 3d ed. New York: Intext, 1973.
3. Sparrow, E. M., and J. L. Gregg. "Laminar Free Convection from a Vertical Plate with Uniform Surface Heat Flux." *Transactions of the A.S.M.E.,* Vol. 78 (February 1956), pp. 435–40.
4. Sparrow, E. M., and J. L. Gregg. "Similar Solutions for Free Convection from a Nonisothermal Vertical Plate." *Transactions of the A.S.M.E.,* Vol. 80 (February 1958), pp. 379–86.
5. Gebhart, B. *Heat Transfer.* 2d ed. New York: McGraw-Hill, 1971.
6. Goldstein, S., ed. *Modern Developments in Fluid Mechanics.* Vol. 2. London: Oxford University Press, 1938.
7. Eckert, E. R. G., and E. Soehngen. "Interferometric Studies on the Stability and Transition to Turbulence of a Free Convection Boundary Layer." *Proceedings of the General Discussion on Heat Transfer,* The Institute of Mechanical Engineering and the American Society of Mechanical Engineering (1951).
8. Eckert, E. R. G., and T. W. Jackson. "Analysis of Turbulent Free Convection Boundary Layer on Flat Plate." *N.A.C.A. Report 1015* (1951).
9. Kays, W. M., and M. E. Crawford. *Convective Heat and Mass Transfer.* 2d ed. New York: McGraw-Hill, 1980.
10. King, W. J. "The Basic Laws and Data of Heat Transmission III—Free Convection." *Mechanical Engineering,* Vol. 54 (1932), pp. 347–53.
11. McAdams, W. H. "Heat Transmission." 3d ed. New York: McGraw-Hill, 1954.
12. Churchill, W. W., and H. H. S. Chu. "Correlating Equations for Laminar and Turbulent Free Convection from a Vertical Plate." *International Journal of Heat and Mass Transfer,* Vol. 18 (1975), pp. 1323–29.
13. Sparrow, E. M., and J. L. Gregg. "Laminar Free Convection Heat Transfer from the Outer Surface of a Vertical Circular Cylinder." *Transactions of the A.S.M.E.,* Vol. 78 (1956), pp. 1823–29.
14. Kyte, J. R., A. J. Madden, and E. L. Piret. "Natural Convection Heat Transfer at Reduced Pressure." *Chemical Engineering Progress,* Vol. 49, No. 12 (December 1953), pp. 653–62.
15. Fishenden, M., and O. A. Saunders. *An Introduction to Heat Transfer.* New York: Oxford University Press, 1957.
16. Fujii, T., and H. Imura. "Natural Convection Heat Transfer from a Plate with Arbitrary Inclination." *International Journal of Heat and Mass Transfer,* Vol. 15 (1972), pp. 755–67.
17. Goldstein, R. J., E. M. Sparrow, and D. C. Jones. "Natural Convection Mass Transfer Adjacent to Horizontal Plates." *International Journal of Heat and Mass Transfer,* Vol. 16 (1973), pp. 1025–35.
18. Lloyd, J. R., and W. R. Moran. "Natural Convection Adjacent to Horizontal Surfaces of Various Plan Forms." *Journal of Heat Transfer,* Vol. 96, No. 4 (1974), pp. 443–47.
19. Al-Arabi, M., and M. K. El-Riedy. "Natural Convection Heat Transfer from Isothermal Horizontal Plates of Different Shapes." *International Journal of Heat and Mass Transfer,* Vol. 19 (1976), pp. 1399–1404.
20. Yousef, W. W., J. D. Tarasuk, and W. J. McKeen. "Free Convection Heat Transfer from Upward Facing Isothermal Horizontal Surfaces." *Journal of Heat Transfer,* Vol. 104, No. 3 (1982), pp. 493–99.
21. Vliet, G. C. "Natural Convection Local Heat Transfer on Constant Flux Inclined Surfaces." *Journal of Heat Transfer,* Vol. 91, No. 4 (1969), pp. 511–16.
22. Fussey, D. E., and I. P. Warneford. "Free Convection from a Downward Facing Inclined Plate." *International Journal of Heat and Mass Transfer,* Vol. 21 (1978), pp. 119–26.
23. Vliet, G. C., and C. K. Liu. "An Experimental Study of Turbulent Natural Convection Boundary Layers." *Journal of Heat Transfer,* Vol. 91, No. 4 (1969), pp. 517–31.
24. Morgan, V. T. "The Overall Convective Heat Transfer from Smooth Circular Cylinders." In *Advances in Heat Transfer,* Vol. 11, T. F. Irvine, Jr. and J. P. Hartnett, eds. New York: Academic Press, 1975.
25. Fand, R. M., E. W. Morris, and M. Lum. "Natural Convection Heat Transfer from Horizontal Cylinders to Air, Water, and Silicone Oils for Rayleigh Numbers Between 3×10^2 and 2×10^7." *International Journal of Heat and Mass Transfer,* Vol. 20 (1977), pp. 1173–84.

26. Churchill, S. W., and H. H. S. Chu. "Correlating Equations for Laminar and Turbulent Free Convection from a Horizontal Cylinder." *International Journal of Heat and Mass Transfer,* Vol. 18 (1975), pp. 1049–53.

27. Yuge, T. "Experiments on Heat Transfer from Spheres Including Combined Natural and Forced Convection." *Journal of Heat Transfer,* Series c., Vol. 82 (August 1960), pp. 214–20.

28. Jakob, M. "Free Heat Convection Through Enclosed Plane Gas Layers." *Transactions of the A.S.M.E.,* Vol. 68 (April 1946), pp. 189–94.

29. Globe, S., and D. Dropkin. "Natural Convection Heat Transfer in Liquids Confined by Two Horizontal Plates and Heated from Below." *Journal of Heat Transfer,* Vol. 81 (February 1959), pp. 24–29.

30. Emery, A. F., and N. C. Chu. "Heat Transfer Across Vertical Layers." *Journal of Heat Transfer,* Vol. 87, No. 1 (February 1965), pp. 110–16.

31. MacGregor, R. K., and A. F. Emery. "Free Convection Between Vertical Plane Layers—Moderate and High Prandtl Number Fluids." *Journal of Heat Transfer,* Vol. 91, No. 3 (1969), pp. 391–403.

32. Randall, K. R., J. W. Mitchell, and M. M. El-Wakil. "Natural Convection Heat Transfer Characteristics of Flat Plate Enclosures." *Journal of Heat Transfer,* Vol. 101, No. 1 (1979), pp. 120–25.

33. Raithby, G. D., and K. G. T. Hollands. "A General Method of Obtaining Approximate Solutions to Laminar and Turbulent Free Convection Problems." In *Advances in Heat Transfer,* Vol. 11, T. F. Irvine, Jr. and J. P. Hartnett, eds. New York: Academic Press, 1975.

34. Sparrow, E. M., and J. L. Gregg. "Buoyancy Effects in Forced-Convection Flow and Heat Transfer." *Journal of Applied Mechanics, Transactions of the A.S.M.E.* (March 1959), pp. 133–34.

35. Acrivos, A. "Combined Laminar Free and Forced Convection Heat Transfer in External Flows." *AIChE Journal,* Vol. 4 (1958), pp. 285–89.

36. Mori, Y. "Buoyancy Effects in Forced Laminar Convection Flow Over a Horizontal Flat Plate." *Journal of Heat Transfer, Transactions of the A.S.M.E.,* Vol. 83, No. 4 (November 1961), pp. 479–82.

37. Metais, V., and E. R. G. Eckert. "Forced, Mixed, and Free Convection Regimes." *Journal of Heat Transfer, Transactions of the A.S.M.E.,* Vol. 86 (May 1964), pp. 295–96.

38. Fand, R. M., and K. K. Keswani. "Combined Natural and Forced Convection Heat Transfer from Horizontal Cylinders to Water." *International Journal of Heat and Mass Transfer,* Vol. 16 (1973), pp. 1175–91.

39. Lloyd, J. R., and E. M. Sparrow. "Combined Forced and Free Convection Flow on Vertical Surfaces." *International Journal of Heat and Mass Transfer,* Vol. 13 (1970), pp. 434–38.

40. ElSherbiny, S. M., G. D. Raithby, and K. G. T. Hollands. "Heat Transfer by Natural Convection Across Vertical and Inclined Air Layers." *Journal of Heat Transfer,* Vol. 104, No. 1 (February 1982), pp. 96–102.

41. Sparrow, E. M., and M. A. Ansari. "A Refutation of King's Rule for Multi-Dimensional External Natural Convection." *International Journal of Heat and Mass Transfer,* Vol. 26, No. 9 (September 1983), pp. 1357–64.

HEAT TRANSFER IN CONDENSATION AND BOILING

7.1 INTRODUCTION

The assumption that a single phase is present has been implicit in all convection problems treated to this point. However, phase changes do occur in some situations, such as the ablation of a spaceship nose cone, the freezing of water or melting of ice, and the boiling of a liquid or the condensation of a vapor. Only these latter two processes involving phase change, condensation and boiling, are dealt with herein. The study of condensation and boiling is important for the following two reasons: (1) The phase change itself may be of primary importance, such as in a steam power plant, refrigerator, or air conditioning system, thus necessitating the use of boilers and condensers in which the prediction of the surface coefficients of heat transfer is important. (2) One of the characteristics of condensation and boiling heat transfer might be used: high heat transfer rates per unit area supported by relatively small temperature differences (in other words, relatively high surface coefficients of heat transfer compared with an "equivalent" single phase convection problem). This particular characteristic of one type of boiling heat transfer has application in the design of the cooling system for boiling water nuclear reactors.

In the sections that follow, condensation heat transfer will be dealt with first and then boiling heat transfer will be discussed. It will be seen that for the most commonly occurring type of condensation, film condensation, the prediction of the surface coefficient of heat transfer can be handled analytically for many important geometries, as long as the film flow is laminar. For the case of turbulent film condensation, experimental correlations will be presented for the surface coefficient of heat transfer. On the other hand, the most common type of boiling heat transfer, nucleate boiling, is so complex and imperfectly understood that, for the most part, experimental correlations are used and are presented herein for the heat transfer coefficient and the maximum heat flux in this type of boiling. Finally, analytical expressions for the surface coefficient will be presented for a type of boiling, film boiling, which bears many resemblances to film condensation.

7.2 CONDENSATION HEAT TRANSFER

When a saturated vapor comes in contact with a surface at a temperature T_w, which is lower than the saturation temperature T_v of the vapor, energy is transferred from the vapor to the surface, causing the vapor to release its latent heat of condensation or vaporization, and to condense to the liquid phase on the colder surface. The condensate may form on the surface as discrete drops, called *dropwise condensation;* as a film covering the entire cold surface, called *film* or *filmwise condensation;* or as a combination of drops and film, called *mixed condensation.* It is found experimentally, for example, that the average surface coefficients of heat transfer in dropwise condensation are two to ten or more times larger than those in an equivalent film condensation situation [1]. Thus, in condenser design, for example, dropwise condensation is preferred. However, for dropwise condensation to occur, it is necessary to treat the surface with a promoter, such as a fatty acid or an oil. The effect of the promoter wears off in a short time and film condensation begins unless a new application of the promoter is made. Hence, the occurrence of dropwise condensation cannot be relied upon; designs are made assuming that the condensation is of the film type. For additional information concerning dropwise condensation, the reader can consult Hampson [1], where some experimental data is presented; Griffith [2], where the mechanism of dropwise condensation and promoters of this type of condensation are discussed in some detail; and Merte, Jr. [3], where an extensive discussion is given of the mechanisms, models, promoters, and experimental data along with experimental correlations for the surface coefficient of heat transfer.

Because of the problems just mentioned in initiating and sustaining dropwise condensation, the usual type of condensation one encounters, and the type designed for, is film condensation. This is the type our attention will be focused on in this chapter. This film referred to is a moving layer of liquid, usually of varying thickness at different points on the condensing surface, flowing over the condensing surface and draining by the action of gravity from the bottom of the surface. When the film flow is laminar, this type of condensation can be mathematically analyzed for many of the practical condensing surface geometries which one must deal with.

7.2a Laminar Film Condensation

The overall method of approach to be used on any general condensing surface geometry will be illustrated in detail for the simplest geometry, the vertical plate. Hence, we will consider the problem of analytically predicting the local and average surface coefficients of heat transfer under steady-state conditions for the vertical flat plate shown in Fig. 7.1 at temperature T_w, which is surrounded by a saturated vapor at temperature T_v. When $T_w < T_v$, laminar film condensation occurs on the plate surface. This situation was analyzed by Nusselt [4], [5]. For this steady film condensation problem, Nusselt made the following assumptions:

1. There is zero shear stress at the liquid/vapor interface, at $y = \delta$ in Fig. 7.1, where δ is the total thickness of the condensate layer at any x.
2. Acceleration effects within the thin liquid film are negligible.

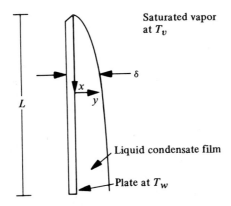

Figure 7.1 Diagram of film condensation on a vertical plate.

3. The temperature at the liquid/vapor interface, at $y = \delta$ for any x, is the saturation temperature T_v of the vapor (that is, all the resistance to heat transfer is concentrated in the condensate film).
4. The subcooling of the liquid in the film is neglected (that is, almost all of the condensed liquid is at a temperature less than T_v and, hence, has had some sensible enthalpy removed, such as $c_p \Delta T$-type terms, as well as the latent heat of condensation).
5. There is a linear temperature distribution from T_w at $y = 0$ to T_v at $y = \delta$ across the liquid film at any x.

In the analysis that follows, we will first attempt to find the velocity field within the moving condensate layer so that it can be used to find the temperature field needed, by virtue of Eq. (5.3), to predict the local surface coefficient of heat transfer. This was done for constant property, single phase, forced convection problems in Chapter 5. This velocity field will also be used to calculate the mass flow rate of the condensate layer, which will be needed in energy balances on that layer.

To find the velocity distribution within the moving condensate layer of Fig. 7.1, the x-momentum equation is applied to a control volume, within the layer, of dimensions dx by dy by one unit, as shown in Fig. 7.2. From the x-momentum theorem, the sum of all the forces on this control volume equals the x-momentum flux out of it minus the x-momentum flux entering it. But Nusselt's assumption 2 indicates that acceleration effects within the liquid are negligible. That is, for a stationary control volume, the momentum flux out in the x direction is equal to the momentum flux in; hence, their difference is zero and the x-momentum equation reduces to

$$\Sigma F_x = 0. \tag{7.1}$$

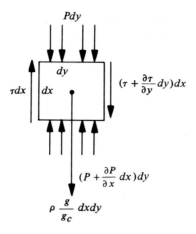

Figure 7.2 The differential control volume with forces for the film condensation analysis shown.

The pressure, shear, and weight forces are shown on the control volume in Fig. 7.2. Thus, from Eq. (7.1),

$$P \, dy - \left(P + \frac{\partial P}{\partial x} \, dx\right) dy - \tau \, dx + \left(\tau + \frac{\partial \tau}{\partial y} \, dy\right) dx$$

$$+ \, \rho \left(\frac{g}{g_c}\right) dx \, dy = 0.$$

Cancelling like terms and dividing by $dx \, dy$ gives

$$\frac{\partial \tau}{\partial y} = \frac{\partial P}{\partial x} - \rho \frac{g}{g_c}, \qquad (7.2)$$

where ρ is the density of the liquid condensate.

To eliminate $\partial P / \partial x$, Eq. (7.1) is applied to a control volume of vapor, of density ρ_v, outside the condensate film, where the shear stress is essentially zero; it is located at the same x as the control volume shown in Fig. 7.2. Thus, in the stationary vapor,

$$\frac{\partial P}{\partial x} = \rho_v \left(\frac{g}{g_c}\right). \qquad (7.3)$$

Since all acceleration effects are negligible and the layer of liquid is thin, $\partial P / \partial x$ in Eqs. (7.2) and (7.3) are the same and can be eliminated. Hence,

$$\frac{\partial \tau}{\partial y} = -(\rho - \rho_v) \frac{g}{g_c}. \qquad (7.4)$$

Since the film flow is laminar and the layer is also assumed to be thin, Newton's law of viscosity (see Chapter 5) gives the local shear stress in the x direction as

$$\tau = \frac{\mu}{g_c} \frac{\partial u}{\partial y}.$$ (7.5)

Substituting the derivative of Eq. (7.5) with respect to y into Eq. (7.4), a differential equation for the velocity profile is

$$\frac{\partial^2 u}{\partial y^2} = -\frac{(\rho - \rho_v)g}{\mu}.$$

Integrating partially with respect to y, where $f_0(x)$ is a function of x, yields

$$\frac{\partial u}{\partial y} = \frac{-(\rho - \rho_v)gy}{\mu} + f_0(x).$$ (7.6)

By assumption 1, at $y = \delta$ for all x, the shear stress and, therefore by Eq. (7.5), $\partial u/\partial y$ are equal to zero. Hence, from Eq. (7.6) at $y = \delta$,

$$0 = \frac{-(\rho - \rho_v)g\delta}{\mu} + f_0(x).$$

Solving for $f_0(x)$ and substituting it into Eq. (7.6) yields

$$\partial u/\partial y = (\rho - \rho_v)(g/\mu)(\delta - y).$$

Integrating partially with respect to y,

$$u = (\rho - \rho_v)(g/\mu)(\delta y - \tfrac{1}{2}y^2) + f_1(x).$$

But at $y = 0$ for all x, $u = 0$; hence, $f_1(x) = 0$ and the velocity distribution is then

$$u = (g/\mu)(\rho - \rho_v)(\delta y - \tfrac{1}{2}y^2).$$ (7.7)

(Note that this is not a complete velocity distribution at this point because of the appearance of the unknown condensate film thickness δ, which is a function of x.)

The mass flow rate across the area dy by 1 in the liquid film at any x is

$$d\omega = \rho u \, dy \, (1).$$

Summing all contributions of this type over the y direction from 0 to δ, the local mass flow rate per unit of width at any x is

$$\omega = \int_0^\delta \rho u \, dy.$$

Substituting the velocity distribution, Eq. (7.7), into the previous equation and integrating yields

$$\omega = \frac{\rho(\rho - \rho_v)g\delta^3}{3\mu}.$$

(7.8)

Thus, the mass flow rate within the condensate layer at any x is given by Eq. (7.8) and the velocity distribution is given by Eq. (7.7). Note that both equations depend upon the unknown local film thickness δ. Therefore, the velocity profile cannot be found from the law of conservation of mass and the momentum theorem by themselves in this film condensation situation. This velocity field is coupled to the temperature field within the film, and hence the energy equation must now be used to find both the velocity field and temperature field. Physically, it is the temperature field that causes the condensation of the vapor in the first place and, as a result, causes the flow of the condensate down the plate. Thus, we can see that film condensation has more in common with single phase *free* convection (where, again, the temperature field causes the velocity field) than with single phase, constant property forced convection.

Next, in order to predict the local and average surface coefficients of heat transfer and the film thickness δ, and to complete the specification of the velocity field and the mass flow rate within the film, the temperature field must be found within the moving condensate. From Eq. (5.3), the definition of the local surface coefficient of heat transfer is

$$h_x = \frac{-k(\partial T/\partial y)_{y=0}}{\Delta T}.$$

[5.3]

Since the "natural" driving temperature difference is the wall temperature minus the vapor saturation temperature, Eq. (5.3) becomes

$$h_x = \frac{-k(\partial T/\partial y)_{y=0}}{T_w - T_v}.$$

(7.9)

From assumption 5, the linear temperature variation in the film at any x, the temperature distribution is

$$T = T_w + (T_v - T_w)y/\delta.$$

(7.10)

Substituting the partial derivative of Eq. (7.10) with respect to y into Eq. (7.9) yields

$$h_x = k/\delta.$$

(7.11)

Hence, once δ is found as a function of x, the local surface coefficient of heat transfer is given by Eq. (7.11). This can be found by making an energy balance on the control volume shown in Fig. 7.3, dx units long in the x direction and δ units thick in the y direction. In

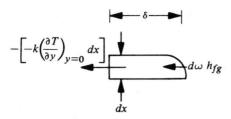

Figure 7.3 Control volume with energy transfer terms for film condensation.

the steady state, with no generation, from Eq. (1.8), $R_{\text{in}} = R_{\text{out}}$. Now, energy is transferred out of the control volume at the plate surface by conduction across the first few stationary molecular layers adjacent to the plate at the rate

$$-\left(-k\left.\frac{\partial T}{\partial y}\right|_{y=0} dx\right). \tag{7.12}$$

Energy enters the control volume between x and $x + dx$ from outside the condensate layer by the transport of the latent heat of condensation h_{fg} by the material which condenses between x and $x + dx$. Since the mass flow rate at any x is given by Eq. (7.8), differentiating it with respect to x yields

$$d\omega = \rho(\rho - \rho_v)(g/\mu)\delta^2 \, d\delta. \tag{7.13}$$

Equation (7.13) represents the rate at which mass enters between x and $x + dx$, since this causes δ to change with x; hence, the energy transfer term becomes

$$d\omega \, h_{fg}. \tag{7.14}$$

Now, energy also crosses the remaining two surfaces of the control volume by conduction, which is considered negligible, and by virtue of the motion of the liquid in the condensate layer, since within this subcooled liquid there is a temperature distribution given by Eq. (7.10). But, assumption 4 neglects this subcooling; hence the only net energy transfer rate terms are Eqs. (7.12) and (7.14). Using Eq. (7.13) in Eq. (7.14) and equating to Eq. (7.12) gives

$$\frac{\rho(\rho - \rho_v)gh_{fg}\delta^2 d\delta}{\mu} = k\left.\frac{\partial T}{\partial y}\right|_{y=0} dx. \tag{7.15}$$

Substituting Eq. (7.10) into the right-hand side of Eq. (7.15) and rearranging yields

$$\delta^3 \, d\delta = \frac{(T_v - T_w)\mu k}{\rho(\rho - \rho_v)gh_{fg}} dx. \tag{7.16}$$

Integrating Eq. (7.16) from $x = 0$, where $\delta = 0$, to x, where the layer thickness is δ, and extracting the fourth root gives

$$\delta = \left[\frac{4\mu k(T_v - T_w)x}{\rho(\rho - \rho_v)gh_{fg}}\right]^{1/4}.$$ (7.17)

Hence, substituting this result into Eq. (7.11), rearranging and noting that $k/k^{1/4} = (k^3)^{1/4}$, the *local surface coefficient of heat transfer in laminar film condensation on a vertical plate* is

$$h_x = \left[\frac{\rho(\rho - \rho_v)gk^3h_{fg}}{4\mu(T_v - T_w)x}\right]^{1/4}.$$ (7.18)

The driving potential difference, $T_v - T_w$, for energy flow across the film is independent of position x along the plate. Therefore, equating the total heat transfer rate to the plate in terms of the *average* surface coefficient of heat transfer h to the local transfer rates integrated over the plate surface leads to the following expression:

$$hL = \int_0^L h_x \, dx.$$

Substituting h_x from Eq. (7.18) and integrating, the *average surface coefficient over the entire plate in laminar film condensation* is

$$h = 0.943 \left[\frac{\rho(\rho - \rho_v)gk^3h_{fg}}{\mu L(T_v - T_w)}\right]^{1/4}.$$ (7.19)

Equation (7.19) can be easily modified to take account of the plate being at an angle ϕ with the horizontal instead of being vertical, $\phi = 90°$. This situation is shown in Fig. 7.4. Equation (7.4) shows that the net shear force is balanced by the difference between the weight of the liquid in the control volume and the hydrostatic pressure gradient (actually a buoyant force); that is,

$$\frac{\partial \tau}{\partial y} = -(\rho - \rho_v)\frac{g}{g_c}.$$ [7.4]

If the plate is inclined, only the component of the right-hand side along the plate can balance $\partial \tau / \partial y$. As shown in Fig. 7.4, this is

$$-(\rho - \rho_v)\frac{g}{g_c}\sin \phi.$$

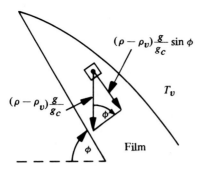

Figure 7.4 Film condensation on an inclined plate.

Hence, the average surface coefficient on an inclined plate is the same as Eq. (7.19) if $(\rho - \rho_v)g$ is replaced by $(\rho - \rho_v)g \sin \phi$. Thus,

$$h = 0.943 \left[\frac{\rho(\rho - \rho_v)g \sin \phi \, k^3 h_{fg}}{\mu L(T_v - T_w)} \right]^{1/4}. \tag{7.20}$$

Equation (7.20) gives the average surface coefficient of heat transfer h over an inclined plate when laminar film condensation occurs on the upper plate surface. Naturally, when the inclination angle ϕ, shown in Fig. 7.4, is 90°, Eq. (7.20) degenerates to the vertical plate result, Eq. (7.19). On the other hand, it is apparent that Eq. (7.20) cannot be correct at very small angles ϕ, such as $\phi = 0°$. However, because of the similarity between the processes of single phase laminar free convection and laminar film condensation, it would appear, from the work of Fujii and Imura [16, Chapter 6], that Eq. (7.20) might be valid for angles ϕ as low as 3°. In any event, it would seem safe to use Eq. (7.20) at least to $\phi = 45°$, which is the angle at which it would seem possible for some condensate to flow over the plate leading edge instead of moving down along the plate.

Relaxation of Some of Nusselt's Assumptions

Next, Nusselt's assumption 4 concerning the neglect of liquid subcooling will be dispensed with, approximately, by employing the integral method of analysis, as developed in Chapters 5 and 6, along with a more nearly correct energy balance on the film. The remaining four assumptions will be retained.

Let h_f be the enthalpy of the saturated liquid at temperature T_v, and h_g the enthalpy of the saturated vapor at temperature T_v. Hence, the enthalpy of the liquid, at any other temperature, per unit mass is

$$h_L = h_f + c_p(T - T_v).$$

Figure 7.5 shows the various energy transfer terms. Note that energy crosses the face at x, through the area dy units by 1 unit, within the liquid film at the rate

$$\rho u[h_f + c_p(T - T_v)]dy,$$

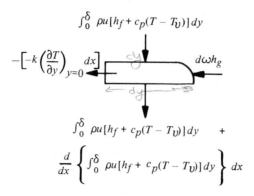

$$\int_0^\delta \rho u [h_f + c_p(T - T_v)] dy$$

$$-\left[-k\left(\frac{\partial T}{\partial y}\right)_{y=0} dx\right] \qquad d\omega h_g$$

$$\int_0^\delta \rho u [h_f + c_p(T - T_v)] dy \qquad +$$

$$\frac{d}{dx}\left\{\int_0^\delta \rho u [h_f + c_p(T - T_v)] dy\right\} dx$$

Figure 7.5 Details of the control volume and the energy transfer terms when liquid subcooling is not neglected.

and the total energy transfer rate across the face at x, from within the film, is the sum of all contributions of this type in the definite integral from $y = 0$ to $y = \delta$. The energy transfer rate across the face at $x + dx$ can be put in terms of that at x by a Taylor series expansion. Hence, Eq. (1.8) becomes

$$d\omega\, h_g + \int_0^\delta \rho u[h_f + c_p(T - T_v)]dy = -\left(-k\left.\frac{\partial T}{\partial y}\right|_{y=0} dx\right)$$
$$+ \int_0^\delta \rho u[h_f + c_p(T - T_v)]\, dy$$
$$+ \frac{d}{dx}\left\{\int_0^\delta \rho u[h_f + c_p(T - T_v)]\, dy\right\} dx.$$

Since the first term on the right is $k(T_v - T_w)\, dx/\delta$,

$$d\omega\, h_g = \frac{k}{\delta}(T_v - T_w)\, dx$$
$$+ \frac{d}{dx}\left\{\int_0^\delta \rho u[h_f + c_p(T - T_v)]\, dy\right\} dx. \qquad (7.21)$$

Using the result for u that led to Eq. (7.19), the first term of the sum operated on by d/dx on the right becomes

$$\int_0^\delta \rho u\, h_f dy = \omega h_f,$$

where h_f is a constant. Thus,

$$\frac{d}{dx}\left(\int_0^\delta \rho u h_f dy\right) dx = d\omega h_f,$$

and, since $d\omega (h_g - h_f) = d\omega h_{fg}$, Eq. (7.21) becomes

$$d\omega \, h_{fg} = \frac{k}{\delta}(T_v - T_w) \, dx + \frac{d}{dx}\left[\int_0^\delta \rho u c_p(T - T_v) \, dy\right] dx. \qquad (7.22)$$

Again using Eq. (7.7) for u and the linear temperature distribution, the second term on the right of Eq. (7.22) becomes

$$\frac{d}{dx}\left[\int_0^\delta \rho c_p \frac{g}{\mu}(\rho - \rho_v)\left(\delta y - \frac{y^2}{2}\right)\left(\frac{y}{\delta} - 1\right)(T_v - T_w) \, dy\right] dx.$$

Integrating and then differentiating, the second term on the right of Eq. (7.22) becomes

$$-\frac{3}{8}\rho(\rho - \rho_v)\frac{g}{\mu} c_p(T_v - T_w)\delta^2 d\delta.$$

Substituting this and Eq. (7.13) into Eq. (7.22) yields

$$\delta^3 \, d\delta = \frac{\mu k(T_v - T_w) \, dx}{\rho(\rho - \rho_v)g \, [h_{fg} + \frac{3}{8} c_p(T_v - T_w)]}.$$

Integrating and then substituting into $h_x = k/\delta$, and determining the average value yields

$$h = 0.943 \left\{\frac{\rho(\rho - \rho_v)g \, k^3 \, [h_{fg} + \frac{3}{8} c_p(T_v - T_w)]}{\mu L \, (T_v - T_w)}\right\}^{1/4}. \qquad (7.23)$$

The effect of subcooling shows up in Eq. (7.23), as contrasted with Eq. (7.19), in the term $\frac{3}{8} c_p(T_v - T_w)$, which is added to the latent heat of condensation h_{fg}.

Rohsenow [6] not only relaxes the subcooling assumption, but also, by use of successive approximations in an integral energy equation, accounts for nonlinearity of the temperature profile. His final result for the *average surface coefficient of heat transfer in laminar film condensation on a vertical plate* is

$$h = 0.943 \left[\frac{\rho(\rho - \rho_v)g \, k^3 \, h_{fg}(1 + 0.68 \, c_p\Delta T/h_{fg})}{\mu L \Delta T}\right]^{1/4}, \qquad (7.24)$$

where $\Delta T = T_v - T_w$ and $0 < c_p\Delta T/h_{fg} < 1$.

In using Eq. (7.24) to predict the value of the average surface coefficient of heat transfer h over a surface, such as a vertical plate or tube, on which a vapor is condensing in a laminar filmwise fashion, the temperature-dependent properties μ, k, ρ, and c_p should be evaluated at the film temperature of

$$T_f = \frac{1}{2} (T_v + T_w),$$

while ρ_v and h_{fg} are evaluated at the vapor temperature.

Sparrow and Gregg [7] include the effects of liquid acceleration as well as subcooling and, by the method of the similarity transformation, determine an exact analytical solution for the local and average surface coefficients in laminar film condensation on a vertical plate. Their results show little difference from those of Rohsenow at Prandtl numbers of the liquid condensate greater than about 1. At lower Prandtl numbers, such as those of the liquid metals, there is considerable deviation.

Hence, for Prandtl numbers of the liquid condensate greater than 1, Eq. (7.24) is the recommended relation for the average surface coefficient of heat transfer in laminar film condensation on a vertical plate rather than the less accurate Eqs. (7.19) and (7.23). If the plate is inclined at angle ϕ to the horizontal, g is replaced by $g \sin \phi$ in Eq. (7.24) as was explained earlier in connection with Eq. (7.20).

Condensation Rate

With the average surface coefficient of heat transfer in laminar film condensation on a vertical plate given by Eq. (7.24), the simplest way to derive an expression for the rate of condensation on the plate surface is through the use of an energy balance on the condensate film. Using a linear temperature distribution within the film and taking account of liquid subcooling, an expression will be derived next for the condensation rate, designated Γ, per unit width of a vertical plate which is L units high.

Figure 7.6 shows the entire film as a control volume and the energy transfer terms. Thus, Eq. (1.8) yields $R_{in} = R_{out}$, or

$$\Gamma h_g = hL(T_v - T_w) + \int_0^{Y_0} \rho u [h_f + c_p(T - T_v)] \, dy, \qquad (7.25)$$

where Y_0 is the value of the condensate thickness δ at $x = L$, and Γ is equal to, by conservation of mass, the mass flow rate of liquid condensate from the bottom of the plate, that is, at $x = L$. Integrating, the last term on the right of Eq. (7.25) gives

$$\int_0^{Y_0} \rho(\rho - \rho_v) \frac{g}{\mu} \left(Y_0 y - \frac{y^2}{2} \right) \left[h_f + c_p \left(\frac{y}{Y_0} - 1 \right) (T_v - T_w) \right] dy$$

$$= \rho(\rho - \rho_v) \frac{g}{3\mu} Y_0^3 [h_f - \tfrac{3}{8} c_p (T_v - T_w)]. \qquad (7.26)$$

But, from Eq. (7.8), the multiplier of the bracket in Eq. (7.26) is Γ, the mass flow of condensate per unit of width at $x = L$. (Note that ω is used as the mass flow rate of condensate at *any x*, not just at $x = L$.) Hence, Eq. (7.25) becomes

$$\Gamma h_g = hL(T_v - T_w) + \Gamma [h_f - \tfrac{3}{8} c_p (T_v - T_w)],$$

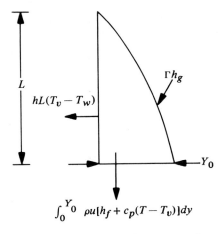

$$\int_0^{Y_0} \rho u[h_f + c_p(T - T_v)]dy$$

Figure 7.6 Energy transfer rates on the entire film.

or

$$\Gamma \left[h_{fg} + \tfrac{3}{8} c_p(T_v - T_w)\right] = hL(T_v - T_w). \tag{7.27}$$

Condensation on Vertical Tubes

When film condensation occurs on either the outside or inside surface of a vertical tube or pipe, the vertical plate relationship, Eq. (7.24), may yield a good approximation to the average surface coefficient of heat transfer. This will be the case if the maximum film thickness Y_0, which occurs at the bottom of the vertical tube, is much smaller than the tube diameter D. If this is so, the vertical tube can be adequately treated as a vertical plate of height L and of width equal to the circumference, πD, of the actual tube. Thus, it is necessary to check the ratio of Y_0 to D, and so an expression for Y_0 is needed. Since the average surface coefficient of heat transfer is already available as Eq. (7.24), it is easiest, as was the case for the condensation rate just treated, to put Y_0 in terms of the average heat transfer coefficient h.

To relate the maximum film thickness Y_0 to the average surface coefficient h, we recall that, for a linear temperature profile in the film, Eq. (7.11) gives

$$h_x = k/\delta. \tag{7.11}$$

Or, since $\delta = Y_0$ at the bottom of the plate where $x = L$, Eq. (7.11) becomes the following:

$$h_{x=L} = k/Y_0. \tag{7.28}$$

Heat Transfer in Condensation and Boiling

Next, $h_{x=L}$ will be related to the average surface coefficient h over the entire plate with the aid of the relation which connects the average and local surface coefficients; this was presented just before Eq. (7.19). After rearrangement, this relation becomes

$$h = \frac{1}{L} \int_0^L h_x dx. \tag{7.29}$$

Examination of Eq. (7.18) indicates that the local surface coefficient is inversely proportional to $x^{1/4}$; that is,

$$h_x = Kx^{-1/4}. \tag{7.30}$$

Equation (7.30) is also the form which the local surface coefficient associated with Eq. (7.24) takes when K represents the multiplier of $x^{-1/4}$, which is independent of x. Substituting Eq. (7.30) into Eq. (7.29) gives

$$h = \frac{1}{L} \int_0^L K x^{-1/4} dx = \frac{K}{L} \frac{4}{3} x^{3/4} \Big|_0^L = \frac{4}{3} \frac{K}{L^{1/4}}.$$

However, from Eq. (7.30), the quantity $K/L^{1/4}$ is the local value of h_x at $x = L$; hence,

$$h = \frac{4}{3} h_{x=L}$$

or

$$h_{x=L} = \frac{3}{4} h. \tag{7.31}$$

Thus, the local surface coefficient of heat transfer at $x = L$ is 3/4 of the average value from $x = 0$ to $x = L$. Combining Eqs. (7.31) and (7.28) gives the relation between the maximum condensate thickness at the bottom of the plate and the average surface coefficient, namely,

$$Y_0 = 4k/3h. \tag{7.32}$$

If $Y_0/D \ll 1$, then the vertical plate relation, Eq. (7.24), will be used in situations in which there is laminar film condensation on either the outside or inside surface of a vertical tube.

Criterion for Laminar Flow Within the Film

The relationships for the surface coefficient of heat transfer developed previously assume that the condensate film is laminar. A criterion is required to determine whether the film flow is laminar or turbulent. Hence, the local film Reynolds number for a flat plate or a vertical tube (if the maximum film thickness is much less than the tube diameter), with an average velocity U_m within the film at any x, is defined as

$$N_{Re_f} = \frac{\rho U_m D_H}{\mu},$$

where ρ is the density of the liquid condensate, and D_H is the local hydraulic diameter, which is defined in Chapter 5 as 4 times the cross-sectional area perpendicular to the flow divided by the wetted perimeter (the portion of the perimeter over which an appreciable shear stress acts).

However,

$$D_H = 4\delta W/W = 4\delta,$$

where W is the plate width and the wetted perimeter is also W, rather than $2W$, since the perimeter at the liquid film/vapor interface does not sustain an appreciable shear stress. As a result,

$$N_{Re_f} = 4 \frac{\rho U_m \delta}{\mu}.$$

But $\rho U_m \delta$ is the condensate flow rate per unit of width and is maximum at the bottom of the plate or vertical tube, where the mass flow rate of condensate per unit of width is Γ. Hence, the film Reynolds number becomes

$$N_{Re_f} = \frac{4\Gamma}{\mu}, \tag{7.33}$$

where Γ is the mass flow rate of condensate, either per unit of width for a plate or per unit of circumference for a vertical tube, at the bottom of the plate or tube. Based on the results presented in [5], [8], and [9], the laminar film condensation relations can be used up to a Reynolds number, defined by Eq. (7.33), of about 1800. For values greater than 1800, an experimental correlation will be subsequently presented.

Thus,

$$\frac{4\Gamma}{\mu} < 1800, \quad Laminar; \quad \frac{4\Gamma}{\mu} > 1800, \quad Turbulent. \tag{7.34}$$

After a value of h is calculated using the assumption of laminar film condensation, Eq. (7.27) is used to determine Γ. Then Eq. (7.34) is used to determine if the film flow is truly laminar.

Even in the absence of film turbulence, experiments show that the actual average surface coefficient of heat transfer is about 20% greater than that predicted by Eq. (7.24). This difference might be attributable to some mixed condensation occurring, ripples at the liquid/vapor interface causing local thinning of the film, and the rate-increasing disturbances that are usually associated with actual equipment. Hence, it is expected that the result given by Eq. (7.24) is conservative; that is, it is lower than the actual surface coefficient as long as the Prandtl number of the liquid is greater than about 1.

EXAMPLE 7.1

A heat transfer surface is idealized as a vertical flat plate which is 0.5 m high and 1.25 m wide at a temperature of 60°C. The surface is in an environment of saturated steam at 90°C. Calculate the rate at which steam is condensed on one side of the surface if the condensation occurs in a filmwise fashion.

Solution
The total rate $\dot{M}$ at which steam condenses is the mass flow rate per meter Γ off the bottom of the plate multiplied by the plate width W; that is,

$$\dot{M} = \Gamma W = 1.25\Gamma. \tag{7.35}$$

If the film flow is laminar—and this will be checked by Eq. (7.34) once Γ is found—Eq. (7.27) can be used to eventually determine the value of Γ. Hence, since $T_v = 90°C$ and $T_w = 60°C$, the film temperature T_f is $T_f = (90 + 60)/2 = 75°C$. From the property tables for the liquid H_2O, we have $c_p = 4185$ J/kg °C, $\rho = 974.8$ kg/m^3, $\mu = 0.384 \times 10^{-3}$ kg/ms, $k = 0.667$ W/m °C, and $N_{Pr} = 2.41$. Also, from the steam tables available in a thermodynamics text such as [10], $h_{fg} = 2283.2 \times 10^3$ J/kg, $\rho_v = 0.424$ kg/m^3.

Hence, with $L = 0.5$ m, Eq. (7.27) becomes

$$\Gamma = \frac{h(0.5)(90 - 60)}{[2{,}283{,}200 + \frac{3}{8}(4185)(90 - 60)]} = 6.44 \times 10^{-6}h. \tag{7.36}$$

The value of h is provided by Eq. (7.24) if the film flow is laminar and if the other conditions for its validity are satisfied. Here, $N_{Pr} = 2.41$; hence, the Prandtl number condition is met and

$$\frac{c_p\Delta T}{h_{fg}} = \frac{4185(90 - 60)}{2{,}283{,}200} = 0.055,$$

so that the inequality involving this group is also fulfilled. Hence, with $g = 9.8$ m/s^2, substituting known values into Eq. (7.24) yields

$$h = 0.943 \left\{ \frac{974.8(974.8 - 0.424)9.8(0.667)^3(2{,}283{,}200)[1 + 0.68(0.055)]}{0.000384(0.5)(90 - 60)} \right\}^{1/4}$$
$$= 5475 \text{ W/m}^2 \text{ °C}.$$

With this, Eq. (7.36) gives

$$\Gamma = 6.44 \times 10^{-6}(5475) = 0.0353 \text{ kg/ms}.$$

Now that Γ has been determined based on the assumption that the film flow is laminar, this assumption is checked by calculating the film Reynolds number, Eq. (7.33).

$$N_{Re_f} = \frac{4\Gamma}{\mu} = \frac{4(0.0353)}{0.000384} = 367.7.$$

Hence, by the criterion given in Eq. (7.34), the flow within the film is laminar. Now, using Eq. (7.35), the condensation rate is found to be

$$\dot{M} = 1.25\Gamma = 1.25(0.0353) = 0.044 \text{ kg/s}.$$

Looking at the value of h (5475 W/m² °C) that occurs in this situation, we note that the surface coefficients in condensation heat transfer are much higher than those for free convection with no phase change, even if the single phase surrounding the plate is liquid water and not vapor. These relatively high heat transfer coefficients, at low temperature differences, in the absence of any forced effect, were referred to earlier in this chapter.

EXAMPLE 7.2

Saturated steam at 180°F is condensing on the outside surface of vertical tubes which are 2 ft high, 1 in. in diameter and are at 140°F. How many tubes are required in this condenser if 1000 lbm/hr of steam is to be condensed?

Solution
The condensation rate per tube is $\dot{M}$ and the unknown number of tubes is n; hence, the total condensation rate of 1000 lbm/hr is

$$n\dot{M} = 1000, \tag{7.37}$$

where

$$\dot{M} = \pi D\Gamma = \frac{1}{12} \pi\Gamma. \tag{7.38}$$

Next, we will assume that the film flow is laminar and that the condensation process on the actual vertical tubes can be approximated adequately as being the same as on a vertical plate. Once the value of h is determined, these two assumptions will be checked by reference to the flow criterion, Eq. (7.34), and to the criterion, $Y_0 \ll D$, associated with Eq. (7.32) for determining when a vertical tube can be idealized as a vertical plate.

Now, $T_v = 180°F$, $T_w = 140°F$, and $T_f = \frac{1}{2}(140 + 180) = 160°F$. Hence, from Appendix A, at $160°F$, $\rho = 61.0$, $\mu = 0.968$, $k = 0.384$, $N_{Pr} = 2.52$, and $c_p = 1.0$. (Note that the value of the Prandtl number of the condensate is always checked, since Eq. [7.24] is restricted to values of the Prandtl number greater than 1.) The steam tables [11] give, at $180°F$, $\rho_v = 0.02$ lbm/ft^3 and $h_{fg} = 990.2$. Substituting known values into Eq. (7.27), with $L = 2$, yields

$$\Gamma \left[990.2 + \tfrac{3}{8}(1)(180 - 140) \right] = h(2)(180 - 140),$$
$$\text{or} \quad \Gamma = 0.0797h. \tag{7.39}$$

Since $c_p \Delta T/h_{fg} = (1)(40)/990.2 = 0.0404$, from Eq. (7.24),

$$h = 0.943 \left\{ \frac{61(61 - 0.02)4.17 \times 10^8(0.384)^3 990.2[1 + 0.68(0.0404)]}{0.968(2)(180 - 140)} \right\}^{1/4}$$
$$= 974 \text{ Btu/hr ft}^2 \text{ °F.}$$

Substituting this into Eq. (7.39) yields

$$\Gamma = 0.0797(974) = 77.6 \text{ lbm/hr ft.}$$

Now that Γ has been found, the criteria referred to earlier can be checked.

The flow is laminar, since from Eq. (7.33), the film Reynolds number has a value less than 1800, namely,

$$N_{Re_f} = \frac{4(77.6)}{0.968} = 321.$$

In addition, these tubes behave as vertical plates, since from Eq. (7.32), the maximum film thickness at the tube bottom is

$$Y_0 = \frac{4(0.384)}{3(974)} = 0.000526 \text{ ft,}$$

and, since $D = 1/12$ ft, $Y_0 \ll D$. Hence, from Eq. (7.38),

$$\dot{M} = 1/12\pi(77.6) = 20.3 \text{ lbm/hr per tube,}$$

and from Eq. (7.37), the required number of tubes is

$$n = 49.3 = 50 \text{ tubes.}$$

7.2b Condensation on Horizontal Tubes

Nusselt [4], in an analysis similar to that for the flat plate, found an analytical solution for the average surface coefficient of heat transfer in laminar film condensation on a horizontal cylinder of diameter D:

$$h = 0.725 \left[\frac{\rho(\rho - \rho_v)gk^3 h_{fg}}{\mu D(T_v - T_w)} \right]^{1/4}. \tag{7.40}$$

As for the vertical plate, ρ_v and h_{fg} are evaluated at the temperature of the vapor, while the remaining properties μ, k, and ρ of the liquid condensate are evaluated at the film temperature. If the effect of liquid subcooling is appreciable—that is, if $c_p\Delta T/h_{fg}$ is significantly greater than zero—h_{fg} in Eq. (7.40) is to be replaced by h_{fg}^1, defined next.

$$h_{fg}^1 = h_{fg}[1 + 0.68 c_p\Delta T/h_{fg}] \tag{7.41}$$

The film Reynolds number given by Eq. (7.33) is used to check for turbulence by interpreting Γ as the mass flow rate per foot at the bottom of the horizontal tube from one side of the tube only. If the total condensate rate per foot at the bottom of the tube is used, the Reynolds number criterion changes from 1800 to 3600. Generally, because of the size of the tubes involved, film condensation on single *horizontal* tubes is almost always laminar.

Vertical Tier of Horizontal Tubes

In some condenser designs, there are n horizontal tubes directly above one another in a vertical tier, as shown in Fig. 7.7. Thus, the condensate from the bottom of the top tube runs down onto the top of the second tube, and so on. Chen [12] has analyzed this situation, including the effect of condensation on the subcooled liquid film between tubes. He developed the following relationship for the average surface coefficient of heat transfer over all n tubes in laminar film condensation:

$$h = 0.725 \left[\frac{\rho(\rho - \rho_v)gk^3 h_{fg}(1 + 0.68\, c_p\Delta T/h_{fg})}{n\mu D(T_v - T_w)} \right]^{1/4}$$
$$\times \left[1 + 0.2 \frac{c_p\Delta T}{h_{fg}}(n-1) \right], \tag{7.42}$$

provided that $c_p\Delta T/h_{fg} \leq 2$, $N_{Pr} \geq 1$.

Condensation Inside Tubes

Condensation occurs inside tubes of the condensers of air-cooled refrigeration and air conditioning units. For a horizontal tube, Eq. (7.40) is also valid for condensation on the inside of the tube as long as the film does not separate from the wall and some means is

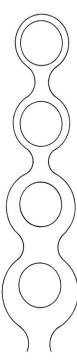

Figure 7.7 Film condensation on a vertical tier of tubes.

provided for condensate drainage at the bottom so that no pool of condensate builds up there.

Generally, however, condensate drainage does not occur through the bottom of the tube, but instead the condensate flows along the bottom of the tube to the tube outlet, thus causing a deep enough condensate thickness to invalidate the use of Eq. (7.40). Chato [13] has analyzed this case and presents the following expressions for the average surface coefficient of heat transfer in laminar film condensation *inside* of almost horizontal tubes for the case in which the interfacial shear stress between the vapor and the condensate can be neglected:

$$h = 0.556 \left[\frac{g\rho(\rho - \rho_v)k^3 h'_{fg}}{\mu D(T_v - T_w)} \right]^{1/4} \tag{7.43}$$

for a tube slope ≤ 0.002 and for $\quad N_{\mathrm{Re}_v} = \dfrac{4\omega_v}{\pi D \mu_v} < 35{,}000.$

The value h'_{fg} is given by Eq. (7.41), ω_v is the inlet mass flow rate of the vapor, and the unsubscripted properties in Eq. (7.43) are to be evaluated at the film temperature T_f of the liquid condensate. For slopes greater than 0.002 and less than 0.6, tabular results needed to determine the value of h are given in the appendix of Chato's article [13]. For inlet vapor Reynolds numbers greater than 35,000 where interfacial shearing stress must be accounted for, results are available in Rohsenow [14].

Condensation on Helical Condensers and Inclined Tubes

Karimi [15] develops a solution to the problem of laminar film condensation on the outside of a condenser formed by a helical circular cylinder. Results for the average surface coefficient of heat transfer are given in graphical form in [15]. However, Karimi notes that these helical circular cylinder results are within about 8.5% of the results for condensation on the outside of an inclined straight circular cylinder, an inclined tube. Hence, only the result for the inclined tube will be presented here. If greater accuracy is needed for a helical circular cylindrical condenser, the graph given in [15] should be consulted.

Thus, the average surface coefficient of heat transfer in laminar, filmwise condensation on the *outside* of an inclined right circular cylinder of diameter D is

$$h = 0.728 \left[\frac{g\rho(\rho - \rho_v)k^3 h_{fg}^1 \cos \phi}{\mu D(T_v - T_w)} \right]^{1/4}. \qquad (7.44)$$

Equation (7.44) is subject to the following additional restrictions:

$$N_{Pr} \geq 1, \quad L/D \geq 25 \tan \phi \qquad (7.45)$$

The value ϕ is the angle between the inclined tube and the horizontal; for a horizontal tube, $\phi = 0$. The value h_{fg}^1 is defined in Eq. (7.41). The values ρ, k, and μ are properties of the liquid condensate which should be evaluated at the film temperature $T_f = (T_w + T_v)/2$. In the second inequality of Eq. (7.45), L is the length of the inclined circular cylinder. This inequality must be satisfied so that the relationship for the average h over the length L is independent of the length L. For inclined cylinders shorter than indicated by this inequality, the very low condensate film thicknesses at the beginning of the cylinder in the starting region, or film development region, give a dependence of the average h on the length L of the cylinder.

EXAMPLE 7.3

One of the vertical tiers in a condenser is made up of 10 horizontal tubes spaced directly below one another. The tubes have an outer diameter of 1.25 cm, a length of 2 m, a surface temperature of 60°C, and are in contact with saturated steam at 120°C. (a) Calculate the condensation rate of steam on this vertical tier. (b) If, instead, the 10 tubes are arranged so that they are single horizontal tubes which do not drip upon one another, find the percentage change in the condensation rate.

Solution

The condensation rate per meter of width, from *one* side of the vertical tier, off the bottom tube in the tier is given by Eq. (7.27) after the plate area $L \times 1$ in that equation is replaced by the surface area per meter of one side of the tier of tubes, $n\pi D/2$. Hence,

$$\Gamma = \frac{hn\pi D(T_v - T_w)/2}{h_{fg} + \frac{3}{8}c_p(T_v - T_w)}. \tag{7.46}$$

If the film flow is laminar, the value of h will be given by Eq. (7.42).

From the steam tables in [10], at $T_v = 120°C$, one finds that $h_{fg} = 2.202 \times 10^6$ J/kg, and $\rho_v = 1.12$ kg/m³. At the film temperature $T_f = (60 + 120)/2 = 90°C$, the properties of the liquid water condensate are found from the property tables in Appendix A to be $c_p = 4206$ J/kg °C, $\rho = 965.4$ kg/m³, $\mu = 0.000318$ kg/ms, $k = 0.676$ W/m °C, and $N_{Pr} = 1.979$. Thus, $c_p\Delta T/h_{fg} = 4206(120 - 60)/2.202 \times 10^6 = 0.115$. Hence, with this and the value of the Prandtl number, the two inequalities associated with Eq. (7.42) are satisfied.

$$h_{fg}(1 + 0.68c_p\Delta T/h_{fg}) = 2.206 \times 10^6[1 + 0.68(0.115)]$$
$$= 2.37 \times 10^6$$

$$1 + 0.2\frac{c_p\Delta T}{h_{fg}}(n - 1) = 1 + 0.2(0.115)(10 - 1) = 1.207$$

Using these results and the needed properties in Eq. (7.42) yields

$$h = 0.725\left[\frac{965.4(965.4 - 1.12)9.8(0.676)^3 2.37 \times 10^6}{10(0.000318)(0.0125)(120 - 60)}\right]^{1/4}(1.207)$$
$$= 6366 \text{ W/m}^2 \text{ °C.}$$

With this value of h, Eq. (7.46) becomes

$$\Gamma = \frac{6366(10)\pi(0.0125)(120 - 60)/2}{2.202 \times 10^6 + \frac{3}{8}(4206)(120 - 60)} = 0.0327 \text{ kg/sm.}$$

Now the film Reynolds number is checked to see if the assumption that the condensate flow is laminar is a correct one.

$$N_{Re_f} = 4\Gamma/\mu = 4(0.0327)/0.000318 = 411$$

Since this value is less than 1800, laminar film flow exists. The value Γ is the condensation rate per meter from one side of the tier. Hence it follows that the total condensation rate $\dot{M}$ is given by, where the width W is 2 m,

$$\dot{M} = 2\Gamma W = 2(0.0327)(2) = 0.1308 \text{ kg/s.} \tag{a}$$

In part b, none of the ten tubes drip upon any other tube, so that the average surface coefficient of heat transfer is simply that for a single tube and is given by either Eq. (7.40) or, equivalently, by Eq. (7.42), with $n = 1$ instead of 10. Hence, using the same numbers as in part a, except $n = 1$ instead of 10, Eq. (7.42) yields

$$h = 9379 \text{ W/m}^2 \text{ °C}.$$

Since the total heat transfer area on the ten tubes is the same in (b) as in (a), the new condensation rate $\dot{M}$ is simply that found in (a) multiplied by the ratio of the heat transfer coefficient in (b) to that of (a), namely,

$$\dot{M} = 0.1308 \left(\frac{9379}{6266}\right) = 0.1927 \text{ kg/s}.$$

Thus the percentage increase in the condensation rate with the new tube arrangement of (b) is

$$\left(\frac{0.1927 - 0.1308}{0.1308}\right) 100 = 47.3\%. \tag{b}$$

Although, from a heat transfer standpoint, it is preferable not to have the tubes dripping on one another, space limitations and requirements often dictate the use of vertical tiers of horizontal tubes. The average h over ten tubes in a vertical tier is less than that for a single tube, since the thicknesses of the film on each tube below the top is greater, on the average, than for the top tube as a result of the accumulation of condensate from the tubes above it. This greater thickness of liquid condensate film offers a greater resistance to heat transfer for the same driving temperature difference than does the thinner film on a single isolated tube. Hence, a lower surface coefficient results than that on a single isolated horizontal tube.

7.2c Turbulent Film Condensation

Earlier, it was pointed out that when the film Reynolds number defined by Eq. (7.33) exceeds a value of about 1800 in a film condensation situation, the condensate flow in the film becomes turbulent and the relations for laminar film condensation are no longer valid. This rarely occurs on horizontal cylinders or even on vertical tiers of horizontal cylinders. However, turbulence does occur on vertical plates and vertical cylinders. In these cases, when the maximum film thickness is much less than the diameter, the average surface coefficient of heat transfer over the surface is given by the experimental correlation of Kirkbride [9] as

$$h\left[\frac{\mu^2}{k^3 \rho(\rho - \rho_v)g}\right]^{1/3} = 0.00762 N_{\text{Re}_f}^{0.4}, \tag{7.47}$$

where $N_{\text{Re}_f} = 4\Gamma/\mu$ from Eq. (7.33).

The use of an energy balance that neglects liquid subcooling permits the rearrangement of Eq. (7.47) to obtain an explicit expression in the average surface coefficient of heat transfer h (note that Eq. [7.47] contains h implicitly in Γ on the right-hand side) for turbulent film condensation over a vertical plate or tube; that is, for $4\Gamma/\mu \geq 1800$,

$$h = 0.000743 \left\{ \frac{L(T_v - T_w)}{\mu h_{fg}} \left[\frac{k^3 \rho(\rho - \rho_v)g}{\mu^2} \right]^{5/6} \right\}^{2/3}. \qquad (7.48)$$

Since, in film condensation, the value of the film Reynolds number cannot ordinarily be calculated until the value of h is found, either laminar or turbulent film flow is assumed and the value of h is calculated based on that assumption. This is then used to determine if the film Reynolds number is in a range which makes the original assumption valid. If it is not, h is recalculated using the other flow regime.

If $\dot{M}$ is the total condensate flow rate at the bottom of a vertical cylinder or plate, the energy balance (with the effect of subcooling neglected) becomes

$$\dot{M}h_{fg} = hA(T_v - T_w), \qquad (7.49)$$

where A is the total surface area of the vertical cylinder or vertical plate on which vapor is being condensed.

EXAMPLE 7.4

Saturated steam at 300°F is to be condensed on the outside of 2-in.-diameter, 10-ft-high vertical tubes. If the outside wall temperature of the tubes is 200°F and if 10,000 lbm/hr of steam is to be condensed, find the number of tubes required.

Solution
From the steam tables [11] at 300°F, $\rho_v = 0.155$ and $h_{fg} = 910$. From Appendix A, for liquid water at the film temperature of 250°F, $\rho = 59$, $\mu = 0.561$, and $k = 0.396$. Hence, since $A = \pi D L n = \frac{1}{6}\pi(10)n$, from Eq. (7.49),

$$(10,000)(910) = 10(\tfrac{1}{6}\pi) \, hn \, (300 - 200), \quad \text{or } n = 17,400/h. \quad (7.50)$$

Now, assume that the vertical plate relations apply, and since the temperature difference and the height are relatively large, the film flow is assumed to be turbulent. Hence, from Eq. (7.48),

$$h = 0.000743 \left\{ \frac{10(300 - 200)}{0.561(910)} \left[\frac{(0.396)^3 59(59 - 0.155)4.17 \times 10^8}{(0.561)^2} \right]^{5/6} \right\}^{2/3}$$
$$= 2690 \text{ Btu/hr ft}^2 \text{ °F.}$$

Substituting into Eq. (7.50) yields

$$n = \frac{17,400}{2690} = 6.46.$$

Hence, 7 tubes will be required, assuming that the use of the correlation for turbulent film condensation is justified. The mass flow rate of condensate per tube is approximately $10,000/7 = 1430$ lbm/hr. Therefore,

$$\Gamma = \frac{1430}{\pi D} = \frac{1430}{\frac{1}{6}\pi} = 2730 \text{ lbm/hr ft.}$$

Hence, from Eq. (7.33),

$$N_{Re_f} = \frac{4(2730)}{0.561} = 19,500,$$

which is greater than 1800. Thus, the film flow is turbulent.

7.2d Effects of Vapor Velocity, Superheat, and Noncondensable Gases on Film Condensation

In the results presented thus far, a zero shear stress at the film edge (i.e., the liquid/vapor interface) has been assumed. However, if there is a vapor velocity parallel to the condensing surface, substantial interfacial shear stresses can be induced. If the vapor velocity is vertically downward, it is expected that the effect of this shear stress is to "drag" the film down the plate more rapidly and cause thinning of the film, which increases the heat transfer rate and the surface coefficient of heat transfer. In a similar fashion, upward flow of the vapor tends to retard the downward motion of the liquid condensate film and decrease the condensing heat transfer rates and the surface coefficients of heat transfer (except for the case in which the upward vapor velocity is so large that the shear force dominates the weight force and causes the liquid condensate to leave at the top of the plate or vertical tube). For film condensation situations with appreciable interfacial shear stress, [14] should be consulted.

If a superheated vapor at temperature T_s (where $T_s > T_v$, the vapor saturation temperature at the pressure of the superheated vapor) is adjacent to a surface at temperature T_w less than T_v, film condensation can occur. It is found that the effect of even very large superheats changes the average surface coefficient of heat transfer in laminar film condensation only slightly [20]. At atmospheric pressure, 100°F of superheat changes the value of h by about 1%, while 200°F of superheat above the saturation temperature of 212°F increases the value of h by about 3%. The surface coefficient h, when superheat is present, is calculated with the ordinary film condensation relationships, using the surface temperature T_w and the saturation temperature T_v of the superheated vapor at the pressure of the superheated vapor, *not* the actual temperature T_s of the superheated vapor.

For extreme cases of superheating, such as 1000°F superheat at atmospheric pressure, which causes an increase in h of roughly 10% over the value with no superheat, a calculation procedure to account for this is given in Sparrow and Eckert [20].

It has long been observed that even relatively small amounts of a noncondensable gas in a vapor, such as small amounts of air in steam, can cause large reductions in the surface coefficient of heat transfer for film condensation situations. The reason for this is that in a condensing situation where the vapor is pure, the condensing process causes some bulk movement of vapor toward the liquid/vapor interface, where it is to be condensed, as a result of the density difference between the liquid and the vapor phase and the fact that mass must be conserved. When the vapor is pure, the flow of the vapor toward the interface proceeds with practically no resistance. Now, if a noncondensable gas is mixed with the vapor, the same bulk motion that carries vapor toward the condensate film also transports the noncondensable toward the film. However, the film is impermeable to the noncondensable, so that its concentration builds up at this liquid/gas mixture interface and establishes a concentration gradient of the noncondensable which, in the steady state, allows the noncondensable to diffuse away from the interface at the same rate at which it arrives with the vapor. But, since the total pressure of the vapor-noncondensable mixture is essentially constant, and the concentration of the noncondensable at the interface is greater than that away from the interface, its partial pressure must be higher. Consequently, the partial pressure of the condensable vapor must be lower at the interface. Since the vapor condenses at a saturation temperature that corresponds to its partial pressure, which is lower at the interface than far away from the interface, the effective vapor saturation temperature is also lower. Thus, there is a smaller driving temperature difference across the film as a result. Thus the effect of the noncondensable gas building up at the interface, in concentrations much greater than its concentration away from the interface, is to reduce the vapor saturation temperature at the interface and, therefore, reduce the heat transfer rates. This lowering of the vapor saturation temperature at the interface also introduces an additional complexity, as far as analysis is concerned, in that the local effective driving temperature difference now varies with position along the surface, since the concentration of the noncondensable gas is greatest where the film is thinnest and least where the film is thickest.

The analysis of laminar film condensation on a vertical plate in the presence of noncondensable gases, performed in [21], as well as the experimental data presented therein, indicates reductions in the heat transfer rate, for film condensation of steam, of more than 50% for mass fractions of the noncondensable air as low as 1.7%. Additional information concerning the effects of a noncondensable gas on film condensation and techniques for accounting for this effect are available in Sparrow and Lin [21] and, particularly, in Rohsenow [14].

7.3 BOILING HEAT TRANSFER

Boiling occurs when heat transfer to the liquid phase of a substance causes some of the liquid to change into the vapor phase. If the boiling occurs with the liquid confined to an open vessel or because a heater is submerged in a bath of essentially stationary liquid, the process is referred to as *pool boiling,* whereas, if the boiling occurs when the liquid

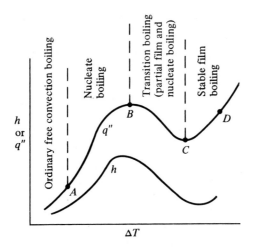

Figure 7.8 Diagram of the heat flux and the surface coefficient of heat transfer versus temperature difference in the various boiling regimes.

is forced through a duct or along a surface, the process is termed *forced convection boiling*. If either pool or forced convection boiling takes place with the bulk of the liquid at the saturation temperature corresponding to the pressure level of the liquid, it is termed *saturated pool, or forced convection, boiling*. If the bulk of the liquid is at a temperature lower than the saturation temperature, the process is called *subcooled, or local boiling*. (Subcooled boiling is often also referred to as *surface boiling*.) The principal concern here is saturated pool boiling. Boiling is one of the most complicated and difficult areas in heat transfer, and the discussion presented here is brief and elementary. For more complete discussions of boiling, see [22] and [23].

7.3a Saturated Pool Boiling

In the pool boiling of a saturated pure liquid, at least four different types of boiling can be distinguished. Figure 7.8 shows the so called "boiling curve" with the four different boiling regimes named. The lower curve is the average surface coefficient of heat transfer between the heating surface and the surrounding saturated liquid, while the upper curve is the heat flux to the saturated liquid from the heating surface. These curves qualitatively show how the surface coefficient of heat transfer and the heat flux vary with the temperature difference between the heater surface and the saturation temperature of the liquid. The experimental results from which the behavior shown in Fig. 7.8 was sketched are contained in [24] and [38].

For the different boiling regimes which result in this boiling curve, consider a pure liquid at its saturation temperature in an open pan which has a heater at its bottom allowing control of the heater surface temperature. When the heater is adjusted so that its temperature is only a few degrees above that of the saturated liquid (anywhere on the curve of q'' versus ΔT before point A in Fig. 7.8), it is found that an ordinary free convection process occurs between the heater surface and the saturated liquid, and the energy is transported to the surface of the saturated liquid by convection currents. It then causes surface evaporation of the saturated liquid. This type of boiling regime, in which no bubbles or vapor pockets are observed in the saturated liquid, but in which a single phase convection process at the heater surface/liquid interface causes surface evaporation at the free surface of the liquid, is called *ordinary free convection boiling.* The average surface coefficients of heat transfer between the heater surface and the saturated liquid can be well approximated by the experimental correlations for single phase free convection heat transfer presented in Chapter 6—Eqs. (6.54–57) for vertical plates, Eq. (6.65) for horizontal plates, and so on.

If the heater temperature is increased so that the value of ΔT in Fig. 7.8 goes beyond point *A,* the formation of vapor bubbles at various points will be observed on the heater surface (at the bottom of the pan). These are the most desirable nucleation sites. The bubbles will grow until the buoyant force and the force on them due to the motion of the surrounding liquid can overcome the weight of the bubble and the surface tension forces, and cause the bubble to break away from the surface. When the bubble breaks away and begins to rise it may, depending upon conditions, either collapse in the liquid before reaching the free surface of the liquid or it may move all the way up to the free surface where it collapses and releases its vapor content to the surroundings. This particular boiling regime, where vapor bubbles originate at various points on the heater surface, is called the *nucleate boiling regime* and is the most commonly recognized. It is seen from Fig. 7.8 that not only are the surface coefficients of heat transfer higher in the nucleate boiling regime than in the ordinary free convection boiling regimes, but they also increase at a much greater rate with an increase in temperature difference ΔT. In the ordinary free convection boiling regime, Eqs. (6.54–55) show that the average surface coefficient of heat transfer h is directly proportional to ΔT to the one-third or one-fourth power. In nucleate boiling, experiment shows that the surface coefficient of heat transfer h can depend upon ΔT to the second through about the fourth power, depending upon the magnitude of ΔT. The much higher heat transfer coefficients in the nucleate boiling regime are believed to be caused by the stirring and agitating effect that the bubble motion has on the surrounding liquid, since most of the energy carried away from the plate is thought to be due to the heat transfer to the liquid near the plate. However, Graham and Hendricks [25] propose that the heat transfer rate, from a surface upon which nucleate boiling occurs, is made up of four different contributions which arise from different portions of the surface. According to this theory, part of the total surface is assumed to be a non-boiling area where ordinary free convection occurs, another part close to the bubble is a region of enhanced free convection, a third part of the surface is being prepared for a bubble by a transient conduction process, and the last contribution is due to evaporation from a microlayer of liquid beneath a bubble. Further research in the area of nucleate boiling is required before a definitive description of the nucleate boiling mechanism can be given.

Referring to the boiling curve in Fig. 7.8, as the heater temperature is further increased past *A,* more and more sites on the heater surface become nucleation points for bubble growth. Upward streaming bubbles, in more or less definite columns, can be observed up to the temperature difference corresponding to point *B* on the curve in Fig. 7.8. At this point, a large fraction of the heater surface is covered by bubbles which are practically touching each other and are making it difficult for the liquid to reach the heater surface. This is called the *maximum heat flux point* or the *burnout flux point* or the *critical heat flux point* or the *departure from nucleate boiling* DNB. As the heater temperature and ΔT are increased just past point B, the heat transfer rate per unit area decreases, as shown in Fig. 7.8. The reason for this decrease is the fact that almost all of the heater surface is covered by a vapor layer or film which tends to insulate the surface because of the relatively low thermal conductivity of the vapor compared with that of the liquid. A large blanket of vapor is observed which suddenly breaks away from the surface, causing liquid to rush to the surface and vaporize almost explosively into another vapor film. Also, at some points, instantaneous nucleate boiling may be observed. The boiling in this regime is called *film boiling,* and the boiling regime between points B and C on the boiling curve in Fig. 7.8 is called *transition boiling,* a regime of partial film and nucleate boiling. Because of the behavior of the vapor film just discussed, this boiling regime is also referred to as the *unstable film boiling regime.* Film boiling, also called the *Leidenfrost effect,* is observed most often when water droplets are thrown into a very hot frying pan and proceed to "dance" along the pan surface as a result of the formation of the insulating film of vapor between the droplet and the pan.

When the heater surface temperature and ΔT are increased past point C in Fig. 7.8, a stable film covering the entire heater surface is observed. This type of boiling is termed *stable film boiling.* As ΔT is further increased the boiling curve begins to rise again. This is due to radiation across the vapor film from the heater surface through the vapor to the liquid, making a substantial contribution to the heat flux from the heater surface.

If the heater surface temperature is controlled by resistors dissipating electrical energy within the heater material, it is difficult to produce the portion of the boiling curve past point B. Consider an electric resistance heater under steady-state conditions operating at point B with the heat transfer rate to the fluid from the heater equal to the rate at which electrical energy is dissipated. Now, to move past point B, the heater surface temperature must be increased. This would be accomplished by increasing the electrical energy dissipated within the heater. But as the curve in Fig. 7.8 shows, the fluid cannot accept this greater amount of energy when ΔT is just past point B, so that the energy dissipated must be partially stored in the heater material, causing the heater surface temperature to rise still further. Again the fluid can accept even less energy at this higher ΔT, so the heater temperature continues to rise until a point on the boiling curve is reached where a steady state is possible. This would be at point D or beyond, where, once again, the fluid can accept energy at the rate dissipated by the heater. For many combinations of saturated liquid and heater surface material, point D is beyond the melting point of the heater surface and destruction of the surface, or *burnout,* will occur if the heater is taken beyond point B without quickly reducing the electrical power supplied to the heater. This leads to the term *burnout flux* used earlier to describe the last point of nucleate boiling, point B. Ordinarily, heat transfer equipment operating in a boiling regime would be designed to operate in the nucleate boiling regime at, or just before, the peak flux point B. In some situations, such as the quenching of certain metals after heat treatment, cryogenic applications involving liquid oxygen or nitrogen, or in equipment used to process chemicals, the film boiling regime cannot be avoided. Hence, expressions for estimating the surface coefficient of heat transfer in film boiling will be presented in a later section.

Although the qualitative shape of the boiling curve in Fig. 7.8 remains essentially the same for different pressure levels, fluid-heating surface material combinations, and heater geometry, these effects do cause the curve to shift and, therefore, give a different boiling curve for each different set of conditions. It is found that the nucleate boiling region, for a given pressure level and fluid-heater surface material, is relatively insensitive to the geometry of the heater surface. Hence, if all other conditions are the same, the same boiling curve would apply to both a relatively thin wire and to a horizontal plate. However, the boiling curve is quite sensitive to changes in pressure level and to the fluid-heater surface material combination. As an example of this sensitivity, the experimental data of Farber and Scorah [24] indicate that, for saturated water at atmospheric pressure boiling on a 0.040-in.-diameter wire with a ΔT of 10°F, the surface coefficient of heat transfer, when a nickel wire is used, is about 200 times higher than that for a chromel C wire under the same conditions.

7.3b Working Relations for Saturated Pool Boiling

Presented in this section will be some of the more widely accepted experimental correlations and analytical and semi-analytical results for the surface heat flux and the surface coefficient of heat transfer in the ordinary free convection, nucleate boiling, and stable film boiling regimes in saturated pool boiling. Included also are expressions for the maximum, or peak, heat flux in nucleate boiling, as well as the temperature difference at which this peak flux occurs.

Ordinary Free Convection Boiling

As indicated earlier, when the surface temperature is only a few degrees or so above the saturation temperature, no vapor bubbles form and the process is one of single phase natural convection from the heater surface to the saturated bulk liquid with consequent evaporation at the liquid free surface. Thus the appropriate expressions for the average surface coefficient of heat transfer are those given in Chapter 6 for the particular heater surface geometry and orientation under consideration. For saturated water at one atmosphere on a horizontal flat plate, this boiling regime may exist for a ΔT, heater surface temperature minus the saturated liquid temperature, of about 4°C or 7°F. However, for higher pressure levels, the ΔT leading to ordinary free convection boiling is considerably smaller than this value (see Rohsenow [22] and Lienhard [26]).

Nucleate Boiling

For nucleate boiling, Rohsenow [27] has correlated experimental data with the equation

$$\frac{c_p(T_w - T_0)}{h_{fg} N_{\mathrm{Pr}}^s} = C_{sf}\left[\frac{q/A}{\mu h_{fg}} \sqrt{\frac{g_c \sigma}{g(\rho - \rho_v)}}\right]^{0.33}, \qquad (7.51)$$

Table 7.1
Constants for Selected Water-Heating Surfaces

Water-Heater Surface Material Combination	C_{sf}
Water-Nickel or Water-Brass	0.006
Water-Platinum or Water-Copper	0.013
Water-Stainless Steel	0.014

From *Boiling Heat Transfer and Two Phase Flow* by L. S. Tong. Copyright © 1965 by John Wiley and Sons, Inc. Used with permission of John Wiley and Sons, Inc.

where T_w is the temperature of the heating surface, T_0 is the saturation temperature of the liquid as determined by the pressure level, h_{fg} is the latent heat of vaporization at the saturation temperature of the liquid, q/A is the heat transfer rate per unit area from the heater to the liquid during nucleate boiling, g is the local acceleration of gravity, and g_c is the unit conversion factor equal to 32.2 lbm ft/lb$_f$ sec^2 in the English Engineering system of units and 1 $\frac{\text{kg m}}{\text{N s}^2}$ in the SI system. The value ρ_v is the vapor density, while σ is the surface tension at the liquid/vapor interface. The properties μ, c_p, N_{Pr}, and ρ are those of the saturated liquid and are to be evaluated at the liquid saturation temperature T_0, as should h_{fg}, σ, and ρ_v. *The exponent s on the Prandtl number in Eq. (7.51) is equal to 1.0 for water and is taken as 1.7 for all other liquids.*

In Eq. (7.51), C_{sf} is an experimentally determined constant which depends upon the liquid-heater surface material combination. For present purposes, the liquid will invariably be water. Table 7.1, which is part of a larger compilation of experimental data presented by Tong [28], contains some common water-heater surface material combinations.

From Rohsenow, the equation for the surface tension of liquid water is derived by using a linear relation to connect its value at 212°F to the value of zero at the critical point of 705°F. That is,

$$\sigma = 58 \times 10^{-4}(1 - 0.00142T) \text{ lbf/ft}, \tag{7.52}$$

where T is in °F.

Or, in SI units, one obtains the following expression for the surface tension of water at the liquid/vapor interface:

$$\sigma = 0.0808 \, [1 - 0.00267T] \, N/m \tag{7.53}$$

when T is in °C.

Experiments indicate that the nucleate boiling heat flux is relatively insensitive to the specific geometry of the heater surface. Hence, Eq. (7.51) should apply to flat plates, cylinders, and other geometries.

Peak Flux on a Horizontal Plate   As mentioned earlier, in order to take advantage of the very high surface coefficients of heat transfer which occur with relatively low driving temperature differences between the surface and the bulk of the fluid, most boiling heat transfer equipment is designed to operate in the nucleate boiling regime. In order to maximize this effect, the operating point should be at or near the peak flux, $(q/A)_{max}$ (near point *B* of Fig. 7.8), and yet not beyond this point, especially for electrically resistive heated surfaces, because of the high surface temperatures that might result. It will be recalled that this *peak heat flux* point *B* in Fig. 7.8 is also called by a number of other names such as *maximum,* or *burnout* flux, *critical heat flux* CHF, *first boiling crisis,* and the *departure from nucleate boiling,* DNB.

Zuber, on the basis of arguments concerning the interference with and blocking of liquid descending toward the heater surface by ascending vapor bubbles, developed an expression for the peak flux which is widely used. An account of his work is available in Rohsenow [22] and in Lienhard [26]. Here we will use the simpler expression given by Rohsenow and Griffith [29] for the peak heat flux in nucleate boiling on a horizontal plate.

For English units,

$$\frac{(q/A)_{max}}{\rho_v h_{fg}} = 143\left(\frac{g}{g_0}\right)^{1/4}\left(\frac{\rho - \rho_v}{\rho_v}\right)^{0.6}, \tag{7.54}$$

where ρ is the density of the liquid at the saturation temperature in lbm/ft^3, ρ_v is the density of the vapor at the liquid saturation temperature in lbm/ft^3, h_{fg} is the latent heat of vaporization at the saturation temperature in Btu/lbm, g/g_0 is the ratio of the local acceleration of gravity g to the acceleration of gravity at sea level (note that $g_0 = 32.2$ ft/sec^2 = 4.17 × 10^8 ft/hr^2), 143 is a *dimensional* constant, and *both sides of Eq. (7.54) have units of ft/hr.*

For SI units, the Rohsenow and Griffith correlation takes the following form:

$$\frac{(q/A)_{max}}{\rho_v h_{fg}} = 0.0121\left(\frac{g}{g_0}\right)^{1/4}\left(\frac{\rho - \rho_v}{\rho_v}\right)^{0.6}. \tag{7.55}$$

In Eq. (7.55), the peak flux, $(q/A)_{max}$, has units of W/m^2, ρ_v is in kg/m^3, and h_{fg} is in J/kg, with 0.0121 being a *dimensional* constant which causes *both sides of the equation to have units of m/s.*

Temperature Difference at Peak Flux Point   Of interest also, in many problems of design and analysis involving boiling heat transfer, is the temperature difference between the heater surface and the saturated liquid at the peak, or maximum, flux point. This can be found by combining Eq. (7.51) with Eq. (7.54) or (7.55) to eliminate the peak flux, $(q/A)_{max}$, and to display explicitly the temperature difference, $(T_w - T_0)_{max}$, at the peak flux point. The maximum temperature difference in nucleate pool boiling of a saturated liquid is

$$\frac{c_p(T_w - T_0)_{max}}{h_{fg} N_{Pr}^s} = C_{sf} \left[\frac{C_0 \rho_v}{\mu} \left(\frac{g}{g_0} \right)^{1/4} \left(\frac{\rho - \rho_v}{\rho_v} \right)^{0.6} \sqrt{\frac{g_c \sigma}{g(\rho - \rho_v)}} \right]^{0.33}. \tag{7.56}$$

In using Eq. (7.56), $s = 1.0$ for water and 1.7 for all other liquids, C_{sf} is given in Table 7.1, and the value of C_0 depends upon the system of units to be used. In the SI system, $C_0 = 0.0121$ m/s, while for the English system, $C_0 = 143$ ft/hr. The individual units to be used in conjunction with either of these values of the dimensional constant C_0 can be ascertained by referring back to the discussion of Eqs. (7.54) and (7.55).

Peak Flux on a Horizontal Cylinder Unlike the nucleate boiling heat flux, which is sensibly independent of heater surface geometry as long as the flux is less than the peak value, the magnitude of the *peak* nucleate boiling flux does exhibit a dependence upon geometry and orientation of the heater surface. A comprehensive collection of peak flux results for other geometries is available in Lienhard [26]. Here, we present only the results for the horizontal cylinder presented by Sun and Lienhard [30].

$$\text{For } 0.15 \le R' \le 3.47, \quad \frac{(q/A)_{max}}{(q/A)_{max_F}} = 0.89 + 2.27 e^{-3.44\sqrt{R'}}. \tag{7.57}$$

$$\text{For } R' > 3.47, \quad \frac{(q/A)_{max}}{(q/A)_{max_F}} = 0.894. \tag{7.58}$$

$$R' = R \sqrt{\frac{g(\rho - \rho_v)}{\sigma}} \tag{7.59}$$

The value R is the cylinder radius, ρ and ρ_v are the density of the saturated liquid and vapor, respectively, g is the local acceleration of gravity, σ is the surface tension between the liquid and its vapor, and $(q/A)_{max_F}$ is the peak heat flux on a horizontal flat plate under the same conditions otherwise. It is given by Eq. (7.54) or (7.55).

EXAMPLE 7.5

Saturated liquid water at 100°C is to be boiled on a horizontal copper heating surface which will be maintained at 120°C. (a) Find the heat transfer rate per unit area and the surface coefficient of heat transfer for this set of conditions. (b) Find the peak, or maximum, flux as well as the temperature difference and the surface coefficient of heat transfer at the peak flux point.

Solution
Since the temperature difference between the heater surface and the saturated liquid, $T_w - T_0 = 120 - 100 = 20$°C, is more than just a few degrees, it is assumed that nucleate boiling is taking place, rather than ordinary free convection boiling. Actually, to insure that the boiling is of the nucleate type and is not transitional or film-type boiling, one

must compute the temperature difference at the peak flux point and check to be sure that the actual difference of 20°C is less than the difference at the peak flux state. Since this is going to be done in part b anyway, we will defer this check, which would ordinarily be made at this point in the analysis. Thus, for the assumed nucleate pool boiling of a saturated liquid, Eq. (7.51) is to be used with $s = 1.0$, since the liquid is water, and, from Table 7.1, $C_{sf} = 0.013$ for the water-copper heating surface combination. From the property tables in Appendix A for liquid water at 100°C, we have $c_p = 4219$ J/kg °C, $\rho = 958.6$ kg/m³, $\mu = 0.283 \times 10^{-3}$ kg/ms, $N_{Pr} = 1.751$. For the vapor at 100°C, from [10], $\rho_v = 0.598$ kg/m³ and $h_{fg} = 2.257 \times 10^6$ J/kg. The surface tension σ is calculated from Eq. (7.53) as $\sigma = 0.0808 [1 - 0.00267(100)] = 0.0592$ N/m. With these, and the local acceleration of gravity $g = 9.8$ m/s², Eq. (7.51) becomes

$$\frac{4219(120 - 100)}{2.257 \times 10^6 (1.751)^1}$$
$$= 0.013 \left[\frac{(q/A)}{0.283 \times 10^{-3}(2.257 \times 10^6)} \sqrt{\frac{(1)(0.0592)}{9.8(958.6 - 0.598)}} \right]^{0.33} .$$

Solving gives $q/A = 1,143,773$ W/m².

To find the surface coefficient of heat transfer, Newton's cooling law is used with the flux just calculated and the driving temperature difference $T_w - T_0 = 20°C$. Thus,

$$h = \frac{(q/A)}{T_w - T_0} = \frac{1,143,773}{20} = 57,189 \text{ W/m}^2 \text{ °C}.$$

We will now find the solution for (b). The peak flux for this horizontal heating surface is given by Eq. (7.55) as

$$\frac{(q/A)_{\max}}{0.598(2.257 \times 10^6)} = 0.0121 \left(\frac{9.8}{9.8}\right)^{1/4} \left(\frac{958.6 - 0.598}{0.598}\right)^{0.6}$$

or $(q/A)_{\max} = 1,367,153$ W/m².

The temperature difference at the peak flux point is found next by inserting the known quantities into Eq. (7.56) with $s = 1.0$, $C_{sf} = 0.013$, and $C_0 = 0.0121$ because of the use of SI units. Hence,

$$\frac{4219(T_w - T_0)_{\max}}{2.257 \times 10^6 (1.751)^1}$$
$$= 0.013 \left[\frac{0.0121(0.598)}{0.283 \times 10^{-3}} \left(\frac{9.8}{9.8}\right)^{1/4} \left(\frac{958.6 - 0.598}{0.598}\right)^{0.6} \right.$$
$$\left. \times \sqrt{\frac{(1)(0.0592)}{9.8(958.6 - 0.598)}} \right]^{0.33} .$$

$$(T_w - T_0)_{\max} = 21.2°C$$

Finally, for the surface coefficient of heat transfer at the peak flux point, we have

$$h = \frac{(q/A)_{\max}}{(T_w - T_0)_{\max}} = \frac{1,367,153}{21.2} = 64,488 \text{ W/m}^2 \text{ °C}.$$

We can see from the result in (b) for the maximum temperature difference, $(T_w - T_0)_{\max}$, that we were justified in assuming nucleate boiling, since the actual temperature difference of 20°C is less than the maximum for nucleate boiling. Of course, the actual ΔT of 20°C is extremely close to the maximum in this case. Hence, to be conservative, it may be worthwhile to use a smaller actual temperature difference to allow for data scatter about correlating curves such as Eqs. (7.51) and (7.55). The magnitude of the surface coefficients of heat transfer and the associated driving temperature differences in Ex. 7.5 attest to the correctness of the statement made earlier about *nucleate* boiling leading to very high heat transfer coefficients with fairly small temperature differences between the surface and the saturated liquid.

7.4 SUBCOOLED NUCLEATE BOILING

To calculate the heat transfer rate per unit area from a heating surface at a temperature T_w greater than the saturation temperature T_0, corresponding to the pressure level, to a liquid whose bulk temperature is T_f, which is lower than T_0, Rohsenow [18] recommends the following equation for nucleate boiling in the subcooled liquid:

$$q/A = h(T_w - T_0) + h_c(T_w - T_f), \tag{7.60}$$

where h_c is the ordinary single phase surface coefficient of heat transfer between the object and the liquid, as if nucleate boiling were not occurring. Note that h_c is found from correlations such as Eqs. (6.56–57) if the process is occurring in a pool, or from the appropriate relation for a single phase forced convection situation in Chapter 5 if the process is one of forced flow over the heating surface. The value of h in Eq. (7.60) refers to the average surface coefficient of heat transfer which would result from saturated nucleate pool boiling alone, and can be estimated from Eq. (7.51) for either pool or forced convection nucleate boiling as long as the flow velocity is not too high.

Generally speaking, the contribution to (q/A) in Eq. (7.60) due to the boiling, the first term on the right of Eq. (7.60), largely overshadows the second term, due to single phase convection, especially for pool boiling. Hence, the flux is not greatly different than would occur if the liquid were saturated rather than subcooled. However, it has been observed that liquid subcooling exerts a substantial influence on the magnitude of the peak flux, as shown in McAdams et al. [31] and in Rohsenow [22].

For subcooled nucleate *pool* boiling, the following simple relationship for the peak flux, due to Ivey and Morris, is presented in Rohsenow [22]:

$$\frac{(q/A)_{\max \text{ sub}}}{(q/A)_{\max}} = 1 + \frac{0.1}{\rho_v h_{fg}} \left(\frac{\rho_v}{\rho}\right)^{1/4} c_p \rho (T_0 - T_{\text{Liq}}). \tag{7.61}$$

In Eq. (7.61), $(q/A)_{\text{max sub}}$ is the peak heat flux in subcooled nucleate pool boiling, while $(q/A)_{\text{max}}$ is the corresponding quantity in saturated nucleate pool boiling and can be calculated from Eqs. (7.54) and (7.55), or (7.57) and (7.58), depending upon heater surface geometry. The value T_0 is the saturation temperature corresponding to the pressure level of the subcooled liquid, and T_{Liq} is the bulk temperature of the subcooled liquid. Other properties are as for Eq. (7.51).

7.5 FILM POOL BOILING

As discussed earlier, the boiling regime beyond point C of the boiling curve in Fig. 7.8 is termed the stable film boiling region. This is a boiling regime which is often amenable to analysis from first principles because of its qualitative similarity to the laminar film condensation process discussed and analyzed earlier in the chapter. An additional complicating aspect, which does not appear in the film condensation process, is radiation heat transfer across the vapor film from the heated surface to the liquid at the liquid/vapor interface.

For predicting the surface coefficient of heat transfer h in stable film boiling on a horizontal tube, Bromley [32], [33] has performed an analysis which yields

$$h = h_{co}\left(\frac{h_{co}}{h}\right)^{1/3} + h_r, \tag{7.62}$$

where h_{co} is the surface coefficient that results if radiation is neglected in the film boiling process. It is given by

$$h_{co} = 0.62\left[\frac{k_v^3\,\rho_v(\rho - \rho_v)g(h_{fg} + 0.4c_{pv}\Delta T)}{\mu_v D\Delta T}\right]^{1/4}, \tag{7.63}$$

if

$$0.8 < \lambda_c/D < 8. \tag{7.64}$$

$$\lambda_c = 2\pi\sqrt{\frac{g_c\sigma}{g(\rho - \rho_v)}} \tag{7.65}$$

The restriction, Eq. (7.64), on the validity of Bromley's result, Eq. (7.63), was not part of his original work. Rather, this is discussed by Jordan [23], and results from the fact that at very small cylinder diameters the vapor film thickness is no longer a small fraction of the cylinder radius, while at very large cylinder diameters a wavy flow pattern develops which is not in accord with Bromley's assumption of a smooth, laminar flow in the film about the cylinder.

In the preceding equations, σ is the surface tension, g_c is the standard gravitational constant, $\Delta T = T_w - T_0$, where T_0 is the saturation temperature of the liquid at the prevailing pressure, all properties subscripted v are those of the vapor and are evaluated at the film temperature T_f, where

$$T_f = \tfrac{1}{2}(T_w + T_0), \tag{7.66}$$

g is the local acceleration of gravity, D is the tube diameter, and ρ is the density of the liquid at its saturation temperature.

In Eq. (7.62), h_r is the radiation surface coefficient which was defined at the end of Chapter 4. Assuming that the emissivity of the liquid is near unity,

$$h_r = \sigma\epsilon \frac{(T_w^4 - T_0^4)}{T_w - T_0}, \tag{7.67}$$

where ϵ is the total hemispheric emissivity of the cylinder surface and σ is the Stefan-Boltzmann constant, *not* the surface tension.

From Eq. (7.62), note that to determine the film boiling surface coefficient of heat transfer h, a trial and error procedure must be used. However, for $0 \leq h_r/h_{co} \leq 10$, Bromley [32] gives the following expression for h which closely approximates Eq. (7.62):

$$h = h_{co} + h_r \left[\frac{3}{4} + \frac{1}{4} \frac{h_r}{h_{co}} \left(\frac{1}{2.62 + h_r/h_{co}} \right) \right]. \tag{7.68}$$

For film boiling on horizontal cylinders for values of λ_c/D outside the range given by Eq. (7.64), Jordan [23] recommends the following relationships:

$$h_{co} = 0.60 \left[\frac{k_v^3 \rho_v(\rho - \rho_v)g(h_{fg} + 0.4c_{pv}\Delta T)}{\mu_v \lambda_c \Delta T} \right]^{1/4} \tag{7.69}$$

if

$$\lambda_c/D < 0.8 \tag{7.70}$$

and

$$h_{co} = 0.16 \left[\frac{k_v^3 \rho_v(\rho - \rho_v)g(h_{fg} + 0.4\,c_{pv}\Delta T)}{\mu_v \lambda_c \Delta T} \right]^{1/4} \left(\frac{\lambda_c}{D} \right)^{0.83} \tag{7.71}$$

if

$$\lambda_c/D > 8. \tag{7.72}$$

Heat Transfer in Condensation and Boiling

These relations, Eqs. (7.69) and (7.71), must, of course, still be combined properly with the expression for radiation across the film in order to give the surface coefficient of heat transfer h in film boiling as shown in Eqs. (7.62) or (7.68).

Film boiling on vertical plates and vertical cylinders is greatly complicated by the fact that the vapor flow in the film usually undergoes a transition to turbulent flow even on relatively short vertical surfaces. An account of the analysis and results for this situation as well as for film boiling on horizontal surfaces is given by Jordan [23]. Additional information on film boiling, such as the minimum flux point, point C in Fig. 7.8, and film boiling from spheres is also available in Rohsenow [22].

EXAMPLE 7.6

A 1-in.-diameter steel bar is quenched in a bath of saturated water at 212°F. Estimate the average surface coefficient of heat transfer between the horizontal steel bar and the water when the bar surface temperature is at 1388°F. The total hemispherical emissivity of the bar surface is $\epsilon = 0.90$.

Solution

Since the temperature difference $T_w - T_0 = 1388 - 212 = 1176$°F is higher than previously encountered in nucleate boiling, it is suspected that the boiling regime may not be the nucleate boiling regime, but the film boiling regime. Reference to Ex. 7.5, or use of Eq. (7.56) if previous results were not available, verifies this, since the temperature difference at the maximum flux point was found to be 21.2°C $= 38.2$°F, for nucleate boiling in saturated water at 100°C $= 212$°F, and this is much lower than the temperature difference of the present problem. Hence, it is assumed that boiling is occurring in the stable film boiling region. In this region the value of the surface coefficient can be estimated using either Eq. (7.62) or Eq. (7.68) if the inequality preceding Eq. (7.68) is satisfied. Since $T_w = 1388 + 460 = 1848$°R and $T_0 = 212 + 460 = 672$°R, Eq. (7.67) yields for the radiation surface coefficient

$$h_r = 0.1714 \times 10^{-8}(0.90)\left(\frac{1848^4 - 672^4}{1848 - 672}\right)$$
$$= 14.93 \text{ Btu/hr ft}^2 \text{ °F.} \tag{7.73}$$

The value h_{co} will be given by Eq. (7.63), (7.69), or (7.71), depending upon the value of the associated inequality involving λ_d/D. The properties of the vapor are evaluated at the vapor film temperature given by Eq. (7.66) as $T_f = \frac{1}{2}(T_w + T_0) = \frac{1}{2}(1388 + 212) = 800$°F. Since the pressure of the vapor is 14.7 psia, properties of *superheated* vapor at 14.7 psia and 800°F are found in the steam tables to be $k_v = 0.029$ Btu/hr ft °F, $\mu_v = 0.0592$ lbm/hr ft, $\rho_v = 0.0196$ lbm/ft³, $\rho = 59.8$ lbm/ft³, $h_{fg} = 970.3$ Btu/lbm, and $c_{pv} = 0.45$. Also, $D = \frac{1}{12}$ ft, $\Delta T = 1388 - 212 = 1176$°, and $g = 4.17 \times 10^8$ ft/hr².

From Eq. (7.52) at $T_0 = 212°F$, one has that $\sigma = 40.5 \times 10^{-4}$ lb$_f$/ft. Thus, using Eq. (7.65), we have

$$\lambda_c/D = \frac{2\pi}{(\frac{1}{12})} \sqrt{\frac{4.17 \times 10^8(40.5 \times 10^{-4})}{4.17 \times 10^8(59.8 - 0.0196)}} = 0.621$$

and $\lambda_c = 0.052$ ft.
This value of λ_c/D satisfies inequality Eq. (7.70) and therefore dictates the use of Eq. (7.69). Inserting the known quantities into Eq. (7.69) yields the following:

$$h_{co} = 0.60 \left\{ \frac{(0.029)^3(0.0196)(59.8 - 0.0196)4.17 \times 10^8}{0.0592(0.052)(1176)} \right.$$
$$\left. \times [970.3 + 0.4(0.45)(1176)] \right\}^{1/4}$$
$$= 26.6 \text{ Btu/hr ft}^2 \text{ °F.} \tag{7.74}$$

Since the ratio

$$\frac{h_r}{h_{co}} = \frac{14.93}{26.6} = 0.561 \tag{7.75}$$

satisfies the inequality above Eq. (7.68), Eq. (7.68) can be used to determine the average surface coefficient of heat transfer between the horizontal bar and the water. Substituting the values given in Eqs. (7.73–74) into Eq. (7.68) yields

$$h = 26.6 + 14.93 \left[\frac{3}{4} + \frac{1}{4}(0.561) \left(\frac{1}{2.62 + 0.561} \right) \right]$$
$$= 38.5 \text{ Btu/hr ft}^2 \text{ °F.}$$

Note that this surface coefficient is low relative to that for nucleate boiling in the same liquid (c.f. Ex. 7.5). Therefore, the statement that heat transfer with phase change yields high values of the surface coefficient of heat transfer at moderate temperature differences does not hold over the entire boiling curve. In boiling processes, it applies principally to the nucleate boiling region.

7.6 FORCED CONVECTION BOILING

Forced convection boiling, or "flow boiling," refers to boiling which occurs as a result of a fluid moving through a duct or over the outside of a tube or some other heater surface geometry, because an external device, such as a pump, causes a forced velocity. This type of boiling is considerably more complicated than pool boiling, especially for the case in which the liquid to be boiled is flowing *inside* a tube or a duct of a different shape. For the case of forced convection boiling inside a tube, the bulk of the fluid has its mass fraction of vapor, the quality x, continually changing along the tube, thus yielding a general two-phase flow problem.

Consider a subcooled liquid entering a vertical tube heated electrically so that the surface heat flux is constant along the tube. Near the tube entrance, single phase forced convection occurs, with both the fluid bulk mean temperature and the tube wall temperature increasing along the tube. When the local wall temperature exceeds the saturation temperature by a great enough amount, bubbles form at the surface, and subcooled nucleate boiling may occur and then change to saturated nucleate boiling when the bulk mean temperature reaches the saturation temperature. Because it contains a moving mixture of liquid and of vapor bubbles, the flow is now termed bubbly flow, and this flow regime persists for relatively low quality, x, flows, quality below about 5 to 10% according to Rohsenow [22]. Above this quality, the high-quality region occurs, in which the flow becomes the "annular" type in which there is an annulus of liquid against the tube wall and a core of vapor which might or might not also contain liquid droplets. Farther down the tube, the wall might become dry of liquid, at and beyond the critical condition, causing an attendant large increase in the wall temperature for a constant flux tube. This condition may be caused by vapor blanketing the surface in analogy with the peak flux point in pool boiling or by the high-velocity vapor in the core entraining the liquid into the core. This region is analogous to film boiling in a pool.

Now that forced convection boiling in a tube has been described qualitatively, a brief discussion of some of the quantitative relations for the flux will be presented. Experimental evidence indicates (see Rohsenow [22]) that in the low quality, bubbly flow regime, the heat flux to the fluid, q'', includes a single phase forced convection component, q_F''. This component is given by Newton's cooling law using values of the surface coefficient of heat transfer presented in Chapter 5 for single phase forced convection. The value q'' also includes a nucleate boiling component, q_B''. As an approximation, q_B'' may be taken as the value given by pool boiling relations such as Eq. (7.51), or it may come from a separate experimental correlation of forced convection nucleate boiling data. Thus we have

$$q'' = q_F'' + q_B''. \tag{7.76}$$

At low surface heat flux q'' or low wall-to-bulk-fluid temperature differences, q_F'' is a significant part of Eq. (7.76). Hence q'' depends strongly on the forced convection velocity in those conditions. At very high values of q'', $q_B'' >> q_F''$ and curves of all different fluid velocities merge into a single boiling curve, $q'' = q_B''$. (Rohsenow's equivalent of Eq. (7.76) also contains another flux, the incipient boiling flux, which is usually small relative to q_F''. For simplicity's sake, this has been dropped from Eq. [7.76].)

For subcooled and low quality (mass fractions of the vapor less than about 10%) forced convection boiling of water, Rohsenow [22] presents the following relation due to Thom and co-workers. For English units,

$$q_B'' = 192.9(T_w - T_s)^2 e^{P/630}. \tag{7.77}$$

In Eq. (7.77), q''_B is in Btu/hr ft², $T_w - T_s$ in °F, and P must be in psia, while 192.9 is a *dimensional* constant with the units of Btu/hr ft² °F².

For SI units,

$$q''_B = 1971.1(T_w - T_s)^2 e^{P/4.344 \times 10^6}. \tag{7.78}$$

In Eq. (7.78), q''_B must have units of W/m², $T_w - T_s$ is in °C, and P is the pressure in N/m², while 1971.1 is a *dimensional* constant with units of W/m² °C².

As mentioned previously, *Eqs. (7.77) and (7.78) are limited to subcooled and low quality (x < 10%) forced convection boiling of water.* The limit on the quality restricts these relations to the bubbly flow regime.

In the bubbly flow regime just discussed, there is no significant dependence of either q''_F or q''_B in Eq. (7.76) on the mass fraction of the vapor, the quality x, as long as the quality is less than about 10%. However, in the annular flow regime, where the quality is higher than about 5 to 10%, both q''_F and q''_B in Eq. (7.76) depend upon the local value of the quality x. A widely accepted correlation for the surface coefficients of heat transfer used to calculate q''_F and q''_B in the annular, higher quality regime is that of Chen [35]. This result can be used in vertical flow through a pipe for qualities up to about 70% or to the point where the annular flow regime ends. These results of Chen, however, require knowledge of the local quality x in order to be used. But the local quality would not ordinarily be known in advance as a function of position along the tube in the annular regime, since the quality is determined by the flux q'' itself. Hence, this is a general two-phase flow problem in which the quality and the flux are mutually coupled and would have to be solved for simultaneously. This problem is beyond the scope of this text; the reader is referred to Rohsenow [22] and Griffith [36]. According to Rohsenow, the critical heat flux in forced convection boiling, which occurs at the end of the annular flow regime, is a more complicated phenomenon and is less well understood than it is in pool boiling. Thus, predictions of the point at which the critical heat flux occurs are not very reliable. Information on the critical flux in forced convection boiling can be found in Rohsenow [22] and in Hahne and Grigull [37].

7.7 SUMMARY OF CONDENSATION AND BOILING EXPRESSIONS

The various expressions presented in this chapter for the surface coefficient of heat transfer and the surface heat flux in condensation and in boiling heat transfer are collected in this section for the reader's convenience. They are presented, in terms of their equation numbers, in Table 7.2.

As with the summary tables of Chapters 5 and 6, the reader is warned that the conditions for validity of the various expressions in Table 7.2 are short and, to some degree, incomplete. Hence, it is recommended that a tentative selection of an expression from Table 7.2 be followed by a re-reading of the restrictions for that expression, which are discussed where the expression is first presented in the chapter.

Table 7.2
Summary of Expressions for the Heat Transfer Coefficient and Heat Flux in Condensation and Boiling

Geometry of Body	Comments and Conditions	Equation Number
Vertical plates and cylinders	Laminar, film condensation, constant T_w. For vertical tubes $Y_0/D \ll 1$; see Eq. (7.32) $$4\Gamma/\mu < 1800, \ 0 < c_p\Delta T/h_{fg} < 1, \ N_{Pr} \geq 1$$	(7.24)
	Turbulent film condensation, constant T_w $$4\Gamma/\mu > 1800$$	(7.48)
Inclined plate	Laminar film condensation on upper surface at constant T_w, or, alternately, replace g in Eq. (7.24) by $g \sin \phi$ $$4\Gamma/\mu < 1800$$	(7.20)
Horizontal and almost horizontal tubes	Laminar film condensation on outside of tubes at constant T_w	
	Single horizontal tube $$4\Gamma/\mu < 1800, \ 0 < c_p\Delta T/h_{fg} < 1, \\ N_{Pr} \geq 1$$	(7.40)
	Vertical tier of n horizontal tubes $$4\Gamma/\mu < 1800, \ c_p\Delta T/h_{fg} \leq 2, \ N_{Pr} \geq 1$$	(7.42)
	Laminar film condensation inside tubes at constant T_w	
	Horizontal tube with drainage through bottom so that no condensate pool forms there $$0 < c_p\Delta T/h_{fg} < 1, \ N_{Pr} \geq 1.$$	(7.40)
	Almost horizontal tube with condensate flow along tube Tube slope ≤ 0.002, $N_{Re_v} < 35,000$.	(7.43)
Inclined tubes and helical condensers	Laminar film condensation on the outside of the constant T_w tube or helical circular cylinder $$N_{Pr} \geq 1, \ L/D \geq 25 \tan \phi$$	(7.44)
Horizontal plates and cylinders and most other geometries	Nucleate pool boiling, relation relatively insensitive to geometry and orientation	(7.51)
Horizontal plate	Peak flux in nucleate pool boiling	
	SI units	(7.55)
	English units	(7.54)
	Temperature difference at the peak flux point	(7.56)
	Peak flux in subcooled nucleate pool boiling	(7.61)

Chapter Seven

Table 7.2
(Continued)

Geometry of Body	Comments and Conditions	Equation Number
Horizontal cylinder	Peak flux in nucleate pool boiling	
	$0.15 \leq R' \leq 3.47$	(7.57)
	$R' > 3.47$	(7.58)
	Peak flux in subcooled nucleate pool boiling	(7.61)
	Film pool boiling	
	$\lambda_c/D \leq 0.8$	(7.69)
	$0.8 \leq \lambda_c/D < 8$	(7.63)
	$\lambda_c D > 8$	(7.71)
Tubes and ducts	Forced convection nucleate boiling of *water* *inside* tubes and ducts. Subcooled or low-quality saturated boiling, $x < 10\%$ in the bubbly flow regime	
	SI units	(7.78)
	English units	(7.77)

7.8 PROBLEMS

7.1 One side of a vertical plate, 0.15 m high and 0.3 m wide, at 80°C is in contact with saturated steam at 1.01×10^5 N/m². Find (a) the average surface coefficient of heat transfer between the plate surface and the steam, and (b) the rate at which condensate flows from the bottom of the plate.

7.2 If all other conditions remain the same, how high must the plate in Prob. 7.1 be before turbulence is encountered in the condensate film?

7.3 Rework Prob. 7.1 if the plate is inclined at 20° to the vertical.

7.4 A vertical condenser tube is 5 cm in diameter, 0.6 m high, and has a surface temperature of 120°C. It is in contact with saturated steam at 2.69×10^5 N/m². If a condenser contains 20 of these tubes, calculate the rate at which steam is condensed.

7.5 A heat transfer surface is idealized as a 1.5-ft-high and 3-ft-wide vertical flat plate at 150°F. The surface is in contact with saturated steam at 200°F. Calculate the total rate at which steam is condensed on one side of the surface. Assume that the condensation is of the laminar film type.

7.6 Find a criterion in terms of $c_p\Delta T/h_{fg}$ which indicates when Eq. (7.19) is within 5% of Eq. (7.24).

7.7 Saturated steam at 14.7 psia is condensing inside a 1-ft-high and 1-in.-diameter vertical tube. The inside tube surface is at 188°F. Estimate the rate of condensate flow out of the bottom of the tube.

7.8   If the plate of Prob. 7.5 is inclined at 60° to the horizontal, determine the total condensation rate.

7.9   Calculate the thickness of the condensate layer at the bottom of one of the tubes of Prob. 7.4.

7.10   A condenser is built utilizing 0.3-m-high, 2.5-cm-diameter vertical tubes whose surface temperature is 60°C. If the condenser is to process 0.063 kg/s of saturated steam at 80°C, calculate the number of tubes needed in the condenser.

7.11   Saturated steam at 40°C condenses on the 30°C outside surface of a vertical tube 1 m high and 7.5 cm in diameter. The tube surface is kept at 30°C by water flowing inside. If the cooling water temperature rise is 6°C, find the mass flow rate of the cooling water through the tube.

7.12   One of the vertical tiers in a condenser consists of 20 horizontal tubes, one above the other, of 1.25 cm diameter, at 60°C. If they are surrounded by saturated steam at 90°C, compute the condensate flow rate from the vertical tier. The tubes are 1 m long.

7.13   Find the average surface coefficient of heat transfer between the steam and the top horizontal tube in the tier of Prob. 7.12.

7.14   A horizontal condenser tube is 1 in. in diameter and 5 ft long, and is at a temperature of 180°F and in contact with saturated steam at 220°F. Find the total condensation rate $\dot{M}$ of the steam.

7.15   One of the vertical tiers in a condenser is made up of 10 horizontal tubes spaced directly below one another. The tubes are of 1/2 in. outside diameter, have a surface temperature of 140°F, and are in contact with 240°F saturated steam. Calculate the condensation rate of steam on this vertical tier if the tubes are 3 ft long.

7.16   A condenser is built with 2.5-cm-diameter, 1.6-m-long horizontal tubes arranged such that each vertical tier contains 10 tubes. If the tube surfaces are at 80°C and will be in contact with saturated steam at 100°C, calculate the number of vertical tiers needed for a total condensation rate of 0.5 kg/s.

7.17   Superheated vapor at 1.43×10^5 N/m² and 130°C surrounds a horizontal 5-cm-diameter tube whose surface temperature is 80°C. Compute the condensate flow rate from the bottom of the 1.6-m-long tube.

7.18   Find an expression for the average surface coefficient of heat transfer between a saturated vapor and the kth tube from the top in a vertical tier consisting of N horizontal tubes at temperature T_w, which is less than the vapor saturation temperature.

7.19   Saturated steam at 440°F is condensing on a 2-in.-diameter, 4-ft-high vertical cylinder whose outside surface is at 300°F. Compute the total rate at which steam condenses.

7.20   A vertical plate condenser is 1 ft wide, at a temperature of 140°F, and is in contact with saturated steam at 200°F. If the steam condensation rate $\dot{M}$ is to be 1000 lbm/hr, find the height of the plate.

7.21   Saturated steam at 110°C is condensing on a horizontal tube of surface temperature 70°C. Calculate the tube diameter which would just make the condensate film turbulent.

7.22   A condensing surface is idealized as a flat plate 1.6 m high and 3 m long which has its surface held at 60°C. It is in contact with saturated steam at 120°C. Compute the rate at which steam is being condensed on one side of this surface.

7.23   A condenser tube, 2 in. in diameter and 10 ft high, is maintained at a surface temperature of 80°F while in contact with saturated steam at 17 psia. If the cooling water passing through the tube is to have a maximum temperature rise of 30°F, calculate (a) the mass flow rate of the cooling water and (b) the rate at which steam is being condensed.

7.24   Consider the inside of the container shown in Fig. 7.9, with the bottom held at constant temperature T_0. Surrounding the container is a vapor at its saturation temperature T_v. Call the instantaneous height of the condensate z, and neglect any condensation on the walls. Make a simple analysis, noting all assumptions, to yield an expression for z as a function of time τ. The cross-sectional area of the container bottom is A, the condensate density is ρ, and latent heat of condensation is h_{fg}.

Figure 7.9

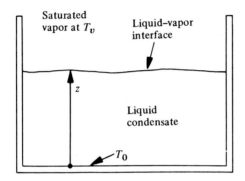

7.25   Find the heat transfer rate per unit area and the surface coefficient of heat transfer for the nucleate pool boiling of saturated water at 212°F on a 242°F horizontal copper heater surface.

7.26   Calculate the maximum heat flux in the nucleate boiling of saturated water at 212°F on a horizontal plate.

7.27   Using Prob. 7.25, find the maximum temperature difference in nucleate boiling of saturated water at 212°F on a horizontal copper heater surface.

7.28   Saturated water at 1.43×10^5 N/m² surrounds a nickel heating rod whose surface temperature is 115°C. Calculate the surface coefficient of heat transfer between the rod and the water.

7.29   Calculate the value of the maximum heat flux in the nucleate pool boiling of saturated water at 1.43×10^5 N/m² on a horizontal 2.5-cm-diameter rod.

7.30   Find the maximum temperature difference in the nucleate boiling of saturated water at 1.43×10^5 N/m² on the nickel heater of Prob. 7.29.

7.31   Nucleate boiling of saturated water at 280°F occurs on the surface of a horizontal stainless steel tube whose surface temperature is 300°F. Estimate the average surface coefficient of heat transfer between the tube and the saturated liquid.

7.32 Estimate the tube surface temperature which would give the maximum heat flux in nucleate boiling in Prob. 7.31 if its diameter is 2 in.

7.33 If the tube wall of Prob. 7.32 is at $T_{w_{max}}$ while the tube is immersed in saturated water at 280°F, estimate the average surface coefficient of heat transfer between the tube wall and the saturated liquid.

7.34 A conventional electric frying pan has a frying surface area of 0.052 m² and dissipates electrical energy at the rate of 1000 W across the surface. The frying pan is filled with nucleate boiling water at atmospheric pressure, and steady-state conditions prevail. Estimate the temperature of the surface in contact with the water.

7.35 Using Eq. (7.51), derive the following expression for the average surface coefficient of heat transfer between a surface and the saturated liquid during nucleate boiling:

$$h = \mu \sqrt{\frac{g(\rho - \rho_v)}{g_c \sigma}} \left(\frac{c_p}{C_{sf} N_{Pr}^s}\right)^3 \frac{(T_w - T_0)^2}{h_{fg}^2}.$$

7.36 Compute (a) the maximum flux and (b) the maximum temperature difference for the nucleate boiling of saturated water at 180°C on a horizontal platinum heating surface.

7.37 A small experimental nuclear reactor has its fuel rods cooled by saturated water at 360°F. If the outside surface temperature of the rod is 375°F, estimate the value of the average surface coefficient of heat transfer between the rod and the water if the outside surface of the rod is clad with stainless steel.

7.38 A boiler is designed to produce 1.25 kg/s of steam during the saturated pool boiling of 140°C water. If the heating surface temperature is 150°C, estimate the total surface area required for the stainless steel heater.

7.39 A kettle, whose 1-ft-diameter bottom surface is at 250°F, contains saturated water at 212°F. If the water level initially is 8 in. above the kettle bottom, estimate the time needed for the water to boil completely away. The kettle bottom is made of stainless steel.

7.40 At a certain time during the quenching of a long, 2.5-cm-diameter brass bar in saturated water at 100°C, the bar's surface temperature is 107°C. Estimate the temperature gradient at the bar surface at this instant of time.

7.41 Calculate the percentage change in the maximum heat flux obtainable in nucleate boiling of water if the saturation pressure is changed from 10^5 N/m² to 2×10^5 N/m² on a horizontal plate heater.

7.42 A 5-cm-diameter steel shaft at 1200°C is quenched in a 100°C water bath at atmospheric pressure. If the shaft is approximately horizontal in the bath, calculate the surface coefficient of heat transfer between the shaft surface and the water bath if the bar surface has a total hemispherical emissivity of 0.90.

7.43 Long brass rods, 1/2 in. in diameter, are cooling in saturated water at 212°F. At the instant of time when the rod surface temperature is 1000°F, calculate (a) the average surface coefficient of heat transfer between the rods and the water and (b) the rate at which the rod temperature is changing with time. For the brass surface, $\epsilon = 0.65$.

REFERENCES

1. Hampson, H. "The Condensation of Steam on a Metal Surface." *General Discussion on Heat Transfer, The Institution of Mechanical Engineers and the American Society of Mechanical Engineers* (1951), pp. 58–61.

2. Griffith, P. "Dropwise Condensation." In *Handbook of Heat Transfer,* edited by W. M. Rohsenow and J. P. Hartnett. New York: McGraw-Hill, 1973.

3. Merte, H., Jr. "Condensation Heat Transfer." In *Advances in Heat Transfer,* Vol. 9, edited by T. F. Irvine, Jr. and J. P. Hartnett. New York: Academic Press, 1973.

4. Nusselt, W. "Die Oberflächenkondensation des Wasserdampfes." *Zeitschrift des Vereines Deutscher Ingenieure,* Vol. 60 (July 1916), pp. 541–46 and 569–75.

5. Monrad, C. C., and W. L. Badger. "The Condensation of Vapors." *Industrial Engineering Chemistry,* Vol. 22, No. 10 (October 1930), pp. 1103–12.

6. Rohsenow, W. M. "Heat Transfer and Temperature Distribution in Laminar Film Condensation." *Transactions of the A.S.M.E.* (November 1956), pp. 1645–48.

7. Sparrow, E. M., and J. L. Gregg. "A Boundary Layer Treatment of Laminar Film Condensation." *Transactions of the A.S.M.E., Journal of Heat Transfer,* Vol. 81, No. 1 (February 1959), pp. 13–18.

8. Cooper, C. M., T. B. Drew, and W. H. McAdams. "Isothermal Flow of Liquid Layers." *Industrial Engineering Chemistry,* Vol. 26, No. 4 (1934), pp. 428–31.

9. Kirkbride, C. G. "Heat Transfer by Condensing Vapor on Vertical Tubes." *Transactions of the AIChE,* Vol. 36 (1934), pp. 170–86.

10. Van Wylen, G. J., and R. E. Sonntag. *Fundamentals of Classical Thermodynamics.* SI version, 2d ed. New York: Wiley, 1976.

11. Van Wylen, G. J., and R. E. Sonntag. *Fundamentals of Classical Thermodynamics.* 2d ed. New York: Wiley, 1973.

12. Chen, M. M. "An Analytical Study of Laminar Film Condensation: Part 2—Single and Multiple Horizontal Tubes." *Transactions of the A.S.M.E., Journal of Heat Transfer,* Vol. 83, No. 1 (February 1961), pp. 55–60.

13. Chato, J. C. "Laminar Condensation Inside Horizontal and Inclined Tubes." *ASHRAE Journal* (February 1962), pp. 52–60.

14. Rohsenow, W. M. "Film Condensation." In *Handbook of Heat Transfer,* edited by W. M. Rohsenow and J. P. Hartnett. New York: McGraw-Hill, 1973.

15. Karimi, A. "Laminar Film Condensation on Helical Reflux Condensers and Related Configurations." *International Journal of Heat and Mass Transfer,* Vol. 20 (1977), pp. 1137–44.

16. Keenan, J. H., and F. K. Keyes. *Thermodynamic Properties of Steam.* New York: Wiley, 1958.

17. Bromley, L. A. "Effect of Heat Capacity of Condensate." *Industrial Engineering Chemistry,* Vol. 44, No. 12 (December 1952), pp. 2966–69.

18. Rohsenow, W. M., and H. Y. Choi. *Heat, Mass and Momentum Transfer.* Englewood Cliffs, N.J.: Prentice-Hall, 1961.

19. Karlekar, B. V., and R. M. Desmond. *Heat Transfer.* 2d ed. St. Paul, Minn.: West, 1982.

20. Sparrow, E. M., and E. R. G. Eckert. "Effect of Superheated Vapor and Noncondensable Gases on Laminar Film Condensation." *AIChE Journal,* Vol. 7 (1961), pp. 473–77.

21. Sparrow, E. M., and S. H. Lin. "Condensation Heat Transfer in the Presence of a Noncondensable Gas." *Transactions of the A.S.M.E., Journal of Heat Transfer,* Vol. 86 (August 1964), pp. 430–36.

22. Rohsenow, W. M. "Boiling." In *Handbook of Heat Transfer,* edited by W. M. Rohsenow and J. P. Hartnett. New York: McGraw-Hill, 1973.

23. Jordan, D. P. "Film and Transition Boiling." In *Advances in Heat Transfer,* Vol. 5, edited by T. F. Irvine, Jr. and J. P. Hartnett. New York: Academic Press, 1968.

24. Farber, E. A., and R. L. Scorah. "Heat Transfer to Water Boiling Under Pressure." *Transactions of the A.S.M.E.,* Vol. 70, No. 4 (1948), pp. 369–84.

25. Graham, R. W., and R. C. Hendricks. "Assessment of Convection, Conduction, and Evaporation in Nucleate Boiling," *NASA TN D-3943* (May 1967).

26. Lienhard, J. H. *A Heat Transfer Textbook.* Englewood Cliffs, N.J.: Prentice-Hall, 1981.

27. Rohsenow, W. M. "A Method of Correlating Heat-Transfer Data for Surface Boiling of Liquids." *Transactions of the A.S.M.E.,* Vol. 74 (August 1952), pp. 969–76.

28. Tong, L. S. *Boiling Heat Transfer and Two Phase Flow.* New York: Wiley, 1965.

29. Rohsenow, W. M., and P. Griffith. "Correlation of Maximum Heat Flux Data for Boiling of Saturated Liquids." *Chemical Engineering Progress Symposium Series No. 18,* Vol. 52 (AIChE-ASME Heat Transfer Symposium, Louisville, Ky., 1955), pp. 47–49.

30. Sun, K., and J. H. Lienhard. "The Peak Boiling Heat Flux on Horizontal Cylinders." *International Journal of Heat and Mass Transfer,* Vol. 13 (1970), pp. 1425–39.

31. McAdams, W. H., W. E. Kennel, C. S. Minden, R. Carl, P. M. Picornell, and J. E. Dew. "Heat Transfer at High Rates to Water with Surface Boiling." *Industrial Engineering Chemistry,* Vol. 41, No. 9 (September 1949), pp. 1945–55.

32. Bromley, L. A. "Heat Transfer in Stable Film Boiling." *Chemical Engineering Progress,* Vol. 46, No. 5 (May 1950), pp. 221–27.

33. Bromley, L. A., N. R. LeRoy, and J. A. Robbers. "Heat Transfer in Forced Convection Film Boiling." *Industrial Engineering Chemistry,* Vol. 45 (1953), pp. 2639–46.

34. Lubin, B. T. "Analytical Derivation for Total Heat Transfer Coefficient in Stable Film Boiling from Vertical Plate." *Transactions of the A.S.M.E., Journal of Heat Transfer,* Vol. 91, No. 3 (August 1969), pp. 452–53.

35. Chen, J. C. "Correlation for Boiling Heat Transfer to Saturated Fluids in Convective Flow." *Industrial and Engineering Chemistry, Process Design and Development,* Vol. 5, No. 3 (July 1966), pp. 322–29.

36. Griffith, P. "Two-Phase Flow." In *Handbook of Heat Transfer,* edited by W. M. Rohsenow and J. P. Hartnett. New York: McGraw-Hill, 1973.

37. Hahne, E., and U. Grigull, eds. *Heat Transfer in Boiling.* Washington, D.C.: Hemisphere, 1977.

38. Nukiyama, S. "The Maximum and Minimum Values of the Heat Q Transmitted from Metal to Boiling Water Under Atmospheric Pressure." *Journal Japan Soc. Mech. Engrs.,* Vol. 37 (1934), pp. 367–74. Translation by C. J. Lee in *International Journal of Heat and Mass Transfer,* Vol. 9 (1966), pp. 1419–33.

HEAT EXCHANGERS

8.1 INTRODUCTION

A piece of equipment often needed and encountered in heat transfer practice is the so-called "heat exchanger." Basically, this is a device which allows energy transfer, because of temperature differences, between a hot fluid and a cold fluid. This occurs either across a solid barrier interposed between the two fluids to prevent their mixing, or to and from this solid wall as a result of the alternate passing of hot and cold fluids over the barrier. Figure 8.1 gives a schematic picture of such a general heat exchanger. In general, there are also, as shown in Fig. 8.1, other surfaces whose function it is to confine one or both of the fluids and which have essentially no heat transfer across them. In many heat exchangers, the barrier between the fluids will be a tube wall or a plane wall and the outer containment barriers might be circular cylindrical surfaces or plane surfaces.

Heat exchangers are commonly used in air conditioning and refrigeration systems, space heating systems, power production systems, in the chemical processing industry, and in engines of all types. Specific examples include the radiator of an automobile, where outside air is used to cool the liquid water often employed as the engine coolant; the condenser of a household refrigerator, where energy is being rejected from the refrigerant, such as Freon-12, to the room air; the boiler of a large power plant, where energy is being transferred from hot combustion gases to liquid water in order to generate steam; and a fuel-oil cooler in a jet engine, where lubrication oil is cooled by having it slightly preheat fuel going to the combustor. Sizes of heat exchangers range from relatively small units, such as a fuel-oil cooler which can be held in one's hand, to extremely large units, such as power plant boilers and feedwater preheaters, whose overall size may dwarf a person and which may have 5000 or 10,000 m² of heat transfer area.

In this chapter, previously developed topics in heat transfer, such as the rate equations for convection and conduction heat transfer, the energy balance, and the various expressions for the surface coefficient of heat transfer, will be integrated in order to analyze the performance of existing heat exchangers and to allow some preliminary steps to be taken in the thermal design of a new heat exchanger.

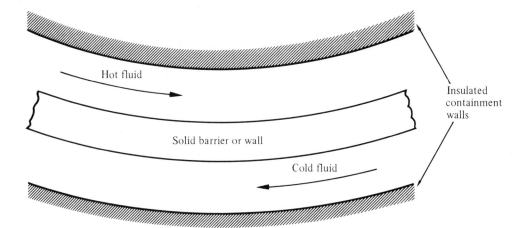

Figure 8.1 Schematic of a general heat exchanger.

8.2 TYPES OF HEAT EXCHANGERS

In a direct contact heat exchanger, as the name implies, the two fluids exchanging energy with one another do so by making actual physical contact. A cooling tower, in which condenser water falls through an atmospheric air column and in which evaporation of part of the water plays a major role, is one example of such an exchanger. A tank in which liquid water is heated by bubbling steam through it is another. This type of exchanger, being less common than some other types, will not be dealt with here. Information concerning the analysis of direct contact exchangers is available in the chapter on cooling towers in Jakob [1].

Excluding direct contact heat exchangers, all other exchangers can be classified as either *regenerators* or *recuperators*.

A regenerator, or regenerative-type heat exchanger, has a solid matrix, such as channels constructed of brickwork, or a package of wire mesh, which has a hot fluid flowing over or through it for a certain time period, thus raising the temperature of the solid and cooling the fluid, and then has the cooler fluid flowing over it for, in general, a different period of time. As the cooler fluid passes over the solid, the solid cools down while transferring its energy to the fluid, causing the fluid temperature to rise. The entire cycle is repeated continually. The alternate passing of the hot and cold fluid through the solid can be accomplished either by switching the hot and cold streams over a stationary solid matrix, as is done in a blast furnace stove, or by moving the solid matrix back and forth from the hot stream to the cold, as is done in the rotary regenerator of a gas turbine power plant. As is evident by its nature, the regenerative heat exchanger never reaches a true steady state. Instead, it ultimately reaches a periodic unsteady state, in which the solid temperature depends upon both position and time.

Because of the dependence upon time, this type of exchanger is more difficult to analyze than the recuperative type. It is also more specialized and not as frequently encountered. An account of regenerator theories is given in Jakob [1] and in Hausen [14]. Other

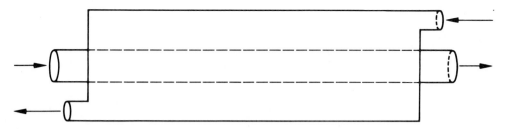

Figure 8.2 Schematic of a simple counterflow heat exchanger.

information related to regenerators is available in Kays and London [2], Fraas and Ozisik [3], Hausen [14], and Afgan and Schlünder [4].

The last class or type of heat exchanger is the recuperative exchanger, sometimes called an ordinary heat exchanger or simply a heat exchanger. In this type, as depicted in Fig. 8.1, there is a solid barrier, the heat transfer surface, between the two fluids, which flow continually over the two sides of the barrier. Thus, in the recuperative exchanger, a true steady state is ultimately reached, and it is this steady-state condition that we will be concerned with in the remainder of the chapter. For unsteady-state performance of a recuperator, what is called heat exchanger dynamics, refer to Kays and London [2] and to Afgan and Schlünder [4]. From this point on, we will be discussing only recuperative exchangers, which we will refer to simply as heat exchangers.

There are a number of important types of recuperative heat exchangers in common usage. The simplest type is the pure counterflow exchanger or the pure parallel flow exchanger, which can be two concentric pipes, as shown in Fig. 8.2. The two fluids flow in opposite directions in these pipes in the simple counterflow exchanger, while both fluids flow in the same direction in a simple parallel flow exchanger. The counterflow exchanger of Fig. 8.2 is also the simplest example of a shell-and-tube heat exchanger, in which one or more tubes carry one of the fluids while the other fluid flows in the opposite direction between the outside surface of the tubes and the outer shell.

However, in most industrial situations, heat exchanger compactness—that is, large surface area for heat transfer to exchanger volume ratios—and economy usually prohibit a pure counterflow design. Instead, either a shell-and-tube design using multiple tube passes and sometimes multiple shell passes, or a crossflow heat exchanger design are used. Figure 8.3(a) shows a one-shell-pass, two-tube-pass shell-and-tube exchanger with transverse baffles. These baffles, if enough are present, direct the shell side fluid over the tubes in a crossflow fashion; this leads to a higher outside surface coefficient of heat transfer, h_o, than would occur if the shell side fluid simply flowed axially along the tubes as it does for the single tube in Fig. 8.2. Note that because of the direction of the tube side fluid in the second tube pass (the three bottom tubes), the overall flow pattern of the tube side fluid relative to the shell side fluid is a combination of counterflow and parallel flow. Figure 8.3(b) shows the shorthand representation which will be used for this type of exchanger. An example of a shell-and-tube heat exchanger with two shell passes and four tube passes is shown in Fig. 8.4.

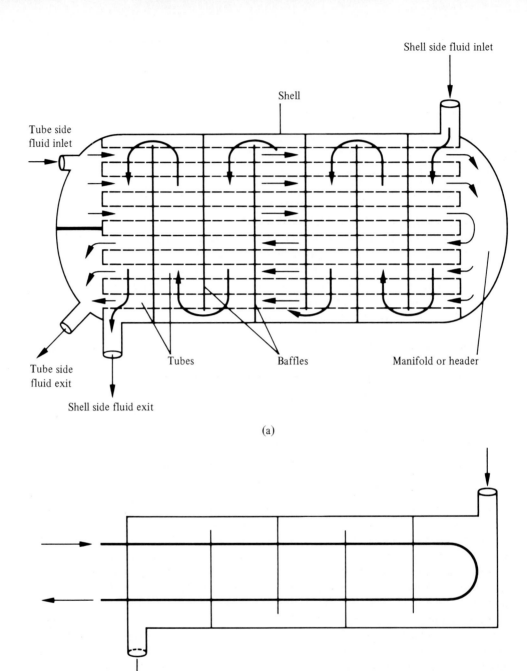

Shell side fluid inlet

Shell

Tube side
fluid inlet

Tube side
fluid exit

Shell side fluid exit

Tubes

Baffles

Manifold or header

(a)

(b)

Figure 8.3 Two representations of a one-shell-pass, two-tube-pass shell-and-tube exchanger.

Chapter Eight

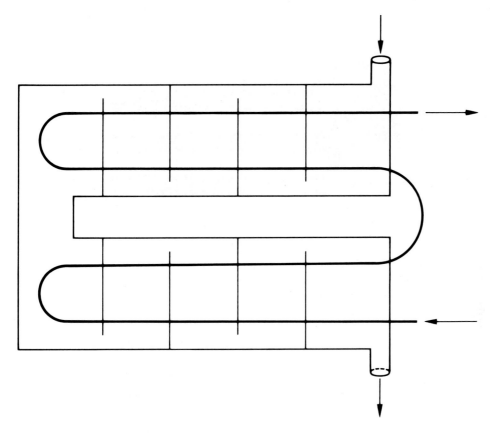

Figure 8.4 Two-shell-pass, four-tube-pass shell-and-tube heat exchanger.

In the crossflow type of heat exchanger, the two fluids involved flow through passages which direct the fluids at right angles to one another, as in Fig. 8.5. The vertical partitions in the crossflow exchanger of Fig. 8.5 prevent fluid 1 from mixing with itself in the transverse section perpendicular to its main flow direction; thus, the temperature of fluid 1 depends upon both x and y. For fluid 2 also, the vertical partitions prevent mixing in the x direction, so that its temperature depends on both the coordinate y in its general flow direction and on the transverse coordinate x. This situation is referred to as both fluids unmixed while in the heat exchanger. When fluids 1 and 2 reach their respective exits from the exchanger, the lack of vertical partitions just beyond the exits allows the transverse turbulent mixing motions to mix the fluids and remove the transverse temperature gradients so that each fluid has essentially a uniform temperature downstream of the actual exchanger exits. An automobile radiator is one example of such a crossflow heat exchanger with both fluids unmixed.

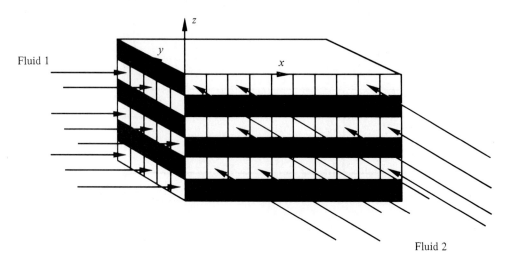

Figure 8.5 Crossflow heat exchanger with both fluids unmixed.

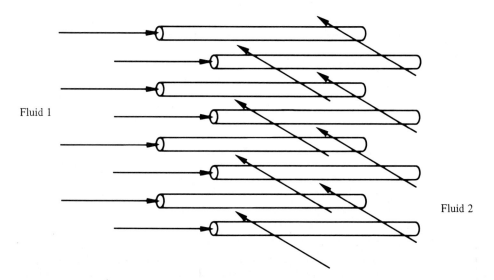

Figure 8.6 Crossflow heat exchanger with fluid 1 unmixed and fluid 2 mixed.

Another commonly occurring crossflow heat exchanger is one in which one fluid is unmixed and the other fluid is mixed. Figure 8.6 depicts this case, in which fluid 1 passing through the tubes is unmixed, while fluid 2, passing over the tubes with no sideways constraints to inhibit turbulent mixing, is the mixed fluid. Figure 8.5 can also represent this case if the vertical partitions for fluid 2 are removed, thus yielding an exchanger in which fluid 1 is unmixed and fluid 2 is mixed. Finally, the case of a crossflow heat exchanger in which both fluids are mixed is illustrated by Fig. 8.7, which is Fig. 8.5 with all vertical partitions removed from the exchanger.

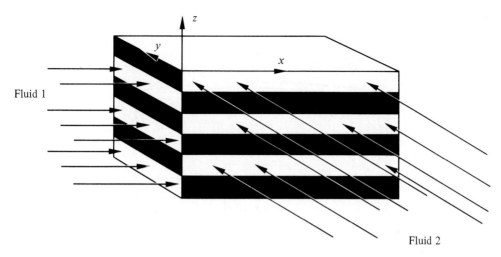

Figure 8.7 Crossflow exchanger with both fluids mixed.

Often, in order to increase the ratio of heat transfer surface area to exchanger volume—that is, to make the exchanger more compact—various types of fins and differently shaped tube passages, including dimpled tubes, are employed in the construction of crossflow heat exchangers. Extensive coverage of compact heat exchangers, including heat transfer and friction data, can be found in Kays and London[2].

8.3 OVERALL COEFFICIENT OF HEAT TRANSMISSION AND FOULING FACTORS

Essential to the fundamental analysis of any heat exchanger is the expression for the heat transfer rate between the two fluids across a differential portion of the solid wall which separates them. The general picture is shown in Fig. 8.8, where at this position in the exchanger, the hotter fluid's temperature is designated T_h while that of the colder fluid is termed T_c. The inside and outside surface coefficients are h_i and h_o, respectively, and their numerical values would be found from the analytical results and experimental correlations presented in Chapters 5, 6, 7, and 9. The basic solid wall of the exchanger is referred to by the subscript w. In general, there will be thin layers of deposits on both sides of this wall, as shown, due to the fouling of the wall by the heat exchanger fluids and dirt and impurities which they might contain. The effect of these fouling deposits will be included by use of experimentally determined fouling factors, r_{fi} and r_{fo}, which are thermal resistances per unit area of the surface containing the deposits. These fouling factors depend upon the total time the surface has been in service, the surface temperature, characteristics of the fluid flowing by the surface, and fluid velocity. When the heat exchanger is first put into service, or just after a periodic cleaning of both surfaces, the fouling factors are zero and heat transfer rates are relatively high. But the continual buildup

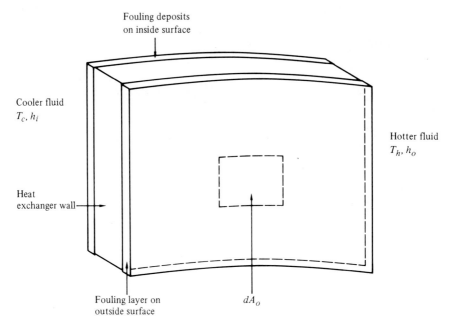

Fouling deposits
on inside surface

Cooler fluid
T_c, h_i

Hotter fluid
T_h, h_o

Heat
exchanger wall

Fouling layer on
outside surface

dA_o

Figure 8.8 General exchanger wall showing fouling layers on both sides.

of these deposits with time as the exchanger becomes fouled or "dirty" can cause a dramatic reduction in the heat transfer rate. Because of this, the Tubular Exchanger Manufacturers Association (TEMA) lists minimum recommended fouling factors [5], and a representative selection of these factors is presented in Table 8.1. For additional information on fouling factors, see [5].

Consider the development of the expression for the heat transfer rate dq across the differential area dA_o located on the outside surface of the heat exchanger wall material. The energy transfer between the two fluids at temperatures T_h and T_c at this location in the exchanger must negotiate the thermal resistances $1/h_o dA_o$ due to the finite outside surface coefficient h_o, the outside fouling layer r_{fo}/dA_o, the finite thickness and thermal conductivity of the wall, r_w/dA_o, the inside fouling layer, r_{fi}/dA_i, and the finite inside surface coefficient, $1/h_i dA_i$. Using the methods developed for steady-state heat transfer across composite barriers in series in Chapter 2, one can write the expression for the heat transfer rate as

$$dq = \frac{T_h - T_c}{\dfrac{1}{h_o dA_o} + \dfrac{r_{fo}}{dA_o} + \dfrac{r_w}{dA_o} + \dfrac{r_{fi}}{dA_i} + \dfrac{1}{h_i dA_i}}. \qquad (8.1)$$

Implicit in Eq. (8.1) is the assumption that the differential areas on the outside of the outer fouling layer and on the inside of the inner fouling layer differ negligibly from dA_o and dA_i, respectively. This will be the case if the layers are thin or are practically planar,

Table 8.1
Selected Values of Typical Approximate Fouling Factors

Fluid Type		Fouling Resistance	
		$m^2 \, °C/W$	$hr \, ft^2 \, °F/Btu$
Seawater	below 52°C (125°F)	0.00009	0.0005
	above 52°C (125°F)	0.00018	0.001
City or well water	below 52°C (125°F)	0.00018	0.001
(such as Great Lakes)	above 52°C (125°F)	0.00035	0.002
Distilled water		0.00009	0.0005
Treated boiler feedwater above 52°C (125°F)		0.00018	0.001
Quenching oil		0.0007	0.004
Fuel oil		0.0009	0.005
Diesel engine exhaust gas		0.0018	0.01
Steam, non–oil bearing		0.00009	0.0005
Industrial air		0.00035	0.002
Refrigerating liquids		0.00018	0.001
Refrigerating vapors condensing from compressors		0.00035	0.002

From *Standards of the Tubular Exchanger Manufacturers Association.* 6th ed. New York: Tubular Exchangers Manufacturers Assoc. (TEMA), 1978.

as is often the case even when the wall is that of a tube, as long as the tube thickness is a small fraction of its inside radius.

Setting dq equal to the product of the overall heat transmission coefficient U_o, based on outside area, with the differential area dA_o and the temperature difference $T_h - T_c$, as was done in Chapter 2, gives

$$dq = U_o dA_o \, (T_h - T_c). \tag{8.2}$$

Equating Eqs. (8.1) and (8.2) and solving for U_o yields

$$U_o = \cfrac{1}{\cfrac{1}{h_o} + r_{fo} + r_w + r_{fi}\cfrac{dA_o}{dA_i} + \cfrac{dA_o}{h_i \, dA_i}}. \tag{8.3}$$

Once the geometry of the heat exchanger wall is selected, r_w and the ratio dA_o/dA_i can be found. If the wall is planar, r_w is simply the wall thickness divided by the wall's thermal conductivity, while $dA_o/dA_i = 1$. For a circular tube wall, the expressions for r_w and dA_o/dA_i were developed in Chapter 2 and are in terms of the tube inside and outside radii and its thermal conductivity.

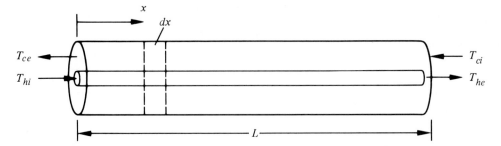

Figure 8.9 Simple counterflow heat exchanger.

Hence, with the surface coefficients determined from Chapters 5, 6, and 7, r_{fi} and r_{fo} found from Table 8.1 or from the TEMA standards [5], and r_w from Chapter 2, a numerical value, or at least an initial estimate of the value, of U_o will be obtained for use in the basic rate equation, Eq. (8.2). The analysis of heat exchangers with the aid of this equation will be the topic of some of the subsequent sections in this chapter.

8.4 ANALYSIS OF SIMPLE COUNTERFLOW AND SIMPLE PARALLEL FLOW HEAT EXCHANGERS

This section will present a detailed analysis of the simple counterflow heat exchanger, which will lead to an expression for the heat transfer rate q in terms of the surface area for heat transfer A, the overall heat transmission coefficient U_o, and the inlet and exit temperatures of both fluids. With this completed, it is a simple matter to write down the equivalent result for the simple parallel flow exchanger.

The analysis of the counterflow exchanger is important in its own right; not only is this type of exchanger sometimes the best solution to the heat exchanger design problem, but it will also be shown that this is thermodynamically the best exchanger. This is so because a counterflow exchanger of infinite surface area will allow one of the fluids to experience the maximum possible temperature change, namely the difference between the cold fluid inlet temperature T_{ci} and the hot fluid inlet temperature T_{hi}. The overall procedure used to analyze the counterflow exchanger is also the one to be used on the more complex exchangers, such as a multi-shell pass, multi-tube pass shell-and-tube exchanger.

Consider the simple counterflow exchanger, consisting of two concentric pipes through which the two fluids move in opposite directions, shown in Fig. 8.9. The hotter fluid enters at $x = 0$ with an inlet temperature T_{hi}, a mass flow rate ω_h, and specific heat c_{ph}. The colder fluid enters at $x = L$ with an inlet temperature T_{ci}, mass flow rate ω_c, and specific heat c_{pc}. The exit temperatures of the hotter and the colder fluids are T_{he} and T_{ce}, respectively. Figure 8.10 shows the qualitative distribution of the local temperatures T_h and T_c of the hot and cold fluids along the x direction. We can see from the figure that T_{he} is closer to T_{ci} than T_{ce} is to T_{hi}. It appears that as the exchanger length L is increased without limit, thus giving an infinite heat transfer surface area, T_{he} will reach T_{ci} and the hotter fluid will have been cooled through the maximum possible temperature range,

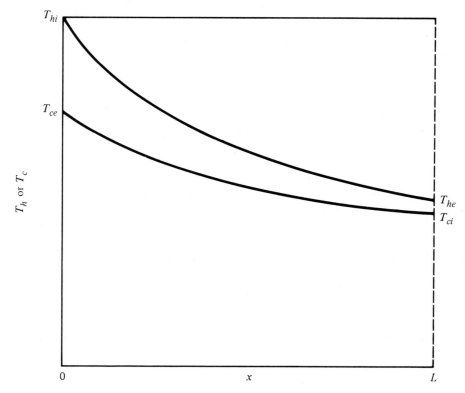

Figure 8.10 Qualitative temperature distributions with x in a counterflow heat exchanger.

$T_{hi} - T_{ci}$. An overall energy balance on the entire exchanger will reveal the condition that determines *which fluid,* the hot fluid or the cold, will undergo the maximum possible temperature change in the ideal counterflow exchanger of infinite area. Thus, applying the energy balance, Eq. (1.8), in the steady state, with no generation, and with negligible heat transfer across the outside of the exchanger gives, for a control volume around the entire exchanger,

$$R_{\text{in}} = R_{\text{out}}, \quad \omega_h c_{ph} T_{hi} + \omega_c c_{pc} T_{ci} = \omega_h c_{ph} T_{he} + \omega_c c_{pc} T_{ce}. \tag{8.4}$$

Define the *capacity rates,* C_c and C_h, as

$$C_c = \omega_c c_{pc}, \quad C_h = \omega_h c_{ph}. \tag{8.5}$$

Using Eq. (8.5) in (8.4) and rearranging gives

$$C_h(T_{hi} - T_{he}) = C_c(T_{ce} - T_{ci}). \tag{8.6}$$

Equation (8.6) tells us what must be true in a counterflow exchanger of infinite area so that the hot fluid exit temperature T_{he} will reach the cold fluid inlet temperature T_{ci}. Replacing T_{he} by T_{ci} and rearranging, Eq. (8.6) becomes

$$\frac{T_{ce} - T_{ci}}{T_{hi} - T_{ci}} = \frac{C_h}{C_c}.$$

Since $T_{hi} - T_{ci}$ is the maximum temperature change of either fluid, it follows that $T_{ce} - T_{ci} < T_{hi} - T_{ci}$, and therefore, from the preceding equation, $C_h < C_c$. Similar reasoning indicates that if the colder fluid experiences the maximum possible temperature rise in an infinite area counterflow exchanger, then it is required that $C_c < C_h$. Hence, *it is the fluid which has the lesser, or minimum, capacity rate* $C = \omega c_p$ *which experiences the maximum possible temperature change in the counterflow exchanger of infinite area.* Note that for a *parallel* flow exchanger of infinite area, *neither* fluid can ever experience the temperature change $T_{hi} - T_{ci}$ (without a phase change), since both fluids are flowing in the same direction and must come to some common temperature T_e *between* T_{hi} and T_{ci}. This statement is also true for other types of exchangers as well, such as the one-shell-pass, two-tube-pass shell-and-tube exchanger.

As indicated earlier, the basic problem is to relate the heat transfer rate q to U_o, the heat transfer surface area A, and the inlet and exit temperatures T_{ci}, T_{hi}, T_{ce}, and T_{he}. The analysis will be carried out for the following conditions:

    a. Steady state
    b. Zero heat transfer across the outside of the exchanger
    c. Constant specific heats c_{pc} and c_{ph}
    d. U does not vary in the x direction
    e. Axial conduction is negligible in the fluids and the wall

The previously derived rate equation, Eq. (8.2), gives the heat transfer rate from the hotter to the cooler fluid across the differential area dA between x and $x + dx$ as

$$dq = U(T_h - T_c)dA. \tag{8.7}$$

Integrating this between $x = 0$ and $x = L$, the other end of the exchanger, where the area is A (the total exchanger heat transfer area), gives, with U independent of x,

$$q = U \int_0^A (T_h - T_c)dA. \tag{8.8}$$

Factoring out the proper average value of the difference in temperature, between the hot and the cold fluids, from the integral operator and calling this temperature difference ΔT, gives

$$q = UA\Delta T \tag{8.9}$$

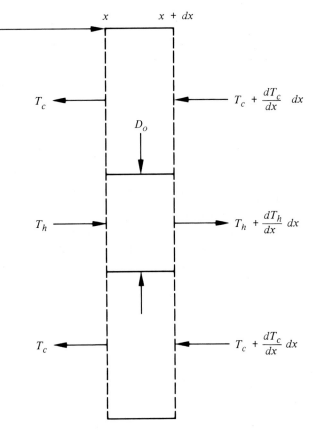

Figure 8.11 Differential control volume for the counterflow heat exchanger of Fig. 8.9.

where

$$\Delta T = \frac{1}{A} \int_0^A (T_h - T_c) dA.$$

The task now is to find out how the appropriate driving potential difference ΔT in Eq. (8.9) is related to the inlet and exit temperatures of the two fluids.

In order to accomplish this, Eqs. (8.9) and (8.7) and energy balances on the two fluids for differential control volumes located at position x down the exchanger will be employed. Toward this end, the control volume shown in Fig. 8.9 between x and $x + dx$ is given in more detail in Fig. 8.11.

At position x, the local bulk mean temperature of the hotter fluid is T_h and that of the colder fluid is T_c with, in general, both of these being unknown functions of x. With the outside diameter of the central tube being D_o, the differential outside surface area for heat transfer is

$$dA = \pi D_o dx,$$

and with this, the basic rate equation, Eq. (8.7), becomes as follows for the heat transfer rate between the hot and cold fluids between positions x and $x + dx$:

$$dq = U(T_h - T_c)\pi D_o dx. \tag{8.10}$$

Next, an energy balance is made on the differential control volume surrounding the hotter fluid. Thus, from Eq. (1.8),

$$R_{\text{in}} = R_{\text{out}}.$$

Energy enters the control volume because of the movement of the hot fluid into the control volume and leaves because of the fluid leaving and because of the heat transfer rate to the colder fluid across the surface area $\pi D_o dx$. Hence,

$$\omega_h c_{ph} T_h = \omega_h c_{ph} \left[T_h + \frac{dT_h}{dx} dx \right] + dq.$$

Using the definitions of Eq. (8.5) and rearranging yields

$$dq = -C_h \, dT_h. \tag{8.11}$$

Following the same procedure, an energy balance on the colder fluid for the differential volume shown between x and $x + dx$ in Fig. 8.11 gives

$$dq = -C_c \, dT_c. \tag{8.12}$$

Solving Eqs. (8.11) and (8.12) for dT_h and dT_c, respectively, and then subtracting dT_c from dT_h leads to

$$dT_h - dT_c = dq \left[\frac{1}{C_c} - \frac{1}{C_h} \right]. \tag{8.13}$$

Substituting for dq from Eq. (8.10) and noting that $dT_h - dT_c = d(T_h - T_c)$, Eq. (8.13) becomes

$$d(T_h - T_c) = \left[\frac{1}{C_c} - \frac{1}{C_h} \right] U(T_h - T_c)\pi D_o dx. \tag{8.14}$$

Equation (8.14) is a differential equation for the local driving temperature difference, $T_h - T_c$, between the two fluids as a function of x and can be solved by separating the variables and integrating. Since we are trying to find, for use in Eq. (8.9), the correct overall driving potential difference ΔT between the two fluids in terms of the inlet and exit temperatures of the fluids, we will integrate between $x = 0$, where $T_h - T_c = T_{hi}$

$- T_{ce}$ as indicated in Fig. 8.9, and $x = L$, where $T_h - T_c = T_{he} - T_{ci}$. Thus, after rearrangement, with these limits Eq. (8.14) becomes

$$\int_{T_{hi} - T_{ce}}^{T_{he} - T_{ci}} \frac{d(T_h - T_c)}{T_h - T_c} = \int_0^L \left[\frac{1}{C_c} - \frac{1}{C_h} \right] U \pi D_o dx. \tag{8.15}$$

Performing the indicated integrations when U is independent of x, and noting that the total outside heat transfer surface area of the exchanger is $A = \pi D_o L$, yields the following equation:

$$\ln \left[\frac{T_{he} - T_{ci}}{T_{hi} - T_{ce}} \right] = UA \left[\frac{1}{C_c} - \frac{1}{C_h} \right]. \tag{8.16}$$

Next, integrating Eqs. (8.11) and (8.12) between the inlet and exit of the hot fluid and the cold fluid, respectively, gives

$$q = C_h(T_{hi} - T_{he}), \tag{8.17}$$

$$q = C_c(T_{ce} - T_{ci}). \tag{8.18}$$

Equations (8.17) and (8.18) could also have been arrived at by applying energy balances to the hot and cold fluid separately for control volumes extending from the inlet, at $x = 0$, and the exit, $x = L$, of the entire exchanger.

Solving for the temperature differences in Eqs. (8.17) and (8.18) and then subtracting the first equation from the second gives

$$q \left[\frac{1}{C_c} - \frac{1}{C_h} \right] = (T_{ce} - T_{ci}) - (T_{hi} - T_{he}). \tag{8.19}$$

But $q = UA\Delta T$, from Eq. (8.9), and using this in Eq. (8.19) results in

$$UA \left[\frac{1}{C_c} - \frac{1}{C_h} \right] \Delta T = (T_{ce} - T_{ci}) - (T_{hi} - T_{he}). \tag{8.20}$$

Finally, using Eq. (8.16) to eliminate $UA \left[\frac{1}{C_c} - \frac{1}{C_h} \right]$ and rearranging leads to the equation for the overall driving potential difference, ΔT, in terms of the inlet and exit temperatures of the two fluids:

$$\Delta T = \Delta T_{\text{LMTD}} = \frac{(T_{he} - T_{ci}) - (T_{hi} - T_{ce})}{\ln \dfrac{T_{he} - T_{ci}}{T_{hi} - T_{ce}}}. \tag{8.21}$$

$$(T_{ce} - T_{ci}) - T_{he} - T_{he}$$

Heat Exchangers

This is called the *logarithmic mean temperature difference* for a simple *counterflow* heat exchanger—hence the use of the letters LMTD as a subscript on ΔT in Eq. (8.21). Furthermore, nothing in Eq. (8.21) requires that the hot fluid enter at $x = 0$, so Eq. (8.21) may be used regardless of which end of the exchanger, $x = 0$ or $x = L$, the hotter fluid enters.

For the case in which one of the fluids is undergoing a phase change at constant temperature T_p as it passes through the exchanger while the remaining fluid changes temperature from T_i at the inlet to T_e at the exit, it can be demonstrated that Eq. (8.21) degenerates to the proper driving potential difference to be used in Eq. (8.9).

One fluid experiencing a phase change at constant temperature T_p

$$\Delta T_{\text{LMTD}} = \frac{T_e - T_i}{\ln\left[\dfrac{T_i - T_p}{T_e - T_p}\right]} \tag{8.22}$$

Equation (8.22) is basically the same as the logarithmic mean temperature difference equation derived in Chapter 5 for the simpler situation in which a fluid was being heated or cooled while passing through a duct at constant wall temperature T_w. This situation was also frequently dealt with in Chapter 5.

Equation (8.21) could also have been derived in the following fashion: Instead of integrating Eq. (8.15) between $x = 0$ and $x = L$, one can integrate between $x = 0$, where $T_h - T_c = T_{hi} - T_{ce}$, and x, where the upper limit is $T_h - T_c$. Then the resulting equation is solved for $T_h - T_c$ as a function of x and results in an exponential dependence upon x. This is then substituted into Eq. (8.8), the integration is performed as indicated there, and then the result is combined with Eq. (8.9) and ΔT is solved for, yielding Eq. (8.21) after rearrangement.

It can also be demonstrated that if a designates one end of the exchanger, regardless of whether or not that is the hot fluid inlet, and if b designates the other end of the exchanger, the logarithmic mean temperature difference for the counterflow exchanger can be written as

$$\Delta T_{\text{LMTD}} = \frac{(T_h - T_c)_a - (T_h - T_c)_b}{\ln\left[\dfrac{(T_h - T_c)_a}{(T_h - T_c)_b}\right]} . \tag{8.23}$$

8.4a Parallel Flow Heat Exchanger

For the parallel flow heat exchanger in which both fluids would flow in the *same* direction in Figs. 8.9 and 8.11, thus giving T_{ci} and T_{hi} at one end of the exchanger and temperatures T_{ce} and T_{he} at the other end of the exchanger, a derivation which is virtually the same as the one which led to Eq. (8.21) leads to the mean temperature difference appropriate to the parallel, or co-current exchanger. This turns out to be the same as Eq. (8.23). Thus, from Eq. (8.23), we can write the following expression for the case of a *parallel flow exchanger:*

$$\Delta T_{\text{LMTD}} = \frac{(T_{hi} - T_{ci}) - (T_{he} - T_{ce})}{\ln \left[\dfrac{T_{hi} - T_{ci}}{T_{he} - T_{ce}} \right]} . \qquad (8.24)$$

8.4b The Counterflow Exchanger as the Best, or Ideal, Heat Exchanger

Direct comparison of Eqs. (8.24) and (8.21) indicates that, for the same set of inlet and exit temperatures,

$$\left[\Delta T_{\text{LMTD}} \right]_{\text{parallel flow exchanger}} < \left[\Delta T_{\text{LMTD}} \right]_{\text{counterflow exchanger}} .$$

Hence, for the same values of U in both exchangers, the parallel flow exchanger requires a greater heat transfer surface area A than does the pure counterflow exchanger. As a matter of fact, there are, for a given set of inlet temperatures, fluid exit temperatures that *cannot* even be reached in a *parallel* flow exchanger, even if it had an infinite amount of heat transfer area. Yet, in a counterflow exchanger, the fluid can reach these temperatures with a *finite* heat transfer area for the exchanger. These comments also apply to other types of exchangers, such as multi-shell-pass and multi-tube-pass shell-and-tube exchangers and crossflow exchangers, when they are compared to a pure counterflow exchanger.

It is considerations such as these, together with the previously mentioned fact that a counterflow exchanger of infinite heat transfer area causes the fluid with the minimum capacity rate to have the maximum possible temperature change that cause one to conclude that, *from the pure thermodynamic standpoint alone,* the counterflow exchanger is the ideal heat exchanger. However, such other factors as allowable pressure drop and the attendant pumping power requirement, compactness, space limitations, and special requirements often mean that the best or most economical exchanger design is a multi-pass shell-and-tube exchanger or a crossflow exchanger [6], [7]. If, however, one of the fluids is undergoing a phase change at constant temperature, then the appropriate ΔT to use in Eq. (8.9) is given as Eq. (8.22), *regardless of the exchanger type,* as is pointed out in Kays and London [2], for instance. Thus, if either one of the fluids undergoes a constant temperature phase change, or if the ratio of the capacity rate of the fluid with the lower capacity rate to that of the fluid with the greater capacity rate ($C_{\text{min}}/C_{\text{max}}$) approaches zero, then all types of recuperative exchangers perform thermodynamically the same as does a pure counterflow exchanger.

8.5 ANALYSIS OF MULTIPASS SHELL-AND-TUBE EXCHANGERS AND CROSSFLOW EXCHANGERS BY THE CORRECTION FACTOR, *F*, METHOD

For multipass shell-and-tube exchangers and for crossflow heat exchangers, the analysis leading to the appropriate mean temperature difference ΔT to use in Eq. (8.9) is considerably more involved, for these more complex exchangers, than it was for the pure counterflow exchanger, where it culminated in the logarithmic mean temperature difference expression, Eq. (8.21). The details of the analysis for some of these more complicated exchangers is given in Jakob [1], and graphical summaries of results are given in Bowman, Mueller, and Nagle [6] and in the TEMA standards [5].

This type of analysis leads to an expression for the ΔT to be used in Eq. (8.9) in terms of the inlet and exit temperatures of the two fluids involved. It is best suited to analysis and design problems in which all four of these temperatures are known, or problems in which all four can be easily found in advance by using the overall energy balance, Eq. (8.6), to relate the unknown inlet or exit temperature to the other known temperatures. For other types of problems in which, for example, both exit temperatures are unknown, this type of analysis, based on having the ΔT of Eq. (8.9) in terms of the four inlet and exit fluid temperatures, is unwieldy and requires an iterative approach. In these cases, the effectiveness-number of transfer units type of analysis, which will be presented in a forthcoming section, is easier to use.

In this section we will discuss the analysis that is based on ΔT, which is called the correction factor-ΔT_{LMTD} method. The results of this type of analysis for ΔT to be used in Eq. (8.9) are presented in terms of a multiple, or correction factor, *F*, which when multiplied by the logarithmic mean temperature difference for a pure counterflow exchanger, ΔT_{LMTD}, will give the correct mean temperature difference ΔT for the more complicated multipass shell-and-tube exchangers and for crossflow exchangers. Thus,

$$\Delta T = F\Delta T_{\text{LMTD}}. \tag{8.25}$$

The value *F* must be less than one, in general, because of the fact that the pure counterflow exchanger is the ideal, or best, exchanger from the thermodynamic viewpoint alone.

The ΔT_{LMTD} on the right side of Eq. (8.25) is to be formed from the inlet and exit temperatures on the actual multipass shell-and-tube or crossflow exchanger treated as if it were a pure counterflow exchanger. That is, ΔT_{LMTD} is evaluated from Eq. (8.21) for the actual exchanger, regardless of where its inlets are relative to its exits and regardless of relative flow directions at the inlets and exits. The correction factor *F* in Eq. (8.25) takes into account the differences between the actual, complex exchanger and the simple pure counterflow exchanger, and gives an actual mean temperature difference, ΔT, which is less than that for the pure counterflow exchanger.

This correction factor *F* for some selected shell-and-tube and crossflow heat exchangers is presented in graphical form as Fig. 8.12 through 8.15. These figures are from Bowman,

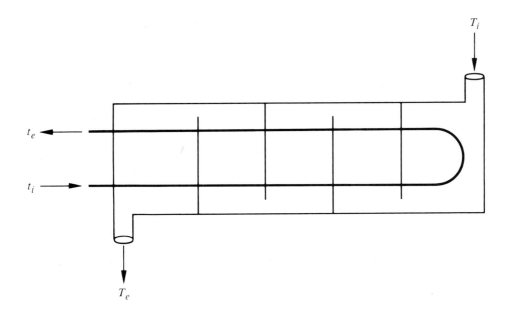

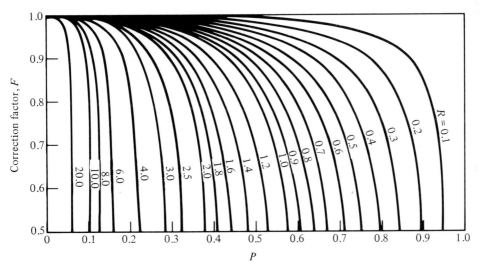

Figure 8.12 Correction factor, *F*, for a one-shell-pass, two-tube-pass exchanger. Generally, also a good approximation for four, six, eight, . . . , tube passes. (Adapted from *A Heat Transfer Textbook* by J. H. Lienhard. Copyright © 1981 by Prentice-Hall, Inc. Used with permission of Prentice-Hall, Inc. From "Mean Temperature Difference in Design" by Bowman/Mueller and Nagle. *Transactions of the A.S.M.E.,* Vol. 62 [1940], pp. 283–94).

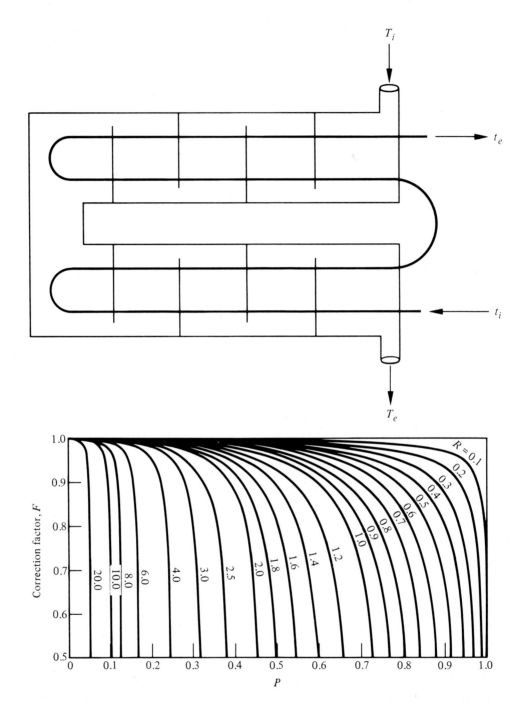

Figure 8.13 Correction factor, *F*, for a two-shell-pass, four-tube-pass exchanger. Generally, also a good approximation for eight, twelve, sixteen, . . . , tube passes. (Adapted from Lienhard, 1981. Used with permission of Prentice-Hall, Inc.)

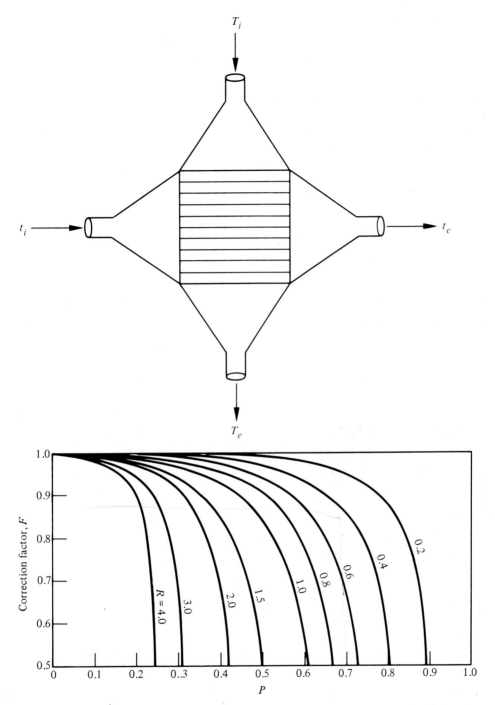

Figure 8.14 Correction factor, F, for a crossflow exchanger with one fluid mixed and the other unmixed. (Adapted from Lienhard, 1981. Used with permission of Prentice-Hall, Inc.)

Heat Exchangers

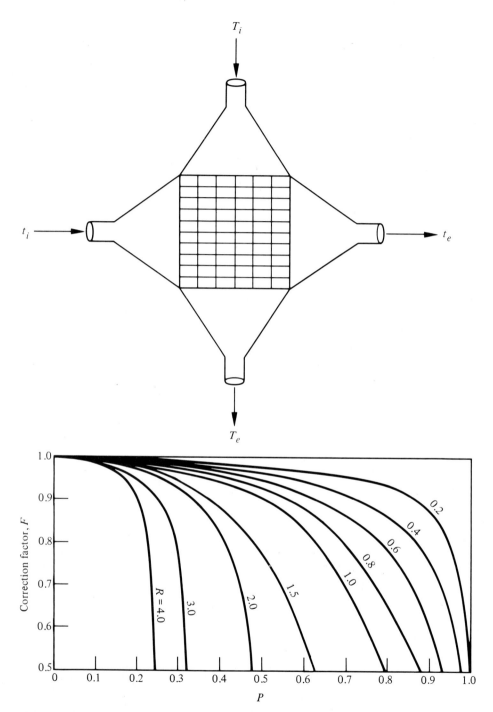

Figure 8.15 Correction factor, F, for a crossflow exchanger with both fluids unmixed. (Adapted from Lienhard, 1981. Used with permission of Prentice-Hall, Inc.)

Mueller, and Nagle [6], as adapted by Lienhard [13]. Additional figures are available in these references for exchangers used less frequently, such as a shell-and-tube exchanger with more than two shell passes.

Notice that Fig. 8.12, which is for a shell-and-tube exchanger with one shell pass and two tube passes, is also approximately valid for greater, even numbers of tube passes, such as 4, 6, 8, etc. [6]. Dodd [8], in a later work, gives more quantitative information about the conditions under which the two-tube-pass results can be used fairly accurately for larger even numbers of tube passes.

Examination of Figs. 8.12 through 8.15 for the factor F indicates that F depends upon the type of exchanger and two temperature ratio parameters, P and R, defined as follows:

$$P = \frac{t_e - t_i}{T_i - t_i},$$

(8.26)

$$R = \frac{T_i - T_e}{t_e - t_i}.$$

(8.27)

As is shown in Figs. 8.12 to 8.14, the lower case letter t *always* refers to the *tube side* fluid, for which t_e is the *exit* temperature and t_i the *inlet* temperature. The uppercase letter T, *without h or c subscripts*, always refers to the shell side fluid, for which T_e is the *exit* temperature and T_i the *inlet* temperature. For the crossflow exchanger with both fluids unmixed, with F given by Fig. 8.15, it is immaterial which fluid is designated shell side or tube side.

In using Figs. 8.12 and 8.13, we assume that the shell-and-tube exchanger possesses enough transverse baffles to justify one of the assumptions made in the analysis which leads to F, that there is no significant transverse temperature variation within the shell side fluid. This point is discussed and investigated in the article by Gardner and Taborek [9].

It is evident from Figs. 8.12 through 8.15 that there are regions of the F versus P curves where the curves are practically vertical lines. Here the slightest change in P can cause an extremely large change in F and can cause severe problems in reliable prediction of the exchanger performance or reliable exchanger design. Thus, it is recommended that one design away from these areas. A rough rule of thumb that is sometimes used is that $F \geq 0.8$ [6], but this can also fail for certain values of the parameter R, as is evident from Figs. 8.12 to 8.15. So, in the last analysis, judgment must be exercised to insure that the exchanger design point is not too close to the almost vertical portions of the curves in those figures.

EXAMPLE 8.1

Lubrication oil leaves an industrial engine at a temperature of 90°C with a mass flow rate of 0.025 kg/s and is to be cooled to 50°C before returning to the engine. To accomplish this, the oil is to flow on the shell side of a one-shell-pass, even number of tube passes heat exchanger constructed of tubes whose outside diameter is $D_o = 1.9$ cm. Water passes through the tubes at a mass flow rate of 0.04 kg/s and with an inlet temperature of 20°C.

The specific heat of the water is 4180 J/kg °C, while that of the oil is 2100 J/kg °C. An estimate has already been made of the overall heat transmission coefficient, based on the outside area of the tubes, as being $U = 250$ W/m² °C. (a) Calculate the needed total outside heat transfer area A of the tubes. (b) If the shell length is not to exceed 0.9 m, how many tube passes are needed?

Solution

The basic relation to be used here—which contains the unknown heat transfer area A—is the overall rate equation, Eq. (8.9), with ΔT given by the product of the correction factor F and the logarithmic mean temperature difference ΔT_{LMTD}, as shown in Eq. (8.25). Hence, combining these two relations gives

$$q = UAF\Delta T_{\text{LMTD}}. \tag{8.28}$$

Note that the oil is the hotter fluid, h, and the water is the colder fluid, c, so that $\omega_h = 0.025$ kg/s, $c_{ph} = 2100$ J/kg °C, $T_{hi} = 90$°C, $T_{he} = 50$°C, $\omega_c = 0.04$ kg/s, $c_{pc} = 4180$ J/kg °C, and $T_{ci} = 20$°C. Therefore, from Eq. (8.5)

$$C_h = \omega_h c_{ph} = 0.025(2100) = 52.5, \checkmark$$

$$C_c = \omega_c c_{pc} = 0.04(4180) = 167.2. \checkmark$$

The exit temperature of the water is needed to find both ΔT_{LMTD} and F in Eq. (8.28). This temperature is found through the overall energy balance on the exchanger, Eq. (8.6), $C_h(T_{hi} - T_{he}) = C_c(T_{ce} - T_{ci})$.
Thus,

$$52.5(90 - 50) = 167.2(T_{ce} - 20),$$

$$T_{ce} = 32.6°C. \checkmark$$

The heat transfer rate q is found from an energy balance on the oil alone (or on the water alone) from Eq. (8.17),

$$q = C_h(T_{hi} - T_{he}) = 52.5(90 - 50) = 2100 \text{ W}. \checkmark$$

The value ΔT_{LMTD} is evaluated as if the actual exchanger were a pure counterflow exchanger. From Eq. (8.21),

$$\Delta T_{\text{LMTD}} = \frac{(50 - 20) - (90 - 32.6)}{\ln\left[\dfrac{50 - 20}{90 - 32.6}\right]} = 42.2. \checkmark$$

Finally, the correction factor F, which takes into proper account the fact that the actual exchanger is not a pure counterflow but a single-shell-pass, even-multiple tube pass exchanger, is found from Fig. 8.12.

In this case, the shell side fluid is the oil, so that $T_i = 90°C$ and $T_e = 50°C$, while the tube side fluid is the water, giving $t_i = 20°C$ and $t_e = 32.6°C$ in the nomenclature of the figure. These temperatures allow the calculation of the parameters P and R as follows. Thus, from Eqs. (8.26) and (8.27), we have

$$P = \frac{t_e - t_i}{T_i - t_i} = \frac{32.6 - 20}{90 - 20} = 0.18,$$

$$R = \frac{T_i - T_e}{t_e - t_i} = \frac{90 - 50}{32.6 - 20} = 3.17.$$

Going to Fig. 8.12 with these values of P and R yields an approximate correction factor F of $F = 0.93$. Using this value of F, along with the previously calculated q and ΔT_{LMTD}, and the given value of U, in Eq. (8.28) yields the required heat transfer surface area.

$$A = \frac{q}{UF\Delta T_{LMTD}} = \frac{2100}{250(0.93)(42.2)} = 0.214 \text{ m}^2. \tag{a}$$

The total length of 1.9-cm diameter tubing needed is L_t and is given by

$$A = \boxed{\pi D_t L_t}$$

$$L_t = \frac{0.214}{\pi(0.019)} = 3.59 \text{ m}.$$

Since the shell length, L_s, is not to exceed 0.9 m, the number n of tube passes needed is

$$n = L_t/L_s = 3.59/0.9 \simeq 4. \tag{b}$$

The number of tube passes calculated in (b) was predicated implicitly on the assumption that only a *single* tube passes from the inlet manifold to the exit manifold. If, instead, one uses two separate tubes, then only two tube passes of each of these separate tubes would be required in the shell length of 0.9 m. (See Fig. 8.3[a] which utilizes, in effect, three separate tubes, each of which makes two passes through the shell.)

EXAMPLE 8.2

Seventy-two thousand lbm/hr of engine oil, whose specific heat $c_p = 0.5$ Btu/lbm °F, is to be cooled from 240°F to 123°F in a heat exchanger by cooling water available at 60°F whose mass flow rate is 40,000 lbm/hr. The oil is to flow inside the exchanger tubes, and the overall heat transmission coefficient based on outside tube area has been calculated to be $U = 33$ Btu/hr ft² °F. In this calculation it was found that the relatively low oil-side surface coefficient of heat transfer is the *controlling resistance* to heat transfer, so that the value of U remains about the same whether the water flows along or across the tubes.

Find the required heat transfer surface area for the following types of exchangers:

a. Simple parallel flow exchanger
b. Simple counterflow exchanger
c. One-shell-pass, two-tube-pass exchanger
d. Two-shell-pass, four-tube-pass exchanger

Solution

The water exit temperature is needed to compute the logarithmic mean temperature difference appropriate to each of the four exchanger types. Hence, since the oil is the hotter fluid and the water the cooler fluid, $T_{hi} = 240°F$, $T_{he} = 123°F$, $T_{ci} = 60°F$, and using Eqs. (8.5),

$$C_h = \omega_h c_{ph} = 72{,}000(0.5) = 36{,}000,$$
$$C_c = 40{,}000(1) = 40{,}000.$$

The energy balance equation for all of the exchangers is Eq. (8.6),

$$C_h(T_{hi} - T_{he}) = C_c(T_{ce} - T_{ci}),$$
$$36{,}000(240 - 123) = 40{,}000(T_{ce} - 60),$$

Thus, $T_{ce} = 165°F$.

Notice that the exit water temperature of 165°F is *greater* than the required exit oil temperature of 123°F. This is not possible in a *parallel* flow exchanger where, of course, both fluids exit at the same end of the exchanger. In such an exchanger both fluids also enter at the same end and then exchange energy as they flow in the same direction along the exchanger. Thus, the Second Law of Thermodynamics, for this type of exchanger, does not allow the originally cooler fluid to ever reach a temperature which is *above* that of the hot fluid *at the same axial position* along the exchanger length. Hence, a simple parallel flow exchanger *cannot* be used, since regardless of its surface area, it cannot do the required cooling job.

Next, the logarithmic mean temperature difference for the pure counterflow exchanger will be calculated, since this is needed in the computations for all three remaining exchangers. From Eq. (8.21),

$$\Delta T_{\text{LMTD}} = \frac{(123 - 60) - (240 - 165.3)}{\ln\left[\dfrac{123 - 60}{240 - 165.3}\right]} = 68.7°F.$$

The required energy transfer rate, q, is found by an energy balance on either fluid; hence,

$$q = \omega_h c_{ph}(T_{hi} - T_{he}) = 72{,}000(0.5)(240 - 123)$$
$$= 4{,}212{,}000 \text{ Btu/hr.}$$

By Eqs. (8.9) and (8.21), this q is related to the needed surface area for the pure counterflow exchanger as follows:

$$q = UA\Delta T_{\text{LMTD}},$$

so

$$4,212,000 = 33\ A(68.7),$$

and $A = 1858\ \text{ft}^2$ for counterflow.

For the one-shell-pass, two-tube-pass and the two-shell-pass, four-tube-pass exchangers, the appropriate rate equation is Eq. (8.28) which, when rearranged, can be written,

$$A = \frac{q}{UF\Delta T_{\text{LMTD}}}. \tag{8.29}$$

The value F will be found from Figs. 8.12 and 8.13 for those two exchangers once the parameters P and R are evaluated from their defining equations, Eqs. (8.26) and (8.27). Since the oil is the tube-side fluid and the water is the shell-side fluid, one has that

$$t_i = T_{hi} = 240°\text{F}, \quad t_e = T_{he} = 123°\text{F},$$
$$T_i = T_{ci} = 60°\text{F}, \quad T_e = T_{ce} = 165.3°\text{F},$$
$$P = \frac{t_e - t_i}{T_i - t_i} = \frac{123 - 240}{60 - 240} = 0.65,$$
$$R = \frac{T_i - T_e}{t_e - t_i} = \frac{60 - 165.3}{123 - 240} = 0.9.$$

Using these values of P and R on Fig. 8.12 for the one-shell-pass, two-tube-pass exchanger, one notes that there is no intersection at $P = 0.65$ with $R = 0.9$, and therefore no solution. Hence, as was the case for the simple parallel flow exchanger, even a one-shell-pass, two-tube-pass exchanger of infinite area cannot perform the task required of it in this problem.

Turning next to the two-shell-pass, four-tube-pass exchanger, Fig. 8.13 yields, at $P = 0.65$ and $R = 0.9$, a correction factor of

$$F \simeq 0.88.$$

Using this in Eq. (8.29) gives the needed area as

$$A = \frac{4,212,000}{33(0.88)(68.7)} = 2111\ \text{ft}^2.$$

Hence, the required areas for the four different exchangers are given in the following list:

a. *Parallel flow.* Exchanger is unable to do the cooling job regardless of its area.
b. *Counterflow.* $A = 1858\ \text{ft}^2.$

c. *One-shell-pass, two-tube-pass.* Exchanger unable to do this cooling job regardless of its area.

d. *Two-shell-pass, four-tube-pass.* $A = 2111 \text{ ft}^2$.

These results also reinforce what was discussed earlier in connection with the pure counterflow exchanger being the "best" or "ideal" exchanger even when it *does not* have infinite heat transfer surface area. Here it is seen that two of the exchangers investigated were unable to perform the required cooling task, even if they could have infinite area. Of the two remaining exchangers that could do the duty called for, the counterflow exchanger requires less area than does the two-shell-pass, four-tube-pass exchanger *as long as both exchangers give essentially the same overall heat transmission coefficient U,* as was the case in this problem.

EXAMPLE 8.3

An existing crossflow heat exchanger has an outside surface area for heat transfer of $A = 3.7 \text{ m}^2$ and is constructed of tubes such that the tube-side fluid would be unmixed and the other outside fluid would be mixed. It is proposed to use this exchanger as a water heater by passing water through the tubes at an inlet temperature of 5°C and a desired exit temperature of 60°C. This is to be accomplished by using air which, after passing over a stove, enters the exchanger at 162°C and is to leave the exchanger at 52°C. An initial rough estimate of the overall heat transmission coefficient based on the outside (air side) area yields $U_o = 20 \text{ W/m}^2 \text{ °C}$. Calculate the mass flow rate of water which can be processed under these conditions.

Solution
The heat transfer rate q between the air and the water is given by the basic rate equation, Eq. (8.9), with ΔT given by Eq. (8.25). Thus,

$$q = U_o A F \Delta T_{\text{LMTD}}.$$

The energy balance on the water alone, Eq. (8.18) since the water is the cold fluid, gives

$$q = C_c(T_{ce} - T_{ci})$$

where the capacity rate, C_c, contains the unknown mass flow rate ω_c. Combining these two equations gives the following:

$$C_c(T_{ce} - T_{ci}) = U_o A F \Delta T_{\text{LMTD}}. \tag{8.30}$$

Since $T_{hi} = 162°C$, $T_{ci} = 5°C$, $T_{he} = 52°C$, and $T_{ce} = 60°C$, the ΔT_{LMTD} calculated as if this were a pure counterflow exchanger is given by Eq. (8.21).

$$\Delta T_{\text{LMTD}} = \frac{(T_{he} - T_{ci}) - (T_{hi} - T_{ce})}{\ln\left[\dfrac{T_{he} - T_{ci}}{T_{hi} - T_{ce}}\right]}$$

$$= \frac{(52 - 5) - (162 - 60)}{\ln\left[\dfrac{52 - 5}{162 - 60}\right]} = 71°C$$

This exchanger is a crossflow type with one fluid, the air, mixed, and the other fluid, the water, unmixed. So F will be found from Fig. 8.14. The water is the tube-side fluid, so

$$t_i = 5°C \quad \text{and} \quad t_e = 60°C.$$

The air is the shell-side fluid, giving

$$T_i = 162°C \quad \text{and} \quad T_e = 52°C.$$

Thus, using Eqs. (8.26) and (8.27), one has the following values of the parameters P and R:

$$P = \frac{t_e - t_i}{T_i - t_i} = \frac{60 - 5}{162 - 5} = 0.35,$$

$$R = \frac{T_i - T_e}{t_e - t_i} = \frac{162 - 52}{60 - 5} = 2.0.$$

With these, Fig. 8.14 yields $F = 0.85$.

With $c_{pc} = 4178$ J/kg °C, from the property tables for water, substitution of known and calculated quantities into Eq. (8.30) gives

$$\omega_c(4178)(60 - 5) = 20(3.7)(0.85)(71), \quad \omega_c = 0.0194 \text{ kg/s.}$$

EXAMPLE 8.4

Eleven thousand lbm/hr of all saturated steam vapor at 8 psia leaves a small steam turbine and now is to be condensed to all saturated liquid. A small shell-and-tube heat exchanger is to be used with city water entering the tubes at 50°F and leaving at 160°F. The estimated overall heat transmission coefficient based on the outside tube area and on clean conditions is about 225 Btu/hr ft² °F. The enthalpy of vaporization of the 8 psia steam is $h_{fg} = 988.4$ Btu/lbm, and its saturation temperature is 182.8°F as found in a standard thermodynamics textbook. (a) Find the mass flow rate of coolant required and the needed heat transfer surface area. (b) After some time of operation has passed, the exchanger becomes fouled or "dirty," and yet is still required to completely condense the 11,000 lbm/hr of steam. Find the needed mass flow rate of coolant and the exit coolant temperature under this condition.

Solution

The required mass flow rate of coolant in the initial, "clean" condition of the exchanger is found by an overall energy balance on the exchanger,

$$w_c c_{pc}(T_{ce} - T_{ci}) = w_h h_{fg},$$

$$w_c(1)(160 - 50) = 11,000(988.4),$$

$$w_c = 98,840 \text{ lbm/hr}.$$

The heat transfer surface area is found by equating the heat transfer rate q as given by the basic rate equation, Eq. (8.9) combined with Eq. (8.25), to the energy transfer rate from the condensing steam, $w_h h_{fg}$, giving,

$$U_o A_o F \Delta T_{\text{LMTD}} = w_h h_{fg}. \tag{8.31}$$

However, in the problem statement, the type of shell-and-tube exchanger, number of shell passes, and number of tube passes was not specified. But since one of the fluids, the steam, is undergoing a phase change at constant temperature, then, as discussed in the text, the logarithmic mean temperature difference with $F = 1$ is to be used regardless of exchanger type, and the form of this temperature difference appropriate to the phase change process is given by Eq. (8.22) as follows:

$$\Delta T_{\text{LMTD}} = \frac{T_e - T_i}{\ln\left[\dfrac{T_i - T_p}{T_e - T_p}\right]}. \tag{8.22}$$

In this problem, $T_p = 182.8°\text{F}$, $T_i = 50°\text{F}$, and $T_e = 160°\text{F}$, so Eq. (8.22) yields

$$\Delta T_{\text{LMTD}} = \frac{160 - 50}{\ln\left[\dfrac{50 - 182.8}{160 - 182.8}\right]} = 62.4°\text{F}.$$

Using this in Eq. (8.31) gives

$$225 \, A_o \, (1)(62.4) = 11,000(988.4), \quad A_o = 774.4 \text{ ft}^2.$$

Also,

$$w_c = 98,840 \text{ lbm/hr} \tag{a}$$

In (b), the exchanger performance is to be studied after it has been in service a long enough period of time for fouling of the surfaces to occur. As discussed in Sec. 8.3, the fouling layers on both the inside and outside of the tubes add resistance to heat transfer and decrease the overall heat transmission coefficient's value relative to its "clean" value of 225 Btu/hr ft² °F. From Table 8.1, the inside fouling resistance due to the city water and the outside fouling resistance due to the steam are found to be $r_{fi} = 0.0015$ and

$r_{fo} = 0.0005$, respectively. Next, these values are incorporated into Eq. (8.3) in order to determine the value of U_o when fouled, U_{of}. Notice that if the exchanger is "clean," so that $r_{fi} = r_{fo} = 0$ in Eq. (8.3), then U_o is given as follows:

$$U_o = \frac{1}{\dfrac{1}{h_o} + r_w + \dfrac{dA_o}{h_i dA_i}} . \tag{8.32}$$

Thus, combining Eq. (8.32) with Eq. (8.3) yields

$$U_{of} = \frac{1}{\dfrac{1}{U_o} + r_{fo} + r_{fi}\dfrac{dA_o}{dA_i}} .$$

If the tube thickness is small relative to its inside radius, $dA_i \approx dA_o$ and we get

$$U_{of} = \frac{1}{\dfrac{1}{225} + 0.0005 + 0.0015} = 155.2 \frac{\text{Btu}}{\text{hr ft}^2 \, °\text{F}} .$$

Using this and Eq. (8.22) in Eq. (8.31) gives

$$155.2(774.4)(1) \frac{(T_e - 50)}{\ln\left[\dfrac{182.8 - 50}{182.8 - T_e}\right]} = 11,000(988.4),$$

$$\ln\left[\frac{132.8}{182.8 - T_e}\right] = \frac{T_e - 50}{90.46} . \tag{8.33}$$

This equation can now be solved iteratively for T_e by using an initial estimate of T_e on the right side and calculating T_e on the left, then using this value as the new estimate on the right until two successive iterates for T_e agree closely enough. As an initial estimate of T_e, one can use the T_e value which the arithmetic average temperature difference,

$$182.8 - \frac{(50 + T_e)}{2} ,$$

would give if it were used in Eq. (8.31) instead of the correct ΔT_{LMTD}. Doing this gives an initial estimate of $T_e = 135°\text{F}$, which when used on the right side of Eq. (8.33) gives $T_e = 130.9°\text{F}$. This, when used on the right side of Eq. (8.33), yields $128.5°\text{F}$. Continuation of this process leads to the new coolant exit temperature of

$$T_e = 124.6°\text{F}.$$

Next, the energy balance on the exchanger yields the new mass flow rate of the coolant, ω_c.

$$\omega_c c_{pc}(T_{ce} - T_{ci}) = \omega_h h_{fg}$$

$$\omega_c(1)(124.6 - 50) = 11,000(988.4)$$

$$\omega_c = 145,743 \text{ lbm/hr.}$$

Also,

$$T_e = 124.6°\text{F.} \tag{b}$$

Thus, the new coolant mass flow rate of 145,743 lbm/hr is required to effect complete condensation of the steam after the exchanger becomes fouled, and this leads to a new exit coolant temperature of 124.6°F.

8.6 EFFECTIVENESS METHOD FOR HEAT EXCHANGER PERFORMANCE ANALYSIS

The correction factor method of analysis, which uses the factor F in conjunction with the logarithmic mean temperature difference ΔT_{LMTD}, is best suited for those calculations in which all exit and inlet temperatures are specified or can be found easily at the onset by an application of a simple energy balance. Thus, one can then calculate the logarithmic mean temperature difference and use it to find the quantity which is generally the unknown in this type of problem, namely, the required heat transfer surface area A. This type of calculation is considered to be one of the steps in a heat exchanger *design* problem. However, the heat transfer engineer is also often confronted with the so-called *performance* calculation, in which a heat exchanger of known area is being used at an off-design point or perhaps under conditions, or with fluids, unrelated to the original design of the exchanger. The problem then becomes one of predicting both outlet fluid temperatures. Since the logarithmic mean temperature difference and the factor F cannot be computed in advance for this type of problem, the correction factor method becomes unwieldy to use; an iterative procedure must be employed to solve for the exit fluid temperatures needed to compute *both* ΔT_{LMTD} and the correction factor F, which depends upon the exit temperatures through the parameters P and R. Example 8.4b gave some indication of the complexity of this type of performance calculation, even though the phase change caused the value of F to be known as unity.

To simplify the work involved in such performance calculations, the effectiveness–number of transfer units method, ϵ-NTU method, will be developed next. This method uses some of the same basic relations from which the correction factor method, F-ΔT_{LMTD}, evolved, but rearranges them in such a way as to eliminate one of the unknown outlet temperatures of the exchanger. This allows the remaining fluid exit temperature to be found directly from the known information in a performance calculation.

8.6a Development of the Effectiveness–Number of Transfer Units Method

In Sec. 8.4, it was shown that the best possible, or ideal, heat exchanger is the simple counterflow exchanger of infinite heat transfer area. This is so because, in this ideal exchanger, the fluid with the smaller capacity rate, $C = \omega c_p$, experiences the maximum possible temperature change, namely, $T_{hi} - T_{ci}$. Thus one has the following expression for this *maximum* possible heat transfer rate, q_{max}:

$$q_{max} = C_{min}(T_{hi} - T_{ci}). \qquad (8.34)$$

The value C_{min} is the smaller of C_c and C_h.

The *actual* heat transfer rate, q, for *any* type of exchanger is given by Eqs. (8.17) and (8.18), when no phase changes occur, as

$$q = C_h(T_{hi} - T_{he}) \qquad [8.17]$$

or

$$q = C_c(T_{ce} - T_{ci}). \qquad [8.18]$$

Since q_{max} represents the heat transfer rate in the best possible exchanger, a heat transfer rate that can only be obtained in the limit of an infinite area counterflow exchanger, the definition of the actual heat exchanger *effectiveness*, ϵ, suggests itself as the ratio of the actual heat transfer rate in any type of exchanger to the maximum possible heat transfer rate in the ideal counterflow exchanger. Thus,

$$\epsilon = \frac{q}{q_{max}}. \qquad (8.35)$$

Using Eqs. (8.17), (8.18), and (8.34) in Eq. (8.35) yields

$$\epsilon = \frac{C_h(T_{hi} - T_{he})}{C_{min}(T_{hi} - T_{ci})}, \qquad (8.36)$$

$$\epsilon = \frac{C_c(T_{ce} - T_{ci})}{C_{min}(T_{hi} - T_{ci})}. \qquad (8.37)$$

In the derivation of expressions for the nondimensional effectiveness ϵ in terms of the exchanger surface area A, overall heat transmission coefficient U, etc., another nondimensional grouping which appears is the *number of transfer units*, NTU. This is a direct measure of the size, the heat transfer area, of the exchanger, and is defined as follows when the overall heat transmission coefficient U is sensibly constant:

$$NTU = \frac{UA}{C_{min}}. \qquad (8.38)$$

In a performance type of calculation in which both exit temperatures T_{he} and T_{ce} are unknown at the start of the problem, the effectiveness ϵ is viewed as the basic dependent variable, since it contains *one* of these unknowns as indicated in Eqs. (8.36) and (8.37). The task now is to develop expressions for ϵ in terms of quantities that do not contain the unknown exit temperatures explicitly. The overall procedure to be followed in order to accomplish this for any type of heat exchanger will be demonstrated in detail next for the case of a counterflow exchanger.

Derivation of ϵ-NTU Relations for the Counterflow Heat Exchanger

As indicated earlier, the same underlying expressions which lead to the logarithmic mean temperature difference ΔT_{LMTD} and to the factor F in the correction factor method also, with different interpretation and emphasis, lead to the ϵ-NTU relations desired. Thus, we begin with Eq. (8.16) for the counterflow exchanger,

$$\ln \left[\frac{T_{he} - T_{ci}}{T_{hi} - T_{ce}} \right] = UA \left[\frac{1}{C_c} - \frac{1}{C_h} \right].$$

[8.16]

Factoring out $-(1/C_h)$ on the right side of Eq. (8.16) and taking antilogs yields

$$\frac{T_{he} - T_{ci}}{T_{hi} - T_{ce}} = e^{-\frac{UA}{C_h}(1 - C_h/C_c)}.$$

(8.39)

Suppose, for now, that the smaller of the two capacity rates C_c and C_h is C_h, so that $C_{\min} = C_h$ and $C_{\max} = C_c$. Hence, when we use the definition of the number of transfer units, Eq. (8.38), Eq. (8.39) is rewritten as shown next.

$$T_{he} - T_{ci} = (T_{hi} - T_{ce}) \, e^{-\text{NTU}(1 - C_{\min}/C_{\max})}$$

(8.40)

With $C_h = C_{\min}$ and $C_c = C_{\max}$, Eqs. (8.36) and (8.37) can now be written as follows:

$$\epsilon = \frac{T_{hi} - T_{he}}{T_{hi} - T_{ci}},$$

(8.41)

$$\epsilon = \frac{C_{\max}}{C_{\min}} \left(\frac{T_{ce} - T_{ci}}{T_{hi} - T_{ci}} \right).$$

(8.42)

Solving Eq. (8.41) for T_{he} gives

$$T_{he} = T_{hi} - \epsilon \, (T_{hi} - T_{ci}).$$

(8.43)

From Eq. (8.42), the following expression for T_{ce} is found:

$$T_{ce} = T_{ci} + \frac{C_{\min}}{C_{\max}} \epsilon \, (T_{hi} - T_{ci}).$$

(8.44)

Equations (8.43) and (8.44) are now used in Eq. (8.40) to eliminate T_{he} and T_{ce} where they appear explicitly in that equation, giving

$$(T_{hi} - T_{ci})(1 - \epsilon)$$
$$= (T_{hi} - T_{ci})\left(1 - \frac{C_{min}}{C_{max}}\epsilon\right)e^{-\text{NTU}(1 - C_{min}/C_{max})}. \tag{8.45}$$

Finally, cancelling the factor $(T_{hi} - T_{ci})$ and solving for ϵ gives the following result:

$$\epsilon = \frac{1 - e^{-\text{NTU}(1 - C_{min}/C_{max})}}{1 - (C_{min}/C_{max})e^{-\text{NTU}(1 - C_{min}/C_{max})}}. \tag{8.46}$$

If the derivation is done over for the case in which C_c happens to be C_{min}, instead of C_h, the same result is obtained, Eq. (8.46), which applies to the counterflow exchanger regardless of which capacity rate is the minimum.

An examination of Eq. (8.46) as well as Eqs. (8.36) and (8.37) shows that since only one of the fluid exit temperatures is contained within ϵ, Eq. (8.46) can be used directly, without the need for the iteration-type solution needed in the correction factor method, to solve for an exit temperature in a performance-type calculation when both exit temperatures are unknown. The remaining fluid exit temperature can then be found by an overall energy balance on the exchanger, or by equating the two expressions for ϵ given in Eqs. (8.36) and (8.37), which yields the energy balance result. Note also that as NTU $\rightarrow \infty$, due to increasing the heat transfer surface area, $\epsilon \rightarrow 1$ as discussed earlier for the *ideal* counterflow heat exchanger.

For convenience of calculation, Eq. (8.46) is available in graphed form in Fig. 8.16.

ε-NTU Relations for Other Types of Heat Exchangers

The same general procedure which culminated in Eq. (8.46) for the counterflow heat exchanger can also be employed for the parallel flow, shell-and-tube, and crossflow exchangers to yield the explicit form of the following implicit relation: $\epsilon = \epsilon(\text{NTU}, C_{min}/C_{max})$. In the case of the crossflow exchanger with one fluid mixed and the other unmixed, the capacity rate ratio $C_{mixed}/C_{unmixed}$ is more convenient to use than is C_{min}/C_{max}. Except for the parallel flow exchanger, the development of the ε-NTU relations is more involved than for the counterflow case because of the more complicated flow geometry of these other exchangers. It will be recalled that this was also the case for the correction factor, F, method considered earlier. The derivations and resulting analytical relations for $\epsilon = \epsilon(\text{NTU}, C_{min}/C_{max})$ are available in Kays and London [2]. Here, only the results in easily used graphical form are presented in Figs. 8.17 through 8.21 for parallel flow exchangers, for certain commonly used multi-tube-pass, multi-shell-pass exchangers, and for some crossflow exchangers.

In all of these figures which plot the effectiveness ϵ versus the number of transfer units, NTU, one observes the increase in ϵ with increasing values of NTU. However, for many of the curves, ϵ changes only slowly with NTU at the higher values of NTU. Thus a point

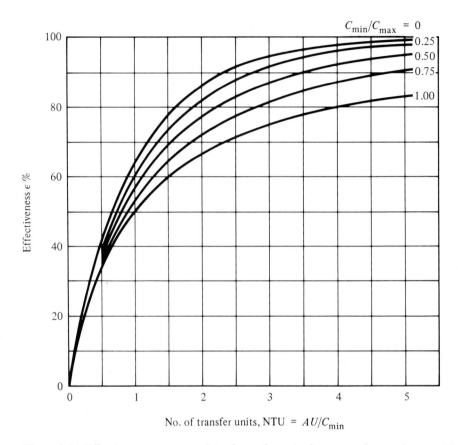

Figure 8.16 Effectiveness versus number of transfer units for counterflow exchanger. (From *Compact Heat Exchangers,* 2d ed., by W. M. Kays and A. L. London. Copyright © 1955, 1964 by the McGraw-Hill Book Company. Used with permission of McGraw-Hill Book Company.)

of diminishing returns is reached where an increase in, say, heat transfer area A causes only a negligible increase in the exchanger effectiveness ϵ and, therefore, would not generally be justified. An extreme example of this behavior is shown in Fig. 8.17 for the parallel flow exchanger where, for $C_{min}/C_{max} > 0.25$, there is virtually no change in ϵ for an increase in the NTU above a value of about 3.0.

By comparing the value of the effectiveness ϵ at, for example, NTU = 5.0 and $C_{min}/C_{max} = 1.0$ for the different types of exchangers represented by Figs. 8.16 to 8.21, the inherent superiority of the counterflow exchanger is immediately evident. We can also see that the ϵ-NTU curve for $C_{min}/C_{max} = 0$, which can be interpreted as a constant temperature phase change of one of the fluids, is the same for all exchanger types. Therefore, for this condition all exchangers exhibit the same performance as does the pure counterflow exchanger.

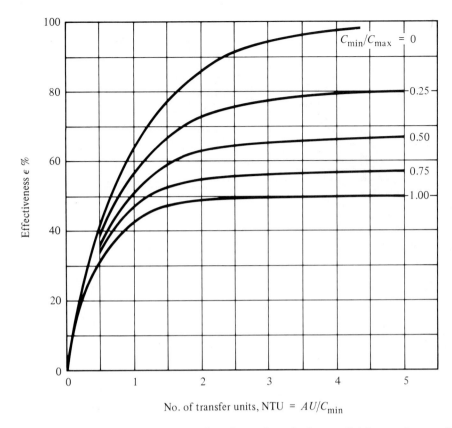

No. of transfer units, NTU $= AU/C_{min}$

Figure 8.17 Effectiveness versus number of transfer units for parallel flow exchanger. (From Kays and London, 1964. Used with permission of McGraw-Hill Book Company.)

EXAMPLE 8.5

Hot engine oil at 100°C with a mass flow rate of 0.03 kg/s and a specific heat of 2100 J/kg °C is to be cooled to 45°C by passing through a single-shell-pass, even number of tube passes exchanger in which the coolant is engine fuel, whose specific heat is 2130 J/kg °C, which enters the exchanger at 15°C and is to leave at 75°C. An overall heat transmission coefficient of $U_o = 200$ W/m² °C is expected and is based on the outside tube area. Find the required heat transfer surface area by use of the ϵ-NTU approach.

Solution
The capacity rate C_c of the cooler engine fuel will be determined first by the overall energy balance, Eq. (8.6), since the ratio of C_{min} to C_{max} is needed in the solution functions of the ϵ-NTU approach. From the given information, we have $T_{hi} = 100$°C, $T_{he} = 45$°C, $C_h = \omega_h c_{ph} = 0.03(2100) = 63$, $T_{ci} = 15$°C, $T_{ce} = 75$°C, and $c_{pc} = 2130$ J/kg °C.

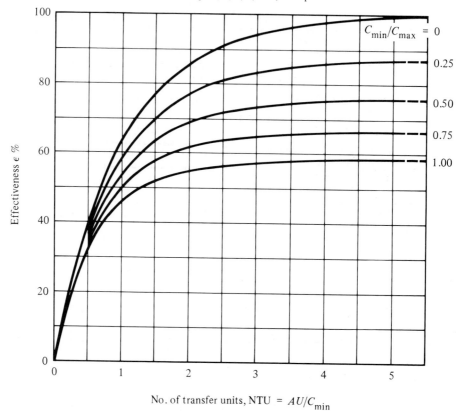

One shell pass, 2, 4, 6, ..., tube passes

No. of transfer units, NTU = AU/C_{min}

Figure 8.18 Effectiveness versus number of transfer units for one shell pass and two, four, six, ..., tube passes. (From Kays and London, 1964. Used with permission of McGraw-Hill Book Company.)

From Eq. (8.6),

$$C_c(T_{ce} - T_{ci}) = C_h(T_{hi} - T_{he})$$
$$C_c(75 - 15) = 63(100 - 45)$$
$$C_c = 57.75.$$

Hence, $C_{min} = 57.75$, $C_{max} = 63$, and $C_{min}/C_{max} = 0.917$.

Next, the effectiveness ϵ of this exchanger is determined by using Eq. (8.37) with C_c cancelling C_{min} in that expression for this problem.

$$\epsilon = \frac{C_c(T_{ce} - T_{ci})}{C_{min}(T_{hi} - T_{ci})} = \frac{75 - 15}{100 - 15} = 0.706$$

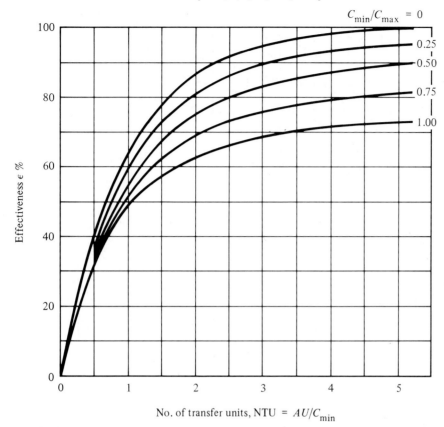

Two shell passes, 4, 8, 12, . . . , tube passes

$C_{min}/C_{max} = 0$

0.25
0.50
0.75
1.00

Effectiveness ϵ %

No. of transfer units, NTU $= AU/C_{min}$

Figure 8.19 Effectiveness versus number of transfer units for two shell passes and four, eight, twelve, . . . , tube passes. (From Kays and London, 1964. Used with permission of McGraw-Hill Book Company.)

Figure 8.18 for the one-shell-pass, even multiple of tube passes exchanger being considered here, with $\epsilon = 0.706 = 70.6\%$ and $C_{min}/C_{max} = 0.917$, yields no solution. This exchanger, even with an infinite amount of surface area giving an infinite number of transfer units, cannot have an effectiveness as high as 70.6% for the capacity rate ratio involved.

Thus, we will go to a *two*-shell-pass, even multiple of four tube passes heat exchanger to see if that one is capable of handling the required heat transfer load.

From Fig. 8.19, one finds that

$$NTU \simeq 3.1 = U_o A_o / C_{min} = 200\, A_o / 57.75$$

$$A_o = 0.895 \text{ m}^2.$$

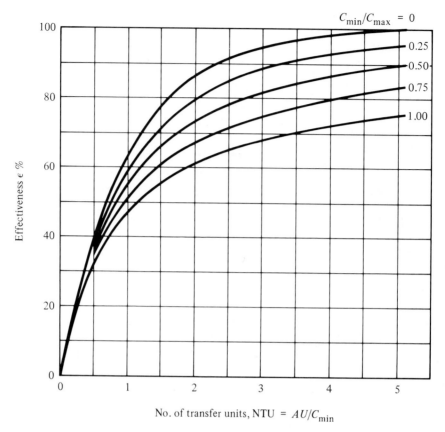

Figure 8.20 Effectiveness versus number of transfer units for a crossflow exchanger with both fluids unmixed. (From Kays and London, 1964. Used with permission of McGraw-Hill Book Company.)

In this problem, since all inlet and exit temperatures were known in advance, the correction factor method could also have been used with approximately the same amount of work, though some practitioners and advocates of the ϵ-NTU method argue that the need to calculate the algebraically messy logarithmic mean temperature difference gives the ϵ-NTU approach an advantage even in this type of "design" problem. The reader can contrast the relative amount of work in the two methods for the "design"-type problem by comparing the problem just solved with the equivalent one, Ex. 8.1(a), solved by the F-ΔT_{LMTD} method.

EXAMPLE 8.6

An already-designed oil cooler is a crossflow heat exchanger with the oil unmixed and the water coolant mixed and a heat transfer surface area of 190 ft². It is proposed to use this exchanger, for a part of the day, at an off-design point where the oil mass flow rate would

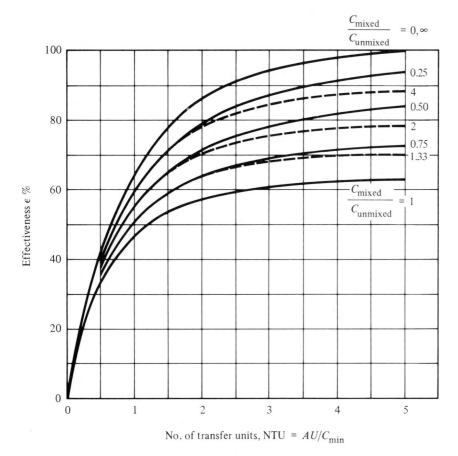

$$\frac{C_{\text{mixed}}}{C_{\text{unmixed}}} = 0, \infty$$

0.25

4

0.50

2

0.75

1.33

$$\frac{C_{\text{mixed}}}{C_{\text{unmixed}}} = 1$$

No. of transfer units, NTU $= AU/C_{\text{min}}$

Figure 8.21 Effectiveness versus number of transfer units for a crossflow exchanger with one fluid mixed and the other unmixed. (From Kays and London, 1964. Used with permission of McGraw-Hill Book Company.)

be doubled to 22,500 lbm/hr entering the exchanger at 230°F while the water coolant mass flow rate remained at 20,000 lbm/hr entering at 60°F. The overall heat transmission coefficient U, based on the given heat transfer area, has been calculated to be 60 Btu/hr ft² °F, while the specific heats of the oil and the water are 0.45 and 1.0 Btu/lbm °F, respectively. Find the exit temperature of both the oil and the water at this off-design condition.

Solution
This problem involves a performance calculation, because both exit temperatures are unknown and would necessitate an iterative, trial-and-error type of solution by the correction factor method.

The ε-NTU method is the appropriate solution technique in this case. We have that

$$\omega_h c_{ph} = C_h = 22{,}500(0.45) = 10{,}125,$$

$$\omega_c c_{pc} = C_c = 20{,}000(1.0) = 20{,}000.$$

Hence, $C_{min} = C_{unmixed} = 10{,}125$ and $C_{max} = C_{mixed} = 20{,}000$.

$$T_{hi} = 230°F \quad \text{and} \quad T_{ci} = 60°F$$

$$NTU = UA/C_{min} = 60(190)/10{,}125 = 1.126$$

$$C_{mixed}/C_{unmixed} = 20{,}000/10{,}125 = 1.975$$

At NTU = 1.126 and the capacity rate ratio of mixed to unmixed fluid of 1.975, Fig. 8.21 for this crossflow exchanger yields a value of effectiveness $\epsilon \approx 56\% = 0.56$. Next the oil exit temperature, T_{he}, is found from the effectiveness expression, Eq. (8.36) with $C_h = C_{min}$ in this problem.

$$0.56 = \frac{10{,}125(230 - T_{he})}{10{,}125(230 - 60)}$$

Therefore, $T_{he} = 135°F$.

Finally, the other exit fluid temperature, T_{ce}, can be found from the overall energy balance, Eq. (8.6), or by equating Eq. (8.36) to (8.37) to yield Eq. (8.6), or by using the numerical value of ϵ directly in Eq. (8.37). This last alternative, Eq. (8.37), gives

$$0.56 = \frac{20{,}000(T_{ce} - 60)}{10{,}125(230 - 60)},$$

$$T_{ce} = 108°F. \quad \textit{Cooling water exit temperature.}$$

The oil exit temperature found previously was $T_{he} = 135°F$.

Note that, when Fig. 8.21 is used, even though the capacity rate ratio is in terms of mixed to unmixed fluids rather than in terms of minimum to maximum capacity rates, the number of transfer units *is still calculated using the capacity rate* C_{min}.

EXAMPLE 8.7

In a water-to-water heat exchange process, 750 kg/s of pressurized liquid water is to be cooled from 288°C to 38°C by 1500 kg/s of cooling water available at 10°C. This is to be accomplished in a multipass shell-and-tube heat exchanger where the initial estimate of the overall heat transmission coefficient based on outside tube surface area is 900 W/m² °C. It is proposed to use a one-shell-pass, even number of tube passes exchanger if it can handle the heat transfer task. If this is not possible, an attempt will be made to use a two-shell-pass, even multiple tube pass unit. Finally, if this still does not do the job, an attempt will be made to use a number of equal size, equal heat transfer area, two-shell-pass exchangers connected in series for the hot fluid and with equal mass flow rates of 10°C cold water entering each exchanger. For the colder water, $c_{pc} = 4180$ J/kg °C, while the hotter water has $c_{ph} = 4350$ J/kg °C. Find the required heat transfer area and the type of exchanger, or number of two-shell-pass exchangers, to be used.

Solution
The capacity rates are found first.

$$C_c = w_c c_{pc} = 1500(4180) = 6,270,000,$$
$$C_h = w_h c_{ph} = 750(4350) = 3,262,500.$$

Hence, $C_{min} = C_h = 3,262,500$ and $C_{min}/C_{max} = 0.52$.

From Eq. (8.36), the effectiveness required of the heat exchanger is found next: $T_{hi} = 288°C$, $T_{he} = 38°C$, $T_{ci} = 10°C$.

$$\epsilon = \frac{C_h(T_{hi} - T_{he})}{C_{min}(T_{hi} - T_{ci})} = \frac{3,262,500(288 - 38)}{3,262,500(288 - 10)} = 0.90$$

Going to Fig. 8.18 with these values of ϵ and C_{min}/C_{max}, we can see that there is no single-pass, even-multiple tube pass exchanger which can handle the heat transfer requirement, since there is no intersection of $\epsilon = 0.90$ with a curve of constant $C_{min}/C_{max} = 0.52$. Similarly, an examination of Fig. 8.19 yields no solution using a two-shell-pass, even multiple of four tube passes exchanger unless this intersection occurs beyond NTU = 5.5, where the curve of C_{min}/C_{max} is practically horizontal. Hence, next we will try to find a solution employing two 2-shell-pass, even multiple of four tube passes exchangers, identical in area, through which the hotter water passes in series and half the coolant mass flow rate, 750 kg/s, enters each of the two exchangers at 10°C.

Call the first exchanger a and the second exchanger b. The inlet temperatures to exchanger a are $T_{hia} = 288°C$ and $T_{cia} = 10°C$. The outlet temperatures T_{hea} and T_{cea} are unknown. The outlet temperature of the hot fluid from a is the inlet temperature of the hot fluid for exchanger b, so $T_{hib} = T_{hea}$. Also, $T_{cib} = 10°C$ and $T_{heb} = 38°C$, while T_{ceb} is unknown.

$$C_{ca} = w_{ca} c_{pc} = C_{cb} = w_{cb} c_{pc} = 750(4180) = 3,135,000 = C_{min}$$

$$(C_{min}/C_{max})a = (C_{min}/C_{max})b = 3,135,000/3,262,500 = 0.96$$

For both exchangers $C_c = C_{min}$, so Eq. (8.37) gives the efficiency expressions as follows, after cancellation of the capacity rates which appear:

$$\epsilon_a = \frac{T_{cea} - T_{cia}}{T_{hia} - T_{cia}} = \frac{T_{cea} - 10}{288 - 10} = \frac{T_{cea} - 10}{278},$$

$$\epsilon_b = \frac{T_{ceb} - T_{cib}}{T_{hib} - T_{cib}} = \frac{T_{ceb} - 10}{T_{hea} - 10}.$$

However, since both exchangers have the same unknown area, the same U, and the same C_{min}, they have the same value of NTU. In addition, since C_{min}/C_{max} is also the same, it follows that so is their effectiveness, that is, $\epsilon_a = \epsilon_b$. Thus, equating the two preceding expressions for ϵ_a and ϵ_b yields

$$\frac{T_{cea} - 10}{278} = \frac{T_{ceb} - 10}{T_{hea} - 10}. \qquad (8.47)$$

Equation (8.47) contains three unknown temperatures, but T_{cea} and T_{ceb} can be put in terms of T_{hea} by overall energy balances on exchangers a and b separately.

Thus, for exchanger a,

$$C_h(288 - T_{hea}) = C_{ca}(T_{cea} - 10),$$

$$3{,}262{,}500(288 - T_{hea}) = 3{,}135{,}000(T_{cea} - 10)$$

or

$$T_{cea} - 10 = 299.7 - 1.04\, T_{hea}. \qquad (8.48)$$

For exchanger b, the energy balance gives

$$C_h(T_{hea} - 38) = C_{cb}(T_{ceb} - 10)$$

or

$$T_{ceb} - 10 = 1.04\, T_{hea} - 39.52. \qquad (8.49)$$

Inserting Eqs. (8.48) and (8.49) into Eq. (8.47) and solving the resulting quadratic equation for T_{hea} yields

$$T_{hea} = 98.3°C.$$

The value T_{cea} is needed in the equation put down earlier for ϵ_a. With T_{hea} now known, T_{cea} is found next from Eq. (8.48) to be

$$T_{cea} = 207.4°C.$$

Thus,

$$\epsilon_a = \frac{T_{cea} - 10}{278} = \frac{207.4 - 10}{278} = 0.71.$$

With $\epsilon_a = 0.71$ and $(C_{min}/C_{max})_a = 0.96$, one finds from Fig. 8.19 that

$$(NTU)_a \cong 3.3 = UA_a/C_{min_a} = 900A_a/3{,}135{,}000, \quad A_a = 11{,}495 \text{ m}^2.$$

Since exchanger b has the same area as exchanger a, two identical heat exchangers of the two-shell-pass, even multiple of four tube passes type are needed with a *total* exchanger surface of 23,000 m².

It is also of interest to see what area would be needed in a pure counterflow exchanger if economy and compactness did not dictate a multi-shell, multi-tube-pass exchanger. Earlier it was found that the effectiveness of a *single* exchanger would have to be 0.9 at a capacity rate ratio $C_{min}/C_{max} = 0.52$. From Fig. 8.16, for a counterflow exchanger,

$$NTU = 3.55 = UA/C_{min} = 900A/3,262,500, \quad A = 12,869 \text{ m}^2.$$

With this first estimate of the required heat transfer surface areas for the two required two-shell-pass, multi-tube-pass exchangers or the single pure counterflow exchanger, along with the multitude of other design requirements such as size, allowable pressure drop, cost, etc., one must decide which of these two options is the better one. Details of such design decisions are discussed in the article by Taborek [7].

EXAMPLE 8.8

Flue gases, whose thermophysical properties are about the same as those of air, are available at 800°F at a mass flow rate of 420,000 lbm/hr. It is proposed to use some of this energy to heat 600,000 lbm/hr of 70°F water in the tubes of a waste heat recovery unit which consists of an existing crossflow heat exchanger. This exchanger is a simple tube bank which has 45 rows of staggered mild steel tubes containing 15 tubes each. The tubes have inside and outside diameters of 0.649 inches and 0.875 inches respectively, are 12 feet long, are placed such that their transverse and longitudinal spacings are both twice the outside diameter, and the exchanger height, H, the transverse dimension for a row, is 26.5 inches. From the TEMA standards [5], the following fouling resistances on the outside and inside tube surfaces are expected: $r_{fo} = 0.01$ hr ft² °F/Btu and $r_{fi} = 0.002$ hr ft² °F/Btu. Make an initial estimate of both the outlet water and flue gas temperatures.

Solution
Since this is a performance type of calculation in which both fluid exit temperatures are unknown, the most straightforward approach is the ϵ-NTU method. The effectiveness ϵ is the basic unknown here, since it contains one of the fluid exit temperatures; hence, the capacity rate ratio and the number of transfer units will have to be found first. We have $\omega_h = 420,000$ lbm/hr, $\omega_c = 600,000$ lbm/hr, $T_{ci} = 70$°F, and $T_{hi} = 800$°F.

Next, in order to evaluate the needed temperature-dependent fluid properties, an initial estimate of the outlet flue gas temperature will be made of $T_{he} \approx 400$°F. Naturally, this must be checked at the end of the calculations, and recalculations must eventually be made, if necessary. With this estimate of T_{he}, the arithmetic average bulk mean air temperature is then $T_{hB_a} = 600$°F, so the property tables give $c_{ph} = 0.25$ and, hence,

$$C_h = \omega_h c_{ph} = 420,000(0.25) = 105,000.$$

An overall energy balance on the heat exchanger, using this estimated outlet air temperature, gives an estimated outlet water temperature as shown next.

$$420,000(0.25)(800 - 400) = 600,000(1)(T_{ce} - 70),$$

or $T_{ce} = 140°F$ (estimate).

With this estimate, the arithmetic average bulk mean temperature of the water is about $100°F$, giving $c_{pc} = 1.0$ lbm °F and

$$C_c = \omega_c c_{pc} = 600,000(1) = 600,000.$$

Therefore $C_{min} = C_h = 105,000$, and since the air, flowing outside the tubes, is mixed and the water, flowing within the tubes, is unmixed, one has

$$C_{mixed}/C_{unmixed} = 105,000/600,000 = 0.175.$$

The total outside surface area A_o for heat transfer in this 675-tube exchanger is given by

$$A_o = n\pi D_o L = 675\pi(0.875/12)12 = 1856 \text{ ft}^2.$$

To find the overall heat transmission coefficient U_o based on the outside tube area, Eq. (8.3) is used, and the inside and outside surface coefficients of heat transfer, h_i and h_o, must be found next.

The mass flow rate of water per tube is given by

$$\omega_t = \omega_c/n = 600,000/675 = 889 \text{ lbm/hr}.$$

At an average, estimated bulk mean temperature of $100°F$, the water properties are $\mu = 1.65$ lbm/hr ft, $k = 0.363$ Btu/hr ft °F, $N_{Pr} = 4.55$. So,

$$N_{Re_{D_i}} = \frac{4\omega_t}{\pi D_i \mu} = \frac{4(889)}{\pi(0.649/12)(1.65)} = 12,684.$$

With this value of estimated Reynolds number, Eq. (5.393) is used from Chapter 5, with an estimated wall temperature of $150°F$ giving $\mu_w = 1.05$ lbm/hr ft, as follows:

$$\frac{h_i(0.649/12)}{0.363} = 0.023(4.55)^{1/3}(12,684)^{0.8}(1.65/1.05)^{0.14}$$

$$h_i = 522 \text{ Btu/hr ft}^2 \text{ °F}.$$

Next, the outside surface coefficient, h_o, will be determined from the information in Chapter 5 on crossflow over tube banks. A check of the transverse spacing between tubes in a single row as well as the diagonal spacing between tubes in successive rows indicates that the minimum area section is in the transverse dimension. This area is given by

$$A_m = L(H - n_{row}D_o) = 12[(26.5/12) - 15(0.875/12)]$$
$$= 13.38 \text{ ft}^2.$$

Hence

$$(\rho U)_{max} = \omega_h/A_m = 420,000/13.38 = 31,390 \text{ lbm/hr ft}^2.$$

With an air film temperature of 375°F, based on the estimated average bulk mean air temperature of 600°F and wall temperature of 150°F, the property tables give $\mu \approx 0.06$ lbm/hr ft, $k = 0.021$ Btu/hr ft °F, $N_{Pr} = 0.683$. Hence,

$$N_{Re_{D_f}} = \frac{(\rho U)_{max}D_o}{\mu} = \frac{31,390(0.875/12)}{0.06} = 38,148.$$

The longitudinal and transverse spacing ratios were given as $S_L/D_o = 2.0$ and $S_T/D_o = 2.0$, so that Table 5.4 yields $B = 0.482$ and $m = 0.556$. These values are now used in Eq. (5.424) as follows:

$$\frac{h_o(0.875/12)}{0.021} = 1.13(0.482)(38,148)^{0.556}(0.683)^{1/3}$$

$$h_o = 48.7 \text{ Btu/hr ft}^2 \text{ °F.}$$

From Chapter 2, we see that r_w is given by, with $k_w = 26$ Btu/hr ft °F for mild carbon steel,

$$r_w = \frac{D_o}{2} \ln\left(\frac{D_o}{D_i}\right)/k_w = \frac{(0.875/12)}{2} \ln\left(\frac{0.875}{0.649}\right)/26 = 0.00042.$$

Since $dA_o/dA_i = \pi D_o dx/\pi D_i dx = D_o/D_i = 0.875/0.649 = 1.35$, Eq. (8.3), with the given fouling factors and the just-computed quantities inserted, yields

$$U_o = \frac{1}{\dfrac{1}{48.7} + 0.01 + 0.00042 + 0.002(1.35) + \dfrac{1.35}{522}}$$
$$= 27.6 \text{ Btu/hr ft}^2 \text{ °F.}$$

With this,

$$\text{NTU} = U_o A_o/C_{min} = 27.6(1856)/105,000 = 0.49.$$

Thus, with NTU = 0.49 and $C_{mixed}/C_{unmixed} = 0.175$, Fig. 8.21 yields

$$\epsilon \approx 38\% = 0.375.$$

From Eq. (8.37),

$$\epsilon = \frac{C_c(T_{ce} - T_{ci})}{C_{min}(T_{hi} - T_{ci})} = \frac{600,000(T_{ce} - 70)}{105,000(800 - 70)} = 0.375,$$

$$T_{ce} = 118°F.$$

From Eq. (8.36),

$$\epsilon = \frac{C_h(T_{hi} - T_{he})}{C_{min}(T_{hi} - T_{ci})} = \frac{105,000(800 - T_{he})}{105,000(800 - 70)} = 0.375,$$

$$T_{he} = 526°F.$$

Thus, the outlet water temperature is 118°F and outlet flue gas temperature is 526°F based on the original estimate of the gas temperature at exit being about 400°F. If it were felt that additional accuracy would be required in this preliminary investigation of feasibility of this existing exchanger, the entire calculation would be redone with the initial estimate of air temperature being 526°F. However, this would only change the average bulk mean temperature of the air by about 63°, to 663°F instead of the previously used 600°F, and the water's average temperature changes by only 6°F. So this will not change the air or water properties significantly, and probably a recalculation is not warranted at this point.

8.7 VARIABLE OVERALL HEAT TRANSMISSION COEFFICIENT, U

So far in this chapter, all analyses have been predicated upon the assumption of a sensibly constant value of U, the overall heat transmission coefficient. As shown in Eq. (8.3), the value of U depends upon, among other quantities, the value of the surface coefficients of heat transfer, h_i and h_o, at the inside and outside heat transfer surface areas, respectively. In general, h_i and h_o can be expected to vary with position, x, down the exchanger because of a number of different factors. First of all, if x is close enough to a point where one of the fluids enters the exchanger, the hydrodynamic and/or the thermal boundary layer may not yet have filled the passage. Hence the value of h will depend upon x because the boundary layer thickness, hydrodynamic or thermal, varies with x as we saw in Chapter 5. In addition, if a point of transition from laminar to turbulent flow occurs within the boundary layer, there will be a concomitant, rather abrupt, change in the surface coefficient with x. The second major reason, and usually the most important one for relatively long heat exchangers, for h_i and h_o and, therefore, for U to vary with x is the temperature dependence of the fluid properties such as viscosity μ, thermal conductivity k, Prandtl number N_{Pr}, etc. The reference temperature, such as the film temperature T_f, at which these properties are evaluated depends upon the local bulk mean temperature of both fluids in the exchanger. These temperatures vary with position x, and therefore also cause wall temperature variation with x, the thermal history effect discussed in Chapter 5. Thus it follows that h_i and h_o depend upon x through the dependence of h on the fluid properties which occur in the various analytical solutions, semi-analytical relations, and experimental correlations which are given in Chapters 5, 6, and 7, and on the thermal history effect.

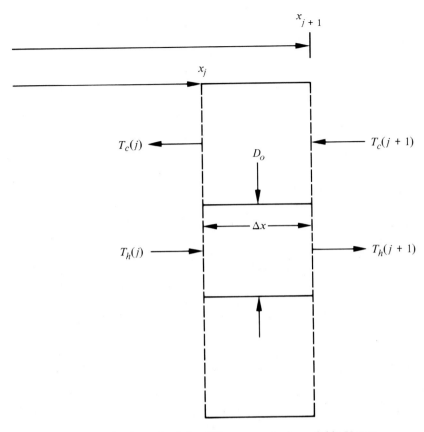

Figure 8.22 Control volume for finite difference analysis, variable U case.

8.7a Finite Difference Method

The most general method for treating the case of variable U is the finite difference approach. To illustrate the overall approach, the counterflow exchanger depicted in Fig. 8.9 will be used. The appropriate modifications to the approach for other types of exchangers will be apparent. The length of the exchanger, be it known or unknown, is broken up into a number of small but finite lengths Δx similar to the differential length dx shown in Figs. 8.9 and 8.11. A control volume including both fluids between position x_j and position x_{j+1}, where $\Delta x = x_{j+1} - x_j$, is taken. This is shown in Fig. 8.22, which is to this finite difference analysis what Fig. 8.11 is to the previously developed differential analysis.

If Δx, though finite, is chosen small enough, the conditions at x_j and at x_{j+1} will be fairly close to one another, so the h_i and h_o values, and hence U, can be determined using conditions at x_j. Once $U(j)$ has been found, the basic heat transfer rate equation gives the approximate heat transfer rate q_j between the fluids across the section of length Δx as the following:

$$q_j = U(j)\pi D_o \Delta x[T_h(j) - T_c(j)]. \tag{8.50}$$

Heat Exchangers

An energy balance on the hot and cold fluids, respectively, yields the following two relations:

$$q_j = \omega_h c_{ph}[T_h(j) - T_h(j + 1)], \tag{8.51}$$

$$q_j = \omega_c c_{pc}[T_c(j) - T_c(j + 1)]. \tag{8.52}$$

Equating the relations, Eqs. (8.51) and (8.50), gives, after some rearrangement,

$$T_h(j + 1) = T_h(j) - \frac{U(j)\pi D_o \Delta x}{\omega_h c_{ph}}[T_h(j) - T_c(j)]. \tag{8.53}$$

Equation (8.53) is the basic finite difference equation to be used to advance the solution, step by step, down the length of the exchanger. The procedure is as follows: At $x = 0$, where the index j is also zero, $T_h(0) = T_{hi}$ and $T_c(0) = T_{ce}$ if the coolant exit temperature is known. If it is not known, an estimate is made of T_{ce}, and this estimate is eventually checked and refined in an iterative process. With $T_h(0)$ and $T_c(0)$ known, $U(0)$ is computed. These, along with the small finite spacing chosen, $\Delta x = x_1 - x_0$, are inserted into Eq. (8.53), which is then solved for $T_h(1)$, the value of T_h at $j = 1$ where $x = x_1$. Then, by equating the energy balances, Eqs. (8.51) and (8.52), $T_c(1)$ is solved for. Now, with $T_h(1)$ and $T_c(1)$ known, the entire procedure is repeated by computing $U(1)$, using $T_h(1)$ and $T_c(1)$ and inserting these into Eq. (8.53) to solve for $T_h(2)$, T_h at $x = x_2$. This procedure continues until the other end of the exchanger is reached if the exchanger area is known, or until T_c reaches T_{ci} if the exchanger area is not known in advance. For the case in which the exchanger area is known but T_{ce} is not known, the check on the original assumed value of T_{ce} is whether or not one gets the actual known inlet temperature, T_{ci}, when the other end of the exchanger is reached. If the calculated T_{ci} does not equal the actual inlet temperature, the entire procedure must be repeated with a new estimate of T_{ce} at $x = 0$ until the T_{ci} is matched at the other end. When this occurs, one has an approximate solution for the two exit temperatures T_{he} and T_{ce} in this performance calculation. In the case of the design-type calculation in which T_{ce} and T_{he} are both known, the value of x—call it x_N—at which $T_h = T_{he}$ is a finite difference approximation to the needed length, and therefore to the heat transfer surface area of the exchanger.

However, regardless of which case is being used, one cannot (as was explained in Chapters 2 and 3) stop the finite difference calculations at this point. Still to be carried out is lattice refinement to insure that the solution is independent of the lattice, or mesh size Δx, that was used. Hence, the next step is to cut the value of Δx in half, doubling the number of increments down the tube, repeat the calculations, and compare the predicted exchanger length or predicted exit fluid temperatures to those arrived at previously. If agreement between the two sets of calculations is not sufficiently close, then the lattice spacing is cut in half once again and the entire procedure repeated until two successive sets of calculations, using successively finer values of Δx, agree satisfactorily. A short computer program would be written to solve for the evolution of T_h and T_c with position down the exchanger and to perform the required lattice refinement.

8.7b Approximate Analytical Method

If the variable overall heat transmission coefficient U varies *linearly* with the local difference in bulk mean temperature $\Delta T = T_h - T_c$ between the two fluids, an exact solution for the heat transfer rate in a pure counterflow exchanger has been developed. This is discussed in Gardner and Taborek [9], and the result is as follows:

$$q = A\Delta[U(T_h - T_c)] \tag{8.54}$$

where

$$\Delta[U(T_h - T_c)] = \frac{U_b(T_h - T_c)_a - U_a(T_h - T_c)_b}{\ln\left[\dfrac{U_b(T_h - T_c)_a}{U_a(T_h - T_c)_b}\right]}. \tag{8.55}$$

As in the case of Eq. (8.23), the subscripts a and b in Eq. (8.55) refer to the two different ends of the counterflow heat exchanger.

Based on Eq. (8.55), an *approximate method* can be used for heat exchanger types *other than counterflow* when the value of U varies linearly with $T_h - T_c$. Equation (8.54) is generalized by the inclusion of the correction factor F previously introduced in Eq. (8.25). Thus,

$$q = FA\Delta[U(T_h - T_c)]. \tag{8.56}$$

The value of F that appears in Eq. (8.56) is the value found for a constant U exchanger from Figs. 8.12 through 8.15, and the $\Delta[U(T_h - T_c)]$ is the one given by Eq. (8.55). In their discussion of this method, Gardner and Taborek [9] indicate that the *approximate* result, Eq. (8.56), generally yields accuracy to within 10% for all but extreme ratios of $(T_h - T_c)_b/(T_h - T_c)_a$ and/or of U_b/U_a, as long as U varies linearly with $T_h - T_c$. Thus, if 10% accuracy is sufficient *and* if U varies in an essentially linear fashion with $T_h - T_c$, Eq. (8.56) can be used instead of the more complicated finite difference approach.

8.8 COMMENTS ON HEAT EXCHANGER DESIGN

Earlier in this chapter, we saw that one aspect of the true heat exchanger design problem was the determination of the heat transfer surface area needed to accomplish the required energy transfer process. Both the correction factor method, $F\text{-}\Delta T_{\text{LMTD}}$, and the effectiveness method, $\epsilon\text{-NTU}$, can be used to find the area A once mass flow rates and inlet and outlet temperatures of both fluids are specified, and, in addition, once an estimate of the overall heat transmission coefficient U is available. However, the determination of U

is coupled to finding the area A in the true design process; tube sizes, for instance, influence fluid velocities and, therefore, the surface coefficients of heat transfer. The latter serve to fix the value of U, as well as to influence the area for heat transfer in a given volume of exchanger.

Thus, the true heat exchanger design problem is not simply the type of problem termed "design" earlier in the chapter, but, like any design venture, is a complex, open-ended problem involving a large number of constraints and design considerations leading to some final choice among a number of alternative designs of a "best" or optimum design solution.

The thermal and fluid mechanical design considerations include the proper energy transfer rate between the fluids to accomplish the desired heating and cooling of the fluids; maximum allowable pressure drops of each fluid which, along with the mass flow rates, determine the pumping power requirements and cost; and fluid velocities, which partially control the magnitude of the overall heat transmission coefficient not only through the surface coefficients directly, but also in connection with the differences in fouling of exchanger surfaces at different velocities. In the true design process, cognizance must be given to the overall size of the exchanger, limits on maximum length or other particular dimensions, corrosion problems, ease of cleaning and maintenance of the unit, use of standard tube diameters and lengths, thermal expansion and contraction, tube vibration, and overall structural integrity. The foregoing, and other considerations and imposed constraints, when coupled with the operating and initial cost considerations, eventually lead to a design which is best for the situation.

An excellent article which presents information on the true heat exchanger design problem is the one by Taborek [7], "Evolution of Heat Exchanger Design Techniques." Although primarily oriented toward shell-and-tube exchangers, this article gives qualitative information which can be adapted to any type of heat exchanger. Other good sources of design information and procedures include Kays and London [2], the TEMA standards [5], Fraas and Ozisik [3], and Kern and Kraus [10]. Valuable practical information concerning heat exchanger design procedures and selection of off-the-shelf exchangers of specific types can be found in the various exchanger manufacturers' catalogs, bulletins, and product manuals [11], [12], and in Walker [15].

8.9 PROBLEMS

8.1 A counterflow heat exchanger is to heat 0.45 kg/s of 4°C air to 27°C by using 0.08 kg/s of stack gases, which behave as air, and which are available at 210°C. For both air streams, $c_p \approx 1000$ J/kg °C, and the overall heat transmission coefficient based on outside tube area has been estimated as $U_o = 12$ W/m² °C. Find the required outside surface area of the tubing.

8.2 Consider a simple parallel flow heat exchanger in which 11,250 lbm/hr of oil is to be cooled in a 1-in. outside diameter tube from 230°F to 100°F by 20,000 lbm/hr of water which enters at 50°F. The specific heat of the oil is about 0.45 Btu/lbm °F, while that of the water is about 1.0 Btu/lbm °F. The overall heat transmission coefficient has been calculated to be $U_o = 60$ Btu/hr ft² °F. Find the water exit temperature and the number of 5-ft-long tubes needed to perform the required task.

8.3 Water at a mass flow rate of 2.5 kg/s is available at 70°C to raise the temperature of 2.0 kg/s of cool water from 5°C to 60°C. The expected heat transmission coefficient $U = 1200$ W/m² °C based on outside tubing area, and $c_p = 4190$ J/kg °C. Find the surface area needed by a parallel flow exchanger and that needed by a counterflow exchanger.

8.4 In a steam superheater, 20,000 lbm/hr of steam, $c_p \approx 0.46$ Btu/lbm °F, at 300°F enters the passages of a crossflow exchanger and is to leave at 600°F. Ninety-six thousand lbm/hr of spent combustion gases, $c_p = 0.24$ Btu/lbm °F, enter the exchanger at 1000°F. The overall heat transmission coefficient, based on the gas-side area, already has been determined to be $U = 73$ Btu/hr ft² °F. If both fluids are unmixed, calculate the required air-side area.

8.5 Ammonia enters a two-shell-pass, four-tube-pass exchanger at 65°C with a mass flow rate of 0.8 kg/s and $c_p = 4800$ J/kg °C. While passing through the tubes, the ammonia is cooled to 20°C by 1.5 kg/s of water, $c_p = 4180$ J/kg °C, which enters at 10°C on the shell side. An overall heat transmission coefficient of $U = 30$ W/m² °C has been calculated. Find the needed heat transfer surface area, and compare it to the area which a pure counterflow exchanger would need.

8.6 A single-shell-pass, two-tube-pass heat exchanger has 9000 lbm/hr of water, $c_p = 1.0$ Btu/lbm °F, entering the shell side at 100°F; the water is to leave at 160°F. To heat the water, 15,000 lbm/hr of oil, $c_p = 0.45$ Btu/lbm °F, enters the tubes at 230°F. The overall heat transmission coefficient, based on the outside tube area, has been estimated to be $U = 60$ Btu/hr ft² °F. Calculate the needed outside surface area of the tubing.

8.7 By appropriate integration of Eq. (8.14) and use of that result in Eq. (8.8), derive Eq. (8.21) by this alternate procedure.

8.8 By use of the basic rate equation, Eq. (8.7), and energy balances, develop the expression for the appropriate driving temperature difference for a simple parallel flow heat exchanger, Eq. (8.24).

8.9 Develop expressions for T_h and T_c as functions of position x and known quantities for a pure counterflow heat exchanger. (*Hint:* begin with Eq. [8.14].)

8.10 A shell-and-tube heat exchanger using two shell passes and eight tube passes has 0.6 kg/s of an oil, $c_p = 2000$ J/kg °C, entering the shell at 120°C which is to leave at 40°C. Water at 20°C enters the tubes with a mass flow rate of 0.5 kg/s. The overall heat transmission coefficient based on "clean" surfaces is $U = 300$ W/m² °C referenced to the outside tube area, and fouling factors of 0.0009 and 0.0001 m² °C/W are expected on the oil and water sides, respectively. The ratio of outside to inside tube area is 1.2. Compute the needed outside area for the exchanger to perform properly when in the fouled condition.

8.11 A truck radiator is to be designed as a crossflow heat exchanger with both fluids unmixed. Four hundred forty lbm/hr of water at 250°F enters and is to leave at 135°F, while 4500 lbm/hr of air at 90°F enters. For the water and air, $c_p = 1.0$ and 0.24 Btu/lbm °F, respectively, and the overall heat transmission coefficient based on the air-side area is estimated to be $U = 20$ Btu/hr ft² °F. (a) Calculate the required heat transfer surface area for this crossflow exchanger. (b) Find the area that would be needed for a simple parallel flow exchanger.

8.12 An ammonia condenser is to be designed as a single-shell-pass, six-tube-pass exchanger in which 0.05 kg/s of saturated ammonia vapor at 50°C is to be condensed to 50°C liquid. The latent heat of condensation is $h_{fg} = 1,052,000$ J/kg. Water, $c_p = 4180$ J/kg °C, at 15°C flows through the tubes and is to leave at 40°C. Based on outside tube area, an overall heat transmission coefficient of $U = 1200$ W/m² °C is expected. Find the mass flow rate of water needed and the outside heat transfer area.

8.13 Lubricating oil from an engine enters a crossflow heat exchanger at a rate of 2000 lbm/hr at a temperature of 200°F and leaves at 110°F. The coolant is air, which enters the exchanger at 80°F at a mass flow rate of 16,700 lbm/hr. The specific heats are $c_{p_{air}} = 0.24$, and $c_{p_{oil}} = 0.4$ Btu/lbm °F. The overall heat transmission coefficient based on the air side area has already been calculated to be $U = 15$ Btu/hr ft² °F. (a) If both fluids are unmixed, compute the required air side area. (b) What area would be needed for a pure counterflow exchanger? (c) In the counterflow exchanger of part b, suppose someone hooked up the oil inlet and exit backwards so that it became a parallel flow exchanger. What would the oil exit temperature be?

8.14 Seven and a half kg/s of water is to be heated from 10°C to 65°C as it passes through the tubes of a crossflow heat exchanger. Twelve and a half kg/s of air flows over the outside of the tubes with an inlet temperature of 260°C. For the air and water, $c_p = 1000$ and 4180 J/kg °C, respectively. The overall heat transmission coefficient based on air-side area has been estimated to be $U = 25$ W/m² °C. (a) Find the heat transfer area required. (b) What area would be needed for a pure counterflow exchanger?

8.15 In a proposed one-shell-pass even multiple of two tube passes heat exchanger for a nuclear aircraft, 300 lbm/s of 1400°F liquid metal enters the tubes and is to leave at 950°F, while 200 lbm/s of air at 600°F enters on the shell side. For the air, $c_p = 0.26$ Btu/lbm °F, while for the liquid metal, $c_p = 0.23$ Btu/lbm °F. The overall heat transmission coefficient based on the air side area is $U = 60$ Btu/hr ft² °F. What heat transfer surface area is needed?

8.16 A shell-and-tube heat exchanger with one shell pass and two tube passes is to be used as a small hot water supply system. Water, $c_p = 4180$ J/kg °C, enters the 1.25-cm outside diameter tubes at 10°C and is to leave at 60°C when its mass flow rate is 0.1 kg/s. Steam, $c_p = 1990$ J/kg °C, enters the shell at 205°C and is to leave at 140°C. The overall heat transmission coefficient based on the outside tube area is 500 W/m² °C. (a) Calculate the length of tubing needed. (b) If the maximum exchanger length is 1 m, how many separate tubes are required?

8.17 Find the outside tube surface area needed in a two-shell-pass, twelve-tube-pass exchanger whose function it is to condense 800,000 lbm/hr of steam which enters the shell side as saturated vapor at 180°F and leaves as saturated liquid. The latent heat of condensation is 990.2 Btu/lbm. Cooling water, $c_p = 1.0$ Btu/lbm °F, enters the tubes at 80°F and leaves at 130°F. An overall heat transmission coefficient of $U = 600$ Btu/hr ft² °F is expected.

8.18 The feasibility of using a simple counterflow heat exchanger is to be investigated for the following conditions: It is desired to heat 0.15 kg/s of water from 10°C to 100°C while it is passing along the outside surface of the single tube whose outer and inner diameters are 2.5 cm and 2.25 cm, respectively, by using high-pressure water which enters the tube at 250°C and is to leave at 70°C. For the water, $c_p = 4180$ J/kg °C, and the tube's thermal conductivity is $k = 110$ W/m °C. Find the length of the tube needed for a shell diameter of 5 cm.

8.19 A boiler's feedwater heater is constructed as a crossflow heat exchanger in which 25,000 lbm/hr of water, $c_p = 1.0$ Btu/lbm °F, enters tubes of 0.625-in. outside diameter and 0.51-in. inside diameter at 170°F and is to leave at 270°. Steam at 300°F condenses on the outside surface of the tubes, and its latent heat of condensation is $h_{fg} = 910.4$ Btu/lbm. If the water velocity in the tubes is limited to 7 ft/sec, find the length and number of tubes required when the tube thermal conductivity is $k = 60$ Btu/hr ft °F.

8.20 A heat exchanger is needed to cool 1 kg/s of engine oil, $c_p = 2000$ J/kg °C, from 170°C to 30°C by using water, $c_p = 4180$ J/kg °C, which is available at 10°C and is to leave at 90°C. An overall heat transmission coefficient of $U = 225$ W/m² °C is expected, based on outside tube area. It is desired to use a single-shell-pass, even multiple tube pass exchanger if it can handle the heat transfer task. If not, a two-shell-pass, even multiple tube pass exchanger is the next preference. If neither of these is suitable, then a pure counterflow exchanger will be used. Find the required outside heat transfer surface area and specify the type of exchanger to be used.

8.21 A two-shell-pass, sixteen-tube-pass exchanger has been designed to cool 109,000 lbm/hr of oil, $c_p = 0.55$ Btu/lbm °F, from 210°F to 120°F by using 80,000 lbm/hr of water, $c_p = 1.0$ Btu/lbm °F, which enters the tubes at 60°F. Another exchanger may have to be built to cool the same mass flow of oil from 210°F down to 90°F using the same water flow rate and inlet temperature. If all dimensions except the length are to be the same in both exchangers, calculate the ratio of the length of the new exchanger to that of the existing one.

8.22 Work Ex. 8.6 by use of the correction factor method rather than the ϵ-NTU approach used when the problem was solved in the main body of the chapter.

8.23 A feasibility study is being made for a heat exchanger in a gas turbine power plant. Both the hot and cold fluids are air with the same mass flow rate of 40,000 lbm/hr and the same specific heat, $c_p = 0.24$ Btu/lbm °F. The heat exchanger preliminary design calls for a crossflow exchanger with both fluids unmixed and an effectiveness, ϵ, of 70%. (a) If the overall heat transmission coefficient based on the hot gas side has been estimated to be about 50 Btu/hr ft² °F, calculate the surface area needed for the hot gas side. (b) What area is required for a pure counterflow exchanger if all other quantities are the same as in part a?

8.24 Rework Prob. 8.3 by use of the ϵ-NTU method.

8.25 Water at 40°F flowing at the rate of 6000 lbm/hr enters the shell side of a two-shell-pass, even multiple of four tube passes exchanger, while oil enters at 200°F on the tube side with a mass flow rate of 10,000 lbm/hr and a specific heat $c_p = 0.45$ Btu/lbm °F. The overall heat transmission coefficient U_o is found to be 70 Btu/hr ft² °F based on the total outside tube area of 150 ft². Compute the exit temperatures of both fluids.

8.26 Suppose the mass flow rate of the water in Ex. 8.3 is doubled, keeping the inlet temperatures and air mass flow rate the same. Find the new exit temperature of the water if the value of U_o does not change significantly.

8.27 If at an off-design condition, the heat exchanger described in Ex. 8.1 has oil entering at 110°C instead of at 90°C, with the mass flow rates of oil and coolant and the coolant inlet temperature being the ones given in that example, find the outlet temperatures of both fluids if the value of U remains unchanged.

8.28 Derive the analytical expression relating the effectiveness, ϵ, to the number of transfer units, NTU, for the parallel flow heat exchanger.

8.29 An air intercooler in a compressor system consists of a counterflow heat exchanger of area 320 ft², and the rough estimate of the overall heat transmission coefficient based on that area is $U = 25$ Btu/hr ft² °F. The water coolant enters at 60°F with a mass flow rate of 4000 lbm/hr, while the air enters at 500°F and is to leave at 100°F. For the water and air, respectively, the specific heat values are $c_p = 1.0$ Btu/lbm °F and $c_p = 0.24$ Btu/lbm °F. Calculate the mass flow rate of air which can be processed in this intercooler.

8.30 The two-shell-pass, even number of tube passes heat exchanger whose area was found to be 0.895 m² in Ex. 8.5 is also to be used in a different situation where the mass flow rate of the oil is 0.02 kg/s and its inlet temperature is 150°C. However, the engine fuel still enters at 15°C with a mass flow rate of 0.027 kg/s. The new conditions give a slightly lower overall heat transmission coefficient of $U_o = 180$ W/m² °C. Calculate the outlet temperatures expected for both fluids.

8.31 The steam superheater whose area was to be found in Prob. 8.4 must also operate for a portion of each day with an increased mass flow rate of steam of 30,000 lbm/hr entering at 300°F. The mass flow rate and entering temperature of the spent combustion gases remain at 96,000 lbm/hr and 1000°F, respectively, and since the gas side surface coefficient of heat transfer is the controlling resistance to heat transfer, the overall heat transmission coefficient's value remains at $U = 73$ Btu/hr ft² °F. Find the exit temperature of the steam.

8.32 A single-shell-pass, two-tube-pass heat exchanger unit has an outside tube surface area of 3.5 m² and is to be used to condense 50°C saturated ammonia vapor, with a latent enthalpy of condensation $h_{fg} = 1,052,000$ J/kg, on the shell side. If 1.5 kg/s of water, $c_p = 4180$ J/kg °C, enters the tubes at 10°C and the estimated overall heat transmission coefficient is $U_o = 1350$ W/m² °C, calculate the condensation rate of the ammonia and the exit water temperature.

8.33 A crossflow heat exchanger with an outside tube area of 850 ft² is to be used as a steam condenser. The latent heat of condensation of the 8 psia, 182.8°F steam to be completely condensed on the shell side is 988.4 Btu/lbm. Cooling water is available at 50°F but is to leave the tubes at a temperature no higher than 140°F. An estimate of the overall heat transmission coefficient leads to $U_o = 175$ Btu/hr ft² °F. Find the mass flow rate of the coolant and the steam condensation rate.

8.34 A water heater has been designed as a crossflow heat exchanger whose outside tube surface area is 150 m². Fifteen kg/s of water enters the tubes at 10°C, while 25 kg/s of 300°C air enters the exchanger and flows over the outside of the tubes in a mixed condition. The overall heat transmission coefficient is found to be $U_o = 22$ W/m² °C when fouling is taken into account by use of fouling resistances of 0.0004 and 0.0002 m² °C/W on the air and water sides, respectively. The ratio of outside to inside tube surface area is 1.10. (a) What is the exit temperature of the water in the fouled condition? (b) What water exit temperature results when the exchanger surfaces are clean?

8.35 Using the ϵ-NTU method, solve Ex. 8.4, which was worked earlier by the correction factor method.

8.36 A crossflow heat exchanger is available for the final cooling of 1.25 kg/s of an oil, $c_p = 1880$ J/kg °C, which flows over the outside of the tubes, whose area is 14 m², in a mixed condition and which must leave the exchanger at 37°C. The coolant available is water, $c_p = 4180$ J/kg °C, which enters the tubes at 7°C with a maximum mass flow rate of 0.75 kg/s. An estimate of the overall heat transmission coefficient based on the area given earlier is $U_o = 400$ W/m² °C. Find the highest temperature at which this oil can enter the exchanger.

8.37 The last stage of heating and cooling of two air streams, $c_p = 0.24$ Btu/lbm °F, is to occur in an existing crossflow exchanger of heat transfer area $A = 168$ ft² in which both streams are unmixed. The hotter stream, whose mass flow rate is 40,000 lbm/hr, must exit at 110°F, while the cooler stream, whose flow rate is 35,000 lbm/hr, must exit at 90°F. An overall heat transmission coefficient of $U = 50$ Btu/hr ft² °F is expected. Find the required inlet temperatures for both air streams.

8.38 An existing quench oil cooler is a one-shell-pass, two-tube-pass exchanger of outside tube area $A_o = 0.9$ m². It operates with 0.022 kg/s of water, $c_p = 4180$ J/kg °C, entering the tube side at 70°C, while 232°C oil, $c_p = 2000$ J/kg °C, enters the shell side. At an off-design operating condition, the exchanger must process 0.5 kg/s of the quenching oil. At this condition, an overall heat transmission coefficient of $U_o = 220$ W/m² °C is expected based on clean exchanger conditions. However, the estimated fouling factors are 0.0007 and 0.0002 m² °C/W on the quench oil and water sides, respectively. The ratio of outside to inside tube area is 1.15. Find the exit temperature of the quenching oil.

8.39 It is proposed to replace a burned-out heat exchanger in a furnace with a new crossflow unit of area 58 ft². The combustion gases enter at 1000°F flowing at a rate of 500 lbm/hr in a mixed condition, while the air to be heated enters tubes at 60°F and at a mass flow rate of 8000 lbm/hr. The specific heat value $c_p = 0.24$ Btu/lbm °F for both the combustion gases and the air. If the overall heat transmission coefficient is expected to be $U_o = 5$ Btu/hr ft² °F, calculate the exit temperatures of both the air and the stack gases, as well as the heat transfer rate to the air.

8.40 In a test of a two-shell-pass, four-tube-pass heat exchanger, it is found that 4 kg/s of water which enters at 20°C leaves at 92°C when 2 kg/s of water enters the tubes at 220°C. The exchanger area is $A = 12.2$ m², and both streams of water have a specific heat of about 4190 J/kg °C. Find the value of the overall heat transmission coefficient.

8.41 In an experiment to determine fouling resistance, a crossflow exchanger of outside tube area $A_o = 27$ ft² is found to raise the temperature of 4000 lbm/hr of fuel oil, $c_p = 0.5$ Btu/lbm °F, from 60°F to 200°F as it flows unmixed through the tubes, while 2175 lbm/hr of steam, $c_p = 0.46$ Btu/lbm °F, enters at 500°F and flows in a mixed condition over the outside of the tubes when all surfaces are clean. The exchanger is operated for a long period of time, during which the steam side is kept free from fouling by periodic cleanings. It is found that the exit temperature of the fuel oil changes to 161°F after a long time has passed. If the ratio of outside to inside tube surface area is 1.20, find the value of the fouling resistance on the fuel oil side.

8.42 In a shell-and-tube exchanger used as a water heater, 1.2 kg/s of 10°C water is to be heated to 60°C by passing 0.9 kg/s of 95°C water on the shell side. For water, $c_p = 4190$ J/kg °C, and the overall heat transmission coefficient based on outside tube area is around 1450 W/m² °C. The first choice for the exchanger would be a one-shell-pass, two-tube-pass unit if it can perform the heat transfer duty. If it cannot, an attempt will be made to use a two-shell-pass, four-tube-pass unit. If that is not able to perform the task, a pure counterflow exchanger will be used. (a) Find the outside tube surface area needed. (b) What is the highest mass flow rate of the 95°C water that can be tolerated without the exit temperature of the other water ever exceeding 80°C? Assume only a negligible change in U_o at this new operating point.

8.43 In a household heating unit, air, $c_p = 0.24$ Btu/lbm °F, enters a crossflow exchanger at 60°F and is to leave at 85°F after flowing over the outside of tubes in a mixed state, with the required energy transfer rate to this air being 120,000 Btu/hr. Water, $c_p = 1.0$ Btu/lbm °F, enters the tubes of the exchanger at 200°F and is to leave at 70°F. The estimated overall heat transmission coefficient based on outside tube area is $U_o = 7$ Btu/hr ft² °F. (a) Find the tube surface area required. (b) For the same mass flow rates and air inlet temperature, what is the lowest the water inlet temperature can be without the exit air temperature dropping below 75°F for the exchanger analyzed in part a?

8.44 A crossflow heat exchanger is to be designed to utilize the energy in the exhaust gases of an industrial diesel engine for a space heating application. Exhaust gases at 370°C enter the tubes at 0.25 kg/s. They are to leave at 95°C when 15°C air flows over the tubes in a mixed condition and leaves the exchanger at 26°C. Both specific heats are about $c_p = 1090$ J/kg °C, while the value of the overall heat transmission coefficient is estimated to be about $U_o = 15$ W/m² °C. (a) Compute the outside tube surface area needed. (b) If, for the exchanger of part a, the mass flow rate of the exhaust gases is reduced to one-half of the previous value, with both inlet temperatures and the air mass flow rate being the same as in part a, calculate the new exit air temperature.

8.45 A recuperator for a stationary gas turbine power plant uses the turbine's exhaust gases to preheat the intake air. This unit is to be designed as a crossflow heat exchanger with both fluids unmixed. The intake air and exhaust gas mass flow rates are equal at 90,000 lbm/hr with the same specific heat, $c_p = 0.24$ Btu/lbm °F. The exhaust gases enter at 550°F and are to leave at 200°F, while the intake air enters at 80°F. The overall heat transmission coefficient is estimated to be about $U = 6$ Btu/hr ft² °F. (a) How much heat transfer surface area is needed for this crossflow unit? (b) At a higher steady-state power level for the exchanger of part a, the inlet exhaust gas temperature is 660°F. If the mass flow rate and intake air inlet temperature are the same as in part a, calculate the outlet temperatures at this new condition.

8.46 A Freon-12 refrigeration system condenser consists of a single-shell-pass, four-tube-pass exchanger in which 0.3 kg/s of F-12 condenses on the shell side at 80°C because of water, $c_p = 4180$ J/kg °C, entering the tubes at 20°C and leaving at 55°C. The latent enthalpy of condensation of the Freon is $h_{fg} = 93,373$ J/kg, and the overall heat transmission coefficient based on outside tube surface area is $U_o = 620$ W/m² °C. (a) Find the needed heat transfer surface area. (b) If the condenser of part a must sometimes operate off-design, where the mass flow rate of 80°C Freon to be completely condensed is 0.45 kg/s with the same water inlet temperature, find the new mass flow rate and exit temperature of the water. Assume that the value of U_o does not change very much.

8.47 A crossflow heat exchanger is to be used as an ammonia condenser in which 600 lbm/hr of 115°F ammonia enters the tubes and is completely condensed. The value h_{fg} = 460.9 Btu/lbm is the latent enthalpy of condensation. Air flows over the outside of the tubes at an inlet temperature of 90°F and an exit temperature of 105°F. The estimate of the overall heat transmission coefficient based on outside tube area is U_o = 8 Btu/hr ft² °F. (a) What is the heat transfer area of the exchanger? (b) With the exchanger designed as in part a, there is the potential for an air mass flow rate 60% greater than in part a, with the same air inlet temperature and ammonia temperature. What new ammonia condensation rate would this allow, and what would the new exit temperature of the air be? It is expected that this will increase the value of U_o to about 10.5 Btu/hr ft² °F.

8.48 A two-shell-pass, four-tube-pass heat exchanger has 0.03 kg/s of lubrication oil, c_p = 2220 J/kg °C, flowing through the shell side with an inlet temperature of 160°C and a required exit temperature of 55°C. Ethylene glycol enters the tube side at 35°C and is to leave at 45°C and has a specific heat c_p = 2475 J/kg °C. The available estimate of the overall heat transmission coefficient referenced to the outside of the tubes is U_o = 170 W/m² °C. (a) Find the required heat transfer area. (b) If the exchanger, designed according to the constraints of part a, must also operate at times with 0.06 kg/s of oil entering at 180°C, calculate the exit temperatures of both fluids at this other operating point.

8.49 Nitrogen gas, c_p = 0.253 Btu/lbm °F, whose mass flow rate is 70,000 lbm/hr is to be heated from 200°F to 800°F in a crossflow heat exchanger with both fluids unmixed, by air which enters at 1250°F and which is to leave at 350°F. For this air, c_p = 0.255 Btu/lbm °F. Based on the air side, the expected overall heat transmission coefficient is U_a = 6.5 Btu/hr ft² °F. (a) Specify the exchanger surface area. (b) This same exchanger will also be utilized part of the time to do the initial heating of 5000 lbm/hr of hydrogen gas, c_p = 3.46 Btu/lbm °F, which enters at 80°F. The air inlet conditions remain the same as in part a. Find the exit temperature of the hydrogen gas if the expected U_a changes to 7.5 Btu/hr ft² °F.

8.50 A small steam condenser is to be designed as a one-shell-pass, four-tube-pass exchanger with 50°C steam condensing at the rate of 0.17 kg/s on the shell side. Its latent heat of condensation is h_{fg} = 2,388,700 J/kg. Water, c_p = 4180 J/kg °C, enters the tubes at 10°C and is to leave at 38°C. A rough estimate gives the value of the overall heat transmission coefficient based on outside tube area of U_o = 2000 W/m² °C. (a) Specify the heat transfer surface area needed. (b) The feasibility of using this same exchanger as an oil cooler is also to be studied. If 1.5 kg/s of oil, c_p = 2000 J/kg °C, enters the shell at 80°C with the water mass flow rate and inlet temperature being the same as for part a, what are the oil and water exit temperatures if the value of U_o is now 340 W/m² °C?

REFERENCES

1. Jakob, M. *Heat Transfer*. Vol. 2. New York: Wiley, 1957.
2. Kays, W. M., and A. L. London. *Compact Heat Exchangers*. 2d ed. New York: McGraw-Hill, 1964.
3. Fraas, A. P., and M. N. Ozisik. *Heat Exchanger Design*. New York: Wiley, 1965.
4. Afgan, N., and E. U. Schlünder, eds. *Heat Exchangers: Design and Theory Sourcebook*. New York: McGraw-Hill, 1974.
5. *Standards of the Tubular Exchanger Manufacturers Association*. 6th ed. New York: Tubular Exchangers Manufacturers Association (TEMA), 1978.

6. Bowman, R. A., A. C. Mueller, and W. M. Nagle. "Mean Temperature Difference in Design." *Transactions of the A.S.M.E.,* Vol. 62 (May 1940), pp. 283–94.

7. Taborek, J. "Evolution of Heat Exchanger Design Techniques." *Heat Transfer Engineering,* Vol. 1, No. 1 (July–September 1979), pp. 15–29.

8. Dodd, R. "Temperature Efficiency of Heat Exchangers with One Shell Pass and Even Numbers of Tube Passes." *Transactions of the Institute of Chemical Engineers,* Vol. 60, No. 6 (1982), pp. 364–68.

9. Gardner, K., and J. Taborek. "Mean Temperature Difference: A Reappraisal." *AIChE Journal,* Vol. 23, No. 6 (November 1977), pp. 777–86.

10. Kern, D. Q., and A. D. Kraus. *Extended Surface Heat Transfer.* New York: McGraw-Hill, 1972.

11. *PLATECOIL, Catalog No. 5-63, Product Data Manual.* Platecoil Division, Tranter Manufacturing Inc., Lansing, Michigan, 1966.

12. *Type "WU" Heat Exchangers.* Bulletin C-130.1, revision 6. Bell & Gossett, ITT, Fluid Handling Division, 8200 N. Austin Ave., Morton Grove, Illinois, 1976.

13. Lienhard, J. H. *A Heat Transfer Textbook.* Englewood Cliffs, N.J.: Prentice-Hall, 1981.

14. Hausen, H. *Heat Transfer in Counterflow, Parallel Flow, and Cross Flow.* New York: McGraw-Hill, 1983.

15. Walker, G. *Industrial Heat Exchangers: A Basic Guide.* Washington, D.C.: Hemisphere, 1982.

ADDITIONAL TOPICS IN HEAT TRANSFER

9.1 INTRODUCTION

This chapter discusses several additional topics in heat transfer which are special in the sense that the average practicing engineer may not encounter these nearly as often as the topics previously presented. These additional topics are (1) *heat transfer in high-speed flow,* knowledge of which is important in the design and analysis of gas turbine power plants, jet engines, high-speed aircraft, and missiles, (2) *liquid metal heat transfer,* whose applications include power cycles and nuclear power reactors, (3) *differential similarity,* which makes possible the intelligent design of experiments that minimize the amount of data taking and maximize the generality of the results by finding the appropriate non-dimensional groups directly from the governing partial differential equations and their boundary conditions, and (4) *transpiration and film cooling,* which find application in gas turbine blades and vanes, combustors, and rocket nozzles.

9.2 HEAT TRANSFER IN HIGH-SPEED FLOW

Chapter 5 on forced convection heat transfer was limited to low-speed flow situations, where the Eckert number is much less than unity (see Eq. [5.200] and Eq. [5.286] and the associated discussions). Hence, the viscous dissipation terms, such as the last terms on the right-hand side of Eqs. (5.199) and (5.285), could be neglected. These viscous dissipation terms represent the rate at which mechanical energy is irreversibly degraded to random thermal energy as a result of the work done by the viscous shearing stresses on the moving fluid. When they are large enough, their result shows up as a heating effect throughout the fluid in the thermal boundary layer. The heating effect of viscous dissipation, also called *aerodynamic heating,* is roughly analogous to the heating effect when sandpapering a block of wood, which is a result of the dissipation of the mechanical work required to move the sandpaper across the wood surface. In a moving fluid, this heating effect occurs at a rate which varies with position throughout the thermal boundary layer.

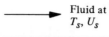

Figure 9.1 Fluid flowing over an insulated flat plate.

To understand some of the effects of viscous dissipation in high-speed flow, consider first *low-speed*, steady, constant property flow of a fluid at temperature T_s and velocity U_s, over a flat plate, which is perfectly insulated at its upper surface, as shown in Fig. 9.1. For steady-state conditions on the plate surface and in the absence of any significant net radiant loss from the surface, the application of the energy balance, Eq. (1.8), to any part of the upper surface of the plate leads to the result that the plate temperature equals the undisturbed free stream temperature; hence, $T_w = T_s$. In fact, the temperature everywhere within the fluid equals T_s, since there is no unbalanced thermal loading in this low-speed flow situation to cause a temperature distribution within the moving fluid.

Now, when U_s is large enough to cause appreciable viscous dissipation, the last term on the right of Eq. (5.285) can no longer be neglected. Thus, appreciable amounts of mechanical energy are degraded by the action of the viscous shearing stresses, resulting in a heating effect which produces a nonconstant temperature distribution as a function of position within the fluid. Since the plate is perfectly insulated, its temperature becomes greater than T_s in order to maintain the zero temperature derivative condition at the plate surface. This temperature, which a perfectly insulated surface achieves in a flow with appreciable viscous dissipation, and with negligible radiation effects, is called the *aerodynamic recovery temperature* T_r. Note that T_r is also called the *recovery temperature* and the *adiabatic wall temperature*. A knowledge of this aerodynamic recovery temperature is essential, not only for finding the highest temperature that a surface over which a high-speed fluid is moving will achieve, but also in the calculation of convection heat transfer rates when the surface is held at some temperature T_w other than T_r; that is, when the surface is not insulated.

9.2a High-Speed Flow Between Parallel Plates

Figure 9.2 shows two parallel plates spaced a distance b apart, between which a constant property fluid flows in a steady, laminar fashion at high enough speed to produce appreciable viscous dissipation effects. The lower plate is perfectly insulated and the upper plate is held at a known temperature T_1. Both the velocity and temperature profiles are fully developed. (See Secs. 5.3b–4a.) The temperature profile has been developed to the point where $\partial T/\partial x = 0$ for all values of x and y.

We want to predict the temperature distribution as a function of y within the moving fluid, and we especially want to predict the temperature of the lower, insulated plate under these conditions, since this temperature is the aerodynamic recovery temperature, T_r. Once this aspect of high-speed flow has been dealt with—namely, the prediction of the temperature attained by a perfectly insulated surface—the other question of heat transfer in high-speed flow will be addressed by analysis of the geometry shown in Fig. 9.2 when

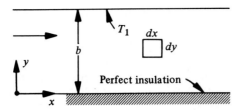

Figure 9.2 Differential control volume of a fluid in high-speed flow between two parallel plates.

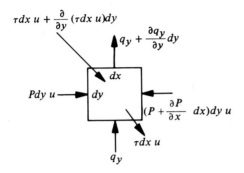

Figure 9.3 Energy transfer terms across a control surface in high-speed flow.

the lower plate is held at a known temperature T_w which is *not* equal to the aerodynamic recovery temperature T_r.

The starting point for analysis of both of these facets of high-speed flow is the governing differential equation for the temperature distribution within such a flow. It is instructive and worthwhile to derive this equation from first principles, an energy balance on the differential control volume shown in Fig. 9.2. In this way we can focus our attention on the shear work terms which cause the viscous dissipation of mechanical energy forms which is responsible for the differences between high-speed and low-speed flows.

The law of conservation of energy, Eq. (1.8), with steady-state conditions prevailing and no generation, reduces to

$$R_{\text{in}} = R_{\text{out}}. \tag{9.1}$$

The control volume is shown in Fig. 9.3 with the various energy transfer terms into and out of the control volume. The x-conduction terms are not considered, since $\partial T/\partial x = 0$ for all x and y, and no energy crosses the upper or lower faces by virtue of mass crossing these faces because the y component of velocity v is zero for all x and y. This is a consequence of the fully developed velocity field; hence, $\partial u/\partial x = 0$ for all x and y. Energy does enter the left-hand face and leaves the right-hand face because of the motion of the mass in the x direction. However, these energy transfer terms in and out are equal, and cancel each other since $\partial T/\partial x = 0$. Hence, the only energy transfer terms across the control surface are the ones shown in Fig. 9.3; that is, the y-conduction terms, the rate at which the shear stresses do work on the control surface, and the rate at which the pressure does work on the control surface. The rate at which the material in the control volume transmits shear work to the slower-moving fluid layers under the lower control surface is $\tau dx\, u$, while the term involving the shear stress τ at the upper surface is the rate at which shear work is being done on the control volume as a result of faster-moving fluid particles above the upper control surface. Substituting into Eq. (9.1), with the appropriate symbols from Fig. 9.3 yields

$$P dy\, u + q_y + \tau dx\, u + \frac{\partial}{\partial y}(\tau u)\, dy dx = \tau dx\, u + q_y + \frac{\partial q_y}{\partial y}\, dy$$
$$+ P dy\, u + \frac{\partial P}{\partial x}\, dx dy\, u. \quad \textbf{(9.2)}$$

Since by Fourier's law of conduction and Newton's law of viscosity,

$$q_y = -k\, dx\, \frac{\partial T}{\partial y} \quad \text{and} \quad \tau = \mu\, \frac{\partial u}{\partial y},$$

respectively, Eq. (9.2) becomes

$$\mu\, \frac{\partial}{\partial y}\left(u\, \frac{\partial u}{\partial y}\right) = -k\, \frac{\partial^2 T}{\partial y^2} + u\, \frac{\partial P}{\partial x}. \quad \textbf{(9.3)}$$

Since $\partial T/\partial x = 0$ for all x and y, Eq. (9.3) is an ordinary differential equation in T. After expansion of the left-hand side,

$$\mu\, u\, \frac{\partial^2 u}{\partial y^2} + \mu\left(\frac{\partial u}{\partial y}\right)^2 = -k\, \frac{d^2 T}{dy^2} + u\, \frac{\partial P}{\partial x}. \quad \textbf{(9.4)}$$

Using the x-momentum equation for this flow situation from Chapter 5, Eq. (5.36), yields

$$\mu\, \frac{\partial^2 u}{\partial y^2} = \frac{\partial P}{\partial x}. \quad \textbf{(9.5)}$$

Multiplying both sides of Eq. (9.5) by the local x component of velocity u, and combining the result with Eq. (9.4) yields

$$k\, \frac{d^2 T}{dy^2} = -\mu\left(\frac{\partial u}{\partial y}\right)^2. \quad \textbf{(9.6)}$$

Note that the term on the right-hand side of Eq. (9.6) is the viscous dissipation function which was negligible in the low-speed flow problems solved in Chapter 5. It cannot be neglected here, because by definition it is significant in a high-speed flow problem. Notice that Eq. (9.6) could also have been arrived at by proper interpretation of Eq. (5.285) once v and $\partial T/\partial x$ were set equal to zero.

Temperature Distribution When the Lower Plate Is Insulated

The governing differential equation for the temperature distribution is Eq. (9.6) subject to the boundary conditions at $y = 0$, $dT/dy = 0$, and at $y = b$, $T = T_1$. The velocity distribution must be known so that the term on the right-hand side of Eq. (9.6) can be determined. Since the flow is constant property, the velocity field is independent of the temperature field and can be solved for separately, using the continuity and x-momentum equations. Thus, from Chapter 5, where this has already been done,

$$u = 6\, U_m \left(\frac{y}{b} - \frac{y^2}{b^2} \right). \qquad \text{[5.44]}$$

Differentiating Eq. (5.44),

$$\frac{\partial u}{\partial y} = 6\, U_m \left(\frac{1}{b} - \frac{2y}{b^2} \right).$$

Therefore,

$$\left(\frac{\partial u}{\partial y} \right)^2 = \frac{36\, U_m^2}{b^4} (b^2 - 4yb + 4y^2).$$

Substituting this into Eq. (9.6), rearranging, and defining $B = 36\, U_m^2 \mu / k b^4$ yields

$$\frac{d^2 T}{dy^2} = -B(b^2 - 4yb + 4y^2).$$

Integrating,

$$\frac{dT}{dy} = -B(b^2 y - 2y^2 b + \tfrac{4}{3} y^3) + C_1. \qquad (9.7)$$

From the first boundary condition, C_1 must equal 0 in Eq. (9.7), which yields

$$\frac{dT}{dy} = -B(b^2 y - 2y^2 b + \tfrac{4}{3} y^3).$$

Integrating again,

$$T = -B(\tfrac{1}{2} b^2 y^2 - \tfrac{2}{3} y^3 b + \tfrac{1}{3} y^4) + C_2. \qquad (9.8)$$

Substituting the second boundary condition into Eq. (9.8) yields

$$T_1 = -B(\tfrac{1}{2} b^4 - \tfrac{2}{3} b^4 + \tfrac{1}{3} b^4) + C_2, \quad \text{or} \quad C_2 = T_1 + \tfrac{1}{6} B b^4.$$

Substituting into Eq. (9.8), the temperature distribution in the moving fluid, when appreciable viscous dissipation is present, is

$$T(y) = T_1 - B(\tfrac{1}{2} b^2 y^2 - \tfrac{2}{3} by^3 + \tfrac{1}{3} y^4 - \tfrac{1}{6} b^4). \qquad (9.9)$$

Note that for low-speed flow, that is, flow without appreciable viscous dissipation, the right-hand side of Eq. (9.6) is nearly zero. Hence, $B = 0$, and Eq. (9.9) becomes

$$T(y) = T_1. \qquad (9.10)$$

Thus, when the temperature profile is caused solely by the viscous dissipation, the distortion of the temperature profile caused by appreciable viscous dissipation is seen immediately by comparing Eqs. (9.9) and (9.10).

The temperature of the insulated plate—that is, the aerodynamic recovery temperature T_r—is also required when viscous dissipation effects are not negligible. This is determined from Eq. (9.9) by setting $y = 0$, which yields

$$T_r = T_1 + \frac{1}{6} Bb^4.$$

From the definition of B,

$$T_r - T_1 = \frac{6\mu U_m^2}{k}.$$

Multiplying the numerator and denominator of the right-hand side by $2c_p$,

$$T_r - T_1 = 12 \left(\frac{\mu c_p}{k} \right) \frac{U_m^2}{2c_p}. \qquad (9.11)$$

Since $\mu c_p / k = N_{\text{Pr}}$, the Prandtl number, Eq. (9.11) in nondimensional form becomes

$$\frac{T_r - T_1}{U_m^2 / 2c_p} = 12 N_{\text{Pr}}. \qquad (9.12)$$

The denominator of the left-hand side of Eq. (9.12) is the temperature rise obtained when the fluid moving at the average velocity U_m is decelerated to zero velocity adiabatically. This is often called the *dynamic temperature rise*. The group on the left-hand side of Eq. (9.12) is called the *aerodynamic recovery factor*, or *recovery factor*, r_c; that is,

$$r_c = \frac{T_r - T_1}{U_m^2 / 2c_p}. \qquad (9.13)$$

Rearranging Eq. (9.12) for the flow between parallel plates with appreciable viscous dissipation yields

$$T_r = T_1 + 12N_{Pr}(U_m^2/2c_p).$$ (9.14)

Note that for low-speed flow, as U_m becomes small, $T_r \to T_1$. Hence, the results above support the qualitative arguments concerning the effect of viscous dissipation that were advanced at the beginning of this section with regard to a flat plate immersed in an external boundary layer–type flow.

Also note, by examining Eqs. (9.9) and (9.14), that the highest fluid temperature resulting from viscous dissipation appears at the insulated surface and is the aerodynamic recovery temperature T_r. Physically, the work dissipated by the shear stresses must leave the control volume if steady-state conditions prevail. However, it cannot leave in the x direction, because $\partial T/\partial x = 0$, and it cannot leave at the insulated boundary. Hence, it must be conducted to the upper surface at the temperature T_1. As a result, T_r must be high enough to ensure that all conduction is toward the upper surface.

Temperature Distribution and Heat Transfer When the Lower Plate Is at T_w

The more generally encountered case for the parallel plate configuration shown in Fig. 9.2 occurs when the lower plate, instead of being perfectly insulated, is held at some temperature T_w which is *not* equal to the aerodynamic recovery temperature T_r which results when the plate is perfectly insulated. In order to find, for this case, the temperature profile within the moving fluid and, more importantly, the heat transfer rate to or from the surface in the presence of viscous dissipation caused by the high-speed flow, the starting point is the governing differential equation for the temperature distribution in the moving fluid, Eq. (9.6). The difference between the present constant lower wall temperature case and the case of the perfectly insulated lower wall, which was just analyzed, is in the boundary condition to be applied at $y = 0$. Thus, the appropriate boundary conditions are the following ones:

$$\begin{aligned} y &= 0 & T &= T_w \\ y &= b & T &= T_1 \end{aligned}$$ (9.15)

Since only the condition at $y = 0$ is different, one can take advantage of some of the work done previously for the perfectly insulated lower plate case by integrating Eq. (9.7) to give the following result:

$$T = -B(\tfrac{1}{2} b^2 y^2 - \tfrac{2}{3} by^3 + \tfrac{1}{3} y^4) + C_1 y + C_2,$$

where C_1 and C_2 are determined by applying the boundary conditions, yielding

$$\begin{aligned} T = &-B(-\tfrac{1}{6} yb^3 + \tfrac{1}{2} b^2 y^2 - \tfrac{2}{3} by^3 + \tfrac{1}{3} y^4) \\ &+ (T_1 - T_w)y/b + T_w. \end{aligned}$$ (9.16)

Now, the definition of the local surface coefficient of heat transfer is

$$h_x = \frac{-k \left.\dfrac{\partial T}{\partial y}\right|_{y=0}}{\Delta T},$$

[5.3]

where ΔT is the appropriate driving potential difference.
  From Eq. (9.16),

$$\left.\frac{\partial T}{\partial y}\right|_{y=0} = \frac{1}{6} Bb^3 + \frac{T_1 - T_w}{b}.$$

Using this in Eq. (5.3), with the definition of B, gives

$$h_x = \frac{\dfrac{k}{b}\left(T_w - T_1 - \dfrac{6\mu U_m^2}{k}\right)}{\Delta T}.$$

(9.17)

Now, if ΔT is selected as $T_w - T_1$, the local surface coefficient of heat transfer h_x would depend upon the driving temperature difference explicitly, since $T_w - T_1$ would still appear on the right-hand side of Eq. (9.17). *Since the appropriate, or natural, driving temperature difference is chosen, whenever possible, in such a way as to make h_x independent of ΔT*, the term in parentheses in Eq. (9.17) is rearranged so that

$$h_x = \frac{k}{b} \frac{\left[T_w - \left(T_1 + \dfrac{12 N_{\mathrm{Pr}} U_m^2}{2 c_p} \right) \right]}{\Delta T}.$$

(9.18)

Using Eq. (9.14), the quantity in parentheses in Eq. (9.18) is the temperature that the lower surface would have if it were perfectly insulated—that is, the aerodynamic recovery temperature T_r. Hence, Eq. (9.18) becomes

$$h_x = \frac{k}{b} \frac{T_w - T_r}{\Delta T}.$$

Thus, for use in Newton's cooling law, the natural driving temperature difference ΔT should be chosen as the actual wall temperature T_w minus the aerodynamic recovery temperature T_r. Then, h_x will not depend upon the driving temperature difference itself. Hence,

$$h_x = k/b,$$

and by Newton's cooling law, the local convective heat transfer rate is

$$q_x'' = h_x(T_w - T_r).$$

(9.19)

For low-speed flow, where viscous dissipation effects are negligible, Eq. (9.18) reduces to

$$h_x = \frac{k}{b} \frac{T_w - T_1}{\Delta T}.$$

For the low-speed case, then, ΔT is chosen as $T_w - T_1$, so that

$$h_x = k/b.$$

Thus, if the driving temperature difference in Newton's cooling law is chosen to be $T_w - T_r$ for flow with appreciable viscous dissipation, the surface coefficient of heat transfer to be used is the same as that for the low-speed situation. This observation considerably simplifies the problem of calculating the heat transfer rate in a high-speed flow, since once the simpler problem of predicting the aerodynamic recovery temperature T_r is solved, this result used in Eq. (9.19), along with the low-speed surface coefficient of heat transfer as given by the analyses and the experiments reported in Chapter 5, allows the calculation of the heat transfer rate in the presence of appreciable viscous dissipation, at least for the parallel plate duct case just considered. One would hope that this simplicity would also carry over to more general and more difficult-to-analyze high-speed flow situations, such as external boundary layer–type flow over a body. This does turn out to be the case, as the results of the next section will show.

9.2b High-Speed Flow Over Bodies Immersed in the Flow

In order to solve for the quantities of interest (namely, the aerodynamic recovery temperature and the local surface heat flux) in external boundary layer flow over a body when there is appreciable viscous dissipation in the fluid, one can, and usually does, start with the partial differential form of the thermal energy equation, Eq. (5.285). However, a rather simple solution for a special case of high-speed flow suggests itself when the total energy equation, Eq. (5.283), is transformed so that the dependent variable is the stagnation temperature T_0 rather than the temperature T.

After considerable rearrangement, the total energy equation, Eq. (5.283), can be written as

$$\rho c_p \left(u \frac{\partial T_0}{\partial x} + v \frac{\partial T_0}{\partial y} \right) = \frac{\partial}{\partial y} \left[k \frac{\partial T_0}{\partial y} + \frac{k}{c_p} (N_{Pr} - 1) \frac{\partial}{\partial y} \left(\frac{u^2}{2} \right) \right], \qquad (9.20)$$

where

$$T_0 = T + \frac{u^2}{2c_p}, \qquad (9.21)$$

where T_0 is the *stagnation temperature* or the *total temperature* and is the temperature which a fluid at temperature T and velocity u will reach if adiabatically decelerated to zero velocity. In other words, it is the sum of the static and dynamic temperatures.

By studying Eq. (9.20), we can see that a vastly simplified form occurs when the Prandtl number of the fluid is unity, since this causes the second term in the right-hand bracket to vanish and leads to the following:

$$\rho c_p \left(u \frac{\partial T_0}{\partial x} + v \frac{\partial T_0}{\partial y} \right) = k \frac{\partial^2 T_0}{\partial y^2}.$$

(9.22)

Next, let us attempt to solve Eq. (9.22) for steady, laminar, constant property, two-dimensional planar boundary layer–type flow of a fluid, whose Prandtl number is unity, over a flat plate such as the one depicted in Fig. 9.1. A solution for the aerodynamic recovery temperature, T_r, will be sought for two reasons. Firstly, T_r is important in its own right, since it is the temperature which a perfectly insulated body attains in a high-speed flow in the absence of any net radiant loss. Since, depending upon the free stream velocity's magnitude, T_r could be a very high temperature, it is important to know its value to see if surfaces which are effectively insulated, such as an uncooled turbine vane or a portion of the skin of an aircraft, might actually have to be cooled so that they can retain suitable strength and structural integrity. Secondly, based upon the results of the parallel plate duct problem, it is expected that T_r will be needed in Newton's cooling law in order to calculate the surface heat flux in a high-speed flow over an uninsulated surface.

The boundary conditions, when the plate is insulated, for the partial differential equation, Eq. (9.22), are as follows:

As $y \to \infty$, $T_0 \to T_{0s}$, where, by Eq. (9.21), the free stream stagnation temperature T_{0s} is

$$T_{0s} = T_s + \frac{U_s^2}{2c_p}.$$

Since the plate is perfectly insulated, the boundary condition at $y = 0$ is that $\partial T / \partial y = 0$. From Eq. (9.21),

$$T = T_0 - \frac{u^2}{2c_p}.$$

(9.23)

Differentiating both sides of Eq. (9.23) partially with respect to y yields

$$\frac{\partial T}{\partial y} = \frac{\partial T_0}{\partial y} - \frac{u \, \partial u / \partial y}{c_p}.$$

(9.24)

Now, at $y = 0$, $\partial T / \partial y = 0$ and $u = 0$; hence, from Eq. (9.24), $\partial T_0 / \partial y = 0$.

The initial condition on x is that at $x = 0$, $T = T_s$, and since $u = U_s$ at $x = 0$, the boundary condition, with the aid of Eq. (9.21), becomes that at $x = 0$, $T_0 = T_{0s}$.

Hence, the three boundary conditions that must be satisfied by the solution to Eq. (9.22) are

$$
\begin{array}{lll}
\text{at } x = 0, & T_0 = T_{0s} & \text{for } 0 < y < \infty, \\
\text{at } y = 0, & \partial T_0 / \partial y = 0 & \text{for } x > 0, \\
\text{at } y \to \infty, & T_0 \to T_{0s} & \text{for } x > 0.
\end{array}
$$

One solution to Eq. (9.22) is seen to be, by inspection,

$$T_0 = K,$$

where K is a constant. Normally such a solution to a partial differential equation is a *trivial* solution, called trivial because it is unable to satisfy the boundary conditions on the equation. However, *this solution satisfies all the boundary conditions if K is chosen to be T_{0s}*. Hence, the solution to Eq. (9.22) which satisfies the three boundary conditions is

$$T_0 = T_{0s}. \tag{9.25}$$

Using Eq. (9.21), the temperature distribution is

$$T = T_{0s} - \frac{u^2}{2c_p}. \tag{9.26}$$

Since from the solution of $u = u(x, y)$ in Chapter 5, u is a known function of x and y for steady constant property flow over a flat plate, $T = T(x, y)$ from Eq. (9.26). The aerodynamic recovery temperature is the temperature of the insulated surface in a high-speed flow with appreciable viscous dissipation and is found by setting $y = 0$ in Eq. (9.26). Since $u = 0$ at $y = 0$, the aerodynamic recovery temperature is

$$T_r = T_{0s}. \tag{9.27}$$

The recovery factor r_c, defined by Eq. (9.13), after using Eqs. (9.21) and (9.27) in Eq. (9.13) and replacing U_m by U_s, and T_1 by T_s, is

$$r_c = \frac{T_r - T_s}{U_s^2/2c_p} = 1 \tag{9.28}$$

as long as $N_{\text{Pr}} = 1$.

Now that the solution for the aerodynamic recovery temperature T_r is in hand as Eq. (9.27), we next turn to the prediction of the local surface heat flux on the plate of Fig. 9.1 when the plate surface, rather than being insulated, is held at a constant known temperature T_w. In order to find the surface heat flux in this high-speed flow, the local surface coefficient of heat transfer must be predicted. Hence, the temperature distribution within the moving fluid must be predicted. We are still working with the restriction that the Prandtl number of the fluid is 1, so that, once again, the governing partial differential equation, with stagnation temperature T_0 as the dependent variable, is given by Eq. (9.22). The boundary conditions, in terms of the stagnation temperature T_0, are

$$
\begin{aligned}
&\text{at } x = 0 \quad \text{and } y > 0, &&T_0 = T_{0s}, \\
&\text{at } y = 0 \quad \text{and } x > 0, &&T_0 = T_w, \\
&\text{as } y \to \infty \text{ and } x > 0, &&T_0 \to T_{0s}.
\end{aligned}
$$

Equation (9.22), with these boundary conditions, looks exactly like Eq. (5.288) and its boundary conditions. Therefore, since T_0 plays the same role in Eq. (9.22) as T in Eq. (5.288), ϕ is defined as

$$\phi = \frac{T_0 - T_w}{T_{0s} - T_w}.$$

Using this, Eq. (9.22) and the boundary conditions become

$$\rho c_p \left(u \frac{\partial \phi}{\partial x} + v \frac{\partial \phi}{\partial y} \right) = k \frac{\partial^2 \phi}{\partial y^2}, \tag{9.29}$$

where

$$
\begin{array}{lll}
\text{at } x = 0 & \text{and } y > 0, & \phi = 1, \\
\text{at } y = 0 & \text{and } x > 0, & \phi = 0, \\
\text{as } y \to \infty & \text{and } x > 0, & \phi \to 1.
\end{array}
$$

Equation (9.29) and the boundary conditions are exactly the same as those following Eq. (5.288), since the velocity fields are also the same because of the constant property condition. (See Eqs. [5.290–92].) Hence, the solution for ϕ is the same known function of the similarity variable η; that is,

$$\eta = y\sqrt{U_s/\nu x}. \tag{9.30}$$

As a result,

$$\phi = \frac{T_0 - T_w}{T_{0s} - T_w} = \phi(\eta),$$

or

$$T_0 = T_w + (T_{0s} - T_w)\phi(\eta). \tag{9.31}$$

From Eq. (5.3), the local surface coefficient of heat transfer is

$$h_x = \frac{-k(\partial T/\partial y)_{y=0}}{\Delta T}. \tag{9.32}$$

Since $T = T_0 - u^2/2c_p$ from Eq. (9.23),

$$\frac{\partial T}{\partial y} = \frac{\partial T_0}{\partial y} - \frac{u \, \partial u/\partial y}{c_p}.$$

Since $u = 0$ at $y = 0$,

$$\left.\frac{\partial T}{\partial y}\right|_{y=0} = \left.\frac{\partial T_0}{\partial y}\right|_{y=0}.$$

Using this in Eq. (9.32),

$$h_x = \frac{-k(\partial T_0/\partial y)_{y=0}}{\Delta T}. \tag{9.33}$$

Now, $(\partial T_0/\partial y)_{y=0}$ is determined by taking the partial derivative of Eq. (9.31) with respect to y and evaluating the result at $y = \eta = 0$. That is,

$$\left.\frac{\partial T_0}{\partial y}\right|_{y=0} = (T_{0s} - T_w)\left.\frac{d\phi}{d\eta}\right|_{\eta=0}\left.\frac{\partial \eta}{\partial y}\right|_{y=0}. \tag{9.34}$$

From Eq. (9.30),

$$\left.\frac{\partial \eta}{\partial y}\right|_{y=0} = \sqrt{\frac{U_s}{\nu x}}. \tag{9.35}$$

Substituting Eqs. (9.34) and (9.35) into (9.33) and rearranging,

$$h_x = \frac{(T_w - T_{0s})[k\sqrt{U_s/\nu x}\,(d\phi/d\eta)_{\eta=0}]}{\Delta T}. \tag{9.36}$$

The bracketed quantity in the numerator does not contain any temperature differences; hence, if the driving temperature difference ΔT in Newton's cooling law is taken as $T_w - T_s$, as in low-speed flow, the resulting h_x would depend explicitly upon the temperature differences $T_w - T_{0s}$ and $T_w - T_s$. In addition, the resulting h_x could assume not only positive values as in low-speed flow (no appreciable viscous dissipation), but also negative values and zero, depending upon the magnitudes of the temperatures T_w, T_s, and T_{0s}. *Thus, the driving temperature difference in Newton's cooling law which suggests itself "naturally" from Eq. (9.36) is $\Delta T = T_w - T_{0s}$, since this choice makes h_x independent of the driving temperature difference.* From Eq. (9.27), T_{0s} is the temperature of the plate if it is perfectly insulated—that is, the aerodynamic recovery temperature T_r. Hence, from Eq. (9.36), with $\Delta T = T_w - T_{0s} = T_w - T_r$, the local surface coefficient of heat transfer is

$$h_x = k\sqrt{\frac{U_s}{\nu x}}\left.\frac{d\phi}{d\eta}\right|_{\eta=0}.$$

Comparing this with the steps preceding Eq. (5.296), *it is seen that the use of $T_w - T_r$ as the driving temperature difference permits the use of the ordinary low-speed surface coefficient of heat transfer from Eq. (5.296), as long as $N_{Pr} = 1$.*

The analysis of the flow between parallel plates and of the boundary layer flow of a fluid, whose Prandtl number is unity, over a flat plate indicates that the convection heat transfer rate in steady, constant property, high-speed flow of a fluid over a surface at temperature T_w can be calculated from Newton's cooling law,[1]

$$q'' = h_x(T_w - T_r). \tag{9.37}$$

Here h_x is the ordinary low-speed surface coefficient of heat transfer and can be found for either laminar or turbulent boundary layer flow over any body from the analytical relations and experimental correlations in Chapter 5. The aerodynamic recovery temperature T_r must be used instead of the free stream temperature T_s in the driving temperature difference. This is the temperature the surface of the body attains if it is perfectly insulated, with no radiant loss, in the same high-speed stream. The aerodynamic recovery temperature T_r is found from either analytical results or experimental correlations for the recovery factor r_c, defined by

$$r_c = \frac{T_r - T_s}{U_s^2/2c_p}. \tag{9.38}$$

For low-speed flows, T_r approaches T_s, and the procedure just described reduces to that previously used for low-speed flow problems in Chapter 5.

Eckert [1], [2] recommends

$$r_c = \sqrt{N_{Pr}} \quad \text{and} \quad r_c = N_{Pr}^{1/3} \tag{9.39}$$

for *laminar flow over a flat plate* and for *turbulent flow over a flat plate,* respectively. The recovery factor changes for shapes other than a flat plate, but Eq. (9.39) can be used as approximations for other shapes, as long as regions of separated flow are avoided. Eckert also shows that the same relations, Eqs. (9.38) and (9.39), can be used even with property variation if the temperature-dependent fluid properties, which appear in these equations and also in the equation for the low-speed surface coefficient of heat transfer, are evaluated at the reference temperature T^*, defined as

$$T^* = T_s + 0.5(T_w - T_s) + 0.22(T_r - T_s). \tag{9.40}$$

For extremely high flow velocities, usually in the supersonic Mach number range, there is considerable property variation across the boundary layer and, perhaps, dissociation of the gas. Under these conditions, Eckert recommends the use of a heat transfer coefficient

---

1. This can also be verified for the more general case in which the Prandtl number is not unity.

and recovery factor based upon an enthalpy difference rather than a temperature difference. (For further details of this procedure, see [2].)

For the more general case of heat transfer in high-speed flow over a *non-isothermal* surface, the reader can consult Rubesin and Inouye [3].

EXAMPLE 9.1

A flat plate, 0.3 m long, has a 38°C, 1-atm pressure air stream flowing over its upper surface with a speed of 396 m/s. (a) If the plate were perfectly insulated, calculate what its surface temperature would be. (b) If the plate is actually being held at a temperature, T_w, of 93°C, find the heat transfer rate to the plate and compare it to the value obtained if, erroneously, this were treated as a low-speed problem; that is, treated as if the effects of viscous dissipation were negligible.

Solution

In part a, the perfectly insulated surface in a high-speed stream reaches the aerodynamic recovery temperature T_r in the absence of any net radiant loss. From Eq. (9.38), since $T_s = 38°C$ and $U_s = 396$ m/s, the recovery factor relationship gives

$$r_c = \frac{T_r - 38}{(396)^2/2c_p^*} = \frac{T_r - 38}{78{,}408/c_p^*}. \tag{9.41}$$

In the case of air at 38°C, the specific heat c_p and the Prandtl number N_{Pr} are relatively weak functions of temperature. Hence, a good first estimate of T_r can be made. That is, c_p is about 1004.6 and N_{Pr} is about 0.70 over a wide temperature range. Now, r_c is either $N_{Pr}^{1/2}$ or $N_{Pr}^{1/3}$, depending upon whether the flow is laminar or turbulent. However, the cube and square root of 0.70 are 0.888 and 0.8265, respectively; therefore, r_c cannot vary through a large range regardless of the value of T^*. A similar statement holds for c_p^*. Thus, using $r_c = 0.888$ and $c_p^* = 1004.6$, $T_r = 107.3°C$ is used as the first estimate for T_r in Eq. (9.40). With this we can calculate T^*, from Eq. (9.40), as

$$T^* = 38 + 0.5(107.3 - 38) + 0.22(107.3 - 38) = 87.9°C.$$

At $T^* = 87.9°C$, the property tables in Appendix A yield $\mu^* = 0.211 \times 10^{-4}$ kg/ms, $c_p^* = 1008.8$ J/kg °C, and $N_{Pr}^* = 0.694$. By the perfect gas equation of state,

$$\rho^* = \frac{P}{RT^*} = \frac{1.013 \times 10^5}{287(87.9 + 273)} = 0.978 \text{ kg/m}^3.$$

With $U_s = 396$ m/s and the plate length $L = 0.3$ m, the plate Reynolds number is

$$N_{Re_L} = \rho^* U_s L/\mu^* = \frac{0.978(396)(0.3)}{0.211 \times 10^{-4}} = 5.51 \times 10^6.$$

Since the Reynolds number is greater than 300,000, the boundary layer flow over the plate is turbulent, and from Eq. (9.39), for the turbulent case,

$$r_c = (N_{Pr}^*)^{1/3} = (0.694)^{1/3} = 0.8853.$$

Substituting known values of r_c and c_p^* into Eq. (9.41), the surface temperature of the plate is found to be

$$T_r = 106.8°C.$$

Since this value of T_r is virtually identical with that used as a first estimate—107.3°C—to calculate T^*, there is no need to recalculate T^*, and the surface temperature of part a is 106.8°C. Here, it is seen that the effects of viscous dissipation, aerodynamic heating, are appreciable since the adiabatic plate is at a temperature 69°C *higher* than it would be in a low-speed flow of 38°C air over it.

We will now answer part b. The heat transfer rate per unit area for high-speed flow is

$$q'' = h_x(T_w - T_r). \tag{9.37}$$

Since T_w and T_r are not dependent upon x here, and if the flow is assumed to be turbulent over the entire plate, Eq. (9.37) can be modified to give the total heat transfer rate from the plate of length L in the x direction and width W. Thus,

$$q = \bar{h}LW(T_w - T_r), \tag{9.42}$$

where $\bar{h}$ is the average surface coefficient over the plate surface.

The reference temperature T^* for the fluid properties is given by Eq. (9.40). The recovery temperature T_r was calculated in part a. (Note that although T^* is different here from that used in part a, this difference is not expected to have much effect on T_r by virtue of the reasoning presented in part a.) Hence, since $T_s = 38°C$, $T_w = 93°C$, and $T_r = 106.8°C$, Eq. (9.40) gives

$$T^* = 38 + 0.5(93 - 38) + 0.22(106.8 - 38) = 80.6°C.$$

From the property tables we can see that the properties of air at 80.6°C are sufficiently close to those at 87.9°C, used in part a, to neglect the slight change. Hence, the values of the Reynolds and Prandtl numbers are those of part a, namely, $N_{Re_L}^* = 5.51 \times 10^6$, and $N_{Pr}^* = 0.694$. From the property tables, $k^* = 0.031$ W/m °K. From Chapter 5, for turbulent flow over an isothermal flat plate, $\bar{h}$ is given by

$$\bar{h}L/k = 0.036\, N_{Re_L}^{0.8} N_{Pr}^{1/3}. \tag{5.383}$$

Substituting known quantities into this equation,

$$\frac{\overline{h}(0.3)}{0.031} = 0.036(5.51 \times 10^6)^{0.8}(0.694)^{1/3},$$

or

$$\overline{h} = 834 \ \text{W/m}^2 \ {}^{\circ}\text{K}.$$

With this, Eq. (9.42) yields

$$q = \overline{h} \ LW \ (T_w - T_r) = 834(0.3)(1)(93 - 106.8) = -3453 \ \text{W}.$$

The minus sign indicates that the convective heat flow is *from* the fluid *to* the plate, even though the plate surface temperature of 93°C is greater than the static temperature of the stream, $T_s = 38$°C. The energy dissipated within the boundary layer by the irreversible action of the viscous shear stresses causes a heat flow to the wall even though the wall is at a higher temperature than the static temperature of the free stream. This is not too surprising if the driving temperature difference in Newton's cooling law for high-speed flow situations is noted. The effective free stream temperature in that driving temperature difference is not T_s, but is rather the aerodynamic recovery temperature T_r. The latter reduces to T_s for low-speed flows, where the Eckert number is much smaller than unity. That is, $T_r \rightarrow T_s$ when $N_{\text{Ec}} = U_s^2/c_p \, (T_w - T_s) << 1$. (The temperature difference used in the denominator of the Eckert number is always a positive quantity; that is, its absolute value $|T_w - T_s|$ must be used.)

If the flow in this situation is treated as low speed, then

$$q_e = \overline{h}LW(T_w - T_s) = 834(0.3)(1)(93 - 38) = 13{,}761 \ \text{W}.$$

In this case, the use of the low-speed relations predicts a heat transfer rate q_e which is not only in error, but which also predicts the wrong direction of the heat flow. As can be seen, at least in this example, the error incurred in not properly accounting for the high-speed flow effects is enormous.

Part a of this example illustrates one of the problems encountered in the design of a supersonic or hypersonic aircraft or missile; that is, the high temperatures that the outside surfaces of such vehicles attain if they are adiabatic with negligible radiant losses. The plate described in part a could be an idealization of an actual aircraft wing.

EXAMPLE 9.2

Mechanical energy is stored in an experimental car by means of a rotating flywheel. The flywheel is shaped like a relatively thin disk of 2 ft diameter and rotates at 18,000 revolutions per minute. The composite material of which the flywheel is constructed has an effective thermal conductivity so small that the flywheel is essentially insulated. Estimate the maximum temperature on the flywheel surface when spinning in 70°F air. Assume that the boundary layer flow is turbulent.

Solution

The speed at any radius in the flywheel is $u = r\omega$. The maximum speed occurs at the outer edge where $r = r_0 = 1$ ft; hence,

$$u = r_0\omega = (1)\omega,$$

$$\omega = 18{,}000 \text{ rev/min} \times 2\pi \text{ rad/rev} \times \frac{1}{60} \text{ (min/sec)} = 1881 \text{ rad/sec.}$$

Since the flywheel is effectively insulated and has a velocity relative to the surrounding air, its temperature at any point will be the local aerodynamic recovery temperature T_r. Hence, to determine r_c, approximate the material at the outer edge of the flywheel as a flat plate over which 70°F air moves at the tip speed of 1881 ft/sec in a turbulent fashion. Therefore, from Eq. (9.39) combined with Eq. (9.38), the recovery factor is

$$\frac{T_r - T_s}{U_s^2/2c_p^*} = (N_{\text{Pr}}^*)^{1/3}. \tag{9.43}$$

Since the Prandtl number and the specific heat are both weak functions of the temperature for air, let $N_{\text{Pr}}^* = 0.690$ and $c_p^* = 0.242$, which corresponds to a T^* of 250°F. Thus, from Eq. (9.43),

$$\frac{T_r - 70}{\dfrac{(1881)(1881)}{2(32.2)(778)(0.242)}} = (0.690)^{1/3}, \quad \text{or} \quad T_r = 328°F.$$

Since $T_w = T_r$, from Eq. (9.40) a revised estimate of T^* is

$$T^* = 70 + 0.5(328 - 70) + 0.22(328 - 70) = 256°F.$$

At $T^* = 256°F$, the values of c_p and N_{Pr} are sufficiently close to their values at 250°F so that no further calculations are required. Thus, the estimated maximum temperature in the flywheel is $T_r = 328°F$.

9.3 LIQUID METAL HEAT TRANSFER

Liquid metals are an important class of heat transfer fluids, since their physical and chemical characteristics make them particularly suitable for use as coolants in nuclear reactors and as working substances in power cycles. Their favorable characteristics include high thermal conductivity (relative to water or air, for example), existence of the liquid state at relatively high temperatures and over a wide range of temperatures, moderate values of viscosity, and moderate values of the vapor pressure, even at high temperatures. Some of their less desirable characteristics include their corrosive action, the tendency of some to violently react with water, and the fact that some of the liquid metals, and/or their vapors, are toxic. Properties of selected liquid metals are presented in Appendix A. Perhaps the most striking of the liquid metal properties is the low value of the Prandtl number.

This ranges from about 0.003 to approximately 0.07, whereas for heat transfer fluids such as air, water, and oils, the Prandtl number has values from about 0.69 (for air) to about 10,000 (for oils). Recall that the Prandtl number N_{Pr} is the ratio of the molecular diffusivity of momentum to the molecular diffusivity of heat, and as such, controls to a great extent, in many circumstances, the ratio of the thickness of the thermal boundary layer to the thickness of the hydrodynamic boundary layer, δ_t/δ. It should also be recalled that certain expressions for the surface coefficient of heat transfer between a surface and the adjacent fluid, such as Eqs. (5.302), (5.333), and (5.353), were derived using the assumption that $\delta_t/\delta \leq 1$, which, by equations such as Eq. (5.335), means that the Prandtl number of the fluid should be close to or greater than unity.

Hence, these *analytical* results for the surface coefficient of heat transfer which were developed in Chapter 5 do *not* apply to very low Prandtl number fluids, specifically the liquid metals. In addition, most of the experimental correlations and semi-analytical results given in Chapters 5 and 6 for forced and free convection, respectively, are not valid at low Prandtl numbers.

However, some of the experimental correlations in these chapters do apply down to a Prandtl number of zero and, therefore, should also be applicable to calculation of liquid metal heat transfer. Thus, for very low Prandtl numbers, Eqs. (5.303) and (5.304) can be used for laminar flow over an isothermal plate, while Eq. (5.354) is valid for laminar flow over a constant flux plate and Eq. (5.387) is appropriate for turbulent flow over a constant temperature plate. Equation (5.415) can be employed at a stagnation point for a low Prandtl number fluid. For free convection in the liquid metals, Eqs. (6.56) and (6.57a) are for laminar and turbulent heat transfer, respectively, to constant temperature vertical plates and cylinders. Equations (6.85) and (6.86) are suitable for isothermal horizontal cylinders.

In this chapter, the discussion of liquid metal heat transfer will focus upon the analytical simplification that can be effected for liquid metals because of their ultra-low Prandtl numbers. We will also discuss a few more experimental correlations, specifically for liquid metals, which complement the results of Chapters 5 and 6 for low Prandtl number fluids. Because the correlations to be presented here evolve primarily from liquid metal data, they may be considered more reliable than their counterparts, when there are any, in Chapter 5 or 6.

9.3a Forced Convection Analysis of Liquid Metal Flow Over a Flat Plate

Results discussed in Chapter 5 suggest that the ratio δ_t/δ is inversely proportional to the cube root of the Prandtl number for external boundary layer–type flow of a fluid over a solid surface; that is,

$$\delta_t/\delta \sim 1/N_{Pr}^{1/3} . \tag{9.44}$$

A very low Prandtl number for a liquid metal, when substituted into Eq. (9.44), shows that the thermal boundary layer thickness is much greater than the hydrodynamic boundary layer thickness. Hence, Fig. 9.4 shows the physical situation for external flow over a body, with $\delta_t \gg \delta$, and where T_s and U_s are the free stream values of temperature and

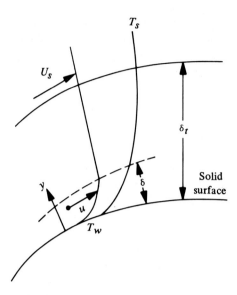

Figure 9.4 Diagram showing the large thickness of the thermal boundary layer relative to the velocity boundary layer, which is characteristic of liquid metal flow.

velocity. Note that the local x component u of velocity within the thermal boundary layer is essentially constant at the value U_s for all values of y in the range $\delta < y < \delta_t$. *When $N_{Pr} \rightarrow 0$, $u = U_s$ for all $y > 0$, or when the Prandtl number is very small, as it is for liquid metals, $u = U_s$ for all y is a reasonable approximation to use.* This, $u \approx U_s$ for all y, is the analytical simplification mentioned earlier for liquid metal heat transfer which will now be employed to predict the local and average surface coefficients of heat transfer.

Let us consider steady, laminar, constant property, two-dimensional planar, thin boundary layer–type flow of a liquid metal over a flat plate with constant wall temperature T_w. The undisturbed free stream temperature and velocity are T_s and U_s, respectively. In order to predict the local and average surface coefficients of heat transfer, the approximate integral method, introduced and developed in detail in Chapter 5, will be used. Conduction in the general flow direction, the x direction, will be neglected. This is the case in which $N_{Re_L} N_{Pr} > 50$ as indicated by Grosh and Cess [7] and by Stein [8]. Thus, from Chapter 5, we have the governing equation, the integral form of the thermal energy equation for low-speed flow, given as the following:

$$\frac{d}{dx} \int_0^{\delta_t} u(T_s - T)\, dy = \alpha \left. \frac{\partial T}{\partial y} \right|_{y=0}.$$ **[5.312]**

For the flow of a liquid metal over the plate, $u \approx U_s$ for all y. Hence, Eq. (5.312) becomes

$$\frac{d}{dx} \int_0^{\delta_t} (T_s - T)\, dy = \frac{\alpha}{U_s} \left. \frac{\partial T}{\partial y} \right|_{y=0}.$$ **(9.45)**

The boundary conditions on the actual temperature profile are the same as those in Chapter 5 for the corresponding analysis of a moderate to high Prandtl number fluid, $N_{Pr} > 0.6$. Hence, the approximating sequence for the temperature distribution function is the same as the one used in that chapter, namely,

$$T = T_w + (T_s - T_w) \left(\frac{3}{2} \frac{y}{\delta_t} - \frac{1}{2} \frac{y^3}{\delta_t^3} \right). \qquad [5.325]$$

Partially differentiating T with respect to y,

$$\frac{\partial T}{\partial y} \bigg|_{y=0} = \frac{3(T_s - T_w)}{2\delta_t}, \qquad (9.46)$$

and, rearranging Eq. (5.325),

$$T_s - T = (T_s - T_w) \left(1 - \frac{3}{2} \frac{y}{\delta_t} + \frac{1}{2} \frac{y^3}{\delta_t^3} \right). \qquad (9.47)$$

Substituting Eqs. (9.46) and (9.47) into Eq. (9.45),

$$\frac{d}{dx} \int_0^{\delta_t} \left(1 - \frac{3}{2} \frac{y}{\delta_t} + \frac{1}{2} \frac{y^3}{\delta_t^3} \right) dy = \frac{3\alpha}{2U_s\delta_t}.$$

Integrating and inserting the limits yields

$$\delta_t \frac{d\delta_t}{dx} = \frac{4\alpha}{U_s}.$$

Separating the variables, integrating, and noting that $\delta_t = 0$ at $x = 0$,

$$\delta_t = \sqrt{8\alpha x / U_s}. \qquad (9.48)$$

Now, from Eq. (5.3), the local surface coefficient of heat transfer is

$$h_x = \frac{-k \dfrac{\partial T}{\partial y} \bigg|_{y=0}}{T_w - T_s}. \qquad [5.3]$$

Substituting Eq. (9.48) into Eq. (9.46) and then substituting the result into Eq. (5.3) yields

$$h_x = \frac{1.5\,k}{\sqrt{8\,\alpha x / U_s}}. \qquad (9.49)$$

Using $N_{\text{Pr}} = \nu/\alpha$ and $N_{\text{Re}_x} = U_s x/\nu$ in rearranging Eq. (9.49), the local Nusselt number for the flow of a liquid metal over a flat plate is

$$N_{\text{Nu}_x} = h_x x/k = 0.531 \, N_{\text{Re}_x}^{1/2} \, N_{\text{Pr}}^{1/2} \,. \tag{9.50}$$

The product of the Reynolds and Prandtl numbers is called the *Peclet number* N_{Pe_x}:

$$N_{\text{Pe}_x} = N_{\text{Re}_x} \, N_{\text{Pr}}. \tag{9.51}$$

Hence, in terms of the Peclet number, Eq. (9.50) is

$$N_{\text{Nu}_x} = 0.531 \, N_{\text{Pe}_x}^{1/2} \,. \tag{9.52}$$

With this, the procedure shown in Chapter 5 for arriving at the average surface coefficient over the entire plate, Eq. (5.302), can be employed on Eq. (9.52). Thus, the average Nusselt number over a plate of length L for a low Prandtl number fluid is

$$\overline{N}_{\text{Nu}_L} = 1.062 \, N_{\text{Pe}_L}^{1/2} \,, \tag{9.53}$$

when the wall temperature is constant at the value T_w.

Next, using the same basic integral procedure which led to Eqs. (9.52) and (9.53) for the constant temperature surface plate, the following equations can be developed for the local surface coefficient and the wall temperature distribution in *laminar flow of a liquid metal over a plate with a constant, known flux* q_w''. The details are left to the reader as an exercise in one of the problems at the end of the chapter.

$$N_{\text{Nu}_x} = 0.75 \, N_{\text{Re}_x}^{1/2} \, N_{\text{Pr}}^{1/2} = 0.75 \, N_{\text{Pe}_x}^{1/2} \tag{9.54}$$

The wall temperature distribution is

$$T_w = T_s + \frac{4}{3} \frac{q_w''}{k} \left(\frac{x}{N_{\text{Pe}_x}^{1/2}} \right). \tag{9.54a}$$

Equations (9.54) and (9.54a) apply *when the surface heat flux is constant at the value* q_w''.

Equations (9.52–54a) were derived on the assumption that the flow of the low Prandtl number fluid is laminar. It was shown in Chapter 5 that if the flow is turbulent, the local heat flux in the y direction, at any value of y, can be written as

$$q''/\rho c_p = -(\alpha + \epsilon_H)\partial T/\partial y,$$

where the thermal diffusivity α represents the molecular transport of heat in the y direction (the conduction), and the eddy diffusivity of heat ϵ_H is a measure of the transport of heat due to the mixing motions of the turbulent eddies and is treated as an "apparent conduction" or "turbulent conduction" term. For the turbulent flow of fluids such as air and water, generally, in the turbulent core, $\epsilon_H \geq \alpha$. For liquid metals, however, with their relatively high thermal conductivity k and thermal diffusivity α, it is found that in turbulent flow at low and moderate Peclet numbers, $\epsilon_H \ll \alpha$. That is, the turbulent transport of heat is relatively small compared with the molecular transport and, hence, the heat transfer coefficient can be predicted using the laminar flow relations as long as the Peclet number is only moderately above the value which corresponds to the onset of turbulent flow.

For a complete discussion of the prediction of liquid metal heat transfer, from the theoretical, analytical, and mathematical standpoints, including turbulent flow, the reader may wish to consult Stein [8].

EXAMPLE 9.3

A heat transfer surface is idealized as a flat plate which is 0.15 m long in the flow direction and is 0.3 m wide. Energy at the rate of 2900 W must be transferred across this constant temperature surface. If liquid sodium at 204°C is flowing over the plate at a speed of 0.76 m/s, calculate the needed surface temperature T_w of the plate.

Solution

Since energy is transferred at the rate of 2900 W by forced convection from the plate surface to the liquid sodium, Newton's cooling law yields

$$2900 = \overline{h}_L\,(0.15)(0.3)(T_w - 204) \quad \text{or,}$$
$$\overline{h}_L\,(T_w - 204) = 64{,}444. \tag{9.55}$$

Next, the plate Reynolds number must be checked to determine if Eq. (9.53) can be used. The unknown wall temperature T_w is also implicit in $\overline{h}_L$, since it depends on temperature-dependent fluid properties such as μ and N_{Pr}. As a first estimate, all the fluid properties will be evaluated at the free stream temperature, T_s, of 204°C. Then a revised estimate will be made based on the value of T_w which is found. Thus the properties of sodium at 204°C are found from Appendix A to be $\rho = 900$ kg/m^3, $\mu = 0.43 \times 10^{-3}$ kg/ms, $k = 80.3$ W/m °C, and $N_{Pr} = 0.0072$. Thus, the Reynolds number is found to be, with $U_s = 0.76$ m/s,

$$N_{Re_L} = \frac{900(0.76)(0.15)}{0.43 \times 10^{-3}} = 238{,}000.$$

Hence, the flow is laminar, and Eq. (9.53) can be used. Since $N_{Pe_L} = N_{Re_L}\,N_{Pr}$ $= 238{,}000(0.0072) = 1714$, one has

$$\frac{\overline{h}_L(0.15)}{80.3} = 1.062(1714)^{1/2}, \quad \text{or} \quad \overline{h}_L = 23{,}537 \text{ W/m}^2 \text{ °C}.$$

Using this in Eq. (9.55) yields,

$$T_w = 206.7°C.$$

Since this is so close to the value of 204°C used for property determination, no re-calculation is necessary, and the required plate surface temperature is 207°C. (Note that the magnitude of the average surface coefficient, almost 24,000 W/m² °C, is much higher than those found in equivalent physical situations with air or water as the fluid. Indeed, this forced convection coefficient, with a liquid metal as the fluid, is of the same order of magnitude as surface coefficients in processes involving water undergoing a phase change, as described in Chapter 7. The relatively high values of h associated with single phase convection between liquid metals and solid surfaces provide one of the reasons for selecting a liquid metal as a heat transfer fluid.)

9.3b Experimental Correlations for Liquid Metals

Mentioned at the beginning of Sec. 9.3 were a number of experimental correlations for the surface coefficient of heat transfer, given in Chapters 5 and 6, which are also valid at low enough Prandtl numbers to be used for liquid metal heat transfer calculations. In this section, experimental correlations will be presented which were developed on the basis of liquid metal heat transfer data specifically. These are considered more reliable than any counterparts they may have in Chapters 5 and 6. Thus, the correlations to be presented next represent an extension of, and complement to, the ones listed in Sec. 9.3.

Forced Convection of Liquid Metals Within Circular Tubes

For fully developed, turbulent flow of liquid metals through a circular pipe *at constant wall temperature,* Seban and Shimazaki [9] show that the following expression fits their analytical results quite well:

$$N_{Nu_D} = 5 + 0.025 \, (N_{Pe_D})^{0.8}, \tag{9.56}$$

where the temperature-dependent fluid properties are evaluated at the average bulk mean temperature. Equation (9.56) should be used only if $L/D > 60$ and $N_{Re_D} > 2000$ or $N_{Pe_D} > 100$.

If the liquid metal flow is fully developed and turbulent through a circular tube with *a constant heat flux condition* at the boundary, Lyon [10] has presented an analytical result similar in form to Eq. (9.56); however, Lubarsky and Kaufman [11], from an extensive review of experimental correlations, represent the data for this case by

$$N_{Nu_D} = 0.625 \, N_{Pe_D}^{0.4}. \tag{9.57}$$

Like Eq. (9.56), Eq. (9.57) should be used when $L/D > 60$ and $N_{Pe_D} > 100$. The properties are evaluated at the average bulk mean temperature of the fluid.

In many practical situations, the high values of the surface coefficient of heat transfer between a pipe wall and the liquid metal cause relatively small temperature differences between the wall and the liquid metal. This is the reason for not including the wall temperature explicitly when evaluating the temperature-dependent fluid properties for use in Eqs. (9.56–57). However, for those situations in which there is an appreciable difference between the wall temperature and the local bulk mean temperature, Viskanta and Touloukian [12] suggest that the Nusselt number be evaluated from the appropriate equation, using the wall temperature to evaluate all temperature-dependent fluid properties. Then this Nusselt number is multiplied by the ratio of the thermal diffusivity at the wall temperature to the thermal diffusivity at the average bulk mean temperature to yield the actual Nusselt number, in which k is evaluated at the average bulk mean temperature. Mathematically,

$$hD/k_m = (N_{Nu_D})_w(\alpha_w/\alpha_m), \tag{9.58}$$

where the subscript w indicates the wall temperature and the subscript m indicates the average bulk mean temperature.

Forced Convection of Liquid Metals Over a Sphere

Witte [13] has correlated experimental data for the flow of liquid sodium over a sphere. In his work, he shows that the existing experimental correlations for moderate to high Prandtl number fluids, such as those given in Chapter 5 for the sphere, are not adequate for the liquid metals. Hence, Witte's result, which is shown next, is recommended and is expected to be valid for other liquid metals as well as sodium.

$$N_{Nu_D} = 2 + 0.386\, N_{Pe_D}^{0.5} \tag{9.59}$$

for $3.6 \times 10^4 < N_{Re_D} < 1.5 \times 10^5$. The Peclet number N_{Pe_D} in Eqs. (9.59), (9.56), and (9.57) is defined as shown in Eq. (9.51) except for the use of the diameter D as the characteristic dimension; that is,

$$N_{Pe_D} = N_{Re_D}\, N_{Pr}. \tag{9.60}$$

Free Convection of Liquid Metals Over a Horizontal Cylinder

For laminar natural convection of liquid metals over the outside of horizontal, isothermal cylinders, Hyman, Bonilla, and Ehrlich [14] recommend the following experimental correlation:

$$N_{Nu} = 0.53 \left[\left(\frac{N_{Pr}}{0.952 + N_{Pr}} \right) N_{Gr} N_{Pr} \right]^{1/4} \tag{9.61}$$

The characteristic dimension to be used in the Grashof and Nusselt numbers is the cylinder diameter, while the temperature-dependent fluid properties are evaluated at the film temperature.

EXAMPLE 9.4

A 44% K and 56% Na liquid is to be cooled from 1300°F to 700°F in a 1-in.-diameter tube whose wall is to be maintained at 680°F. If the mass flow rate is 20,000 lbm/hr, estimate the required length L of tubing when (a) the properties are treated as constant at the values appropriate to the average bulk mean fluid temperature, and (b) the correction for temperature-dependent fluid properties given by Eq. (9.58) is employed.

Solution

An energy balance on the entire tube of unknown length L yields

$$\omega c_p(T_{B_2} - T_{B_1}) = h\pi DL \, \Delta T_L,$$

where ΔT_L is the logarithmic mean temperature difference given by

$$\Delta T_L = \frac{T_{B_2} - T_{B_1}}{\ln[(T_w - T_{B_1})/(T_w - T_{B_2})]}$$

$$= \frac{700 - 1300}{\ln[(680 - 1300)/(680 - 700)]} = -175.$$

(Cf. Chapter 5.) The average bulk mean temperature is $T_{B_a} = \frac{1}{2}(1300 + 700) = 1000°F$. The property tables in Appendix A, for 44% K and 56% Na at 1000°F, give $\rho = 48.8$, $\mu = 0.43$, $k = 16.4$, $c_p = 0.248$, and $N_{Pr} = 0.0065$. Hence, the energy balance becomes

$$20{,}000(0.248)(700 - 1300) = h\pi(1/12) \, L(-175),$$
$$\text{or} \quad hL = 65{,}100. \tag{9.62}$$

Since $\omega = \rho \frac{1}{4} \pi D^2 U_m$,

$$20{,}000 = 48.8(0.785)(1/12)^2 U_m, \quad \text{or} \quad U_m = 75{,}500 \text{ ft/hr}.$$

Hence, the Reynolds and Peclet numbers are

$$N_{Re_D} = \frac{48.8(75{,}500)(1/12)}{0.43} = 714{,}000,$$
$$N_{Pe_D} = 714{,}000(0.0065) = 4640.$$

Thus, Eq. (9.56) yields

$$\frac{h(1/12)}{16.4} = 5 + 0.025(4640)^{0.8}, \quad \text{or} \quad h = 5220 \text{ Btu/hr ft}^2 \, °\text{F}.$$

Substituting this into Eq. (9.62), the required length of tubing is

$$L = \frac{65,100}{5220} = 12.5 \text{ ft.} \tag{a}$$

(A check of the value of L/D indicates it is greater than 60 and, therefore, the use of Eq. [9.56] was justified.)

There is little change in c_p between $1000°F$ and $680°F$; hence, Eq. (9.62) remains as in part a:

$$hL = 65,100. \tag{9.62}$$

The values of the properties at $T_w = 680°F$ are $\rho = 51.3$, $\mu = 0.56$, $N_{Pr} = 0.009$, and $\alpha_w = 1.23 \text{ ft}^2/\text{hr}$. From the property tables at $T_{B_a} = 1000°F$, $\alpha_m = 1.35$.

Since $20,000 = 51.3(0.785)(1/12)^2 U_m$, $U_m = 71,800 \text{ ft/hr}$. Hence, the Reynolds and Peclet numbers are

$$(N_{Re_D})_w = \frac{51.3(71,800)(1/12)}{0.56} = 548,000,$$

$$(N_{Pe_D})_w = 548,000(0.009) = 4940.$$

Thus, from Eq. (9.56),

$$(N_{Nu_D})_w = 5 + 0.025(4940)^{0.8} = 27.3.$$

Substituting these values into Eq. (9.58),

$$\frac{h(1/12)}{16.4} = (27.3)\frac{1.23}{1.35}, \quad \text{or} \quad h = 4910 \text{ Btu/hr ft}^2 \, °F.$$

Hence, from Eq. (9.62), the required tube length is

$$L = \frac{65,100}{4910} = 13.27 \text{ ft.} \tag{b}$$

Hence, the percentage error in the calculation in part a is

$$\left(\frac{13.27 - 12.50}{13.27}\right) \times 100\% = 5.8\%.$$

Therefore, even in this rather extreme example involving very large temperature differences between the surface and the liquid metal, the use of the average bulk mean temperature for calculating properties gives less than a 6% error in the tube length and less than a 7% error in the average surface coefficient of heat transfer h.

9.4 Differential Similarity

In certain convection heat transfer situations, analytical solutions are often not possible as a result of a lack of appropriate mathematical techniques. In other cases, analytical solutions are difficult or are simply not the optimum approach. Thus, experiments are conducted and the data correlated to predict the outcome in these situations. Experimental correlations have been used extensively to determine the surface coefficients of heat transfer in physical situations that are not amenable to analysis, such as turbulent flow through a tube bundle. (See Chapters 5, 6, and 9.) However, when an experiment is conducted, it must be designed so that maximum generality of the results is achieved with a minimum expenditure of time, effort, and money in acquiring the data. This is frequently done by combining the variables involved into a lesser number of nondimensional groups, and then systematically varying the nondimensional groups which will serve as the new independent variables, while measuring the values of the nondimensional groups which are serving as the new dependent variables. These nondimensional groups can be found in many ways: for example, by the use of Buckingham's pi theorem, which is usually stressed in fluid mechanics. However, the technique for determining these nondimensional groups discussed herein is *differential similarity*. This method requires that both the governing partial differential equations and the boundary conditions be written down even when they cannot be solved. Then, all of the dependent and independent variables, in *both* the equations and the boundary conditions, are made nondimensional, usually by dividing them by known values that have some meaning in the problem. Then, *both* equations and boundary conditions are transformed into nondimensional form. Finally, the nondimensional dependent variables can be written as an unknown function of all the nondimensional independent variables and of all nondimensional parameters that appear *both* in the governing nondimensional equations and in the nondimensional boundary conditions on those equations.

To illustrate the overall technique of differential similarity, we would like to find the nondimensional groups upon which the velocity field and the local and average skin friction coefficient depend in steady, laminar, low-speed, constant property, two-dimensional planar boundary layer flow over a flat plate of length L at zero angle of attack to the flow at undisturbed velocity U_s.

The first step is to write down the governing partial differential equations for the velocity field. In this constant property flow, they are the continuity equation and x-momentum equation, given in Sec. 5.3e as

$$\frac{\partial u}{\partial x} + \frac{\partial v}{\partial y} = 0, \quad \rho\left(u\,\frac{\partial u}{\partial x} + v\,\frac{\partial u}{\partial y}\right) = \mu\,\frac{\partial^2 u}{\partial y^2},$$

with the boundary conditions

$$\begin{aligned}
&\text{at } 0 < x < L \text{ and } y = 0, && u = 0 \text{ and } v = 0,\\
&\text{as } y \to \infty, && u \to U_s,\\
&\text{at } x = 0 \text{ and } 0 < y < \infty, && u = U_s.
\end{aligned}$$

Chapter Nine

Next, all the dependent and independent variables in the equations *and boundary conditions* are changed to nondimensional form by dividing them by an appropriate known value that has some significance; thus, all space coordinates are divided by the plate length L, and all velocities by the free stream value U_s. The nondimensional dependent and independent variables are denoted, whenever possible, by the upper case of the letter which symbolizes their dimensional values. Thus, the nondimensional dependent and independent variables are

$$X = \frac{x}{L}, \quad Y = \frac{y}{L}, \quad U = \frac{u}{U_s}, \quad V = \frac{v}{U_s}.$$

Since the dimensional variables x, y, u, and v are constant multiples of the corresponding nondimensional variables (i.e., $x = LX$, $u = U_sU$, etc.), the partial differential equations become

$$\frac{U_s}{L}\frac{\partial U}{\partial X} + \frac{U_s}{L}\frac{\partial V}{\partial Y} = 0, \quad \rho\left(\frac{U_s^2}{L}U\frac{\partial U}{\partial X} + \frac{U_s^2}{L}V\frac{\partial U}{\partial Y}\right) = \frac{\mu U_s}{L^2}\frac{\partial^2 U}{\partial Y^2}.$$

These two equations are now put in nondimensional form by dividing by U_s/L in the first equation and by $\rho U_s^2/L$ in the second equation; thus,

$$\frac{\partial U}{\partial X} + \frac{\partial V}{\partial Y} = 0, \quad U\frac{\partial U}{\partial X} + V\frac{\partial U}{\partial Y} = \left(\frac{\mu}{\rho U_s L}\right)\frac{\partial^2 U}{\partial Y^2}.$$

In this last equation, the nondimensional parameter which appears as the coefficient of $\partial^2 U/\partial Y^2$ is the reciprocal of the plate Reynolds number N_{Re_L}; hence, the nondimensional governing equations become

$$\frac{\partial U}{\partial X} + \frac{\partial V}{\partial Y} = 0, \tag{9.63}$$

$$U\frac{\partial U}{\partial X} + V\frac{\partial U}{\partial Y} = \left(\frac{1}{N_{\mathrm{Re}_L}}\right)\frac{\partial^2 U}{\partial Y^2}. \tag{9.64}$$

The boundary conditions in nondimensional form are

$$\begin{aligned}
&\text{at } Y = 0 \text{ and } 0 < X < 1, &&U = 0 \text{ and } V = 0,\\
&\text{as } Y \to \infty, &&U \to 1,\\
&\text{at } X = 0 \text{ and } 0 < Y < \infty, &&U = 1.
\end{aligned}$$

Note that the nondimensional dependent variables U and V depend upon the nondimensional independent variables X and Y, the value of the nondimensional parameter in Eq. (9.64), the Reynolds number, and the values of any nondimensional parameters that may appear in the boundary conditions. Here, no such parameters are present in the boundary conditions, since the nondimensional boundary conditions are pure numbers which do not change in problems of this type.

Hence,

$$U = f_1(X, Y, N_{\text{Re}_L}), \quad V = f_2(X, Y, N_{\text{Re}_L}), \tag{9.65}$$

where f_1 and f_2 are functions determined by experiment. (If an analytical solution is available, the results can be arranged in the general nondimensional form shown in Eq. [9.65].) *Thus, in an experiment to measure the velocity field, Eq. (9.65) suggests that values of U and V be measured at various nondimensional stations X and Y for different values of the Reynolds number.*

The fundamental reason, of course, for attempting to find the nondimensional groups such as X, Y, U, V, etc., is that there are fewer of these than of the original dimensional quantities. For example, in the situation at hand, the governing equations and boundary conditions in the original dimensional variables u and v indicate that the original dimensional dependent variables depend on the dimensional independent variables x and y, as well as on the dimensional parameters U_s, L, ρ, and μ; that is,

$$u = f_3(x, y, \rho, U_s, L, \mu), \quad v = f_4(x, y, \rho, U_s, L, \mu).$$

Thus, it might be thought that ρ, U_s, L, and μ should all be independently varied, rather than just the one group which contains all of them, the Reynolds number. Hence, the application of differential similarity effects a substantial savings in both acquiring and presenting the experimental data.

Next, let us use Eq. (9.65) to find the nondimensional groups upon which the local and average skin friction coefficient depend. In most practical engineering problems, these are needed more often than the details of the full velocity field.

From Sec. 5.3, the nondimensional local and average skin friction coefficients are

$$C_{f_x} = \frac{\tau_w}{\frac{1}{2}\rho U_s^2}, \tag{9.66}$$

$$C_{f_L} = \frac{\overline{\tau_w}}{\frac{1}{2}\rho U_s^2} = \frac{1}{L}\int_0^L C_{f_x}\, dx. \tag{9.67}$$

From Newton's law of viscosity,

$$\tau_w = \mu \left.\frac{\partial u}{\partial y}\right|_{y=0}$$

Since $u = UU_s$ and $y = YL$,

$$\tau_w = \frac{\mu U_s}{L}\left.\frac{\partial U}{\partial Y}\right|_{Y=0} \tag{9.68}$$

Since $N_{Re_L} = \rho U_s L/\mu$, substituting Eq. (9.68) into Eq. (9.66) yields

$$C_{f_x} = \frac{2}{N_{Re_L}} \frac{\partial U}{\partial Y}\bigg|_{Y=0} \qquad (9.69)$$

It was previously shown that

$$U = f_1(X, Y, N_{Re_L}).$$

Hence,

$$\frac{\partial U}{\partial Y} = g(X, Y, N_{Re_L}), \qquad \frac{\partial U}{\partial Y}\bigg|_{Y=0} = g(X, N_{Re_L}), \qquad (9.70)$$

where $g = \partial f_1/\partial Y$. (Note that the derivative is not a function of Y since it is evaluated at a specific value, $Y = 0$, of Y.)

Substituting Eq. (9.70) into Eq. (9.69) yields

$$C_{f_x} = \frac{2}{N_{Re_L}} g(X, N_{Re_L}). \qquad (9.71)$$

Equation (9.71) shows that the local skin friction coefficient depends upon the plate Reynolds number and the nondimensional position X away from the leading edge of the plate. In the analytical solution to this problem, in Sec. 5.3, it was found that the local skin friction coefficient depends upon the local length Reynolds number only. *Hence, differential similarity does not necessarily establish the simplest possible form for a nondimensional variable of interest, but it does establish an implicit functional relationship that is simpler than the one that would be expected if the nondimensional groups had not been determined.*

From Eq. (9.67), the average skin friction coefficient, with $dx/L = dX$, is

$$C_{f_L} = \int_0^1 C_{f_x}\, dX. \qquad (9.72)$$

Substituting Eq. (9.71) into Eq. (9.72) yields

$$C_{f_L} = \int_0^1 \frac{2}{N_{Re_L}} g(X, N_{Re_L})\, dX. \qquad (9.73)$$

After integration, the results on the right of Eq. (9.73) will not contain X, since the integration between fixed limits has removed any ultimate dependence on X; hence, Eq. (9.73) can be written as

$$C_{f_L} = F(N_{Re_L}).$$

Thus, the average skin friction coefficient depends only upon the plate Reynolds number.

The discussion of the use of differential similarity to find the appropriate nondimensional groups will now be completed by considering the heat transfer aspect of the fluid flow problem just dealt with.

If the fluid has undisturbed free stream temperature T_s while flowing in a steady, laminar, constant property, low-speed, boundary layer fashion over an *isothermal* flat plate, we would like to find the nondimensional groups upon which the temperature distribution depends, and also the nondimensional groups which determine the average surface coefficient of heat transfer.

From Eq. (5.287), the governing partial differential equation for the temperature field is

$$\rho c_p \left(u \frac{\partial T}{\partial x} + v \frac{\partial T}{\partial y} \right) = k \frac{\partial^2 T}{\partial y^2}, \tag{9.74}$$

with the boundary conditions

$$
\begin{aligned}
&\text{at } x = 0 \text{ and } 0 < y < \infty, &\quad T &= T_s, \\
&\text{at } y = 0 \text{ and } 0 < x < L, &\quad T &= T_w. \\
&\text{as } y \rightarrow \infty, &\quad T &\rightarrow T_s.
\end{aligned}
\tag{9.75}
$$

In general, the governing partial differential equations and boundary conditions for the velocity field $u = u(x, y)$ and $v = v(x, y)$ should also be considered. However, for constant property problems the velocity field can be found independent of the temperature field, and the nondimensional velocity field has previously been found as Eq. (9.65).

All dependent and independent variables are made nondimensional in Eqs. (9.74–75) by dividing all variables with the dimension of length by the plate length L, all velocities by the free stream value U_s, and the temperature T minus the wall temperature T_w by the difference between the free stream temperature T_s and the wall temperature T_w. Hence, the nondimensional variables become

$$X = \frac{x}{L}, \quad Y = \frac{y}{L}, \quad U = \frac{u}{U_s}, \quad V = \frac{v}{U_s}, \quad \phi = \frac{T - T_w}{T_s - T_w}. \tag{9.76}$$

Using these values in Eqs. (9.74–75) yields

$$\frac{\rho c_p U_s}{L} \left(U \frac{\partial \phi}{\partial X} + V \frac{\partial \phi}{\partial Y} \right) = \frac{k}{L^2} \frac{\partial^2 \phi}{\partial Y^2}, \tag{9.77}$$

$$
\begin{aligned}
&\text{at } X = 0 \text{ and } 0 < Y < \infty, &\quad \phi &= 1, \\
&\text{at } Y = 0 \text{ and } 0 < X < 1, &\quad \phi &= 0, \\
&\text{as } Y \rightarrow \infty, &\quad \phi &\rightarrow 1.
\end{aligned}
$$

Equation (9.77) is made nondimensional by dividing both sides by $\rho c_p U_s / L$. Hence,

$$U \frac{\partial \phi}{\partial X} + V \frac{\partial \phi}{\partial Y} = \frac{k}{\rho U_s L c_p} \frac{\partial^2 \phi}{\partial Y^2}. \tag{9.78}$$

The nondimensional group which is the coefficient on the right of Eq. (9.78) is rearranged as

$$\frac{k}{\rho U_s L c_p} = \left(\frac{\mu}{\rho U_s L}\right)\left(\frac{k}{\mu c_p}\right) = \frac{1}{N_{Re_L} N_{Pr}}.$$

Hence, Eq. (9.78) becomes

$$U \frac{\partial \phi}{\partial X} + V \frac{\partial \phi}{\partial Y} = \frac{1}{N_{Re_L} N_{Pr}} \frac{\partial^2 \phi}{\partial Y^2}. \tag{9.79}$$

The nondimensional dependent variable ϕ depends on the nondimensional independent variables X and Y, the nondimensional parameters N_{Re_L} and N_{Pr}, the nondimensional groups that U and V depend on (from Eq. [9.65], they are X, Y, and N_{Re_L}), and, finally, on any nondimensional parameters in the boundary conditions. However, we can see that none appear in the boundary conditions here. Hence, in implicit functional form, the nondimensional temperature is

$$\phi = F(X, Y, N_{Re_L}, N_{Pr}). \tag{9.80}$$

From Chapter 5, the average surface coefficient of heat transfer over the length of the plate, for an isothermal plate, is

$$\overline{h} = \frac{1}{L} \int_0^L h_x dx.$$

Since $dx/L = dX$,

$$\overline{h} = \int_0^1 h_x dX. \tag{9.81}$$

Also, from Chapter 5,

$$h_x = \frac{-k(\partial T/\partial y)_{y=0}}{T_w - T_s}.$$

Changing T and y to the nondimensional variables,

$$h_x = \frac{k}{L} \frac{\partial \phi}{\partial Y}\bigg|_{Y=0}$$

Substituting this into Eq. (9.81) and rearranging so that both sides are nondimensional yields

$$\frac{\overline{h}L}{k} = \int_0^1 \left.\frac{\partial \phi}{\partial Y}\right|_{Y=0} dX, \tag{9.82}$$

and, from Eq. (9.80),

$$\left.\frac{\partial \phi}{\partial Y}\right|_{Y=0} = G(X, N_{\text{Re}_L}, N_{\text{Pr}}). \tag{9.83}$$

(Note that setting $Y = 0$ after differentiation removes the dependency of the partial derivative on Y because it is now being evaluated at $Y = 0$, a specific Y.) Substituting Eq. (9.83) into Eq. (9.82), the integration over X removes the dependency of the result on X. Hence, Eq. (9.82) becomes

$$\overline{h}L/k = f(N_{\text{Re}_L}, N_{\text{Pr}}). \tag{9.84}$$

Hence, the method of differential similarity predicts that the average Nusselt number depends on the plate Reynolds number and the Prandtl number. The actual functional dependence f is determined by experiment or by a solution to the equations. Equation (9.84) predicts the nondimensional groups which are relevant, and, by referring to the analytical solution of this problem in Chapter 5, we can see that these are the same nondimensional groups as were formed in the analytical solution.

So far, even though we have stressed the importance of putting the *boundary conditions* into nondimensional form and then examining these nondimensional boundary conditions for nondimensional groups, no nondimensional groups that were not pure numbers have appeared as a result of the boundary conditions. Suppose, instead of the isothermal plate just discussed, we have instead a plate with an unheated starting length x_0; that is, the wall temperature is equal to the stream temperature T_s from the leading edge to the coordinate x_0, and is held at the constant value T_w from x_0 to L. Let us see what change this demands in the expression for the average surface coefficient of heat transfer over the heated section.

The velocity field in a constant property flow is unaffected by a change in the thermal boundary conditions; hence, U and V still depend only on X, Y, and N_{Re_L}. The partial differential form of the thermal energy equation also remains unchanged, since the condition changed in this problem involves a boundary condition; hence, Eq. (9.79) applies here. The new boundary conditions are

$$
\begin{array}{lll}
\text{at } x = 0 \text{ and } 0 < y < \infty, & T = T_s, \\
\text{at } y = 0 \text{ and } 0 < x < x_0, & T = T_s, \\
\text{at } y = 0 \text{ and } x_0 < x < L, & T = T_w, \\
\text{as } y \to \infty, & T \to T_s.
\end{array}
$$

Using the quantities defined in Eq. (9.76), these boundary conditions in nondimensional form are

$$
\begin{aligned}
\text{at } X = 0 \text{ and } 0 < Y < \infty, && \phi &= 1, \\
\text{at } Y = 0 \text{ and } 0 < X < x_0/L, && \phi &= 1, \\
\text{at } Y = 0 \text{ and } x_0/L < X < 1, && \phi &= 0, \\
\text{as } Y \to \infty, && \phi &\to 1.
\end{aligned}
$$

Now ϕ depends on the nondimensional independent variables X and Y of the governing partial differential equation, the groups X, Y, and N_{Re_L} which U and V depend on, the nondimensional parameters N_{Re_L} and N_{Pr} in the partial differential equation, *and any nondimensional parameters in the boundary conditions*. The boundary conditions yield the nondimensional parameter x_0/L. Therefore, the nondimensional temperature distribution is

$$
\phi = F(X, Y, N_{\mathrm{Re}_L}, N_{\mathrm{Pr}}, x_0/L). \tag{9.85}
$$

When we use the same procedure as was used for Eq. (9.84) but with Eq. (9.85), the implicit functional relationship for the average Nusselt number is

$$
\overline{h}L/k = f(N_{\mathrm{Re}_L}, N_{\mathrm{Pr}}, x_0/L). \tag{9.86}
$$

Thus, the nondimensional boundary conditions yield an additional group which $\overline{h}L/k$ depends on, namely, the group x_0/L.

At the beginning of this section, it was stated that the method of differential similarity, or other methods of obtaining the appropriate nondimensional groups, should be the first step in the intelligent design of experiments to measure values of the surface coefficient of heat transfer or, for that matter, to measure any quantity of interest. With this in mind, we can see that Eq. (9.86) provides us with the information about how the data should be taken when we want to measure the values of the average surface coefficient of heat transfer on a flat plate with an unheated starting length, and present a comprehensive experimental correlation of the results. The value of the Prandtl number and nondimensional unheated starting length x_0/L are fixed, and the Reynolds number is varied. Then, the average Nusselt number at each different Reynolds number is measured. While we keep the same unheated starting length, the Prandtl number is then changed to a new value and fixed, the Reynolds number varied, and the average Nusselt numbers measured at each Reynolds number. Again, the Prandtl number is changed, keeping the original unheated starting length, and the procedure is repeated. After a sufficiently wide range of Prandtl numbers has been encompassed, the entire procedure is repeated for each new value of dimensionless unheated starting length x_0/L. The data is plotted with the Nusselt number as the ordinate, the Reynolds number as the abscissa, and the Prandtl number as the parameter, at one value of x_0/L. A similar graph is plotted at each different value of x_0/L. Next, the curves on the graphs are examined to determine if an experimental equation can be devised to fit the data over a wide range of all the parameters concerned.

When this is possible, the equation often has the form

$$N_{\mathrm{Nu}_L} = C(N_{\mathrm{Re}_L})^a (N_{\mathrm{Pr}})^b (x_0/L)^d,$$

where C, a, b, and d are frequently constants determined by forcing the equation to fit the experimental data.

EXAMPLE 9.5

Consider steady, laminar free convection of a fluid at temperature T_s along a vertical flat plate of length L and surface temperature T_w. Using differential similarity, find the groups which the nondimensional velocity field, temperature field, and average Nusselt number depend on.

Solution

From Chapter 6, the governing partial differential equations for the velocity and temperature fields are

$$\frac{\partial u}{\partial x} + \frac{\partial v}{\partial y} = 0, \tag{6.1}$$

$$\rho \left(u \frac{\partial u}{\partial x} + v \frac{\partial u}{\partial y} \right) = \mu \frac{\partial^2 u}{\partial y^2} + g\beta\rho(T - T_s), \tag{6.16}$$

$$\rho c_p \left(u \frac{\partial T}{\partial x} + v \frac{\partial T}{\partial y} \right) = k \frac{\partial^2 T}{\partial y^2}. \tag{6.3}$$

The boundary conditions are

$$
\begin{array}{lll}
\text{at } x = 0 \text{ and } 0 < y < \infty, & u = 0 \text{ and } T = T_s, \\
\text{at } y = 0 \text{ and } 0 < x < L, & u = v = 0 \text{ and } T = T_w, \\
\text{as } y \rightarrow \infty, & u \rightarrow 0 \text{ and } T \rightarrow T_s.
\end{array}
$$

The dependent and independent variables are made nondimensional by dividing all lengths by L, and by dividing the temperature minus T_w by the undisturbed fluid temperature T_s minus T_w. All velocities should be divided by a reference velocity that has meaning in this problem; however, the free stream velocity is zero and, therefore, cannot be used. The velocity in a pure free convection situation is caused by the net upward force per unit volume $g\beta\rho(T - T_s)$, or, when used as a reference force, $g\beta\rho(T_w - T_s)$. A measure of the work done by this force on a unit volume of fluid, from $x = 0$ to $x = L$, is the product of the force and the length; that is,

$$\text{work} = g\beta\rho(T_w - T_s)L.$$

If no viscous restraining forces are present, the work changes the kinetic energy of the unit volume of material from 0 at $x = 0$ to $\frac{1}{2}\rho U_0^2$ at $x = L$; that is,

$$\frac{1}{2}\rho U_0^2 = g\beta\rho(T_w - T_s)L. \tag{9.87}$$

Even in the presence of the restraining viscous forces, Eq. (9.87) is a *measure* of the level of an induced reference free convection velocity U_0, in terms of the variables which cause this velocity. Hence for the reference velocity, let

$$U_0 = \sqrt{g\beta\Delta TL} . \tag{9.88}$$

Also, let $\Delta T = T_w - T_s$, or if the flow is downward—that is, $T_w < T_s$—let $\Delta T = T_s - T_w$, so that ΔT is always positive. Thus, the nondimensional independent and dependent variables are

$$X = \frac{x}{L}, \quad Y = \frac{y}{L}, \quad U = \frac{u}{U_0}, \quad V = \frac{v}{U_0}, \quad \phi = \frac{T - T_w}{T_s - T_w}.$$

Substituting these values into Eqs. (6.1), (6.16), and (6.3), and making the equations nondimensional by dividing the dimensional quantities into both sides of the respective equations, the nondimensional partial differential equations are

$$\frac{\partial U}{\partial X} + \frac{\partial V}{\partial Y} = 0, \tag{9.89}$$

$$U\frac{\partial U}{\partial X} + V\frac{\partial U}{\partial Y} = \frac{\mu}{\rho U_0 L}\frac{\partial^2 U}{\partial Y^2} + 1 - \phi, \tag{9.90}$$

$$U\frac{\partial \phi}{\partial X} + V\frac{\partial \phi}{\partial Y} = \frac{k}{\rho U_0 L c_p}\frac{\partial^2 \phi}{\partial Y^2}. \tag{9.91}$$

In nondimensional form, the boundary conditions are

at $X = 0$ and $0 < Y < \infty$, $\quad U = 0$ and $\phi = 1$,
at $Y = 0$ and $0 < X < 1$, $\quad U = V = 0$ and $\phi = 0$,
as $Y \to \infty$, $\quad U \to 0$ and $\phi \to 1$.

Since the temperature and velocity fields are mutually coupled and a simultaneous solution of Eqs. (9.89–91) is required to determine them, their nondimensional values depend on all the nondimensional independent variables and parameters that appear in Eqs. (9.89–91) (note that no nondimensional groups appear in the boundary conditions); that is,

$$U = f_1\left(X, Y, \frac{\mu}{\rho U_0 L}, \frac{k}{\rho U_0 L c_p}\right),$$

$$V = f_2\left(X, Y, \frac{\mu}{\rho U_0 L}, \frac{k}{\rho U_0 L c_p}\right), \tag{9.92}$$

$$\phi = f_3\left(X, Y, \frac{\mu}{\rho U_0 L}, \frac{k}{\rho U_0 L c_p}\right).$$

The last nondimensional group in Eq. (9.92) can be rewritten as

$$\frac{k}{\rho U_0 L c_p} = \left(\frac{\mu}{\rho U_0 L}\right) \frac{k}{\mu c_p} = \frac{\mu}{\rho U_0 L} \frac{1}{N_{Pr}}.$$

Since $\mu/\rho U_0 L$ already appears, the last group may be replaced by the Prandtl number. Thus, the third equation of Eqs. (9.92) becomes

$$\phi = f_4(X, Y, \mu/\rho U_0 L, N_{Pr}).$$

Using Eq. (9.88) yields

$$\frac{\mu}{\rho U_0 L} = \frac{\mu}{\rho L \sqrt{g\beta \Delta T L}} = \frac{1}{\sqrt{\dfrac{g\beta \rho^2 \Delta T L^3}{\mu^2}}}.$$

The nondimensional group under the radical is the Grashof number

$$N_{Gr} = \frac{g\beta \rho^2 \Delta T L^3}{\mu^2}.$$

Using this, the nondimensional velocity and temperature fields are

$$U = f_5(X, Y, N_{Gr}, N_{Pr}), \quad V = f_6(X, Y, N_{Gr}, N_{Pr}),$$
$$\phi = f_7(X, Y, N_{Gr}, N_{Pr});$$

and using the procedure which led to Eq. (9.86),

$$\overline{N}_{Nu} = \overline{h} L/k = f_8(N_{Gr}, N_{Pr}).$$

This result indicates that the same nondimensional groups used in Chapter 6 are characteristic of free convection solutions and correlations.

9.5 TRANSPIRATION AND FILM COOLING

Certain surfaces in rocket, jet, and gas turbine engines, such as rocket nozzles, blades, vanes, combustors, and afterburners, must operate under such severe thermal loadings (that is, with high-temperature gases flowing over them, often at relatively high speed) that more sophisticated cooling schemes than simple convection cooling may be required to keep the temperature of the surface low and the thermal gradients within the material small. Two of the more important advanced cooling schemes are *transpiration cooling* and *film cooling*. The differences between convection, transpiration, and film cooling of a surface are shown in Fig. 9.5(a–c), where the upper surface, in contact with the hot gas flowing over it, is the surface to be protected from high temperatures and high thermal

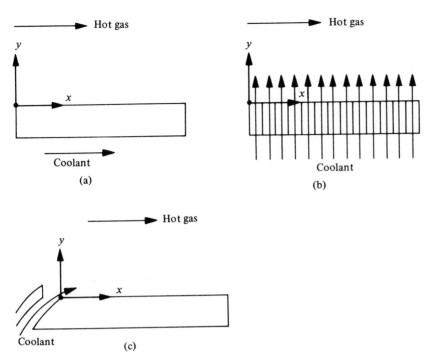

Figure 9.5 (a) Ordinary convection cooling, (b) transpiration cooling, and (c) film cooling of a plate.

gradients. Figure 9.5(a) shows convection cooling, in which a coolant is forced along the underside of the material, or even perpendicular to the underside of the material (impingement cooling) to keep the upper surface temperature low. If the surface coefficient of heat transfer between the upper surface and the hot gas is very high, as it would be for high gas speeds, the amount of coolant needed and the coolant speeds required may not permit an economical convection cooling design. Here transpiration and film cooling have an advantage, since the coolant eventually reaches the surface to be protected. In transpiration cooling, depicted in Fig. 9.5(b), the object to be protected is constructed of a porous material through which the coolant is forced, cooling the material as it passes through it, and then leaves at the surface to be protected and "blows" into the boundary layer. This blowing of the coolant into the boundary layer distorts the temperature field in that the local surface coefficient of heat transfer is reduced from the value it would normally have for a hot gas flowing over a solid plate under the same conditions. In film cooling, as shown in Fig. 9.5(c), a layer of coolant is discharged from a slot, usually parallel to the upper surface and in the direction of the hot gas flow. This layer of coolant tends to insulate the surface from the hot gas, but its effectiveness continually decays as x increases because of the mixing between the film flow and the hot gas. As a result, it affords considerable thermal protection in a local region just downstream of the injection point. This decay of the film's effectiveness can be overcome by adding additional downstream slots, spaced a finite distance apart.

Additional Topics in Heat Transfer

819

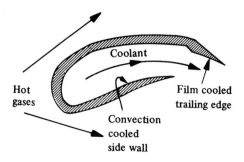

Figure 9.6 Film cooling of the thin trailing edge of a turbine blade.

In general, the coolant used in transpiration and film cooling schemes might be a gas different from the hot gas flowing over the upper surface, or it might even be a liquid. (If a liquid is used, there is the possibility of taking advantage of a change in phase of the coolant at the surface, from the liquid to the gas phase, and, hence, utilizing its latent heat of vaporization in the cooling process.) Only transpiration and film cooling involving injection of the same type of gas as flows over the surface is considered herein, as the greatest amount of analytical and experimental results are available for these situations. Furthermore, these schemes are most often used because of the requirements of weight and simplicity in rocket and jet engines.

Note that actual advanced cooling scheme designs often incorporate two or more of the schemes shown in Figs. 9.5(a–c), as, for example, in a turbine blade, where a side wall can be convectively cooled and then the coolant exhausted through a slot so that it film cools the thin trailing edge of the blade. This is shown in Fig. 9.6.

9.5a Transpiration Cooling

The local surface coefficient of heat transfer h_x on a flat plate with constant surface velocity transpiration cooling, when the flow is laminar, is related to the local surface coefficient h_{0x} on a flat plate with no transpiration by

$$\frac{h_x}{h_{0x}} = \frac{B}{e^B - 1},$$

(9.93)

where

$$B = \frac{\rho v_0 c_p}{h_{0x}},$$

where ρ and c_p are the density and specific heat, respectively, of the injected fluid, and v_0 is the constant injection velocity at the plate surface [22]. The local surface coefficient h_{0x} for steady, laminar, constant property flow over a solid flat plate can be estimated using the relations in Chapter 5. In reality, *Eq. (9.93) can be used for either laminar or turbulent flow,* as is shown in a Couette-type analysis by Kays and Crawford [23]. This leads to Eq. (9.93) for the case of turbulent flow.

A limit on the validity of Eq. (9.93) occurs when the injection velocity v_0 is large enough to "blow" the boundary layer off the surface and lead to a more complex flow not of the thin boundary layer type, in a phenomenon somewhat reminiscent of separation of a boundary layer, which was discussed in Chapter 5. Kays and Crawford provide the following rule of thumb for blowoff of the boundary layer:

Blowoff occurs when

$$\frac{v_0}{U_s} \geq 0.01. \tag{9.94}$$

The results above, for laminar flow, are for a constant blowing velocity v_0. Next an exact solution for a case in which the blowing velocity depends upon x, $v_w(x)$, will be presented.

The local surface coefficient of heat transfer in steady, laminar, constant property, low-speed, boundary layer–type flow over a flat plate, at constant temperature, which is being transpiration-cooled by the same constant property fluid, has been predicted in an exact analytical fashion using the method of the similarity transformation [19]. For a proper similarity variable to exist, the transpiration surface velocity v_w must be inversely proportional to the square root of the distance from the leading edge; that is,

$$v_w \sim 1/\sqrt{x} \ .$$

From Hartnett and Eckert [19], for blowing with $N_{Pr} = 0.7$, the results can be represented by a straight line with a maximum error of about 5%; that is,

$$\frac{N_{Nu_x}}{N_{Re_x}^{1/2}} = 0.29 - 0.48 \left(\frac{v_w}{U_s} N_{Re_x}^{1/2} \right), \tag{9.95}$$

for $0 \leq (v_w/U_s)N_{Re_x}^{1/2} \leq 0.50$. [Note that near $(v_w/U_s)N_{Re_x}^{1/2} = 0.61$, Hartnett and Eckert [19] predict $N_{Nu_x} = 0$, i.e., the boundary layer blown off the surface, and the assumption of a thin boundary layer is no longer valid.]

Surface Temperature of a Transpiration-Cooled Plate

In general, the upper surface temperature $T_w(x)$ of the transpiration-cooled plate, depicted in Fig. 9.5(b), will not be known in advance and depends upon the energy exchange between it, the coolant, and the external fluid stream.

Consider steady, constant property flow of a fluid at undisturbed temperature T_s over the upper surface of the porous plate of Fig. 9.5(b) when coolant fluid of the same type, but of undisturbed temperature T_c below the plate bottom, is forced upward through the

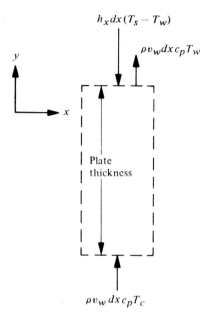

$$h_x dx (T_s - T_w)$$

$$\rho v_w dx c_p T_w$$

$$\rho v_w dx c_p T_c$$

Figure 9.7 The control volume and energy transfer terms for the transpiration cooling situation.

plate with local y component of velocity $v_w(x)$. The upper surface temperature distribution will be derived for the case of negligible net radiant loss and negligible x conduction within the plate by use of an energy balance on an appropriate portion of the plate. Thus, a control volume dx long, consisting of both the porous plate material and the coolant flowing through it, is chosen with the control surface shown as a dashed line in Fig. 9.7. Energy at the rate $\rho v_w dx\, c_p T_c$ crosses the lower control surface into the control volume by virtue of the motion of the coolant, which is at the temperature T_c. The y conduction across the lower control surface is considered negligible, because the lower control surface is taken well below the lower plate surface in the undisturbed coolant at temperature T_c. Because of the large surface area to volume ratio of the porous material, generally the coolant temperature and the upper surface temperature are virtually identical at $T_w(x)$. Hence, at the upper control surface, energy at the rate $\rho v_w dx\, c_p T_w$ leaves the control volume because of the mass blown into the boundary layer, and also enters the control volume at the rate $h_x dx(T_s - T_w)$ by convection from the free stream. Hence, using Eq. (1.8) in the steady state with no generation and radiation neglected, $R_{\text{in}} = R_{\text{out}}$. Therefore,

$$\rho v_w dx\, c_p T_c + h_x dx(T_s - T_w) = \rho v_w dx\, c_p T_w.$$

Solving for the surface temperature T_w of the transpiration-cooled plate yields

$$T_w = \frac{T_s + (\rho c_p v_w T_c / h_x)}{1 + (\rho c_p v_w / h_x)}. \tag{9.96}$$

The blowing velocity v_w required to keep the surface at a particular temperature when the hot gas temperature is T_s and the coolant inlet temperature is T_c is obtained from Eq. (9.96), which can be arranged as

$$\frac{T_w - T_c}{T_s - T_w} = \frac{h_x}{\rho c_p v_w}.$$ (9.97)

Equations (9.96–97) show that if v_w is a constant over the plate surface, the plate surface temperature must vary with x, since h_x is a function of x. However, if the surface blowing velocity varies inversely as the square root of x along the surface, Eq. (9.96) or Eq. (9.97), together with Eq. (9.95), show that the plate surface temperature is independent of x.

References [4], [19], [20], [22], and [23] give detailed discussions of transpiration cooling. Laminar transpiration cooling, with both constant and variable fluid properties, for both the flat plate and plane stagnation point, is discussed in Hartnett and Eckert [19] along with the laminar recovery factor for high-speed flow over a transpiration-cooled flat plate. Kays and Crawford [23] deal with transpiration cooling for both laminar and turbulent flow, including cases in which the coolant gas is different from the gas in the free stream. The surface temperature distribution of transpiration cooled materials is discussed in [20–21]. Mickley et al. [22] suggest the use of Eq. (9.93) for turbulent flow over transpiration-cooled plates, as well as for laminar flow.

Finally, some of the disadvantages and problems associated with transpiration cooling should be noted. Many of the highly porous materials lack strength and machinability. In addition, many of the porous materials exhibit a tendency to oxidize rather quickly, causing the pores to be blocked by oxide film, and causing the coolant flow through the material to be reduced, resulting in material failure. Because of these problems, there has been a shift toward investigating a cooling scheme classified between transpiration and film cooling which involves only a small number of pores. This scheme calls for a large number of small-diameter holes to be drilled into conventional turbine blade alloys. These holes could be on the order of five diameters apart. Esgar, Colladay, and Kaufman [21] call this scheme "full coverage film cooling." The relatively large holes used reduce the possibility of their being closed by oxide film and make this a promising cooling scheme when the problems involved in drilling very small holes are solved.

EXAMPLE 9.6

Air flows over a flat plate with a speed U_s of 30 m/s and a pressure of 10^5 N/m². The plate is porous and cooling air is forced through it with a constant speed of 0.015 m/s for all values of x. If the film temperature is 27°C, find the ratio of the surface coefficient with transpiration to that which would occur without transpiration at a point $x = 0.15$ m from the leading edge. In addition, find the value of h_x at this point.

Solution

Equation (9.93), in conjunction with Eq. (5.297) for h_{0x} if the flow is laminar or with Eq. (5.382) if it is turbulent, will be used because the blowing velocity is constant over the surface.

From the property tables, at the given temperature and pressure, $\rho = 1.177$ kg/m³, $\mu = 1.983 \times 10^{-5}$ kg/ms, $k = 0.0262$ W/m °C, $c_p = 1005.7$ J/kg °C, $N_{Pr} = 0.708$. Hence,

$$N_{Re_x} = \rho U_s x / \mu = 1.177(30)(0.15)/1.983 \times 10^{-5} = 2.67 \times 10^5.$$

Thus, the flow is laminar, and Eq. (5.297) can be used to calculate h_{0x}.

$$h_{0x}(0.15)/0.0262 = 0.332(2.67 \times 10^5)^{1/2}(0.708)^{1/3}, \quad \text{or}$$

$$h_{0x} = 26.7 \text{ W/m}^2 \text{ °C}.$$

From Eq. (9.93),

$$B = \rho v_0 c_p / h_{0x} = 1.177(0.015)(1005.7)/26.7$$

$$B = 0.665$$

$$\frac{h_x}{h_{0x}} = \frac{0.665}{e^{0.665} - 1} = 0.704$$

and

$$h_x = 0.704(26.7) = 18.8 \text{ W/m}^2 \text{ °C}.$$

Thus a 30% decrease in the surface coefficient from the flat plate value results when a transpiration velocity at the surface which is only 0.05% of the free stream velocity is used.

9.5b Film Cooling

Film cooling consists of interposing a film of coolant between the surface to be protected and the hot gas stream. This coolant can be injected along the surface through a slot, as shown in Figs. 9.5(c) and 9.6. As a result of mixing between the hot gas stream and the coolant, the protection this film affords the surface continually decreases in the downstream direction. Eventually, the film offers no protection far downstream from the slot, unless it is replenished from other slots downstream of the first slot. One of the most important characteristics of film cooling is its ability to provide thermal protection when the passing of a coolant through the surface, or within the material just below the surface, is not possible because of the size and shape of the object. An example is the trailing edge of the turbine blade, shown in Fig. 9.6, which may be so thin (for aerodynamic reasons) that coolant holes cannot be cast or drilled into it. Thus, it is protected from the hot gas by injecting the coolant on the surface from some upstream portion of the blade.

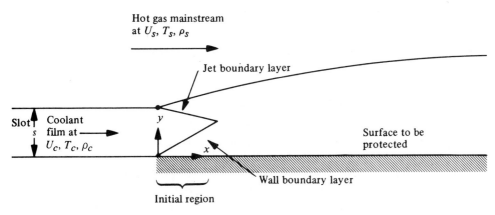

Hot gas mainstream
at U_s, T_s, ρ_s

Jet boundary layer

Slot
s

Coolant
film at
U_c, T_c, ρ_c

y

x

Surface to be
protected

Wall boundary layer

Initial region

Figure 9.8 Diagram of and nomenclature for the tangential film cooling of a surface from a slot.

In order to define and explain the parameters used in film cooling, and the technique used to calculate the heat transfer rate in the presence of a film, consider Fig. 9.8. The surface is a perfectly insulated flat plate, called the *surface to be protected,* which is to be kept at reasonably low temperatures by the coolant, which is injected tangential to the plate at $x = 0$ from a slot of height s. The coolant velocity, temperature, and density at the injection point are designated U_c, T_c, and ρ_c, respectively, while the corresponding quantities in the hot main gas stream are U_s, T_s, and ρ_s. (Probably the first work done in "film cooling" by Weighardt [24] was actually film heating, in connection with the de-icing of wing tips. Most of the relations developed for film cooling are equally applicable to film heating.)

Now, if there were no film, the perfectly insulated surface would attain the static temperature T_s of the hot free stream (neglecting the net radiant loss) for low-speed flow, or the aerodynamic recovery temperature T_r for high-speed flow. However, when the film is injected over the surface, say for a low-speed flow, the temperature of the adiabatic surface attains values less than T_s. The temperature which a perfectly insulated surface reaches when it has a flow of coolant (i.e., the film) forced over it in the absence of radiation is defined as the *film adiabatic wall temperature* T_{aw}. In general, T_{aw} is a function of x; that is, $T_{aw} = T_{aw}(x)$. In the initial region shown in Fig. 9.8, the mixing region between the mainstream and the coolant (i.e., the jet boundary layer) has not yet intersected the developing boundary layer along the wall surface, and therefore there is a core of undisturbed fluid which is still at the temperature T_c. Hence, for the initial region, $T_{aw} = T_c$, approximately. For values of x past those in the initial region, the local film adiabatic wall temperature T_{aw} rises continually and approaches T_s (or T_r if the flow is high speed) at very large values of x unless the film is replenished by injection from another downstream slot. Very often the surfaces to be protected by a film are adiabatic, or essentially adiabatic (such as the trailing edge of the turbine blade in Fig. 9.6, because of the relatively low thermal conductivity of most of the high-temperature turbine blade alloys). Hence it is desirable to predict the film adiabatic wall temperature as a function of x. Now, T_{aw} is treated in the nondimensional form, called the *film effectiveness,* or *adiabatic film effectiveness, η;* that is,

$$\eta = \frac{T_{aw} - T_s}{T_c - T_s}. \tag{9.98}$$

If there is appreciable viscous dissipation (i.e., high-speed flow), T_s in Eq. (9.98) is replaced by the aerodynamic recovery temperature T_r. Because of the complexity of the hydrodynamic and thermal picture in Fig. 9.8, especially near the slot, most theoretical and semitheoretical approaches have been limited to a region far downstream of the slot where, if $U_c < U_s$, the flow behaves as an ordinary turbulent boundary layer. Perhaps the best analytical prediction of film effectiveness, even relatively close to the slot, is the integral method of Nicoll and Whitelaw [31]. In the initial region and just downstream of it, the numerical finite difference methods show considerable promise. Since it is difficult to calculate $\eta = \eta(x)$ from Nicoll and Whitelaw [31], a simpler result from the less rigorous analysis of Stollery and El-Ehwany [29] is presented. The *blowing parameter m* is defined by

$$m = \rho_c U_c / \rho_s U_s, \tag{9.99}$$

and the slot or coolant Reynolds number by

$$N_{Re_c} = \rho_c U_c s / \mu_c. \tag{9.100}$$

From [29], with these definitions, the film adiabatic wall effectiveness for flow over a flat plate becomes

$$\eta = 3.09 \left(\frac{ms}{x} \right)^{0.8} \left(N_{Re_c} \frac{\mu_c}{\mu_s} \right)^{0.2}, \tag{9.101}$$

or

$$\eta = 3.09 \, (\overline{x})^{-0.8},$$

where $\overline{x} \equiv (x/ms)[N_{Re_c}(\mu_c/\mu_s)]^{-1/4} > 5$ and $U_c/U_s < 1.5$.

Equation (9.101), when compared with the vast amount of experimental data for η, tends to give a lower, and therefore more conservative, value of η than the data indicate. There is a large amount of scatter in the data from different investigators when plotted on the same graph due to differences in slot geometry, starting length of the free stream boundary layer, and property variation with temperature. Hence, the value of η computed from Eq. (9.101) is to be considered a first estimate which, in many cases, will also be a conservative one.

Equation (9.101) permits an estimate of the plate surface temperature as a function of x, at least for the larger values of x, when the plate, the surface to be protected, is perfectly insulated. If the plate is not insulated, but instead is held at some known constant temperature T_w as a partial result of internal convection cooling in addition to the film, it is desirable to know the effect of the film on the calculation of the local heat transfer rate per unit area. In a fashion similar to that used earlier in dealing with heat transfer in high-speed flow, Seban, Chen, and Scesa [28] show that the difference between the film adiabatic wall temperature T_{aw} and the actual wall temperature T_w is the appropriate or "natural" driving temperature difference to use in Newton's cooling law. Hence, the heat transfer rate per unit area is calculated from

$$q'' = h_x(T_w - T_{aw}). \tag{9.102}$$

Thus, in film cooling situations, the film adiabatic wall temperature T_{aw} is the effective temperature potential of the free stream, rather than the static temperature T_s for low-speed flow without a film, or the aerodynamic recovery temperature T_r for high-speed flow without a film.

Seban and Back [26] and Seban[27] indicate that the local surface coefficient of heat transfer h_x is very close to the value for the body without a film, if x/s is greater than about 30 and if the blowing parameter m is less than unity. Thus, under these conditions, the relations of Chapter 5 for the forced convection coefficients may be used. If these conditions are not satisfied, see [26–27].

In some design situations, injection of the film parallel to the surface to be protected (i.e., tangential injection) may not be possible, and the injected coolant may have to enter the stream at some other angle, possibly even normal to the surface. For this condition, see Seban, Chen, and Scesa [28]. It is shown that normal injection reduces the effectiveness by about 50% near the slot (compared with tangential injection), and that this reduction is not as great at greater distances from the slot.

To alleviate the structural problems which may arise in transpiration cooling a porous material, particular in gas turbine blades and vanes, rows of film cooling holes can be used at various downstream locations, on the surface to be protected, in order to continually replenish the spent film with cooler fluid to maintain a high cooling effectiveness of the film. This is called "full coverage film cooling," and material relevant to it can be consulted in [35] through [38].

Finally, from Hartnett, Birkebak, and Eckert [25], the temperature-dependent fluid properties in the local length Reynolds number, Prandtl number, and Nusselt number should be evaluated in film cooling problems at the reference temperature

$$T^* = 0.5(T_w - T_s) + T_s + 0.22(T_{aw} - T_s). \tag{9.103}$$

Additional Topics in Heat Transfer

EXAMPLE 9.7

The insulated film-cooled flat plate shown in Fig. 9.8 is 0.4 ft long and has a coolant film slot height of 0.01 ft. The hot mainstream is air at 1000°F and 30 psia moving with a speed of 250 ft/sec, while the coolant is air at 700°F and 30 psia, and moves with a speed of 100 ft/sec. (a) Calculate the highest temperature on the plate surface. (b) If the plate is not insulated and, instead, an internal coolant causes the plate temperature at the downstream end to be 600°F with the coolant film flowing over the plate, calculate the heat transfer rate per unit area at $x = 0.4$ ft, and calculate the percent reduction in the heat flux to the plate at this point as a consequence of the film flow over the outside surface of the plate.

Solution

Since $T_c = 700°F$, $T_s = 1000°F$, $U_s = 250$ ft/sec, $U_c = 100$ ft/sec $= 3.6 \times 10^5$ ft/hr, $\rho_c = 30(144)/53.3(1160) = 0.0698$ lbm/ft³, $\rho_s = 30(144)/53.3(1460) = 0.0555$ lbm/ft³, and $s = 0.01$ ft,

$$m = \frac{\rho_c U_c}{\rho_s U_s} = \frac{0.0698(100)}{0.0555(250)} = 0.503.$$

Now, from the property tables for air in Appendix A, $\mu_c = 0.0765$ lbm/hr ft and $\mu_s = 0.0884$ lbm/hr ft. Also, $U_c/U_s = 100/250 = 0.4$. Since this is less than 1.5, part of the criterion for the use of Eq. (9.101) is satisfied.

Now the point of interest on the plate is the point at which the surface temperature is highest. Since effectiveness decreases with increasing x, the location of this point is at the downstream edge, that is, at $x = 0.4$ ft. Thus,

$$x/ms = \frac{0.4}{0.503(0.01)} = 79.5.$$

Since

$$\frac{\mu_c}{\mu_s} = \frac{0.0765}{0.0884} = 0.866$$

and

$$N_{Re_c} = \frac{\rho_c U_c s}{\mu_c} = \frac{0.0698(3.6 \times 10^5)(0.01)}{0.0765} = 3280,$$

then $N_{Re_c}\mu_c/\mu_s = 3280(0.866) = 2840$, so that

$$\frac{x}{ms}\left(N_{Re_c}\frac{\mu_c}{\mu_s}\right)^{-1/4} = \frac{79.5}{(2840)^{1/4}} = 10.89 = \bar{x}.$$

Since the parameter $\overline{x}$ is greater than 5 and $U_c/U_s < 1.5$, Eq. (9.101) can be used to calculate the value of the adiabatic film effectiveness η at $x = 0.4$ ft. Thus,

$$\eta = 3.09(\overline{x})^{-0.8} = \frac{3.09}{(10.89)^{0.8}} = 0.456.$$

From Eq. (9.98),

$$\eta = \frac{T_{aw} - T_s}{T_c - T_s} = \frac{T_{aw} - 1000}{700 - 1000} = 0.456,$$

and the highest temperature on the plate surface is

$$T_{aw} = 863°F. \tag{a}$$

Since the effectiveness in the initial region, just downstream of the slot, is about unity, the plate temperature there is equal to the coolant temperature of 700°F. Hence, with the film, the surface temperature of this adiabatic plate varies from 700°F at $x = 0$ to a maximum of 863°F at the downstream edge, at $x = 0.4$ ft. If the plate was not film-cooled, its temperature over the entire surface would be the free stream static temperature of 1000°F. (A check shows that viscous dissipation effects are negligible.)

Since $T_{aw} = 863°F$ at $x = 0.4$ ft from part a, the heat transfer rate to the plate, from Eq. (9.102), is

$$q'' = h_x(T_w - T_{aw}) = h_x(600 - 863) = -263 \, h_x. \tag{9.104}$$

Now $x/s = 0.40/0.01 = 40$, and since this is greater than 30, the ordinary low-speed surface coefficients given in Chapter 5 can be used.

The reference temperature for the fluid properties is evaluated from Eq. (9.103) as

$$T^* = 0.5(600 - 1000) + 1000 + 0.22(863 - 1000) = 770°F.$$

At this temperature, the property tables in Appendix A give $\mu^* = 0.079$, $k^* = 0.03$, $c_p^* = 0.25$, and $N_{Pr}^* = 0.684$. Also, $\rho^* = 30(144)/53.3(770 + 460) = 0.0658$ lbm/ft^3. Thus,

$$N_{Re_x} = \frac{0.0658(250 \times 3600)(0.4)}{0.079} = 3 \times 10^5.$$

Even with this relatively low length Reynolds number, the flow over most of the plate is expected to be turbulent as a result of the mixing occurring in the jet boundary layer of the initial region. Thus, from Eq. (5.382), the local surface coefficient is

$$h_x(0.4)/0.03 = 0.0288(3 \times 10^5)^{0.8}(0.684)^{1/3},$$

or

$$h_x = 45.8 \, \text{Btu/hr ft}^2 \, °F.$$

Hence, from Eq. (9.104),

$$q'' = -45.8(263) = -12,060 \text{ Btu/hr ft}^2. \qquad \textbf{(b)}$$

If there were no film, the local surface coefficient would again be 45.8 (neglecting the slight difference in the reference temperature at which the fluid properties are evaluated and assuming that the flow is turbulent). However, the driving temperature difference is $T_w - T_s$, instead of $T_w - T_{aw}$, so that the heat flux at $x = 0.4$ ft without a film is

$$q'' = 45.8(600 - 1000).$$

Thus, the percent reduction, due to the presence of the film, in the heat flux at $x = 0.4$ ft is

$$\left[\frac{(600 - 1000) - (600 - 863)}{600 - 1000} \right] (100) = 34.2\%. \qquad \textbf{(b)}$$

9.6 PROBLEMS

9.1 Estimate the surface temperature of the wing of a fighter aircraft which is flying at 760 m/s in 0°C air.

9.2 A perfectly insulated 3.8-cm-long flat plate moves with a speed of 30 m/s in the positive x direction, while 15°C air is moving over the plate with a speed of 75 m/s in the negative x direction. Calculate (a) the surface temperature of the plate if the air pressure is 10^5 N/m² and (b) the direction and value of the heat transfer rate for one side of the plate, if the plate surface temperature must be held at 25°C.

9.3 Air enters an axial compressor stage at a static temperature T_s and at an absolute speed V_s which, together, determine that the stagnation temperature of the incoming air is T_{0s}. The compressor disk in this stage is rotating at a high angular speed ω_0 and conditions are such that the disk behaves as if it were essentially adiabatic. A test engineer claims that temperatures greater than T_{0s} were measured on the rotating disk. Is this possible? If so, explain how. (Note that the Prandtl number of air is less than one.)

9.4 Oil is flowing in steady fashion between two stationary parallel plates at an average speed of 1.8 m/s. If one of the plates is insulated and the other is at 40°C, calculate the temperature of the insulated plate. The oil properties are $N_{Pr} = 3000$ and $c_p = 1940$ J/kg °C.

9.5 A perfectly insulated flat plate, 1 ft long, is immersed in a 100°F, 15 psia air stream moving at 1300 ft/sec. Calculate the surface temperature of the plate.

9.6 Suppose that the surface of the plate of Prob. 9.5 is held at 250°F (rather than being perfectly insulated). (a) Determine the heat transfer rate from one side of the plate and (b) compare it with the value obtained if, erroneously, this were treated as a low-speed problem—that is, treated as if the effects of viscous dissipation were negligible.

9.7 Rework Prob. 9.5 if the plate is held at $T_w = 200°$F. Neglect changes in the air properties due to the slightly different value of the reference temperature.

9.8 Find the maximum angular speed at which a 0.3-m-radius disk, made of a low thermal conductivity material, can be rotated in 20°C air if the disk temperature is not to exceed 80°C.

9.9 An uncooled turbine vane can be idealized as a perfectly insulated 1-in.-long flat plate. If the combustion gases, which have the same properties as air, pass over the vane at 500 ft/sec and are at a static temperature of 700°F and a pressure of 30 psia, calculate the surface temperature of the turbine vane.

9.10 Consider two parallel plates, with a fluid between them, separated by a distance b. The lower plate is insulated, while the upper plate is held at the known temperature T_1 and moves with a known speed V_s. This causes the fully developed velocity distribution in the fluid to be

$$u = V_s(y/b),$$

where y is measured from the lower, stationary, insulated plate. The fluid flow is laminar and constant property, but the fluid velocities are high enough to cause appreciable viscous dissipation. If the fluid temperature depends only upon y, find an expression for the aerodynamic recovery factor

$$r_c = \frac{T_r - T_1}{V_s^2/2c_p},$$

where T_r is the temperature of the insulated plate.

9.11 Air at 10^5 N/m² and 30°C is flowing with a velocity of 150 m/s over the upper surface of a flat plate which is 0.6 m long in the direction of flow. The plate surface is held at 40°C. Neglecting radiation, compute the convection heat transfer rate between the plate and the air.

9.12 Figure 9.9 shows a turbine blade from a jet engine. Its leading edge (the area around the stagnation point) can be approximated as a 1/4-in.-diameter circular cylinder. The combustion gases are at a static temperature of 1020°F and have an absolute velocity of 1500 ft/sec as they leave the stationary nozzles (vanes) and move toward the blade, whose tangential velocity is 500 ft/sec. The gases enter the blading smoothly. Assuming that the properties of the gas are those of air, find the heat transfer rate per unit area at the stagnation point for a maximum blade temperature of 900°F. The nozzle inclination angle is 20°, and $r_c = 1$ at a stagnation point.

Figure 9.9

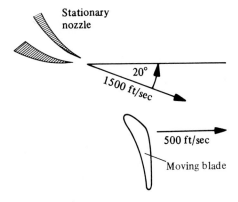

9.13   A 2-in.-diameter pipe has a surface temperature of 90°F. Air at 110°F and 14.7 psia blows over the pipe in crossflow with a speed of 900 ft/sec. Compute the heat transfer rate per unit area in the vicinity of the stagnation point where the recovery factor is unity.

9.14   Liquid lead at 482°C passes over a 0.15-m-long flat plate at a speed of 0.1 m/s. If the plate temperature is 371°C and the plate is 0.6 m wide, calculate (a) the average surface coefficient of heat transfer and (b) the heat transfer rate to the plate.

9.15   A flat plate heat exchanger dissipates energy at the constant rate per unit area of 35,000 Btu/hr ft² over its surface. The coolant is liquid bismuth at 800°F flowing over the 18-in.-long plate at a speed of 0.6 ft/sec. Find the value of the maximum plate temperature.

9.16   Consider steady, laminar, two-dimensional flow of a liquid metal in the vicinity of a stagnation point, where the free stream velocity variation is $U_s = ax$. Using integral methods and the limiting case of $N_{Pr} \rightarrow 0$, derive the expression

$$N_{Nu_x} = 0.75\, N_{Pe_x}^{1/2}$$

for the local Nusselt number when T_w is a constant.

9.17   Derive equations for the local surface coefficient of heat transfer and the wall temperature distribution when a low Prandtl number fluid (such as a liquid metal) flows in a steady, laminar, low-speed, constant property manner over a flat plate which supplies a known constant heat flux q_w'' to the fluid. Axial conduction can be neglected.

9.18   Sodium, at an average bulk mean temperature of 371°C, flows in a laminar fashion through a 1.25-cm-diameter tube with a constant flux of 30,000 W/m² at the tube wall. Calculate (a) the average surface coefficient of heat transfer if it can be assumed that the velocity is practically constant at its average value and that the flow is fully developed thermally, and (b) the maximum tube wall temperature if the maximum bulk mean temperature of the sodium is 386°C.

9.19   A liquid metal flows over a flat plate in a steady, boundary layer–type fashion. Can there ever be a portion of the plate in which $\delta_t << \delta$ when a pressure gradient is absent? Explain.

9.20   Liquid mercury at 100°C enters a 2.5-cm-diameter tube at an average speed of 0.3 m/s. If a constant flux condition is maintained at the wall, estimate the value of the surface coefficient of heat transfer.

9.21   Liquid zinc, flowing at 100,000 lbm/hr, is to be heated from 900°F to 1100°F in a 2-in.-diameter, 10-ft-long tube. Calculate the required tube wall temperature.

9.22   Liquid potassium at 500°F flows at 1.1 ft/sec over a 1-ft-long, 1-ft-wide flat plate at 530°F. Compute the heat transfer rate from the plate to the potassium.

9.23   A 2.5-cm-diameter, 1.8-m-long coolant tube in a reactor has a wall flux maintained at 250,000 W/m². Liquid potassium enters the tube at 149°C. If the maximum tube temperature is not to exceed 250°C, estimate the required mass flow rate of the potassium coolant.

9.24   The wall of a heat exchanger is idealized as a 1-ft-long and 1-ft-wide flat plate, with a constant flux of 50,000 Btu/hr ft² across it. If a liquid metal mixture of 44% potassium and 56% sodium at 400°F passes over the plate at a speed of 2.3 ft/sec, calculate the highest temperature on the plate surface.

9.25 Sodium, flowing at 0.28 kg/s, at 371°C, enters a 3.75-cm-diameter cooling tube in a nuclear reactor. Energy released by the fission process causes a constant heat flux to the tube of 300,000 W/m² along the length of the 1.8-m tube. If the flow is fully developed, find the maximum wall temperature of the tube.

9.26 Show what, for the limiting case of $N_{Pr} \rightarrow 0$, the turbulent free convection relation, Eq. (6.57a), reduces to.

9.27 Liquid lead at 454°C is in contact with a 0.6-m-high vertical plate whose wall temperature is maintained at 508°C. If the thermal coefficient of volume expansion is equal to 0.1152 $\times$ 10^{-3} °C^{-1}, estimate the value of the average surface coefficient of heat transfer between the plate and the lead.

9.28 The lead described in Prob. 9.27 is in contact with a 2.5-cm-diameter horizontal tube held at 508°C. Calculate the heat transfer rate from the tube if it is 3 m long.

9.29 A 1-in.-diameter horizontal tube has an outside surface temperature of 425°F and is surrounded by liquid sodium at 375°F. The thermal coefficient β of volume expansion is 0.150 $\times$ 10^{-3} °F^{-1}. Calculate the value of the average surface coefficient of heat transfer between the outside tube surface and the sodium.

9.30 Consider steady, laminar, boundary layer–type, constant property flow with appreciable viscous dissipation over a perfectly insulated flat plate. The undisturbed values of velocity and static temperature are U_s and T_s, respectively. If the nondimensional temperature is

$$\sigma = \frac{T - T_s}{U_s^2/2c_p},$$

find the groups which this nondimensional temperature depends on.

9.31 (a) Find the groups which the nondimensional temperature depends on in steady, laminar, constant property flow with appreciable viscous dissipation over a flat plate at constant wall temperature T_w. The free stream temperature and velocity are T_s and U_s. (b) Also find the groups which the average Nusselt number depends on.

9.32 Consider steady, laminar, constant property, boundary layer–type flow at an upstream velocity U_s, over a circular cylinder in crossflow. The cylinder of radius R is rotating slowly at the angular speed ω_0 rad/sec. Find the groups which the nondimensional velocity distribution depends on.

9.33 Suppose that a plate is porous and fluid is blown into the boundary layer from the plate surface at a known constant velocity v_c in the y direction. Use differential similarity to deduce the nondimensional groups which the nondimensional velocity field depends on.

9.34 Suppose that there is a two-dimensional, steady, constant property flow, but not of the thin boundary layer type, over some stationary solid body. The undisturbed flow velocity far upstream of the body is U_0, while a characteristic linear dimension in the x direction on the body itself has the value L. If there are no body forces, find the nondimensional groups which the nondimensional velocity field depends on.

9.35 For laminar, steady, constant property flow over a flat plate at temperature T_w with transpiration—fluid being blown into the boundary layer at the plate surface at the constant speed v_c in the y direction—find the nondimensional groups which the average Nusselt number depends on. The free stream properties of the fluid flowing over the plate are U_s and T_s, while the fluid entering the boundary layer at the plate surface enters at temperature T_w.

9.36 Consider steady, laminar, constant property, low-speed flow of a fluid at temperature T_s over a flat plate whose surface temperature is $T_w = a + bx$, where a and b are known constants. Find the groups which the nondimensional temperature distribution depends on.

9.37 The governing partial differential equations for the velocity and temperature fields in steady, laminar, low-speed flow of a fluid at T_s over a flat plate at temperature T_w, when all fluid properties are constant except for the viscosity, which is given by $\mu = aT^b$ where a and b are known constants, are

$$\frac{\partial u}{\partial x} + \frac{\partial v}{\partial y} = 0, \quad \rho\left(u\frac{\partial u}{\partial x} + v\frac{\partial u}{\partial y}\right) = \frac{\partial}{\partial y}\left(\mu\frac{\partial u}{\partial y}\right),$$

$$\rho c_p\left(u\frac{\partial T}{\partial x} + v\frac{\partial T}{\partial y}\right) = k\frac{\partial^2 T}{\partial y^2}.$$

Find the groups which the nondimensional temperature field T/T_s depends on. The plate length is L and free stream velocity is U_s.

9.38 Consider steady, laminar, constant property, two-dimensional boundary layer–type flow of a fluid at temperature T_s and speed U_s over a constant temperature flat plate, when free convection effects and viscous dissipation effects are important. If the plate is of length L, find the groups which the average Nusselt number depends on. Discuss possible criteria for neglecting free convection and viscous dissipation effects.

9.39 Consider a long solid cylinder of outer radius R which is initially at the constant temperature T_0 and is then suddenly exposed to a fluid at temperature T_s and known constant surface coefficient of heat transfer h. If the thermal properties of the solid cylinder are constant, deduce the nondimensional groups which the nondimensional temperature excess ratio $\phi = (T - T_s)/(T_0 - T_s)$ depends on. (Note that a nondimensional time can be defined by dividing time by R^2 and multiplying it by the thermal diffusivity.)

9.40 A turbine vane is idealized as a 2.5-cm-long flat plate. If the plate Reynolds number is 300,000 and the average film temperature is 500°C, estimate the percent reduction in the local surface coefficient at the end of the plate if transpiration cooling is used with a constant surface velocity of the coolant equal to 0.06 m/s. The combustion products, behaving like air, flow over the vane at 600,000 N/m².

9.41 Calculate the constant surface velocity v_0 of coolant required to reduce the local surface coefficient by 50% in Prob. 9.40.

9.42 Repeat Prob. 9.41, where $v_w = c/\sqrt{x}$, and c is a constant, and calculate the value of v_w required at the end of the plate.

9.43 Air flows over a flat plate with a speed U_s of 100 ft/sec and a pressure of 14.7 psia. The plate is porous and cooling air is forced through it with a speed of 0.05 ft/sec at the surface. Estimate the ratio of the surface coefficient with transpiration to that without transpiration at $x = 0.5$ ft from the leading edge. The film temperature is 100°F.

9.44 Air flows over a flat plate with a speed U_s of 100 ft/sec and a pressure of 14.7 psia. The plate is porous and cooling air is forced through it such that v_w is inversely proportional to the square root of x and has the value 0.05 ft/sec at 1/2 ft from the leading edge. If the film temperature is 100°F, find the ratio of the local surface coefficient of heat transfer with transpiration to that with no transpiration at $x = 0.5$ ft.

9.45 Hot air at 650°C and 10^5 N/m² flows over a 0.15-m-long flat plate with a speed of 60 m/s. If the plate surface temperature is 450°C as a result of forcing 250°C cooling air through the plate such that its surface velocity is $v_w = c/\sqrt{x}$, calculate (a) the surface velocity at the end of the plate, (b) the percent reduction in the local surface coefficient at the end of the plate, and (c) the total mass flow rate of coolant if the plate is 0.3 m wide.

9.46 Air at 1200°F and 40 psia flows over a 1-ft-long insulated flat plate at a speed of 200 ft/sec. The plate is film-cooled by 600°F coolant air issuing from a 0.01-ft-high slot at the plate leading edge, with a pressure of 40 psia and a speed of 50 ft/sec. Estimate the value of the highest surface temperature on the plate.

9.47 Rework Prob. 9.46 where the free stream speed is 1000 ft/sec and the coolant speed in the slot is 250 ft/sec.

9.48 If the plate of Prob. 9.46 is held at the constant temperature 600°F instead of being perfectly insulated, compute (a) the heat flux at $x = 0.75$ ft and (b) the percent reduction of the flux due to the presence of the coolant film.

9.49 Air at 400°C and 136,000 N/m² is flowing over a 0.15-m-long flat plate at 180 m/s. The plate is essentially insulated and is film-cooled by coolant air at 100°C and 136,000 N/m², which is injected tangentially onto the plate surface from a slot 0.15 cm high located at the plate leading edge. If the maximum plate temperature is not to exceed 300°C, calculate (a) the velocity of the coolant required at the slot exit and (b) the mass flow rate of coolant air per meter of width.

REFERENCES

1. Eckert, E. R. G. "Engineering Relations for Friction and Heat Transfer to Surfaces in High Velocity Flow." *Journal of the Aero. Sciences* (August 1955), pp. 585–87.
2. Eckert, E. R. G. "Engineering Relations for Heat Transfer and Friction in High Velocity Laminar and Turbulent Boundary Layer Flow Over Surfaces with Constant Pressure and Temperature." *Transactions of the A.S.M.E.*, Vol. 78 (August 1956), pp. 1273–83.
3. Rubesin, M. W., and M. Inouye. "Forced Convection External Flows." In *Handbook of Heat Transfer*, edited by W. M. Rohsenow and J. P. Hartnett. New York: McGraw-Hill, 1973.
4. Eckert, E. R. G., and R. M. Drake, Jr. *Analysis of Heat and Mass Transfer*. New York: McGraw-Hill, 1972.
5. Holman, J. P. *Heat Transfer*. 5th ed. New York: McGraw-Hill, 1981.
6. Karlekar, B. V., and R. M. Desmond. *Heat Transfer*. 2d ed. St. Paul, Minn.: West, 1982.
7. Grosh, R. J., and R. D. Cess. "Heat Transfer to Fluids with Low Prandtl Numbers for Flow Across Plates and Cylinders of Various Cross Section." *Transactions of the A.S.M.E.*, Vol. 80 (1958), pp. 667–76.
8. Stein, R. P. "Liquid Metal Heat Transfer." In *Advances in Heat Transfer*, Vol. 3, edited by T. F. Irvine, Jr., and J. P. Hartnett. New York: Academic Press, 1966.
9. Seban, R. A., and T. T. Shimazaki. "Heat Transfer to a Fluid Flowing Turbulently in a Smooth Pipe with Walls at Constant Temperature." *Transactions of the A.S.M.E.*, Vol. 73 (1951), pp. 803–9.

10. Lyon, R. N. "Liquid Metal Heat Transfer Coefficients." *Chemical Engineering Progress,* Vol. 47, No. 2 (1951), pp. 75–79.

11. Lubarsky, B., and S. J. Kaufman. "Review of Experimental Investigations of Liquid Metal Heat Transfer." *NACA Report 1270* (1956).

12. Viskanta, R., and Y. S. Touloukian. "Heat Transfer to Liquid Metals with Variable Properties." *Journal of Heat Transfer, Transactions of the A.S.M.E.,* Vol. 82, No. 4 (1960), pp. 333–40.

13. Witte, L. C. "An Experimental Study of Forced Convection Heat Transfer from a Sphere to Liquid Sodium." *Journal of Heat Transfer,* Vol. 90, No. 1 (February 1968), pp. 9–12.

14. Hyman, C. S., C. F. Bonilla, and S. W. Ehrlich. "Natural Convection Transfer Processes: I. Heat Transfer to Liquid Metals and Non-Metals at Horizontal Cylinders." *Chemical Engineering Progress Symposium Series,* Vol. 49, No. 5 (1953), pp. 21–31.

15. Skupinski, E., J. Tortel, and L. Vautrey. "Determination des Coefficients de Convection d'un Alliage Sodium-Potassium dans un Tube Circulaire." *International Journal of Heat and Mass Transfer,* Vol. 8 (1965), pp. 937–51.

16. Lienhard, J. H. *A Heat Transfer Textbook.* Englewood Cliffs, N. J.: Prentice-Hall, 1981.

17. *Liquid Metals Handbook.* 2d ed. (revised). Atomic Energy Commission, Dept. of the Navy (January 1954).

18. *Liquid Metals Handbook, Sodium-NaK Supplement.* 3d ed. Atomic Energy Commission, Dept. of the Navy (June 1955).

19. Hartnett, J. P., and E. R. G. Eckert. "Mass Transfer Cooling in a Laminar Boundary Layer with Constant Fluid Properties." *Transactions of the A.S.M.E.,* Vol. 79 (1957), pp. 247–54.

20. Nealy, D. A., and P. W. McFadden. "Some Boundary Layer–Porous Wall Coupling Effects in Laminar Flows with Surface Injection." *A.S.M.E. Paper No. 70—GT-1* (1970).

21. Esgar, J. B., R. S. Colladay, and A. Kaufman. "An Analysis of the Capabilities and Limitations of Turbine Air Cooling Methods." *NASA TN D-5992* (September 1970).

22. Mickley, H. S., R. C. Ross, A. L. Squyers, and W. E. Stewart. "Heat, Mass, and Momentum Transfer for Flow Over a Flat Plate with Blowing or Suction." *NASA TN 3203* (1954).

23. Kays, W. M., and M. E. Crawford. *Convective Heat and Mass Transfer.* 2d ed. New York: McGraw-Hill, 1980.

24. Wieghardt, K. "Hot Air Discharge for De-icing." *Report F-TS-919–RE Wright-Patterson Air Force Base,* Dayton, Ohio (1946).

25. Hartnett, J. P., R. C. Birkebak, and E. R. G. Eckert. "Velocity Distributions, Temperature Distributions, Effectiveness, and Heat Transfer for Air Injected Through a Tangential Slot into a Turbulent Boundary Layer." *Journal of Heat Transfer, Transactions of the A.S.M.E.,* Vol. 83, No. 3 (August 1961), pp. 293–306.

26. Seban, R. A., and L. H. Back. "Velocity and Temperature Profiles in Turbulent Boundary Layers with Tangential Injection." *Journal of Heat Transfer, Transactions of the A.S.M.E.,* Vol. 84 (February 1962), pp. 45–54.

27. Seban, R. A. "Heat Transfer and Effectiveness for a Turbulent Boundary Layer with Tangential Fluid Injection." *Journal of Heat Transfer, Transactions of the A.S.M.E.,* Vol. 82, No. 4 (November 1960), pp. 303–12.

28. Seban, R. A., H. W. Chen, and S. Scesa. "Heat Transfer to a Turbulent Boundary Layer Downstream of an Injection Slot." *A.S.M.E. Paper No. 57-A-36* (1957).

29. Stollery, J. L., and A. A. M. El-Ehwany. "A Note on the Use of a Boundary Layer Model for Correlating Film Cooling Data." *International Journal of Heat and Mass Transfer,* Vol. 8 (1965), pp. 55–65.

30. Stollery, J. L., and A. A. M. El-Ehwany. "On the Use of Boundary Layer Model for Correlating Film Cooling Data." *International Journal of Heat and Mass Transfer,* Vol. 10 (1967), pp. 101–5.

31. Nicoll, W. B., and J. H. Whitelaw. "The Effectiveness of the Uniform Density, Two-Dimensional Wall Jet." *International Journal of Heat and Mass Transfer,* Vol. 10 (1967), pp. 623–39.

32. Sucec, J. "Numerical Prediction of the Temperature Distribution in the Initial Region of a Two-Dimensional Turbulent Film Cooling Configuration." *International Journal for Numerical Methods in Engineering,* Vol. 2 (1970), pp. 297–314.

33. Williams, J. J., and W. H. Giedt. "The Effect of Gaseous Film Cooling on the Recovery Temperature Distribution in Rocket Nozzles." *A.S.M.E. Paper No. 70-HT/SpT-42* (1970).

34. Mayle, R. E., F. C. Kopper, M. F. Blair, and D. A. Bailey. "Effect of Streamline Curvature on Film Cooling." *Journal of Engineering for Power,* Vol. 99 (January 1977), pp. 77–82.

35. Afejuka, W. O., N. Hay, and D. Lampard. "The Film Cooling Effectiveness of Double Rows of Holes." *Journal of Engineering for Power,* Vol. 102 (1980), pp. 601–6.

36. Hampel, H., R. Friedrich, and S. Wittig. "Full Coverage Film Cooled Blading in High Temperature Gas Turbines: Cooling Effectiveness, Profile Loss, and Thermal Efficiency." *Journal of Engineering for Power,* Vol. 102 (1980), pp. 957–63.

37. Crawford, M. E., W. M. Kays, and R. J. Moffat. "Full Coverage Film Cooling—Part I: Comparison of Heat Transfer Data with Three Injection Angles." *Journal of Engineering for Power,* Vol. 102 (1980), pp. 1000–1005.

38. Crawford, M. E., W. M. Kays, and R. J. Moffat. "Full Coverage Film Cooling—Part II: Heat Transfer Data and Numerical Simulation." *Journal of Engineering for Power,* Vol. 102 (1980), pp. 1006–12.

Appendix A

SELECTED THERMOPHYSICAL PROPERTIES OF SUBSTANCES

Table A.1.a
Thermal Conductivity k, Specific Heat c_p, Density ρ, and Thermal Diffusivity α, of Metals and Alloys (SI Units)

Metal	Properties at 20°C				Thermal Conductivity k, W/m °K									
	ρ (kg/m³)	c_p (J/kg °C)	k (W/m °C)	α (m²/s)	−100 C / −148 F	0 C / 32 F	100 C / 212 F	200 C / 392 F	300 C / 572 F	400 C / 752 F	600 C / 1112 F	800 C / 1472 F	1000 C / 1832 F	1200 C / 2192 F
Aluminum														
Pure	2,707	0.896 × 10³	204	8.418 × 10⁻⁵	215	202	206	215	228	249				
Al-Cu (Duralumin) 94–96 Al, 3–5 Cu, trace Mg	2,787	0.883	164	6.676	126	159	182	194						
Al-Mg (Hydronalium) 91–95 Al, 5–9 Mg	2,611	0.904	112	4.764	93	109	125	142						
Al-Si (Silumin) 87 Al, 13 Si	2,659	0.871	164	7.099	149	163	175	185						
Al-Si (Silumin, copper bearing) 86.5 Al, 1 Cu	2,659	0.867	137	5.933	119	137	144	152	161					
Al-Si (Alusil) 78–80 Al, 20–22 Si	2,627	0.854	161	7.172	144	157	168	175	178					
Al-Mg-Si 97 Al, 1 Mg, 1 Si, 1 Mn	2,707	0.892	177	7.311		175	189	204						
Lead	11,373	0.130	35	2.343	36.9	35.1	33.4	31.5	29.8					
Iron														
Pure	7,897	0.452	73	2.034	87	73	67	62	55	48	40	36	35	36
Wrought iron (C < 0.5%)	7,849	0.46	59	1.626		59	57	52	48	45	36	33	33	33
Cast iron (C ≈ 4%)	7,272	0.42	52	1.703										
Steel (C max ≈ 1.5%) Carbon steel C ≈ 0.5%	7,833	0.465	54	1.474		55	52	48	45	42	35	31	29	31
1.0%	7,801	0.473	43	1.172		43	43	42	40	36	33	29	28	29
1.5%	7,753	0.486	36	0.970		36	36	36	35	33	31	28	28	29

Material	ρ	c	λ	a	T1	T2	T3	T4	T5	T6	T7	T8	T9	T10
Nickel steel Ni ≈ 0%	7,897	0.452	73	2.026										
10%	7,945	0.46	26	0.720										
20%	7,993	0.46	19	0.526										
30%	8,073	0.46	12	0.325										
40%	8,169	0.46	10	0.279										
50%	8,266	0.46	14	0.361										
60%	8,378	0.46	19	0.493										
70%	8,506	0.46	26	0.666										
80%	8,618	0.46	35	0.872										
90%	8,762	0.46	47	1.156										
100%	8,906	0.448	90	2.276										
Invar Ni = 36%	8,137	0.46	10.7	0.286										
Chrome steel Cr = 0%	7,897	0.452	73	2.026	87	73	67	62	55	48	40	36	35	36
1%	7,865	0.46	61	1.665		62	55	52	47	42	36	33	33	
2%	7,865	0.46	52	1.443		54	48	45	42	38	33	31	31	
5%	7,833	0.46	40	1.110		40	38	36	36	33	29	29	29	
10%	7,785	0.46	31	0.867		31	31	31	29	29	28	28	29	
20%	7,689	0.46	22	0.635		22	22	22	22	24	24	26	29	
30%	7,625	0.46	19	0.542										
Cr-Ni (chrome-nickel): 15 Cr, 10 Ni	7,865	0.46	19	0.526										
18 Cr, 8 Ni (V2A)	7,817	0.46	16.3	0.444		16.3	17	17	19	19	22	26	31	
20 Cr, 15 Ni	7,833	0.46	15.1	0.415										
25 Cr, 20 Ni	7,865	0.46	12.8	0.361										
Ni-Cr (nickel-chrome): 80 Ni, 15 Cr	8,522	0.46	17	0.444		14.0	15.1	15.1	16.3	17	19	22		
60 Ni, 15 Cr	8,266	0.46	12.8	0.333										
40 Ni, 15 Cr	8,073	0.46	11.6	0.305										
20 Ni, 15 Cr	7,865	0.46	14.0	0.390										
Cr-Ni-Al: 6 Cr, 1.5 Al, 0.55 Si (Sicromal 8)	7,721	0.490	22	0.594										
24 Cr, 2.5 Al, 0.55 Si (Sicromal 12)	7,673	0.494	19	0.501										

Continued on next page

Table A.1.a (Continued)

Metal	Properties at 20°C				Thermal Conductivity k, W/m °K									
	ρ (kg/m^3)	c_p $(J/kg\ °C)$	k $(W/m\ °C)$	α (m^2/s)	−100 C −148 F	0 C 32 F	100 C 212 F	200 C 392 F	300 C 572 F	400 C 752 F	600 C 1112 F	800 C 1472 F	1000 C 1832 F	1200 C 2192 F
Manganese steel														
Mn = 0%	7,897	0.494	73	1.863										
1%	7,865	0.46	50	1.388										
2%	7,865	0.46	38	1.050		38	36	36	36	35	33			
5%	7,849	0.46	22	0.637										
10%	7,801	0.46	17	0.483										
Tungsten steel														
W = 0%	7,897	0.452	73	2.026										
1%	7,913	0.448	66	1.858										
2%	7,961	0.444	62	1.763		62	59	54	48	45	36			
5%	8,073	0.435	54	1.525										
10%	8,314	0.419	48	1.391										
20%	8,826	0.389	43	1.249										
Silicon steel														
Si = 0%	7,897	0.452	73	2.026										
1%	7,769	0.46	42	1.164										
2%	7,673	0.46	31	0.888										
5%	7,417	0.46	19	0.555										
Copper														
Pure	8,954	0.3831×10^3	386	11.234×10^{-5}	407	386	379	374	369	363	353			
Aluminum bronze 95 Cu, 5 Al	8,666	0.410	83	2.330										
Bronze 75 Cu, 25 Sn	8,666	0.343	26	0.859										
Red Brass 85 Cu, 9 Sn, 6 Zn	8,714	0.385	61	1.804		59	71							
Brass 70 Cu, 30 Zn	8,522	0.385	111	3.412	88		128	144	147	147				
German silver 62 Cu, 15 Ni, 22 Zn	8,618	0.394	24.9	0.733	19.2		31	40	45	48				
Constantan 60 Cu, 40 Ni	8,922	0.410	22.7	0.612	21		22.2	26						

Table — thermophysical properties of metallic solids. Density ρ (kg/m³), specific heat c_p (kJ/kg·°C), thermal conductivity k (W/m·°C) at 20°C, thermal diffusivity $\alpha \times 10^{5}$ (m²/s), and thermal conductivity k (W/m·°C) at the temperatures indicated.

Substance	ρ	c_p	k	α	−100°C	0°C	100°C	200°C	300°C	400°C	600°C	800°C	1000°C	1200°C
Magnesium Pure	1,746	1.013	171	9.708	178	171	168	163	157					
Mg-Al (electrolytic) 6–8% Al, 1–2% Zn	1,810	1.00	66	3.605		52	62	74	83					
Mg-Mn 2% Mn	1,778	1.00	114	6.382	93	111	125	130						
Molybdenum	10,220	0.251	123	4.790	138	125	118	114	111	109	106	102	99	92
Nickel Pure (99.9%)	8,906	0.4459	90	2.266	104	93	83	73	64	59	55			
Impure (99.2%)	8,906	0.444	69	1.747		69	64	59	55	52				
Ni-Cr 90 Ni, 10 Cr	8,666	0.444	17	0.444		17.1	18.9	20.9	22.8	24.6				
80 Ni, 20 Cr	8,314	0.444	12.6	0.343		12.3	13.8	15.6	17.1	18.9	22.5			
Silver Purest	10,524	0.2340	419	17.004	419	417	415	412	362	360				
Pure (99.9%)	10,524	0.2340	407	16.563	419	410	415	374						
Tungsten	19,350	0.1344	163	6.271		166	151	142	133	126	112	76		
Zinc, pure	7,144	0.3843	112.2	4.106	114	112	109	106	100	93				
Tin, pure	7,304	0.2265	64	3.884	74	65.9	59	57						

From *Analysis of Heat and Mass Transfer* by E. R. G. Eckert and R. M. Drake. Copyright © 1972 by the McGraw-Hill Book Company. Used with permission of McGraw-Hill Book Company.

Selected Thermophysical Properties of Substances

Table A.1.b
Thermal Conductivity k, Specific Heat c_p, Density ρ, and Thermal Diffusivity α of Metals and Alloys (English Units)

Material	k (Btu/hr ft °F) 32°F	212°F	572°F	932°F	c_p (Btu/lbm °F) 32°F	ρ (lbm/cu ft) 32°F	α (ft²/hr) 32°F
Metals							
Aluminum	117	119	133		0.208	169	3.33
Bismuth	4.9	3.9			0.029	612	0.28
Copper, pure	224	218	212		0.091	558	4.42
Gold	169	170			0.030	1203	4.68
Iron, pure	35.8	36.6			0.104	491	0.70
Lead	20.1	19	18		0.030	705	0.95
Magnesium	91	92			0.232	109	3.60
Mercury	4.8				0.033	849	0.17
Nickel	34.5	34	32		0.103	555	0.60
Silver	242	238		155	0.056	655	6.6
Tin	36	34			0.054	456	1.46
Zinc	65	64	59	207	0.091	446	1.60
Alloys							
Admiralty metal	65	64					
Brass, 70% Cu, 30% Zn	56	60	66		0.092	532	1.14
Bronze, 75% Cu, 25% Sn	15				0.082	540	0.34
Cast iron							
Plain	33	31.8	27.7		0.11	474	0.63
Alloy	30	28.3	27	24.8	0.10	455	0.66
Constantan, 60% Cu, 40% Ni	12.4	12.8			0.10	557	0.22
18–8 Stainless steel,							
Type 304	8.0	9.4	10.9	12.4	0.11	488	0.15
Type 347	8.0	9.3	11.0	12.8	0.11	488	0.15
Steel, mild, 1% C	26.5	26	25	22	0.11	490	0.49

Table A–1 "Thermal conductivity k, specific heat c, density ρ, and thermal diffusivity α of metals and alloys" (p. 634) in *Principles of Heat Transfer*, 3d ed., by Frank Kreith (Intext Co.). Copyright © 1958, 1963, 1973 by Harper & Row, Publishers, Inc. By permission of Harper & Row, Publishers, Inc.

Appendix A

Table A.2.a
Physical Properties of Some Nonmetals (SI Units)

Material	T (°C)	ρ (kg/m³)	c_p (J/kg °C)	k (W/m °C)	α (m²/s)
Aerogel, silica	120	136.2		0.022	
Asbestos	−200	469.3		0.074	
	0	469.3		0.156	
	0	576.7	0.816×10^3	0.151	
	100	576.7	0.816	0.192	
	200	576.7		0.208	
	400	576.7		0.223	
	−200	696.8		0.156	
	0	696.8		0.234	
Brick, dry	20	1,762–1,810	0.84	0.38–0.52	$0.028\text{–}0.034 \times 10^{-5}$
Bakelite	20	1,273.5	1.59	0.232	0.0114
Cardboard, corrugated				0.064	
Clay	20	1,457.7	0.88	1.279	0.101
Concrete	20	1,906–2,307	0.88	0.81–1.40	0.049–0.070
Coal, anthracite	20	1,201–1,506	1.26	0.26	0.013–0.015
Powdered	30	737	1.30	0.116	0.013
Cotton	20	80	1.30	0.059	0.194
Cork, board	30	160		0.043	
Expanded scrap	20	44.9–118.5	1.88	0.036	0.015–0.044
Ground	30	150.6		0.043	
Diatomaceous earth	38	320.4		0.062	
	871	320.4		0.142	
Earth, coarse gravelly	20	2,050	1.84	0.52	0.0139
Felt, wool	30	330.0		0.05	
Fiber, insulating board	21	237.1		0.048	
Red	20	1,289.5		0.47	
Glass plate	20	2,707	0.8	0.76	0.034
Glass, borosilicate	30	2,227		1.09	
Wool	20	200.2	0.67	0.040	0.028
Granite				1.7–4.0	
Ice	0	913	1.93	2.22	0.124
Marble	20	2,499–2,707	0.808	2.8	0.139
Rubber, hard	0	1,198.2		0.151	
Sandstone	20	2,162–2,307	0.71	1.63–2.1	0.106–0.126
Silk	20	57.7	1.38	0.036	0.044
Wood, oak radial	20	609–801	2.39	0.17–0.21	0.0111–0.0121
Fir (20% moisture) radial	20	416.5–421.3	2.72	0.14	0.0124

From *Analysis of Heat and Mass Transfer* by E. R. G. Eckert and R. M. Drake. Copyright © 1972 by the McGraw-Hill Book Company. Used with permission of McGraw-Hill Book Company.

Table A.2.b
Physical Properties of Some Nonmetals (English Units)

Material	Average Temperature (°F)	k (Btu/hr ft °F)	c_p (Btu/lbm °F)	ρ (lbm/cu ft)	α (ft²/hr)
Insulating materials					
Asbestos	32	0.087	0.25	36	~0.01
	392	0.12		36	~0.01
Cork	86	0.025	0.04	10	~0.006
Cotton, fabric	200	0.046			
Diatomaceous earth, powdered	100	0.030	0.21	14	~0.01
	300	0.036			
	600	0.046			
Molded pipe covering	400	0.051		26	
	1600	0.088			
Glass wool					
Fine	20	0.022			
	100	0.031			
	200	0.043		1.5	
Packed	20	0.016			
	100	0.022		6.0	
	200	0.029			
Hair felt	100	0.027		8.2	
Kaolin insulating					
Brick	932	0.15		27	
	2102	0.26			
Firebrick	392	0.05		19	
	1400	0.11			
85% magnesia	32	0.032		17	
	200	0.037		17	
Rock wool	20	0.017		8	
	200	0.030			
Rubber	32	0.087	0.48	75	0.0024

Building Materials					
Brick					
Fire clay	392	0.58	0.20	144	0.02
	1832	0.95			
Masonry	70	0.38	0.20	106	0.018
Zirconia	392	0.84		304	
	1832	1.13			
Chrome brick	392	0.82		246	
	1832	0.96			
Concrete					
Stone	~70	0.54	0.20	144	0.019
10% moisture	~70	0.70		140	~0.025
Glass, window	~70	~0.45	0.2	170	0.013
Limestone, dry	70	0.40	0.22	105	0.017
Sand					
Dry	68	0.20		95	
10% H_2O	68	0.60		100	
Soil					
Dry	70	~0.20			~0.01
Wet	70	~1.5	0.44		~0.03
Wood					
Oak perpendicular to grain	70	0.12	0.57	51	0.0041
parallel to grain	70	0.20	0.57	51	0.0069
Pine perpendicular to grain	70	0.06	0.67	31	0.0029
parallel to grain	70	0.14	0.67	31	0.0067
Ice	32	1.28	0.46	57	0.048

Table A–2 "Physical properties of some nonmetals" (p. 635) in *Principles of Heat Transfer*, 3d ed., by Frank Kreith (Intext Co.). Copyright © 1958, 1963, 1973 by Harper & Row Publishers, Inc. By permission of Harper & Row Publishers, Inc.

Table A.3.a
Property Values of Engine Oil, Unused (SI Units)

T (°C)	ρ (kg/m³)	c_p (J/kg °C)	ν (m²/s)	k (W/m °C)	α (m²/s)	N_{Pr}	β (°K⁻¹)
0	899.12	1.796×10^3	0.00428	0.147	0.911×10^{-7}	47,100	0.70×10^{-3}
20	888.23	1.880	0.00090	0.145	0.872	10,400	
40	876.05	1.964	0.00024	0.144	0.834	2,870	
60	864.04	2.047	0.839×10^{-4}	0.140	0.800	1,050	
80	852.02	2.131	0.375	0.138	0.769	490	
100	840.01	2.219	0.203	0.137	0.738	276	
120	828.96	2.307	0.124	0.135	0.710	175	
140	816.94	2.395	0.080	0.133	0.686	116	
160	805.89	2.483	0.056	0.132	0.663	84	

From *Analysis of Heat and Mass Transfer* by E. R. G. Eckert and R. M. Drake. Copyright © 1972 by the McGraw-Hill Book Company. Used with permission of McGraw-Hill Book Company.

Table A.3.b
Property Values of Engine Oil, Unused (English Units)

T (°F)	ρ (lb/ft³)	c_p (Btu/lbm °F)	ν (ft²/sec)	k (Btu/hr ft °F)	α (ft²/hr)	N_{Pr}	β (1/°R)
32	56.13	0.429	0.0461	0.085	3.53×10^{-3}	47,100	0.39×10^{-3}
68	55.45	0.449	0.0097	0.084	3.38	10,400	
104	54.69	0.469	0.0026	0.083	3.23	2,870	
140	53.94	0.489	0.903×10^{-3}	0.081	3.10	1,050	
176	53.19	0.509	0.404	0.080	2.98	490	
212	52.44	0.530	0.219	0.079	2.86	276	
248	51.75	0.551	0.133	0.078	2.75	175	
284	51.00	0.572	0.086	0.077	2.66	116	
320	50.31	0.593	0.060	0.076	2.57	84	

From *Heat and Mass Transfer*, 2d ed., by E. R. G. Eckert and R. M. Drake. Copyright © 1959 by the McGraw-Hill Book Company. Used with permission of the McGraw-Hill Book Company.

Appendix A

Table A.4.a
Properties of Air at Atmospheric Pressure (SI Units)

T (°K)	ρ (kg/m³)	c_p (kJ/kg °C)	μ (kg/ms)	ν (m²/s)	k (W/m °C)	α (m²/s)	N_{Pr}	a^* (1/m³ °C) $\times 10^{-6}$
100	3.6010	1.0266	0.6924×10^5	1.923×10^6	0.009246	0.02501×10^4	0.770	20396.0
150	2.3675	1.0099	1.0283	4.343	0.013735	0.05745	0.753	2622.0
200	1.7684	1.0061	1.3289	7.490	0.01809	0.10165	0.739	641.8
250	1.4128	1.0053	1.488	9.49	0.02227	0.13161	0.722	282.9
300	1.1774	1.0057	1.983	15.68	0.02624	0.22160	0.708	87.61
350	0.9980	1.0090	2.075	20.76	0.03003	0.2983	0.697	45.21
400	0.8826	1.0140	2.286	25.90	0.03365	0.3760	0.689	25.13
450	0.7833	1.0207	2.484	28.86	0.03707	0.4222	0.683	16.31
500	0.7048	1.0295	2.671	37.90	0.04038	0.5564	0.680	9.285
550	0.6423	1.0392	2.848	44.34	0.04360	0.6532	0.680	6.182
600	0.5879	1.0551	3.018	51.34	0.04659	0.7512	0.680	4.234
650	0.5430	1.0635	3.177	58.51	0.04953	0.8578	0.682	3.009
700	0.5030	1.0752	3.332	66.25	0.05230	0.9672	0.684	2.188
750	0.4709	1.0856	3.481	73.91	0.05509	1.0774	0.686	1.638
800	0.4405	1.0978	3.625	82.29	0.05779	1.1951	0.689	1.246
850	0.4149	1.1095	3.765	90.75	0.06028	1.3097	0.692	0.9734
900	0.3925	1.1212	3.899	99.3	0.06279	1.4271	0.696	0.7674
950	0.3716	1.1321	4.023	108.2	0.06525	1.5510	0.699	0.6132
1000	0.3524	1.1417	4.152	117.8	0.06752	1.6779	0.702	0.4961
1100	0.3204	1.160	4.44	138.6	0.0732	1.969	0.704	0.3270
1200	0.2947	1.179	4.69	159.1	0.0782	2.251	0.707	0.2272
1300	0.2707	1.197	4.93	182.1	0.0837	2.583	0.705	0.1605
1400	0.2515	1.214	5.17	205.5	0.0891	2.920	0.705	0.1160
1500	0.2355	1.230	5.40	229.1	0.0946	3.262	0.705	0.0878
1600	0.2211	1.248	5.63	254.5	0.100	3.609	0.705	0.0672
1700	0.2082	1.267	5.85	280.5	0.105	3.977	0.705	0.0518
1800	0.1970	1.287	6.07	308.1	0.111	4.379	0.704	0.0407
1900	0.1858	1.309	6.29	338.5	0.117	4.811	0.704	0.0319
2000	0.1762	1.338	6.50	369.0	0.124	5.260	0.702	0.0253
2100	0.1682	1.372	6.72	399.6	0.131	5.715	0.700	0.0206
2200	0.1602	1.419	6.93	432.6	0.139	6.120	0.707	0.0167
2300	0.1538	1.482	7.14	464.0	0.149	6.540	0.710	0.0139
2400	0.1458	1.574	7.35	504.0	0.161	7.020	0.718	0.0116
2500	0.1394	1.688	7.57	543.5	0.175	7.441	0.730	0.0097

$*g\beta^2 c_p/\mu k$ for standard gravity and standard atmospheric pressure.
From *Applied Heat Transfer* by J. P. Todd and H. B. Ellis. New York: Harper & Row, 1982.

Table A.4.b
Properties of Air at Atmospheric Pressure (English Units)

T (°F)	μ (lbm/hr ft)	k (Btu/hr ft °F)	c_p (Btu/lbm °F)	N_{Pr}	$a^* \times 10^{-6}$ (1/ft³ °F)
−100	0.0319	0.0104	0.239	0.739	10.22
−50	0.0358	0.0118	0.239	0.729	5.4
0	0.0394	0.0131	0.240	0.718	3.13
50	0.0427	0.0143	0.240	0.712	1.94
100	0.0459	0.0157	0.240	0.706	1.26
150	0.0484	0.0167	0.241	0.699	0.86
200	0.0519	0.0181	0.241	0.693	0.59
250	0.0547	0.0192	0.242	0.690	0.42
300	0.0574	0.0203	0.243	0.686	0.312
400	0.0626	0.0225	0.245	0.681	0.180
500	0.0675	0.0246	0.248	0.680	0.111
600	0.0721	0.0265	0.250	0.680	0.072
700	0.0765	0.0284	0.254	0.682	0.049
800	0.0806	0.0303	0.257	0.684	0.0346
900	0.0846	0.0320	0.260	0.687	0.0251
1000	0.0884	0.0337	0.263	0.690	0.0187

$*g\beta\rho^2 c_p/\mu k$ for standard gravity and standard atmospheric pressure.

From *Heat Transfer*, 2d ed., by B. Gebhart. Copyright © 1961, 1971 by the McGraw-Hill Book Company. Used with permission of the McGraw-Hill Book Company.

Appendix A

Table A.5.a
Properties of Water (Saturated Liquid) (SI Units)

Temp. (°C)	c_p (kJ/kg °C)	ρ (kg/m³)	μ (kg/ms)	k (W/m °C)	$a = \dfrac{g\beta\rho^2 c_p}{\mu k}$ (1/m³·°C)	N_{Pr}	r (kJ/kg)	β (1/°C)
0	4.194	1000.0	1.79×10^{-3}	0.566	0.0	13.26	2501	0.00000
10	4.202	1000.0	1.31	0.585	5.37×10^9	9.41	2478	0.00010
20	4.190	998.0	1.01	0.602	13.5	7.03	2454	0.00020
30	4.179	996.0	0.803	0.619	23.7	5.42	2431	0.00029
40	4.177	992.6	0.656	0.633	36.9	4.33	2407	0.00038
50	4.178	988.1	0.536	0.644	53.3	3.48	2383	0.00046
60	4.183	983.3	0.475	0.654	67.6	3.04	2359	0.00053
70	4.187	977.8	0.408	0.664	85.4	2.573	2334	0.00059
80	4.197	971.8	0.359	0.671	103	2.245	2309	0.00064
90	4.206	965.4	0.318	0.676	122	1.979	2283	0.00069
100	4.219	958.6	0.283	0.682	146	1.751	2257	0.00074
110	4.233	951.3	0.253	0.685	173	1.563	2230	0.00080
120	4.251	943.4	0.229	0.685	203	1.421	2203	0.00086
130	4.270	935.1	0.211	0.685	230	1.315	2174	0.00091
140	4.294	926.4	0.196	0.685	258	1.229	2145	0.00096
150	4.321	917.3	0.185	0.684	287	1.169	2114	0.00102
160	4.350	907.8	0.174	0.681	320	1.111	2083	0.00108
170	4.383	897.7	0.164	0.679	354	1.059	2050	0.00114
180	4.420	887.3	0.154	0.676	390	1.007	2015	0.00120
190	4.461	876.4	0.147	0.671	428	0.977	1979	0.00135
200	4.506	865.0	0.139	0.667	470	0.939	1941	0.00138
210	4.558	853.0	0.133	0.661	516	0.917	1901	0.00142
220	4.616	840.8	0.127	0.654	570	0.896	1859	0.00154
230	4.684	827.1	0.121	0.648	633	0.875	1814	0.00161
240	4.761	813.7	0.116	0.638	706	0.866	1767	0.00170
250	4.850	799.4	0.107	0.627	792	0.866	1716	0.00188
260	4.955	783.7	0.107	0.616	896	0.861	1663	0.00198
270	5.079	768.0	0.103	0.598	1.02×10^{12}	0.875	1605	0.00214
280	5.224	750.8	0.098	0.581	1.21	0.881	1544	0.00239
290	5.399	732.1	0.094	0.560	1.42	0.906	1477	0.00263
300	5.610	712.3	0.091	0.530	1.67	0.963	1405	0.00288

Table B–4 "Properties of water (saturated liquid)" (pp. 489–90) (after Kreith and Moore) in *Applied Heat Transfer* by James P. Todd and Herbert B. Ellis. Copyright © 1982 by James P. Todd and Herbert B. Ellis. By permission of Harper & Row, Publishers, Inc.

Selected Thermophysical Properties of Substances

Table A.5.b
Properties of Water (English Units)

T (°F)	Saturation Pressure (psia)	ρ (lbm/ft³)	μ (lb/hr ft)	k (Btu/hr ft °F)	c_p (Btu/lbm °F)	N_{Pr}	a^* (1/ft³ °F)
40	0.122	62.43	3.74	0.326	1.0041	11.5	
50	0.178	62.41	3.16	0.334	1.0013	9.49	
60	0.256	62.36	2.72	0.341	0.9996	7.98	154×10^6
70	0.363	62.30	2.37	0.347	0.9987	6.81	238
80	0.507	62.22	2.08	0.353	0.9982	5.89	330
90	0.698	62.12	1.85	0.358	0.9980	5.15	435
100	0.949	62.00	1.65	0.363	0.9980	4.55	547
110	1.28	61.86	1.49	0.367	0.9982	4.06	670
120	1.69	61.71	1.35	0.371	0.9985	3.64	804
130	2.22	61.55	1.24	0.375	0.9989	3.29	949
140	2.89	61.38	1.13	0.378	0.9994	3.00	1100
150	3.72	61.20	1.05	0.381	1.0000	2.74	1270
160	4.74	61.00	0.968	0.384	1.0008	2.52	1450
170	5.99	60.80	0.900	0.386	1.0017	2.33	1640
180	7.51	60.58	0.839	0.388	1.0027	2.17	1840
190	9.34	60.36	0.785	0.390	1.0039	2.02	2050
200	11.53	60.12	0.738	0.392	1.0052	1.89	2270
212	14.696	59.83	0.686	0.394	1.0070	1.76	2550
220	17.19	59.63	0.655	0.394	1.0084	1.67	2750
240	24.97	59.11	0.588	0.396	1.0124	1.51	3270
260	35.43	58.86	0.534	0.397	1.0173	1.37	3830
280	49.20	57.96	0.487	0.397	1.0231	1.26	4420
300	67.01	57.32	0.449	0.396	1.0297	1.17	5040
320	89.66	56.65	0.418	0.395	1.0368	1.10	5660
340	118.01	55.94	0.393	0.393	1.0451	1.04	6300
360	153.0	55.19	0.371	0.391	1.0547	1.00	6950
380	195.8	54.38	0.351	0.388	1.0662	0.97	7610
400	247.3	53.51	0.333	0.384	1.0800	0.94	8320
420	308.8	52.61	0.316	0.379	1.0968	0.92	9070
440	381.6	51.68	0.301	0.374	1.1168	0.90	9900
460	466.9	50.70	0.285	0.368			
480	566.1	49.67	0.272	0.362			

*$g\beta\rho^2 c_p/\mu k$ for standard gravity and low pressure.

From *Heat Transfer*, 2d ed., by B. Gebhart. Copyright © 1961, 1971 by the McGraw-Hill Book Company. Used with permission of the McGraw-Hill Book Company.

Table A.6.a
Physical Properties of Some Liquid Metals (SI Units)

Metal, Composition, and Melting Point	T (°C)	ρ kg/m³	c_p (J/kg °K) $\times 10^3$	ν (m²/s) $\times 10^{-7}$	k W/m °K	α (m²/s) $\times 10^{-5}$	N_{Pr}
Bismuth (271.1°C)	316	10,011	0.1444	1.617	16.4	1.138	0.0142
	538	9,739	0.1545	1.133	15.6	1.035	0.0110
	760	9,467	0.1645	0.8343	15.6	1.001	0.0083
Lead (327.2°C)	371	10,540	0.159	2.276	16.1	1.084	0.024
	482	10,412	0.155	1.849	15.6	1.223	0.017
	704	10,140		1.347	14.9		
Mercury (−38.9°C)	77	13,490	0.138	1.0	9.2	0.49	0.02
	127	13,368	0.137	0.95	9.8	0.58	0.016
Potassium (63.9°C)	149	807.3	0.80	4.608	45.0	6.99	0.0066
	427	741.7	0.75	2.397	39.5	7.07	0.0034
	704	674.4	0.75	1.905	33.1	6.55	0.0029
Sodium (97.8°C)	93	929.1	1.38	7.516	86.2	6.71	0.011
	204	900	1.34	4.8	80.3	6.66	0.0072
	371	860.2	1.30	3.270	72.3	6.48	0.0051
	704	778.5	1.26	2.285	59.7	6.12	0.0037
NaK(56/44, 19°C)	93	887.4	1.130	6.522	25.6	2.552	0.026
	371	821.7	1.055	2.871	27.5	3.17	0.0091
	704	740.1	1.043	2.174	28.9	3.74	0.0058
NaK(22/78, −11.1°C)	93	849.0	0.946	5.797	24.4	3.05	0.019
	399	775.3	0.879	2.666	26.7	3.92	0.0068
	760	690.4	0.883	2.118			
PbBi(44.5/55.5, 125°C)	149	10,524	0.147	1.496	9.05	0.586	0.189
	371	10,236	0.147	1.171	11.86	0.790	
	649	9,835					

Adapted by permission from *Heat Transfer*, 2d ed., by Bhalchandra V. Karlekar and Robert M. Desmond. Copyright © 1982 by West Publishing Company. All rights reserved. Also from *Liquid Metals Handbook*, Atomic Energy Commission and Department of the Navy. NAVEXOS P-733 (rev.) 2d ed. Revised June 1954, 3d ed. Sodium and Potassium Supplement, 1955.

Selected Thermophysical Properties of Substances

Table A.6.b
Physical Properties of Some Liquid Metals (English Units)

Metal	Temperature (°F)	Thermal Conductivity (Btu/hr ft °F)	Density (lbm/ft³)	Heat Capacity (Btu/lbm °F)	Absolute Viscosity* (lbm/ft sec)	Kinematic Viscosity (ft²/sec)	Thermal Diffusivity (ft²/hr)	c_p/k (hr ft/lbm)	N_{Pr}
Bismuth	600	9.5	625	0.0345	1.09×10^{-3}	1.74×10^{-6}	0.44	3.6×10^{-3}	0.014
	800	9.0	616	0.0357	0.90	1.5	0.41	4.0	0.013
	1000	9.0	608	0.0369	0.74	1.2	0.40	4.1	0.011
	1200	9.0	600	0.0381	0.62	1.0	0.39	4.2	0.0094
	1400	9.0	591	0.0393	0.53	0.9	0.39	4.4	0.0084
Gallium	85(m.p.)	19.5	381	0.082	1.39	3.6	0.61	4.3	0.022
	200		378	0.082	1.05	2.8			
	500		370	0.082	0.73	2.0			
	800		363		0.58	1.6			
	1200		355		0.47	1.3			
	1600		348		0.44	1.2			
Lead	700	9.3	658	0.038	1.61	2.45	0.37	4.1	0.024
	850	9.0	652	0.037	1.38	2.12	0.37	4.1	0.020
	1000	8.9	646	0.037	1.17	1.81	0.37	4.2	0.017
	1150	8.7	639	0.037	1.02	1.60	0.37	4.3	0.016
	1300	8.6	633		0.92	1.45			
Lithium	400	22	31.6	1.0	0.40	13	0.70	0.046	0.065
	600		31.0	1.0	0.34	11			
	800		30.5	1.0	0.37	12			
	1200		29.4	1.0	0.29	9.9			
	1800		27.6	1.0	0.28	10			
Mercury	50	4.7	847	0.033	1.07	1.2	0.17	7.1	0.027
	200	6.0	834	0.033	0.84	1.0	0.22	5.5	0.016
	300	6.7	826	0.033	0.74	0.90	0.25	4.9	0.012
	400	7.2	817	0.032	0.67	0.82	0.27	4.5	0.011
	600	8.1	802	0.032	0.58	0.72	0.31	4.0	0.0084
Potassium	300	26.0	50.4	0.19	0.25	5.0	2.7	7.4	0.0066
	500	24.7	48.7	0.19	0.16	3.3	2.7	7.5	0.0043
	800	22.8	46.3	0.18	0.12	2.6	2.7	8.0	0.0035
	1100	20.6	43.8	0.18	0.10	2.3	2.6	8.8	0.0032
	1300	19.1	42.1	0.18	0.090	2.1	2.5	9.6	0.0031

Appendix A

Substance	Temp				(×10⁻³)	(×10⁻⁶)		(×10⁻³)	
Sodium	200	49.8	58.0	0.33	0.47	8.1	2.6	6.7	0.011
	400	46.4	56.3	0.32	0.29	5.1	2.6	6.9	0.0072
	700	41.8	53.7	0.31	0.19	3.5	2.5	7.3	0.0050
	1000	37.8	51.2	0.30	0.14	2.7	2.4	8.0	0.0040
	1300	34.5	48.6	0.30	0.12	2.5	2.4	8.7	0.0038
Tin	500	19	433	0.0580	1.22	2.82	0.76	3.1	0.013
	700	19.4	428	0.0603	0.98	2.3	0.75	3.1	0.011
	850	19	425	0.0621	0.85	2.0	0.72	3.3	0.010
	1000	19	421	0.0639	0.76	1.8	0.71	3.4	0.0093
	1200	19	417	0.0662	0.67	1.6	0.69	3.5	0.0084
Zinc	600	35.4	435	0.123	2.10	4.88	0.66	3.48	0.027
	850	33.7	431	0.119	1.72	4.02	0.66	3.52	0.022
	1000	33.2	428	0.116	1.39	3.29	0.67	3.50	0.017
	1200	32.8	422	0.113	0.983	2.4	0.69	3.44	0.014
	1500	32.6	408	0.107			0.74	3.29	
Na, 56 wt % K, 44 wt %	200	14.8	55.4	0.270	0.390	7.04	0.994	18.2	0.026
	400	15.3	53.8	0.261	0.244	4.54	1.09	17.0	0.015
	700	15.9	51.3	0.252	0.158	3.08	1.23	15.9	0.0090
	1000	16.4	48.8	0.248	0.119	2.43	1.35	15.2	0.0065
	1300	16.7	46.2	0.249	0.108	2.34	1.45	14.9	0.0058
Na, 22 wt % K, 78 wt % (near eutectic)	200	14.1	53.0	0.226	0.330	6.20	1.17	16.1	0.019
	400		51.3	0.217	0.216	4.21			
	750	15.4	48.4	0.210	0.139	2.87	1.52	14.1	0.0068
	1100		45.5	0.208	0.107	2.35			
	1400		43.1	0.211	0.0981	2.28			
Pb, 44.5 wt % Bi, 55.5 wt % (eutectic)	300	5.23	657	0.035	1.18	1.83	0.227	6.69	0.024
	550	6.20	646	0.035	1.03	1.62	0.274	5.65	0.019
	700	6.85	639	0.035	0.800	1.29	0.306	5.11	
	1100		620		0.772	1.26			
	1200		614						

*The absolute viscosity μ in lbm/hr ft is found by multiplying the value given in the table by 3600 sec/hr.

From *Liquid Metals Handbook*. Atomic Energy Commission and Department of the Navy. NAVEXOS P-733 (rev.) 2d ed. Revised June 1954.

TOTAL HEMISPHERICAL EMISSIVITIES

Table B.1
Selected Values of the Total Hemispheric Emissivity ϵ

Material and Surface Condition	Temperature (°F)	ϵ
Metals		
Aluminum		
Highly polished plate 98.3%	440	0.039
pure	1070	0.057
Commercial sheet	212	0.09
Heavily oxidized	200	0.20
	940	0.31
Brass		
Highly polished	500	0.03
	700	0.032
Oxidized	500	0.61
	1000	0.59
Copper		
Polished	100	0.04
	1000	0.18
Oxidized	600	0.47
	1250	0.70
Gold		
Polished	100	0.02
	1000	0.02
Inconel X		
Stably oxidized	600	0.89
	2000	0.925

Continued on next page

Table B.1 *(Continued)*

Material and Surface Condition	Temperature (°F)	ϵ
Iron and Steel		
Iron, oxidized	250	0.74
	1000	0.84
	2000	0.885
Mild steel, polished	100	0.10
	1000	0.35
Molten cast iron or molten mild steel	2000	0.29
Stainless steel 301	100	0.16
	1000	0.25
	1500	0.49
Lead, gray oxidized	100	0.28
Nickel, oxidized	100	0.35
	1000	0.66
Titanium, Ti-75A	200	0.10
	800	0.19
Tungsten filament	1000	0.11
	2000	0.16
	6000	0.39
Nonmetallic Materials		
Asbestos, board	100	0.96
Brick, red, rough	70	0.93
Firebrick	1832	0.75
Coal soot	68	0.95
Concrete, rough	32–200	0.94
Glass	68	0.93
Ice, smooth	32	0.92
Paints, oil, all colors	68	0.89–0.97
Paints, aluminum	68	0.40–0.70
Paper	203	0.89
Plaster	68	0.92
Porcelain, glazed	72	0.92
Rubber, hard	68	0.92
Water	100	0.96
Wood	68	0.8–0.9

This short table of total hemispheric emissivities was prepared using selected entries from the following source, which is one of the most complete and extensive compilations of such data to date: *Thermal Radiation Properties Survey,* 2d ed. by G. G. Gubareff, J. E. Janssen, and R. H. Torborg. Minneapolis, Minnesota: Honeywell Research Center, Minneapolis-Honeywell Regulator Company, 1960.

TABLES OF THE BESSEL AND ERROR FUNCTIONS

Table C.1
Selected Values of the Bessel Functions of the First and Second Kinds, Orders Zero and One

x	$J_0(x)$	$J_1(x)$	$Y_0(x)$	$Y_1(x)$
0.0	1.00000	0.00000	$-\quad\infty$	$-\quad\infty$
0.2	+0.99002	0.09950	−1.0811	−3.3238
0.4	+0.96039	+0.19603	−0.60602	−1.7809
0.6	+0.91200	+0.28670	−0.30851	−1.2604
0.8	+0.84629	+0.36884	−0.08680	−0.97814
1.0	+0.76520	+0.44005	+0.08825	−0.78121
1.2	+0.67113	+0.49830	+0.22808	−0.62113
1.4	+0.56686	+0.54195	+0.33790	−0.47915
1.6	+0.45540	+0.56990	+0.42043	−0.34758
1.8	+0.33999	+0.58152	+0.47743	−0.22366
2.0	+0.22389	+0.57672	+0.51038	−0.10703
2.2	+0.11036	+0.55596	+0.52078	+0.00149
2.4	+0.00251	+0.52019	+0.51042	+0.10049
2.6	−0.09680	+0.47082	+0.48133	+0.18836
2.8	−0.18503	+0.40971	+0.43591	+0.26355
3.0	−0.26005	+0.33906	+0.37685	+0.32467
3.2	−0.32019	+0.26134	+0.30705	+0.37071
3.4	−0.36430	+0.17923	+0.22962	+0.40101
3.6	−0.39177	+0.09547	+0.14771	+0.41539
3.8	−0.40256	+0.01282	+0.06450	+0.41411
4.0	−0.39715	−0.06604	−0.01694	+0.39792
4.2	−0.37656	−0.13864	−0.09375	+0.36801
4.4	−0.34226	−0.20278	−0.16333	+0.32597
4.6	−0.29614	−0.25655	−0.22345	+0.27374
4.8	−0.24042	−0.29850	−0.27230	+0.21357

Continued on next page

Table C.1 *(Continued)*

x	$J_0(x)$	$J_1(x)$	$Y_0(x)$	$Y_1(x)$
5.0	−0.17760	−0.32760	−0.30851	+0.14786
5.2	−0.11029	−0.34322	−0.33125	+0.07919
5.4	−0.04121	−0.34534	−0.34017	+0.01013
5.6	+0.02697	−0.33433	−0.33544	−0.05681
5.8	+0.09170	−0.31103	−0.31775	−0.11923
6.0	+0.15065	−0.27668	−0.28819	−0.17501
6.2	+0.20175	−0.23292	−0.24830	−0.22228
6.4	+0.24331	−0.18164	−0.19995	−0.25955
6.6	+0.27404	−0.12498	−0.14523	−0.28575
6.8	+0.29310	−0.06252	−0.08643	−0.30019
7.0	+0.30007	−0.00468	−0.02595	−0.30267
7.2	+0.29507	+0.05432	+0.03385	−0.29342
7.4	+0.27859	+0.10963	+0.09068	−0.27315
7.6	+0.25160	+0.15921	+0.14243	−0.24280
7.8	+0.25541	+0.20136	+0.18722	−0.20389
8.0	+0.17165	+0.23464	+0.22352	−0.15806
8.2	+0.12222	+0.25800	+0.25011	−0.10724
8.4	+0.06916	+0.27079	+0.26622	−0.05348
8.6	+0.01462	+0.27275	+0.27146	−0.00108
8.8	−0.03923	+0.26407	+0.26587	+0.05436
9.0	−0.09033	+0.24531	+0.24994	+0.10431
9.2	−0.13675	+0.21471	+0.22449	+0.14911
9.4	−0.17677	+0.18163	+0.19074	+0.18714
9.6	−0.20898	+0.13952	+0.15018	+0.21706
9.8	−0.23227	+0.09284	+0.10453	+0.23789
10.0	−0.24594	+0.04347	+0.05567	+0.24902

Reprinted with permission of Macmillan Publishing Company from *Heat Transfer*, 2d ed., by Alan J. Chapman. Copyright © 1967 by Alan J. Chapman.

Table C.2
Selected Values of the Modified Bessel Functions of the First and Second Kinds, Orders Zero and One

x	$I_0(x)$	$I_1(x)$	$(2/\pi)K_0(x)$	$(2/\pi)K_1(x)$
0.0	1.000	0.0000	$+\infty$	$+\infty$
0.2	1.0100	0.1005	1.1158	3.0405
0.4	1.0404	0.2040	0.70953	1.3906
0.6	1.0920	0.3137	0.49498	0.82941
0.8	1.1665	0.4329	0.35991	0.54862
1.0	1.2661	0.5652	0.26803	0.38318
1.2	1.3937	0.7147	0.20276	0.27667
1.4	1.5534	0.8861	0.15512	0.20425
1.6	1.7500	1.0848	0.11966	0.15319
1.8	1.9896	1.3172	0.92903×10^{-1}	0.11626

Continued on next page

Tables of the Bessel and Error Functions

Table C.2 *(Continued)*

x	$I_0(x)$	$I_1(x)$	$(2/\pi)K_0(x)$	$(2/\pi)K_1(x)$
2.0	2.2796	1.5906	0.72507	0.89041×10^{-1}
2.2	2.6291	1.9141	0.56830	0.68689
2.4	3.0493	2.2981	0.44702	0.53301
2.6	3.5533	2.7554	0.35268	0.41561
2.8	4.1573	3.3011	0.27896	0.32539
3.0	4.8808	3.9534	0.22116	0.25564
3.2	5.7472	4.7343	0.17568	0.20144
3.4	6.7848	5.6701	0.13979	0.15915
3.6	8.0277	6.7028	0.11141	0.12602
3.8	9.5169	8.1404	0.8891×10^{-2}	0.9999×10^{-2}
4.0	11.3019	9.7595	0.7105	0.7947
4.2	13.4425	11.7056	0.5684	0.6327
4.4	16.0104	14.0462	0.4551	0.5044
4.6	19.0926	16.8626	0.3648	0.4027
4.8	22.7937	20.2528	0.2927	0.3218
5.0	27.2399	24.3356	0.2350	0.2575
5.2	32.5836	29.2543	0.1888	0.2062
5.4	39.0088	35.1821	0.1518	0.1653
5.6	46.7376	42.3283	0.1221	0.1326
5.8	56.0381	50.9462	0.9832×10^{-3}	0.1064
6.0	67.2344	61.3419	0.7920	0.8556×10^{-3}
6.2	80.7179	73.8859	0.6382	0.6879
6.4	96.9616	89.0261	0.5146	0.5534
6.6	116.537	107.305	0.4151	0.4455
6.8	140.136	129.378	0.3350	0.3588
7.0	168.593	156.039	0.2704	0.2891
7.2	202.921	188.250	0.2184	0.2331
7.4	244.341	227.175	0.1764	0.1880
7.6	294.332	274.222	0.1426	0.1517
7.8	354.685	331.099	0.1153	0.1424
8.0	427.564	399.873	0.9325×10^{-4}	0.9891×10^{-4}
8.2	515.593	483.048	0.7543	0.7991
8.4	621.944	583.657	0.6104	0.6458
8.6	750.461	705.377	0.4941	0.5220
8.8	905.797	852.663	0.4000	0.4221
9.0	1093.59	1030.91	0.3239	0.3415
9.2	1320.66	1246.68	0.2624	0.2763
9.4	1595.28	1507.88	0.2126	0.2236
9.6	1927.48	1824.14	0.1722	0.1810
9.8	2329.39	2207.13	0.1396	0.1465
10.0	2815.72	2670.99	0.1131	0.1187

Reprinted with permission of Macmillan Publishing Company from *Heat Transfer*, 2d ed., by Alan J. Chapman. Copyright © 1967 by Alan J. Chapman.

Table C.3
The Error Function Erf $(X) = \dfrac{2}{\sqrt{\pi}} \displaystyle\int_0^X e^{-t^2}\, dt$

X	Erf (X)	X	Erf (X)	X	Erf (X)
0.00	0.00000	0.64	0.63459	1.28	0.92973
0.02	0.02256	0.66	0.64938	1.30	0.93401
0.04	0.04511	0.68	0.66378	1.32	0.93807
0.06	0.06762	0.70	0.67780	1.34	0.94191
0.08	0.09008	0.72	0.69143	1.36	0.94556
0.10	0.11246	0.74	0.70468	1.38	0.94902
0.12	0.13476	0.76	0.71754	1.40	0.95229
0.14	0.15695	0.78	0.73001	1.42	0.95538
0.16	0.17901	0.80	0.74210	1.44	0.95830
0.18	0.20094	0.82	0.75381	1.46	0.96105
0.20	0.22270	0.84	0.76514	1.48	0.96365
0.22	0.24430	0.86	0.77610	1.50	0.96611
0.24	0.26570	0.88	0.78669	1.55	0.97162
0.26	0.28690	0.90	0.79691	1.60	0.97635
0.28	0.30788	0.92	0.80677	1.65	0.98038
0.30	0.32863	0.94	0.81627	1.70	0.98379
0.32	0.34913	0.96	0.82542	1.75	0.98667
0.34	0.36936	0.98	0.83423	1.80	0.98909
0.36	0.38933	1.00	0.84270	1.85	0.99111
0.38	0.40901	1.02	0.85084	1.90	0.99279
0.40	0.42839	1.04	0.85865	1.95	0.99418
0.42	0.44747	1.06	0.86614	2.00	0.99532
0.44	0.46623	1.08	0.87333	2.10	0.99702
0.46	0.48466	1.10	0.88021	2.20	0.99814
0.48	0.50275	1.12	0.88679	2.30	0.99886
0.50	0.52050	1.14	0.89308	2.40	0.99931
0.52	0.53790	1.16	0.89910	2.50	0.99959
0.54	0.55494	1.18	0.90484	2.60	0.99976
0.56	0.57162	1.20	0.91031	2.70	0.99987
0.58	0.58792	1.22	0.91553	2.80	0.99992
0.60	0.60386	1.24	0.92051	2.90	0.99996
0.62	0.61941	1.26	0.92524	3.00	0.99998

The selected values of the error function used in this short table were taken from the following two sources, which provide a more complete and extensive compilation: *Handbook of Mathematical Functions with Formulas, Graphs, and Mathematical Tables,* by Milton Abramowitz; Irene A. Stegun, ed., U.S. Department of Commerce, National Bureau of Standards Applied Mathematics Series 55, 1964; *Tables of the Error Function and its Derivative,* U.S. Department of Commerce, National Bureau of Standards Applied Mathematics Series 41, 1954.

ABBREVIATIONS AND CONVERSION FACTORS FOR SI AND ENGLISH UNITS

Table D.1

Abbreviations of Units in the English and SI Systems

Quantity	Engineering System		SI System	
	Unit	*Abbreviation*	*Unit*	*Abbreviation*
Length	feet	ft	meter	m
Area	square feet	ft²	square meter	m²
Volume	cubic feet	ft³	cubic meter	m³
Mass	pound mass	lbm	kilogram	kg
Time	second	sec	second	sec
	hour	hr	hour	hr
Force	pound force	lbf	Newton	N
Energy	British Thermal Unit	Btu	Joule (1 Nm)	J
Power		Btu/hr	watt (1 J/sec)	W
Temperature	degrees Fahrenheit	°F	degrees Celsius	°C
	degrees Rankine	°R	degrees Kelvin	°K

Table D.2
Conversion Factors—English and SI Units

Quantity	Conversion Equivalents
Length	1 ft = 0.3048 m
Area	1 ft^2 = 0.0929 m^2
Volume	1 ft^3 = 0.0283 m^3
Mass	1 lbm = 0.4536 kg
Force	1 lbf = 4.448 N
Energy	1 Btu = 1055 J
Pressure	1 lbf/ft^2 = 47.9 N/m^2
Power	1 Btu/hr = 0.293 W
Temperature, T	$T(°F) = \dfrac{9}{5} T(°C) + 32$
	$T(°R) = \dfrac{9}{5} T(°K)$
Mass flow rate	1 lbm/hr = 0.000126 kg/s
Heat flux	1 Btu/hr ft^2 = 3.154 W/m^2
Specific heat	1 Btu/lbm °F = 4186 J/kg °C
Density	1 lbm/ft^3 = 16.018 kg/m^3
Absolute viscosity	1 lbm/hr ft = 0.000413 kg/ms
Thermal diffusivity	1 ft^2/hr = 2.5806 × 10^{-5} m^2/s
Thermal conductivity	1 Btu/hr ft °F = 1.73 W/m °C
Surface coefficient of heat transfer	1 Btu/hr ft^2 °F = 5.678 W/m^2 °C

REFERENCES

1. *Liquid Metals Handbook.* 2d ed., revised. Atomic Energy Commission and Department of the Navy (June 1954).
2. Pratt and Whitney Aircraft. *Aeronautical Vest Pocket Handbook.* 11th ed. East Hartford, Conn.: Pratt and Whitney Aircraft, May 1966.
3. "Metrification in Scientific Journals." *American Scientist,* Vol. 56, No. 2 (1968), pp. 159–64.
4. Sears, F. W., and M. W. Zemansky. *University Physics.* 4th ed. Reading, Mass.: Addison-Wesley, 1970.

INDEX

H

Heat exchangers, 721
  counterflow, 723, 730, 754
  crossflow, 723, 725, 738, 755
  design of, 771–72
  direct contact, 722
  fouling factors, 727
  parallel flow, 723, 736, 757
  recuperative, 723
  regenerative, 722
  shell and tube, 723, 738, 755
Heat flux, 5
  in nucleate boiling, 702
  in subcooled forced convection
    boiling, 712–13
Heat transfer
  coefficient of, 8. *See also*
    Surface coefficient of heat
    transfer
  in condensation and boiling,
    673
  in conduction, 33
  definition of, 1
  in film cooling, 818, 827
  in forced convection, 414
  in free convection, 612
  in high-speed flow, 781
  integral methods in, 538
  in liquid metals, 798
  mechanism in turbulent flow,
    444
  in mixed convection, 654
  quasi-one-dimensional. *See*
    Conduction, steady quasi-
    one-dimensional
  in radiation, 306
  roughness effect on, 573
  shape factors for conduction,
    100, 102–3
  in transpiration cooling, 818
Heisler's charts, 193
Hemispheric emissivity, 308, 310
High-speed flow, 781
Hydraulic diameter, 506
Hydrodynamic boundary layer,
  455

I

Implicit finite difference
  equations, 251, 258
Inclined tubes, 693
Initial condition, 35

Injection into the boundary
  layer, 818, 824
Inner layer, 448
Insulation, critical thickness of,
  59
Integral methods, 466
  equations
    approximate solution to, 473
    derivation of, 466
    energy, 554
    laminar flow over flat plate,
      475
    momentum, 473, 619
    turbulent flow over flat plate,
      486
  in free convection, 621
  in heat transfer, 538
  in laminar film condensation,
    683
Intensity of radiation, 322–24
Internal resistance to heat
  transfer, 74, 76
Inviscid flow, 456
Irradiation, 314

J

j factor, 562

K

Kantorovich profile, 474
Kinematic viscosity, 435. *See
  also* Viscosity
Kirchhoff's law, 318

L

Laboratory black body, 315
Laminar film condensation, 674
Laminar flow, 425
Laminar sublayer, 448, 450
Latent heat of condensation, 679
Lattice points, 117
Law
  of conservation of energy, 12
  of conservation of mass, 418,
    469
  of corresponding corners, 341
  Fourier's, 3, 5
  Kirchhoff's, 318
  Newton's, for convective heat
    transfer, 8
  Newton's, for shear stress, 425
  one-seventh power, 447

Planck's, distribution of
  radiant emission, 308
Stefan-Boltzmann, 11
of the wake, 448
of the wall, 451
Wien's displacement, 309
Leidenfrost effect, 701
Linear vector space, 474
Liquid metal heat transfer, 798
Liquid metal properties, A15
LMTD, 506
Logarithmic mean temperature
  difference, 506
  for counterflow, 736
  for one fluid changing phase,
    736
  for parallel flow, 737
Low-speed flow, 495
Lumped parameter method, 227
  for body with convection, 232
  for body with convection and
    generation, 233
  for body with convection to a
    finite amount of fluid, 240
  for body with convection to a
    fluid with harmonically
    varying temperature, 237
  for body with radiation, 235
Lumped temperature
  distribution, 73

M

Matrix method of stability
  analysis, 260
Maximum heat flux, 701, 704
Mean beam length, 389, 393
Mechanical energy balance, 495,
  530
Metals, properties of, A2
Method Classification Scheme,
  for radiation heat transfer,
  342
Method of undetermined
  coefficients, 238
Mixed condensation, 674
Mixed free and forced
  convection, 654
Mixing cup temperature, 490
Momentum equation, integral
  form, 473, 619
Momentum theorem, 418

Temperature distribution, 416
  for laminar flow over constant
    flux flat plate, 546
  for laminar flow over
    isothermal flat plate, 542
  for laminar, fully developed
    flow in isothermal tube,
    502
  for laminar high-speed flow
    between parallel plates,
    786–87
  for laminar high-speed flow
    over flat plate, 791
  for surface of constant flux flat
    plate, 546, 565
  for surface of transpiration-
    cooled plate, 822
Thermal boundary layer, 527
Thermal coefficient of volume
  expansion, 616, A10, A13
Thermal conductivity, 3, 4
  of air, A11
  of engine oil, unused, A10
  of liquid metals, A15
  of metals and alloys, A2
  of nonmetals, A7
  of selected materials, 4
  variable, solutions for, 45, 279
  of water, A13
Thermal diffusivity, 34
  of air, A11
  of engine oil, unused, A10
  of liquid metals, A15
  of metals and alloys, A2
  of nonmetals, A7
Thermal energy equation, 495
  for boundary layer flow, 531
  for flow in tubes, 495
  integral form, 537

Thermal entrance length, 506
Thermal history, 505, 547, 559
Thermal radiation, 11
Thermal resistance, to heat
  transfer, 5, 53
Total energy equation for
  boundary layer flow, 530
Total hemispheric emissivity, 11
  selected values of, A18
Total temperature, 789
Transient. See Conduction,
  unsteady-state
Transition boiling, 701
Transmissivity, 314
Transpiration cooling, 818
Truncation error, of finite
  difference equations, 118
Tube banks, 586
Turbulent film condensation,
  695
Turbulent flow, 436
Turbulent Prandtl number, 514
Turbulent shear stress, 439

U

Undetermined coefficients,
  method of, 238
Unheated starting length, 538
Universal velocity distribution,
  451
Unstable film boiling, 701
Unstable finite difference
  equations, 259

V

Vapor velocity, effect on film
  condensation, 697
Variable generation, 68
Variable thermal conductivity,
  solutions for, 45, 279
Velocity boundary layer, 455
Velocity distribution
  for flat plate, laminar flow,
    480, 481
  in laminar film condensation,
    677
  in laminar flow inside parallel
    plate duct, 429
  in laminar flow inside pipe,
    434
  one-seventh power law, 447
  in turbulent flow in pipe, 446
  universal, 451
  in viscous sublayer, 450
View factor. See Angle factor
Viscosity
  of air, A11
  of engine oil, unused, A10
  of liquid metals, A15
  of water, A13
Viscous dissipation, 495, 531,
  781
Viscous layer, 448
Viscous stress, normal, 423, 460
Viscous sublayer, 448, 450
von Kàrmàn's analogy, 519

W

Wake region, in turbulent flow,
  448
Wavelength, distribution of
  emitted radiant energy, 308
Wien's displacement law, 309